LIFE IN THE UNIVERSE

Second Edition

JEFFREY BENNETT

University of Colorado at Boulder

SETH SHOSTAK

SETI Institute

PEARSON

Addison
Wesley

San Francisco   Boston   New York   Capetown
Hong Kong   London   Madrid   Mexico City   Montreal   Munich
Paris   Singapore   Sydney   Tokyo   Toronto

Vice President and Editorial Director: Adam Black, Ph.D.
Senior Acquisitions Editor: Lothlórien Homet
Development Manager: Michael Gillespie
Development Editor: Debbie Hardin
Media Producer: Matthew Phillips
Associate Editor: Deb Greco
Senior Administrative Assistant: Cathy Glenn
Director of Marketing: Christy Lawrence
Managing Editor: Corrine Benson
Production Supervisor: Nancy Tabor
Production Management: Elm Street Publishing Services
Composition: Thompson Type
Illustrators: John & Judy Waller, Scientific Illustrations
Photo Research: Kristin Piljay
Director, Image Resource Center: Melinda Patelli
Rights and Permissions Manager: Zina Arabia
Cover and Text Design: Marilyn Perry
Cover Printer: Phoenix Color Corporation
Printer and Binder: QuebecorWorld, Dubuque
Cover Image: Getty Images
Cover Illustration: Graham Johnson of www.fiVth.com

Library of Congress Cataloging-in-Publication Data

Bennett, Jeffrey O.
    Life in the universe. — 2nd ed. / Jeffrey Bennett, Seth Shostak.
      p. cm.
    Includes bibliographical references and index.
    ISBN 0-8053-4753-4
    1. Exobiology—Textbooks.   2. Life—Origin—Textbooks.
    I. Shostak, G. Seth.   II. Title.

QH327.B45 2007
576.8'39—dc22

                2006025657

ISBN 0-8053-4753-4

PEARSON

Addison
Wesley

1 2 3 4 5 6 7 8 9 10—QWD—09 08 07 06

www.aw-bc.com

DEDICATION

All this world is heavy with the promise of greater things, and a day will come, one day in the unending succession of days, when beings, beings who are now latent in our thoughts and hidden in our loins, shall stand upon this earth as one stands upon a footstool, and shall laugh and reach their hands amidst the stars.

H. G. Wells, 1902

THE QUEST TO UNDERSTAND LIFE ON EARTH and the prospects for life elsewhere in the universe touches on the most profound questions of human existence. It sheds light on our origins, teaches us to appreciate how and why our existence on Earth became possible, and inspires us to wonder about the incredible possibilities that may await us in space. We dedicate this book to all who wish to join in this quest, with the sincere hope that knowledge will help our species act wisely and responsibly.

Brief Contents

Detailed Contents

5 The Nature of Life on Earth 148

6 The Origin and Evolution of Life on Earth 190

PART III

LIFE IN THE SOLAR SYSTEM

7 Searching for Life in Our Solar System 235

Preface

TO THE READER

Few questions have so inspired humans through the ages as the mystery of whether we are alone in the universe. Many ancient Greek philosophers were confident that intelligent beings could be found far beyond Earth. When the first telescopes were trained on the Moon in the seventeenth century, some eminent astronomers interpreted lunar features as proof of an inhabited world. Barely a century ago, belief in a civilization on Mars became so widespread that the term *martian* became synonymous with *alien*. But despite this historical interest in the possibility of extraterrestrial life, until quite recently few scientists devoted any effort to understanding the issues surrounding it, let alone to making a serious search for life.

In the past couple of decades, however, a remarkable convergence of biology, geology, astronomy, and other sciences has suddenly placed the issue of extraterrestrial life at the forefront of research. Advances in our understanding of the origin of life on Earth are helping us predict the conditions under which life might arise in other places. Discoveries of microbes thriving under extreme conditions (at least by human standards) on Earth have raised hopes that life might survive even in some of the harsh environments found elsewhere in our own solar system. Proof that planets exist around other stars—first obtained in the 1990s—has given added impetus to the study of the conditions that might allow for life in other star systems. Technological advances are making it possible for us to engage in unprecedented, large-scale scrutiny of the sky for signals from other civilizations, spurring heightened interest in the search for extraterrestrial intelligence (SETI). Perhaps most important, scientists have found the interdisciplinary study of issues related to the search for life beyond Earth to have intrinsic value, independent of whether the search is ultimately successful.

Given the intense research efforts being undertaken by the scientific community and the long-standing public fascination with the search for life, it should be no surprise that the study of life in the universe—also known as *astrobiology*—has become one of the most publicly visible sciences. Colleges, too, have recognized the growing importance of this discipline, and many have begun to institute astrobiology courses. This book aims to serve such courses by offering a comprehensive introduction to the broad science of life in the universe.

Although this book is a text, it is designed to be of interest to *anyone* with a desire to learn about the current state of research in astrobiology. No special scientific training or background is assumed, and all necessary scientific concepts are reviewed as they arise. If you have a basic high school education and a willingness to learn, you are capable of understanding every topic covered in this book. We wish you well in your efforts.

Jeffrey Bennett
Seth Shostak

TO CURRENT OR PROSPECTIVE INSTRUCTORS

The rest of this preface is aimed primarily at current or prospective instructors teaching courses on life in the universe. Students and general readers might still find it useful, because it explains some of the motivation behind the pedagogical features and organization of this book and may thereby help you get the most from your reading.

Why Teach a Course on Life in the Universe?

By itself, the rapid rise of research interest in astrobiology might not be enough to justify the creation of new courses for nonscience majors. But the subject has at least three crucial features that together make a strong case for adding it to the standard science offerings:

1. For students who take only one or a few required science courses, the interdisciplinary nature of the study of life in the universe offers a broader understanding of the range of scientific research than can a course in any single discipline.
2. Public fascination with UFOs and alien visitation offers a unique opportunity to use life in the universe courses as vehicles for teaching about the nature of science and how to distinguish true science from pseudoscience.
3. The science of life in the universe considers many of the most profound questions we can ask, including: What is life? How did life begin on Earth? Are we alone? Could we colonize other planets or other star systems? Students are virtually always interested in these questions, making it easy to motivate even those students who study science only because it is required.

These features probably also explain the growing number of life in the universe courses being offered at colleges around the world as well as at the high school level. It's worth noting that, besides being fascinating to students, a course on life in the universe can be a great experience for instructors. The interdisciplinary nature of the subject means that no matter what your specific scientific background, you are sure to learn something new when you teach an astrobiology course at any level.

Using This Book for Your Course

As courses in astrobiology began to appear, instructors faced an immediate challenge: The interdisciplinary nature of the subject made it difficult to decide where emphasis should be placed. Over time, however, a general consensus emerged in favor of a rough balance between the different disciplines that contribute to the study of life in the universe. This book was written to serve that consensus course, and the success of our first edition gives us confidence that we achieved that goal. We have maintained the same interdisciplinary approach for this second edtion, while also responding to feedback from the many users of our first edition and updating the book with new developments in the science. With an interdisciplinary course goal in mind, we now turn our attention to a few details that should help instructors use this book effectively.

Course Types   This book is designed primarily for use in courses for nonscience majors, such as core course requirements in natural science or elective follow-on courses for students who lack the preparation needed for more technical offerings in astrobiology. It can also be used at the senior high school level, especially for integrated science courses that seek to break down the traditional boundaries separating individual science disciplines.

Overall Structure   We've developed this book with a four-part structure that matches the content of most courses on life in the universe. The table of contents gives more detail; a brief outline of the structure follows:

Part I.   Introducing Life in the Universe (Chapters 1–3). Chapter 1 offers a brief overview of the topic of life in the universe and why this science has moved to the forefront of research. Chapter 2 discusses the nature of science based on the assumption that this is many students' first real exposure to how scientific thinking differs from other modes of thinking. Chapter 3 presents fundamental astronomical and physical concepts necessary for understanding the rest of the course material.

Part II.   Life on Earth (Chapters 4–6). This is the first of three parts devoted to in-depth study of astrobiology issues. Here we discuss the current state of knowledge about life on Earth. Chapter 4 discusses the geological conditions that have made Earth habitable. Chapter 5 explores the nature of life on Earth. Chapter 6 discusses current ideas about the origin and subsequent evolution of life on Earth.

Part III.   Life in the Solar System (Chapters 7–10). We next use what we've learned about life on Earth in Part II to explore the possibilities for life elsewhere in our solar system. Chapter 7 discusses the environmental requirements for life and then offers a brief tour of various worlds in our solar system, exploring their potential habitability. Chapters 8 and 9 focus on the places that seem most likely to offer possibilities for life: Mars (Chapter 8) and the jovian moons Europa, Ganymede, Callisto, Titan, Enceladus, and Triton (Chapter 9). Chapter 10 discusses how habitability evolves over time in the solar system, with emphasis on comparing the past and present habitability of Venus and Earth; this chapter also introduces the concept of a habitable zone around a star, setting the stage for the discussion of life beyond our solar system in Part IV.

Part IV.   Life Among the Stars (Chapters 11–13). This final set of chapters deals with the question of life beyond our solar system. Chapter 11 focuses on the types of stars that seem suitable as "suns" for habitable planets, and then discusses the methods of detection and results of recent discoveries of extrasolar planets; it also covers the question of whether we should expect Earth-like planets to be rare or common. Chapter 12 covers the search for extraterrestrial intelligence (SETI). Chapter 13 discusses the challenge of and prospects for interstellar travel, and then uses these ideas to investigate the Fermi paradox ("Where is everybody?"), the potential solutions to the paradox, and the implications of the considered solutions.

Pace of Course Coverage   Although the chapters are not all of equal length, it should be possible to cover them at an average rate of approximately one chapter per week in a typical 3-hours-per-week college course. Thus, the 13 chapters in this book should provide about the right amount of material for a typical one-semester college course. If you are teaching a one-quarter course, you might need to be selective in your coverage, perhaps dropping some topics entirely. If you are teaching a yearlong course, as might be the case at the high school level, you can spread out the material to cover it at an average rate of about one chapter every 2 weeks.

New for the Second Edition

Astrobiology is a fast-moving field, and there have been many new developments since we wrote the first edition. Here, briefly, is a list of the major changes you'll find in this second edition:

- *New Structure Centered on Learning Goals:* We have rewritten the book to focus on clear, succinct "Learning Goals" that appear at the beginning of each chapter and around which the chapter is organized. This structure makes the key "big picture" goals much more obvious to students, while at the same time making it much easier for students to organize their studying. We have previously used this same structure in our astronomy texts with the same lead author (*The Cosmic Perspective* and *The Essential Cosmic Perspective*), and it has proven extremely useful both to students and instructors.

- *Updating the Science:* In the process of revising the book with this new pedagogy, we have carefully reviewed every sentence and every word to ensure that the book is as clear and streamlined as possible. We have also thoroughly updated the book with the many recent discoveries in astrobiology, including:
  - new discoveries about the potential habitability of the early Earth, including evidence for early oceans and continents within 100 million years after Earth's formation
  - new discoveries about the genetic relationships between the three domains of life and about the early development of life on Earth
  - the latest results from the rovers and orbiters studying Mars
  - the latest *Cassini* and *Huygens* results from Titan and Enceladus

- new discoveries of extrasolar planets and their implications to our understanding of solar system formation
- and much more

- *Annotated Figures:* Key figures in each chapter now include the research-proved technique of "annotation"—carefully crafted text placed on the figure (in blue) to guide students through interpreting graphs, following process figures, and translating among different representations.

- *Interactive Figures™ and Interactive Photos™:* Key figures and photos include an icon directing students to interactive versions online. They're also available for in-class presentation on the Media Manager CD-ROM for instructors.

- *Expanded Exercise Sets:* We have significantly expanded the end-of-chapter exercise sets with additional multiple-choice and quantitative problems, in most cases providing at least twice as many exercises as we had in the first edition. The sets are now organized as follows:
  - Review Questions: Questions that students should be able to answer from the reading alone.
  - Does It Make Sense? (or similar title): A set of short statements for which students are expected to determine whether each statement makes sense, and to explain why or why not. These exercises are generally easy once students understand a particular concept, but quite difficult otherwise; thus, they are an excellent probe of comprehension.
  - Quick Quiz: New to this edition, this short multiple-choice quiz allows students to check their progress.
  - Short-Answer/Essay Questions: Questions that go beyond the Review Questions in asking for conceptual interpretation.
  - Quantitative Problems: Problems that require some mathematics, usually based on topics covered in the Cosmic Calculations boxes.
  - Discussion Questions: Open-ended questions for class discussion.
  - Web Projects: Projects designed for independent research on the Web.

- *Optional Mathematics:* We have added Cosmic Calculations boxes to each chapter (except introductory Chapter 1). These boxes offer mathematics that can be used optionally, and form the basis for many of the Quantitative Problems at the end of each chapter.

- *Changes to Chapter Organization:* Users of the first edition will notice some significant changes to the chapter structure, including the following:
  - Chapter 1 is a chapter new to this edition, offering a short introduction to astrobiology before we move into more detailed discussions in subsequent chapters.

- Chapters 4 and 5 are now in reverse order from how they appeared in the first edition, and are significantly reorganized and rewritten.

- Chapter 13 combines material that formerly appeared in two separate chapters of the first edition.

- In addition, note that ALL chapters have been heavily reorganized and rewritten to fit our new pedagogy and the changes in science that have occurred since the publication of the first edition.

Supplements and Resources

In addition to the book itself, a number of supplements are available to help you as an instructor. The following is a brief summary; contact your local Addison-Wesley representative for more information.

- *Life in the Universe Media Manager* CD-ROM (ISBN 0-8053-0641-2). This CD-ROM provides all Bennett adopters with the largest library of purpose-built, in-class presentation materials for teaching introductory astrobiology. It includes jpegs of all of the figures and photos from the book, Interactive Figures™, Tutorials, and Test Bank questions.

- *Life in the Universe Companion Web Site* (http://www.aw-bc.com/bennett/). This open access Web site provides students with an out-of-class study aid. It includes Interactive Figures™ and Interactive Photos™ from the text, Interactive Tutorials, Movies, weblinks, and more. The site also provides resources to help instructors set up and design a course, including links to courses currently taught at other schools and advice from experts in the field.

- *Life in the Universe Activities Manual,* Second Edition, by Ed Prather, Erika Offerdahl, and Tim Slater (ISBN 0-8053-1712-0). Revised in conjunction with the main text, this manual provides creative projects that explore a wide range of concepts in astrobiology. It can be used as a laboratory component for a life in the universe course or as a source for group activities in the classroom.

Acknowledgments

A textbook may carry the names of its authors, but it is the result of the hard work of a long list of committed individuals. We could not possibly name everyone who has had a part in this book, but we would like to call attention to a few people who have played particularly important roles. First, we thank the friends and family members who put up with us during the long hours that we worked on this book. Without their support, this book would not have been possible.

At Addison Wesley, we offer special thanks to our current editors Adam Black, Lothlórien Homet, and Deb Greco (who put in countless hours to make this book meet its schedule). Many others have also helped make this book happen, including Joan Marsh, Mary Douglas, Margot Otway, Claire Masson, Michael Gillespie, Debbie Hardin, our accuracy checker Stephen Mojzsis, Brandi Nelson, Angel Chavez, and our artists John and Judy Waller.

We've also been fortunate to be able to draw on the expertise of several other Addison Wesley authors, in some cases drawing ideas and artwork directly from their outstanding texts. For her gracious help, we thank Jane Reece for permission to use content from *Biology,* Seventh Edition (Campbell and Reece), *Essential Biology,* Second Edition (Campbell, Reece, and Simon), and *Biology Concepts & Connections,* Fifth Edition (Campbell, Reece, Taylor and Simon); and the authors of *The Cosmic Perspective* astronomy texts. And very special thanks go to Bruce Jakosky, who was our coauthor on the first edition and provided much of the vision around which this book has been built.

Finally, we thank the many people who have carefully reviewed portions of the book (in either the first or second edition) in order to help us make it both as scientifically up-to-date and as pedagogically useful as possible:

Wayne Anderson, Sacramento City College
Timothy Barker, Wheaton College
Wendy Hagen Bauer, Wellesley College
Raymond Bigliani, Farmingdale State University of New York
Sukanta Bose, Washington State University
Greg Bothun, University of Oregon
Paul Braterman, University of North Texas
Juan Cabanela, Haverford College
Christopher Churchill, New Mexico State University
Leo Connolly, San Bernardino State
Manfred Cuntz, University of Texas at Arlington
Steven J. Dick, U.S. Naval Observatory
James Dilley, Ohio University
Jack Farmer, Arizona State University
Steven Federman, University of Toledo
Eric Feigelson, Penn State University
Richard Frankel, California Polytechnic State University
Rica S. French, MiraCosta College
Tracy Furutani, California Polytechnic State University
Bob Garrison, University of Toronto
Harold Geller, George Mason University
Perry A. Gerakines, University of Alabama at Birmingham
Donna H. Gifford, Pima Community College
Bob Greeney, Holyoke Community College
Bruce Hapke, University of Pittsburgh
William Hebard, Babson College
Beth Hufnagel, Anne Arundel Community College
James Kasting, Pennsylvania State University
Laura Kay, Barnard College
Jim Knapp, Holyoke Community College
David W. Koerner, Northern Arizona University
Karen Kolehmainen, California State University, San Bernardino
Kristin Larson, Western Washington University
James Lattimer, Stony Brook University

About the Authors

JEFFREY BENNETT

Jeffrey Bennett holds a B.A. (1981) in biophysics from the University of California, San Diego, and an M.S. and Ph.D. (1987) in astrophysics from the University of Colorado, Boulder. He has taught at every level from preschool through graduate school, including more than 50 college classes in astronomy, physics, mathematics, and education. He served 2 years as a visiting senior scientist at NASA headquarters, where he created NASA's "IDEAS" grant program, started a program to fly teachers aboard NASA's airborne observatories (including the hopefully soon-to-be-flying SOFIA), and worked on numerous educational programs for the Hubble Space Telescope and other space science missions. He also proposed the idea for and helped develop the Voyage scale model solar system, which opened in 2001 on the National Mall in Washington, D.C. (He is pictured here with the model Sun.) In addition to this astrobiology textbook, he has written college-level textbooks in astronomy, mathematics, and statistics, along with two books for the general public. He is also author of the award-winning children's books *Max Goes to the Moon* and *Max Goes to Mars*. When not working, he enjoys participating in masters swimming and in the daily adventures of life with his wife, Lisa; his children, Grant and Brooke; and his dog, Cosmo. You can read more about his projects on his personal Web site, www.jeffreybennett.com.

SETH SHOSTAK

Seth Shostak earned his B.A. in physics from Princeton University (1965) and a Ph.D. in astronomy from the California Institute of Technology (1972). He is currently a senior astronomer at the SETI Institute in Mountain View, California, where he helps press the search for intelligent cosmic company. For much of his career, Seth conducted radio astronomy research on galaxies and investigated the fact that these massive objects contain large amounts of unseen mass. He has worked at the National Radio Astronomy Observatory in Charlottesville, Virginia, as well as at the Kapteyn Astronomical Institute in Groningen, the Netherlands (where he learned to speak bad Dutch). Seth also founded and ran a company that produced computer animation for television. He has written several hundred popular articles on various topics in astronomy, technology, film, and television and teaches courses at the California Academy of Sciences and elsewhere. A frequent fixture on the lecture circuit, Seth gives approximately 70 talks annually at both educational and corporate institutions. He is a Distinguished Speaker for the American Institute of Aeronautics and Astronautics and a frequent commentator on astronomical matters for radio and television. His book *Sharing the Universe* (Berkeley Hills Books, 1998) details the rationale and expectations of the scientific search for intelligent life. When he's not trying to track down aliens, Seth can often be found in the darkroom, where he continues to hope that something interesting will develop.

How to Succeed in Your Astrobiology Course

USING THIS BOOK

Each chapter in this book is designed to make it easy for you to study effectively and efficiently. To get the most out of each chapter, you might wish to use the following study plan:

- A textbook is not a novel, and you'll learn best by reading the elements of this text in the following order:

  1. Start by reading the Learning Goals and the introductory paragraphs at the beginning of the chapter so that you'll know what you are trying to learn.
  2. Next, get an overview of key concepts by studying the illustrations and reading their captions. The illustrations highlight almost all of the major concepts, so this "illustrations first" strategy gives you an opportunity to survey the concepts before you read about them in depth.
  3. Read the chapter narrative, but save the boxed features (Special Topics, Cosmic Calculations) to read later. As you read, make notes on the pages to remind yourself of ideas you'll want to review later. Avoid using a highlight pen; underlining with pen or pencil is far more effective, because it forces you to take greater care and therefore helps keep you alert as you study. Be careful to underline selectively—it won't help you later if you've underlined everything.
  4. Make a second pass through the chapter, this time reading the boxed features and rereading any material that is not yet fully clear to you.
  5. Use the Chapter Summary to make sure you have correctly interpreted key points. The best way to use the summary is to try to answer the Learning Goal questions for yourself before reading the short, given answers.

- After completing the reading as above, try the end-of-chapter Review Questions; if you don't know an answer, look back through the chapter until you figure it out. Then test your understanding a little more deeply by trying the "Does It Make Sense?" and Quick Quiz questions.

- If your course has a quantitative emphasis, work through all of the examples in the Cosmic Calculations before trying the Quantitative Problems for yourself. Remember that you should always try to answer questions qualitatively before you begin plugging numbers into a calculator. For example, make an order of magnitude estimate of what your answer should be so that you'll know your calculation is on the right track, and be sure that your answer makes sense and has the appropriate units.

THE KEY TO SUCCESS: STUDY TIME

The single most important key to success in any college course is to spend enough time studying. A general rule of thumb for college classes is that you should expect to study about 2 to 3 hours per week *outside* of class for each unit of credit. For example, based on this rule of thumb, a student taking 15 credit hours should expect to spend 30 to 45 hours each week studying outside of class. Combined with time in class, this works out to a total of 45 to 60 hours spent on academic work—not much more than the time a typical job requires, and you get to choose your own hours. Of course, if you are working while you attend school, you will need to budget your time carefully.

As a rough guideline, your studying time in astrobiology might be divided as shown in the table. If you find that you are spending fewer hours than these guidelines suggest, you can probably improve your grade by studying longer. If you are spending more hours than these guidelines suggest, you may be studying inefficiently; in that case, you should talk to your instructor about how to study more effectively.

If Your Course Is	Time for Reading the Assigned Text (per week)	Time for Homework Assignments (per week)	Time for Review and Test Preparation (average per week)	Total Study Time (per week)
3 credits	2 to 4 hours	2 to 3 hours	2 hours	6 to 9 hours
4 credits	3 to 5 hours	2 to 4 hours	3 hours	8 to 12 hours
5 credits	3 to 5 hours	3 to 6 hours	4 hours	10 to 15 hours

GENERAL STRATEGIES FOR STUDYING

- Don't miss class. Listening to lectures and participating in discussions is much more effective than reading someone else's notes. Active participation will help you retain what you are learning.

- Take advantage of resources offered by your professor, whether it be e-mail, office hours, review sessions, online chats, or simply finding opportunities to talk to and get to know your professor. Most professors will go out of their way to help you learn in any way that they can.

- Budget your time effectively. One or 2 hours each day is more effective, and far less painful, than studying all night before homework is due or before exams.

- If a concept gives you trouble, do additional reading or studying beyond what has been assigned. And if you still have trouble, ask for help: You surely can find friends, peers, or teachers who will be glad to help you learn.

- Working together with friends can be valuable in helping you understand difficult concepts. However, be sure that you learn *with* your friends and do not become dependent on them.

- Be sure that any work you turn in is of *collegiate quality:* neat and easy to read, well organized, and demonstrating mastery of the subject matter. Although it takes extra effort to make your work look this good, the effort will help you solidify your learning and is also good practice for the expectations that future professors and employers will have.

PREPARING FOR EXAMS

- Study the Review Questions, and rework problems and other assignments; try additional questions to be sure you understand the concepts. Study your performance on assignments, quizzes, or exams from earlier in the term.

- Study your notes from lectures and discussions. Pay attention to what your instructor expects you to know for an exam.

- Reread the relevant sections in the textbook, paying special attention to notes you have made on the pages.

- Study individually *before* joining a study group with friends. Study groups are effective only if every individual comes prepared to contribute.

- Don't stay up too late before an exam. Don't eat a big meal within an hour of the exam (thinking is more difficult when blood is being diverted to the digestive system).

- Try to relax before and during the exam. If you have studied effectively, you are capable of doing well. Staying relaxed will help you think clearly.

1
A Universe of Life?

LEARNING GOALS

Gazing at the night sky, we can't help but be amazed at the glittering stars that silently beckon us from the depths of space. Stars are suns, much like our own Sun, each shining brightly for millions or billions of years. Many have planets, some of which may be much like the planets in our own solar system. Among these countless worlds, it may seem hard to imagine that Earth could be the only home for life. But while the possibility of life beyond Earth might seem quite reasonable, we do not yet have sufficient evidence to know whether it actually exists.

Learning whether the universe is full of life holds great significance for the way we view ourselves and our planet. If life is rare or nonexistent elsewhere, we will view our planet with added wonder. If life is common, we'll know that Earth is not quite as special as it may seem. If civilizations are common, we'll be forced to accept that we ourselves are just one of many intelligent species throughout the universe. The profound implications of finding—or not finding—extraterrestrial life make the question of life beyond Earth an exciting topic of study.

The primary purpose of this book is to give you the background needed to understand new and exciting developments in the human quest to find life beyond Earth. To do that, in coming chapters we will focus in some detail on the scientific issues that frame the search for life. First, however, let's start with a brief introduction to the subject, so that you'll understand why it has become such a hot topic of scientific research.

1.1 THE POSSIBILITY OF LIFE BEYOND EARTH

Aliens are everywhere, at least if you follow the popular media (Figure 1.1). Starships on television, such as the *Enterprise* or *Voyager,* are on constant prowl throughout the galaxy, seeking out new life and hoping it speaks English (or something close enough to English for the "universal translator"). In *Star Wars,* aliens from many planets gather at bars to share drinks and stories, and presumably to marvel at the fact that they have greater similarity in their level of technology than do different nations on Earth. Closer to home, movies like *Independence Day, Men in Black,* and *War of the Worlds* feature brave Earthlings battling evil aliens, while tabloids like the *Weekly World News* routinely carry headlines about the latest alien landings. Even serious newspapers and magazines run occasional articles about UFO sightings or about claims that the U.S. government is hiding frozen alien corpses at "Area 51." Given this intense media interest, it's probably not surprising that close to half of all Americans claim to believe that we have already been visited by beings from other worlds.

Figure 1.1

Aliens have become a part of modern culture, as illustrated in this movie poster.

Scientists are interested in aliens too, although most scientists remain deeply skeptical about reports of aliens on Earth (for reasons we'll discuss later in the book). Scientists are therefore searching for life elsewhere, looking for evidence of life on other worlds in our solar system, trying to learn whether we should expect to find life on planets orbiting other stars, and searching for signals broadcast by other civilizations. Indeed, the study of life in the universe is one of the most rapidly growing fields of active scientific research, largely because of its clear significance: The discovery of life of any kind beyond Earth would forever change our perspective on how we fit into the universe as a whole, and would undoubtedly teach us much more about life here on Earth as well.

THINK ABOUT IT   Check television and movie listings to see what is currently showing that involves aliens of some sort. Do you think any of the shows portray aliens in a scientifically realistic fashion? Explain.

• What are we searching for?

When we say we are searching for *life* in the universe, just what is it that we are looking for? Is it the kind of intelligent life we see portrayed in science fiction TV shows and films? Is it something more akin to the plants and animals we see in parks and zoos? Is it tiny, bacteria-like microbes? Or could it be something else entirely?

The simple answer is "all of the above." When we search for **extraterrestrial life,** or life beyond Earth, we are looking for any sign of life, be it simple, complex, or intelligent. We don't care if it looks exactly like life we are familiar with on Earth, or if it is dramatically different. However, we can't really answer the question of what we are looking for unless we know what life *is*.

Unfortunately, defining life is no simple matter, not even here on Earth where we have bountiful examples of it. Ask yourself: What common attributes make us think that a bacterium, a beetle, a mushroom, a tumbleweed, a maple tree, and a human are all alive, while we don't think the same of a crystal, a cloud, an ocean, or a fire? If you spend just a little time thinking about this question, you'll begin to appreciate its difficulty. For example, you might say that life can move, but the same is true of clouds and oceans. You might say that life can grow, but so can crystals. Or you might say that life can reproduce and spread, but so can fire. We will explore in Chapter 5 how scientists try to answer this question and come up with a general definition of life, but for now it should be clear that this is a complicated question that affects how we search for life in the universe.

Because of this definitional difficulty, the scientific search for extraterrestrial life in the universe generally presumes a search for life that is at least somewhat Earth-like and that we could therefore recognize based on what we know from studying life on Earth. Science fiction fans will object that this search is far too limited, and they may be right—but we have to start somewhere, so we begin with what we understand.

• Is it reasonable to imagine life beyond Earth?

The scientific search for life in the universe is a relatively recent development in human history, but the idea of extraterrestrial life is not. Many ancient cultures told stories about beings living among the stars and, as we'll discuss in Chapter 2, the ancient Greeks engaged in serious philosophical debate about the possibility of life beyond Earth.

Until quite recently, however, all these ideas remained purely speculative, because there was no way to study the question of life beyond Earth scientifically. It was always possible to *imagine* extraterrestrial life, but there was no scientific reason to think that it could really exist. Indeed, the relatively small amounts of data that might have shed some light on the question of life beyond Earth were often misinterpreted. Prior to the twentieth century, for example, some scientists guessed that Venus might harbor a tropical paradise—a guess that was based on little more than the fact that Venus is covered by clouds and closer than Earth to the Sun. Mars was the subject of even more intense debate, largely because a handful of scientists thought they saw long, straight canals on the surface [Section 8.1]. The canals, which don't really exist, were cited as evidence of a martian civilization.

Today, we have enough telescopic and spacecraft photos of the planets and large moons in our solar system to be quite confident that no civilization has ever existed on any of them. The prospect of large animals or plants seems almost equally unlikely. Nevertheless, scientific interest in life beyond Earth has exploded in the past few decades. Why? We'll spend most of the rest of the book answering this question, but we can summarize the key points briefly. First, although large, multicellular life in our solar system seems unlikely anywhere but on Earth, new discoveries in both planetary science and biology have given us some reason to think that simpler life—perhaps tiny microbes—might yet exist on other planets or moons that orbit our Sun. Second, while we've long known that the universe is full of *stars*, we've only recently gained concrete evidence telling us that it is also full of *planets*, which means there are far more places where we could potentially search for life. Third, advances both in scientific understanding and in technology now make it possible to study the question of life in the universe through established techniques of science, something that was not possible just a few decades ago. For example, we now understand enough about biology to explore the conditions that might make it possible for life to exist on other worlds, and we know enough about planets to consider which ones might be capable of harboring life. Indeed, while at present we have only limited ability to actually search for life beyond Earth, we are rapidly developing the spacecraft technology needed to search for microbes on other worlds of our solar system and the telescopic technology needed to look for signs of life among the stars.

The bottom line is that while it remains possible that life exists only on Earth, we now have plenty of scientific reasons to think that life might be widespread and that we could detect it if it is. In the rest of this chapter, we'll briefly discuss the scientific context of the search and the places where we might conduct the search, so that you will have a sense of what we will cover in more depth in the rest of the book.

1.2 THE SCIENTIFIC CONTEXT OF THE SEARCH

Almost every field of scientific research has at least some bearing on the search for life in the universe. Even seemingly unrelated fields such as mathematics and computer science play important roles. For example, we use mathematics to do the many computations that help us understand all other areas of science, and we use computers to simulate everything from the formation of stars and planets to the way in which the molecules of life interact with one another. However, three particular areas play an especially important role in framing the context of the scientific search for life: astronomy; planetary science, which includes geology and atmospheric science; and biology. In this section, we'll take a brief look at the context contributed by each of these three sciences.

• How does astronomy help us understand the possibilities for extraterrestrial life?

For most of human history, our conception of the universe was quite different from what it is today. Earth was widely assumed to be the center of the universe. Planets were lights in the sky, named for ancient gods, and no one had reason to think they could be *worlds* on which we might search for life. Stars were simply other lights in the sky, distinguished from the planets only by the fact that they remained fixed in the patterns of the constellations, and few people even considered the possibility that our Sun could be one of the stars. Moreover, with the Sun and planets presumed to be orbiting around Earth, there was certainly no reason to think that stars could have planets of their own, let alone planets on which there might be life.

When you consider the dominance that this Earth-centered, or **geocentric**, view of the universe held for thousands of years, it becomes obvious that astronomy plays a key role in framing the context of the modern search for life. We will discuss in Chapter 2 how and why the human view of the cosmos changed dramatically about 400 years ago, and we'll consider the modern astronomical context in some detail in Chapter 3. But the point should already be clear: We now know that Earth is but one tiny world orbiting one rather ordinary star in a vast cosmos, and this fact opens up countless possibilities for life on other worlds.

Astronomy provides context to the search for life in many other ways as well, but one more is important enough to mention right now: By studying distant objects, we have learned that the physical laws that operate in the rest of the universe are the same as those that operate right here on Earth. This tells us that if something happened here, it is possible that the same thing could have happened somewhere else, at least in principle. We are not the center of the universe in location, and we have no reason to think we are "central" in any other way, either.

• How does planetary science help us understand the possibilities for extraterrestrial life?

Planetary science is the name we give to the study of almost everything having to do with planets. It includes the study of planets themselves, as well as the study of moons orbiting planets, the study of how planets form, and the study of other objects that may form in association with planets (such as asteroids and comets). Planetary science helps set the context for the search for life in the universe in several different ways, but two are especially important.

First, by learning how planets form, we develop an understanding of how common we might expect planets to be. Until just about the middle of the last century, we really had no basis for assuming that many other

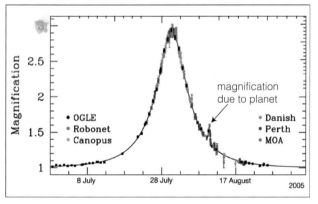

Figure 1.2

The painting (above) shows an artist's impression of what an extrasolar planet discovered in early 2006 may look like. It was the smallest extrasolar planet yet discovered, with a mass about five times that of Earth. Its odd-sounding name, OGLE-2005-BLG-390Lb, is a designation indicating that it was discovered by a technique based on Einstein's general theory of relativity (gravitational lensing, discussed in Chapter 11). The graph shows the data that led to the planet's discovery.

stars would have their own planets. Some scientists thought this likely, while others did not, and we lacked the data needed to distinguish between the two possibilities [Section 3.5]. But during the latter half of the twentieth century, a growing understanding of the processes by which our own solar system formed—much of it based on evidence obtained through human visits to the Moon and spacecraft visits to other planets—gradually made it seem more likely that other stars might similarly be born with planetary systems.

Still, as recently as 1995, no one was sure whether planets existed around other stars. That was the year in which the first strong evidence was obtained for the existence of **extrasolar planets**, or planets orbiting stars other than our Sun.* Since that time, additional discoveries of extrasolar planets have poured in at an astonishing rate, so that the known extrasolar planets now far outnumber the planets of our own solar system (Figure 1.2). Based on the statistics of these discoveries, it now seems likely that many or most stars have planets and, as we'll discuss in Chapter 11, it seems reasonable to imagine that life—and possibly even civilizations—could exist on at least some of these planets or their moons.

The second way in which planetary science shapes the context for the search for life is by helping us understand how planets work. For example, by studying planets and comparing them to one another, we have learned why some planets are rocky like Earth while others, like Jupiter, contain vast amounts of hydrogen and helium gas. We've also learned why Venus is *so* much hotter than Earth despite the fact that, in the scheme of our solar system, it is only slightly closer to the Sun. Similarly, we can now explain why the Moon is desolate and barren even though it orbits the Sun at essentially the same distance as Earth, and we have a fairly good idea of why Mars is cold and dry today, when evidence shows that it was warmer and wetter in the distant past.

This understanding of how planets work gives us deeper insight into the nature of planetary systems in general. More important to our purposes, it also helps us understand what to look for as we search for **habitable worlds**—worlds that contain the basic necessities for life. After all, given that there are far more worlds in the universe than we can ever hope to study in detail, we can improve our odds of success in finding life by constraining the search to those worlds that are the most promising. Be sure to note that when we ask whether a world is habitable, we are asking whether it offers environmental conditions under which life *could* arise or survive, not whether it actually harbors life.

Also keep in mind that when we say a world is habitable, we do not necessarily mean that plants, animals, or people could survive there. For much of Earth's history, nearly all life was microscopic, and even today, the total mass of microbes on Earth is greater than that of all plants and animals combined. The search for habitable worlds is primarily a search for places where microbes of some kind might survive, though we might find larger organisms as well.

• How does biology help us understand the possibilities for extraterrestrial life?

Astronomy, planetary science, and other science disciplines play important roles in shaping the context for the search for life in the universe,

---

*There was one earlier discovery, but it was of objects orbiting what we call a *neutron star,* not an "ordinary" star like our Sun.

but since we are searching for *life,* the context of biology is especially important. Just as you wouldn't look for a house to buy without knowing something about real estate, it would make no sense to search for life if we didn't know something about how life functions. The key question about the biological context of the search revolves around whether we should expect biology to be rare or common in the universe.

Wherever we have looked in the universe, we have found clear evidence that the same laws of nature are operating. We see galaxies sprinkled throughout space, and we see that the same stellar processes that occur in one place also occur in others. In situations in which we can observe orbital motions, we find that they agree with what we expect from the law of gravity. These and other observations make us confident that the basic laws of physics that we've discovered here on Earth also hold throughout the universe.

We can be similarly confident that the laws of chemistry are universal. Observations of distant stars show that they are made of the same chemical elements that we find here in our own solar system, and that interstellar gas clouds contain many of the same molecules we find on Earth. This provides conclusive evidence that atoms come in the same types and combine in the same ways throughout the universe.

— The universality of physics and chemistry is what makes us confident that we will find planets and other worlds, including many that are at least potentially habitable, throughout our Milky Way Galaxy and throughout the universe. Could biology also be universal? That is, could the biological processes we find on Earth be common throughout the universe? If the answer is yes, then the search for life elsewhere should be exciting and fruitful. If the answer is no, then life may be a rarity.

Because we haven't yet observed biology anywhere else in the universe, we can't yet know whether biology is universal. However, evidence from Earth gives us reason to think that it might be. Laboratory experiments suggest that chemical constituents found on the early Earth would have combined readily into complex organic (carbon-based) molecules, including many of the building blocks of life [Section 6.2]. Indeed, scientists have found organic molecules in meteorites (chunks of rock that fall to Earth from space) and, through spectroscopy [Section 3.4], in clouds of gas between the stars. The fact that such molecules form even under the extreme conditions of space suggests that they form quite readily and may be common on many worlds.

Of course, the mere presence of organic molecules does not necessarily mean that life will arise, but the history of life on Earth gives us some reason to think that the step from chemistry to biology is not especially difficult. As we'll discuss in Chapter 6, geological evidence tells us that life on Earth arose quite early in Earth's history, at least on a geological time scale. This early origin of life on Earth suggests—but certainly does not prove—that the same process might occur on other worlds. If the transition from chemistry to biology was exceedingly improbable, we might expect that it would have required much more time. Thus, the early origin of life on Earth makes it seem reasonable to think that life would emerge just as quickly on other worlds with similar conditions.

THINK ABOUT IT Microbial life on Earth predates intelligent life like us by at least 3–4 billion years. Do you think this fact tells us anything about the likelihood of finding *intelligent* life, as opposed to finding any life, on extrasolar planets? Explain.

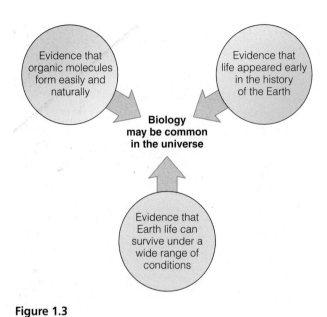

Figure 1.3

Three lines of evidence that give us at least some reason to think that biology may be common in the universe.

If life really can be expected to emerge under the right conditions, the only remaining question is the prevalence of those "right" conditions. Here, too, recent discoveries give us reason to think that biology could be common. In particular, biologists have found that microscopic life can survive and prosper under a much wider range of conditions than was believed only a few decades ago [Section 5.5]. For example, we now know that life exists in extremely hot water near deep-sea volcanic vents, in the frigid conditions of Antarctica, and inside rocks buried a kilometer or more beneath the Earth's surface. Indeed, if we were to export these strange organisms from Earth to other worlds in our solar system—perhaps to Mars or Europa—it seems possible that at least some of them would survive. This suggests that the range of "right" conditions for life may be quite broad, in which case it might be possible to find life even on planets that are significantly different in character from Earth.

In summary, we have no reason to think that life ought to be rare and several reasons to expect that it may be quite common (Figure 1.3). If life is indeed common, studying it will give us new insights into life on Earth, even if we don't find other intelligent civilizations. These enticing prospects have captured the interest of scientists from many disciplines and from around the world, giving birth to a new science devoted to the study of, and search for, life in the universe.

1.3   PLACES TO SEARCH

The study of life in the universe involves fundamental research in all the scientific areas we have already mentioned, and others as well. Indeed, as you'll see throughout this book, the study of life in the universe goes far beyond simply searching for living organisms. Still, all of this study is driven by the possibility that life exists elsewhere, so before we dive into any details, it's worth a quick overview of the places and methods we use to search for life beyond Earth.

• Where should we search for life in the universe?

The search for life in the universe takes place on several different levels. First, and in many ways foremost, it is a search for life right here on Earth. As we discussed earlier, we are still learning about the places and conditions under which life exists on Earth, and many scientists are busy searching for new species of life on our own world. After all, the more we know about life here, the better we'll be able to search for it elsewhere.

Searching Our Own Solar System   Turning our attention to places besides Earth, the first place to search for life is on other worlds in our own solar system. Our solar system has a lot of worlds: It has the planets orbiting the Sun, moons orbiting planets, and huge numbers of smaller objects such as asteroids and comets.

Figure 1.4 shows a portrait of some of our best current views of the planets in our solar system. Note that it is *not* to scale, since its purpose is to show each planet as we know it today from spacecraft or through telescopes; you can turn to Figure 3.3 to see the sizes correctly scaled. Note also that we have included both Pluto and the recently discovered Eris (which is slightly larger than Pluto), though these two objects are now considered "dwarf planets" (see Special Topic 1.1).

The photos alone make clear how different Earth is from every other planet in our solar system. Ours is the only planet with oceans of liquid

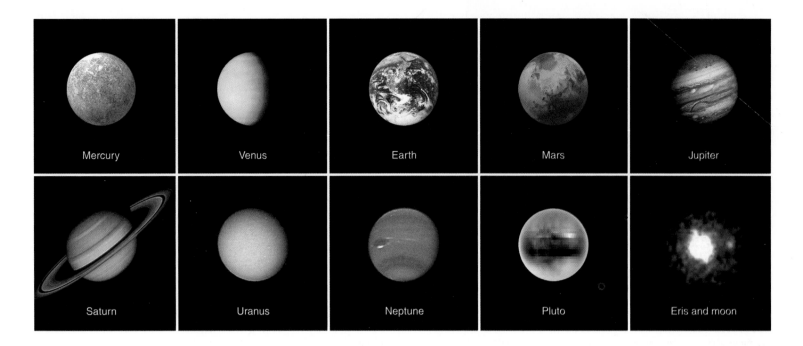

Mercury	Venus	Earth	Mars	Jupiter
Saturn	Uranus	Neptune	Pluto	Eris and moon

water on its surface, a fact that provides an instant clue about why Earth is home to so much life: Water is crucial to all life on Earth. Indeed, as we'll discuss in Chapters 5 and 7, we have good reason to think that liquid water is always a requirement for life, though it's possible that a few other liquids might work in place of water.

Given that we are primarily looking for life that is at least somewhat Earth-like, the need for water or some other liquid places constraints on where we might find life. Among the planets, Mars is by far the most promising candidate. As we'll discuss in detail in Chapter 8, strong evidence tells us that the now-barren surface of Mars (Figure 1.5) once had flowing water, making it seem reasonable to imagine life having arisen on Mars at that time. Mars still has significant amounts of water ice, so it is even possible that life exists on Mars today, perhaps hidden away in places where volcanic heat keeps underground water liquid. Past or present life seems much less likely on any of the other planets, though we can't rule

Figure 1.4

A "family portrait" of the planets that orbit our Sun, shown in order of increasing average distance from the Sun; the photos are *not* shown to scale. Pluto and Eris are included, though they are now considered "dwarf planets."

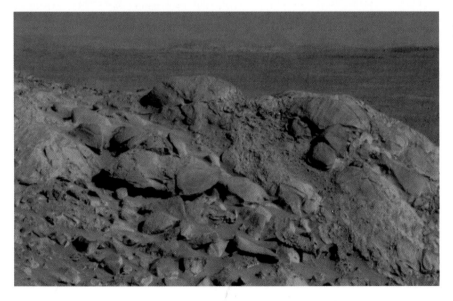

Figure 1.5

The surface of Mars, photographed by NASA's *Spirit* rover from a perch in the Columbia Hills. The martian surface is dry and barren today, but strong evidence points to liquid water on its surface in the distant past.

Figure 1.6

This photograph shows Jupiter and two of its moons: Io is the moon in front of Jupiter's Great Red Spot, and Europa is to the right. Scientists suspect that Europa has a deep ocean beneath its surface of ice, making it a prime target in the search for life in our solar system.

it out completely; we'll discuss these dim planetary prospects for life in Chapter 7.

Aside from the planets, the most promising abodes for life in the solar system are a few of the large moons. At least six moons are potential candidates for life, with the best candidate being Jupiter's moon Europa (Figure 1.6). Current evidence strongly suggests that Europa hides a deep ocean of liquid water under its icy crust. Indeed, if we are interpreting the evidence correctly, the Europan ocean may have considerably more water than all of Earth's oceans combined [Section 9.2]. Because we suspect that life on Earth got started in the deep oceans [Section 6.1], Europa may well have all the conditions needed both for life to have arisen and for its ongoing survival. Two other moons of Jupiter—Ganymede and Callisto—also show some evidence of subsurface oceans, though the evidence is less strong and other considerations (primarily availability of energy) give them poorer prospects for harboring life. A fourth candidate for a life-bearing moon is Titan, which orbits Saturn and is the only moon in the solar system with a substantial atmosphere. Titan is far too cold for liquid water on its surface, but it apparently has liquid methane. It may also have liquid water deep underground. At the least, studies of Titan [Section 9.4] show that it has interesting organic chemistry, even if it does not have life. The fifth and sixth moons for which we have found evidence

SPECIAL TOPIC 1.1: HOW MANY PLANETS ARE THERE IN OUR SOLAR SYSTEM?

Everyone in college today grew up in an era when we were taught that there are nine planets in our solar system. Well, as you've probably heard, astronomers recently voted to demote Pluto from being a regular planet to being a "dwarf planet," leaving only eight official planets in our solar system. But don't despair if you're a Pluto fan. Counting planets is about words, not about science.

A little background: Pluto was discovered in 1930, at a time when astronomers were searching for an object thought to be causing slight perturbations to the orbits of other planets around the Sun. Neptune itself had been discovered in just this way in 1846, after scientists used perturbations in Uranus's orbit to predict the existence and location of an "eighth planet." Neptune was found as soon as astronomers pointed telescopes to the calculated position, which is why we sometimes say that Neptune was discovered with physics and mathematics, and only confirmed with a telescope.

Over the ensuing decades, a few scientists claimed to see ongoing orbital discrepancies and embarked on a search for a "ninth planet" that might be causing them. Pluto was found during this search, though about 12 full-moon-widths away from the predicted position. And though hailed as a planet upon discovery, its status gradually became suspect, as we learned that its mass is much less than 1% that of Earth, and its orbit is much more tilted and elongated than that of any of the other planets. Worse, reanalysis of past observations suggested that the claimed orbital discrepancies had simply been errors in measurement, making Pluto a solution to a nonexistent problem.

In 1951, by analyzing comet orbits, astronomer Gerard Kuiper (rhymes with "piper") predicted the existence of a "comet belt" beyond Neptune—now called the *Kuiper belt*—analogous to the asteroid belt between Mars and Jupiter. Telescope technology caught up with this prediction in the 1990s, and astronomers confirmed the existence of vast numbers of "Kuiper belt objects" orbiting the Sun in the general vicinity of Pluto. Scientifically, it became obvious that Pluto was much more like a large member of this group than like a small version of any

of the other eight planets. Still, as long as Pluto was the largest known Kuiper belt object, most astronomers were content to leave its planetary status intact.

Then, in July 2005, astronomer Michael Brown announced the discovery of a Kuiper belt object—given the name Eris after a Greek goddess who bred discord among humans—that is slightly larger than Pluto. This discovery was a fatal blow to the status quo of nine planets; after all, if Pluto is big enough to count as a planet, then Eris must belong to the club too. But what of the dozens of Kuiper belt objects only a little smaller than Pluto, and of future discoveries yet to be made? How would we draw the line between "planet" and "large Kuiper belt object"?

Official decisions on astronomical names and definitions rest with the International Astronomical Union (IAU), an organization made up of professional astronomers from around the world. At a contentious meeting in August 2006, the IAU voted to define a "planet" as an object that: (1) orbits a star (but is itself neither a star nor a moon); (2) is massive enough for its own gravity to give it a nearly round shape; and (3) has cleared the neighborhood around its orbit. By this definition, Pluto gets kicked out of the planetary club, because it shares its orbital neighborhood with other Kuiper belt objects. Pluto, Eris, and other objects that are large enough to be round but have not cleared their orbital neighborhoods are designated "dwarf planets." The myriad smaller objects that orbit the Sun, including most asteroids and comets, become members of a class called "small solar system bodies."

The important thing to remember is that this new definition was established by vote, making it politics, not science. The politics may yet change again (some astronomers have already voiced objections to the new definition), but Pluto will remain the same. Moreover, many people are likely to keep thinking of Pluto as a planet regardless of what the IAU says, much as most people still think of Europe and Asia as separate continents, although technically both are part of Eurasia. The bottom line is that while it's good to know the current official definitions, it's better to understand the science behind them.

of subsurface liquids—possibly including liquid water—are Saturn's moon Enceladus and Neptune's moon Triton.

Searching Among the Stars In terms of numbers, there are many more places to look for life on planets and moons around other stars than in our own solar system. However, the incredible distances to the stars [Section 3.2] make searches of these worlds much more difficult. All stars are so far away that we will need great leaps in technology to have any hope at all of sending spacecraft to study their planets up close; for example, with current spacecraft, the journey to even the nearest stars would take close to 100,000 years.

With visits out of reach, telescopic searches represent our only hope of finding life on extrasolar worlds. Current telescope technology is barely able to detect the existence of extrasolar planets; indeed, so far, we can detect only rather large planets, not planets as small as Earth. But the technology is advancing rapidly. We may be capable of detecting Earth-size planets by 2010 or sooner, and within a couple of decades we may have telescopes that are able to obtain crude pictures and spectra of planets and moons around other stars. As a result, one important area of research lies in trying to figure out the photographic or spectral "signatures" that would tell us we are looking at a world with life.

• Could aliens be searching for us?

So far we have talked about searching for life that is not searching for us; that is, life that we could identify only by seeing it with our spacecraft or telescopes. But if life really is common in the universe, there could be other places like Earth where life has evolved to become intelligent enough to be interested in searching for life beyond its home world. In that case, it is possible that other civilizations might actually be broadcasting signals that we could detect. The **search for extraterrestrial intelligence**, or **SETI**, which we'll discuss in Chapter 12, focuses on the search for such signals from alien civilizations (Figure 1.7). Although we don't know whether the search will meet with success, we can be sure that the unambiguous receipt of an alien message would be one of the most significant discoveries in human history—not to mention the fact that it would also probably answer many of our other questions about life in the universe.

1.4 THE NEW SCIENCE OF ASTROBIOLOGY

We have seen that the study of life in the universe is a multidisciplinary field of scientific research, involving scientists with training in many different specialties. Nevertheless, because it has become a prominent and important area of study, it would be good to give the science of life in the universe its own name. A number of different names are in use, including "exobiology" and "bioastronomy," but in this book we follow the lead of NASA and call it **astrobiology.** This term is meant to invoke the combination of astronomy (the study of the universe) and biology (the study of life), so *astrobiology* literally means "the study of life in the universe."

• How do we study the possibility of life beyond Earth?

Because astrobiology is a young science, scientists are still working to decide where to focus their research efforts. One major player in this effort

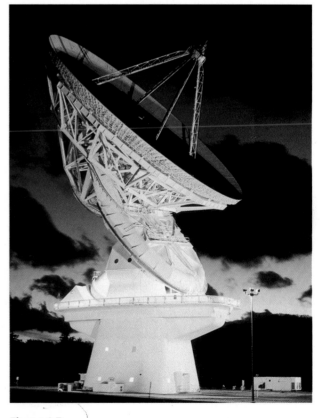

Figure 1.7

This 140-foot radio telescope in West Virginia was used in 1996 to search for signals from extraterrestrial civilizations.

is the NASA Astrobiology Institute, a collaboration involving scientists from NASA and more than a dozen other research institutions across the United States. Similar efforts are under way in other countries, including the United Kingdom, Sweden, France, Spain, Russia, and Australia. These collaborations are among the most interdisciplinary in any area of science, bringing together astronomers, biologists, geologists, chemists, and many others seeking to understand the prospects of finding life beyond Earth.

Although different groups focus their efforts in different areas, most astrobiology research is concentrated in the following three areas:

1. Studying the conditions conducive to the origin and ongoing existence of life.
2. Looking for such conditions on other planets in our solar system and around other stars.
3. Looking for the actual occurrence of life elsewhere.

Notice that astrobiology therefore includes much more than simply searching for extraterrestrial life or civilizations. At a fundamental level, astrobiology research seeks to reveal the connections between living organisms and the places where they reside. In this sense, finding no life (on Mars, for example) is just as significant a result as finding life, because either way we learn about the conditions that can lead to the presence of life, about how life evolves in conjunction with planets, and about whether life is likely to be rare or common throughout the universe.

In the rest of this book, we will focus on the same three areas listed above. After discussing the scientific context of the search in greater detail in Chapters 2 and 3, we'll turn our attention in Chapters 4 through 6 to the nature, origin, and evolution of life on Earth. This study of the history of life on our planet will help us understand the conditions under which we might expect to find life elsewhere. Next we'll discuss prospects for life elsewhere in our solar system in Chapters 7 through 10, and then discuss the prospects for finding life—including intelligent life—beyond our solar system in Chapters 11 through 13. Along the way, we'll learn what science can currently say about the future of life on Earth, we'll consider possible futures for our own species, and we'll discuss the philosophical implications of the search for—and potential discovery of—life beyond Earth.

◯ THE BIG PICTURE
Putting Chapter 1 into Context

In this chapter, we have offered a brief overview of the ideas we will cover in more depth in the rest of the book, primarily so that you will have a sense of what to expect in the rest of your study of life in the universe. As we will in every chapter, we conclude with a brief "big picture" recap of how these ideas fit into the overall goals of the scientific study of life in the universe:

- Despite the abundance of aliens in popular media, we don't yet have any convincing evidence for life beyond Earth. Nevertheless, current understanding of astronomy, planetary science, and biology gives us good reason to think that it is at least reasonable to imagine that life may be widespread, and the discovery of life of any kind would have profound significance to our understanding of life in the universe.

- It's conceivable that life may exist on any of several worlds in our own solar system, but it's extremely unlikely that any of this life is intelligent. However, we find many more possibilities when we consider life on planets or moons around other stars. And, through the search for extraterrestrial intelligence (SETI), it is even possible that we could receive a signal from an advanced civilization.

- The prospect that life may be common in the universe has given rise to the new science of astrobiology, an exciting and interdisciplinary topic of research that focuses both on understanding the possibility of finding life elsewhere and on the actual search for life beyond Earth.

SUMMARY OF KEY CONCEPTS

1.1 THE POSSIBILITY OF LIFE BEYOND EARTH

- **What are we searching for?**

The search for **extraterrestrial life** is in principle a search for *any* kind of life. However, the difficulty of clearly defining life means that it's easier to focus the search on life that is at least somewhat similar to life here on Earth. This still opens a wide range of possibilities, from bacteria-like microbes to complex plants and animals.

- **Is it reasonable to imagine life beyond Earth?**

People have long considered the possibility of life beyond Earth, but only recently have we been able to examine this possibility through the lens of science. While we have no evidence at this time of actual life beyond Earth, our scientific understanding of the possibilities makes it reasonable to think that life could exist elsewhere.

1.2 THE SCIENTIFIC CONTEXT OF THE SEARCH

- **How does astronomy help us understand the possibilities for extraterrestrial life?**

Astronomy tells us that we live on just a tiny planet orbiting one rather ordinary star in a vast cosmos, and that the same physical laws that operate here also operate throughout the universe. Together these ideas suggest that there could be many other worlds with life.

- **How does planetary science help us understand the possibilities for extraterrestrial life?**

Planetary science helps us understand both how common we should expect other worlds to be and the conditions under which we might expect those worlds to be capable of harboring life. Based on current understanding of how planets form, we expect planets to be common around other stars—an idea that has been confirmed by recent discoveries of **extrasolar planets**. By learning how planets work, we learn the conditions that might make a **habitable world**, meaning a world that has the basic necessities for life, even if it does not actually have life.

- **How does biology help us understand the possibilities for extraterrestrial life?**

With no life currently known to exist beyond Earth, we do not know whether biology should be common on worlds with habitable conditions. However, three lines of evidence suggest that life could be common: (1) The fact that life arose quickly on Earth suggests that it might occur on any world that has the "right" conditions. (2) We know from observations of meteorites and interstellar clouds that organic molecules are common throughout the galaxy, suggesting that we'll find them on many other worlds. (3) The fact that life on Earth survives even under some seemingly extreme conditions suggests that life is hardy enough to survive in many other places as well.

1.3 PLACES TO SEARCH

• Where should we search for life in the universe?

The search begins right here on Earth, as we seek to learn more about the life on our own planet. Elsewhere in our solar system we can search many planets and moons, but current understanding suggests that the most promising candidates for life are Mars and Jupiter's moon Europa. In the future, as telescope technology improves, we should be able to conduct telescopic searches for life around other stars.

• Could aliens be searching for us?

If life is common in the universe, civilizations might also be common, in which case other civilizations might be conducting their own searches and broadcasting signals of their existence. We look for such signals from alien civilizations through the **search for extraterrestrial intelligence**, or **SETI**.

1.4 THE NEW SCIENCE OF ASTROBIOLOGY

• How do we study the possibility of life beyond Earth?

The science of life in the universe, or astrobiology, focuses on three major areas: (1) studying the conditions conducive to the origin and ongoing existence of life; (2) looking for such conditions on other planets in our solar system and around other stars; and (3) looking for the actual occurrence of life elsewhere. Together, these studies should help us understand the connections between living organisms and the places where they reside.

EXERCISES AND PROBLEMS

REVIEW QUESTIONS
Short-Answer Questions Based on the Reading

1. Why are scientists interested in the possibility of life beyond Earth?

2. People have long been interested in life beyond Earth. What is different today that makes this possibility seem scientifically reasonable?

3. What do we mean by a *geocentric universe?* In general terms, contrast a geocentric view of the universe with our modern view of the universe.

4. What are *extrasolar planets?* In what way does their discovery make it seem more reasonable to imagine finding life elsewhere?

5. What do we mean by a *habitable* world? Does a habitable world necessarily have life?

6. What do we mean by the "universality" of physics and chemistry? Although we don't know yet whether biology is similarly universal, what evidence makes it seem that it might be?

7. Besides Earth, what worlds in our solar system seem most likely to have life? Why?

8. Could we actually detect life on an extrasolar planet and moon with current technology? Explain.

9. What is the *search for extraterrestrial intelligence* (SETI)?

10. What do we mean by *astrobiology?* What other terms are sometimes used to describe it? What are the major areas of research in astrobiology?

TEST YOUR UNDERSTANDING

Quick Quiz

Choose the best answer to each of the following. Explain your reasoning with one or more complete sentences.

11. An *extrasolar planet* is: (a) a planet that is larger than our Sun; (b) a planet that orbits a star other than our Sun; (c) a planet located in another galaxy.

12. A *habitable planet* is: (a) a planet that has oceans like the Earth; (b) a planet that has life of some kind; (c) a planet that may or may not have life, but that has environmental conditions under which it seems that life could arise or survive.

13. By a *geocentric* view of the universe, we mean: (a) the ancient idea that Earth resided at the center of the universe; (b) the idea that Earth is the only planet with life in the universe; (c) a view of the universe shaped by current understanding of geological science.

14. According to current scientific understanding, life on Earth: (a) was exceedingly improbable; (b) arose quite soon after conditions allowed it; (c) may have been inevitable, but took billions of years to arrive.

15. The correct order for the eight official planets in our solar system, from closest to farthest from the Sun, is: (a) Mercury, Venus, Earth, Mars, Saturn, Jupiter, Neptune, Uranus; (b) Mercury, Venus, Earth, Mars, Jupiter, Uranus, Neptune, Saturn; (c) Mercury, Venus, Earth, Mars, Jupiter, Saturn, Uranus, Neptune.

16. Today, the research known as SETI is conducted primarily by: (a) scanning the skies for signals from alien civilizations; (b) sending spacecraft to the planets; (c) using telescopes to observe extrasolar planets.

17. If we sent one of our current spacecraft to a nearby star (besides the Sun), the trip would take about: (a) a decade; (b) a century; (c) 100,000 years.

18. Scientists today are interested in searching for life on Mars because: (a) We see clear evidence of a past civilization on Mars. (b) Mars contains frozen water ice at its polar caps. (c) Evidence suggests that Mars had liquid water on its surface in the distant past.

19. Based on current evidence, the object in our solar system most likely to have a deep, subsurface ocean of liquid water is: (a) Mars; (b) Europa; (c) Titan.

20. Based on the way scientists view the study of astrobiology, failure to find life on any other world would mean: (a) The whole subject has been a waste of time. (b) We must have done something wrong, since life has to exist beyond Earth. (c) We have learned important lessons about the conditions that made life on Earth possible.

INVESTIGATE FURTHER
In-Depth Questions to Increase Your Understanding

Discussion Questions

21. *Aliens Among Us.* Take an informal poll of your friends or classmates. How many believe we have already been visited by aliens? On what basis do they hold their beliefs? How strong are their convictions on this issue? In light of your findings and what you've learned in this chapter, discuss whether public interest in aliens visiting Earth has any bearing on the scientific study of astrobiology.

22. *Conducting the Search.* Given the large number of possible places to look for life, how would you prioritize the search? In other words, where would you look first for life on other worlds in our own solar system, and how would you come up with a search strategy for other star systems? Explain your priorities and strategies clearly.

23. *Funding for Astrobiology.* Imagine that you are a member of Congress, so it is your job to decide how much government funding goes to research in astrobiology. What factors would influence your decision? Do you think you would increase or decrease such funding from the current level? Explain.

WEB PROJECTS

24. *Astrobiology News.* Go to NASA's Astrobiology home page and read some of the recent news from astrobiology research. Choose one recent news article, and write a one- to two-page summary of the research and how it fits into astrobiology research in general.

25. *The NASA Astrobiology Institute.* Go to the home page for the NASA Astrobiology Institute (NAI) and learn more about how it is organized and the type of research it supports. Also learn whether your school or any nearby institutions participate in the NAI. In one page or less, describe the NAI and its work and discuss the particular contributions of any institutions located near you.

26. *International Astrobiology.* Search the Web for information on astrobiology efforts outside the United States. Learn about the effort in one particular country or group of countries. What areas of research are emphasized? How do the researchers involved in the effort collaborate with other international astrobiology efforts? Write a one- to two-page report on your findings.

27. *The Search for Extraterrestrial Intelligence.* Go to the home page for the SETI Institute. Learn more about how SETI is funded and how the institute does its work. Summarize your findings in about one page.

2
The Science of Life in the Universe

LEARNING GOALS

2.1 THE ANCIENT DEBATE ABOUT LIFE BEYOND EARTH

- How did attempts to understand the sky start us on the road to science?
- Why did the Greeks argue about the possibility of life beyond Earth?

2.2 THE COPERNICAN REVOLUTION

- How did the Copernican revolution further the development of science?
- How did the Copernican revolution alter the ancient debate on extraterrestrial life?

2.3 THE NATURE OF MODERN SCIENCE

- How can we distinguish science from nonscience?
- What is a scientific theory?

2.4 THE PROCESS OF SCIENCE IN ACTION: UNDERSTANDING GRAVITY

- What is gravity?
- Do we really understand gravity?

All our science, measured against reality, is primitive and childlike—and yet is the most precious thing we have.
Albert Einstein

Extraterrestrial life may sound like a modern idea, but stories of life beyond Earth actually reach far back into ancient times. Many of these stories concerned mythical or supernatural beings, such as gods and others, living among the constellations, but some were not so different from the ideas we consider today. Nevertheless, the present-day search for life in the universe differs from all these ancient speculations in an important way: While ancient people could do little more than guess about the possibility of finding life elsewhere, we can now study this possibility with the powerful methods of modern science.

Given that we don't yet know of any life beyond Earth, you might wonder how we can make a science of life in the universe. The answer is that we use science to help us understand the conditions under which we might expect to find life, the likely characteristics of life elsewhere, and the methods we can use to search for it. Without this science, the search for life would be a shot in the dark. With it, we can focus our efforts in a way that will help us know where to look and how to interpret our results—whether or not we find life.

In this chapter, we will discuss how we moved from ancient speculation to the methods of modern science that we use to search for life in the universe. In doing so, we will learn about the nature of science and explore the question of how knowledge obtained through science differs from knowledge obtained in other ways.

2.1 THE ANCIENT DEBATE ABOUT LIFE BEYOND EARTH

More than 2,300 years ago, scholars of ancient Greece were already engaged in a lively debate over the possibility of life existing beyond Earth. Some scholars argued that there *must* be life elsewhere, while others, especially Aristotle, argued just the opposite. This impassioned debate may in some ways seem a historical curiosity, but the mere fact that it occurred tells us that a major change in human thinking was already under way.

Deeper in the past, our ancestors looked at the sky and attributed what they saw to the arbitrary actions of mythological beings, an idea still reflected in the fact that the planets carry the names of mythological gods. In contrast, the Greek scholars sought rational explanations for what they could observe in the universe around them. As far as we know, these Greek efforts marked the first attempts to understand the universe through methods closely resembling the ones we use in science today. Thus, if we want to understand how modern science works—and how we can use it to study the possibility of life beyond Earth—we must begin by peering more than two millennia into the past, to see how observations of the sky started humanity on the road to modern science and kindled interest in the question of whether the universe is ours alone.

Figure 2.1

This photograph, taken at Arches National Park with a 6-hour exposure, shows daily paths of stars in the sky. Notice that stars near the pole star (Polaris) make complete daily circles, while those farther from the pole star rise in the east and set in the west. Ancient people were quite familiar with patterns of motion like these.

• How did attempts to understand the sky start us on the road to science?

Imagine living in ancient times, looking up at the sky without the benefit of any of our modern knowledge. What would you see?

Every day, the Sun rises in the east and sets in the west, though its precise path through the sky varies with the seasons. Every night, the stars likewise move across the sky, with different constellations prominent at different times of year (Figure 2.1). The Moon goes through its monthly phases, from new to full and back again, while the planets gradually meander among the stars in seemingly mysterious ways. All the while, the ground beneath you feels steady and solid. It would be quite natural to assume—as did people of many early cultures—that Earth is a flat, motionless disk surmounted by a domelike sky across which the heavenly bodies move.

The story of how we progressed from this simple, intuitive view of Earth and the heavens to our modern understanding of Earth as a tiny planet in a vast cosmos is in many ways the story of the development of science itself. Our ancestors were surely curious about many aspects of the world around them, but astronomy held special interest. The Sun clearly plays a central role in our lives, governing daylight and darkness and the progression of the seasons. The Moon's connection to the tides would have been obvious to people living near the sea. The evident power of these celestial bodies probably explains why they attained prominent roles in many early religions and may be one reason why it seemed so important to know the sky. Careful observations of the sky also served practical needs by enabling ancient peoples to keep track of the time and the seasons—crucial skills for agricultural societies.

As civilizations rose, astronomical observations became more careful and elaborate. In some cases, the results were recorded in writing. The ancient Chinese kept detailed records of astronomical observations beginning some 5,000 years ago. By about 2,500 years ago, written records allowed the Babylonians (in the region of modern-day Iraq) to predict eclipses with great success. Halfway around the world (and a few centuries later), the Mayans of Central America independently developed the same ability.

These ancient, recorded observations of astronomy represent databases of facts—the raw material of science. But in most cases for which we have historical records, it appears that these facts were never put to use for anything beyond meeting immediate religious and practical needs. One clear exception was ancient Greece, where scholars began to use these facts in an attempt to understand the architecture of the cosmos.

Early Greek Science   Greece gradually rose as a power in the Middle East beginning around 800 B.C., and was well established by about 500 B.C. Its geographical location placed it at a crossroads for travelers, merchants, and armies of northern Africa, Asia, and Europe. Building on the diverse ideas brought forth by the meeting of these many cultures, ancient Greek philosophers began to move human understanding of nature from the mythological to the rational.

We generally trace the origin of Greek science to the philosopher Thales (c. 624–546 B.C.; pronounced "*THAY-lees*"). Among his many accomplishments, Thales was the first person known to have addressed the question, "What is the universe made of?" without resorting to supernatural explanations. His own guess—that the universe fundamentally consisted of water and that Earth was a flat disk on an infinite ocean—was not widely accepted even in his own time, but his mere asking of the question helped set the stage for all later science. For the first time, someone had suggested that the world was inherently understandable and not just the result of arbitrary or incomprehensible events.

The scholarly tradition begun by Thales was carried on by many others, perhaps most famously by Plato (428–348 B.C.) and his student Aristotle (384–322 B.C.). Each Greek philosopher introduced new ideas, sometimes in contradiction to the ideas of others. None of these ideas rose quite to the level of modern science, primarily because the Greeks tended to rely more on pure thought and intuition than on observations or experimental tests. Nevertheless, with hindsight we can see at least three major innovations in Greek thought that helped pave the way for modern science.

First, the Greek philosophers developed a tradition of trying to understand nature without resorting to supernatural explanations. For example, although earlier Greeks might simply have accepted that the Sun moves across the sky because it is pulled by the god Apollo in his chariot—an idea whose roots were already lost in antiquity—the philosophers sought a natural explanation that caused them to speculate anew about the construction of the heavens. They were free to think creatively because they were not simply trying to prove preconceived ideas, and they recognized that new ideas should be open to challenge. As a result, they often worked communally, debating and testing each other's ideas. This tradition of challenging virtually every new idea remains one of the distinguishing features of scientific work today.

Second, the Greeks developed mathematics in the form of geometry. They valued this discipline for its own sake, and they understood its power, using geometry to solve both engineering and scientific problems. Without their mathematical sophistication, they would not have gone far in their attempts to make sense of the cosmos. Like the Greek tradition of challenging ideas, the use of mathematics to help explore the implications of new ideas remains an important part of modern science.

Third, while much of their philosophical activity consisted of subtle debates grounded only in thought and thus was not scientific in the modern sense, the Greeks also saw the power of reasoning from observations. They understood that an explanation about the world could not be right if it disagreed with observed facts. This willingness to discard explanations that simply don't work is also a crucial part of modern science.

The Geocentric Model

Perhaps the greatest Greek contribution to science came from the way they synthesized all three innovations into the idea of creating **models** of nature, an idea that is still central to modern science. Scientific models differ somewhat from the models you may be familiar with in everyday life. In our daily lives, we tend to think of models as miniature physical representations, such as model cars or airplanes. In contrast, a scientific model is a conceptual representation whose purpose is to explain and predict observed phenomena. For example, a model of Earth's climate uses logic, mathematics, and known physical laws in an attempt to represent the way in which the climate works. Its purpose is to explain and predict climate changes, such as the changes that may occur with global warming. Just as a model airplane does not faithfully represent every aspect of a real airplane, a scientific model may not fully explain all our observations of nature. Nevertheless, even the failings of a scientific model can be useful, because they often point the way toward building a better model.

north celestial pole

Figure 2.2

The early Greek geocentric model consisted of a central Earth surrounded by the celestial sphere, which is shown here marked with modern constellation borders and a few reference points and circles. We still use the idea of the celestial sphere when making astronomical observations, but we no longer imagine that it reflects reality.

TABLE 2.1 *The Seven Days of the Week and the Astronomical Objects They Honor*

The correspondence between objects and days is easy to see in French and Spanish. In English, the correspondence becomes clear when we look at the names of the objects used by the Teutonic tribes who lived in the region of modern-day Germany.

Object	Teutonic Name	English	French	Spanish
Sun	Sun	Sunday	dimanche	domingo
Moon	Moon	Monday	lundi	lunes
Mars	Tiw	Tuesday	mardi	martes
Mercury	Woden	Wednesday	mercredi	miércoles
Jupiter	Thor	Thursday	jeudi	jueves
Venus	Fria	Friday	vendredi	viernes
Saturn	Saturn	Saturday	samedi	sábado

THINK ABOUT IT Conceptual models aren't just important in science; they often affect day-to-day policy decisions. For example, government economists use models to predict future tax revenues, and politicians use the results in making tax-and-spending decisions. Describe at least two other cases in which models affect our daily lives.

In astronomy, the Greeks constructed conceptual models of the universe in an attempt to explain what they observed in the sky, an effort that quickly led them past simplistic ideas of a flat Earth under a dome-shaped sky to a far more sophisticated view of the cosmos. One of the first crucial steps was taken by a student of Thales, Anaximander (c. 610–547 B.C.). In an attempt to explain the way the sky appears to turn around the pole star each day (see Figure 2.1), Anaximander suggested that the heavens must form a complete sphere—the **celestial sphere**—around Earth (Figure 2.2). Moreover, based on how the sky varies with latitude, he realized that Earth's surface must be curved rather than flat, leading him to guess that Earth is cylinder-shaped.

We do not know precisely when other Greeks first realized that Earth is round, but this idea was being taught as early as about 500 B.C. by the famous mathematician Pythagoras (c. 560–480 B.C.). He and his followers imagined Earth as a sphere floating at the center of the celestial sphere. Much of their motivation for adopting a spherical Earth probably was philosophical: The Pythagoreans had a mystical interest in mathematical perfection, and they considered a sphere to be geometrically perfect. More than a century later, Aristotle cited observations of Earth's curved shadow on the Moon during lunar eclipses as evidence for a spherical Earth. Thus, Greek philosophers adopted a **geocentric model** of the universe (recall that *geocentric* means "Earth-centered"), with a spherical Earth at the center of a great celestial sphere.

Incidentally, this shows the error of the widespread myth that Columbus proved Earth to be round when he sailed to America in 1492. Not only were scholars of the time well aware of Earth's round shape, they even knew Earth's approximate size: Earth's circumference was first measured (fairly accurately) in about 240 B.C. by the Greek scientist Eratosthenes. In fact, a likely reason why Columbus had so much difficulty finding a sponsor for his voyages was that he tried to argue a point on which he was dead wrong: He claimed the distance by sea from western Europe to eastern Asia to be much less than the scholars knew it to be. Indeed, when he finally found a patron in Spain and left on his journey, he was so woefully underprepared that the voyage would almost certainly have ended in disaster if the Americas hadn't stood in his way.

The Mystery of Planetary Motion If you watch the sky closely, you'll notice that while the patterns of the constellations seem not to change, the Sun, the Moon, and the five planets visible to the naked eye (Mercury, Venus, Mars, Jupiter, and Saturn) gradually move among the constellations from one day to the next. Indeed, the word *planet* comes from the Greek for "wanderer," and it originally referred to the Sun and Moon as well as to the five visible planets. Our seven-day week is directly traceable to the fact that seven "planets" are visible in the heavens (Table 2.1).

The wanderings of these objects convinced the Greek philosophers that there had to be more to the heavens than just a single sphere surrounding Earth. The Sun and Moon each move steadily through the constellations, with the Sun completing a circuit around the celestial sphere each year and the Moon completing each circuit in about a month (think "moonth"). The Greeks could account for this motion by adding separate spheres for the Sun and Moon, each nested within the sphere of the stars, and allowing these spheres to turn at different rates from the sphere of the stars. But the five visible planets posed a much greater mystery.

If you observe the position of a planet (such as Mars or Jupiter) relative to the stars over a period of many months, you'll find not only that its speed and brightness vary considerably but that its direction of motion sometimes also changes. While the planets usually move eastward relative to the constellations, sometimes they reverse course and go backward (Figure 2.3). These periods of **apparent retrograde motion** (*retrograde* means "backward") last from a few weeks to a few months, depending on the planet.

This seemingly erratic planetary motion was not so easy to explain with rotating spheres, especially because the Greeks generally accepted a notion of "heavenly perfection," enunciated most clearly by Plato, which demanded that all heavenly objects move in perfect circles. How could a planet sometimes go backward when moving in a perfect circle? The Greeks came up with a number of ingenious ideas that preserved Earth's central position, culminating with a complex model of planetary motion described by the astronomer Ptolemy (c. A.D. 100–170; pronounced "TOL-e-mee"); we refer to Ptolemy's model as the **Ptolemaic model** to distinguish it from earlier geocentric models. This model reproduced retrograde motion by having planets move around Earth on small circles that turned around larger circles. A planet following this circle-on-circle motion traces a loop as seen from Earth, with the backward portion of the loop mimicking apparent retrograde motion (Figure 2.4).

The circle-on-circle motion may itself seem somewhat complex, but Ptolemy found that he also had to use many other mathematical tricks, including putting some of the circles off-center, to get his model to agree with observations. Despite all this complexity, he achieved remarkable success: His model could correctly forecast future planetary positions to within a few degrees of arc—roughly equivalent to holding your hand at arm's length against the sky. Indeed, the Ptolemaic model generally worked so well that it remained in use for the next 1,500 years. When Arabic scholars translated Ptolemy's book describing the model in around A.D. 800, they gave it the title *Almagest*, derived from words meaning "the greatest compilation."

An Alternate Model

In about 260 B.C., the Greek scientist Aristarchus (c. 310–230 B.C.) offered a radical departure from the conventional wisdom: He suggested that Earth goes around the Sun, rather than vice versa. Little of Aristarchus's work survives to the present day, so we do not know exactly how he came up with his Sun-centered idea. We do know that he made measurements that convinced him that the Sun is much larger than Earth, so perhaps he simply concluded that it was more natural for the smaller Earth to orbit the larger Sun. In addition, he almost certainly recognized that a Sun-centered system offers a much more natural explanation for apparent retrograde motion.

You can see how the Sun-centered system explains retrograde motion with a simple demonstration (Figure 2.5a). Find an empty area (such as a sports field or a big lawn), and mark a spot in the middle to represent the Sun. You can represent Earth, walking counterclockwise around the Sun, while a friend represents a more distant planet (such as Mars,

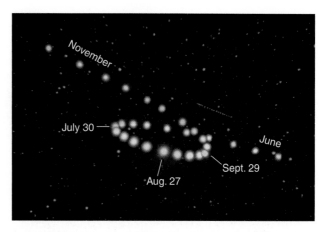

Figure 2.3

This composite of 29 photographs, each taken at 5-to-8-day intervals, shows Mars between early June and late November 2003; notice how it usually moves eastward (left) relative to the stars, but reverses course during its apparent retrograde motion. (The white dots in a line just right of center are the planet Uranus, which by coincidence was in the same part of the sky.)

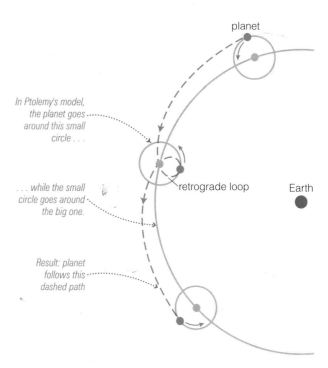

Figure 2.4

This diagram shows how the Ptolemaic model accounted for apparent retrograde motion. Each planet is assumed to move around a small circle that turns on a larger circle. The resulting path (dashed) includes a loop in which the planet goes backward as seen from Earth.

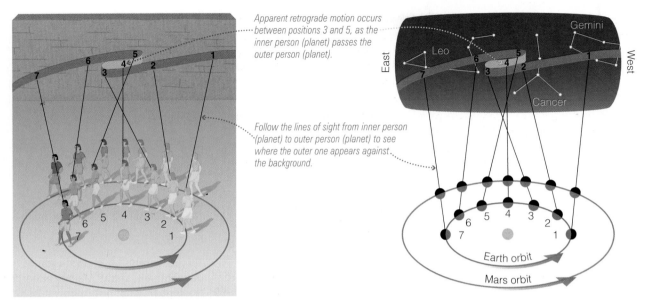

Apparent retrograde motion occurs between positions 3 and 5, as the inner person (planet) passes the outer person (planet).

Follow the lines of sight from inner person (planet) to outer person (planet) to see where the outer one appears against the background.

Gemini

Leo

Cancer

East   West

Earth orbit

Mars orbit

a The retrograde motion demonstration: Watch how your friend (in red) usually appears to you (in blue) to move forward against the background of the building in the distance but appears to move backward as you catch up to and pass him or her in your "orbit."

b This diagram shows how the idea from the demonstration applies to planets. Follow the lines of sight from Earth to Mars in numerical order. Notice that Mars appears to move westward relative to the distant stars as Earth passes it by in its orbit (roughly from points 3 to 5 in the diagram).

Figure 2.5   Interactive Figure

Apparent retrograde motion—the occasional "backward" motion of the planets relative to the stars—has a simple explanation in a Sun-centered solar system.

Jupiter, or Saturn) by walking counterclockwise around the Sun at a greater distance. Your friend should walk more slowly than you, because more distant planets orbit the Sun more slowly. As you walk, watch how your friend appears to move relative to buildings or trees in the distance. Although both of you always walk in the same direction around the Sun, your friend will appear to move backward against the background during the part of your "orbit" at which you catch up to and pass him or her. To understand the apparent retrograde motions of Mercury and Venus, which are closer to the Sun than is Earth, simply switch places with your friend and repeat the demonstration. The demonstration applies to all the planets. For example, because Mars takes about 2 years to orbit the Sun (actually, 1.88 years), it covers about half its orbit during the 1 year in which Earth makes a complete orbit. If you trace lines of sight from Earth to Mars from different points in their orbits, you will see that the line of sight usually moves eastward relative to the stars but moves westward during the time when Earth is passing Mars in its orbit (Figure 2.5b). Like your friend in the demonstration, Mars never actually changes direction. It only appears to change direction from our perspective on Earth.

Despite the elegance of this Sun-centered model for the universe, Aristarchus had little success in convincing his contemporaries to accept it. Some of the reasons for this rejection were purely philosophical and not based on any hard evidence. However, at least one major objection was firmly rooted in observations: Aristarchus's idea seemed inconsistent with observations of stellar positions in the sky.

To understand the inconsistency, imagine what would happen if you placed the Sun rather than Earth at the center of the celestial sphere, with Earth orbiting the Sun some distance away. In that case, Earth would be closer to different portions of the celestial sphere at different times of year. When we were closer to a particular part of the sphere, the stars on that part of the sphere would appear more widely separated than they would when we were farther from that part of the sphere, just as the spacing between the two headlights on a car looks greater when you are closer to the car. This would create annual shifts in the separations of stars—but the Greeks observed no such shifts. They knew that there were only two possible ways to account for the lack of an observed shift: Either Earth was at the center of the universe or the stars were so far away as to make the shift undetectable by eye. To most Greeks, it seemed unreasonable to imagine that the stars could be *that* far away, which inevitably led them to conclude that Earth must hold a central place.

This argument about stellar shifts still holds when we allow for the reality that stars lie at different distances rather than all on the same sphere: As Earth orbits the Sun, we look at particular stars from slightly different positions at different times of year, causing the positions of nearby stars to shift slightly relative to more distant stars (Figure 2.6). Although such shifts are much too small to measure with the naked eye—because stars really are very far away [Section 2.2]—they are easily detectable with modern telescopes. Thus, these annual shifts in stellar position, called **stellar parallax**, now provide concrete proof that Earth really does go around the Sun.

The Roots of Modern Science

Although the Greeks ultimately rejected the correct idea—that Earth orbits the Sun—we have seen that they did so for reasons that made good sense at the time. Not all of their reasons would pass the test of modern science; for example, their preference for motion in perfect circles came only from their cultural ideas of aesthetics and not from any actual data. But they also went to a lot of effort to ensure that their models were consistent with observations, and in that way they laid the foundation of modern science. And while Aristarchus may not have won the day in his own time, his idea remained alive in books. Some 1,800 years after he first proposed it, Aristarchus's Sun-centered model apparently came to the attention of a Polish astronomer named Nicholas Copernicus (1473–1543), who took the idea and ran with it in a way that led directly to the development of modern science. We'll return to this story shortly.

• Why did the Greeks argue about the possibility of life beyond Earth?

Almost from the moment that Thales asked his question of what the universe was made of, the Greeks realized that the answer would have bearing on the possibility of life elsewhere. This might seem surprising in light of their geocentric beliefs, because they didn't think of the planets or stars as worlds in the way we think of them today. Instead, the Greeks generally considered the "world" to include both Earth and the heavenly spheres that they imagined to surround it, and they were at least open to the possibility that other such "worlds" might exist.

As we noted earlier, Thales guessed that the world consisted fundamentally of water, with Earth floating on an infinite ocean, but his student Anaximander imagined a more mystical element that he called *apeiron,* meaning "infinite." Anaximander suggested that all material things arose from and returned to the apeiron, which allowed him to imagine that worlds might be born and die repeatedly through eternal time. So even though he made no known claim of life existing elsewhere in the present, Anaximander essentially suggested that other Earths and other beings might exist at other times.

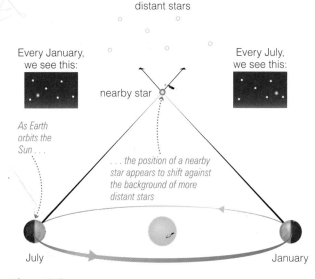

Figure 2.6

If Earth orbits the Sun, then over the course of each year we should see nearby stars shift slightly back and forth relative to more distant stars (*stellar parallax*). The Greeks could not detect any such shift, and used this fact to argue that Earth must be at the center of the universe. Today, we *can* detect stellar parallax with telescopic observations, proving that Earth does orbit the Sun. (This figure is greatly exaggerated; the actual shift is far too small to detect with the naked eye.)

Other Greeks took the debate in a slightly different direction, and eventually a consensus emerged in favor of the world's having been built from four elements: fire, water, earth, and air. However, two distinct schools of thought emerged concerning the nature and extent of these elements:

- The *atomists* held that both Earth and the heavens were made from an infinite number of indivisible atoms of each of the four elements.

- The *Aristotelians* (after Aristotle) held that the four elements—not necessarily made from atoms—were confined to the realm of Earth, while the heavens were made of a distinct fifth element, often called the *aether* (or *ether*) or the *quintessence* (literally, "the fifth essence").

The differences in the two schools of thought led to two fundamentally different conclusions about the possibility of extraterrestrial life.

THINK ABOUT IT Look up the words *ethereal* and *quintessence* in the dictionary. How do their definitions relate to the Aristotelian idea that the heavens were composed of an element distinct from the elements of Earth? Explain.

The atomist doctrine was developed largely by Democritus (c. 470–380 B.C.), and his views show how the idea led almost inevitably to belief in extraterrestrial life. Democritus argued that the world—both Earth and the heavens—had been created by the random motions of infinite atoms. Because this idea held that the number of atoms was infinite, it was natural to assume that the same processes that created our world could also have created others. This philosophy on life beyond Earth was clearly described in the following quotation from a later atomist, Epicurus (341–270 B.C.):

> There are infinite worlds both like and unlike this world of ours ... we must believe that in all worlds there are living creatures and plants and other things we see in this world.*

Aristotle had a different view. He believed that each of the four elements had its own natural motion and place. For example, he believed that the element earth moved naturally toward the center of the universe, an idea that offered an explanation for the Greek assumption that Earth resided in a central place. The element fire, he claimed, naturally rose away from the center, which explained why flames jut upward into the sky. These incorrect ideas about physics, which were not disproved until the time of Galileo and Newton almost 2,000 years later, caused Aristotle to reject the atomist idea of many worlds. If there was more than one world, there would be more than one natural place for the elements to go, which would be a logical contradiction. Aristotle concluded:

> The world must be unique.... There cannot be several worlds.

Interestingly, Aristotle's philosophies were not particularly influential until many centuries after his death. His books were preserved and valued—in particular, by Islamic scholars of the late first millennium—but they were unknown in Europe until they were translated into Latin

---

*From Epicurus's "Letter to Herodotus"; the authors thank David Darling for finding this quotation and the one from Aristotle, both of which appear in Darling's book *The Extraterrestrial Encyclopedia*, Three Rivers Press, 2000.

in the twelfth and thirteenth centuries. St. Thomas Aquinas (1225–1274) integrated Aristotle's philosophy into Christian theology. At this point, the contradiction between the Aristotelian notion of a single world and the atomist notion of many worlds became a subject of great concern to Christian theologians. Moreover, because the atomist view held that our world came into existence through random motions of atoms, and hence without the need for any intelligent Creator, atomism became associated with atheism. The debate about extraterrestrial life thereby became intertwined with debates about religion. Even today, the theological issues are not fully settled, and echoes of the ancient Greek debate between the atomists and the Aristotelians still reverberate in our time.

2.2 THE COPERNICAN REVOLUTION

Greek ideas gained great influence in the ancient world, in large part because the Greeks proved to be as adept at politics and war as they were at philosophy. In about 330 B.C., Alexander the Great began a series of conquests that expanded the Greek Empire throughout the Middle East. Alexander had a keen interest in science and education, perhaps because he grew up with Aristotle as his personal tutor. Alexander established the city of Alexandria in Egypt, which soon became home to the greatest library the world had ever seen. The Library of Alexandria remained the world's preeminent center of research for some 700 years. At its peak, the library may have held more than a half million books, all handwritten on papyrus scrolls. When the library was finally destroyed during a time of anti-intellectual fervor in the fifth century A.D., most of the ancient Greek writings were lost forever.

Much more would have been lost if not for the rise of a new center of intellectual achievement in Baghdad (in present-day Iraq). While European civilization fell into the Dark Ages, scholars of the new religion of Islam sought knowledge of mathematics and astronomy in hopes of better understanding the wisdom of Allah. The Islamic scholars translated and thereby saved many of the remaining ancient Greek works. Building on what they learned from the Greek manuscripts, they went on to develop the mathematics of algebra as well as many new instruments and techniques for astronomical observation.

The Islamic world of the Middle Ages was in frequent contact with Hindu scholars from India, who in turn brought ideas and discoveries from China. Hence, the intellectual center in Baghdad achieved a synthesis of the surviving work of the ancient Greeks, the Indians, the Chinese, and the contributions of its own scholars. This accumulated knowledge spread throughout the Byzantine Empire (the eastern part of the former Roman Empire). When the Byzantine capital of Constantinople (modern-day Istanbul) fell in 1453, many Eastern scholars headed west to Europe, carrying with them the knowledge that helped ignite the European Renaissance. The stage was set for a dramatic rethinking of humanity and our place in the universe.

• How did the Copernican revolution further the development of science?

In 1543, Nicholas Copernicus published *De Revolutionibus Orbium Coelestium* ("Concerning the Revolutions of the Heavenly Spheres"), launching what we now call the **Copernican revolution**. In his book, Copernicus revived Aristarchus's radical suggestion of a Sun-centered solar system and described the idea with enough mathematical detail to make it a valid competitor to the Earth-centered, Ptolemaic model. Over the next century and a half, philosophers and scientists (who were often one and the same) debated and tested the Copernican idea. Many of the ideas that now form the foundation of modern science first arose as this debate played out. Indeed, the Copernican revolution had such a profound impact on philosophy that we cannot understand modern science without first understanding the key features of this revolution.

Copernicus—The Revolution Begins

By the time of Copernicus's birth in 1473, tables of planetary motion based on the Ptolemaic model had become noticeably inaccurate. However, few people were willing to undertake the difficult calculations required to revise the tables. Indeed, the best tables available were already two centuries old, having been compiled under the guidance of the Spanish monarch Alphonso X (1221–1284). Commenting on the tedious nature of the work involved, the monarch is said to have complained that "If I had been present at the creation, I would have recommended a simpler design for the universe."

Copernicus began studying astronomy in his late teens. He soon became aware of the inaccuracies of the Ptolemaic predictions and began a quest for a better way to predict planetary positions. He adopted Aristarchus's Sun-centered idea, probably because he was drawn to its simple explanation for the apparent retrograde motion of the planets (see Figure 2.5). As he worked out the mathematical details of his model, Copernicus also discovered simple geometric relationships that allowed him to calculate each planet's orbital period around the Sun and its relative distance from the Sun in terms of Earth–Sun distance. The success of his model in providing a geometric layout for the solar system further convinced him that the Sun-centered idea must be correct. Despite his own confidence in the model, Copernicus was hesitant to publish his work, fearing that the idea of a moving Earth would be considered absurd.* However, he discussed his system with other scholars, including high-ranking officials of the Church, who urged him to publish a book. Copernicus saw the first printed copy of his book on the same day that he died— May 24, 1543.

Publication of the book spread the Sun-centered idea widely, and many scholars were drawn to its aesthetic advantages. Nevertheless, the Copernican model gained relatively few converts over the next 50 years, and for a good reason: It didn't work all that well. The primary problem was that while Copernicus had been willing to overturn Earth's central place in the cosmos, he had held fast to the ancient belief that heavenly motion must occur in perfect circles. This incorrect assumption forced him to add numerous complexities to his system (including circles on circles much like those used by Ptolemy) to get it to make any reasonable predictions. In the end, his complete model was no more accurate and no less complex than the Ptolemaic model, and few people were willing to throw out thousands of years of tradition for a new model that worked just as poorly as the old one.

---

*Indeed, in the Preface of *De Revolutionibus,* Copernicus offered a theological defense of the Sun-centered idea: "Behold, in the middle of the universe resides the Sun. For who, in this most beautiful Temple, would set this lamp in another or a better place, whence to illumine all things at once?"

Tycho—A New Standard in Observational Data   Part of the difficulty faced by astronomers who sought to improve either the Ptolemaic or the Copernican model was a lack of quality data. The telescope had not yet been invented, and existing naked-eye observations were not particularly accurate. In the late 1500s, Danish nobleman Tycho Brahe (1546–1601), usually known simply as Tycho (commonly pronounced "*TIE-koe*"), set about correcting this problem.

Tycho was an eccentric genius who, at age 20, had lost part of his nose in a sword fight with another student over who was the better mathematician. Taking advantage of his royal connections, he built large naked-eye observatories that worked much like giant protractors, and over a period of three decades he used them to measure planetary positions to within 1 minute of arc ($\frac{1}{60}$ of $1°$)—which is less than the thickness of a fingernail held at arm's length.

Kepler—A Successful Model of Planetary Motion   Tycho never came up with a fully satisfactory explanation for his observations (though he made a valiant attempt), but he found someone else who did. In 1600, he hired a young German astronomer named Johannes Kepler (1571–1630). Kepler and Tycho had a strained relationship,* but in 1601, as he lay on his deathbed, Tycho begged Kepler to find a system that would make sense of his observations so "that it may not appear I have lived in vain."

Kepler was deeply religious and believed that understanding the geometry of the heavens would bring him closer to God. Like Copernicus, he believed that planetary orbits should be perfect circles, so he worked diligently to match circular motions to Tycho's data. After years of effort, he found a set of circular orbits that matched most of Tycho's observations quite well. Even in the worst cases, which were for the planet Mars, Kepler's predicted positions differed from Tycho's observations by only about 8 arcminutes.

Kepler surely was tempted to ignore these discrepancies and attribute them to errors by Tycho. After all, 8 arcminutes is barely one-fourth the angular diameter of the full moon. But Kepler trusted Tycho's careful work. The small discrepancies finally led Kepler to abandon the idea of circular orbits—and to find the correct solution to the ancient riddle of planetary motion. About this event, Kepler wrote,

> *If I had believed that we could ignore these eight minutes [of arc], I would have patched up my hypothesis accordingly. But, since it was not permissible to ignore, those eight minutes pointed the road to a complete reformation in astronomy.*

Kepler's decision to trust the data over his preconceived beliefs marked an important transition point in the history of science. Once he abandoned perfect circles, he was free to try other ideas and he soon hit on the correct one: Planetary orbits take the shapes of the special types of ovals known as *ellipses.* He then used his knowledge of mathematics to put his new model of planetary motion on a firm footing, expressing the key features of the model with what we now call **Kepler's laws of planetary motion**:

- **Kepler's first law:** *The orbit of each planet about the Sun is an ellipse with the Sun at one focus* (Figure 2.7). In essence, this law tells us that

---

*For a particularly moving version of the story of Tycho and Kepler, see *Cosmos,* by Carl Sagan, Episode 3.

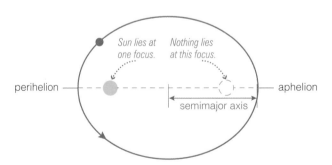

Figure 2.7   Interactive Figure

Kepler's first law states that the orbit of each planet about the Sun is an ellipse with the Sun at one focus. The ellipse shown here is more "stretched out" than the orbits of planets in our solar system, most of which are *almost* (but not quite!) perfect circles.

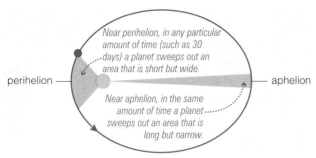

Near perihelion, in any particular amount of time (such as 30 days) a planet sweeps out an area that is short but wide.

Near aphelion, in the same amount of time a planet sweeps out an area that is long but narrow.

perihelion ⎯ ⎯ aphelion

The areas swept out in 30-day periods are all equal.

Figure 2.8 Interactive Figure

Kepler's second law tell us that a planet sweeps out equal areas in equal times as it orbits the Sun, which means it moves fastest near perihelion and slowest near aphelion.

This zoomout box makes it easier to see the data points for the inner planets.

The straight line tells us that the square of each planet's orbital period equals the cube of its average distance from the Sun.

Figure 2.9

This graph shows that Kepler's third law ($p^2 = a^3$) does indeed hold true; for simplicity, the graph shows only the planets known in Kepler's time.

a planet's distance from the Sun varies during its orbit. It is closest at the point called **perihelion** (from the Greek for "near the Sun") and farthest at the point called **aphelion** (from the Greek for "away from the Sun"). The *average* of a planet's perihelion and aphelion distances is the length of its **semimajor axis**. We will refer to the semimajor axis simply as the planet's average distance from the Sun.

- **Kepler's second law:** *As a planet moves around its orbit, it sweeps out equal areas in equal times.* As shown in Figure 2.8, this means the planet moves a greater distance when it is near perihelion than it does in the same amount of time near aphelion, which also means it moves faster when it is nearer to the Sun and slower when it is farther from the Sun.

- **Kepler's third law:** *More distant planets orbit the Sun at slower average speeds, obeying the precise mathematical relationship*

$$p^2 = a^3$$

where p is the planet's orbital period in years and a is its average distance (semimajor axis) from the Sun in astronomical units; one **astronomical unit** (**AU**) is defined as Earth's average distance from the Sun, or about 149.6 million kilometers. Figure 2.9 shows the $p^2 = a^3$ law graphically.

Kepler published his first two laws in 1610 and his third in 1618. Together, they made a model that could predict planetary positions with far greater accuracy than Ptolemy's Earth-centered model. Indeed, Kepler's model has worked so well that we now see it not just as an abstract idea, but as something that reveals a deep, underlying truth about planetary motion.

Galileo—Answering the Remaining Objections   The success of Kepler's laws in matching Tycho's data provided strong evidence in favor of Copernicus's placement of the Sun, rather than Earth, at the center of the solar system. Nevertheless, many scientists still voiced reasonable objections to the Copernican view. There were three basic objections, all rooted in the 2,000-year-old beliefs of Aristotle:

- First, Aristotle had held that Earth could not be moving because, if it were, objects such as birds, falling stones, and clouds would be left behind as Earth moved along its way.

- Second, the idea of noncircular orbits contradicted the view that the heavens—the realm of the Sun, Moon, planets, and stars—must be perfect and unchanging.

- Third, no one had detected the stellar parallax that should occur if Earth orbits the Sun.

Galileo Galilei (1564–1642), nearly always known by only his first name, answered all three objections.

Galileo defused the first objection with experiments that almost single-handedly overturned the Aristotelian view of physics. In particular, he used experiments with rolling balls to demonstrate that a moving object remains in motion *unless* a force acts to stop it (an idea now codified in Newton's first law of motion). This insight explained why objects that share Earth's motion through space—such as birds, falling stones, and clouds—

should *stay* with Earth rather than falling behind as Aristotle had argued. This same idea explains why passengers stay with a moving airplane even when they leave their seats.

The notion of heavenly perfection was already under challenge by Galileo's time, because Tycho had observed a supernova and proved that comets lie beyond the Moon; these observations showed that the heavens *do* sometimes undergo change. But Galileo drove the new idea home after he built a telescope in late 1609.* Through his telescope, Galileo saw sunspots on the Sun, which were considered "imperfections" at the time. He also used his telescope to prove that the Moon has mountains and valleys like the "imperfect" Earth by noticing the shadows cast near the dividing line between the light and dark portions of the lunar face (Figure 2.10). If the heavens were not perfect, then the idea of elliptical orbits (as opposed to "perfect" circles) was not so objectionable.

The third objection—the absence of observable stellar parallax—had been a particular concern of Tycho's. Based on his estimates of the distances of stars, Tycho believed that his naked-eye observations were sufficiently precise to detect stellar parallax if Earth did in fact orbit the Sun. Refuting Tycho's argument required showing that the stars were more distant than Tycho had thought and therefore too distant for him to have observed stellar parallax. Although Galileo didn't actually prove this fact, he provided strong evidence in its favor. For example, he saw with his telescope that the Milky Way resolved into countless individual stars. This discovery helped him argue that the stars were far more numerous and more distant than Tycho had believed.

In hindsight, the final nails in the coffin of the Earth-centered universe came with two of Galileo's earliest discoveries through the telescope. First, he observed four moons clearly orbiting Jupiter, not Earth. Soon thereafter, he observed that Venus goes through phases in a way that proved that it must orbit the Sun and not Earth (Figure 2.11).

---

*Contrary to a common belief, Galileo did *not* invent the telescope. It was invented in 1608 by Hans Lippershey. However, Galileo took what was little more than a toy and turned it into a scientific instrument.

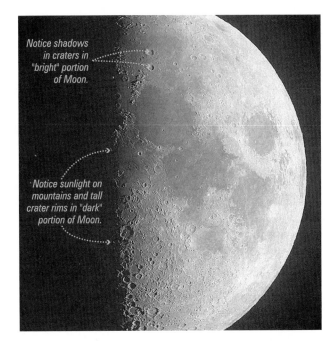

Figure 2.10

The shadows cast by mountains and crater rims near the dividing line between the light and dark portions of the lunar face prove that the Moon's surface is not perfectly smooth.

Figure 2.11

Galileo's telescopic observations of Venus proved that it orbits the Sun rather than Earth.

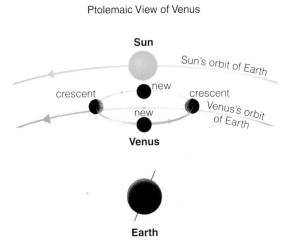

Ptolemaic View of Venus

a In the Ptolemaic system, Venus orbits Earth, moving around a small circle on its larger orbital circle; the center of the small circle lies on the Earth-Sun line. Thus, if this view were correct, Venus's phases would range only from new to crescent.

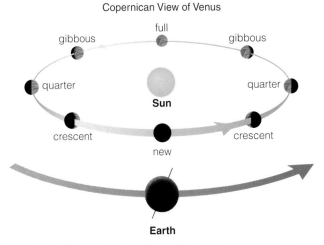

Copernican View of Venus

b In reality, Venus orbits the Sun, so from Earth we can see it in many different phases. This is just what Galileo observed, allowing him to prove that Venus really does orbit the Sun.

COSMIC CALCULATIONS 2.1

Kepler's Third Law

When Kepler discovered his third law ($p^2 = a^3$), he knew only that it applied to the orbits of planets about the Sun. In fact, it applies to any orbiting object as long as the following two conditions are met:

1. The object orbits the Sun *or* another star of precisely the same mass.
2. We use units of *years* for the orbital period and *AU* for the orbital distance.

(Newton extended the law to *all* orbiting objects; see Cosmic Calculations 7.1.)

Example 1: The largest asteroid, Ceres, orbits the Sun at an average distance (semimajor axis) of 2.77 AU. What is its orbital period?

Solution: Both conditions are met, so we solve Kepler's third law for the orbital period p and substitute the given orbital distance, $a = 2.77$ AU:

$$p^2 = a^3 \quad \Rightarrow \quad p = \sqrt{a^3} = \sqrt{2.77^3} \approx 4.6$$

Ceres has an orbital period of 4.6 years.

Example 2: A planet is discovered orbiting every three months around a star of the same mass as our Sun. What is the planet's average orbital distance?

Solution: The first condition is met, and we can satisfy the second by converting the orbital period from months to years: $p = 3$ months $= 0.25$ year. We now solve Kepler's third law for the average distance a:

$$p^2 = a^3 \quad \Rightarrow \quad a = \sqrt[3]{p^2} = \sqrt[3]{0.25^2} \approx 0.40$$

The planet orbits its star at an average distance of 0.40 AU, which is nearly the same as Mercury's average distance from the Sun.

Together, these observations offered clear proof that Earth is *not* the center of everything.*

Although we now recognize that Galileo won the day, the story was more complex in his own time, when Catholic Church doctrine still held Earth to be the center of the universe. On June 22, 1633, Galileo was brought before a Church inquisition in Rome and ordered to recant his claim that Earth orbits the Sun. Nearly 70 years old and fearing for his life, Galileo did as ordered and his life was spared. However, legend has it that as he rose from his knees, he whispered under his breath, *Eppur si muove*—Italian for "And yet it moves." (Given the likely consequences if Church officials had heard him say this, most historians doubt the legend.)

The Church did not formally vindicate Galileo until 1992, but the Church had given up the argument long before that. Galileo's book *Dialogue Concerning the Two Chief World Systems* was removed in 1824 from the Church's index of banned books. Today, Catholic scientists are at the forefront of much astronomical research, and official Church teachings are compatible not only with Earth's planetary status but also with the theories of the Big Bang and the subsequent evolution of the cosmos and of life.

THINK ABOUT IT Although the Catholic Church today teaches that science and the Bible are compatible, not all religious denominations hold the same belief. Do *you* think that science and the Bible are compatible? Defend your opinion.

Newton—The Revolution Concludes

Kepler's model worked so well and Galileo so successfully defused the remaining objections that by about the 1630s, scientists were nearly unanimous in accepting the validity of Kepler's laws of planetary motion. However, no one yet knew *why* the planets should move in elliptical orbits with varying speeds. The question became a topic of great debate, and a few scientists even guessed the correct answer—but they could not prove it, largely because the necessary understanding of physics and mathematics didn't exist yet. This understanding finally came through the remarkable work of Sir Isaac Newton (1642–1727), who invented the mathematics of calculus and used it to explain and discover many fundamental principles of physics.

In 1687, Newton published a famous book usually called *Principia*, short for *Philosophiae Naturalis Principia Mathematica* ("Mathematical Principles of Natural Philosophy"). In it, he laid out precise mathematical descriptions of how motion works in general, ideas that we now describe as **Newton's laws of motion**. For reference, Figure 2.12 illustrates the three laws of motion, although we will not make much use of them in this book. (Be careful not to confuse *Newton's* three laws, which apply to all motion, with *Kepler's* three laws, which describe only the motion of planets moving about the Sun.) Newton continued on in *Principia* to describe his universal law of gravitation (see Section 2.4), and then used mathematics to prove that Kepler's laws are natural consequences of the laws of motion and gravity.

---

*While these observations proved that Earth is not the center of everything, they did not by themselves prove that Earth orbits the Sun; direct proof of that fact did not come until later, with measurements of stellar parallax and of an effect known as the *aberration of starlight* that also occurs only because of Earth's motion. Nevertheless, the existence of Jupiter's moons showed that moons can orbit a moving planet like Jupiter, which overcame some critics' complaints that the Moon could not stay with a moving Earth, and the proof that Venus orbits the Sun provided clear validation of Kepler's model of Sun-centered planetary motion.

Newton's first law of motion:
An object moves at constant velocity unless a net force acts to change its speed or direction.

Example: A spaceship needs no fuel to keep moving in space.

Newton's second law of motion:
Force = mass × acceleration

Example: A baseball accelerates as the pitcher applies a force by moving his arm. (Once the ball is released, the force from the pitcher's arm ceases, and the ball's path changes only because of the forces of gravity and air resistance.)

Newton's third law of motion:
For any force, there is always an equal and opposite reaction force.

Example: A rocket is propelled upward by a force equal and opposite to the force with which gas is expelled out its back.

Figure 2.12

Newton's three laws of motion.

In essence, Newton had created a new model for the inner workings of the universe in which motion is governed by clear laws and the force of gravity. The model explained so much about the nature of motion in the everyday world, as well as about the movements of the planets, that the geocentric idea could no longer be taken seriously.

Looking Back at Revolutionary Science Fewer than 150 years passed between Copernicus's publication of *De Revolutionibus* in 1543 and Newton's publication of *Principia* in 1687, such a short time in the scope of human history that we call it a revolution. A quick look back shows that the revolution not only caused a radical change in human perspective on our place in the universe—shifting Earth from a central role to being just one of many worlds—but also altered our ideas about how knowledge should be acquired. For example, while previous generations had tolerated inaccuracies in the predictions of the Ptolemaic model, Copernicus and his followers felt compelled to find models of nature that could actually reproduce what they observed.

The eventual success of Kepler's model also led to a new emphasis on understanding *why* nature works as it does. Past generations had relied almost solely on their cultural senses of aesthetics in guessing that the world was built with perfect circles and spheres and indivisible atoms, and they seemed content to accept these guesses even without any evidence of their reality. By Newton's time, guessing was no longer good enough: Instead, you had to present hard evidence, backed by rigorous mathematics, to convince your colleagues that you'd hit on something that truly brought us closer to understanding the nature of the universe.

• How did the Copernican revolution alter the ancient debate on extraterrestrial life?

The Copernican revolution did not deal directly with the question of life in the universe, but it had a major effect on the way people thought about the issue. You can see why by thinking back to the ancient Greek debate.

Recall that while the atomists believed that there were many worlds, Aristotle held that this world *must* be unique and located in the center of everything, largely because his ideas of physics convinced him that all the "earth" in the universe would have naturally fallen to the center. The Copernican revolution therefore proved that Aristotle was wrong: Earth is not the center of the universe, after all.

Of course, the fact that Aristotle was wrong did *not* mean that the atomists had been right, but many of the Copernican-era scientists assumed that they had been. Galileo suggested that lunar features he saw through his telescope might be land and water much like that on Earth. Kepler agreed and went further, suggesting that the Moon had an atmosphere and was inhabited by intelligent beings. Kepler even wrote a science fiction story, *Somnium* ("The Dream"), in which he imagined a trip to the Moon and described the lunar inhabitants. Giordano Bruno was so convinced of the existence of extraterrestrial life that he battled authorities until they finally had him burned at the stake (see Special Topic 2.1).

Later scientists took the atomist belief even further. William Herschel (1738–1822), most famous as co-discoverer (with his sister Caroline) of the planet Uranus, assumed that all the planets were inhabited. In the late nineteenth century, when Percival Lowell (1855–1916) imagined seeing canals on Mars [Section 8.1], it's quite likely that he was still being influenced by the philosophical ruminations of people who had lived more than 2,000 years earlier.

If this debate about extraterrestrial life shows anything, it's probably this: It's possible to argue almost endlessly, as long as there are no actual

SPECIAL TOPIC 2.1: GEOCENTRISM AND THE CHURCH

The case of Galileo is often portrayed as having exposed a deep conflict between science and religion. However, the history of the debate over geocentrism shows that the reality was much more complex, with deep divisions even within the Church hierarchy.

Perhaps the clearest evidence for a more open-minded Church comes from the case of Copernicus, whose revolutionary work was supported by many Church officials. A less-well-known and even earlier example concerns Nicholas of Cusa (1401–1464), who published a book arguing for a Sun-centered solar system in 1440, more than a century before Copernicus's book. This Nicholas even weighed in on the subject of extraterrestrial life, writing:

Rather than think that so many stars and parts of the heavens are uninhabited and that this earth of ours alone is peopled ... we will suppose that in every region there are inhabitants, differing in nature by rank and allowing their origin to God ...

Church officials were apparently so untroubled by these radical ideas that they ordained Nicholas as a priest in the same year his book was published, and he later became a Cardinal. (Copernicus probably was not aware of this earlier work by Nicholas of Cusa.)

Many other scientists received similar support within the Church. Indeed, for most of his life, Galileo counted Cardinals—and even the pope who later excommunicated him—among his friends. Some historians suspect that Galileo got into trouble less for his views than for the way he portrayed them. For example, in 1632—just a year before his famous trial—he published a book in which two fictional characters debated the geocentric and Sun-centered views. He named the character taking the geocentric position Simplicio—essentially "simpleminded"—and someone apparently convinced the pope that the character was meant to be him. Moreover, as described by the noted contemporary author Isaac Asimov:

The book was all the more damaging to those who felt themselves insulted, because it was written in vigorous Italian for the general public (and not merely for the Latin-learned scholars) and was quickly translated into other languages—even Chinese!

If it was personality rather than belief that got Galileo into trouble, he was not the only one. The Italian philosopher Giordano Bruno (1548–1600), who had once been a Dominican monk, became an early and extreme supporter not only of the Copernican system but also of the idea of extraterrestrial life. In his book *On the Infinite Universe and Worlds*, published in 1584, Bruno wrote,

[It] is impossible that a rational being ... can imagine that these innumerable worlds, manifest as like to our own or yet more magnificent, should be destitute of similar or even superior inhabitants.

Note that Bruno was so adamant in his beliefs that he claimed that no "rational being" could disagree with him, so it's unsurprising that he drew the wrath of conservative Church officials. Bruno was branded a heretic and burned at the stake on February 17, 1600.

Perhaps the main lesson to be drawn from these stories is that while science has advanced dramatically in the past several centuries, people remain much the same. The Church was never a monolithic entity, and just as different people today debate the meaning of words in the Bible or other religious texts, Church scholars also held many different opinions at the time of the Copernican revolution. The political pendulum swung back and forth—or perhaps even chaotically—between the geocentric and Copernican views. Even when the evidence became overwhelming, a few diehards never gave in, and only the passing of generations finally ended the antagonism that had accompanied the great debate.

facts to get in the way. With hindsight, it's easy for us to see that everything from the musings of the ancient Greeks to Lowell's martian canals were based more on hopes and beliefs than on any type of real evidence.

Nevertheless, the Copernican revolution really did mark a turning point in the debate about extraterrestrial life. For the first time, it was possible to test one of the ancient ideas—Aristotle's—and its failure caused it to be discarded. And while the Copernican revolution did not tell us whether the atomists had been right about life, it did make clear that the Moon and the planets really are other *worlds,* not mere lights in the sky. That fact alone makes it plausible to imagine life elsewhere, even if we still do not have the data necessary to determine whether such life actually exists.

2.3   THE NATURE OF MODERN SCIENCE

The story of how our ancestors gradually figured out the basic architecture of the cosmos exhibits many features of what we now consider "good science." For example, we have seen how models were formulated and tested against observations, and then modified or replaced if they failed those tests. The story also illustrates some classic mistakes, such as the apparent failure of anyone before Kepler to question the belief that orbits must be circles. The ultimate success of the Copernican revolution led scientists, philosophers, and theologians to reassess the various modes of thinking that played a role in the 2,000-year process of discovering Earth's place in the universe. Now, let's examine how the principles of modern science emerged from the lessons learned in the Copernican revolution.

• How can we distinguish science from nonscience?

Perhaps surprisingly, it turns out to be quite difficult to define the term *science* precisely. The word comes from the Latin *scientia,* meaning "knowledge," but not all knowledge is science. For example, you may know what music you like best, but your musical taste is not a result of scientific study.

Approaches to Science   One reason science is difficult to define is that not all science works in the same way. For example, you've probably heard it said that science is supposed to proceed according to something called the "scientific method." As an illustration of this method in its most idealized form, consider what you would do if your flashlight suddenly stopped working. In hopes of fixing your flashlight, you might *hypothesize* that the batteries have died. In other words, you've created a tentative explanation, or **hypothesis**, for the flashlight's failure. A hypothesis is sometimes called an *educated guess*—in this case, it is "educated" because you already know that flashlights need batteries. Your hypothesis then allows you to make a simple prediction: If you replace the batteries with new ones, the flashlight should work. You can test this prediction by replacing the batteries. If the flashlight now works, you've confirmed your hypothesis. If it doesn't, you must revise or discard your hypothesis, usually in favor of some other one that you can also test (such as that the bulb is burned out). Figure 2.13 illustrates the basic flow of this process.

The scientific method can be a useful idealization, but real science rarely progresses in such an orderly way. Scientific progress sometimes occurs when someone goes out and looks at nature in a general way, hoping to learn something new and unexpected, rather than conducting a careful

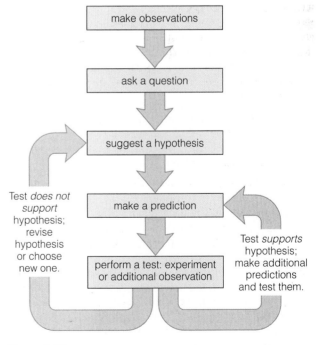

Figure 2.13

This diagram illustrates what we often call the "scientific method."

set of experiments.* This was the case with Galileo, who wasn't looking for anything in particular when he pointed his telescope at the sky and made his first startling discoveries. We still often approach science in this way today, such as when we build new telescopes or send missions to other worlds. For example, we did not know that the *Voyager 1* and *2* spacecraft would find evidence of a subsurface ocean on Europa when we sent them flying past Jupiter. We sent them just to see what was out there, and in the process we gained new and important scientific knowledge.

Another case in which we cannot use the idealized scientific method comes with attempts to understand past events, such as the history of Earth or the origin and evolution of life on Earth. We cannot repeat or vary the past, so we must instead rely on careful study of evidence left behind by past events. For example, we learn about early life on Earth not by observing it directly but by piecing together its story from an examination of fossils and other evidence that we can find today. Nevertheless, we can still apply at least some elements of the scientific method. For example, when scientists first proposed the idea that a massive impact may have been responsible for the death of the dinosaurs [Section 6.4], they were able to predict some of the other types of evidence that should exist if their hypothesis was correct. These predictions allowed other scientists to plan observations that might uncover this evidence, and when they succeeded—such as in discovering an impact crater of the right age—support for the impact hypothesis grew much stronger.

A further complication in describing how science works comes from the fact that scientists are human beings, so their intuitions and personal beliefs inevitably influence their work. Copernicus, for example, adopted the idea that Earth orbits the Sun not because he had carefully tested this idea but because he believed it made more sense than the prevailing view of an Earth-centered universe. As we have seen, while his intuition guided him to the right general idea, he erred in the specifics because he still clung to Plato's ancient belief that heavenly motion must be in perfect circles.

Given the great variety of ways in which it is possible to approach science, how can we identify what is science and what is not? To answer this question, we must look a little deeper at the distinguishing characteristics of scientific thinking.

Hallmarks of Science
One way to define scientific thinking is to list the criteria that scientists use when they judge competing models of nature. Historians and philosophers of science have examined (and continue to examine) this issue in great depth, and different experts express somewhat different viewpoints on the details. Nevertheless, everything we now consider to be science shares the following three basic characteristics, which we will refer to as the *hallmarks of science* (Figure 2.14):

- Modern science seeks explanations for observed phenomena that rely solely on natural causes.

- Science progresses through the creation and testing of models of nature that explain the observations as simply as possible.

- A scientific model must make testable predictions about natural phenomena that would force us to revise or abandon the model if the predictions do not agree with observations.

Hallmarks of Science

Seeks explanations for observed phenomena that rely solely on natural causes.

Progresses through creation and testing of models of nature that explain the observations as simply as possible.

Science

Makes testable predictions about natural phenomena. If predictions do not agree with observations, model must be revised or abandoned.

Figure 2.14

Hallmarks of science.

---

*Science done in this way is sometimes called *discovery science*, to distinguish it from the *hypothesis-driven science* in which observations or experiments are conducted to test a specific hypothesis.

Each of these hallmarks is evident in the story of the Copernican revolution. The first shows up in the way Tycho's exceptionally careful measurements of planetary motion motivated Kepler to come up with a better explanation for those motions. The second is evident in the way several competing models were compared and tested, most notably those of Ptolemy, Copernicus, and Kepler. We see the third in the fact that each model could make precise predictions about the future motions of the Sun, Moon, planets, and stars in our sky. When a model's predictions failed, the model was modified or ultimately discarded. Kepler's model gained acceptance in large part because its predictions were so much better than those of the Ptolemaic model in matching Tycho's observations.

Occam's Razor   The criterion of simplicity in the second hallmark deserves further explanation. Remember that the original model of Copernicus did *not* match the data noticeably better than Ptolemy's model. Thus, if scientists had judged Copernicus's model solely on the accuracy of its predictions, they might have rejected it immediately. However, many scientists found elements of the Copernican model appealing, such as the simplicity of its explanation for apparent retrograde motion. They therefore kept the model alive until Kepler found a way to make it work.

In fact, if agreement with data were the sole criterion for judgment, we could imagine a modern-day Ptolemy adding millions or billions of additional circles to the geocentric model in an effort to improve its agreement with observations. A sufficiently complex geocentric model could in principle reproduce the observations with almost perfect accuracy—but it still would not convince us that Earth is the center of the universe. We would still choose the Copernican view over the geocentric view because its predictions would be just as accurate yet would follow from a much simpler model of nature. The idea that scientists should prefer the simpler of two models that agree equally well with observations is called *Occam's razor,* after the medieval scholar William of Occam (1285–1349).

Verifiable Observations   The third hallmark of science forces us to face the question of what counts as an "observation" against which a prediction can be tested. Consider the claim that aliens are visiting Earth in UFOs. Proponents of this claim say that many thousands of eyewitness observations of UFO encounters provide evidence that it is true. But should these personal testimonials count as *scientific* evidence? On the surface, the answer may not be obvious, because all scientific studies involve eyewitness accounts on some level. For example, only a handful of scientists have personally made detailed tests of Einstein's theory of relativity, and it is their personal reports of the results that have convinced other scientists of the theory's validity. However, there's an important difference between personal testimony about a scientific test and an observation of a UFO: The first is verifiable by anyone, at least in principle, while the second is not.

Understanding this difference is crucial to understanding what counts as science and what does not. Even though you may never have conducted a test of Einstein's theory of relativity yourself, there's nothing stopping you from doing so. It might require several years of study before you have the necessary background to conduct the test, but you could then confirm the results reported by other scientists. In other words, while you may currently be trusting the eyewitness testimony of scientists, you always have the option of verifying their testimony for yourself.

In contrast, there is no way for you to verify someone's eyewitness account of a UFO. Without hard evidence such as clear photographs or pieces of the UFO, there is nothing that you could evaluate for yourself, even in principle. (And in those cases where "hard evidence" for UFO sightings has been presented, scientific study has never yet found the evidence to be strong enough to support the claim of alien spacecraft [Section 12.6].) Moreover, scientific studies of eyewitness testimony show it to be notoriously unreliable. For example, different eyewitnesses often disagree on what they saw even immediately after an event has occurred. As time passes, memories of the event may change further. In some cases in which memory has been checked against reality, people have reported vivid memories of events that never happened at all. This explains something that virtually all of us have experienced: disagreements with a friend about who did what and when. Since both people cannot be right in such cases, at least one person must have a memory that differs from reality.

Because of its demonstrated unreliability, eyewitness testimony alone should *never* be used as evidence in science, no matter who reports it or how many people offer similar testimony. It can be used in support of a scientific model only when it is backed up by independently verifiable evidence that anyone could in principle check for him- or herself. (Indeed, for much the same reason, eyewitness testimony alone is usually insufficient for a conviction in criminal court; at least some other evidence, such as motive, is required.)

Science, Nonscience, and Pseudoscience

It's important to realize that science is not the only valid way of seeking knowledge. For example, suppose you are shopping for a car, learning to play drums, or pondering the meaning of life. In each case, you might make observations, exercise logic, and test hypotheses. Yet these pursuits clearly are not science, because they are not directed at developing testable explanations for observed natural phenomena. As long as nonscientific searches for knowledge make no claims about how the natural world works, they do not conflict with science. In other words, just because something is not science does not make it wrong.

However, you will often hear claims about the natural world that seem to be based on observational evidence but do not treat evidence in a truly scientific way. Such claims are often called **pseudoscience,** which literally means "false science." To distinguish real science from pseudoscience, a good first step is to check whether a particular claim exhibits all three hallmarks of science. Consider the example of people who claim a psychic ability to "see" the future. If they truly can see the future, psychics should be able to make specific, testable predictions. In this sense, "seeing" the future sounds scientific, because we can test it. However, numerous studies have tested the predictions of "seers" and have found that their predictions come true no more often than would be expected by pure chance. If the seers were scientific, they would admit that this evidence undercuts their claim of psychic abilities. Instead, they generally make excuses. For example, they might say that the predictions didn't come true because of some type of "psychic interference." Making testable claims but then ignoring the results of the tests marks as pseudoscience the claimed ability to see the future.

Objectivity in Science

The idea that science is objective, meaning that all people should be able to find the same results, is important to the validity of science as a means of seeking knowledge. However, there is a difference between the overall objectivity of science and the objectivity of individual scientists. In particular, because science is practiced by human beings, individual scientists bring their personal biases and beliefs to their scientific work.

Personal bias can influence the way a scientist proposes or tests a model. For example, most scientists choose their research projects based on personal interests rather than on some objective formula. In some extreme cases, scientists have even been known to cheat—either deliberately or subconsciously—to obtain a result they desire. For example, consider Percival Lowell's claims of mapping artificial canals on Mars. Because no such canals actually exist, he must have been allowing his beliefs about extraterrestrial life to influence the way he interpreted blurry telescopic images—in essence, a form of cheating, though probably not intentional.

Bias can sometimes show up even in the thinking of the scientific community as a whole. Some valid ideas may not be considered by any scientist because the ideas fall too far outside the general patterns of thought, or **paradigm,** of the time. Einstein's theory of relativity provides an example. Many scientists in the decades before Einstein had gleaned hints of the theory but did not investigate them, at least in part because the ideas seemed too outlandish.

The beauty of science is that it encourages continued testing by many people. Even if personal biases affect some results, tests by others should eventually uncover the mistakes. Similarly, if a new idea is correct but falls outside the accepted paradigm, sufficient testing and verification of the idea should eventually force a change in the paradigm. Thus, although individual scientists rarely follow the idealized scientific method, the collective action of many scientists over many years should keep science objective as a whole.

• What is a scientific theory?

The most successful scientific models explain a wide variety of observations in terms of just a few general principles. When a powerful yet simple model makes predictions that survive repeated and varied testing, scientists elevate its status and call it a **theory.** Some famous examples are Isaac Newton's theory of gravity, Charles Darwin's theory of evolution, and Albert Einstein's theory of relativity.

The Meaning of the Term *Theory* The scientific meaning of the word *theory* is quite different from its everyday meaning, in which we equate a theory more closely with speculation or a hypothesis. In everyday life, someone might get a new idea and say, for example, "I have a new theory about why people enjoy the beach." Without the support of a broad range of evidence that others have tested and confirmed, this "theory" is really only a guess. In contrast, Newton's theory of gravity qualifies as a scientific theory because it uses simple physical principles to explain a great many observations and experiments.

Despite its success in explaining observed phenomena, a scientific theory can never be proved true beyond all doubt, because ever more sophisticated observations may eventually disagree with its predictions. However, anything that qualifies as a scientific theory must be supported by a large, compelling body of evidence.

In this sense, a scientific theory is not at all like a hypothesis or any other type of guess. We are free to change a hypothesis at any time, because it has not yet been carefully tested. In contrast, we can discard or replace a scientific theory only if we have a better way of explaining the evidence that supports it.

Again, the theories of Newton and Einstein offer great examples. A vast body of evidence supports Newton's theory of gravity, but by the late 1800s scientists had begun to discover cases where its predictions did not perfectly match observations. These discrepancies were explained only when Einstein developed his general theory of relativity in the early twentieth century, which was able to match the observations. Still, the many successes of Newton's theory could not be ignored, and Einstein's theory would not have gained acceptance if it had not been able to explain these successes equally well. It did, and that is why we now view Einstein's theory as a broader theory of gravity than Newton's theory. As we will discuss in the next section, some scientists today are seeking a theory of gravity that will go beyond Einstein's. If any new theory ever gains acceptance, it will have to match all the successes of Einstein's theory as well as work in new realms where Einstein's theory does not.

THINK ABOUT IT   When people claim that something is "only a theory," what do you think they mean? Does this meaning of *theory* agree with the definition of a theory in science? Do scientists always use the word *theory* in its "scientific" sense? Explain.

The Quest for a Theory of Life in the Universe   We do not yet have a theory of life in the universe, because we do not yet have the data to distinguish between many different hypotheses, which range from the hypothesis of no life anyplace else to the hypothesis that civilizations are abundant in our own galaxy. But thanks to the historical process that gave us the principles of modern science, we have a good idea of what we need to do if we ever hope to verify one of those hypotheses and turn it into a broad-based theory of life in the universe. That is why we can now make a modern science of astrobiology: not because we actually understand it yet but because we now know how to choose appropriate research projects to help us learn about the possibility of finding life elsewhere and how to go out and search for life that might exist within our solar system or beyond.

◤ **MOVIE MADNESS: CINEMA ALIENS**

Aliens have invaded Hollywood.

Helped by new computer graphics technology that has made it easier for artists to create weird-looking extraterrestrials, the silver screen is now crawling with cosmic critters eager to chow down on us, abduct us, or just tick us off as they total our planet. A few, like lovable little E.T., have kinder intentions: That big-eyed, wrinkled little extraterrestrial came 2 million light-years (the distance from the Andromeda galaxy) merely to pick some plants and hang with the kids.

Why are aliens infesting the local multiplex? Partly it's because after the collapse of the Soviet Union, Hollywood had to hunt around for a new source of bad guys. But our own space program has also convinced many among the popcorn-eating public that visiting other worlds will be a walk in the park for any advanced species. The movie moguls quietly ignore the fact that traveling between the stars is enormously more difficult than checking out the planets of your own solar system.

The aliens won't do it just to abduct you for unauthorized breeding experiments.

But the really big problem with Hollywood aliens, other than the fact that they seldom wear any clothes, is that they are inevitably portrayed as being close to our own level of development. This is so that we can take them on in aerial dogfights or challenge them to a manly light-saber duel. But the reality is somewhat different. As we'll discuss in Chapter 13, if we ever make contact with real aliens, their culture will probably be thousands, millions, or even billions of years beyond ours.

Of course, an invasion by hostile aliens with a million-year head start on *Homo sapiens* wouldn't make for an interesting movie. It would be Godzilla versus the chipmunks. But you don't mistake the movies for reality, do you?

2.4 THE PROCESS OF SCIENCE IN ACTION: UNDERSTANDING GRAVITY

We've completed our overview of the nature of modern science and its historical development. We've discussed the general process by which science advances, a process that is crucial to all sciences but is particularly important in astrobiology where, for example, widespread belief in aliens sometimes makes it difficult to separate fact from fiction. Because of its importance, we will continue to focus on the process of science throughout the book. In addition, in the final numbered section of this and all remaining chapters, we will take one topic and explore it in more depth, using it to illustrate some aspect of the process of science in action.

In this chapter, we will focus on gravity. Gravity is obviously important to life in the universe. On a simple level, life would float right off its planet without gravity. On a deeper level, stars and planets could never have been born in the first place without gravity, so we presume that life could not start in a universe in which gravity were absent or in which it worked significantly differently than it does in our universe.

Even more important to our purposes, gravity provides a great illustration of the sometimes surprising distinction that scientists make between "facts" and "theories." Gravity is clearly a fact: Things really do fall down when you drop them, and planets really do orbit the Sun. But scientifically, gravity is also a theory, because we use detailed, mathematical models of gravity to explain *why* things fall down and *why* planets orbit. Gravity is not unique in this way; for example, scientists make the same type of distinction when they talk about the fact of atoms being real and the atomic theory used to explain them, and when they talk about evolution having really occurred and the theory used to explain it. We can gain deeper insight into the way words often carry dual meanings by exploring how models of gravity have changed through time.

• What is gravity?

The true nature of gravity has presumably never changed, but human ideas about it have. In ancient Greece, Aristotle imagined gravity as an inherent property of heavy objects. For example, he claimed that earth and water had gravity, which made them fall toward our central world, while air and fire had "levity," which made them rise up. Aristotle's idea successfully explained a few observed facts, such as that rocks fall to the ground, but it didn't really make any specific predictions, such as how long it would take a rock to reach the ground if dropped from a tall cliff. Still, no one came up with a much better idea for nearly 2,000 years.

The first real breakthrough in human understanding of gravity came in 1666 when Newton (by his own account) saw an apple fall to Earth and suddenly realized that the gravity making the apple fall was the same force that held the Moon in orbit around Earth. He soon worked out a mathematical model to explain his observations, though he didn't share his results widely until 1687, when he published *Principia*.

Newton expressed the force of gravity mathematically with his **universal law of gravitation**. Three simple statements summarize this law:

- Every mass attracts every other mass through the force called *gravity*.

- The strength of the gravitational force attracting any two objects is *directly proportional* to the product of their masses. For example,

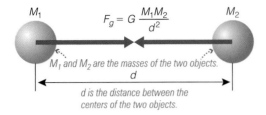

The **universal law of gravitation** tells us the strength of the gravitational attraction between the two objects.

M_1 $F_g = G \dfrac{M_1 M_2}{d^2}$ M_2

M_1 and M_2 are the masses of the two objects.

d

d is the distance between the centers of the two objects.

Figure 2.15

The universal law of gravitation is an *inverse square law*, which means the force of gravity declines with the *square* of the distance d between two objects.

doubling the mass of *one* object doubles the force of gravity between the two objects.

- The strength of gravity between two objects decreases with the *square* of the distance between their centers. That is, the gravitational force follows an **inverse square law** with distance. For example, doubling the distance between two objects weakens the force of gravity by a factor of 2^2, or 4.

These three statements tell us everything we need to know about Newton's universal law of gravitation. Mathematically, all three statements can be combined into a single equation, usually written like this:

$$F_g = G \frac{M_1 M_2}{d^2}$$

where F_g is the force of gravitational attraction, M_1 and M_2 are the masses of the two objects, and d is the distance between their centers (Figure 2.15). The symbol G is a constant called the **gravitational constant**, and its numerical value has been measured to be $G = 6.67 \times 10^{11}$ m^3/(kg $\times$ s^2).

THINK ABOUT IT   How does the gravitational force between two objects change if the distance between them triples? If the distance between them drops by half?

Newton showed that this law explained a great many facts that other scientists had by then discovered. For example, he showed that, when combined with his laws of motion, the universal law of gravitation explained the orbits of all the planets around the Sun (Kepler's laws), including Earth, as well as the orbit of the Moon around Earth. He also showed that it explained Galileo's observation that, absent air resistance, all objects fall to the ground at the same rate, regardless of their mass. Soon thereafter, Sir Edmund Halley (1656–1742) used the law to calculate the orbit of a comet that was seen in 1682. His calculations showed that the comet would return in 1758. Although he did not live to see it, Halley's Comet returned right on schedule, which is why it now bears his name.

These and other successes of Newton's universal law of gravitation made many eighteenth-century scientists think the mystery of gravity had been solved. But a problem came up a few decades after the 1781 discovery of the planet Uranus: Observations of Uranus showed its orbit to be slightly inconsistent with the orbit expected according to Newton's laws. Some scientists began to wonder if Newton's law of gravity might not be quite so exact as they had imagined. However, in the summer of 1846, French astronomer Urbain Leverrier suggested that the inconsistency could be explained by a previously unseen "eighth planet" orbiting the Sun beyond Uranus.* He used Newton's universal law of gravitation to predict the precise location in the sky where he thought the planet must be located. He sent a letter to Johann Galle of the Berlin Observatory, suggesting a search for the eighth planet. On the night of September 23, 1846, Galle pointed his telescope to the position suggested by Leverrier. There, within 1° of its predicted position, he saw the planet Neptune. It

---

*The same idea had been put forward a few years earlier in England by a student named John Adams, but he did not succeed soon enough in convincing anyone to search for the planet; Leverrier was apparently unaware of Adams's work.

was a stunning triumph for Newton's laws, and gave scientists far more confidence in the idea that the law of gravity truly was universal.

Today, we can see Newton's universal law of gravitation in action throughout the universe, in the orbits of extrasolar planets around their stars, of stars around the Milky Way Galaxy, and of galaxies in orbit about each other. There seems no reason to doubt the universality of the law. However, we also now know that Newton's law does not tell the entire story of gravity. Moreover, while Newton's law gives us a useful description of *how* gravity works, it still doesn't really tell us what it *is*.

• Do we really understand gravity?

Not long after Leverrier's success in predicting the existence of Neptune, astronomers discovered another slight discrepancy between a planetary orbit and the prediction made with Newton's law of gravity. This time, it involved the planet Mercury. Leverrier again set to work on the problem, suggesting it might be solved if there were yet another unseen planet, this one orbiting the Sun closer than Mercury. He even gave it a name—Vulcan. But searches turned up no sign of this planet, and we now know that it does not exist. So why was there a discrepancy in Mercury's orbit? Albert Einstein (1879–1955) provided the answer when he published his general theory of relativity in 1915.

To understand what Einstein did, we need to look a little more deeply at Newton's conception of gravity. According to Newton's theory, every mass exerts a gravitational attraction on every other mass, no matter how far away it is. If you think about it, this idea of "action at a distance" is rather mysterious. For example, how does Earth "feel" the Sun's attraction and know to orbit it? Newton himself was troubled by this idea. A few years after publishing his law of gravity in 1687, Newton wrote:

> That one body may act upon another at a distance through a vacuum, . . . and force may be conveyed from one to another, is to me so great an absurdity, that I believe no man, who has . . . a competent faculty in thinking, can ever fall into it.*

This type of "absurdity" was troubling to Einstein, whose scientific career can in many ways be viewed as a quest to find simple principles underlying mysterious laws. Although we will not go into the details, Einstein discovered that he could explain the mysterious action at a distance by assuming that all objects reside in something known as four-dimensional *spacetime*. Massive objects curve this spacetime, and other objects simply follow the curvature much like marbles following the contours of a bowl. Figure 2.16 uses a two-dimensional analogy to illustrate the idea, showing how planetary orbits are the straightest paths allowed by the structure of spacetime near the Sun. Thus, Einstein removed the mystery of "action at a distance" by telling us that gravity arises from the way in which masses affect the basic structure of the universe; in other words, he told us that gravity *is* "curvature of spacetime."

When Einstein worked out the mathematical details of his theory, he found that it gave *almost* precisely the same answers as Newton's universal law of gravitation for relatively weak sources of gravity, but more substantially different answers when gravity was stronger. Thus, he got essentially

*Letter from Newton, 1692–1693, as quoted in J. A. Wheeler, *A Journey into Gravity and Spacetime,* Scientific American Library, 1990, p. 2.

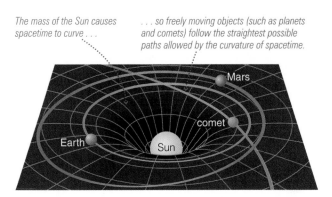

The mass of the Sun causes spacetime to curve . . .

. . . so freely moving objects (such as planets and comets) follow the straightest possible paths allowed by the curvature of spacetime.

Figure 2.16

According to Einstein's general theory of relativity, the Sun curves spacetime much like a heavy weight curves a rubber sheet, and planets simply follow this curvature in their orbits.

the same answers as Newton for the orbits of planets far from the Sun, where the effects of the Sun's gravity are weaker, but a slightly different answer for Mercury, where the effects of the Sun's gravity are much stronger. Einstein's answer matched the observed orbit of Mercury, giving him confidence that he had discovered an underlying truth about the nature of gravity.

Note that Einstein did *not* show that Newton's theory was wrong. After all, Newton's law of gravity had already proved valid in countless situations throughout the universe. Instead, he showed that Newton's theory was only an *approximation* to a more exact theory of gravity—the general theory of relativity. Under most circumstances the approximation is so good that we can barely tell the difference between the two ways of viewing gravity, but in cases of strong gravity, Einstein's theory works and Newton's fails.

Over the past century, scientists have had ample opportunity to test Einstein's theory, both through observations of the distant cosmos and through experimental tests in laboratories and orbiting spacecraft. To date, it has passed every test with flying colors. So does that mean we really do understand gravity today?

Most scientists doubt it, because there is at least one known "hole" in Einstein's theory. According to general relativity, the curvature of spacetime must become infinitely great when gravity becomes infinitely strong, as it would at the centers of the objects known as *black holes*. However, another scientific theory, also well tested, gives a very different answer for what should happen in such places: The theory of quantum mechanics, which successfully explains the workings of atoms, gives an answer that directly contradicts the answer given by relativity. Scientists are working hard to reconcile this discrepancy between two otherwise successful theories, but no one yet knows how it will turn out.

The bottom line is that we currently have a successful theory of gravity, known as Einstein's general theory of relativity, that appears to work throughout the universe. Newton's older universal law of gravitation is now considered an approximation to Einstein's theory. But we probably still don't understand gravity fully, and until we plug the known holes in Einstein's theory, we cannot predict how a more complete theory will alter our view of the role of gravity in the universe.

THINK ABOUT IT   Do you think that the known discrepancy between general relativity and quantum mechanics means that Einstein's theory of gravity is *wrong*? Defend your opinion.

THE BIG PICTURE
Putting Chapter 2 into Context

In this chapter, we've explored the development and nature of science, and how thoughts about life in the universe changed with the development of science. As you continue your studies, keep in mind the following "big picture" ideas:

- The questions that drive research about life in the universe have been debated for thousands of years, but only recently have we begun to acquire data that allow us to address the questions scientifically. In particular, the fundamental change in human perspective that came with the Copernican revolution had a dramatic impact on the question of life in the universe, because it showed that planets really are other *worlds* and not mere lights in the sky.

- The ideas that underlie modern science—what we've called the "hallmarks of science"—developed gradually, and largely as a result of the attempt to understand Earth's place in the universe. Science always begins by assuming that the world is inherently understandable and that we can learn how it works by observing it and by examining the processes that affect it. All of science, therefore, is based on observations of the world around us.

- Science is not the only valid way in which we can seek knowledge, but it has proved enormously useful, having driven the great progress both in our understanding of nature and in the development of technology that has occurred in the past 400 years.

SUMMARY OF KEY CONCEPTS

2.1 THE ANCIENT DEBATE ABOUT LIFE BEYOND EARTH

- **How did attempts to understand the sky start us on the road to science?**

The development of science began with Greek attempts to create **models** to explain observations of the heavens. Although most Greek philosophers favored a **geocentric model,** which we now know to be incorrect, their reasons for this choice made sense at the time. One of the primary difficulties of the model was that it required a complicated explanation for the **apparent retrograde motion** of the planets, with planets going around small circles on larger circles that went around Earth, rather than the much simpler explanation that we find with a Sun-centered model.

- **Why did the Greeks argue about the possibility of life beyond Earth?**

Some Greek philosophers (the *atomists*) held that our world formed among an infinite number of indivisible atoms, and this infinity implied the existence of other worlds. In contrast, Aristotle and his followers (the *Aristotelians*) argued that all earth must have fallen to the center of the universe, which rationalized the belief in a geocentric universe and the belief that the heavens were fundamentally different from Earth. This implied that Earth must be unique, in which case no other worlds or other life could exist.

2.2 THE COPERNICAN REVOLUTION

- **How did the Copernican revolution further the development of science?**

During the Copernican revolution, scientists began to place much greater emphasis on making sure that models successfully reproduced observations, and learned to trust data even when it contradicted deeply held beliefs. This willingness to let data drive the development of models led Kepler to develop what we now call **Kepler's laws of planetary motion,** and later led to the deeper understanding that came with **Newton's laws of motion** and the **law of universal gravitation**.

- **How did the Copernican revolution alter the ancient debate on extraterrestrial life?**

The Copernican revolution showed that Aristotle's Earth-centered beliefs had been incorrect, effectively ruling out his argument for Earth's uniqueness. Many scientists of the time therefore assumed that the atomists had been correct, and that other worlds and life are widespread. However, the data didn't really support this view, which is why we still seek to learn whether life exists elsewhere.

2.3 THE NATURE OF MODERN SCIENCE

• How can we distinguish science from nonscience?
Science generally exhibits these three hallmarks: (1) Modern science seeks explanations for observed phenomena that rely solely on natural causes. (2) Science progresses through the creation and testing of models of nature that explain the observations as simply as possible. (3) A scientific model must make testable predictions about natural phenomena that would force us to revise or abandon the model if the predictions do not agree with observations.

• What is a scientific theory?
A scientific **theory** is a simple yet powerful model that explains a wide variety of observations in terms of just a few general principles, and has attained the status of a theory by surviving repeated and varied testing.

2.4 THE PROCESS OF SCIENCE IN ACTION: UNDERSTANDING GRAVITY

• What is gravity?

According to Newton's **universal law of gravitation,** gravity is a force that causes every mass to attract every other mass. The strength of the force is proportional to the product of the masses and inversely proportional to the *square* of the distance between their centers. But while this statement describes the force of gravity, it still doesn't really tell us what gravity *is*.

• Do we really understand gravity?
Einstein's **general theory of relativity** explains the mysterious "action at a distance" of Newton's law, effectively telling us what gravity is ("curvature of spacetime"). This theory improves on Newton's, because it agrees much better with observations in cases where gravity is strong; thus, we see Newton's theory of gravity as an approximation to Einstein's more general theory. Einstein's theory works extremely well, but we know of at least one place where it appears to break down, so we cannot claim a complete understanding of gravity.

EXERCISES AND PROBLEMS

REVIEW QUESTIONS
Short-Answer Questions Based on the Reading

1. Describe at least three characteristics of Greek thinking that helped pave the way for the development of modern science.

2. What do we mean by a *model* of nature? Summarize the development of the Greek *geocentric model,* from Thales through Ptolemy.

3. What is *apparent retrograde motion,* and why was it so difficult to explain with the geocentric model? What is its real explanation?

4. Who first proposed the idea that Earth is a planet orbiting the Sun, and when? Why didn't this model gain wide acceptance in ancient Greece?

5. Briefly describe and contrast the different views of the atomists and the Aristotelians on the subject of extraterrestrial life.

6. What was the *Copernican revolution,* and how did it change the human view of the universe? Briefly describe the major players and events in the Copernican revolution.

7. Why didn't Copernicus's model gain immediate acceptance? Why did some scientists favor it, despite this drawback?

8. Describe each of *Kepler's laws of planetary motion.* In what sense did these laws provide us with a far more accurate model of planetary motion than either the models of Ptolemy or Copernicus?

9. Briefly describe three reasonable objections to the Sun-centered model that still remained even after Kepler's work, and then describe how Galileo's work overcame each of these objections.

10. How did Newton's discoveries about the laws of motion and the universal law of gravitation put the Sun-centered model on an even stronger footing?

11. How did the Copernican revolution affect scholarly thought regarding the question of life beyond Earth?

12. What is the difference between a *hypothesis* and a *theory* in science?

13. Describe each of the three hallmarks of science and give an example of how we can see each one in the unfolding of the Copernican revolution.

14. What is Occam's razor? Give an example of how it applies.

15. Why doesn't science accept personal testimony as evidence? Explain.

16. In what sense is gravity both a fact and a theory? Explain clearly.

17. What is *Newton's universal law of gravitation?* Write it in equation form, and clearly explain what the equation tells us. What do we mean when we say that the law is an *inverse square law?*

18. How did Einstein's general theory of relativity change our view of gravity? Why do we say that Newton's law of universal gravitation is still a valid approximation to Einstein's theory of gravity?

TEST YOUR UNDERSTANDING

Science or Nonscience?

Each of the following statements makes some type of claim. Decide in each case whether the claim could be evaluated scientifically or whether it falls into the realm of nonscience. Explain clearly; not all of these have definitive answers, so your explanation is more important than your chosen answer.

19. David Beckham is the best soccer player of his generation.

20. Several kilometers below its surface, Europa has an ocean of liquid water.

21. My house is haunted by ghosts, who make the creaking noises I hear each night.

22. There is no liquid water on the surface of Venus today.

23. Bacteria from Earth can survive on Mars.

24. Children born when Jupiter is in the constellation Taurus are more likely to be musicians than other children.

25. Aliens can manipulate time so that they can abduct people and perform experiments on them without the people ever realizing they were taken.

26. Newton's law of gravity explains the orbits of planets around other stars just as well as it explains the orbits of planets in our own solar system.

27. God created the laws of motion that were discovered by Newton.

28. A huge fleet of alien spacecraft will land on Earth and introduce an era of peace and prosperity on January 1, 2020.

Quick Quiz

Choose the best answer to each of the following. Explain your reasoning with one or more complete sentences.

29. In the Greek geocentric model, the retrograde motion of a planet occurs when: (a) Earth is about to pass the planet in its orbit around the Sun. (b) The planet actually goes backward in its orbit around Earth. (c) The planet is aligned with the Moon in our sky.

30. Which of the following was *not* a major advantage of Copernicus's Sun-centered model over the Ptolemaic model? (a) It made significantly better predictions of planetary positions in our sky. (b) It offered a more natural explanation for the apparent retrograde motion of planets in our sky. (c) It allowed calculation of the orbital periods and distances of the planets.

31. Earth is closer to the Sun in January than in July. Therefore, in accord with Kepler's second law: (a) Earth travels faster in its orbit around the Sun in July than in January. (b) Earth travels faster in its orbit around the Sun in January than in July. (c) Earth has summer in January and winter in July.

32. According to Kepler's third law: (a) Mercury travels fastest in the part of its orbit in which it is closest to the Sun. (b) Jupiter orbits the Sun at a faster speed than Saturn. (c) All the planets have nearly circular orbits.

33. Tycho Brahe's contribution to astronomy included: (a) inventing the telescope; (b) proving that Earth orbits the Sun; (c) collecting data that enabled Kepler to discover the laws of planetary motion.

34. Galileo's contribution to astronomy included: (a) discovering the laws of planetary motion; (b) discovering the law of gravity; (c) making observations and conducting experiments that dispelled scientific objections to the Sun-centered model.

35. Which of the following is *not* true about scientific progress? (a) Science progresses through the creation and testing of models of nature. (b) Science advances only through strict application of the scientific method. (c) Science avoids explanations that invoke the supernatural.

36. Which of the following is *not* true about a scientific theory? (a) A theory must explain a wide range of observations or experiments. (b) Even the strongest theories can never be proved true beyond all doubt. (c) A theory is essentially an educated guess.

37. How did the Copernican revolution alter perceptions of the ancient Greek debate over extraterrestrial life? (a) It showed that Aristotle's argument for a unique Earth was incorrect. (b) It showed that the atomists were correct in their belief in an infinite cosmos. (c) It proved that extraterrestrial life must really exist.

38. When Einstein's theory of gravity (general relativity) gained acceptance, it demonstrated that Newton's theory had been: (a) wrong; (b) incomplete; (c) really only a guess.

INVESTIGATE FURTHER
In-Depth Questions to Increase Your Understanding

Short-Answer/Essay Questions

39. *Greek Models.* As we discussed in this chapter, the Greeks actually considered both Earth-centered and Sun-centered models of the cosmos.
    a. Briefly describe the pros and cons of each model as they were seen in ancient times, and explain why most Greeks preferred the geocentric model.
    b. Suppose you could travel back in time and show the Greeks *one* observation from modern times. If your goal was to convince the Greeks to accept the Sun-centered model, what observation would you choose? Do you think it would convince them? Explain.

40. *Copernican Players.* Using a bulleted list format, write a one-page summary of the major roles that Copernicus, Tycho, Kepler, Galileo, and Newton played in overturning the ancient belief in an Earth-centered universe, along with a brief description of how each individual's work contributed to the development of modern science.

41. *What Makes It Science?* Read ahead and choose a single idea in the modern view of the cosmos that is discussed in Chapter 3, such as "The universe is expanding," "The universe began with a Big Bang," "We are made from elements manufactured by stars," or "The Sun orbits the center of the Milky Way Galaxy."
    a. Briefly describe how the idea you have chosen is rooted in each of the three hallmarks of science discussed in this chapter. (That is, explain how it is based on observations, how our understanding of it depends on a model, and how the model is testable.)
    b. No matter how strongly the evidence may support a scientific idea, we can never be certain beyond all doubt that the idea

is true. For the idea you have chosen, describe an observation that might cause us to call the idea into question. Then briefly discuss whether you think that, overall, the idea is likely or unlikely to hold up to future observations. Defend your opinion.

42. *Atomists and Aristotelians.* The ancient Greek arguments about the possible existence of extraterrestrial life continued for centuries. Write a short summary of the arguments, and then write a one- to two-page essay in which you describe how the Greek debate differs from the current scientific debate about extraterrestrial life.

43. *UFO Reports.* Thousands of people have reported sighting UFOs that they claim are alien spacecraft. Do these reports qualify as scientific evidence? Why or why not?

44. *Testing UFOs.* Consider at least one claim that you've heard about alien visitation (such as a claim about the Roswell crash, about an alien abduction, or about aliens among us). Based on what you've heard, can the claim be tested scientifically? If so, how? If not, why not? Do you think the claim should be considered more seriously or more skeptically? Defend your opinion.

45. *Science or Nonscience?* Find a recent news report from "mainstream" media (i.e., not from a supermarket tabloid) that makes some type of claim about extraterrestrial life. Analyze the report and decide whether the claim is scientific or nonscientific. Write two or three paragraphs explaining your conclusion.

46. *Tabloid Astrobiology.* Find a recent article from a supermarket tabloid that makes a claim about alien visitation to Earth. Analyze the article and write two or three paragraphs explaining whether the claim is scientific.

47. *Influence on History.* Based on what you have learned about the Copernican revolution, write a one- to two-page essay about how you believe it altered the course of human history.

48. *The Theory of Gravity.* How does the *fact* of gravity—for example, that things really do fall down—differ from what we think of as the *theory* of gravity? Briefly explain how and why Einstein's theory of gravity supplanted Newton's theory of gravity, and why we expect that we'll eventually find a theory that is even more general than Einstein's.

49. *Discovery of Neptune.*
    a. In what sense was Neptune discovered by mathematics, rather than by a telescope? How did this discovery lend further support to Newton's theory of gravity? Explain.
    b. According to the idea known as *astrology,* the positions of the planets among the constellations, as seen from Earth, determine the courses of our lives. Astrologers claim that they must carefully chart the motions of *all* the planets to cast accurate predictions (horoscopes). In that case, say skeptics, astrologers should have been able to predict the existence of Neptune long before it was predicted by astronomers, since they should have noticed inaccuracies in their predictions. But they did not. Do you think this fact tells us anything about the validity of astrology? Defend your opinion in a one- to two-page essay.

50. *Biographical Research: Post-Copernican Viewpoints on Life in the Universe.* Many seventeenth- and eighteenth-century writers expressed interesting opinions on extraterrestrial life. Each individual listed below wrote a book that discussed this topic; book titles (and original publication dates) follow each name. Choose one or more individuals and research their arguments about

extraterrestrial life. (You can find many of these books online in their entirety.) Write a one- to two-page summary of the person's arguments, and discuss which (if any) parts of these arguments are still valid in the current debate over life on other worlds.

   Bishop John Wilkins, *Discovery of a World in the Moone* (1638).
   René Descartes, *Philosophical Principles* (1644).
   Bernard Le Bovier De Fontenelle, *Conversations on the Plurality of Worlds* (1686).
   Richard Bentley, *A Confutation of Atheism from the Origin and Frame of the World* (1693).
   Christiaan Huygens, *Cosmotheros, or, Conjectures Concerning the Celestial Earths and Their Adornments* (1698).
   William Derham, *Astro-Theology: Or a Demonstration of the Being and Attributes of God from a Survey of the Heavens* (1715).
   Thomas Wright, *An Original Theory or New Hypothesis of the Universe* (1750).
   Thomas Paine, *The Age of Reason* (1793).

51. *Research: Religion and Life Beyond Earth.* Choose one religion (your own or another) and investigate its beliefs with regard to the possibility of life on other worlds. If scholars of this religion have made any definitive statements about this possibility, what did they conclude? If there are no definitive statements, discuss whether the religious beliefs are in any way incompatible with the idea of extraterrestrial life. Report your findings in a short essay.

Quantitative Problems

Be sure to show all calculations clearly and state your final answers in complete sentences.

52. *Sedna Orbit.* The recently discovered object Sedna orbits our Sun beyond Pluto, at an average distance (semimajor axis) of 509 AU. What is its orbital period?

53. *Eris Orbit.* The dwarf planet named Eris, which is larger than Pluto, orbits the Sun every 560 years. What is its average distance (semimajor axis) from the Sun? How does its average distance compare to that of Pluto?

54. *New Planet Orbit.* A newly discovered planet orbits a distant star with the same mass as the Sun at an average distance of 112 million kilometers. Find the planet's orbital period.

55. *Halley Orbit.* Halley's comet orbits the Sun every 76.0 years. (a) Find its semimajor axis distance. (b) Halley's orbit is an extremely eccentric (stretched-out) ellipse, so that at perihelion it is only about 90 million kilometers from the Sun, compared to more than five billion kilometers at aphelion. Does Halley's comet spend most of its time near its perihelion distance, near its aphelion distance, or halfway in between? Explain.

56. *Newton's Universal Law of Gravitation.*
    a. How does quadrupling the distance between two objects affect the gravitational force between them?
    b. Suppose the Sun were somehow replaced by a star with twice as much mass. What would happen to the gravitational force between Earth and the Sun?
    c. Suppose Earth were moved to one-third of its current distance from the Sun. What would happen to the gravitational force between Earth and the Sun?

Discussion Questions

57. *Science and Religion.* Science and religion are often claimed to be in conflict. Do you believe this conflict is real and hence irreconcilable, or is it a result of misunderstanding the differing natures of science and religion? Defend your opinion.

58. *The Impact of Science.* The modern world is filled with ideas, knowledge, and technology that developed through science and application of the scientific method. Discuss some of these things and how they affect our lives. Which of these impacts do you think are positive? Which are negative? Overall, do you think science has benefited the human race? Defend your opinion.

59. *Absolute Truth.* An important issue in the philosophy of science is whether science deals with absolute truth. We can think about this issue by imagining the science of other civilizations. For example, would aliens necessarily discover the same laws of physics that we have discovered, or would the laws they observe depend on the type of culture they have? How does the answer to this question relate to the idea of absolute truth in science? Overall, do you believe that science is concerned with absolute truth? Defend your opinion.

WEB PROJECTS

60. *The Galileo Affair.* In recent years, the Vatican has devoted a lot of resources to learning more about the trial of Galileo and understanding past actions of the Church in the Galileo case. Learn more about such studies, and write a short report about the current Vatican view of the case.

61. *Pseudoscience.* Choose a pseudoscientific claim that has been in the news recently, and learn more about it and how scientists have "debunked" it. Write a short summary of your findings.

62. *UFOlogy.* You can find an amazing amount of material about UFOs on the Web. Search for some such sites. Choose one that looks particularly interesting or entertaining and, in light of what you have learned about science, evaluate the site critically. Write a short review of the site.

63. *Gravitational Lensing.* Go to the Hubble Space Telescope Web site to find out what astronomers mean by "gravitational lensing," and locate at least two pictures that show examples of this phenomenon. How does the existence of gravitational lensing support Einstein's general theory of relativity, and what does it tell us about the idea that gravity works the same way throughout the universe?

3
The Universal Context of Life

LEARNING GOALS

3.1 THE UNIVERSE AND LIFE
- What major lessons does modern astronomy teach us about our place in the universe?

3.2 THE STRUCTURE, SCALE, AND HISTORY OF THE UNIVERSE
- What does modern science tell us about the structure of the universe?
- What does modern science tell us about the history of the universe?
- How big is the universe?

3.3 THE NATURE OF WORLDS
- How do other worlds in our solar system compare to Earth?
- Why do worlds come in different types?
- Should we expect habitable worlds to be common?

3.4 A UNIVERSE OF MATTER AND ENERGY
- What are the building blocks of matter?
- What is energy?
- What is light?

3.5 THE PROCESS OF SCIENCE IN ACTION: CHANGING IDEAS ABOUT THE FORMATION OF THE SOLAR SYSTEM
- How did the nebular model win out over competing models?
- Why isn't the nebular model set in stone?

Do there exist many worlds, or is there but a single world? This is one of the most noble and exalted questions in the study of Nature.

Saint Albertus Magnus (c. 1206–1280)

The study of life in the universe brings together many different fields of research, each contributing to different aspects of our understanding. Biology helps us understand the nature of life, so that we know what we are searching for. Chemistry and biochemistry help us understand how life works, and how it might have arisen in the first place. Planetary science, which includes geology and atmospheric science, helps us understand the conditions that might allow life to arise and survive on other worlds. Physics teaches us about the fundamental laws of nature that both enable and constrain the possibilities for life elsewhere. That is why we will study aspects of all of these sciences as we continue our survey of astrobiology in this book.

The universe is the canvas on which all these sciences come together, so to gain a true appreciation of the scientific search for life, we first need to understand a few fundamental ideas about the universe itself. In this chapter, we'll explore some of the most important ideas, including modern understanding of the overall nature of the universe, current theory about how planets are born, and basic properties of the matter and energy that make up all living things.

3.1 THE UNIVERSE AND LIFE

In Chapter 2, we saw how and why people long assumed that Earth was at the center of the universe and that the Sun, Moon, planets, and stars belonged to an entirely separate realm known as "the heavens." This Earth-centered (geocentric) view of the universe gave our planet a unique place in the cosmos and implied a clear distinction between Earth and anyplace else. Although the geocentric belief did not prevent people from speculating about inhabitants of the heavens—recall the debate between the atomists and the Aristotelians—it certainly limited the possibilities.

The possibilities changed dramatically with the Copernican revolution, when Earth finally lost its central place through the work of Copernicus, Tycho Brahe, Kepler, and Galileo. This new view of Earth's place in the universe, which gained acceptance only about 400 years ago, showed that other planets really are other worlds and not just wandering lights in the night sky. The realization that the Sun is a star added even more possibilities, because it became reasonable to imagine planets around other stars.

We still do not know whether any other planet has life, either in our own solar system or elsewhere. However, we have learned a great deal about the universe in the past 400 years, including much about its size, content, and history. As we will discuss in this chapter, these new discoveries have given us good reason to think that it's worth some scientific effort to search for life beyond Earth.

• What major lessons does modern astronomy teach us about our place in the universe?

The study of astronomy is so old that we cannot even pinpoint when it began, but for most of human history this study focused almost exclusively on observing the motions of visible objects in the sky. These observations were enough to meet immediate practical needs, such as being able to tell the time and the seasons from the Sun's path through the sky, or being able to predict the tides from the position and phase of the Moon.

The human realm of astronomy expanded when the Greeks began their attempts to *explain* the observed motions by seeking to learn the architecture of the cosmos, an effort that bore fruit some 2,000 years later when Kepler finally succeeded in describing the laws by which the planets orbit the Sun. Even then, however, science had advanced only far enough to say *how* the planets move around the Sun, not *why* they move as they do.

The key change in human understanding that allowed the emphasis to shift from "how" to both "how and why" occurred when Newton discovered that the planets are held in their orbits about the Sun by the same force of gravity that makes an apple fall to Earth. With this discovery, Newton delivered the final, shattering blow to the Aristotelian conception that Earth must by necessity be unique. The heavens could no longer be considered a separate realm made from different material (the *ether* or *quintessence*) and operating under different laws from Earth. Newton ended the ancient distinction between Earth and the heavens, finally making it possible to think of both as part of one *universe*.

The modern science of astronomy begins where Newton left off, as we seek to use his and other discoveries about the laws of nature to understand both the history and workings of the objects we see in the sky. Today, powerful telescopes enable us to study objects whose existence was not even contemplated in Newton's time, and experimental techniques allow us to probe the inner workings of the cosmos at a level far deeper than Newton probably could have imagined. Almost everything we have learned through modern astronomy has at least some importance to the study of life in the universe. For our purposes in this book, however, three ideas are especially important in framing the universal context for everything else we will study:

- *The universe is vast and old.* Its vastness implies an enormous number of worlds on which life might possibly have arisen, and its old age means there has been plenty of time for life to begin and evolve.

- *The elements of life are widespread.* Observations show that the basic chemical elements that make up Earth and life are present throughout the universe. Thus, at minimum, the raw ingredients of life should be found on many other worlds.

- *The same physical laws that operate on Earth operate throughout the universe.* Every experiment and observation made to date has given additional support to Newton's conclusion that the laws of nature are the same everywhere. In that case, it is reasonable to think that the same processes that made life possible on Earth have also made life possible on other worlds.

Together, these ideas reinforce the primary lesson of the Copernican revolution: *We are not the center of the universe.* Our planet may be special to us, but it is just one planet orbiting one rather average star in a universe that is certainly home to many similar stars and likely home to many similar planets. The apparent ordinariness of our circumstances is a major reason that it seems plausible to imagine a universe full of life.

3.2 THE STRUCTURE, SCALE, AND HISTORY OF THE UNIVERSE

We will devote the rest of this chapter to understanding the three important ideas listed on the previous page. They are all interrelated, so we cannot simply go through them one at a time. Instead, we'll focus on key features of the universe as we understand it today, so that you can see for yourself how the major ideas emerge from this modern picture. In this section, we'll focus on the structure and history of the universe, as well as the all-important issue of scale; after all, an understanding of the scale of space and time is the only way to appreciate the true vastness and age of the universe.

• What does modern science tell us about the structure of the universe?

The Copernican revolution displaced Earth from the center of the universe, but it was not until much later that we learned conclusively that the Sun holds no central place either. Indeed, it's not quite yet a century since careful studies of stellar orbits proved that our Sun and other stars orbit a point quite far away from us in the Milky Way, and it's been only about 85 years since Edwin Hubble (for whom the Hubble Space Telescope is named) proved that the Milky Way is just one of many galaxies in the universe. Let's begin our brief survey of the universe by examining the current state of knowledge about its general makeup.

Our Cosmic Address   We can understand how Earth fits into the overall structure of the universe by considering what we might call our "cosmic address," illustrated in Figure 3.1. Earth is a planet in our **solar**

KEY ASTRONOMICAL DEFINITIONS

This box summarizes a few key astronomical definitions that we will use throughout the book.

star: Our Sun and other ordinary stars are large, glowing balls of gas that generate heat and light through nuclear fusion in their cores.

planet: A moderately large object that orbits a star and shines primarily by reflecting light from its star. More precisely, according to a definition approved in 2006, an object can be considered a planet only if it: (1) orbits a star; (2) is large enough for its own gravity to make it round, and (3) has cleared most other objects from its orbital path. An object that meets the first two criteria but has *not* cleared its orbital path, like Pluto, is designated a *dwarf planet.*

extrasolar planet: A planet orbiting a star other than our Sun.

habitable planet (or habitable world): A planet (or other type of world) with environmental conditions under which life could potentially arise or survive.

moon (or satellite): An object that orbits a planet. The term *satellite* is also used more generally to refer to any object orbiting another object.

asteroid: A relatively small and rocky object that orbits a star.

comet: A relatively small and ice-rich object that orbits a star.

solar system: The Sun and all the material that orbits it, including the planets. The term *solar system* technically refers only to our own star system (because solar means "of the Sun"), but it is sometimes applied to other star systems.

star system: A star (sometimes more than one star) and any planets and other materials that orbit it. (Roughly half of all star systems contain two or more stars.)

galaxy: A great island of stars in space, containing from a few hundred million to a trillion or more stars, all held together by gravity and orbiting a common center.

universe (or cosmos): The sum total of all matter and energy; that is, all galaxies and everything within and between them.

Universe

approx. size: 10^{21} km

Local Supercluster

Local Group

approx. size: 3×10^{19} km

Milky Way Galaxy

approx. size: 10^{18} km

Solar System
(not to scale)

approx. size: 10^{10} km

Earth

approx. size: 10^4 km

Figure 3.1

Our cosmic address.

system, which consists of the Sun and all the objects that orbit it: the planets and their moons, along with countless smaller objects including rocky asteroids and icy comets.

Our Sun is a star, just like the stars we see in our night sky. The Sun and all the stars we can see with the naked eye make up only a small part of a huge, disk-shaped collection of stars called the **Milky Way Galaxy**. A **galaxy** is a great island of stars in space, containing from a few hundred million to a trillion or more stars. The Milky Way Galaxy is relatively large, containing more than 100 billion stars. Our solar system is located about halfway out through the disk of the galaxy, where it orbits the galactic center once every 230 million years.

Billions of other galaxies are scattered through space. Some galaxies are fairly isolated, but many others are found in groups. Our Milky Way, for example, is a member of a group of about 40 galaxies that we call the *Local Group*. Groups of galaxies with more than several dozen members are often called *galaxy clusters*.

Observations show that galaxies and galaxy clusters are not evenly distributed throughout space, but instead appear to be loosely arranged in giant chains and sheets; you can see this structure in the background of the painting in Figure 3.1. The galaxies and galaxy clusters are more tightly packed in some places than in others, forming giant structures called *superclusters*. Not surprisingly, the supercluster to which our Local Group belongs is called the *Local Supercluster*. Between the vast groupings of galaxies lie huge voids containing few, if any, galaxies.

Together, all these structures make up what we call our **universe**. In other words, the universe is the sum total of all matter and energy, encompassing the superclusters and voids and everything within them.

Figure 3.2

This photo shows the pedestals housing the Sun (the gold sphere on the nearest pedestal) and the inner planets in the Voyage scale model solar system (Washington, D.C.). The model planets are encased in the sidewalk-facing disks visible at about eye level on the planet pedestals. The building at the left is the National Air and Space Museum.

> **THINK ABOUT IT** Some people think that our tiny physical size in the vast universe makes us insignificant. Others think that our ability to learn about the wonders of the universe gives us significance despite our size. What do *you* think?

The Scale of the Solar System The levels of structure illustrated in Figure 3.1 are easy to list: planet, star system, galaxy, galaxy group, supercluster, and universe. However, if you look at the size labels shown for each level of structure, you'll see that this simple hierarchy hides vast jumps in scale from one level to the next. An understanding of this scale is crucial to the study of life in the universe, because it tells us of the great challenges that confront us as we search for life on other worlds. Let's begin by considering the scale of our own solar system.

Illustrations and photo montages often make our solar system look like it is crowded with planets and moons, but the reality is far different. One of the best ways to develop perspective on cosmic sizes and distances is to imagine our solar system shrunk down to a scale on which you could walk through it. The Voyage scale model solar system in Washington, D.C., makes such a walk possible (Figure 3.2). The Voyage model shows the Sun and the planets, and the distances between them, at *one ten-billionth* of their actual sizes and distances.

Figure 3.3a shows the Sun and planets at their correct sizes on the Voyage scale. Notice that, on this scale, the Sun is about the size of a large grapefruit, Jupiter is about the size of a marble, and Earth is about the size of the ballpoint in a pen. Comparing the sizes immediately reveals some key facts about our solar system. For example, the Sun is far larger than any of the planets; in mass, the Sun outweighs all the planets combined by a factor of more than 1,000. The planets also vary considerably in size:

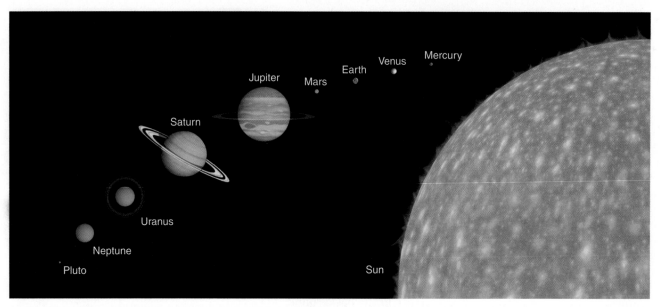

a This painting shows the Sun and planets at their scaled sizes. (Distances are not to scale in this painting.)

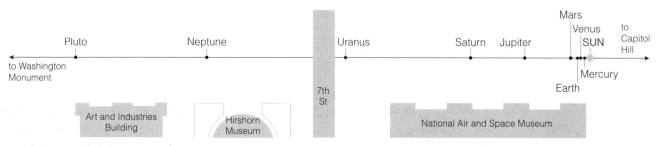

b This map shows the locations of the Sun and planets in the Voyage model; the distance from the Sun to Pluto is about 600 meters (1/3 mile). Planets are lined up in the model, but in reality each planet orbits the Sun independently and a perfect alignment essentially never occurs.

Figure 3.3

The Voyage model represents the solar system at *one ten-billionth* of its actual size. Pluto is included in the Voyage model, which was built before the International Astronomical Union decided that Pluto would be considered a "dwarf planet" rather than a true planet.

The entire Earth could be swallowed up by the storm on Jupiter known as the Great Red Spot (visible near Jupiter's lower left in the painting).

While Figure 3.3a shows the sizes of the Sun and planets at one ten-billionth of their actual sizes, it does not show the distances between them to scale. To see that, we must look at a map of where the planets are located in the Voyage model, as shown in Figure 3.3b. By combining the sizes shown in Figure 3.3a with the locations shown in Figure 3.3b, we can begin to get a true sense of the scale of our solar system. For example, the ballpoint–size Earth is located about 15 meters (16.5 yards) from the grapefruit-size Sun. To picture Earth's orbit around the Sun, imagine placing a grapefruit on a table and then holding a ballpoint as you walk in a circle of radius 15 meters around the grapefruit—and taking a year to complete one walk around.

Perhaps the most striking feature of our solar system when we view it to scale is its emptiness. The Voyage model shows the planets along a straight path, but in reality each planet orbits the Sun independently, although all the orbits lie in nearly the same plane. Thus, to show the full extent of our planetary system on the Voyage scale, we'd need to draw each planet's orbit around the model Sun. Fitting all these orbits would require an area measuring more than a kilometer (0.6 mile) on a side, which is larger than most college campuses. Spread over this large area, only the grapefruit-size Sun, the planets, and a few moons would be big enough to notice with your eyes. The rest of it would look virtually empty (that's why we call it *space!*).

Seeing the solar system to scale can help us understand why the search for life in the solar system is only just beginning. The Moon, the only other world on which humans have ever stepped, lies only about 4 centimeters (an inch and a half) away from Earth in the model. Thus, on this scale, the palm of your hand can cover the entire region of the universe in which humans have so far traveled. Our robotic spacecraft have traveled much farther, but these journeys are long and difficult. For example, the trip to Mars is some 200 times as far as the trip to the Moon. And while you can walk from the Sun to Pluto in just a few minutes on the Voyage scale, the *New Horizons* spacecraft that is currently making the real journey (launched January 2006) will have been in space nearly a decade by the time it finally flies past Pluto.

Distances to the Stars The scale of our solar system is literally astronomical, yet it is still tiny compared to the distances that separate the stars. To begin to think about stellar distances, imagine that you start at the Voyage model Sun in Washington, D.C. You walk the roughly 600-meter distance to Pluto (just over one-third mile) and then decide to keep going to find the nearest star besides the Sun. How far would you have to go?

Amazingly, you would need to walk to California. That is, on the same scale that allows you to walk from the Sun to Pluto in minutes, you would have to cross the entire United States just to find the nearest stars besides the Sun. If this answer seems hard to believe, you can check it for yourself. We usually measure the distances to stars in units of **light-years**; 1 light-year is the *distance* that light can travel in 1 year, which is about 10 trillion kilometers (see Cosmic Calculations 3.1). Thus, on the 1-to-10-billion Voyage scale, a light-year becomes about 1,000 kilometers (because 10 trillion ÷ 10 billion = 1,000 km). The nearest star system to our own, a three-star system called Alpha Centauri, is about 4.4 light-years away. This distance becomes about 4,400 kilometers (2,700 miles) on the 1-to-10-billion scale, or roughly equivalent to the distance across the United States.

The tremendous distances to the stars give us some perspective on the technological challenge of searching for life in other star systems. For example, because the largest star of the Alpha Centauri system is roughly the same size and brightness as our Sun, viewing it in the night sky is somewhat like being in Washington, D.C., and seeing a bright grapefruit in San Francisco (neglecting the problems introduced by the curvature of Earth). It may seem remarkable that we can see this star at all, but the blackness of the night sky allows the naked eye to see it as a faint dot of light. It looks much brighter through powerful telescopes, but we still cannot see any features of the star's surface.

Now, consider the difficulty of seeing *planets* orbiting nearby stars. It is equivalent to looking from Washington, D.C., and trying to see ballpoints or marbles orbiting grapefruits in California or beyond. When you consider this challenge, it is all the more remarkable to realize that we now have technology capable of finding such planets, at least in some cases [Section 11.2]. However, most of the known extrasolar planets have been detected indirectly by virtue of their gravitational effects on their stars, rather than by actual photographic imaging. We'll need additional leaps in technology before we are able to study extrasolar planets in sufficient detail to learn whether they actually harbor life.

COSMIC CALCULATIONS 3.1

How Far Is a Light-Year?

One light-year (ly) is defined to be the distance that light can travel in 1 year. This distance is fixed because light always travels through space at the *speed of light,* which is 300,000 kilometers per second (186,000 miles per second).

It's easy to calculate the distance represented by a light-year if you recall that

$$\text{distance} = \text{speed} \times \text{time}$$

For example, if you travel at a speed of 50 kilometers per hour for 2 hours, you will travel 100 kilometers. To find the distance represented by 1 light-year, we therefore need to multiply the speed of light by 1 year:

$$1 \text{ light-year} = (\text{speed of light}) \times (1 \text{ yr})$$

Because we are given the speed of light in units of kilometers per second but the time as 1 year, we must carry out the multiplication while converting 1 year into seconds. You can find a review of unit conversions in Appendix C; here, we show the result for this particular case:

$$1 \text{ light-year} = \left(300{,}000\, \frac{\text{km}}{\text{s}}\right) \times (1 \text{ yr})$$

$$= \left(300{,}000\, \frac{\text{km}}{\text{s}}\right) \times$$

$$\left(1 \text{ yr} \times 365\, \frac{\text{day}}{\text{yr}} \times 24\, \frac{\text{hr}}{\text{day}} \times 60\, \frac{\text{min}}{\text{hr}} \times 60\, \frac{\text{s}}{\text{min}}\right)$$

$$= 9{,}460{,}000{,}000{,}000 \text{ km}$$

That is, 1 light-year is equivalent to 9.46 trillion kilometers, or almost 10 trillion kilometers. Be sure to note that a light-year is a unit of *distance*, not time.

The vast distances to the stars offer a sobering lesson about interstellar travel. Although science fiction shows and movies like *Star Trek* and *Star Wars* make such travel look easy, the reality is far different. Consider the *Voyager 2* spacecraft. Launched in 1977, *Voyager 2* flew by Jupiter in 1979, Saturn in 1981, Uranus in 1986, and Neptune in 1989. It is now bound for the stars at a speed of close to 50,000 kilometers per hour—about 100 times as fast as a speeding bullet. But even at this speed, *Voyager 2* would take about 100,000 years to reach Alpha Centauri if it were headed in that direction (which it's not). Clearly, convenient interstellar travel remains well beyond our present technology, if it is possible at all (see Chapter 13).

The Scale of the Galaxy   We have examined the scale of our solar system and the distances to nearby stars, but we must make another huge jump in scale to begin to comprehend our Milky Way Galaxy. As we saw in Figure 3.1, the galaxy is some 100,000 light-years in diameter. Our solar system is located about 28,000 light-years from the galactic center, and the galaxy contains more than 100 billion stars. The 1-to-10-billion scale does not work well for visualizing such an enormous structure, because only a handful of the nearest stars could even fit on Earth on this scale.

Let's reduce our solar system scale even more, this time by another factor of one billion (making it a scale of 1 to 10^{19}). On this new scale, each light-year becomes 1 millimeter, and the 100,000-light-year diameter of the Milky Way Galaxy becomes 100 meters, or about the length of a football field. Visualize a football field with a scale model of our galaxy centered over midfield. Our entire solar system is a microscopic dot located around the 20-yard line. The 4.4-light-year separation between our solar system and Alpha Centauri becomes just 4.4 millimeters on this scale—smaller than the width of your little finger. If you stood at the position of our solar system in this model, millions of star systems would lie within reach of your arms.

Another way to get a handle on the size of the galaxy is to think about light-travel times. Light travels extremely fast by earthly standards. If you could circle Earth at the speed of light of 300,000 kilometers per second, you could complete almost eight circuits in just 1 second. But despite this awesome speed, light requires years to cross the vast chasms between the stars. That is why we measure interstellar distances in light-years. For example, when we say that Alpha Centauri is 4.4 light-years away, we mean that its light takes 4.4 years to reach us. This fact has an astonishing implication: It means that we cannot see what Alpha Centauri looks like today, but can see only what it looked like 4.4 years ago, when the light that is now reaching our eyes and telescopes first left on its journey. It also has an important implication to the possibility of carrying on a conversation with any beings who might happen to live in the Alpha Centauri system. We generally transmit messages over long distances with radio waves, which are a form of light and hence travel at the speed of light (see Section 3.4). Thus, if we sent a radio message to Alpha Centauri, the message would take 4.4 years to get there, and any reply would take the same 4.4 years to travel to us. You'd need a lot of patience for a conversation in which it would be almost 9 years from the time you said, "Hello, how are you?" until you heard the reply, "Fine, thanks, and you?"

The effect becomes more dramatic at greater distances. Consider the Orion Nebula, a giant cloud of gas and dust (meaning tiny solid particles) in which new stars and planets are currently being born (Figure 3.4). The Orion Nebula lies about 1,500 light-years away. Thus, we see the Orion

Figure 3.4

The Orion Nebula, located about 1,500 light-years away, photographed by the Hubble Space Telescope. The inset shows its location within the constellation Orion.

Nebula as it looked about 1,500 years ago—about the time of the fall of the Roman Empire. If we were to receive a radio message from aliens in the Orion Nebula, it would have to have been sent some 1,500 years ago. If we sent a message in return, we couldn't expect to hear a reply for at least 3,000 years.

The Orion Nebula is still quite nearby relative to the scale of the galaxy: Using our football-field-size scale model for the galaxy, the nebula lies only about 1.5 meters from Earth. It takes light 100,000 years to cross the 100,000 light-year diameter of the Milky Way Galaxy. Given that we are located about 28,000 light-years from the galactic center, a signal now reaching us from the far outer edge of the galaxy would have been sent more than 70,000 years ago. The *search for extraterrestrial intelligence,* or SETI, which listens for signals from alien civilizations, is in essence a search to hear from civilizations that used radio technology some decades, centuries, or millennia in the past.

The number of star systems in the Milky Way Galaxy is no less remarkable than its size. Imagine that you are having difficulty falling asleep at night, perhaps because you are contemplating the vastness of our galaxy. Instead of counting sheep, you decide to count stars. The Milky Way has more than 100 billion stars (perhaps as many as a trillion). If you are able to count about one star each second, on average, how long would it take you to count 100 billion stars? Clearly, the answer is 100 billion (10^{11}) seconds, but how long is that? You can get the answer by dividing 100 billion seconds by 60 seconds per minute, 60 minutes per hour, 24 hours per day, and 365 days per year. If you do this calculation, you'll find that 100 billion seconds is more than 3,000 years. In other words, you would need thousands of years just to *count* the stars in the Milky Way Galaxy, let alone to study them or search their planets for signs of life. And this assumes you never take a break—no sleeping, no eating, and absolutely no dying!

The Content of the Universe   The Milky Way Galaxy is just one of billions of galaxies in the universe, and we'll discuss the overall extent of the universe shortly. First, however, it's worth briefly discussing the content of the universe. We've said that the universe is the sum total of all matter and energy, but what exactly is this? Until a few decades ago, astronomers assumed that the matter of the universe was primarily found in stars and galaxies, while the energy of the universe took the form of light. It now seems that this "visible" matter and energy is just the tip of the iceberg in a universe that remains far more mysterious.

Just as planets orbit the Sun, stars orbit the center of the Milky Way Galaxy. The more massive the galaxy, the stronger its gravity and the faster stars should be orbiting. By carefully studying stellar orbits, astronomers have been able to put together a map of the distribution of matter in the Milky Way. The surprising result is that while most of the matter that we can see consists of stars and gas clouds in the galaxy's relatively flat *disk,* most of the mass lies unseen in a much larger, spherical *halo* that surrounds the disk (Figure 3.5). We don't know the nature of this unseen mass in the halo, so we call it **dark matter** to indicate that we have not detected any light coming from it, even though we have detected its gravitational effects. Studies of other galaxies suggest that they also are made mostly of dark matter. In fact, most of the mass in the universe seems to be made of this mysterious dark matter, which means that its gravity must have played a key role in assembling galaxies.

Figure 3.5

This painting shows an edge-on view of the Milky Way Galaxy. Most visible stars reside within the galaxy's central bulge and thin *disk,* which runs horizontally in this figure. However, careful study of galactic rotation suggests that most of the mass lies in the galactic *halo*—a large, spherical region that surrounds and encompasses the disk. Because this mass emits no light that we have detected, we call it *dark matter.*

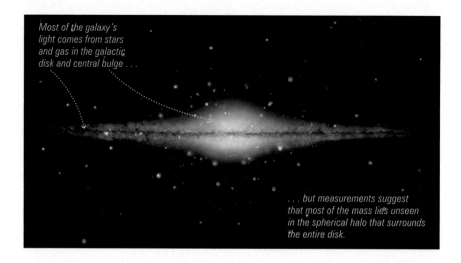

Most of the galaxy's light comes from stars and gas in the galactic disk and central bulge . . .

. . . but measurements suggest that most of the mass lies unseen in the spherical halo that surrounds the entire disk.

Evidence for the existence of dark matter has been building for several decades. More recently, scientists have gathered evidence for an even greater mystery: The universe seems to contain a mysterious form of energy—nicknamed **dark energy** by analogy to dark matter—that is pushing galaxies apart even while their gravity tries to draw them together. As is the case with dark matter, scientists have good reason to think that dark energy exists but lack any real understanding of its nature.

In recent years, scientists have been able to conduct a sort of census of the matter and energy in the universe. The results show that dark energy and dark matter are by far the main ingredients of the universe. The ordinary matter—atoms and molecules—that makes up stars and planets and life apparently represents no more than a few percent of all the matter and energy in the universe.

Because they appear to be the dominant constituents of the universe but we don't know much about them, dark matter and dark energy are arguably the biggest mysteries in astronomy today. However, they do not appear to affect the general evolution of stars, planets, or life, so they seem unlikely to affect our study of life in the universe. Nevertheless, as we seek to answer questions about life elsewhere, the mysteries of dark matter and dark energy should remind us that nature may still hold surprises that no one has foreseen.

• What does modern science tell us about the history of the universe?

We have surveyed the structure and scale of our vast universe, finding that Earth seems to hold a rather ordinary place. Modern understanding of the history of the universe further reinforces the idea that we live on a world that ought to be similar to many others, giving us additional reason to think that life might exist elsewhere.

Figure 3.6 gives a quick overview of the history of the universe as we understand it today, showing how matter has evolved from simple beginnings to the complexity of life on Earth today. Let's briefly examine the history shown in the figure.

The Big Bang and the Expanding Universe   The history begins in the upper left corner of the figure, where the three cubes represent one

Birth of the Universe. *The expansion of the universe began with the hot and dense Big Bang. The cubes show how one region of the universe has expanded with time. The universe continues to expand, but on smaller scales gravity has pulled matter together to make galaxies.*

Galaxies as Cosmic Recycling Plants. *The early universe contained only two chemical elements: hydrogen and helium. All other elements were made by stars and recycled from one stellar generation to the next within galaxies like our Milky Way.*

Life Cycles of Stars. *Many generations of stars have lived and died in the Milky Way.*

Stars are born in clouds of gas and dust; planets may form in surrounding disks.

Stars shine with energy released by nuclear fusion, which ultimately manufactures all elements heavier than hydrogen and helium.

Massive stars explode when they die, scattering the elements they've produced into space.

Earth and Life. *By the time our solar system was born, about 4½ billion years ago, about 2% of the original hydrogen and helium had been converted into heavier elements. Thus, we are "star stuff," because we and our planet are made from elements manufactured in stars that lived and died long ago.*

Figure 3.6

Our cosmic origins.

of the most crucial facts about the universe: The entire universe is **expanding**. That is, telescopic observations, first made by Edwin Hubble, show that average distances between galaxies are increasing with time (see Special Topic 3.1). The fact that distances between galaxies are increasing implies that these distances must have been smaller in the past. If we go back far enough, there must have been a time when galaxies—or the materials from which they were made—were all on top of one another, marking the beginning of the universal expansion. The idea that expansion implies a precise moment of beginning for the universe was first enunciated in 1927 by the astronomer and Catholic priest Georges Lemaître. Today, we call this beginning the **Big Bang**, and measurements of the expansion rate tell us that it occurred about 14 billion years ago. (More precisely, current data suggest that the universe is 13.7 billion years old, give or take a few hundred million years.)

Although Lemaître based his idea solely on the logic of working the current expansion backward in time, today the Big Bang is supported by at least two other lines of strong, independent evidence. First, we have detected radiation left over from the Big Bang. Just as compressing gas inside a car engine (the *piston* compresses gas in a *cylinder*) makes the gas much hotter and denser, the universe must have been much hotter and denser if it was smaller in the past. Thus, if the Big Bang really occurred, the universe should have begun with its matter compressed to extremely high temperature and density. Calculations show that as this matter expanded and cooled with time, it should have left behind a faint "glow" of radiation that we could detect with radio telescopes. This radiation, known as the *cosmic microwave background,* has indeed been detected and carefully studied. Its characteristics (such as its spectrum and distribution) precisely match the characteristics expected if it really is the leftover radiation from the Big Bang, providing strong support for the idea that the Big Bang really happened about 14 billion years ago.

The second line of evidence comes from the overall chemical composition of the universe. Calculations that run the expansion backward allow scientists to predict exactly when and how the chemical elements should have been born in the early universe. The calculations clearly predict that if the Big Bang occurred, then the chemical composition of the universe should be about three-fourths hydrogen and one-fourth helium (by mass). Observations show that this is indeed a close match to the overall chemical composition of the universe. This excellent agreement between prediction and observation gives additional strong support to the Big Bang theory. Note also that the prediction means the universe was born without any elements heavier than hydrogen and helium (except a trace of lithium)—which means the early universe lacked the elements that make life on Earth. As we'll discuss shortly, these elements were made later.

The universe as a whole has continued to expand ever since the Big Bang, but on smaller scales the force of gravity has drawn matter together. Structures such as galaxies and clusters of galaxies occupy regions where gravity has won out against the overall expansion. That is, while the universe as a whole continues to expand, individual galaxies and their contents do *not* expand. This idea is also illustrated by the three cubes in Figure 3.6, which represent how one region of the universe has changed with time. Notice that as the region as a whole grows larger, the matter within it has clumped into galaxies and galaxy clusters. Most galaxies, including our own Milky Way, probably formed within a few billion years after the Big Bang.

At the dawn of the last century, many astronomers assumed that the universe as a whole was permanent and largely unchanging. However, thanks to work started in the 1920s by Edwin Hubble, we now know that the universe is expanding. That is, average distances between galaxies in the universe are increasing with time, and space itself is growing to account for these larger distances.

Hubble discovered the universal expansion by observing many galaxies and the speeds at which they appear to move relative to Earth. His observations revealed two key facts:

1. Virtually every galaxy in the universe (except those within the Local Group) is moving away from us.
2. The more distant the galaxy, the faster it appears to be racing away.

Figure 1 uses a simple analogy to show how these observations lead to the conclusion that the universe is expanding. Imagine that you make a raisin cake in which the distance between adjacent raisins is

1 centimeter. You place the cake in an oven, where it expands as it bakes. After 1 hour, you remove the cake, which has expanded so that the distance between adjacent raisins has increased to 3 centimeters. From the outside, the expansion of the cake is fairly obvious. But what would you see if you lived in the cake, as we live in the universe?

Pick any raisin (it doesn't matter which one), call it the Local Raisin, and identify it in the pictures of the cake both before and after baking. Figure 1 shows one possible choice for the Local Raisin, with three nearby raisins labeled. The accompanying table summarizes what you would see if you lived within the Local Raisin. Notice, for example, that Raisin 1 starts out at a distance of 1 centimeter before baking and ends up at a distance of 3 centimeters after baking, which means it moves a distance of 2 centimeters away from the Local Raisin during the hour of baking. Hence, its speed as seen from the Local Raisin is 2 centimeters per hour. Raisin 2 moves from a distance of 2 centimeters before baking to a distance of 6 centimeters after baking, which means it moves a distance of 4 centimeters away from the Local Raisin during the hour. Hence, its speed is 4 centimeters per hour, or twice as fast as the speed of Raisin 1. Generalizing, the fact that the cake is expanding means that all raisins are moving away from the Local Raisin, with more distant raisins moving away faster. Hubble's discovery that galaxies are moving in much the same way as the raisins in the cake, with most moving away from us and more distant ones moving away faster, implies that the universe in which we live is expanding much like the raisin cake.

Hubble's original measurements of the universal expansion were fairly crude, but they have been greatly improved on since his time. Today, we know the rate of expansion to within about 10%, and we even have measurements that show roughly how the expansion rate has changed with time. Just as knowing a car's speed and its current distance from home can allow you to determine how long it's been since the car left, knowing the rate of expansion and the current separations of galaxies allows astronomers to determine how long it's been since the expansion began. The answer—about 14 billion years—must represent the age of the universe.

Measurements of the expansion rate are also responsible for one of the biggest mysteries in astronomy: the mystery of *dark energy*, discussed briefly in this chapter. If you throw a ball upward, gravity makes it slow down as it rises. In much the same way, we would expect the mutual gravitational attraction of all the galaxies in the universe to slow the expansion rate with time. However, measurements seem to show just the opposite: The expansion rate has been *increasing* with time, at least for the past few billion years. No one knows what is causing this acceleration of the expansion, but it must be some type of energy that can push galaxies apart. That is where the idea of the dark energy comes from, even though we do not yet know what it is.

Figure 1

An expanding raisin cake offers an analogy to the expanding universe. Someone living in one of the raisins inside the cake could figure out that the cake is expanding by noticing that all the other raisins are moving away, with more distant raisins moving away faster. In the same way, we know that we live in an expanding universe because all galaxies outside our Local Group are moving away from us, with more distant ones moving faster.

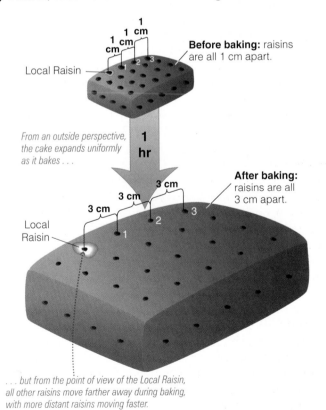

Local Raisin

Before baking: raisins are all 1 cm apart.

From an outside perspective, the cake expands uniformly as it bakes . . .

1 hr

After baking: raisins are all 3 cm apart.

Local Raisin

. . . but from the point of view of the Local Raisin, all other raisins move farther away during baking, with more distant raisins moving faster.

Distances and Speeds as Seen from the Local Raisin

Raisin Number	Distance Before Baking	Distance After Baking (1 hour later)	Speed
1	1 cm	3 cm	2 cm/hr
2	2 cm	6 cm	4 cm/hr
3	3 cm	9 cm	6 cm/hr
⋮	⋮	⋮	⋮

Hydrogen fusion

4 ¹H → 1 ⁴He + energy

Figure 3.7

The hydrogen fusion reaction: Four hydrogen nuclei (protons, in red) fuse to make one helium nucleus (two protons and two neutrons). The helium nucleus has slightly less mass than the four hydrogen nuclei combined (by about 0.7%); this "lost" mass is converted to energy in accord with Einstein's formula $E = mc^2$. This diagram shows the overall fusion reaction; in reality, this reaction proceeds in several steps, with only two nuclei fusing at a time.

Figure 3.8

The Crab Nebula is the remnant gas from a massive star whose explosion (supernova) was witnessed on Earth in A.D. 1054. The glowing gas is moving outward at high speed from the center, confirming its explosive origin. (The central object, a *pulsar*, offers further confirmation.) In a few tens of thousands of years, the gas will have fully dispersed, mixing the elements forged in the exploded star with other gas in the Milky Way Galaxy.

Stellar Lives and Galactic Recycling Within galaxies like the Milky Way, gravity drives the collapse of clouds of gas and dust to form stars. Stars are not living organisms, but they nonetheless go through "life cycles," as illustrated in the lower right of Figure 3.6. A star is born when gravity compresses the material in a cloud to the point where the center becomes dense enough and hot enough to generate energy by **nuclear fusion**, the process in which lightweight atomic nuclei smash together and stick (or fuse) to make heavier nuclei. Planets may be born at the same time. In much the same way that spinning a ball of dough causes it to spread out into a flat pizza, the natural spin of a contracting interstellar cloud keeps some of its gas spread away from its center while shaping it into a flattened disk. (We'll discuss the process in more detail in Section 3.3.) The planets of our own solar system formed in such a disk, which is why they all ended up orbiting the Sun in nearly the same plane.

Once a star is born, it shines with energy released by the nuclear fusion in its core. During most of a star's life, nuclear fusion combines hydrogen nuclei to make helium nuclei (Figure 3.7). It takes four hydrogen nuclei to make one helium nucleus (the process involves several steps); energy is released because a helium nucleus has slightly less mass than the four hydrogen nuclei. This means that a small amount of the mass of the hydrogen has disappeared and become energy in accord with Einstein's famous formula expressing the equivalence of matter and energy, $E = mc^2$ (where E is the energy, m is the mass, and c is the speed of light). Indeed, that is how our Sun shines today—with energy generated deep in its core by the fusion of hydrogen into helium.

A star lives until it exhausts all its usable fuel for fusion. The rate at which a star burns through its fuel depends on its mass: More massive stars, with much denser and hotter cores, burn through their fuel at far greater rates than less massive stars. In essence, more massive stars have their engines running hotter and therefore faster. This more than compensates for the fact that larger stars have more fuel to burn. The most massive stars live only a few million years, while stars like our Sun live 10 billion years and lower-mass stars can live hundreds of billions of years.

In its final death throes, when its fuel for fusion has been exhausted, a star blows much of its content back out into space. In particular, massive stars die in titanic explosions called *supernovae* (Figure 3.8). The returned matter mixes with other matter floating between the stars in the galaxy, eventually becoming part of new clouds of gas and dust from which new generations of stars can be born. Thus, galaxies function as cosmic recycling plants, reusing material expelled from dying stars to make new generations of stars and planets.

We Are Star Stuff The recycling of stellar material is connected to our existence in an even deeper way. As noted earlier, the Big Bang theory predicts that the universe should have been born containing only the simplest chemical elements: hydrogen and helium (and a trace of lithium). Living things, and Earth itself, are made primarily of other elements, such as carbon, nitrogen, oxygen, and iron. Where did these other elements come from? Apparently, all these elements were manufactured by stars.

We cannot see inside stars, but we can use the laws of physics to predict what must happen under the high-temperature and high-density conditions found in stellar cores. These types of calculations tell us that stars spend most of their lives generating energy by fusing hydrogen into helium. Toward the ends of their lives, stars like our Sun can fuse helium into carbon: Fusing three helium nuclei together makes one carbon nucleus. The Sun will stop there, but the cores of more massive stars can continue on to create many other heavy elements. For example, they can fuse carbon into oxygen and silicon, oxygen into neon and sulfur, and so on up to iron. Still other elements can be produced by nuclear reactions that accompany stellar death. All these manufactured elements then disperse into space after the star dies.

At least three lines of observational evidence confirm this theoretical prediction. First, stars of different ages show the expected pattern in the proportions of elements heavier than helium that they contain. The oldest stars contain hardly anything at all besides hydrogen and helium (heavier elements make up less than about 0.1% of their mass), just as we would expect for objects born before there had been time for stars to make much else. Younger stars, like our Sun, contain higher proportions (up to about 2%) of their mass in the form of elements heavier than hydrogen and helium, telling us that they were born from gas clouds that contained the elements manufactured and released by earlier generations of stars.

The second line of evidence comes from studies of the overall abundances of chemical elements in the universe today. The theory of nuclear fusion in massive stars makes specific predictions about relative abundances; for example, it says that the elements carbon and oxygen should be more abundant than nitrogen and that neon should be more abundant than fluorine. Figure 3.9 shows the observed relative abundances of the elements. Notice, for example, that nitrogen is indeed less abundant than carbon and oxygen. In fact, detailed calculations predict a pattern of abundances that almost perfectly matches these observations, even including all the up and down wiggles that appear on the graph.

The third line of evidence comes from studies of the gas from exploding stars (such as the Crab Nebula shown in Figure 3.8). Models of massive stars and their deaths allow astronomers to calculate the precise makeup expected for these clouds from recently deceased stars, and again, the observations match the predictions quite well.

The importance of this stellar manufacturing should be clear: Without it, our universe would not contain the chemical elements of which we are made. The recycling of matter and the production of heavier elements had already been taking place in the Milky Way Galaxy for billions of years before the Sun and the planets were born. By that time, stars had converted about 2% of the original hydrogen and helium into heavier elements. Thus, the cloud that gave birth to our solar system was made of about 98% hydrogen and helium and 2% of everything else (by mass). This 2% may sound small, but it was more than enough to make the small rocky planets of our solar system, including Earth. On Earth, some of these elements became the raw ingredients of life, ultimately blossoming into the great diversity of life we see today.

In summary, most of the material from which we and our planet are made was created inside stars that died before the birth of our Sun. We are intimately connected to the stars because we are products of stars. As astronomer Carl Sagan (1934–1996) used to say, we are "star stuff."

Implications for Life in the Universe
The fact that we are made of "star stuff" has important implications for the possibility of finding life elsewhere in the universe. The processes of stellar and galactic recycling operate throughout the Milky Way Galaxy, as well as in every similar galaxy throughout the universe, so we expect the chemical composition of many other star systems to be quite similar to that of our own. Observations confirm this expectation. We find that while there is some variation

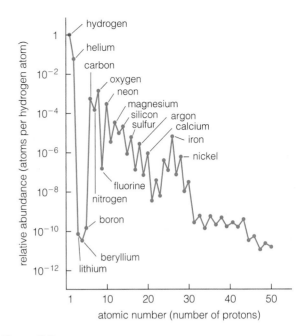

Figure 3.9

This graph shows the observed relative abundances of elements (by number) in the galaxy. For example, the relative abundance of 10^{-4} for nitrogen means that nitrogen is only about $10^{-4} = 0.0001$ times as abundant as hydrogen. The observed abundances agree well with what we expect if stars really have manufactured all the elements except hydrogen and helium. Note that hydrogen and helium are still by far the dominant chemical elements; the overall chemical content of our galaxy is about 98% hydrogen and helium (by mass), and 2% everything else combined.

in the precise proportions—in particular, older stars have much lower proportions of heavy elements—the overall composition of our solar system is typical of what we find for many other stars. We conclude that many and perhaps even most other star systems have the necessary raw materials to build Earth-like planets and life.

THINK ABOUT IT   The oldest stars in the galaxy are generally found in the halo, while younger stars are always found in the disk. Identify these regions in Figure 3.5. What does the difference in heavy-element abundance tell you about which region of the galaxy formed first? Do you think the difference affects the likelihood of finding Earth-like planets or life in the halo versus the disk? (*Note:* We'll discuss this topic further in Chapter 11.)

The Scale of Time   When we discussed the structure of the universe, we found that we had to carefully consider scale to understand how greatly one level differs from the next. In much the same way, it's easy to state that the universe is 14 billion years old, but it requires some deeper thought to begin to grasp the truly astronomical meaning of this age.

You are probably familiar with the use of time lines to represent historical events. We'll use a slight variation on this theme, making a scale for time in which we imagine compressing the entire history of the universe, from the Big Bang to the present, into a single year (Figure 3.10). On this *cosmic calendar*, the Big Bang takes place at the first instant of January 1, and the present is the stroke of midnight on December 31. For a universe that is about 14 billion years old, each month on the cosmic calendar represents a little more than 1 billion years, each day represents about 40 million years, and every second represents more than 400 years.

On this time scale, the Milky Way Galaxy probably formed by early February, though it may not have had its spiral shape right away. Many generations of stars lived and died in the subsequent cosmic months, enriching the galaxy with the "star stuff" from which we and our planet are made.

Figure 3.10

The cosmic calendar compresses the 14-billion-year history of the universe into 1 year, so that each month represents a little more than 1 billion years (more precisely, 1.17 billion years). This cosmic calendar is adapted from a version created by Carl Sagan.

THE HISTORY OF THE UNIVERSE IN 1 YEAR

January 1: The Big Bang

February: The Milky Way forms

September 3: The Earth forms

September 22: Early life on earth

December 17: Cambrian explosion

December 26: Rise of the dinosaurs

December 30: Extinction of the dinosaurs

Our solar system and our planet did not form until early September on this scale, or about $4\frac{1}{2}$ billion years ago in real time [Section 4.2]. By late September, life on Earth was flourishing. However, for most of Earth's history, living organisms remained microscopic in size. On the scale of the cosmic calendar, recognizable animals became prominent only in mid-December. Early dinosaurs appeared on the day after Christmas. Then, in a cosmic instant, the dinosaurs disappeared forever—probably due to the impact of an asteroid or a comet [Section 6.4]. In real time, the death of the dinosaurs occurred some 65 million years ago, but on the cosmic calendar it was only yesterday. With the dinosaurs gone, furry mammals inherited Earth. Some 60 million years later, or around 9 P.M. on December 31 of the cosmic calendar, early hominids (human ancestors) began to walk upright.

Perhaps the most astonishing thing about the cosmic calendar is that the entire history of human civilization falls into just the last half-minute. The ancient Egyptians built the pyramids only about 11 seconds ago on this scale. About 1 second ago, Kepler and Galileo proved that Earth orbits the Sun rather than vice versa. The average college student was born about 0.05 second ago, around 11:59:59.95 P.M. on the cosmic calendar. On the scale of cosmic time, the human species is the youngest of infants, and a human lifetime is a mere blink of an eye.

Like the scale of space, the fantastic scale of time carries important lessons about extraterrestrial life, if it exists. For example, the fact that the universe is so much older than Earth means that there ought to be many worlds that have had plenty of time for life to arise and evolve. If those worlds have had biological histories similar to Earth's, they might have had civilizations millions or even billions of years ago. We'll explore this idea and its astonishing implications in Chapter 13. The scale of time also holds sobering lessons for our own future. Species have come and gone in the months of the cosmic calendar during which life has flourished on Earth, and there's no special reason to think that our fate should be any different. Unless we learn enough about ourselves and our planet to find ways to survive into the next cosmic year, we will end up as little more than a momentary blip in the long history of the universe.

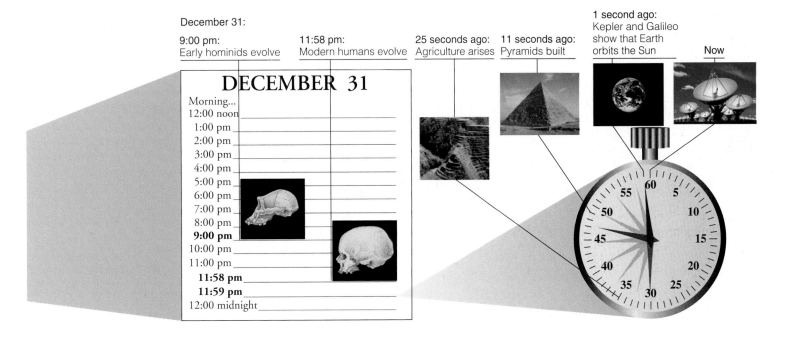

December 31:

9:00 pm: Early hominids evolve

11:58 pm: Modern humans evolve

25 seconds ago: Agriculture arises

11 seconds ago: Pyramids built

1 second ago: Kepler and Galileo show that Earth orbits the Sun

Now

DECEMBER 31

Morning...
12:00 noon
1:00 pm
2:00 pm
3:00 pm
4:00 pm
5:00 pm
6:00 pm
7:00 pm
8:00 pm
9:00 pm
10:00 pm
11:00 pm
11:58 pm
11:59 pm
12:00 midnight

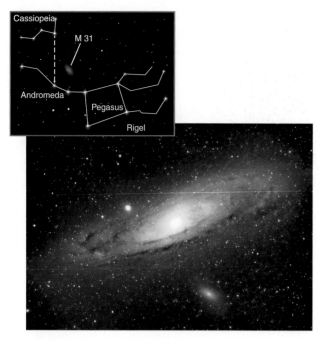

Figure 3.11

The Great Galaxy in Andromeda (M31). When we look at this galaxy—which is faintly visible to the naked eye in the location shown in the inset—we see light that has been traveling through space for 2.5 million years.

• How big is the universe?

We've stated that there are billions of galaxies in the universe, but can we be any more precise? In fact, when we think of the universe as the sum total of *all* matter and energy, we really have no idea how large it is—the universe could well be infinite, in which case it contains an infinite number of galaxies. However, the age of the universe places a fundamental limit on the portion of the universe that we can possibly see, even with the most powerful telescopes imaginable. To understand why, we must think again about the time it takes light to travel vast distances through the universe.

The Observable Universe When we look to great distances, we are also looking far back into the past. Figure 3.11 shows the nearest large galaxy to our own—the Great Galaxy in Andromeda, also known as M31. It is located about 2.5 million light-years away, which means the photo in Figure 3.11 shows this galaxy as it was about 2.5 million years ago, long before modern humans even existed. This might seem like a long time in the past, but it's unlikely that the Andromeda galaxy looks significantly different today: The galaxy is so large that it takes some 200 million years just to rotate once, so in 2 million years it barely changes at all. At much greater distances, however, we begin to see back to a time when the entire universe was significantly younger than it is today.

Consider, for example, a galaxy that is 1 billion light-years away. Its light has taken 1 billion years to reach us, which means we are seeing it as it looked 1 billion years ago—when the universe was only 13 billion years old, rather than its current 14 billion years.* Next, consider a galaxy that is 7 billion light-years away. We see this galaxy as it looked 7 billion years ago—which means we see it as it was when the universe was only half its current age. If we look at a galaxy that is 12 billion light-years away, we see it as it was 12 billion years ago, when the universe was only 2 billion years old. In a universe that is 14 billion years old, we cannot possibly see anything more than 14 billion light-years away. If we wanted to look more than 14 billion light-years away—say, to a distance of 15 billion light-years—we'd be trying to look to a time before the universe existed, when there's nothing to see (Figure 3.12). Thus, our **observable universe**—the portion of the entire universe that we can potentially observe—consists only of objects that lie within 14 billion light-years of Earth.

The concept of the observable universe has at least two important philosophical implications that are worth keeping in mind. First, the fact that we cannot observe anything more than 14 billion light-years away does *not* mean that nothing exists at such distances. In fact, we have good reason to think that the universe goes on far beyond 14 billion light-years, and some evidence suggests that the universe is infinite in extent. It's just that we have no hope of seeing or studying any objects that lie beyond the bounds of our observable universe. Second, notice that by definition, we are the center of our observable universe, since it is defined by a light-travel distance in all directions from *us*. However, being in the center of the *observable* universe is very different from being in the center of *the* universe. The latter would imply a special location, and the

---

*If you think about it, you'll realize that distances to faraway galaxies become difficult to define in an expanding universe, because the galaxies today are significantly farther away than they were when their light left on its journey to us. When we state a distance like 1 billion light-years, we really mean that we are seeing light that has traveled through space for 1 billion years to reach us. Unlike distance, this light-travel time (which astronomers call the *lookback time*) is clearly defined.

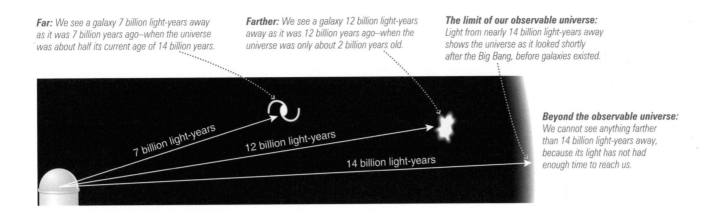

Far: We see a galaxy 7 billion light-years away as it was 7 billion years ago—when the universe was about half its current age of 14 billion years.

Farther: We see a galaxy 12 billion light-years away as it was 12 billion years ago—when the universe was only about 2 billion years old.

The limit of our observable universe: Light from nearly 14 billion light-years away shows the universe as it looked shortly after the Big Bang, before galaxies existed.

7 billion light-years

12 billion light-years

14 billion light-years

Beyond the observable universe: We cannot see anything farther than 14 billion light-years away, because its light has not had enough time to reach us.

former does not. Every observer on any planet around any star in any galaxy must also be at the center of their own observable universe; for example, anyone living in a distant galaxy would say the observable universe extends 14 billion light-years in all directions from them rather than from us. (You may realize that this means that they can see at least some galaxies that we cannot, and vice versa, because their observable universe only partially overlaps ours.) Thus, while the idea that we lie at the center of our own observable universe may sound like a throwback to pre-Copernican times, it's really not, since it still does not give us any special place in the *whole* universe.

Worlds Beyond Imagination   Because the observable universe has a finite size, it must contain a finite number of galaxies. We do not know exactly how many, because there are too many to count and because some galaxies are so faint that we cannot see them even with our most powerful telescopes. Nevertheless, we can *estimate* the number of galaxies in the observable universe by counting the number that we can see in pictures made with our most powerful telescopes. Figure 3.13 shows a remarkable photo, taken by the Hubble Space Telescope, that shows a tiny piece of the sky in great detail. To understand what you are seeing in this photo, imagine holding a grain of sand at arm's length against the

Figure 3.12

The farther away we look in space, the further back we look in time. Thus, the age of the universe puts a limit on the size of the observable universe—the portion of the entire universe that we can observe, at least in principle.

Figure 3.13

This photograph, known as the Hubble Ultra-Deep Field, was made with more than 10 days of exposure time by the Hubble Space Telescope. It shows thousands of galaxies, some more than 12 billion light-years away. The field of view of this image would fit behind a grain of sand held at arm's length against the sky. Nearly every tiny dot in this photo is an entire galaxy of stars.

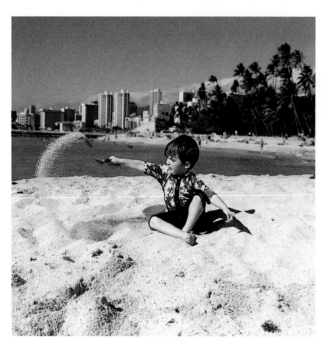

Figure 3.14

The number of stars in the observable universe is comparable to the number of grains of dry sand on all the beaches on Earth.

sky; everything you see in Figure 3.13 would fit within the field of view directly behind that grain of sand. Almost every blob and dot that you see in the photo is an entire galaxy of stars. Even the nearest of these galaxies are millions of light-years away, and the most distant ones (some of the tiniest dots) are more than 12 billion light-years away.

By counting the galaxies in Figure 3.13 and multiplying by the number of such photos it would take to make a montage of the entire sky, astronomers estimate that the observable universe contains about 100 billion galaxies. Just as it would take thousands of years to count the more than 100 billion stars in the Milky Way, it would take thousands of years to count all the galaxies in the observable universe.

Now, let's think about the total number of *stars* in all these galaxies. If we assume 100 billion stars per galaxy—similar to the number in the Milky Way—the total number of stars in the observable universe is roughly 100 billion × 100 billion, or 10,000,000,000,000,000,000,000 (10^{22}). How big is this number? Visit a beach. Run your hands through the fine-grained sand. Imagine counting each tiny grain of sand as it slips through your fingers. Then imagine counting every grain of sand on the beach and continuing on to count *every* grain of dry sand on *every* beach on Earth. If you could actually complete this task, you would find that, roughly speaking, the number of grains of sand is comparable to the number of stars in the observable universe (Figure 3.14).

The total number of *worlds*—by which we mean any reasonably large bodies in space, such as planets, moons, or even large asteroids—may be even greater. If planetary systems are as common as recent discoveries suggest, many or even most stars may have at least a few planets or moons. Clearly, our universe contains worlds beyond imagination.

> **THINK ABOUT IT** Contemplate the fact that there are as many stars as grains of sand on all the beaches on Earth and that each star is a potential sun for a system of planets. Does this vast number of possible homes for life affect your belief about the likelihood of finding life elsewhere? Why or why not?

The incredible size of the universe poses a practical challenge to the search for life beyond Earth: We can no more hope to search all the possible places where we might find life than we could hope to study every grain of dry sand on every beach on Earth. We will therefore confine our discussions of the search for life to the search within the Milky Way Galaxy, and presume that we'd find similar results if we could study other galaxies. Keep in mind that the search is daunting even for our own galaxy, especially when you realize that we have only recently developed the technology needed to discover planets around just a few of the most nearby of our galaxy's more than 100 billion stars.

The Fine-Tuned Universe We have briefly surveyed modern understanding of the universe, and we have discussed a little bit of the evidence that has given us this picture. But why is the universe like this? This may seem a strange question to ask in a science book, but it is one that many scientists are now asking themselves. The interest in this question has been sparked by the realization that our universe appears to be "fine-tuned" for life.

The logic behind the fine-tuning idea, sometimes known as the *anthropic principle*, goes like this: We are here today, able to study the universe and learn about its basic properties. But if any of those properties were much different, we could never have come to exist in the first place. For example, consider the expansion of the universe. If the universe were expanding much more rapidly, all its matter would have flown apart before gravity could have collected it into galaxies, stars, or planets. And if the universe were expanding much more slowly, gravity would have pulled all the matter back together, causing the universe to collapse before there was time for life to get started. The expansion rate had to be "just right" for galaxies to form, a fact that looks even more remarkable when you take into account the recently discovered acceleration of the expansion, which also has to be of just the right value so that it would have accelerated neither too much nor too little by now. Similar considerations apply to many other fundamental properties of the universe, including the ratio of the strengths of different forces (such as gravity and the electromagnetic force), the size scale on which quantum effects become important, and even the fact that we live in three dimensions of space—current theories of physics hold that many more dimensions actually exist, so three was not the only possibility. In every case, a change—perhaps even a very small change—could have prevented us from being here.

There's no doubt that the universe really is fine-tuned in a way that makes our existence possible; the debate is over what this means. Some people argue that it is merely a philosophical question, since we obviously would not be here talking about it if it weren't so, in which case it falls outside the realm of science. Others argue that it implies some "specialness" to humans in particular, with a few going as far as to claim that we should conclude that we *are* unique and hence the only life in the universe. And, of course, many physicists are seeking a natural explanation for the observed fine-tuning. One set of models that physicists have proposed to explain the workings of our universe suggests that in fact there should be a huge number of universes (sometimes called a "multiverse"), each with its own set of parameters such as its number of dimensions, ratio of force strength, and expansion rate. In essence, this viewpoint

MOVIE MADNESS: THE DAY THE EARTH STOOD STILL

There's a galactic club out there, and Earth has been asked to join. Actually, membership is not optional. You *will* join, and you *will* behave.

That's the stern message delivered in *The Day the Earth Stood Still*, a 1951 film classic that suggests that the specter of Cold War nuclear holocaust might best be eliminated with the help of some extraterrestrial meddling. The movie begins as well-intentioned aliens decide to visit the nation's capital and save us from ourselves. Citizens watch in dismay as a glowing saucer cruises by some famous landmarks and eventually sets down on the White House lawn. Help is at hand.

The landed craft has few occupants: Klaatu, a well-spoken alien who resembles a New York lounge lizard, and his robot pal, Gort, who looks like the Tin Man on steroids. These two characters step out of their disk-shaped craft to give us "the word." Klaatu declares, with some pride, that they've traveled 250 million miles to save us. That sounds like a long haul, but it's not: Within that paltry distance all you'll find are the Sun, Moon, Mercury, Venus, and Mars. These aliens supposedly represent some sort of cosmic organization put together to keep the peace, but the low mileage on their odometer suggests that they're just members of the inner-solar-system neighborhood association.

Klaatu wants to speak to the United Nations, but the politicians won't agree to listen, so the debonair alien opts for second best: He'll pitch his spiel to a conclave of the world's top scientists. After a brief demo of extraterrestrial capabilities to get everyone's attention (Klaatu and Gort shut down all earthly machinery for a half-hour—an amazing feat routinely, and annoyingly, managed by your local power company), they get to the point: Stop threatening everyone in the universe with atomic weaponry, or Gort and his robotic kin will wipe out our planet. Apparently Gort is a "death star" on legs.

So it boils down to this: Mom has told you to behave, and if you don't, Dad is going to really punish you. Needless to say, this isn't quite the agenda of any cosmic empire imagined by scientists, nor even that of the galactic federation in *Star Trek*. If there are real aliens out there who've mastered both interstellar travel and the colonization of other worlds, they undoubtedly regard A-bombs as mere firecrackers.

And besides, would anyone really want to trust a bunch of nickel-plated robots to keep the peace? After all, what if they get wet and seize up?

holds that lots of universes exist, most of which are unsuitable for life, and we live in one of the rare universes that is safe for life simply because we can. Other physicists suspect that we'll ultimately find a simpler explanation for why the universe turned out just right, and that once we find this explanation, we'll realize that it had to be this way; as Einstein once put it, perhaps "God had no choice" in setting the parameters of creation. From this viewpoint, our apparent "specialness" may simply be a consequence of our still-incomplete knowledge of physics.

3.3 THE NATURE OF WORLDS

Given that our galaxy contains far more worlds than we can possibly study in detail, we can improve our odds of success in searching for life in the universe by constraining the search in sensible ways. One obvious way is to learn enough about planetary systems in general so that we can make reasonable guesses as to which worlds are most likely to be habitable. In this section, we'll briefly survey how current understanding of planetary science helps us understand the prospects of finding habitable worlds, both in our own solar system and beyond.

Before we begin, it's useful to have a clear "picture" of the solar system that you can refer back to as needed in the coming discussion. Figure 3.15 shows essential features of our solar system as it would appear from somewhere out beyond the orbit of Neptune; remember that the

Figure 3.15

The layout of our solar system as it would appear from beyond Neptune, if we could magnify the sizes of the planets relative to their orbits by about a million times. Each planet's orbital direction is indicated, as is its tilt and direction of rotation (the small circling arrows). Moons are not shown.

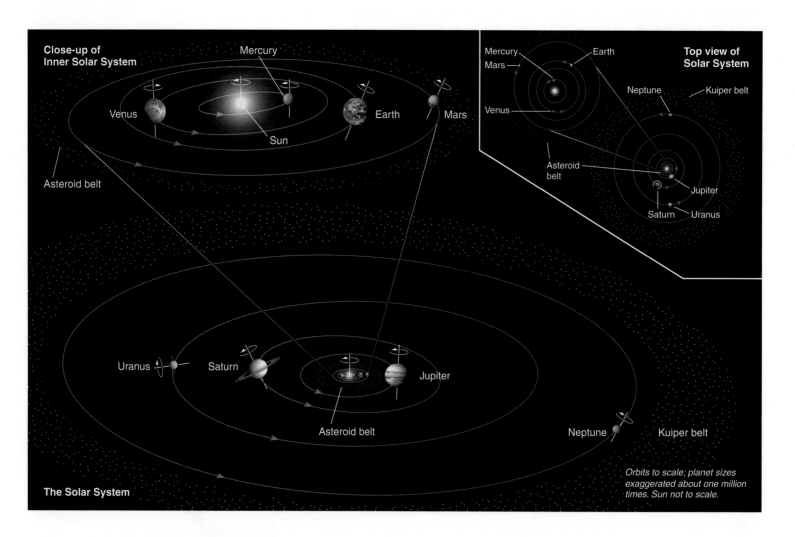

sizes of the planets are highly exaggerated relative to their distances in this figure, because if they were not, the planets would be microscopic on this scale. Notice that, as we saw earlier (see Figure 3.3), the inner planets are bunched much more closely together than the outer planets. Notice also that all the planets orbit the Sun in the same direction and in nearly the same plane and with nearly circular orbits, and that most rotate on their axes in the same direction as well. At the center of the solar system, the Sun rotates in the same direction that the planets orbit. And, although it is not shown, a similar idea holds for most large moons, which orbit nearly in their planet's equatorial plane and in the same direction that their planet rotates.

Figure 3.16

The full moon as seen from Earth. Notice the dark, smooth regions called *maria*.

• How do other worlds in our solar system compare to Earth?

When hearing the term *world*, most people picture a place like Earth, with a solid surface, oceans, and an atmosphere. In centuries past, many scientists also made this assumption. Even the Moon was often thought to have great seas, or *maria* (Latin for "seas"), because the smooth, dark regions reminded seventeenth-century scientists of the smooth surfaces of oceans seen from afar (Figure 3.16). Only as telescopes improved, and as scientists developed techniques for determining the composition of distant worlds from their light, did we learn that the Moon is actually dry and barren, and that Earth's surface oceans and oxygen-rich atmosphere are unique in the solar system.

Today, we know that no other world in our solar system is much like Earth, at least on the surface. No other world has surface oceans of liquid water, an atmosphere rich in oxygen, or a climate so hospitable to life. Nevertheless, planetary scientists have found that we can categorize worlds into groups by their general traits. Understanding these categorizations helps us understand the conditions that might make a world habitable. Let's look at the main categories that we will continue to discuss throughout this book, which are based on the types of objects we find here in our own solar system.

Two Major Types of Planets If you look back at Figure 3.3 or 3.15, you'll see that the eight official planets in our solar system fall into two general groups by size: The four inner planets (Mercury, Venus, Earth, and Mars) are much smaller than the four outer planets (Jupiter, Saturn, Uranus, and Neptune). These size differences reflect basic differences in planetary composition.

The four inner planets are made almost entirely of metal and rock, which makes their average densities several times that of water. They have solid surfaces and their atmospheres (if any) are quite thin compared to the planets themselves. Because Earth is a member of this group, we refer generally to these rocky worlds as **terrestrial planets** (*terrestrial* means "Earth-like").

Jupiter, Saturn, Uranus, and Neptune are quite different in character and composition from the terrestrial planets. Because Jupiter is the largest member of this group, we refer generally to these worlds as **jovian planets** (*jovian* means "Jupiter-like"). Rather than metal and rock, the jovian planets are made largely of hydrogen, helium, and **hydrogen compounds** such as water (H_2O), methane (CH_4), and ammonia (NH_3); this composition gives them much lower average densities than the terrestrial planets. Because hydrogen compounds are generally gases under earthly conditions (except for water, which can be either solid, liquid, or gas on Earth), the jovian planets are often called "gas giants." However, the pressure throughout most of their interiors is so high that these "gases" are not actually in the gas phase; instead, they may be compressed into liquid or into other high-density phases, which makes them behave quite differently than they do on Earth. Moreover, while the jovian planets contain metal and rock deep in their cores, the high pressure means that even these cores are unlikely to resemble the solid surfaces of the terrestrial worlds. Thus, there would be no place to "land" on the jovian planets; if you plunged into one of them, you would continue your descent until you were crushed by the growing pressure.

Keep in mind that there are great differences between the planets even within these two groups, and one of the major goals of planetary research is to understand these differences and why they occur. For example, while all the terrestrial worlds are similar in overall composition, their surfaces look quite different. Among the jovian planets, Uranus and Neptune contain much higher proportions of metal, rock, and hydrogen compounds than do Jupiter and Saturn, which are mostly made from hydrogen and helium. Nevertheless, the importance of the general distinction between terrestrial and jovian planets should be clear: If we assume that life needs a solid surface or oceans in which to get started, the jovian planets seem far less likely to be habitable [Section 7.3]. Thus, at least when we are looking for habitable *planets* in other star systems (we'll discuss moons shortly), we should concentrate on finding planets that are likely to be terrestrial in character.

Dwarf Planets and Small Bodies That Orbit the Sun

Neptune may be the outermost of the official planets, but the solar system continues beyond its orbit. Pluto, generally counted as a ninth planet until 2006 (see Special Topic 1.1), orbits on average at a distance about a third farther from the Sun than Neptune. Thousands of other icy objects share Pluto's region of the solar system. Indeed, we now know that Pluto is not even the largest of these icy objects. A slightly larger one, named Eris, was discovered in 2005, and others may still await discovery. These objects are now officially categorized as *dwarf planets* if they are large enough for their own gravity to make them round,* and as *small bodies* if they are low enough in mass to take any other shape.

Traditionally, astronomers have categorized small bodies in two groups: **asteroids** made mostly of metal and rock and **comets** made mostly of rock and ice. There's little debate over the definition of an asteroid—it's an object that resembles a terrestrial planet in composition but is too small to count as a planet itself. Most asteroids orbit in the region called the **asteroid belt**, which lies between the orbits of Mars and Jupiter. The definition of *comet,* however, has been almost as contentious as the definition of *planet.*

Comets got their name (which comes from the Greek word for "hair") from the objects with long tails that occasionally appear in our skies here on Earth (Figure 3.17). We now know that comets come from the distant reaches of the solar system, and they grow tails only when they come close enough to the Sun for the Sun's heat to convert some of their ice into gas. By studying comet orbits and doing some statistics on the number that we see in the inner solar system in any given period of time,

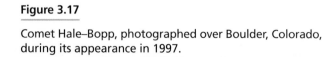

Figure 3.17

Comet Hale–Bopp, photographed over Boulder, Colorado, during its appearance in 1997.

*As of 2006, the objects that clearly meet the roundness criterion and count as dwarf planets are Pluto, Eris, and the asteroid Ceres. However, pending further observations, dozens and perhaps hundreds of additional objects may also qualify as dwarf planets.

astronomers have concluded that comets come from two vast "reservoirs": the **Kuiper belt** (Kuiper rhymes with "piper"), which occupies the region of the solar system beyond Neptune and in which we find both Pluto and Eris, as well as many similar but smaller objects; and a much more distant, spherically shaped region called the **Oort cloud** (Oort rhymes with "court"). Because the objects we see as comets in our sky must be only a tiny fraction of the similar objects in the Kuiper belt and Oort cloud, we've generally called these objects "comets" as well. The statistics tell us that there must be millions of small comets in the Kuiper belt and as many as a trillion in the Oort cloud.

Figure 3.18 shows the relative locations of the asteroid belt, Kuiper belt, and Oort cloud. Both in location and in composition, there is no doubt that both Pluto and Eris are members of the Kuiper belt, along with many other moderately large objects. In that sense, these objects are just unusually large comets, since they have the same general composition as much smaller comets that occasionally fall inward from the Kuiper belt toward the Sun. However, because of their large sizes, many astronomers are uncomfortable referring to Pluto and its cousins as "comets" and prefer the general term **Kuiper belt objects**, adding that they are "dwarf planets" if they are round.

Could any of the dwarf planets or small bodies be habitable? It's doubtful: Except perhaps for a short time when a comet enters the inner solar system, neither asteroids, comets, nor other Kuiper belt objects are likely to have any liquids of any kind. In the case of the asteroids, they may contain water chemically bound with their rock, but this water is not in liquid form (though some of it may have been in the distant past). In the case of comets, they generally reside so far from the Sun that their water is perpetually frozen, and their other ices melt or become gaseous only rarely and in small quantities. Nevertheless, asteroids and comets are important to life in a different way: They occasionally crash into planets, and these collisions can have a profound effect on any living organisms that inhabit the planet at the time.

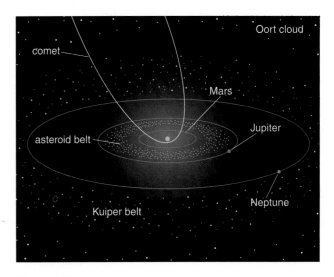

Figure 3.18

The small objects of our solar system reside in three major regions: the asteroid belt, the Kuiper belt, and the Oort cloud. The figure is *not* to scale; the comet orbit shown is typical of that of a comet entering the inner solar system from the Oort cloud.

THINK ABOUT IT   Suppose we discover a new object in our solar system that is as large as Earth in diameter, but has a density and composition similar to that of Pluto and orbits at twice Pluto's distance from the Sun. How would you classify this object? Explain.

Moons   Moons are objects that orbit planets or other objects, such as asteroids or dwarf planets. For example, Pluto is orbited by three moons: The largest, Charon, is about half the size of Pluto; two smaller moons (Nix and Hydra) were discovered in 2005.

Moons represent another clear distinction between the terrestrial and jovian planets. The terrestrial planets have few moons: Mercury and Venus have no moons at all, Earth has one relatively large moon, and Mars has two very small moons—so small that scientists suspect they were once asteroids that were somehow captured into martian orbit. In contrast, moons are common around the jovian planets: The four jovian planets of our solar system together have at least 150 moons; Jupiter alone has more than 60. Most of these moons are tiny (just a few kilometers across), but a few are large enough that we would probably consider them planets if they orbited the Sun independently. For example, Jupiter's moon Ganymede and Saturn's moon Titan are larger than the planet Mercury, while Jupiter's moons Io, Europa, and Callisto and Neptune's moon Triton are all larger than Pluto and the recently discovered Eris. These moons are planetlike in almost every way except for their orbits. Some of them even have active geology or atmospheres. For example, Io is the most volcanically active world in the solar system, Europa must occasionally have water or ice flowing across its surface, and Titan has an atmosphere thicker than Earth's.

Like the planets, moons vary in composition in a way that correlates with their locations in the solar system. Our Moon is made mostly of rock, much like the rocky composition of the terrestrial planets (though with some important differences, too [Section 4.6]). In contrast, the moons of the jovian planets generally contain large amounts of ice, including water ice (H_2O). When you combine this icy composition with the fact that some of the large jovian moons have internal heat and geological activity, you can see why scientists suspect that some of these moons may hide oceans beneath their icy surfaces. Thus, large moons around jovian planets offer a second category (after terrestrial planets) of potentially habitable worlds. Indeed, the prospects for some of these moons being habitable seem so good that we'll devote an entire chapter—Chapter 9—to discussing them.

• Why do worlds come in different types?

Knowing the different types of worlds in our solar system helps us understand their prospects of habitability, but we can gain even further insight by asking *why* worlds come in these different types. To answer this question, we must investigate the origin of our solar system, which in turn will give us insight into the prospects of finding habitable worlds in other star systems.

Our current theory of the solar system's formation developed gradually over many decades and, like any theory, is still subject to modification; we will discuss some of this history and ongoing debate in Section 3.5. Here, we will outline basic features of this **nebular theory**, so named because it starts with the idea that our solar system was born from the gravitational collapse of an interstellar cloud, or *nebula,* of gas and dust (*nebula* is Latin for "cloud"). The particular cloud that gave birth to our own solar system is usually called the **solar nebula**.

Contraction and Disk Formation   The solar nebula presumably began as a large, diffuse cloud, roughly spherical in shape. We do not know precisely what caused this cloud to start collapsing under its own gravity (though some evidence points to a nearby supernova as the trigger), but observations of similar clouds that exist today show that they can indeed collapse and give birth to stars. Once the collapse began, the laws of physics ensured that the solar nebula would heat up, spin faster, and flatten into a disk as it shrank in size (Figure 3.19).

The heating occurred because gas particles tend to move faster as they fall inward under gravity, much like a falling brick speeds up as it approaches the ground. The particles in the solar nebula collided with each other as they fell inward, transforming their energy of motion into heat. The cloud became hottest near the center, where the Sun formed. In more technical terms, this heating was a consequence of **the law of conservation of energy**; this law states that energy can be neither created nor destroyed, but only transformed from one form to another. The cloud began with a great deal of *gravitational potential energy*—energy that

The original cloud is large and diffuse, and its rotation is imperceptibly slow. The cloud begins to collapse.

Because of conservation of energy, the cloud heats up as it collapses. Because of conservation of angular momentum, the cloud spins faster as it contracts.

Collisions between particles flatten the cloud into a disk.

The result is a spinning, flattened disk, with mass concentrated near the center and the temperature highest near the center.

it had due to its particles being far from the cloud center—and it is this energy that ultimately was transformed into heat.

The spin of the cloud was a result of another physical law, known as **conservation of angular momentum**. This law essentially states that the total amount of "circling motion" of an object (or set of objects) must be conserved. We won't go into the details here, but you've probably seen this law in action with ice skating: It explains why a spinning skater's rate of spin increases when she pulls in her arms. In much the same way, a shrinking cloud of gas must spin faster as it contracts, as long as it had at least some small rate of spin to begin with. In the case of interstellar clouds, random motions ensure that they almost inevitably have some small overall rotation, though it is often imperceptible when the cloud is large. As the cloud shrinks, its spin therefore becomes noticeable and fast.

The flattening was a consequence of the spin. As particles collide in a spinning cloud, they tend to add to each other's motion when they are moving in the direction of the rotation, but to cancel each other's motion in other directions. Much like a spinning ball of pizza dough, this tends to force all the particles to end up in a flattened, spinning disk. Observations confirm that many young stars are surrounded by such flattened, spinning disks of material (Figure 3.20), just as we should expect from the nebular theory.

Condensation

As we discussed earlier, the overall composition of the galaxy and of our own Sun tells us that the composition of the solar nebula was about 98% hydrogen and helium and 2% of everything else. This 2% was crucial to planet formation, and it existed in essentially three different forms: metals, rock, and hydrogen compounds such as water, methane, and ammonia. Table 3.1 summarizes the overall composition of the solar nebula.

According to the nebular theory, the planet formation process began when tiny solid particles *condensed* from the gas, essentially becoming the "seeds" around which planets grew. You may be familiar with the idea of condensation in weather: Air on Earth contains mostly nitrogen and oxygen, but also has small amounts of water vapor. At the high altitudes of clouds, temperatures are low enough for this water vapor to condense into liquid droplets or solid ice flakes, which can ultimately fall to the ground as rain or snow. Condensation in the solar nebula occurred in much the same way: A few minor constituents of the disk-shaped cloud condensed into solid form even while the abundant hydrogen and helium remained gaseous.

Figure 3.19

This sequence of paintings shows how the gravitational collapse of a large cloud of gas causes it to become a spinning disk of matter. The hot, dense central bulge becomes a star, while planets can form in the surrounding disk.

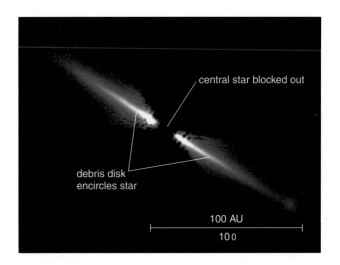

central star blocked out

debris disk encircles star

100 AU

100

Figure 3.20

This Hubble Space Telescope photo shows a flattened, spinning disk around the star AU Microscopii. We see this particular disk edge-on, but see disks in other orientations around other stars. (This particular disk probably represents a relatively late stage in disk development, but the structure is visually similar to earlier stages.)

TABLE 3.1 *Materials in the Solar Nebula*

A summary of the four types of materials present in the solar nebula. The squares in the final column represent the relative proportions of each type (by mass).

	Examples	Typical Condensation Temperature	Relative Abundance (by mass)
Hydrogen and Helium Gas	hydrogen, helium in nebula	do not condense	98%
Hydrogen Compounds	water (H_2O) methane (CH_4) ammonia (NH_3)	<150 K	1.4%
Rock	various minerals	500–1,300 K	0.4%
Metals	iron, nickel, aluminum	1,000–1,600 K	0.2%

The key to understanding condensation in the solar nebula lies in the third column of Table 3.1, which shows typical condensation temperatures. Because temperatures were high in the inner solar system, only materials with high condensation temperatures could become solid—which means only metal and rock could condense within about the present location of the asteroid belt. Farther out, where temperatures were much lower, hydrogen compounds could condense to make solid bits of ice. The result was that the disk remained filled with hydrogen and helium gas everywhere, but in the inner solar system there were solid particles of metal and rock, while the solid particles of the outer solar system were made largely of ice, mixed with smaller amounts of metal and rock. Figure 3.21 summarizes the condensation that occurred at different distances from the Sun.*

Accretion and Terrestrial Planet Formation The solid particles gradually grew larger through a process that we call **accretion**. Accretion began with the microscopic solid particles that condensed from the gas of the solar nebula. These particles orbited the forming Sun with the same orderly, circular paths as the gas from which they condensed. Individual particles therefore moved at nearly the same speed as neighboring particles, so "collisions" were more like gentle touches. Although the particles were far too small to attract each other gravitationally at this point, they were able to stick together through electrostatic forces—the same "static electricity" that makes hair stick to a comb. Small particles thereby began to combine into larger ones. As the particles grew in mass, gravity began to aid the process of their sticking together, accelerating their growth into boulders large enough to count as **planetesimals**, which means "pieces of planets."

The planetesimals grew rapidly at first. As a planetesimal grew larger, it would have both more surface area to make contact with other planetesimals and more gravity to attract them. Many planetesimals probably grew to hundreds of kilometers in size in only a few million years—a long time in human terms, but only about $\frac{1}{1,000}$ the present age of the solar system. Indeed, at this early time in the solar system's history, there may have been dozens or even hundreds of these large planetesimals (sometimes called *protoplanets*) orbiting the Sun between the present-day orbits of Mercury and Mars. When these large planetesimals met, the results were as likely to be shattering collisions as accretion to larger size.

The sizes of accreting planetesimals are too small for us to see them directly in the disks that surround other stars, although we have observed structure in some of these disks that strongly suggests the gravitational influence of large planetesimals or young planets. We can learn even more about the accretion process through computer simulations. A simulation begins with a set of data representing the conditions we observe in interstellar clouds. Then, with the aid of a computer, we apply the laws of physics to predict the changes that should occur over time. These computer simulations successfully reproduce the collapse of large clouds into spinning disks, as well as the basic process of accretion. The simulations cannot yet tell us exactly what to expect in the later stages of accretion,

---

*There is some scientific debate over how much of the solid material was preexisting dust grains from interstellar space, in which case the compositional differences with distance from the young Sun may have been due more to differences in what *vaporized* at different distances than differences in what condensed. But the end result is probably the same in either case: grains of metal and rock in the inner solar system, and icy grains only at larger distances from the Sun.

particularly in the balance between shattering collisions and growth, but the actual solar system speaks for itself: In at least four cases, objects in the inner solar system grew to planetary size, becoming Mercury, Venus, Earth, and Mars. It seems likely that at least a few other Moon- to Mars-size objects might also have been present at early times, but they eventually smashed into one of the four planets that survived. Indeed, as we'll discuss in Chapter 4, we now suspect that our own Moon formed when a Mars-size object smashed into the young Earth.

Jovian Planet Formation   Accretion should have occurred similarly in the outer solar system, but condensation of ices meant that there was more solid material. Thus, the planetesimals that accreted in the outer solar system contained large amounts of ice in addition to metal and rock. The solid objects that reside in the outer solar system today, such as comets, Kuiper belt objects, and the moons of the jovian planets, still show this ice-rich composition. However, the growth of icy planetesimals cannot be the whole story of jovian planet formation, because it would not explain how the jovian planets came to have so much hydrogen and helium gas.

The way in which the jovian planets formed remains a topic of scientific debate, but the leading model holds that they acquired their gas after beginning existence as large, ice-rich planetesimals. According to this model, some of the abundant icy planetesimals of the outer solar system grew to masses many times that of Earth. With these large masses, their gravity became strong enough to capture and hold some of the hydrogen and helium gas that made up the vast majority of the surrounding solar nebula. As the growing planets accumulated gas, their gravity grew stronger still, allowing them to capture even more gas. Ultimately, Jupiter and Saturn captured so much hydrogen and helium that their current compositions bear little resemblance to the icy seeds from which they started. Uranus and Neptune captured less gas, so their compositions more closely resemble the icy planetesimals, with larger amounts of hydrogen compounds and rock and smaller proportions of hydrogen and helium gas than we find in Jupiter and Saturn.

This model also explains the many large moons of the jovian planets. The same processes of heating, spinning, and flattening that made the disk of the solar nebula should also have affected the gas drawn by gravity to the young jovian planets. Thus, each jovian planet probably was surrounded by its own disk of gas, spinning in the same direction that the planet rotated. Moons accreted from ice-rich planetesimals within these disks, which is why these moons tend to orbit in the same direction as their planet rotates, with nearly circular orbits lying close to the equatorial plane of their parent planet.

This model of accretion followed by gas capture explains the observed features of the jovian planets quite well. Nevertheless, some scientists have recently advanced a competing model in which disturbances (technically, instabilities) within the disk of the solar nebula could have led clumps of gas to collapse and form jovian planets without the need for first forming icy planetesimals. Because this model is relatively new, scientists have not yet worked out enough details to know whether it can explain all the characteristics of the jovian planets. Most planetary scientists therefore still favor the accretion-and-gas-capture model, but good practice of science demands that we remain open to considering new evidence and ideas.

Clearing the Disk   As the planets formed, the Sun also formed, so much of the gas in the disk accreted onto the Sun. What happened to the rest

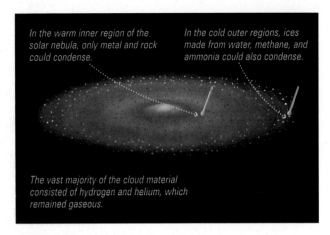

In the warm inner region of the solar nebula, only metal and rock could condense.

In the cold outer regions, ices made from water, methane, and ammonia could also condense.

The vast majority of the cloud material consisted of hydrogen and helium, which remained gaseous.

Figure 3.21

Temperature differences in the disk-shaped cloud that surrounded the young Sun caused different types of material to condense in the inner and outer regions of the solar system, leading to the differences in composition that we see today between terrestrial and jovian planets.

of the gas? Studies of other stars suggest that the young Sun would have had a strong **solar wind**, consisting of particles blown off its surface and out into space. This wind may have swept any remaining gas into interstellar space, effectively ending the era of planet formation. Alternatively, the gas may have been cleared by effects of sunlight itself.

However it occurred, the clearing of the gas sealed the compositional fate of the planets. If the gas had not been cleared soon after the planets formed, it might have continued to cool until hydrogen compounds condensed into ices even in the inner solar system. In that case, the terrestrial planets might have accreted abundant ice, and perhaps hydrogen and helium gas as well, changing their basic nature. At the other extreme, if the gas had been blown out too early, the raw materials of the planets might have been swept away before the planets could fully form. Although these extreme scenarios did not occur in our solar system, they may sometimes occur in other star systems.

Explaining the Worlds   We've seen how condensation and accretion led to the existence of terrestrial and jovian planets, and why the jovian planets tend to have numerous moons. The planets all orbit the Sun in the same direction and nearly the same plane because they were born in a flattened, spinning disk. They have nearly circular orbits, because particles with more elliptical orbits would have tended to suffer more collisions, which would have eventually forced them into more circular patterns. The Sun rotates in the same direction, because it was born at the center of the spinning cloud. The general cloud rotation also explains why most planets rotate in the same direction and most large moons orbit their planets in the same direction; however, because these are smaller-scale motions, it was easier for them to be changed by collisions, which may explain why not all of them follow this pattern.

Similar ideas explain the small bodies that orbit the Sun. While many of the planetesimals that formed early on ended up in one of the eight major planets, many others did not, and some large planetesimals shattered into pieces. Asteroids are the remaining planetesimals (and shattered planetesimals) of the inner solar system; most reside in the asteroid belt because the influence of Jupiter's gravity "herds" them in a way that makes them less likely to suffer collisions than asteroids located in other regions of the inner solar system. Pluto, Eris, and other large Kuiper belt objects remain in the region where they formed, out beyond the orbit of Neptune, along with many smaller comets.

The Oort cloud comets are slightly more difficult to explain, but models suggest that they actually originated as comets orbiting among the jovian planets. When they passed near a large jovian planet, such as Jupiter, they were likely to be flung out to a great distance by the planet's gravity, in much the same way that scientists have taken advantage of Jupiter's gravity to accelerate spacecraft to planets beyond. While it may sound strange for gravity to fling an object away, it's a direct consequence of the law of conservation of energy: When two objects interact through their gravity, their combined energy must remain unchanged, which means that one will lose energy and the other will gain it. For Jupiter and a comet, Jupiter's loss (or gain) of energy would be unnoticeable because it is so much larger, while the comet's gain (or loss) would completely change its orbit.

Figure 3.22 summarizes the scenario we have discussed for the formation of the solar system.

Figure 3.22

A summary of the key features of the nebular theory of how our solar system formed.

A large, diffuse interstellar gas cloud (solar nebula) contracts due to gravity.

Contraction: *As it contracts, the cloud heats, flattens, and spins faster, becoming a spinning disk of dust and gas.*

The Sun will be born in the center.

Planets will form in the disk.

Warm temperatures allow only metal/rock "seeds" to condense in inner solar system.

Condensation: *Hydrogen and helium remain gaseous, but other materials can condense into solid "seeds" for building planets.*

Cold temperatures allow "seeds" to contain abundant ice in the outer solar system.

Terrestrial planets are built from metal and rock.

Accretion: *Solid "seeds" collide and stick together. Larger ones attract others with their gravity, growing bigger still.*

The seeds of jovian planets grow large enough to attract hydrogen and helium gas, making them into giant, mostly gaseous planets; moons form in disks of dust and gas that surround the planets.

Clearing: *The remaining gas is cleared into space, sealing the compositional fates of the planets.*

Terrestrial planets remain in the inner solar system.

Jovian planets remain in the outer solar system.

"Leftovers" from the formation process become asteroids (metal/rock) and comets or Kuiper belt objects (ice/rock).

• Should we expect habitable worlds to be common?

Because the formation of a spinning disk is a consequence of physical laws that operate everywhere, we should expect that most stars should be born surrounded by spinning disks like that in which the planets of our own solar system were born. Observations support this idea, as we have found many young stars to be surrounded by such disks. Once a disk forms, we would expect condensation to occur in much the same way that it began in our solar system, with particles of metal and rock in the hot inner regions and particles of ice in the cold outer regions. Thus, based only on what we see in our solar system, we would expect to find many other solar systems with terrestrial and jovian planets laid out in the same general way as they are in our solar system. We might therefore expect habitable terrestrial planets and habitable jovian moons to be common throughout the galaxy.

Prior to the discoveries of extrasolar planets, astronomers generally assumed that this would indeed be the case, with most other planetary systems laid out much like ours. However, as we'll discuss further in Chapter 11, the reality appears to be more complex. All or nearly all of the extrasolar planets discovered to date are so large that they are almost certainly jovian in their basic composition and nature. This fact alone is not surprising—current technology allows us to detect only fairly massive planets (though we are rapidly gaining the ability to detect smaller ones)—but some of these large planets have surprising orbits. Many of them orbit much closer to their stars than do the jovian planets of our solar system; some orbit their stars much closer than Mercury orbits the Sun. Given that we expect jovian planets to form only in cold regions far from their stars, how can these large planets have ended up so close to their stars? Planetary scientists have a number of ideas that might explain these seemingly anomalous orbits, but so far we lack the evidence needed to know which, if any, of these ideas might be correct.

Because of this uncertainty, at this point we cannot say with any confidence whether solar systems like ours should be rare or common. Nevertheless, given the vast number of stars in the Milky Way Galaxy, even "rare" could mean large total numbers. For example, if only one in a million other star systems is like ours, that would still mean some 100,000 such systems among 100 billion stars (because 100 billion ÷ 1 million = 100,000).

The bottom line is that unless there is something dramatically wrong with our ideas about how planets form, it seems almost inevitable that our galaxy contains many worlds that have liquid water and hence would seem to make suitable homes for life.

3.4 A UNIVERSE OF MATTER AND ENERGY

We have now surveyed modern understanding of the universe on the large scale, and we've considered how this understanding affects the possibilities for extraterrestrial life. The small-scale universe is equally important, because life, like everything else in the universe, is at its most basic level an interplay of the things that we call *matter* and *energy*.

We have already talked a fair amount about matter and energy in passing, as it's hard to have any discussion of science without these concepts. However, for our purposes in this book, it is important for you to be familiar with a few more details that relate to matter and energy as we understand them today. Some of these details may already be familiar to you, but if not, this section should provide the background you'll need for the rest of this book. Note that we will restrict our focus to "ordinary" matter and energy, ignoring the *dark* matter and energy that may well make up most of the content of the universe; although this may seem a rather large restriction, the ordinary matter and energy appear to be all we need to understand stars, planets, and life.

• What are the building blocks of matter?

In Chapter 2, we discussed the ancient Greek idea that matter consists of four elements—fire, water, earth, and air—and the further assumption of some philosophers that these elements come in tiny, indivisible pieces that they called *atoms*, a Greek term meaning "indivisible." Today, we have a similar idea, but there are a lot more elements and fire, water, earth, and air are *not* among them. In addition, we now know that the atoms that make the elements are themselves made from smaller pieces. Let's take a brief look at current understanding of atoms and other microscopic forms of matter.

Atomic Structure   Atoms come in different types, and each type corresponds to a different chemical **element**. Some of the most familiar chemical elements are hydrogen, helium, carbon, oxygen, silicon, iron, gold, silver, lead, and uranium.

Atoms are made of particles that we call **protons**, **neutrons**, and **electrons** (Figure 3.23). Protons and neutrons are found in the tiny **nucleus** at the center of the atom. The rest of the atom's volume contains electrons, which surround the nucleus. Although we can think of electrons as tiny particles, they are not quite like tiny grains of sand and they don't orbit the nucleus the way planets orbit the Sun. Instead, the electrons in an atom form a kind of "smeared out" cloud that surrounds the nucleus and gives the atom its apparent size. The electrons aren't really cloudy, but it is impossible to pinpoint their positions in the atom.

Figure 3.23 also shows several other important features of atoms. First, notice that atoms are incredibly small: Millions could fit end to end across the period at the end of this sentence, and the number of atoms in a single drop of water (typically, 10^{22} to 10^{23} atoms) may exceed the number of stars in the observable universe. At the same time, the electrons give the atom a size far larger than its nucleus; if you imagine an atom on a scale that makes its nucleus the size of your fist, its electron cloud would be many kilometers wide. Nevertheless, most of the atom's mass resides in its nucleus, because protons and neutrons are each about 2,000 times as massive as an electron.

The properties of an atom depend mainly on the amount of **electrical charge** in its nucleus; an object's electrical charge is a measure of how strongly it will interact with other charged particles. We define the electrical charge of a proton as the basic unit of positive charge, which we write as +1. The electron has an electrical charge that is precisely opposite that of a proton, so we say it has negative charge (−1). Neutrons are electrically neutral, meaning that they have no charge. Oppositely charged particles attract one another, and similarly charged particles repel one another. An atom is held together by the attraction between the positively charged protons in the nucleus and the negatively charged electrons that surround the nucleus.*

---

*You may wonder why electrical repulsion doesn't cause the positively charged protons in a nucleus to fly apart from one another. The answer is that an even stronger force, called the *strong force*, overcomes electrical repulsion and holds the nucleus together.

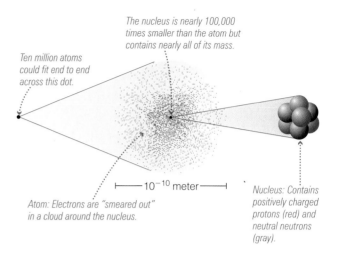

The nucleus is nearly 100,000 times smaller than the atom but contains nearly all of its mass.

Ten million atoms could fit end to end across this dot.

├── 10^{-10} meter ──┤

Atom: Electrons are "smeared out" in a cloud around the nucleus.

Nucleus: Contains positively charged protons (red) and neutral neutrons (gray).

Figure 3.23

The structure of a typical atom. Notice that atoms are extremely tiny: The atom shown in the middle is magnified to about one billion times its actual size, and the nucleus on the right is magnified to about 100 trillion times its actual size.

atomic number = number of protons
atomic mass number = number of protons + neutrons
(A neutral atom has the same number of electrons as protons.)

Hydrogen (^{1}H)   **Helium (^{4}He)**   **Carbon (^{12}C)**

atomic number = 1   atomic number = 2   atomic number = 6
atomic mass   atomic mass   atomic mass
number = 1   number = 4   number = 12
(1 electron)   (2 electrons)   (6 electrons)

*Different **isotopes** of a given element contain the same number of protons, but different numbers of neutrons.*

Isotopes of Carbon

carbon-12   carbon-13   carbon-14

^{12}C ^{13}C ^{14}C
(6 protons   (6 protons   (6 protons
+ 6 neutrons)   + 7 neutrons)   + 8 neutrons)

Figure 3.24

Terminology of atoms.

Most of the atoms in and around you contain the same number of electrons as protons, making them electrically neutral overall. However, atoms often lose or gain electrons, in which case they obtain a net electrical charge. We call such atoms **ions**. A *positive ion* is an atom that has lost one or more electrons so that it has more positive than negative charge overall; a *negative ion* is an atom that has gained one or more electrons, giving it a net negative charge. The net electrical charge of an atom turns out to be exceedingly important to life: Because the nucleus is buried so deeply inside an atom, interactions between atoms are almost exclusively interactions between their electrons. Indeed, these electrical interactions between atoms essentially make up everything that we think of as *chemistry*—and since chemical reactions are the foundation of all the processes that occur in living organisms, we see that the electrical interactions of atoms underlie everything we know about life.

Atomic Terminology   There are several pieces of atomic terminology that we will make use of throughout this book; they are summarized in Figure 3.24. First, each different chemical element contains a different number of protons in its nucleus. This number is its **atomic number**. For example, a hydrogen nucleus contains just one proton, so its atomic number is 1. A helium nucleus contains two protons, so its atomic number is 2. The complete set of the more than 100 known elements is listed in the **periodic table of the elements** (see Appendix D).

The *combined* number of protons and neutrons in an atom is called its **atomic mass number**. The atomic mass number of ordinary hydrogen is 1 because its nucleus is just a single proton. Helium usually has two neutrons in addition to its two protons, giving it an atomic mass number of 4. Carbon usually has six protons and six neutrons, giving it an atomic mass number of 12.

Every atom of a given element contains exactly the same number of protons, but the number of neutrons can vary. For example, all carbon atoms have six protons, but they may have six, seven, or eight neutrons. Versions of an element with different numbers of neutrons are called **isotopes** of the element. Isotopes are named by listing their element name and atomic mass number. For example, the most common isotope of carbon has 6 protons and 6 neutrons, giving it atomic mass number $6 + 6 = 12$, so we call it carbon-12. The other isotopes of carbon are carbon-13 (six protons and seven neutrons give it atomic mass number 13) and carbon-14 (six protons and eight neutrons give it atomic mass number 14). We can also write the atomic mass number of an isotope as a superscript to the left of the element symbol: ^{12}C, ^{13}C, ^{14}C. We read ^{12}C as "carbon-12."

 THINK ABOUT IT   The symbol ^{16}O represents oxygen with an atomic mass number of 16; it is the most common form of oxygen, containing eight protons and eight neutrons. What does the symbol ^{18}O represent?

Molecules   The number of different material substances is far greater than the number of chemical elements because atoms can combine to form **molecules**. Some molecules consist of two or more atoms of the same element. For example, we breathe O_2, oxygen molecules made of two oxygen atoms. Other molecules, such as water, are made up of atoms of two or more different elements. The symbol H_2O tells us that a water molecule contains two hydrogen atoms and one oxygen atom. Substances

oxygen   water   methane   glucose

Organic molecules contain carbon (and usually also contain hydrogen).

Compounds are molecules made from atoms of two or more different elements.

Molecules consist of two or more atoms.

Figure 3.25

Terminology of molecules. (Adapted from Campbell, Reece, Taylor, Simon, *Biology Concepts & Connections*, Fifth Edition, © 2006.)

composed of molecules with two or more different types of atoms are called **compounds**. Thus, water is a compound.

The chemical properties of a molecule are different from those of its individual atoms. For example, molecular oxygen (O_2) behaves differently from atomic oxygen (O), and water behaves differently from pure hydrogen or pure oxygen. Life on Earth is based on the complex chemistry of molecules (compounds) containing carbon, which are called **organic molecules** (or **organic compounds**). In diagrams, molecules are often represented with ball-and-stick models that show how their atoms are arranged (Figure 3.25).

Phases of Matter   Everyday experience tells us that a substance can behave dramatically differently in different *phases*, even though it is still made of the same atoms or molecules. For example, molecules of H_2O can exist in three familiar phases: as **solid** ice, as **liquid** water, and as the **gas** we call water vapor. How can the same molecules look and act so differently in these different phases?

You are probably familiar with the idea of a **chemical bond**, the name we give to the interactions between electrons that hold the atoms in a molecule together. For example, we say that chemical bonds hold the hydrogen and oxygen atoms together in a molecule of H_2O. Similar but much weaker interactions among electrons hold together the many water molecules in a block of ice or a pool of water. Thus, we can think of the interactions that keep neighboring atoms or molecules close together as another type of bond.

If we think in terms of bonds, the phases of solid, liquid, and gas differ in the strength of the bonds between neighboring atoms and molecules. Phase changes occur when one type of bond is broken and replaced by another. Changes in either pressure or temperature (or both) can cause phase changes, but it's easier to think about temperature.

Consider water as an example (Figure 3.26). At low temperatures, water molecules are bound tightly to their neighbors, making the *solid* structure of ice. As long as the temperature remains below freezing, the water molecules in ice remain rigidly held together; we often say that they have a *crystal* structure, meaning that the molecules are arranged in a precise geometrical pattern. However, the molecules within this crystal structure are always vibrating, and higher temperature means greater vibrations.

The **melting point** (0°C at sea level on Earth) is the temperature at which the water molecules finally break the solid bonds of ice. The molecules can then move much more freely among one another, allowing the water to flow as a *liquid*. However, the molecules in liquid water are

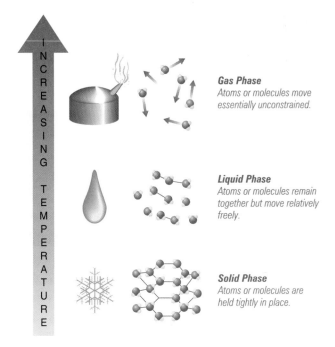

INCREASING TEMPERATURE

Gas Phase
Atoms or molecules move essentially unconstrained.

Liquid Phase
Atoms or molecules remain together but move relatively freely.

Solid Phase
Atoms or molecules are held tightly in place.

Figure 3.26

The basic progression of phase changes in water.

not completely free of one another, as we can tell from the fact that droplets of water can stay intact. Thus, adjacent molecules in liquid water are still held together by a type of bond, though a much looser bond than the one that holds them together in solid ice.

If we continue to heat the water, the average speeds of the water molecules increase, and high enough speeds will ultimately break the bonds between neighboring molecules altogether. The molecules will then be able to move freely, and freely moving particles constitute what we call a *gas*. Above the **boiling point** (100°C at sea level), all the bonds between adjacent molecules are broken so that the water can exist only as a gas.

We see ice melting into liquid water and liquid water boiling into gas so often that it's tempting to think that's the end of the story. However, a little thought should convince you that the reality has to be more complex. For example, you know that Earth's atmosphere contains water vapor that condenses to form clouds and rain. But Earth's temperature is well below the boiling point of water (luckily for us!), so how is it that our atmosphere can contain water in the gas phase?

You'll understand the answer if you remember that temperature is a measure of the *average* motion of the particles in a substance; individual particles may move substantially faster or slower than the average. Thus, even at the low temperatures at which most water molecules are bound together as ice or liquid, a few molecules will always move fast enough to break free of their neighbors and enter the gas phase. In other words, some gas (water vapor) is always present along with solid ice or liquid water. The process by which molecules escape from a solid is called **sublimation**, and the process by which molecules escape from a liquid is called **evaporation**. Higher temperatures lead to higher rates of sublimation or evaporation.

The same basic ideas hold for other substances, but their melting and boiling temperatures differ from that of water. Moreover, although we won't go into detail here, remember that pressure also has an important effect. For example, high pressure can cause a substance to remain in the solid phase even when the temperature is above the low-pressure boiling point.

• What is energy?

At the beginning of this section, we stated that life, like everything else in the universe, is at its most basic level an interplay between matter and energy. We've briefly discussed the makeup of matter, but what is energy?

In essence, energy is what makes matter move. Because this statement is so broad, we often distinguish between many different types of energy. For example, we talk about the energy we get from the food we eat, the energy that makes our cars go, and the energy a lightbulb emits. Fortunately, we can classify all these various types of energy into just three major categories (Figure 3.27):

- Energy of motion, or **kinetic energy** (*kinetic* comes from a Greek word meaning "motion"). Falling rocks, orbiting planets, and the molecules moving in the air around us are all examples of objects with kinetic energy.

- Energy carried by light, or **radiative energy** (the word *radiation* is often used as a synonym for *light*). All light carries energy, which is why light can cause changes in matter. For example, light can alter

Energy can be converted from one form to another.

kinetic energy
(energy of motion)

radiative energy
(energy of light)

potential energy
(stored energy)

Figure 3.27

The three basic categories of energy. Energy can be converted from one form to another, but it can never be created or destroyed, an idea embodied in the law of conservation of energy.

molecules in our eyes—thereby allowing us to see—or warm the surface of a planet.

- Stored energy, or **potential energy**, which might later be converted into kinetic or radiative energy. For example, a rock perched on a ledge has *gravitational* potential energy because it will fall if it slips off the edge, and gasoline contains *chemical* potential energy that can be converted into the kinetic energy of the moving car. Einstein discovered that mass itself is a form of stored energy, sometimes called mass-energy, as described by his famous formula $E = mc^2$ (introduced earlier in the chapter).

It is possible for energy to change from one form to another—indeed, such changes are the primary drivers of life. For example, our bodies take the chemical potential energy stored in food and use it to make molecules move in ways that allow our leg muscles to contract for walking, our blood and skin to create scabs over wounds, and neurons in our brains to fire in ways that make thought possible.

However, while energy can *change* from one form to another, it can be neither created nor destroyed. This idea, which we discussed briefly earlier, is what we call *the law of conservation of energy*. This law helps us understand everything from how the Sun and planets formed (discussed earlier) to the requirements of life. Although some form of stored energy is available almost everywhere, in most cases there is no viable way for life to extract the energy for its own use. Thus, as we'll discuss in later chapters, the availability of a viable energy source is one of the crucial factors that determines the habitability of a world.

• What is light?

As we have just seen, light is essentially a form of energy, one that happens to travel through space at the high speed of 300,000 kilometers per second. We've also discussed the fact that the Sun produces light from the stored energy of mass, which is released through the process of nuclear fusion. In essence, then, nuclear fusion converts a bit of the mass in hydrogen nuclei into light that ultimately flies through space until it hits something, at which point the light can give the energy to something else—our eyes, a telescope, or a plant that uses it for photosynthesis.

Characteristics of Light   In our everyday lives, we often think of light only as the illumination that makes it possible for us to see. In fact, most forms of light are invisible to our eyes.

As discovered by James Clerk Maxwell in the 1860s, light is an **electromagnetic wave;** that is, a wave that can cause charged particles (such as electrons) to move up and down. Like other types of waves (such as water waves, sound waves, or waves on a vibrating string), light is characterized by a **wavelength** and a related **frequency** (Figure 3.28). The wavelength is the distance between adjacent peaks of the wave, and the frequency measures the number of wave peaks passing a given point each second. The unit of frequency, *hertz* (hz), is equivalent to waves (or cycles) per second; for example, 10^3 hertz means that $10^3 = 1,000$ wave peaks pass by a point each second.

All light travels through empty space at the same speed: the speed of light (abbreviated c) of $c = 300,000$ km/s. Thus, wavelength and frequency are inversely related. The longer the wavelength, the lower the frequency,

Wavelength is the distance between adjacent peaks of the electric (and magnetic) field . . .

. . . while frequency is the number of times each second that the electric (and magnetic) field vibrates up and down (or side to side) at any point.

wavelength

All light travels with speed c = 300,000 km/s.

Figure 3.28

Light is an electromagnetic wave, which technically means a wave of vibrating electric and magnetic fields.

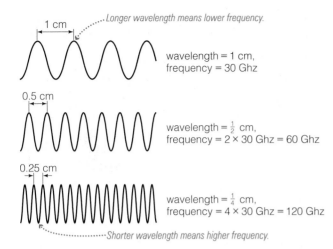

Longer wavelength means lower frequency.

wavelength = 1 cm, frequency = 30 Ghz

wavelength = ½ cm, frequency = 2 × 30 Ghz = 60 Ghz

wavelength = ¼ cm, frequency = 4 × 30 Ghz = 120 Ghz

Shorter wavelength means higher frequency.

Figure 3.29

Because all light travels through empty space at the same speed, light of longer wavelength must have lower frequency, and vice versa. The unit of frequency, hertz, is equivalent to waves (or cycles) per second. (GHz stands for gigahertz, or 10^9 hertz.)

and vice versa (Figure 3.29). This idea can be expressed with a simple formula: wavelength × frequency = speed of light.

Early in the twentieth century, scientists discovered that in addition to its wave properties, light also has properties that make it act like lots of individual particles. These particles of light, called **photons,** can exert pressure, knock individual electrons out of atoms, or cause molecules to start rotating and vibrating. We still characterize photons by their wavelength and frequency, but we can also add a third basic characteristic: the photon's *energy.* The energy of a photon is directly proportional to its frequency: The higher the frequency (or shorter the wavelength), the higher the energy.

The Electromagnetic Spectrum Light can in principle have any wavelength, frequency, or energy. The complete range of possibilities, shown in Figure 3.30, is called the **electromagnetic spectrum.** For convenience, we usually refer to different portions of the electromagnetic spectrum by different names. The **visible light** that we see with our eyes is only a tiny portion of the complete spectrum, with wavelengths from about 400 nm at the blue end of the rainbow to about 700 nm at the red end. (One nanometer [nm] is a billionth of a meter.)

Light with wavelengths somewhat longer than red light is called **infrared,** because it lies beyond the red end of the rainbow. **Radio waves**

Figure 3.30

The electromagnetic spectrum. Notice that wavelength increases as we go from gamma rays to radio waves, while frequency and energy increase in the opposite direction.

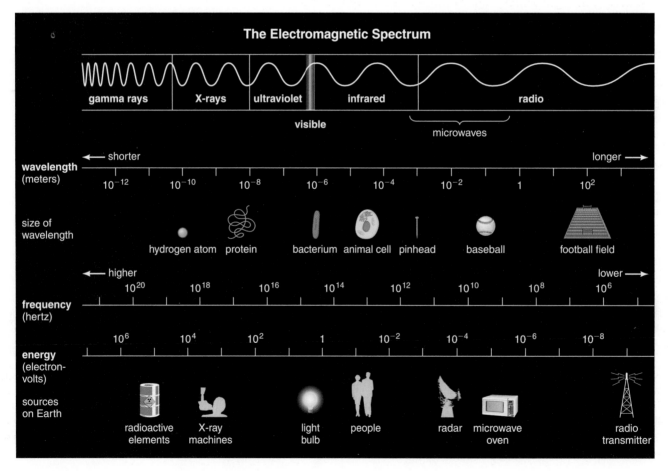

are the longest-wavelength light. Thus, radio waves are a form of light, *not* a form of sound. The region near the border between infrared and radio waves, where wavelengths range from micrometers to centimeters, is sometimes given the name **microwaves**.

On the other side of the spectrum, light with wavelengths somewhat shorter than blue light is called **ultraviolet**, because it lies beyond the blue (or violet) end of the rainbow. Light with even shorter wavelengths is called **X rays**, and the shortest-wavelength light is called **gamma rays**. Notice that visible light is an extremely small part of the entire electromagnetic spectrum: The reddest red that our eyes can see has only about twice the wavelength of the bluest blue, but the radio waves from your favorite radio station are a billion times longer than the X rays used in a doctor's office.

The different energies of different forms of light explain many familiar effects in everyday life. Radio waves carry so little energy that they have no noticeable effect on our bodies. However, radio waves can make electrons move up and down in an antenna, which is how the antenna of your car radio receives the radio waves coming from a radio station. Molecules moving around in a warm object emit infrared light, which is why we sometimes associate infrared light with heat. Receptors in our eyes respond to visible-light photons, making vision possible. Ultraviolet photons carry enough energy to harm cells in our skin, causing sunburn or skin cancer. X-ray photons have enough energy to penetrate through skin and muscle but can be blocked by bones or teeth. That is why doctors and dentists can see our underlying bone structures on photographs taken with X-ray light.

Spectroscopy Because different wavelengths of light contain different amounts of energy, the light we see from a distant object carries information about the processes that must have occurred in the object—or in the space between us and the object—to give the light the energy that it has. This basic idea makes it possible to extract an enormous amount of information from light, and it is the way we have learned almost everything we know about the nature of astronomical objects.

The most obvious way of learning about a distant object from light is to use a telescope to take a picture of it. But there are also other ways to learn from light. For our purposes in this book, one particular way of learning from light is especially important: **spectroscopy**, which involves collecting light through a telescope, then dispersing it into a *spectrum* in much the same way a prism disperses white light into a rainbow of color (Figure 3.31).

Modern technology allows us to see far more detail in an object's spectrum than we can see with a simple prism, and this detail can offer a wealth of information. For example:

- In different parts of the spectrum, an object may emit its own light, absorb incoming light, or reflect incoming light. We can learn a great deal by studying the precise patterns of emission, absorption, and reflection. For example, the planet Mars emits its own infrared light, absorbs blue light from the Sun, and reflects red light from the Sun (hence its red color). Careful study of Mars's spectrum tells us the planet's surface temperature and helps us identify minerals and ices on its surface.

- In many cases, spectra show patterns of bright or dark lines, called **spectral lines** (Figure 3.32), which serve as chemical "fingerprints." Every chemical element, every ion of each element, and every molecule produces its own, unique pattern of spectral lines. Thus,

Figure 3.31

When we pass white light through a prism, it disperses into a rainbow of color that we call a *spectrum.*

Figure 3.32

An astronomical spectrum can show much more detail than we see in light dispersed by a typical prism. Here, we see an idealized spectrum of warm molecular hydrogen gas (H_2). The specific pattern of the many bright lines tells us that they were produced by molecular hydrogen rather than by something else.

careful study of a spectrum can allow us to determine the chemical composition of a distant world. In this way, we can learn the molecular composition of a planet's atmosphere or the chemical composition of a star. Different isotopes of an element also have slightly different spectra, so we can sometimes even determine isotopic ratios in distant worlds.

- By studying spectral lines in detail—for example, how bright or dark they are, how wide they are, and what precise set of atoms and ions is represented in a spectrum—scientists can infer even more information about distant objects. Among other things, spectral lines can tell us an object's temperature, rotation rate, and, in some cases, magnetic field strength. For objects with atmospheres, spectral lines can tell us their atmospheric density and pressure and provide information about atmospheric motions.

Perhaps more than any other astronomical technique, spectroscopy would have seemed like magic to scientists as recently as two centuries ago. Indeed, before scientists discovered and began to learn the meaning of spectral lines in the nineteenth century, many people assumed that we would never be able to discover what distant planets and stars are made of.

The power of spectroscopy is limited only by how much light we can collect, not by distance. Spectroscopy has therefore allowed us to determine the composition of planets, stars, and even very distant galaxies. It is also an indispensable tool in the search for life in the universe, because it allows us to learn about habitability from a distance. Someday, we may make the first conclusive detection of life outside our solar system by finding unmistakable spectral signatures in the atmosphere of some distant planet. But that's a topic for another chapter; let's conclude our study of the universal context of life by turning to an example that will shed some light on the way science has helped us learn all the amazing things we've discussed in this chapter.

3.5 THE PROCESS OF SCIENCE IN ACTION: CHANGING IDEAS ABOUT THE FORMATION OF THE SOLAR SYSTEM

Today, we are quite confident that our solar system was born from the gravitational collapse of an interstellar cloud. But how did scientists come to accept this idea, and why are we so confident that it is correct? As we discussed in Chapter 2, scientific progress generally involves someone proposing a model to explain a variety of observations, and then putting that model to the test. Often, more than one model may be proposed, and the competition between models helps us develop them further until one may clearly win out over the others.

The development of our current model of the solar system's formation—which we now consider a *theory* because the evidence for it is so strong—occurred in precisely this way. Moreover, as is always also the case with scientific theories, the nebular theory is subject to ongoing debate and modification. Because the nebular theory is critically important to the study of life in the universe—after all, it explains how a planet like ours came to exist—we will use its history and ongoing development as this chapter's case study in the process of science in action.

• How did the nebular model win out over competing models?

After the Copernican revolution, and as we learned more about the overall layout and motions of our solar system, it was only natural that scientists would begin to speculate about how our solar system came to be. Around 1755, German philosopher Immanuel Kant became the first person known to propose that our solar system formed from the gravitational collapse of an interstellar cloud of gas. About 40 years later, French mathematician Pierre-Simon Laplace put forth the same idea independently (he apparently was unaware of Kant's proposal). Because an interstellar cloud is usually called a *nebula*, the idea of Kant and Laplace became known as the *nebular hypothesis.*

The Kant/Laplace nebular hypothesis may sound quite similar to our nebular theory today, but there was an important difference: While we now have a detailed model and plenty of evidence to support the modern theory, there was little evidence at the time to support the Kant/Laplace hypothesis. Other scientists therefore put forth other ideas. For example, in 1745, or 10 years before Kant's publication, French scientist Georges Buffon suggested that the planets had been born when a massive object (which he guessed to be a comet) collided with the Sun and splashed out debris that coalesced into the planets. This basic idea came to be the leading competitor to the nebular hypothesis. It took almost 200 years for science to collect enough data to allow us to choose between these two general ideas about how our solar system was born. To understand how one model won out, we must first understand a bit more about how we decide what types of observations a theory must explain.

Observations That a Formation Theory Must Address   The primary goal of a scientific theory is to explain a broad range of diverse observations in terms of just a few fundamental principles. But what observations should we focus on? You might at first guess that we'd want a theory to be able to explain *everything* about a particular topic, but that is neither realistic nor even useful. For example, while we expect any theory of gravity—whether it is Newton's theory, Einstein's theory, or a future theory that is even broader—to explain planetary orbits and Galileo's discovery that mass does not affect an object's rate of fall, we do *not* expect the theory of gravity to explain why a sheet of paper falls more slowly when it is flat than when it is crumpled up into a small ball. The falling paper is certainly affected by gravity, but we recognize that it is also subject to other forces, such as air resistance, and we therefore don't expect gravity alone to explain the paper's motion. Thus, one key to coming up with a successful theory of gravity was realizing the differences between those observations that gravity alone ought to be able to explain, and those in which something else might also be affecting our observations.

THINK ABOUT IT   Drop a rock and a sheet of paper from the same height at the same time. Which one falls faster? Now crumple the sheet of paper into a tight ball, and repeat your experiment. How do your results demonstrate that mass does not affect an object's rate of fall? How do they demonstrate that gravity alone cannot explain everything about falling? What would happen if you could try the same experiments outside on the Moon (where there is no air), and why?

In the case of the solar system, an enormous number of observations might seem at least potentially relevant, from general characteristics of planetary orbits to the shapes of individual asteroids. Historically, as we've learned more about the solar system, scientists at different times have focused on different sets of observations. At the time of Buffon, Kant, and Laplace, before asteroids had been discovered and before we recognized differences between terrestrial and jovian worlds, the focus was almost entirely on explaining the mere existence of planets. Half a century ago, as astronomers began to realize that asteroids and comets are by far the most populous members of the solar system, it became important for the theory to be able to explain their existence and their orbits—and fairly obvious that their general characteristics are much more important than their individual shapes. Today, we expect a formation theory to explain not only our own solar system, but also the observations we've made of planets in other star systems and of disks of material around other stars.

Overall, with our current hindsight, we can group the many known properties of our own solar system (we'll discuss other solar systems shortly) into a list of four major features that a theory of its formation must explain:

1. *Orderly motions of large bodies.* The theory should explain the organized patterns that we see in the orbits and rotations of the larger objects of our solar system. Recall, for example, that all the planets orbit the Sun with nearly circular orbits, all going in the same direction and in nearly the same plane. The orbital direction—counterclockwise as viewed from far above Earth's North Pole—is the same as the direction of the Sun's rotation, the direction of most planet rotations, and the direction in which most large moons orbit their planets.

2. *Two major types of planets.* We must also explain why the major planets divide clearly into two groups, with the small, rocky terrestrial planets close together and close to the Sun, while the large, gas-rich jovian planets are farther apart and farther out.

3. *Small bodies.* The planets are far outnumbered by small bodies—asteroids, comets (in both the Kuiper belt and Oort cloud), and Kuiper belt objects—so we must also explain how these objects formed and came to be in their current orbits.

4. *Exceptions to the rules.* Finally, the generally orderly solar system also has some notable "exceptions to the rules." For example, Earth is unique among the inner planets in having a large moon, and Uranus has an odd, sideways tilt. A successful theory of our solar system must make allowances for such exceptions even as it explains the general rules.

We have already seen how the nebular theory explains the first three of these features: The orderly motions are a consequence of the way the Sun and planets were born in a spinning disk of gas. The two types of planets arose because of the way different materials condensed and accreted at different distances from the Sun. Small bodies are essentially "leftovers" from the birth of the planets. We have not yet discussed the exceptions to the rules, but we suspect that most of them are a result of collisions between some of the "leftovers" and the planets [Section 4.6]. The fact that so many small bodies still exist tells us that at least some large collisions must have been likely in the past, so this idea is fully consistent with the nebular theory. Given all this success, why did it take so long for the nebular theory to gain acceptance? The answer has to do with the way evidence was gradually collected and studied.

The Fall and Rise of the Nebular Model Although Buffon's idea of planets forming in a collision with the Sun always had some supporters, by and large the Kant/Laplace nebular hypothesis was more popular throughout the nineteenth century. By the early twentieth century, however, scientists had found a few aspects of our solar system that the nebular hypothesis did not seem to explain well, at least in its original form. In particular, Laplace had proposed a specific physical mechanism by which he claimed the planets were made. His mechanism basically envisioned the planets forming in successive rings of gas that formed as the cloud contracted and spun faster, but the details are not important here. Instead, the important point is that his mechanism was testable, and as scientists began to put it to the test, they found that it did not work. That is, Laplace's mechanism could not actually build planets as he had thought. This failure meant the model needed to be either modified or discarded.

Some scientists sought to modify the nebular hypothesis by looking for alternative ways to build planets, while others looked for entirely different ideas about how the solar system might have formed. Before too long, a new version of Buffon's old idea began to gain favor. In this new version, instead of a direct collision with the Sun, scientists imagined a *near*-collision between the Sun and another star. According to this *close encounter hypothesis*, the planets formed from blobs of gas that had been gravitationally pulled out of the Sun during the near-collision.

For several decades, the two models battled almost to a draw. Each had at least some features that seemed to agree well with observations, and there was no conclusive evidence that favored one over the other. However, as scientists studied the models in greater depth, they learned to calculate the consequences of each model more precisely. By the mid-twentieth century, these calculations showed that the close encounter hypothesis could *not* account for either the observed orbital motions of the planets or the neat division of the planets into two major categories (terrestrial and jovian). With this clear failure, the close encounter model rapidly lost favor. Moreover, this failure made scientists take more seriously a second problem with the model: It required a highly improbable event. Given the vast separation between star systems in our region of the galaxy, the chance that any two stars would pass close enough to cause a substantial gravitational disruption is so small that it would be difficult to imagine it happening even in the one case needed to make our own solar system. While low probability alone could not rule out the close encounter hypothesis, it certainly did not help the case for a hypothesis that also failed on other grounds. (Today, we can rule out the close encounter hypothesis definitively, because the low probability of forming planets in collisions leads to a testable prediction that we now know to be wrong: If planets are born only in rare events, then planets should be rare around other stars. The fact that we now know planets to be quite common therefore rules out the close encounter hypothesis.)

At the same time that the close encounter hypothesis was losing favor, new ideas about the physics of planet formation led to modifications of the nebular hypothesis. Laplace's mechanism was discarded and replaced by the idea of condensation and accretion, and scientists soon realized that this important modification could indeed allow the nebular model to explain the major features of our solar system. Perhaps even more important, new discoveries about our solar system—such as learning of the existence of the Kuiper belt and Oort cloud and learning more about the differing compositions of planets and moons—fit quite well into the nebular model. By the latter decades of the twentieth century, so much evidence had accumulated in favor of the nebular hypothesis that it achieved the status of a scientific theory—the nebular *theory* of our solar system's birth.

Implications to Life in the Universe   The historical competition between the nebular and close encounter models may sound like scientific trivia, but it had a profound effect on scientific attitudes about life in the universe. The reason is probably clear: If the close encounter hypothesis had been correct, then other planetary systems would have been exceedingly rare. Indeed, it would have been likely that no other habitable worlds would exist in our galaxy, and perhaps even in the universe. In that case, any chance of finding life beyond Earth would have been limited to the other worlds in our own solar system. The fact that this model seemed quite plausible for much of the first half of the twentieth century partially explains why scientific interest in extraterrestrial life waned during that period. Once the close encounter model was discarded and the nebular theory gained acceptance, it became immediately clear that other planetary systems were to be *expected,* making the possibility of life on other worlds seem far more reasonable.

• Why isn't the nebular model set in stone?

Given the strength of evidence for the nebular theory, you might at first expect it to be considered "settled science," with nothing left to learn about it. But if you think back to our discussion of gravity in Chapter 2, you'll realize that even the strongest theories are never set in stone. Just as physicists are engaged in great debate today over what theory of gravity will replace and improve on Einstein's, the nebular theory is one of the hottest topics of debate in modern planetary science. As is usually the case, the scientific debate is being driven by new discoveries.

The new discoveries that are forcing reconsideration of the nebular theory fall into two major categories. First, there are the extrasolar planets with their surprising orbits. Second, there are new observations of young star systems, in which increasingly powerful telescopes are being used to allow us to see the phenomena that accompany stellar birth. The two sets of new evidence are interrelated and must be considered simultaneously by scientists, but let's start by focusing on the implications of the extrasolar planet discoveries.

As we've discussed, scientists developed the nebular model so that it neatly explains why jovian planets in our solar system exist only far from the Sun, while the terrestrial planets exist only close in. The discovery of extrasolar planets that are massive like jovian planets but located in their *inner* solar systems was therefore quite unexpected, and it immediately caused scientists to begin questioning the nebular model. At first, many scientists wondered if the nebular model might be fundamentally flawed, needing to be discarded almost entirely. However, observations of star-forming clouds and young star systems have made scientists increasingly confident in the basic idea that stars and planets are born in collapsing clouds of gas. The issues with the nebular theory, then, must be more in the details of what happens after the cloud forms a spinning disk around a young star.

We still do not know exactly how to reconcile the surprising planetary orbits with the mechanism of planetary formation in the current nebular theory, but scientists have many plausible ideas. For example, one set of models that looks quite promising starts by assuming that jovian planets do indeed form in their outer solar systems, but that gravitational interactions or other processes occurring in the spinning disk can cause them to migrate inward after they form. Scientists are actively trying to model this process with computer simulations, in hopes of seeing whether it really works or whether other ideas are needed to reconcile the nebular theory with observed extrasolar planets.

Observations of young star systems have also revealed plenty of surprises. The basic model of the nebular theory, summarized in Figure 3.22, makes the process of planetary formation look rather smooth and calm, except perhaps in the late stages of accretion when shattering collisions become possible. However, observations of young stars show that their births are actually quite violent. Even as gas is falling inward to make a central star and a rotating disk, we observe huge "jets" of matter being shot outward along the disk's rotation axis (Figure 3.33). Astronomers are working hard to understand exactly what causes these powerful outflows of matter, which offer concrete proof that star formation is a much more complex process than the basic nebular model presumes. Other observations are leading to more detailed understanding of exactly how the central star is born, a process that sheds further light on the formation of the surrounding disk.

Observations of disks themselves also reveal surprises. Some relatively old stars still have disks, suggesting that planets may not be an inevitable outcome of star formation. Other disks show structure indicating that they are quite turbulent, which is one reason that some scientists are exploring the idea of jovian planets being born from instabilities in the disk rather than from condensation and accretion.

The overall point should now be clear: The nebular theory explains so much so well that the basic concept stands on extremely solid ground. However, much remains to be learned about its details, and that makes it an exciting and ongoing topic of scientific research.

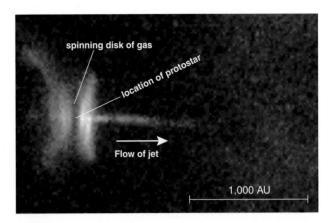

Figure 3.33

This photograph shows jets (red) being shot out along the axis of the disk of gas (green) that surrounds a *protostar*—a star that is still in the process of forming. (The disk is not really split in two as it appears; rather, the central region of the disk is darker and does not show up in this photo.)

THE BIG PICTURE
Putting Chapter 3 into Context

In this chapter, we have explored the universal context with which we conduct the search for life in the universe. As you continue in your studies, keep in mind the following "big picture" ideas:

- We are not the center of the universe, and we have no reason to think that any special circumstances contributed to make our solar system, Earth, or life. This simple idea is one of the major reasons why it seems reasonable to imagine life beyond Earth.

- We are "star stuff" in that we are made of elements that were manufactured in stars. The same elements are available to make planets and life throughout the universe, and we already have both theoretical and observational evidence that planets are common. Thus, while we cannot be sure that life exists elsewhere, we know that the necessary raw materials are available and that many worlds exist on which these raw materials might have given rise to life.

- The universe is vast, and even our own galaxy is so big that we have no hope of studying all of it, at least within the time scales of our lifetimes. Thus, it is important to learn about the general nature of worlds, so that we can come up with sensible ways in which to focus our search for life.

- Galaxies, stars, planets, and life all are ultimately a result of interactions between matter and energy. It is therefore important to have at least some understanding of matter and energy if we wish to understand the processes that make life possible, whether on Earth or beyond.

SUMMARY OF KEY CONCEPTS

3.1 THE UNIVERSE AND LIFE

- **What major lessons does modern astronomy teach us about our place in the universe?**
Three major lessons of modern astronomy are: (1) the universe is vast and old; (2) the elements of life are widespread; (3) the same physical laws that operate on Earth operate throughout the universe.

3.2 THE STRUCTURE, SCALE, AND HISTORY OF THE UNIVERSE

- **What does modern science tell us about the structure of the universe?**

We now know that Earth is a planet orbiting a rather ordinary star among the more than 100 billion stars in the **Milky Way Galaxy**, which in turn is just one among billions of galaxies. The scale of the universe is truly astronomical: If we imagine the Sun the size of a grapefruit, Earth is a pinhead about 15 meters away, while the nearest stars are thousands of kilometers away, and this is still just a tiny part of the cosmic distance scale.

- **What does modern science tell us about the history of the universe?**

The universe began about 14 billion years ago in the **Big Bang**. It has been expanding ever since, except in localized regions where gravity has caused matter to collapse into galaxies and stars. The Big Bang essentially produced only two chemical elements: hydrogen and helium. The rest have been produced by stars and recycled within galaxies from one generation of stars to the next, which is why we are "star stuff." The universe is extremely old: On a cosmic calendar that compresses the history of the universe into 1 year, human civilization is just a few seconds old, and a human lifetime lasts only a fraction of a second.

- **How big is the universe?**
Because light takes time to travel the vast distances across space, the age of the universe limits how far we can see. For a universe that is 14 billion years old, our **observable universe** extends to a distance of 14 billion light-years. This is extremely large: The total number of stars in all the galaxies of the observable universe is comparable to the number of grains of dry sand on all Earth's beaches combined.

3.3 THE NATURE OF WORLDS

- **How do other worlds in our solar system compare to Earth?**
We categorize worlds into several major types. The eight official planets form two major groups: rocky **terrestrial planets** and gas-rich **jovian planets**. Moons are especially common around jovian planets, and many of these moons are made of ice. Small bodies orbiting the Sun include rocky **asteroids**, found mostly in the **asteroid belt**; and ice-rich **comets**, found in the **Kuiper belt** just beyond the orbit of Neptune, and in the more distant **Oort cloud**.

- **Why do worlds come in different types?**

The different types of worlds are consequences of the processes that formed our solar system. According to the **nebular theory**, the solar system began with the gravitational contraction of an interstellar cloud, and laws of nature dictate that such clouds take up the shape of a disk. The Sun formed in the hot and dense center. Terrestrial planets formed in the inner parts of the disk, where high temperatures allowed only metal and rock to condense. Jovian planets and icy moons formed in the cold outer solar system, where ices could condense, and some worlds grew so large that their gravity could capture some of the surrounding gas.

• **Should we expect habitable worlds to be common?**
Our basic understanding of astronomy and planetary science suggests that solar systems like ours, with habitable worlds, should be common. However, some recently discovered extrasolar planets do not fit the expected pattern. Until we can explain this surprising fact, we will not know whether planetary systems like ours are common or rare. However, even if they are comparatively rare, the vast number of star systems in our galaxy makes it seem likely that we'll find many habitable worlds.

3.4 A UNIVERSE OF MATTER AND ENERGY

• **What are the building blocks of matter?**

Ordinary matter is made of **atoms**, which are made of **protons**, **neutrons**, and **electrons**. Atoms of different **chemical elements** have different numbers of protons. **Isotopes** of a particular chemical element all have the same number of protons but different numbers of neutrons. **Molecules** are made from two or more atoms. The appearance of matter depends on its phase: **solid**, **liquid**, or **gas**.

• **What is energy?**
Energy makes matter move, and while it comes in many different forms, we can categorize these forms into three basic categories: energy of motion or **kinetic energy**; stored or **potential energy**; and energy of light or **radiative energy**. The **law of conservation of energy** tells us that energy can change its form but can never be either created or destroyed.

• **What is light?**
Light is an **electromagnetic wave**, but also comes in individual "pieces" called **photons**. Each photon has a precise wavelength, frequency, and energy: The shorter the wavelength, the higher the frequency and energy. In order of decreasing wavelength (or increasing frequency or energy), the forms of light are **radio waves**, **microwaves**, **infrared**, **visible light**, **ultraviolet**, **X rays**, and **gamma rays**.

3.5 THE PROCESS OF SCIENCE IN ACTION: CHANGING IDEAS ABOUT THE FORMATION OF THE SOLAR SYSTEM

• **How did the nebular model win out over competing models?**
For almost 200 years, the nebular hypothesis competed with another idea that proposed a collision or near-collision with a massive object and the Sun as the mechanism for planet formation. The nebular model won out only after it was tested in great detail and the competing ideas failed the test of explaining the observed features of our solar system.

• **Why isn't the nebular model set in stone?**

Like all scientific theories, the nebular theory is subject to ongoing study and modification. In recent years, two sets of observations have forced scientists to revisit the details of the theory: observations of extrasolar planets that have surprising orbits, and observations of young star systems in the process of formation.

EXERCISES AND PROBLEMS

REVIEW QUESTIONS
Short-Answer Questions Based on the Reading

1. List three major ideas of astronomy that help frame the context of the search for life in the universe. Describe each one, along with its importance to astrobiology.

2. Briefly define and describe each of the various levels of structure illustrated in Figure 3.1.

3. Describe the solar system as it looks on the 1-to-10-billion scale used in the text. How far away are other stars on this same scale? How does this model show the difficulty of detecting planets around other stars? What does it tell us about the challenge of interstellar travel?

4. What is a *light-year?* Is it a unit of distance or time? Explain clearly.

5. Briefly describe the scale of the galaxy. How long would it take to count 100 billion stars? Why is the search for extraterrestrial intelligence (SETI) primarily a search for signals broadcast by civilizations in the past, rather than an attempt to carry out two-way radio conversations?

6. What evidence makes scientists think the universe is made mostly of *dark matter* and *dark energy,* and why are these things so mysterious? Are these mysteries likely to have an impact on the question of life in the universe? Explain.

7. What do we mean when we say that the universe is expanding? How does expansion lead to the idea of the Big Bang? Briefly describe the evidence supporting the idea that our universe began with the Big Bang.

8. What do we mean when we say that Earth and life are made from "star stuff"? Explain how this star stuff was made, and briefly describe the evidence supporting this idea.

9. Imagine describing the cosmic calendar to a friend. In your own words, give your friend a feel for how the human race fits into the scale of time.

10. What do we mean by the *observable universe?* How big is it? Answer both in absolute terms (that is, a size in light-years) and by describing a way of putting its vast size into perspective.

11. What do we mean when we say that the universe appears to be "fine-tuned" for life? Briefly describe the possible implications of this idea.

12. Briefly describe the general characteristics of each of the following types of worlds: terrestrial planets, jovian planets, moons of terrestrial planets, moons of jovian planets, asteroids, comets, and large Kuiper belt objects.

13. Describe each of the three key processes that led the *solar nebula* to take the form of a spinning disk. What observational evidence supports this scenario?

14. Briefly explain *why* we think our solar system ended up with rocky worlds in its inner regions and icy or gaseous worlds in its outer regions. How do we explain the small bodies that populate the *asteroid belt, Kuiper belt,* and *Oort cloud?*

15. Why might we expect many other solar systems to be similar to ours? What do discoveries of extrasolar planets tell us about this expectation?

16. Briefly describe the structure of an atom. What determines the atom's *atomic number?* What determines its *atomic mass number?* Under what conditions are two atoms different *isotopes* of the same element? What is a *molecule?*

17. What is the difference between matter in the phases of *solid, liquid,* and *gas?* What are *sublimation* and *evaporation?*

18. Define *kinetic energy, radiative energy,* and *potential energy.* For each type of energy, give at least two examples of objects that either have it or use it. What is the *law of conservation of energy?*

19. What are the characteristics of a *photon* of light? List the different forms of light in order from lowest to highest energy. Would the list be different if you went in order from lowest to highest frequency? From shortest to longest wavelength? Explain.

20. What is *spectroscopy,* and what can we learn from it?

21. Briefly discuss how we decide what types of evidence must be explained by a successful theory of solar system formation. Why don't we expect the theory to explain *everything?*

22. Summarize the four general features of our solar system that a solar system formation theory must address, and how the nebular theory succeeds for each feature.

23. What was the *close encounter hypothesis* for our solar system's formation, and why was it ultimately rejected in favor of the *nebular theory?* How did this rejection affect our understanding of possibilities for extraterrestrial life?

24. Briefly describe how and why recent discoveries are leading scientists to revise the nebular theory.

TEST YOUR UNDERSTANDING

Does It Seem Reasonable?

Suppose that, some day in the future, you heard the following announcements. (These are *not* real discoveries.) In each case, use what you've learned in this chapter to decide whether the announcement seems reasonable or difficult to believe. Explain clearly; because not all of these have definitive answers, your explanation is more important than your chosen answer.

25. Scientists announced today that one of the robotic Mars rovers had driven into the remains of an ancient city on Mars.

26. The *Voyager 2* spaceship, launched in 1977, has just crash-landed on a planet orbiting another star.

27. At a junior high school talent show, 14-year-old Sam Smally read off the names he had given to each of the 100 billion stars in the Milky Way Galaxy.

28. Astronomers have discovered a young solar system located in the Orion Nebula, more than 1,500 light-years away.

29. Astronomers announced that they had just found the largest extrasolar planet yet discovered, and it is made of solid gold.

30. SETI researchers announced today that if they receive a message from a civilization located on the other side of the Milky Way Galaxy, they plan to respond with a message asking the aliens 20 questions about current mysteries in science.

31. Scientists have just found the first strong evidence of life beyond Earth, and the evidence is for life on a moon orbiting the planet Jupiter.

32. Astronomers have discovered another star system that is virtually the reverse of ours: It has all its gaseous planets, icy moons, and comets in its inner regions, and its rocky planets and asteroids in its outer regions.

33. A noted physicist today announced that he has found evidence that gravity operates only on Earth, and nowhere else in the universe.

34. Using new, powerful telescopes, biologists today announced that they had discovered evidence of complex organic molecules in the atmosphere of an extrasolar planet.

Quick Quiz

Choose the best answer to each of the following. Explain your reasoning with one or more complete sentences.

35. The *Milky Way Galaxy* is: (a) another name for our solar system; (b) a small group of stars visible in our night sky; (c) a collection of more than 100 billion stars, of which our Sun is one member.

36. If we represent the solar system on a scale that allows you to walk from the Sun to Pluto in a few minutes, then: (a) the planets would be the size of basketballs and the nearest stars would be a few miles away; (b) the planets would be marble-size or smaller and the nearest stars would be thousands of miles away; (c) the planets would be microscopic and the stars would be light-years away.

37. A television advertisement claiming that a product is "light-years ahead of its time" does not make sense because: (a) It doesn't specify the number of light-years. (b) It uses "light-years" to talk about time, but a light-year is a unit of distance. (c) Light-years can only be used to talk about light.

38. When we say the universe is *expanding,* we mean that: (a) Everything in the universe is growing in size. (b) The average distance between galaxies is growing with time. (c) The universe is getting older.

39. According to observations, the overall chemical composition of our solar system and other similar star systems is approximately: (a) 98% hydrogen and helium, 2% all other elements combined; (b) 98% ice, 2% metal and rock; (c) 100% hydrogen and helium.

40. The age of our solar system is about: (a) one-third of the age of the universe; (b) three-fourths of the age of the universe; (c) 2 billion years less than the age of the universe.

41. The total number of stars in the observable universe is roughly equivalent to: (a) the number of grains of dry sand on all the beaches on Earth; (b) the number of dry grains of sand on Miami Beach; (c) infinity.

42. How many of the planets orbit the Sun in the same direction that Earth does? (a) a few; (b) most; (c) all.

43. Which of the following is *not* a general difference between terrestrial planets and jovian planets? (a) Terrestrial planets are much smaller and less massive than jovian planets. (b) Terrestrial planets are made largely of metal and rock while jovian planets also contain abundant hydrogen compounds such as methane, ammonia, and water. (c) Terrestrial planets have oceans of liquid water and jovian planets do not.

44. Some nitrogen atoms have seven neutrons and some have eight neutrons. These two forms of nitrogen are: (a) ions of each other; (b) phases of each other; (c) isotopes of each other.

INVESTIGATE FURTHER
In-Depth Questions to Increase Your Understanding

Short-Answer/Essay Questions

45. *Our Cosmic Origins.* Write one to three paragraphs summarizing why we could not be here if the universe did not contain both stars and galaxies.

46. *Perspective on Space and Time.* Come up with your own idea, different from any given in this chapter, to give perspective to some aspect of space or time, such as giving perspective to the size of our solar system, or to the Earth–Sun distance, or to the age of Earth, or to the time scale of civilization, or so on. Your goal should be to explain the size or time you have chosen in a way that will make sense to people who have not studied astronomy. Write up your explanation in the form of a short essay.

47. *Alien Technology.* Some people believe that Earth is regularly visited by aliens who travel here from other star systems. For this to be true, how much more advanced than our own would the space travel technology of the aliens have to be? Write one to two paragraphs to give a sense of the technological difference. (*Hint:* The ideas of scale in this chapter can help you contrast the distance the aliens travel easily with the distances we are now capable of traveling.)

48. *Common Levels of Technology.* In *Star Wars,* aliens from many worlds share approximately the same level of technological development. Does this seem plausible? Explain clearly. (*Hint:* Consider the scale of time and the amount of time for which our own civilization has so far existed.)

49. *Patterns of Motion.* In one or two paragraphs, explain why the existence of orderly patterns of motion in our solar system should suggest that the Sun and the planets all formed at one time from one cloud of gas, rather than as individual objects at different times.

50. *Two Kinds of Planets.* The jovian planets differ from the terrestrial planets in a variety of ways. Using phrases or sentences that members of your family would understand, explain why the jovian planets differ from the terrestrial planets in each of the following: composition, size, density, distance from the Sun, and number of satellites.

51. *Pluto and Eris.* How does the nebular theory explain the origin of objects like Pluto and Eris? How was their formation similar to jovian and terrestrial planets, and how was it different?

52. *Atomic Terminology Practice.*
    a. The most common form of iron has 26 protons and 30 neutrons in its nucleus. State its atomic number, atomic mass number, and number of electrons if it is electrically neutral.
    b. Consider the following three atoms: Atom 1 has seven protons and eight neutrons; atom 2 has eight protons and seven neutrons; atom 3 has eight protons and eight neutrons. Which two are *isotopes* of the same element?
    c. Consider fluorine atoms with nine protons and ten neutrons. What are the atomic number and atomic mass number of this fluorine? Suppose we could add a proton to this fluorine nucleus. Would the result still be fluorine? Explain. What if we added a neutron to the fluorine nucleus?
    d. The most common isotope of uranium is ^{238}U, but the form used in nuclear bombs and nuclear power plants is ^{235}U. Given that uranium has atomic number 92, how many neutrons are in each of these two isotopes of uranium?

53. *Origin of Your Energy.* Suppose you have just thrown a ball, and it is now in mid-flight so that it has energy of motion. Trace back the origin of that energy in as much detail as you can; for example, it got its energy from the throwing motion of your arm, but where did your arm get this energy? If possible, trace the energy all the way back to the Big Bang.

54. *Your Microwave Oven.* A *microwave oven* emits microwaves that have just the right wavelength needed to cause energy level changes in water molecules. Use this fact to explain how a microwave oven cooks your food. Why doesn't a microwave oven make a plastic dish get hot? Why do some clay dishes get hot in the microwave? Why do dishes that aren't themselves heated by the microwave oven sometimes still get hot when you heat food on them?

55. *A Strange Star System.* Suppose that we discovered a star system with ten planets, in which nine orbit the star in the same direction but one travels in the opposite direction. Would this observation be consistent with what we would expect according to the nebular theory? Do you think this one observation would be enough to make us discard the nebular theory, or would we just seek to revise it? Defend your opinion.

56. *Oort Cloud Impact.* Most of our solar system's comets appear to be tucked "safely away" in the distant Oort cloud, where they are highly unlikely to ever come in and hit Earth. However, we also learned that they are thought to have ended up so far away largely through the action of Jupiter. Suppose that Jupiter did not exist, and these trillion comets were located much nearer to the Sun. How would you expect the impact rate to have been different? Explain. (*Note:* We'll discuss the possible importance of this impact rate difference in Chapter 11.)

Quantitative Problems

Be sure to show all calculations clearly and state your final answers in complete sentences.

57. *Distances by Light.* Just as a light-year is the distance that light can travel in 1 year, we define a light-second as the distance that light can travel in 1 second, a light-minute as the distance that light can travel in 1 minute, and so on. Calculate the distance in both kilometers and miles represented by each of the following: (a) 1 light-second; (b) 1 light-minute; (c) 1 light-hour; (d) 1 light-day.

58. *Communication with Mars.* We use radio waves, which travel at the speed of light, to communicate with robotic spacecraft. How long does it take a message to travel from Earth to a spacecraft on Mars when: (a) Mars is at its closest distance to Earth; (b) Mars is at its farthest distance from Earth. (*Data:* The distance from Earth to Mars ranges between about 56 and 400 million kilometers.)

59. *Scale of the Solar System.* The real diameters of the Sun and Earth are approximately 1.4 million kilometers and 12,800 kilometers, respectively. The Earth–Sun distance is approximately 150 million kilometers. Calculate the sizes of Earth and the Sun, and the distance between them, on a scale of 1 to 10 billion. Show your work clearly.

60. *Moon to Stars.* How many times greater is the distance to Alpha Centauri (4.4 light-years) than the distance to the Moon? What does this tell you about the relative difficulty of sending astronauts to other stars compared to sending them to the Moon?

61. *Galaxy Scale.* Consider the 1 to 10^{19} scale on which the disk of the Milky Way Galaxy fits on a football field. On this scale, how far is it from the Sun to Alpha Centauri (real distance: 4.4 light-years)? How big is the Sun itself on this scale? Compare the Sun's size on this scale to the size of a typical atom (real diameter: about 10^{-10} meter).

62. *Counting Stars.* Suppose there are 400 billion stars in the Milky Way Galaxy. How long would it take to count them if you could count continuously at a rate of one per second? Show your work clearly.

63. *Interstellar Travel.* Our fastest current spacecraft travel away from Earth at a speed of roughly 50,000 km/hr. At this speed, how long would it take to travel the 4.4-light-year distance to Alpha Centauri (the nearest star system to our own)? Show your work clearly. (*Hint:* Recall that a light-year is approximately 9.5×10^{12} km.)

64. *Faster Trip.* Suppose you wanted to reach Alpha Centauri in 100 years. (a) How fast would you have to go, in km/hr? (b) How many times faster is the speed you found in (a) than the speeds of our fastest current spacecraft (around 50,000 km/hr)?

65. *Planet Probabilities.* Suppose that one in ten million stars is orbited by an Earth-like planet. If there are 100 billion stars in the Milky Way Galaxy, how many Earth-like planets are there in the galaxy? If there are 100 billion galaxies in the observable universe, how many Earth-like planets are there in the observable universe?

66. *What Are the Odds?* The fact that all the planets orbit the Sun in the same direction is cited as support for the nebular hypothesis. Imagine that there's a different hypothesis in which planets can be created orbiting the Sun in either direction. Under this hypothesis, what is the probability that ten planets would end up traveling the same direction? (*Hint:* It's the same probability as that of flipping a coin ten times and getting ten heads or ten tails.)

Discussion Questions

67. *The Changing Limitations of Science.* In 1835, French philosopher Auguste Comte stated that science would never allow us to learn the composition of stars. Although spectral lines had been seen in the Sun's spectrum by that time, not until the mid-1800s (primarily through the work of Foucault and Kirchhoff) did scientists recognize that spectral lines give clear information about chemical composition. Why might our present knowledge have seemed unattainable in 1835? Discuss how new discoveries can change the apparent limitations of science. Today, other questions seem beyond the reach of science, such as the question of *why* there was a Big Bang. Do you think such questions will ever be answerable through science? Defend your opinion.

68. *Lucky to Be Here?* Considering the overall process of solar system formation, do you think it was likely for a planet like Earth to have formed? Could random events in the early history of the solar system have prevented our being here today? What implications do your answers have for the possibility of Earth-like planets around other stars? Defend your opinions.

69. *Perpetual Motion Machines.* Every so often, someone claims to have built a machine that can generate energy perpetually from nothing. Why isn't this possible according to the known laws of nature? Why do you think claims of perpetual motion machines sometimes receive substantial media attention?

WEB PROJECTS

70. *Dark Matter and Dark Energy.* Look for recent discoveries that might shed light on the possible nature of dark matter or dark energy. Choose one such discovery, and write a short report on its implications to our understanding of the universe.

71. *Tour of the Solar System.* Take the virtual tour of the Voyage scale model solar system posted on the text Web site. Write a few paragraphs about which planet is your personal favorite, and why.

72. *Star Birth.* Search the Internet for recent images from the Hubble Space Telescope and other telescopes that show young star systems in the process of formation. Choose five to ten favorite images, and create a photojournal with a page for each picture, along with a short description of the picture and what it may tell us about the process of star and planet formation.

4
The Habitability of Earth

A Rock, A River, A Tree
Hosts to species long since departed
Marked the mastodon,
The dinosaur, who left dry tokens
Of their sojourn here
On our planet floor,
Any broad alarm of their hastening doom
Is lost in the gloom of dust and ages.

Maya Angelou, from *On the Pulse of the Morning*

Some people think that the vast numbers of worlds in our galaxy make it inevitable that we live in a galaxy full of life. Others think that life beyond Earth will prove to be rare or nonexistent. But no matter what you think, one fact seems indisputable: So far, we have no convincing evidence for the existence of life anyplace except right here on Earth.

This simple fact means that the scientific search for life in the universe must necessarily begin with the study of life on Earth. After all, unless we can understand why life exists here, we have little hope of understanding the prospects for finding life elsewhere. Our goal in this and the next two chapters is to understand why life thrives on our planet, so that in later chapters we can apply this understanding to the search for life on other worlds.

In this chapter, we will focus on understanding the physical conditions that make our planet habitable. We'll explore the role that geology plays in Earth's habitability, and discuss how, when, and why Earth became a suitable home for life. With this understanding under our belts, we'll then be ready to turn our attention in Chapters 5 and 6 to life itself.

4.1 GEOLOGY AND LIFE

It's easy to take for granted the qualities that make Earth so suitable for human life: a moderate temperature, abundant water, a protective atmosphere, and a relatively stable environment. But we need look only as far as our neighboring worlds of the inner solar system—the Moon and the three other terrestrial planets—to see how fortunate we are (Figure 4.1). The Moon and Mercury are airless and barren, with surfaces covered by the craters of past impacts. Venus was long hidden from view by its thick, global cloud cover, but we now know that it is a searing hothouse, with surface temperatures higher than that of a pizza oven and surface pressure nearly as great as that a kilometer deep in Earth's oceans. Mars has many features that look almost Earth-like and might indicate a hospitable past, but today Mars has an atmosphere so thin that a visiting astronaut would require a full space suit at all times, and the only surface water is frozen as ice.

Why is it that Earth alone appears to have a habitable surface among the five inner worlds of our solar system? Our distance from the Sun is

Mercury's surface looks much like that of the Moon, with many impact craters.

Cloud-penetrating radar was used to obtain this view of a twin-peaked volcano on Venus.

We see land, ocean, and snow-capped peaks in this view of Earth from orbit.

The Moon's surface is heavily cratered.

Mars lacks surface liquid water today, but features that look like dry riverbeds suggest a more hospitable past.

Figure 4.1

The inner worlds of our solar system, shown to scale, along with sample close-ups from orbiting spacecraft.

clearly an important factor, but that cannot be the whole story. After all, the Moon shares our distance from the Sun, yet lacks all of the qualities that make Earth habitable. Comparing Earth and the Moon in Figure 4.1 suggests that Earth's much larger size is important, but that can't be the whole story either, since hothouse Venus is only slightly smaller than Earth.

Scientists suspect that Earth owes its habitability primarily to a combination of its size and distance from the Sun, and perhaps to a few other factors as well. We'll devote most of this chapter to exploring how these factors have led to Earth's habitability, but first let's look a little more closely at the surprising ways in which Earth's geology is crucial to our existence.

• How is geology crucial to our existence?

Geology is a word with multiple meanings. Taken literally, it is the study of Earth (because *geo* means "Earth"), but scientists commonly extend the meaning to encompass the study of any world with a solid surface. We also use the word *geology* to describe the processes and features that shape worlds; for example, when we speak of Earth's geology, we can mean anything from the composition of our planet to the volcanoes and other processes that rework the surface.

When we consider life on relatively short time scales, such as decades or centuries, we generally don't need to think about how life interacts with our planet's geology. Except for organisms that obtain their energy directly from chemical interactions with their environment—an important class of organisms that we'll discuss further in Chapter 5—we can understand most short-term biology by considering interactions between species; only rarely are these interactions affected directly by something like a volcanic eruption. However, over longer time scales, geology and life are deeply intertwined. In fact, it is Earth's geology that has made our planet habitable, ultimately allowing not only the existence of life but also the long-term evolution of life into complex forms that include us.

Geology is important to life on Earth in many ways, but three aspects of Earth's geology stand out as being especially important:

- **Volcanism**. A volcanic eruption can be a spectacular sight, but volcanoes are important to our existence on a much deeper level: Volcanic activity releases gases trapped in Earth's interior, and these gases were the original source of Earth's atmosphere and oceans. In addition, volcanism releases heat that, we suspect, helped lead to the origin of life on our planet.

- **Plate tectonics**. Earth's surface has been shaped largely by the movement and recycling of rock between the surface and the interior. This process, called *plate tectonics,* is best known for gradually re-arranging the continents of Earth, but its most profound relevance to life involves Earth's climate: According to modern understanding, plate tectonics is largely responsible for the long-term climate stability that has allowed life to evolve and thrive for some 4 billion years.

- **Earth's magnetic field**. Compass needles point north because our planet has a global magnetic field generated deep in its interior. You may know that the magnetic field has at least a few biological effects—for example, some birds use the magnetic field to help guide their migrations—but its deeper significance is to our atmosphere. The magnetic field shields Earth's atmosphere from the energetic particles of the solar wind [Section 3.3], and without this shielding, it's likely that a significant portion of our planet's atmosphere would by now have been stripped away into space.

Because these factors seem so important to life on Earth, we'd like to understand the likelihood of finding them on other worlds. To do so, we must understand how they work and how they came about, a task to which we'll devote most of this chapter. First, however, it's important to understand how we learn about these and other processes that shape our planet, which ultimately comes down to methods for reconstructing Earth's history from the clues we find written in rocks and fossils.

4.2 RECONSTRUCTING THE HISTORY OF EARTH AND LIFE

Human recorded history dates back only a few thousand years on a planet that has existed for about $4\frac{1}{2}$ billion years. To put this fact in perspective, imagine making a timeline to represent Earth's history. On a timeline the length of a football field, human civilization would appear only in a tiny sliver at the end of the line—a sliver no thicker than a piece of paper. How, then, can we possibly know anything about the long history that preceded human civilization?

The answer is that this history is recorded in rocks and **fossils**—relics of organisms that lived and died long ago—that contain clues we can read to learn about the past. Reading this history is not as easy as reading words on a page, but with proper scientific tools it can be read reliably. Our task in this section is to explore a few of the key scientific ideas that have helped us put together a chronology of geology and life on our planet.

a This solidified lava is an example of igneous rock.

b Metamorphic rock has gone through transformations that often give a contorted appearance.

c Sedimentary rock tends to build up in layers like those visible here.

What can we learn from rocks and fossils?

Recall that matter is found in three basic phases: solid, liquid, and gas. The atoms and molecules of liquids or gases are in constant motion, remixing so rapidly that we can't possibly learn much about how they were arranged in the past. But solid objects offer a different story: Because atoms and molecules are essentially locked in place in a solid, they preserve information about the time at which they first became locked together—that is, at the time the object solidified.

Many solid objects preserve past records, including ancient bones and archaeological artifacts such as pieces of pottery or cloth. But when we seek to learn about Earth's history, we must look to the rocks and fossils that make up what we call the **geological record**. (Some people use the term *fossil record* synonymously, but the latter technically refers only to relics of life.) To see how we read this record, let's begin by discussing how we classify the rocks that preserve past history.

Types of Rocks   Geologists classify rocks into three basic types according to how they are made (Figure 4.2):

- **Igneous rock** is made from molten rock that cools and solidifies.

- **Metamorphic rock** is rock that has been structurally or chemically transformed by high pressure or heat that was not quite high enough to melt it.

- **Sedimentary rock** is made by the gradual compression of *sediments*, such as sand and silt at the bottoms of seas and swamps.

Note that rock can change from one type to another. For example, an igneous rock may be transformed by high pressure or heat into a metamorphic rock, and both igneous and metamorphic rock may be eroded into sediments and become part of a sedimentary rock. Sedimentary rock may then be carried deep underground, where it can melt and then resolidify as igneous rock. In fact, each of the three rock types can be transformed into the others, an idea often described as the *rock cycle* (Figure 4.3).

Because rock can be recycled among the three types, a rock's type does not necessarily tell us much about its composition. In fact, individual rocks of any of the three types usually contain a mixture of different crystals in close contact. Each individual crystal represents a **mineral**, which is the word we use to describe a crystal of a particular chemical composition and

Figure 4.2

Samples of the three basic rock types.

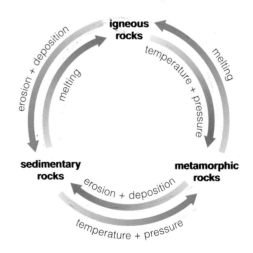

Figure 4.3

The rock cycle describes how rocks can change from one to another of the three basic types. Not shown are loops that transform rock within the individual types; for example, igneous rocks can melt and then reform as new igneous rocks.

structure. Geologists have identified more than 3,000 distinct mineral types, but we often group them by their primary constituents. For example, all minerals that contain substantial amounts of silicon and oxygen are called *silicates;* familiar silicates include quartz and feldspar. Similarly, *carbonates,* such as limestone, are minerals containing large amounts of carbon and oxygen.

Overall, a rock's type—igneous, metamorphic, or sedimentary—tells us *how* it was made, while its mineral composition tells us *what* it is made of. However, because rocks of any type may contain different mixes of minerals, geologists use many more names to subclassify rocks. For example, basalt and granite are names given to two of the most common and important subtypes of igneous rock: **Basalt** is a dark, dense igneous rock that is commonly produced by undersea volcanoes and that is rich in iron and magnesium-based silicate minerals. **Granite**, which is much lighter in color and less dense than basalt, is an igneous rock common in mountain ranges; it gets its name from its grainy appearance and it is composed largely of quartz and feldspar minerals.

Sedimentary Strata   Sedimentary rock is particularly important to our study of Earth's history not only because it may contain intact fossils, but also because it forms in a way that tends to produce a record of time. The sediments that make sedimentary rock are produced primarily by erosion on land. Wind, water, and ice can all help break up solid rock into small pieces, some smaller than a millimeter across, and these small pieces (or *grains*) comprise sediments. Sediments can be carried away by rivers and deposited on floodplains or in the oceans. Over millions of years, sediments pile up on the seafloor, and the weight of the upper layers compresses underlying layers into rock. Fossils can be made when remains of living organisms are buried along with the sediments. Aquatic organisms may be buried in sediments simply because they settle to the bottom of the sea. Some land organisms form fossils when their remains are swept into bodies of water. In other cases, remains of land organisms may be buried in place by windblown silt and later compressed by sediments deposited on top of them when changing sea levels rise over them.

Sediments deposited at different times tend to look different as a result of changes in the rate of sedimentation, in the composition or grain size of sediments settling to the bottom, or in the type of organisms leaving fossils. As a result, sedimentary rock tends to be marked by distinct layers, or **strata** (singular, *stratum*). We can view these strata in sedimentary rocks that have been exposed by, for example, the gradual action of a river carving through the rock over millions of years or by a cut made through a mountain to make way for a road. Figure 4.4 summarizes the process by

Figure 4.4

Formation of sedimentary rock. Each stratum, or layer, represents a particular time and place in Earth's history and is characterized by fossils of organisms that lived in that place and time. (Adapted from Campbell, Reece, *Biology,* Seventh Edition, © 2005.)

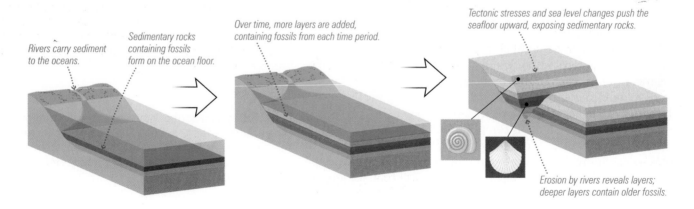

Rivers carry sediment to the oceans.

Sedimentary rocks containing fossils form on the ocean floor.

Over time, more layers are added, containing fossils from each time period.

Tectonic stresses and sea level changes push the seafloor upward, exposing sedimentary rocks.

Erosion by rivers reveals layers; deeper layers contain older fossils.

which sedimentary rock forms. Figure 4.5 shows part of the Grand Canyon, which was carved by the Colorado River. The rock layers of the Grand Canyon walls record more than 500 million years of Earth's history.

Because sedimentary rock builds up gradually over time, at any particular location the more deeply buried layers generally are older. This allows geologists to determine the *relative* ages of rocks and fossils buried in sediments. For example, because fossils of dinosaurs appear only in layers older than those in which we find fossils of primates, we conclude that dinosaurs lived before primates.

No single location contains strata from the entire history of Earth, but geologists have put together a fairly detailed geological record by comparing sedimentary strata from many sites around the world. Scientists correlate the strata from different sites by looking for layers with similar fossils. For example, suppose an upper stratum at one location contains fossils of the same type as those found in a lower stratum at another location. In that case, the first location must contain more ancient strata than the second location (Figure 4.6).

Rock Analysis You can often tell a rock's type simply from its appearance. For example, a piece of recently solidified lava is obviously igneous, while a rock with an embedded seashell is probably sedimentary. Scientists, too, often start by analyzing a rock's appearance and considering where it was found, as both of these offer clues to the rock's history. But if you really want to learn about a rock, you need to analyze it in the laboratory.

Scientists can analyze rocks in a variety of ways, but three types of analysis are particularly important in reconstructing a rock's history and hence the history of our planet.

- **Mineralogical analysis** generally means identifying the minerals present in a rock.

- **Chemical analysis** generally means determining the elemental or molecular composition of a rock or mineral. For example, chemical analysis will tell you the percentages of a rock that consist of iron, silicon, carbon, or other elements.

- **Isotopic analysis** generally means determining the ratio of different *isotopes* [Section 3.4] of particular elements in a rock. For example, oxygen has three stable isotopes: oxygen-16 (eight protons and eight neutrons) is by far the most common, but it is always

Figure 4.5

The walls of the Grand Canyon are exposed sedimentary rock in which the strata record more than 500 million years of Earth's history.

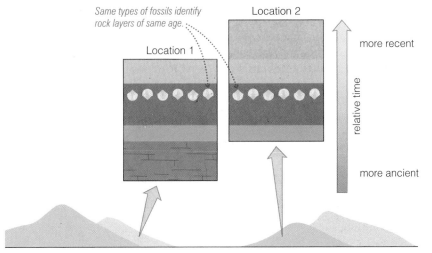

Figure 4.6

In this diagram, we imagine comparing sedimentary strata at two locations. We find that the fossils found in a particular layer near the top at Location 1 are of the same type as those found in a lower layer at Location 2. We conclude that the two sets of strata represent overlapping time periods, with the strata at Location 1 going farther back in time.

mixed with small amounts of oxygen-17 (eight protons and nine neutrons) and oxygen-18 (eight protons and ten neutrons); isotopic analysis can tell us the relative amounts of the three oxygen isotopes in a rock.

These three types of analysis are often used in tandem and each can provide clues about a rock's history; these techniques can also be applied to fossils and other solids. The importance of chemical analysis is probably fairly obvious, since it is always useful to know what an object is made of. Mineralogical analysis can tell us about the temperature and pressure conditions under which a rock formed. For example, while graphite and diamond are both minerals made of nearly pure carbon (they differ in their crystal structures), diamond forms only under much higher pressure conditions than graphite.

Isotopic analysis can be particularly illuminating, because measurements show that isotopes tend to exist in particular overall ratios in nature. For example, the overall ratio of oxygen-16 to oxygen-18 in nature is about 2,000 to 1. Thus, if we find a rock that has more than 1 in 2,000 of its oxygen atoms in the form of oxygen-18, we know that something must have happened to cause the rock to become enriched with this heavier oxygen isotope. Even more important, some isotopes turn out to be **radioactive**, meaning that their nuclei are unstable and tend to change over time in a predictable way. This peculiar property of certain isotopes means that they can serve as natural clocks that allow us to learn *when* a rock (or other solid object) formed.

• How do we learn the age of a rock or fossil?

The most reliable method for measuring the age of a rock, fossil, or other solid object is known as **radiometric dating**. This method relies on careful measurement of an object's proportions of various atoms and isotopes.

A **radioactive isotope** has a nucleus that can undergo spontaneous change, or **radioactive decay**, such as breaking apart or having one of its protons turn into a neutron. For example, carbon comes in two stable isotopes: carbon-12 (98.9% of carbon atoms), with 6 protons and 6 neutrons, and carbon-13 (1.1% of carbon atoms), with 6 protons and 7 neutrons. This stable carbon is sometimes found mixed with much smaller amounts of the radioactive isotope carbon-14, which has 6 protons and 8 neutrons. Other elements, such as uranium, are always unstable—and therefore radioactive—no matter which isotope we are dealing with. When a radioactive nucleus undergoes decay, we say that the original nucleus (before decay) is the *parent* nucleus and the changed nucleus (after decay) is the *daughter* isotope.

Radioactive decay can occur in a variety of ways. Sometimes a large atomic nucleus ejects a helium nucleus, which consists of two protons and two neutrons. (This process is often called *alpha decay*.) In that case, the remaining daughter nucleus has a lower atomic mass than its parent nucleus. For example, uranium-238 decays by ejecting a helium nucleus, leaving thorium-234 as its daughter. However, this isotope is not the final daughter of uranium-238, because thorium-234 is also radioactive. Through a chain of individual decays, eight of which involve the emission of a helium nucleus, uranium-238 ultimately decays into lead-206, which is stable and decays no further (Figure 4.7a).

In other cases, radioactive decay occurs when a nucleus spontaneously emits or absorbs an electron, causing one of its neutrons to turn

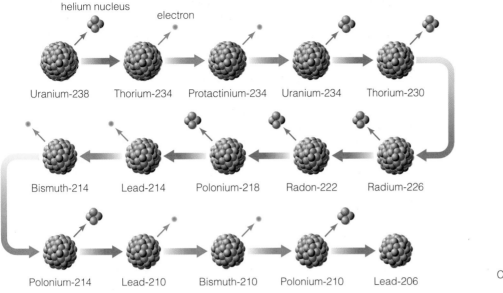

helium nucleus
electron

Uranium-238 → Thorium-234 → Protactinium-234 → Uranium-234 → Thorium-230

Bismuth-214 ← Lead-214 ← Polonium-218 ← Radon-222 ← Radium-226

Polonium-214 → Lead-210 → Bismuth-210 → Polonium-210 → Lead-206

half-life for decay chain = 4.470 billion years

a Uranium-238 decays through a chain of individual decay processes, eight of which involve the emission of a helium nucleus, ultimately leaving lead-206 as its stable daughter isotope. The half-life for the decay chain as a whole is 4.47 billion years.

Figure 4.7

Examples of radioactive decay.

Key:

• electron

● neutron

● proton

electron

Carbon-14 → Nitrogen-14

half-life = 5,730 years

b Carbon-14 decays by emitting an electron from its nucleus, which changes one neutron into a proton to make nitrogen-14 as its stable daughter. The half-life of carbon-14 is 5,730 years.

into a proton, or a proton to turn into a neutron. (The processes are called *beta decay* and *electron capture*, respectively; an absorbed electron is one of the atom's own electrons that gets captured by the nucleus.) In these cases, the parent and daughter nuclei will both have the same atomic mass number (the same total number of protons and neutrons), but they will represent different elements because they have different numbers of protons. For example, carbon-14 decays by emitting an electron as one of its neutrons becomes a proton, so its daughter isotope has seven protons and seven neutrons and therefore is nitrogen-14 (Figure 4.7b).

Regardless of the decay process, radioactive decay always occurs at a specific and measurable rate that is different for every radioactive isotope. Thus, the basic idea behind radiometric dating is to determine the age of a rock (or other solid object) from the ratio of parent and daughter atoms within it, which depends only on the decay rate and the length of time over which the decay has been occurring. To fully understand how this technique works, we must explore the nature of radioactive decay in a bit more detail.

The Probabilistic Nature of Radioactive Decay   Radioactive decay is governed by the laws of quantum physics, which means it is a probabilistic process, much like coin tossing. If you toss a single coin, you cannot determine for sure whether it will land heads or tails. All you can say is that it has a 50% chance of landing heads and a 50% chance of landing tails. However, while the probabilistic nature of coin tossing means you cannot predict the outcome for a single coin toss, it allows you to predict the outcome accurately for many coin tosses. For example, if you toss a fair coin one million times, you can be quite certain that heads will come up close to 500,000 times (but you would not expect the number to be *exactly* 500,000).

To convince yourself that probability can be used to make predictions involving large numbers of atoms or coins, toss a coin 100 times (or toss 100 coins all at once) and record your results. Do you get a result close to 50, as we would expect? Compare your results with those of other students, and discuss what they tell you about the nature of probability.

In the case of radioactive decay, the relevant probability describes the chance of a single nucleus decaying within a specific amount of time (such as within a year). However, because atoms are so tiny, we nearly always deal with such huge numbers of them that we can ignore the individual probabilities and focus instead on the rate at which large numbers of the radioactive nuclei decay.

For example, suppose we study a sample of 1 microgram of a radioactive substance, which despite its small weight (about one-millionth the weight of a paper clip) contains trillions of individual atoms. If we find that 1% of the atoms in the sample undergo radioactive decay within 1 year—meaning that at the end of the year we have 0.99 microgram of the parent substance remaining—we can conclude that any individual atom has a 1% probability of decaying in a 1-year period.

Note that, like coin tosses, results from the past do not affect results for the future. Just as a coin landing heads on one toss does not change the 50% probability of its landing heads on the next toss, the past history of the radioactive sample does not affect the future probability of decay. In our example, we would find that 1% of the atoms in the sample decay in any 1-year period, regardless of how we obtained the sample, how many atoms it contains, or how long we have been studying it. We can therefore reliably and reproducibly measure the decay rate for any radioactive substance by studying a sample of the substance in a laboratory.

The Concept of a Half-Life   Although the probability of decay in a 1-year period is a perfectly good way to describe a substance's decay rate, we usually describe the decay rate by the substance's **half-life**—the time it would take for half the atoms in a sample of the substance to decay. That is, the amount of radioactive substance in any size sample drops by half with each half-life.

Consider the radioactive decay of potassium-40, which undergoes spontaneous change into argon-40 when its nucleus absorbs an electron to change one of its protons into a neutron. Laboratory measurements show that this decay process has a half-life of 1.25 billion years. Suppose a small piece of rock contained 1 microgram of potassium-40 and no argon-40 when it formed (solidified) long ago. The half-life of 1.25 billion years means that half the original potassium-40 would have decayed into argon-40 by the time the rock was 1.25 billion years old, so at that time the rock would have contained $\frac{1}{2}$ microgram of potassium-40 and $\frac{1}{2}$ microgram of argon-40. Half of this remaining potassium-40 would then decay by the end of the next 1.25 billion years, so after 2×1.25 billion $= 2.5$ billion years, the rock would have contained $\frac{1}{4}$ microgram of potassium-40 and $\frac{3}{4}$ microgram of argon-40. After three half-lives, or 3×1.25 billion $= 3.75$ billion years, only $\frac{1}{8}$ microgram of potassium-40 would have remained, while $\frac{7}{8}$ microgram had become argon-40. Figure 4.8 summarizes the gradual decrease in the amount of potassium-40 and the corresponding rise in the amount of argon-40.

Briefly describe how it is possible for us to know that potassium-40 has a half-life of 1.25 billion years, even though we have been capable of studying potassium-40 in the laboratory for only a few decades.

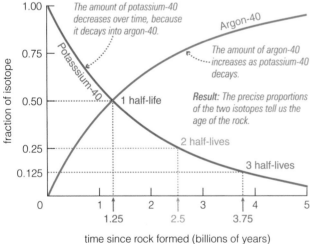

Figure 4.8

Potassium-40 is radioactive, decaying into argon-40 with a half-life of 1.25 billion years. The red line shows the decreasing amount of potassium-40, and the blue line shows the increasing amount of argon-40. The remaining amount of potassium-40 drops by half with each successive half-life.

The Essence of Radiometric Dating   The potassium-40 example shows the essence of radiometric dating. Suppose you find a rock (or a mineral within a rock) that contains equal amounts of potassium-40 and argon-40. As long as you can be confident that the rock contained no argon-40 when it formed (which you can, as we'll discuss shortly) and has lost none during its history, you can conclude that the rock is one half-life, or 1.25 billion years, old. If instead the rock contains seven times as much argon-40 as potassium-40—that is, only $\frac{1}{8}$ of the original potassium-40 remains—you can conclude that it is three half-lives, or 3.75 billion years, old. More generally, you can get the age of any such rock from the relative amounts of potassium-40 and argon-40 and the graphs in Figure 4.8.

Radiometric dating with other isotopes works the same way. By comparing the amounts of parent and daughter isotopes in the rock (or other object), we can determine the fraction of parent isotopes that has decayed since the rock formed.* Then, based on laboratory measurements of the decay rate (half-life), we can determine the rock's age. Table 4.1 lists several of the parent–daughter isotope pairs commonly used in radiometric dating. Notice that the half-lives vary dramatically, so different pairs are useful for dating materials of different ages. For example, carbon-14 is useful only for dating objects less than about 50,000 years old, while uranium-238 is useful for dating the oldest rocks on Earth.

The major issue that sometimes arises in radiometric dating concerns knowing a rock's original composition. If we're going to determine the rock's age by comparing the amounts of a radioactive isotope and its daughter product, we must have some way of knowing how much of the daughter was originally present, if any. In the case of potassium-40, this is quite easy. The daughter product, argon-40, is a gas that does not combine with other elements and can become trapped in rock only under special circumstances. Thus, argon-40 generally cannot be part of a rock when it solidifies, so any argon-40 trapped inside a rock must have come from radioactive decay of potassium-40. Moreover, since the argon-40 atoms become trapped within the rock at the time they are produced by the decay of potassium-40, we can be similarly confident that no argon has escaped from the rock, unless the rock has undergone heating or some other transformation that could have released the trapped gas.

Radiometric dating with other substances is not always as easy, but detailed study of several isotope ratios or several different mineral grains within a single rock often allows geologists to determine a rock's original composition. Still, if a rock has undergone partial melting or a major shock during its history, or if water has removed some atoms from the rock, we may find somewhat different ages when we date the rock with different parent–daughter isotope pairs. Nevertheless, decades of experience in working with these complications have given geochemists a precise understanding of how they affect radiometric dating. As a result, it is nearly always possible to know whether an age is correct or uncertain.

Note that, while radiometric dating is an extremely powerful technique, it works only for dating solids that have not undergone significant change since they formed. Thus, for rocks, we get unambiguous ages only for igneous rocks. Metamorphic rocks are more challenging to date, because they may have undergone change that has altered their isotope ratios from what they would be due to decay alone. Still, we can often place

TABLE 4.1 *Selected Isotopes Used for Radiometric Dating of Rocks and Fossils*

Some of the isotopes decay in several stages of parent–daughter pairs, and only the final daughter product is shown. (Source: Berkeley Laboratory Isotopes Project)

Parent Isotope	Daughter Isotope	Half-Life
Rubidium-87	Strontium-87	49.4 billion years
Lutetium-176	Hafnium-176	37.1 billion years
Thorium-232	Lead-208	14.0 billion years
Uranium-238	Lead-206	4.47 billion years
Potassium-40	Argon-40	1.25 billion years
Uranium-235	Lead-207	704 million years
Aluminum-26	Magnesium-26	717,000 years
Carbon-14	Nitrogen-14	5,730 years

---

*The specific methodology varies in some cases. Carbon-14, for example, is continually produced in Earth's atmosphere by interactions between nitrogen-14 and high-energy particles coming from the Sun. An estimate of the original carbon-14 content of a fossil is based on the atmospheric production rate rather than on a parent–daughter isotope ratio.

COSMIC CALCULATIONS 4.1

Radiometric Dating

The amount of a radioactive substance decays by half with each half-life, so we can express the decay process with a simple formula relating the current amount of a radioactive substance in a rock to the original amount:

$$\frac{\text{current amount}}{\text{original amount}} = \left(\frac{1}{2}\right)^{t/t_{half}}$$

where t is the time since the rock formed and t_{half} is the half-life of the radioactive material. We can solve this equation for t by taking the logarithm of both sides and rearranging the terms:

$$t = t_{half} \times \frac{\log_{10}\left(\frac{\text{current amount}}{\text{original amount}}\right)}{\log_{10}\left(\frac{1}{2}\right)}$$

Example: You chemically analyze a small sample of a meteorite. Potassium-40 and argon-40 are present in a ratio of approximately 0.85 unit of potassium-40 atoms to 9.15 units of gaseous argon-40 atoms. (The units are unimportant, because only the relative amounts of the parent and daughter materials matter.) How old is the meteorite?

Solution: Because no argon gas could have been present in the meteorite when it formed (see discussion in text), the 9.15 units of argon–40 must originally have been potassium–40 that has decayed with its half-life of 1.25 billion years. The sample must therefore have started with 0.85 + 9.15 = 10 units of potassium-40 (the original amount), of which 0.85 unit remains (the current amount). The formula now reads:

$$t = 1.25 \text{ billion yr} \times \frac{\log_{10}\left(\frac{0.85}{10}\right)}{\log_{10}\left(\frac{1}{2}\right)}$$

$$= 1.25 \text{ billion yr} \times \left(\frac{-1.07}{-0.301}\right)$$

$$= 4.45 \text{ billion yr}$$

This meteorite solidified about 4.45 billion years ago.

fairly clear age constraints on metamorphic rocks, such as a minimum time since a particular rock formed, and learn much about the changes they have undergone. A sedimentary rock cannot be dated as a whole because it is a compressed mix of rock grains containing minerals of different ages; we can sometimes date mineral grains found within a sedimentary rock, but this tells us when the grains originally formed as opposed to when the sediments were deposited or compressed into a solid rock. To estimate ages of sedimentary rocks, we therefore generally rely on dating igneous rocks buried above them or intruding into them. Similar ideas apply when we use radiometric dating for fossils and other artifacts: A radiometric age will be reliable if we can be confident in knowing precisely how much of the original parent isotope has decayed, and any uncertainty in decay fraction will mean a corresponding uncertainty in the age.

The Reliability of Radiometric Ages Scientists today have great confidence in the reliability of ages measured through radiometric dating. The underlying processes that govern radioactive decay are well understood, and the measurements needed to determine half-lives are straightforward when done with care. Perhaps most important, in many cases scientists can use several different radioactive isotopes to measure the ages of rocks or fossils. The ages from different parent-daughter isotope pairs almost always agree for a particular rock, fossil, or mineral grain, and when they don't it is usually possible to determine a cause for the difference. This agreement shows that the technique of radiometric dating is highly reliable. Ages of different rocks and fossils from particular strata of sedimentary rock also agree, further confirming the reliability of the technique.

Additional checks come from the fact that, at least in some cases, we can estimate ages in ways that are independent of radiometric dating. For example, if radiometric ages are correct, then they should confirm the ordering of the relative ages in sedimentary strata—and they do. A more precise check can be made for relatively young archaeological artifacts: Some ancient Egyptian artifacts have dates printed on them, and the dates agree with ages found with radiometric dating; in other cases, we can confirm radiometric ages by comparing them to ages that we can obtain from tree ring data. We can even confirm the $4\frac{1}{2}$-billion-year radiometric age for the solar system as a whole by comparing it to an age based on detailed study of the Sun. Theoretical models of the Sun, along with observations of other stars, show that stars slowly expand and brighten as they age. The model ages are not nearly as precise as radiometric ages, with uncertainties of up to a billion years or so, but the current state of the Sun agrees with what we expect for a star of its mass and age.

Taken together, our theoretical understanding of radioactive decay and the various ways of confirming ages found by radiometric dating leave no room for doubt about the reliability of the technique. Today, scientists routinely use radiometric dating to determine ages of rocks and fossils, and the overall uncertainty in radiometric age dates is usually less than about 1–2%.

• What does the geological record show?

Today, we have a detailed understanding of many of the events that mark the geological record, because geologists have carefully studied and dated tens of thousands of rocks and fossils. But before we get into the specifics, it's important to remember that while radiometric dates can be quite precise even very far back in time, our understanding of Earth's prevailing conditions becomes less certain as we look deeper into the past. Any individual rock or fossil may give us only limited information about the conditions under which it formed, so a fuller understanding requires studying many rocks and fossils from the same time period. However, the geological record is more sparse as we look further back. Earth's surface undergoes continual change through volcanism, plate tectonics, and erosion, making older rocks comparatively rarer than younger ones. For fossils left by life, the problem is compounded by other factors, best understood if we think about how fossils form.

Fossil Formation   Although we tend to think of fossils as "remains" of living organisms, most fossils contain little or no organic matter. Figure 4.9 summarizes some of the ways in which fossils are made. In general, when an organism dies and gets buried in sediments, minerals

Figure 4.9

A gallery of fossils.

a A dinosaur bone preserved in sandstone in Dinosaur National Monument, which straddles Utah and Colorado.

b A 190-million-year-old petrified (stone) tree in Arizona.

c These 375-million-year-old impressions are casts of dead organisms (called ammonites) made when minerals filled the empty space left after the organism decayed.

d This 40-million-year-old leaf still retains organic material, including DNA.

e An insect preserved in hardened tree resin (often called *amber*).

f These tusks belong to a whole 23,000-year-old mammoth discovered in Siberian ice in 1999.

g This boy is standing in a 150-million-year-old dinosaur track in Colorado.

dissolved in groundwater gradually replace organic material. Mineral-rich portions of organisms, such as bones, teeth, and shells, may be left behind, becoming fossils like those of the dinosaur bones displayed in many museums (Figure 4.9a). In some cases, the mineral replacement is complete and organisms literally turn to stone; the "stone trees" of Arizona's Petrified Forest formed in this way (Figure 4.9b). In many other cases, the organisms themselves decay, but in doing so they leave an empty mold that fills with minerals dissolved in water. The minerals may then make a cast in the shape of the dead organism (Figure 4.9c).

More rarely, some of the organic material from a dead organism may be preserved well enough to allow at least some study. Some fossil plant leaves are still green and well enough preserved for their cells to be studied with microscopes, even though they died millions of years ago (Figure 4.9d). In other rare cases, whole organisms may be preserved in tree resin (Figure 4.9e) or frozen in ice (Figure 4.9f). One of the most interesting types of fossil is left not by a dead organism but by the activity of an organism while it was alive. For example, *coprolites* are rocks that consist of petrified excrement, which can allow us to learn about an animal's diet. In other cases, scientists have found fossilized dinosaur footprints, made when mineral processes preserved impressions left by a dinosaur as it walked through soft soil or mud (Figure 4.9g). Such fossil tracks provide clues about how dinosaurs walked and can help scientists hypothesize about dinosaur behavior.

THINK ABOUT IT   Molecules of DNA tend to break down rapidly and easily after an organism dies, making it difficult or impossible to identify even fragments of DNA in most fossils. What types of fossils do you think are most likely to yield intact DNA? What types are least likely to yield intact DNA? Explain.

Although fossils can be made in numerous ways, only a tiny fraction of living organisms leave behind any kind of fossil at all. The vast majority of dead organisms decay—becoming food for living organisms in the soil—long before any mineral replacement can occur. If you've ever read about criminologists exhuming the skeletons of people who have been dead for just a few years, you know that even bones and teeth usually decay quite rapidly after death.

The tiny fraction of organisms that have left fossils in relatively recent times in Earth's history still means a substantial fossil record. But fossils become rarer as we look deeper into the geological record, primarily for two reasons. First, fossils can suffer the same fates as rocks, so older fossils are more likely to have been destroyed over time by volcanism, erosion, or other geological processes. The second reason comes from the nature of past life itself. Studies of the geological record show that large plants and animals—which make the most easily discovered fossils—are relatively recent arrivals in Earth's history. For the first 90% or so of Earth's history, any living organisms were all microscopic, which means their fossilized remains must also be microscopic. Clearly, it is much more difficult to find and identify microscopic fossils than huge fossils of dinosaur bones. This fact is especially important in astrobiology, since we generally focus more on the origin and early evolution of life than on large plants and animals. Indeed, as we'll discuss in Chapter 6 , the difficulty of finding and studying microfossils (and other potential traces of life in very old rocks) has led to significant scientific controversies about exactly when life took hold on Earth.

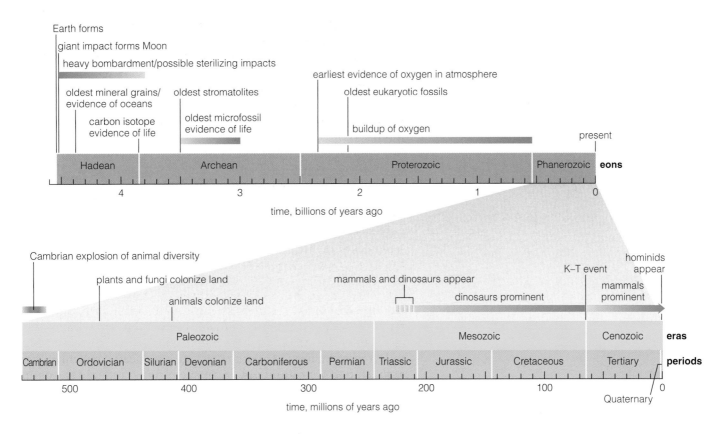

Earth forms
giant impact forms Moon
heavy bombardment/possible sterilizing impacts
earliest evidence of oxygen in atmosphere
oldest mineral grains/ evidence of oceans
oldest stromatolites
oldest eukaryotic fossils
carbon isotope evidence of life
oldest microfossil evidence of life
buildup of oxygen
present

| Hadean | Archean | Proterozoic | Phanerozoic | **eons** |

time, billions of years ago

Cambrian explosion of animal diversity
plants and fungi colonize land
mammals and dinosaurs appear
hominids appear
K–T event
animals colonize land
dinosaurs prominent
mammals prominent

| Paleozoic | Mesozoic | Cenozoic | **eras** |

| Cambrian | Ordovician | Silurian | Devonian | Carboniferous | Permian | Triassic | Jurassic | Cretaceous | Tertiary | **periods** |

Quaternary

time, millions of years ago

The Geological Time Scale

Although the geological record gives us far more information about recent events than the distant past, geologists have nevertheless been able to piece together a fairly detailed history of our planet going nearly all the way back to the time of its formation. To help organize this history, scientists divide it into a set of distinct intervals that make up what we call the **geological time scale**. Figure 4.10 shows the names of the various intervals on a timeline, along with numerous important events that we will discuss later in this chapter and in Chapter 6. It is not necessary to memorize the names of the geological time intervals for the purposes of this book. However, the names are commonly used in books and articles about Earth or astrobiology, so it's useful to be familiar with them.

The first major division of geological time is into a set of four **eons**: the Hadean, Archean, Proterozoic, and Phanerozoic. We can understand these names by working backward in time and looking at their Greek roots. The Phanerozoic eon extends from the present back to about 540 million years ago; its name comes from the Greek for "visible life" because it is marked by the presence of fossils visible to the naked eye. The Proterozoic eon, which extends from 540 million to about 2.5 billion years ago, means the eon of "earlier life" because it shows fossils of single-celled organisms that lived before the Phanerozoic eon. The Archean eon extends from 2.5 to about 3.85 billion years ago and is named for "ancient life"; it got this name after the discovery of fossils from the first half of Earth's history. The Hadean eon got its name at a time when it was presumed that the early Earth would have had "hellish" conditions during this early time (Hades was the Greek mythological name for the underworld); however, as we'll discuss in the next section, recent evidence suggests that the Hadean may not have been quite that bad.

The fact that the geological record is much richer for more recent times is reflected in the more detailed naming system used for these times.

Figure 4.10

The geological time scale. Notice that the lower timeline is an expanded view of the last portion (Phanerozoic) of the upper timeline. The eons, eras, and periods are defined by changes in the rocks and fossils present in the geological record. The absolute ages come from radiometric dating. (Some of the events listed on the timeline are discussed later in this chapter, and others are discussed in Chapter 6.)

The most recent, or Phanerozoic, eon is subdivided into three major **eras**: the Paleozoic, Mesozoic, and Cenozoic. These names also have Greek roots and mean, respectively, "old life," "middle life," and "recent life." The three eras are further subdivided into **periods**. The periods do not follow any consistent naming scheme. For example, the Cambrian period* gets its name from the Roman name for Wales (in Great Britain), the Jurassic period gets its name from rocks found in the Jura mountains of Europe, and the Tertiary period simply means "third" period. The recent geologic periods are even further subdivided into *epochs* and *ages,* but these are not shown in Figure 4.10.

Note that the eons, eras, and periods do not have uniform lengths. For example, the Proterozoic eon is about three times the length of time of the Phanerozoic eon, and the Paleozoic era is longer than the Mesozoic and Cenozoic eras combined. The divisions in the geological time scale are determined not by duration but by specific changes in the geological record, such as where certain species disappear and other new ones appear.

The Age of Earth Figure 4.10 starts with Earth's birth, but we cannot read the precise time at which that occurred from the geological record. The oldest intact Earth rocks date to about 4.0 billion years ago. Apparently, rocks that formed prior to this time have been remelted or reshaped so much that we cannot obtain an age through radiometric dating. So how do we know the age of our planet?

We have a variety of ways of looking back to earlier times. The most direct way is based on studies of tiny mineral grains of zirconium silicate, or *zircons* for short (Figure 4.11). Although they are found embedded in much younger sedimentary rocks, radiometric dating (based on analysis of uranium and hafnium isotopes found within them) shows that some zircons solidified as much as 4.4 billion years ago. Moreover, their isotopic content suggests that they formed at a time when continents had already begun to form, suggesting that Earth's crust had separated from interior material by about 4.5 billion years ago.

Moon rocks also help us constrain Earth's age. Some Moon rocks brought back by the *Apollo* astronauts are considerably older than any intact Earth rocks, reflecting the fact that volcanism and other geological processes have done far less reshaping of the Moon's surface than they have of Earth's surface. Radiometric dating with isotopes of uranium and lead places the ages of the oldest Moon rocks at more than 4.4 billion years, telling us that Earth and the Moon must have formed before this time. In fact, the leading model for the Moon's formation holds that it formed as a result of a *giant impact* that blasted material out of Earth's outer layers; we'll discuss this model in detail in Section 4.6. Earth must therefore be at least slightly older than the Moon, which means older than 4.4 billion years.

While the zircons and lunar rocks set a minimum age for Earth, we can set a maximum age by dating the formation of the solar system as a whole. We do this by studying meteorites. Some meteorites are younger than the solar system as a whole, because they are fragments of asteroids that formed and later shattered in collisions. However, a large number of meteorites have a chemical structure that suggests they were among the

Figure 4.11

This image shows a tiny zircon crystal—about 0.2 millimeter across—that was found embedded in a rock formation in Western Australia. Radiometric dating shows it to be nearly 4.4 billion years old. The image was made with a technique called cathodoluminescence, in which an electron beam is focused on the crystal, causing it to emit visible light; the colors in the image are not real, but instead correspond to differing intensities in the emitted light.

*The Cambrian is the earliest period in the Phanerozoic eon and was once thought to be the first period in which fossil organisms could be found. For that reason, the entire time before the Phanerozoic eon—that is, the Hadean, Archean, and Proterozoic eons—is often called the *Precambrian.*

first pieces of solid material to condense in the early history of the solar system. These meteorites also all have about the same age, offering further evidence that they represent material from the very beginning of the solar system; these date to 4.57 billion years ago (with an uncertainty of less than about 0.02 billion years).

The meteorite data allow scientists to use other techniques to further constrain the ages of both Earth and the Moon. These techniques are somewhat complex in their details, but they are based on comparisons of isotope ratios in meteorites, Earth rocks, and Moon rocks. The results show that both Earth and the Moon had formed within about 50 to 70 million

KEY GEOLOGICAL DEFINITIONS

This box summarizes the definitions of a few of the most basic geological terms used in this chapter and throughout the book.

ROCKS, MINERALS, FOSSILS:

minerals: The basic pieces of solid rock; a particular mineral is distinguished from other minerals by its chemical composition or crystal structure (or both).

rocks: Intact solids that may contain a variety of minerals. We classify rocks into three basic types by their formation process: *Igneous rock* is of volcanic origin, made when molten rock cools and solidifies. *Sedimentary rock* is made by the gradual compression of sediments, which may contain bits of other rock types as well as fossils. *Metamorphic rock* was once either igneous or sedimentary but has since been transformed (but not melted) by high heat or pressure. Rocks of any of the three types may be subclassified by the minerals they contain.

fossil: Any relic left behind by living organisms that died long ago.

TERMS RELATED TO GEOLOGICAL TIME:

geological record: The information about Earth's past that is recorded in rocks and fossils; the latter record is sometimes called the *fossil record.*

geological time scale: The time scale used to measure the history of Earth.

radiometric dating: The method of determining the age of a rock or fossil from study of radioactive isotopes contained within it.

half-life: The time it takes for half of the atoms to decay in a sample of a radioactive substance.

TERMS RELATED TO EARTH'S GEOLOGICAL HISTORY:

differentiation: A process in which materials separate by density. In Earth, differentiation led to a dense *core* made mostly of iron and nickel, a rocky *mantle* made mostly of *silicates* (minerals rich in silicon and oxygen), and a *crust* made of the lowest-density rocks.

heavy bombardment: The period of time during which the planets were heavily bombarded by leftover planetesimals, starting from the time the planets first formed and likely ending some 3.8–4.0 billion years ago.

outgassing: The process of releasing gases trapped in a planetary interior into the atmosphere.

lithosphere: The layer of cooler, more rigid rock that sits above the warmer, softer mantle rock below. It encompasses both the crust and the uppermost portion of the mantle, extending to a depth of about 100 kilometers. On Earth, the lithosphere is broken into a set of large *plates.*

seafloor crust: The relatively dense, thin, young crust found on Earth's seafloors, composed largely of the igneous rock called *basalt.*

continental crust: The thicker, lower-density crust that makes up Earth's continents. It is made when remelting of seafloor crust allows lower-density rock to separate, and typically consists of granite. Continental crust ranges in age from very young to as old as about 4.0 billion years.

plate tectonics: The geological process in which lithospheric plates move around the surface of Earth. It acts like a conveyor belt, with new seafloor crust erupting and spreading outward from mid-ocean ridges and then being recycled back into the mantle by subduction at ocean trenches. It also explains *continental drift,* because plates carry the continents with them as they move.

TERMS RELATED TO CLIMATE AND CLIMATE REGULATION:

greenhouse effect: The effect that makes a planet's surface warmer than it would be in the absence of an atmosphere. It is caused by the presence of *greenhouse gases*—such as carbon dioxide (CO_2), water vapor (H_2O), and methane (CH_4)—that can absorb infrared light emitted by the planetary surface (after the surface is heated by sunlight).

carbon dioxide cycle (CO_2 cycle): The cycle that keeps the amount of carbon dioxide in Earth's atmosphere small and nearly steady and hence keeps Earth habitable. Over time, this cycle has locked up most of Earth's carbon dioxide in *carbonate* rocks (rocks rich in carbon and oxygen) such as limestone.

ice ages: Periods of time during which Earth becomes unusually cold, so that water from the oceans freezes out as ice and covers a substantial portion of the continents.

Snowball Earth: Refers to periods of extreme ice ages that may have occurred several times before about 580 million years ago.

global warming: Usually refers to the current warming of Earth caused by human input of greenhouse gases into the atmosphere.

years (0.05 to 0.07 billion years) after the oldest meteorites formed. We therefore conclude that our planet accreted quickly in the early solar system, and the giant impact that formed the Moon happened quite early in Earth's history as well. By about 4.5 billion years ago, our planet had its Moon and must have been essentially at its current mass and size, ready for its geology to begin shaping the features that would ultimately make it our home.

4.3   THE HADEAN EARTH AND THE DAWN OF LIFE

We know comparatively little about the Hadean eon, which constitutes a little more than the first half-billion years of Earth's history. Nevertheless, it was clearly an important time in the history of our planet, and evidence that we'll discuss in Chapter 6 shows that life arose during or not long after this period of time. So if we want to understand when and how our planet became habitable and gave birth to life, we need to start back in this earliest of Earth's time periods.

• How did Earth get an atmosphere and oceans?

Before Earth could have life, it needed an atmosphere and liquid water oceans. Neither existed when Earth first formed. Our planet was too small and too warm to capture significant amounts of hydrogen and helium gas as it accreted within the solar nebula [Section 3.3], and no other gases were present in large enough quantities to make a substantial atmosphere. In fact, the presence of Earth's atmosphere and oceans once posed a mystery to solar system formation theory, because the planetesimals that condensed at Earth's distance from the Sun should have been made only of rock and metal, with no gaseous or icy content at all. How, then, did Earth obtain the water and gases that make up our oceans and atmosphere?

◤ MOVIE MADNESS: FLINTSTONES AND DINOSAURS

The Paleolithic (early stone age) began roughly 2 million years ago. You can be sure that for any of your ancestors unlucky enough to be born in those dim and distant times, life was generally short and brutish.

That's reality. But the impression you'd get from watching Hollywood's stone age sitcom *The Flintstones* is that our predecessors enjoyed modern lifestyles and contemporary situations. The only real difference between their existence and yours was that all the Flintstones' technology was rock-based, with domesticated dinosaurs supplying the power. Cars rolled on stone wheels, smiling dinos served as cranes and derricks, and air travel was facilitated by obliging pterodactyls. Fred Flintstone and his buddy Barney were smooth-skinned and beardless (try shaving with a rock razor), and regularly took showers using elephants to provide the spray, although it's a puzzle how those bathroom pachyderms avoided being preyed on by the dinosaurs.

Needless to note, most of this doesn't make sense. *The Flintstones* is time out of joint from top to bottom, and the most serious offense is mingling our Paleolithic ancestors with dinosaurs. After all, the genus *Homo* has been around for only a few million years, and *Homo sapiens*

have been strutting our stuff for only the last 100 thousand. Dinosaurs, on the other hand, disappeared about 65 million years ago.

The Flintstones may be harmless Hollywood entertainment, but some non-Hollywood folks have seriously argued that dinosaurs and humans really were contemporaries. In the 1960s, a claim was made that fossilized human footprints could be seen alongside dinosaur prints in the Paluxy riverbed near Glen Rose, Texas. If true, this would indicate that our reading of the geological record is in serious error. But careful investigation by paleontologists has shown that the supposed "man tracks" were, in fact, just more dinosaur prints in which the small, three-pronged toe marks had been filled in by mud or eroded away.

Humans never met dinosaurs, and we could never have used them to power society. Fortunately, our predecessors eventually learned to work with metals, invent engines, and learn about the world through science. It was these developments that give us a lifestyle that our stone age ancestors could never have imagined.

Models of planetary formation now show that while Earth formed primarily from "local" planetesimals of rock and metal, it should also have incorporated some planetesimals from farther out in the solar system; these planetesimals were flung inward by gravitational interactions with other planetesimals and forming planets. Some of these planetesimals came from far enough away—probably from the region currently occupied by the asteroid belt—that they contained ices or rock chemically bound with molecules of water or other common gases. As these planetesimals became part of the forming Earth, their gaseous content became trapped on or within our planet.

The young Earth therefore contained trapped gases, including water, held under pressure in the interior in much the same way that the gas in a carbonated beverage is trapped in a pressurized bottle. When molten rock erupts onto the surface as lava, the release of pressure violently expels the trapped gas in a process we call **outgassing.** Outgassing probably released most of the water vapor that condensed to form our oceans as well as most of the gas that formed our atmosphere.

Some water and gas may also have been supplied directly to the surface by impacts after Earth formed. This process may have begun to create an atmosphere even before Earth was fully formed—perhaps when it was only one-third its current radius. It is difficult to determine the relative contributions of this process and outgassing, but recent studies of comet composition suggest that outgassing was more important to the formation of the atmosphere and oceans.

Volcanism is the major source of outgassing, as you can probably guess from looking at any photo of a volcanic eruption (Figure 4.12). Some outgassing also occurs as a result of impacts—the heat of an impact can melt rock and allow gas to escape. Studies of present-day volcanoes show that the primary gases released by outgassing are water vapor (H_2O), carbon dioxide (CO_2), nitrogen (N_2), sulfur-bearing gases (H_2S or SO_2), and hydrogen (H_2). Because Earth's overall composition has not changed much since its formation, we expect that the same gases were released in early times.* The water vapor condensed as rain to fill Earth's oceans (along with water that bubbled out of the ground at volcanic vents). The gases that remained airborne made up Earth's early atmosphere.

Isotopic analysis of the oldest zircon grains (see Figure 4.11) suggests that substantial amounts of water—and probably oceans—were already present on Earth at the time these grains solidified, some 4.4 billion years ago. If so, much of Earth's atmospheric gas must have been released quite early, perhaps as a side effect of Earth's interior melting or of the giant impact thought to have formed the Moon. Taken alongside the evidence noted earlier suggesting that continents had also begun to form when the zircons solidified, we are led to the intriguing idea that Earth may have had early continents, oceans, and an atmosphere within only about 100 million years after the planet first formed.

The composition of the early atmosphere was very different from that of the atmosphere today. The early atmosphere was probably dominated by carbon dioxide[†] (for reasons we will discuss shortly), while today's atmosphere is dominated by nitrogen (about 78% of the atmosphere)

Figure 4.12

This photo shows the eruption of Mount St. Helens (in Washington State) on May 18, 1980. Eruptions are accompanied by a tremendous amount of outgassing. (The gas itself is generally invisible, so what you see is dust and ash expelled along with the gas.)

---

*Some present-day volcanoes release recycled gases, while others release "juvenile" gases (gases that have not previously been outgassed). The outgassing composition is determined from those releasing juvenile gases, since these should be indicative of the gases originally trapped in Earth's interior.

[†]Significant amounts of hydrogen may also have been present, a controversial idea that we'll discuss in Chapter 6.

and contains only trace amounts of carbon dioxide (less than 0.1%). More important to us, the early atmosphere contained essentially no molecular oxygen (O_2), while the present atmosphere is about 21% oxygen. Thus, we could not have breathed the atmosphere on the early Earth. Earth's present oxygen atmosphere is almost certainly a result of photosynthesis by living organisms, as we will discuss further in Chapter 6. Life therefore must have arisen in a nearly oxygen-free environment.

• Could life have existed during Earth's early history?

The Hadean eon got its name because of its presumed hellish conditions, but the recent zircon evidence tells a different story. If the evidence is being interpreted correctly, it means that Earth may have been habitable within 100 million years after its formation. Many modern-day microbes survive just fine in the absence of oxygen, and it seems likely that such organisms could have thrived almost from the moment the oceans and atmosphere first formed.

However, while the Hadean might have been reasonably balmy at most times, those calm periods must have been frequently interrupted by great violence: Volcanic eruptions were probably more frequent and larger than those of today and, even more significantly, the young Earth was bombarded by planetesimals left over from the birth of the solar system.

The Heavy Bombardment   The accretion of the planets did not end suddenly. The young planets must have shared the solar system with vast numbers of "leftover" planetesimals. Some of these leftovers still survive as asteroids and comets, but many more must have crashed into the planets. The vast majority of these collisions occurred in the first few hundred million years of our solar system's history, during the period we call the **heavy bombardment.** On planets and moons with solid surfaces, such collisions leave **impact craters** as visible scars. A small telescope reveals the presence of numerous impact craters on the Moon (see Figure 4.1).

Earth must have experienced even more impacts than the Moon, both because it presents a larger target and because its stronger gravity tends to attract more objects. However, while the Moon still shows scars of craters formed throughout much of its history, craters on Earth tend to get erased with time by erosion, volcanic eruptions, and plate tectonics. Thus, if we wish to learn about the past cratering rate on Earth, we must look to the evidence recorded on the Moon.

Figure 4.13 shows a map of the Moon's entire surface. Notice that some regions are much more crowded with craters than others. The most heavily cratered regions are the **lunar highlands**, where craters are so abundant that we see overlapping crater boundaries and craters on top of other craters. Radiometric dating of Moon rocks from the lunar highlands shows them to be more ancient than rocks from other regions of the Moon; the highlands are the sources of the Moon rocks that date to more than 4.4 billion years ago, and even their youngest rocks are generally at least 4.0 billion years old. Thus, we conclude that the many craters of the lunar highlands formed more than 4 billion years ago. Clearly, the Moon was being heavily pelted from space during the first half-billion years of the solar system's history, and the same must have been true for Earth.

In contrast, we see relatively few craters in the regions known as the **lunar maria** (see Figure 3.16). The maria are huge impact craters that are smooth because they were covered over by molten lava. The large impacts fractured the lunar crust, creating cracks through which molten

Lunar highlands are ancient and heavily cratered.   *Lunar maria are huge impact basins that were flooded by lava. Only a few small craters appear on the maria.*

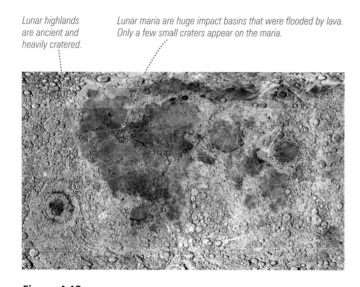

Figure 4.13

This flat map shows the entire surface of the Moon in the same way that a flat map of Earth represents the entire globe. The "lunar highlands" are heavily cratered, and radiometric dating shows that these portions of the lunar surface are quite ancient—some highland rocks date back to some 4.4 billion years ago. The lunar maria are about a half-billion or more years younger, and their relatively few craters tell us that the impact rate had dropped dramatically by the time the maria formed.

lava escaped at some later time; radiometric dating of rocks from the maria shows that the lava flows occurred between about 3.9 and 3.0 billion years ago. The lava flooded the crater basins, covering the existing craters. Thus, the craters we now see within the maria must be the result of impacts that occurred *after* the lava flows, and the relatively small number of these craters tells us that impacts have been relatively rare since the time the maria solidified.

More detailed studies of the ages of lunar craters, as well as analysis of mineral grains in meteorites and old zircons in Earth rocks, can in principle tell us more precisely when the impact rate dropped. Although we might expect that the heavy bombardment would have tapered off gradually, substantial evidence points to a **late heavy bombardment**, around 3.9 billion years ago, that may have been responsible for many of the Moon's large craters; some evidence also suggests that the time between 4.5 billion years ago and the late heavy bombardment may have been relatively quiet. We do not know why the impact rate might have spiked upward with the late heavy bombardment, but some scientists hypothesize that it may have been due to "planetary migration" in which gravitational interactions caused the orbits of the young planets to change. We suspect that similar migration may explain the surprising orbits of some extrasolar planets [Section 11.3], and planetary movement could have disrupted the orbits of asteroids and led to a period of many impacts. In any event, the lunar record shows that the heavy bombardment came to an end by about 3.8 billion years ago.

Every object in the solar system must have been similarly pelted during the heavy bombardment, so all solid surfaces should originally have been as crowded with craters as the lunar highlands. So in addition to helping us understand the early history of Earth and the Moon, these ideas allow us to estimate surface ages for worlds throughout the solar system. Those surface regions that still have abundant craters, such as much of Mercury's surface (see Figure 4.1), must therefore have undergone little change during the last 4 billion years or so. Surfaces with fewer craters must be correspondingly younger, indicating that their original craters have been erased in some way. Thus, thanks to the reconstruction of cratering rates made possible by studies of Moon rocks, planetary scientists can use simple counts of craters on other worlds to estimate the ages of the surfaces. These age estimates are not nearly as precise as radiometric dating—indeed, ages from crater counts are sometimes uncertain by as much as a billion years or so—but until we begin collecting rock samples from other worlds, they are the only age evidence we have to go by.

THINK ABOUT IT   Crater abundance varies greatly in different regions of Mars. For example, the "southern highlands" are quite crowded with craters, while the "Tharsis Bulge" has volcanoes but few impact craters. Which region has the older surface? Does this imply any difference in the number of impacts that occurred in the two regions? Why or why not?

Possible Sterilizing Impacts

We have found that the Hadean Earth must have endured large impacts at a rate much higher than Earth has experienced since. What does this mean for the possibility of life during the Hadean? Remember that the Hadean lasted more than 600 million years, which means that even with "frequent" large impacts, individual events were typically separated by thousands or millions of years. There was never a time when impacts occurred so rapidly that they could have done the equivalent of hitting every living organism on the head. Given the evidence of continents and oceans all the way back to 4.4 to 4.5 billion years ago, there seems no reason why life could not have arisen during the Hadean. Thus, the question of life during the Hadean seems less a question of whether it could have *existed* than whether it could have *survived* the occasional large impacts.

The effects of impacts depend primarily on their sizes. As we'll discuss in Chapter 6, the impact of an object about 10 to 20 kilometers across is thought to have precipitated a series of global changes that caused the extinction of the dinosaurs some 65 million years ago. Fortunately for us, impacts of that size are extremely rare today. But the lunar evidence tells us that far larger impacts occurred during the heavy bombardment.

To understand the effects of very large impacts, scientists calculate the amount of energy they would release and then model how the energy would heat the planet. Such calculations tell us that the impact of an asteroid more than about 350–400 kilometers across would provide enough heat to completely vaporize Earth's oceans and raise the global surface temperature to as high as 2,000°C (3,600°F). It's unlikely that any life could survive such effects, so we conclude that an impact of that size would be a **sterilizing impact**—one that would kill off all life on the planet. Slightly smaller impacts could have vaporized the top few hundred meters of the oceans. These impacts probably would have been sufficient to kill off most life, although life near the ocean bottom (such as near deep-sea vents) or deep underground might have been able to survive.

Impacts large enough to have sterilized much or all of the planet should have been rare even in the Hadean, and we have no way to know exactly how many occurred or when they occurred. Nevertheless, studies of lunar craters and probability arguments suggest they should have happened at least a half-dozen to a dozen times during Earth's first few hundred million years of existence, with the last one probably occurring some time between about 4.2 and 3.8 billion years ago. We therefore reach an important conclusion about life on Earth: While conditions may have allowed life to arise during the Hadean, perhaps even as far back as 4.4 to 4.5 billion years ago, most or all of this life would have been wiped out by major impacts. If an impact fully sterilized the planet, life would have had to start all over again; indeed, it is possible that life arose and was subsequently extinguished multiple times on the early Earth. If an impact was just short of sterilizing, it still would have wiped out all but the forms of life adapted to live in the "safe" environments of the deep ocean or underground. Either way, only after the last major impact was life able to have free rein, and begin the long evolution that ultimately blossomed into the great variety of life on Earth today.

4.4 GEOLOGY AND HABITABILITY

Although no sterilizing impacts have occurred since the end of the heavy bombardment, smaller impacts still occasionally occur; as the case of the dinosaurs shows, these can have great effect on our planet and its life. Nevertheless, if we focus on geology, impact craters only rarely have a significant effect. Instead, virtually every known feature of Earth's surface (with the exception of the relatively few impact craters) has been shaped by one or more of the following three geological processes: volcanism, plate tectonics, and erosion. Why are these processes so important on Earth?

Remarkably, all three of these processes that continually reshape our planet's surface are directly attributable to internal heat. The connection is most obvious for volcanism, since we actually see molten rock rising up from Earth's interior, but it applies equally to plate tectonics. The connection to internal heat is a little more subtle for erosion by wind, water, and ice, but remember that these things exist only because our planet has oceans and an atmosphere, and that the oceans and atmosphere were produced by volcanic outgassing.

Even more important, as we noted in Section 4.1, volcanism and plate tectonics have played major roles in Earth's long-term habitability. Thus, if we hope to understand Earth's surface habitability, we must first understand what our planet is like on the inside and how the interior conditions drive these crucial geological processes. That is our task in this section; as we'll see, internal heat is also a major factor in creating Earth's magnetic field, which we identified in Section 4.1 as a third important ingredient in long-term habitability.

• What is Earth like on the inside?

Our deepest drills have barely pricked Earth's surface, reaching only a few kilometers down into the nearly 6,400 kilometers to Earth's center. Nevertheless, we have managed to learn a lot about our planet's interior structure. One set of clues comes from the nature of surface rocks. For example, the fact that surface rocks are considerably less dense than Earth's overall average density tells us that much denser material must reside in the interior than on the surface. Other clues come from precise measurements of Earth's gravitational field strength in different locations and studies of Earth's magnetic field.

We learn much more about Earth's internal structure from the study of **seismic waves**—waves that propagate much like sound waves both through Earth's interior and along its surface after an earthquake. The precise speed and direction of seismic waves depend on the composition, density, pressure, temperature, and phase (solid or liquid) of the material they pass through. After an earthquake, geologists record the arrival times and the strengths of seismic waves at stations distributed all around the world. At each station, the arriving seismic waves tell us something about the average state of the material through which they have passed. By comparing data from many different stations, scientists have pieced together a fairly detailed picture of Earth's internal structure.

Earth's Interior Structure   Seismic studies reveal that Earth's interior is divided into three major layers by density (Figure 4.14):

- **Core:** The highest-density material, consisting primarily of the metals nickel and iron, resides in the central core. Earth's core has two distinct regions: a solid **inner core**, surrounded by a molten (liquid) **outer core**.

- **Mantle:** Rocky material of moderate density—mostly silicate minerals rich in silicon and oxygen—forms the thick mantle that surrounds the core and makes up most of Earth's volume.

- **Crust:** The lowest-density rock, including igneous rocks such as granite and basalt, forms the thin crust that essentially represents Earth's outer skin.

The molten lava from volcanoes gives many people the incorrect impression that Earth is molten inside, but in fact the lava rises upward from a fairly narrow zone of rock in the upper mantle that is only partially molten. The vast majority of Earth's interior is solid rock, and only the

Figure 4.14

Earth's interior structure, determined from seismic studies; the layering by core-mantle-crust is based on density, while the identification of the lithosphere is based on rock strength. The circular arrows represent the general pattern of mantle convection.

outer core is fully molten. However, the interior rock is not equally strong throughout, and differences in rock strength are often more important than differences in density for understanding many geological processes.

The idea that rock can vary in strength may seem a bit surprising, since in our everyday lives we tend to think of rock as the ultimate in strength and stability (hence the phrase "solid as a rock"). However, the strength of rock depends on its temperature and the surrounding pressure, and under the conditions found in Earth's mantle, even "solid" rock can flow gradually. The popular toy Silly Putty provides a good analogy. The putty can feel pretty solid, especially when it is cold; you can even form it into a ball and bounce it. But if you put a pile of it on a table or inside its "egg" container, after a few days you'll see that it has flowed slowly outward.

In terms of rock strength, geologists define Earth's outer layer as the relatively cool and rigid rock, called the **lithosphere** (*lithos* is Greek for "stone"), that "floats" on warmer, softer rock beneath. The lithosphere encompasses the crust and the upper part of the mantle, as shown in Figure 4.14. Beneath the lithosphere, the higher temperatures allow rock to deform and to flow much more easily. In fact, the mantle rock flows with a characteristic pattern called **convection,** in which hot material expands and rises while cooler material contracts and falls. You are probably familiar with convection in other situations, as it can occur any time a substance is strongly heated from below. For example, you can see convection in a pot of soup on a hot stove, and convection is important in weather because the warm air near the ground tends to rise while the cool air above tends to fall. In the mantle, convection is driven by heat from the core. This heat causes rock near the base of the mantle to expand, giving it a tendency to rise because it becomes lighter and less dense than the rock above it. Meanwhile, rock near the top of the mantle (just below the lithosphere) can cool as its heat flows to the surface (by conduction), causing it to contract and sink. The ongoing process of convection creates *convection cells,* indicated by the circular arrows in Figure 4.14. Keep in mind that mantle convection involves the flow of solid rock, so it is quite slow: Typically, mantle rock flows at a rate of only perhaps 10 centimeters per year, slow enough that it would take about 100 million years for a particular piece of rock to be carried from the base to the top of the mantle.

Differentiation and Internal Heat   Earth's interior layering tells us that it at one time underwent the process known as **differentiation,** in which materials separate according to their density. Earth must have undergone differentiation quite early in its history; as we'll discuss in Section 4.6, the giant impact that formed the Moon must have occurred after differentiation, and we already discussed the fact that this event happened quite early.* For differentiation to have occurred so rapidly, our planet must have been molten or nearly molten throughout its interior. The melting allowed material to separate by density, much as oil separates from water when you mix them and let the mixture sit.

The heat that caused rock to melt came from three main sources. First, the impacts of accretion created heat that melted the outer layers of the young Earth. Second, as denser materials sank through the molten outer layers, their gravitational potential energy was converted into ther-

---

*A second line of evidence for early differentiation comes from isotopic comparisons between meteorites and lead ore on Earth's surface, leading to the conclusion that most lead must have sunk to the core by the time our planet was about 30 million years old.

mal energy that added further heat to the interior. Third, heat is continually released by the radioactive decay of elements within Earth. This heat source is still important today, but it was even more important in early times, when there was more radioactive material to decay. Once the outer layers began to melt due to the first heat source (accretion), the second and third heat sources (sinking of dense materials and radioactive decay) ensured that our planet would completely melt and differentiate fully.

All the terrestrial worlds in our solar system underwent similar melting and differentiation, and the evidence we discussed earlier suggests that differentiation occurred when the planets were very young. Since that time, they have never again been hot enough to melt fully, and they have all been slowly cooling with time. However, different worlds have cooled at different rates.

In general, two factors determine a world's cooling rate. The first is size: Large worlds tend to retain their internal heat much longer than do smaller worlds. You can see why by picturing a large world as a smaller world wrapped in extra layers of rock. The extra rock acts as insulation, making it take longer for interior heat to escape into space. If you now add the fact that the larger world contains more heat in the first place, it's clear that a large world will take much longer to cool than a small world. The Moon's relatively small size probably allowed it to cool substantially within a billion years or so after its formation, while Earth's interior still remains hot enough to keep iron molten in the outer core.

THINK ABOUT IT Give an example from everyday life of a small object cooling faster than a larger one and of a small object warming up more quickly than a larger one. How do these examples relate to the issue of geological activity on Earth and the Moon?

The second general factor in the cooling rate is ongoing heat deposition: If a world has a source of ongoing internal heat, it will tend to cool more slowly with time. For the terrestrial worlds, the only significant source of ongoing heat is radioactive decay. Because many radioactive materials have long half-lives, they can continue to add heat to the interior for billions of years. Over time, radioactive decay has contributed several times as much heat to Earth's interior as accretion and differentiation. As we'll discuss in Chapter 9, some moons of jovian planets have other sources of ongoing heat deposition (such as *tidal heating*) that can keep them much hotter than we would otherwise expect given their relatively small sizes.

• How does plate tectonics shape Earth's surface?

Our discussion of internal heat explains most of the differences in geological activity that we see between Earth and the other terrestrial worlds. The Moon and Mercury have essentially no active, internally driven geology, because their small sizes allowed them to cool long ago. Mars is large enough to have had significant volcanism in the past, but its interior must now be much cooler than it was in its heyday. Venus is only slightly smaller than Earth, so it probably retains nearly as much internal heat; confirmation of this idea comes from the fact that Venus has few impact craters, indicating that lava flows or other processes that recycle surface material have erased craters from the more distant past. However,

Earth appears to be geologically distinct from Venus and all the other terrestrial worlds in one crucial way: Earth is the only planet with ongoing plate tectonics.

The word *tectonics* comes from the Greek word for "builder," *tekton;* you'll notice the same root in the word *architect,* which means "master builder." In geology, tectonics refers to "building" performed on planetary surfaces—that is, any surface reshaping that results from stretching, compression, or other forces acting on the lithosphere. Tectonic processes have operated to some extent on all the terrestrial worlds, but it is most important on Earth, where it operates by the distinctive mechanism of plate tectonics.

The Meaning of Plate Tectonics

The term *plate tectonics* refers to the scientific theory that explains much of Earth's surface geology as a result of the slow motion of *plates*—fractured pieces of the lithosphere—driven by the underlying convection of the mantle. According to the theory, the lithosphere fractured because of stresses generated by mantle convection, and the resulting plates now essentially "float" over the mantle, gradually moving over, under, and around each other as convection moves Earth's interior rock. Because the term *plate tectonics* refers to the theory, it is generally considered to be singular rather than plural, despite the *-s* on *tectonics*.

Earth's lithosphere is broken into about a dozen plates, with known boundaries shown in Figure 4.15. Except during earthquakes, the motions of the plates are barely noticeable on human time scales—a few centimeters per year, which is about the rate at which your fingernails grow. However, geologists can now measure plate motions by comparing readings taken with the global positioning system (GPS) on either side of plate boundaries.

Evidence for Plate Tectonics

The GPS measurements offer the most direct evidence of plate motion. However, the overall theory of plate tectonics rests on three additional significant lines of evidence: evidence of past continental motion, evidence that plates spread apart on seafloors, and a difference between the nature of Earth's crust on the seafloors and the continents.

Figure 4.15

This relief map shows plate boundaries (solid yellow lines), with arrows to represent directions of plate motion. Color represents elevation, progressing from blue (lowest) to red (highest).

We usually trace the origin of the theory of plate tectonics to a slightly different idea proposed in 1912 by German meteorologist and geologist Alfred Wegener: *continental drift*, the idea that continents gradually drift across the surface of Earth. Wegener got his idea in part from the puzzlelike fit of continents such as South America and Africa (Figure 4.16). This fit had been noted by earlier mapmakers, but Wegener took the idea further. He noted that similar types of distinctive rocks and rare fossils were found in eastern South America and western Africa, suggesting that these two regions once had been close together.

Figure 4.16

The puzzle-like fit of South America and Africa.

Despite these strong hints, no one at Wegener's time knew of a mechanism that could allow the continents to push their way through the solid rock beneath and around them. Wegener suggested that Earth's gravity and tidal forces from the Sun and Moon were responsible, but other scientists quickly showed that these forces were too weak to move continents around. As a result, Wegener's idea of continental drift was rejected by all but a few geologists for decades after he proposed it, even though his evidence of a "continental fit" for Africa and South America ultimately proved correct. Today, far more extensive fossil evidence makes it clear that the continents really were arranged differently in the past, and our understanding of how rock can flow in the mantle allows us to understand the real reasons that the continents move around.

Scientists began to recognize the mechanism of continental motion through the discovery of *mid-ocean ridges* in the mid-1950s. Mantle material erupts onto the ocean floor along these ridges, such as the mid-Atlantic Ridge shown in Figure 4.15, while the existing seafloor spreads outward to either side. This **seafloor spreading** helped explain how the continents could move apart with time. Because this idea is quite different from Wegener's original notion of continents plowing through the solid rock beneath them, geologists no longer use the term *continental drift* and instead consider continental motion within the context of plate tectonics.

The third line of evidence for plate tectonics (in addition to continental motion and seafloor spreading) comes from the fact that Earth's surface has two very different types of crust (Figure 4.17). **Seafloor crust** is made primarily of the relatively high-density igneous rock called *basalt*, which commonly erupts from volcanoes like those along mid-ocean ridges and in Hawaii. Seafloor crust is typically only 5–10 kilometers thick, and radiometric dating shows that it is quite young—the average age is about 70 million years, and even the oldest seafloor crust is less than about 200 million years old. **Continental crust** is made of lower-density rock, such as granite, and its rock spans a wide range of ages from the very young all the way back to the oldest rocks found on Earth. Continental crust is much thicker than seafloor crust—typically 20–70 kilometers thick—but it sticks up only slightly higher because its weight presses it down farther onto the mantle below.

The two types of crust make it clear that Earth's surface must undergo continual change. New seafloor crust continually emerges at sites of seafloor spreading, while the wide age range of continental crust tells us that the continents have gradually built up over time.

The Mechanism of Plate Tectonics

Today, we understand that plates move in concert with underlying mantle convection, driven by the heat released from Earth's interior. Over millions of years, the movements involved in plate tectonics act like a giant conveyor belt for Earth's lithosphere

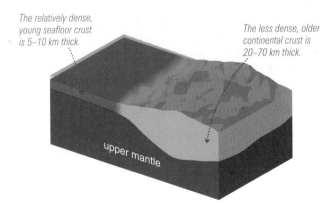

The relatively dense, young seafloor crust is 5–10 km thick.

The less dense, older continental crust is 20–70 km thick.

upper mantle

Figure 4.17

Earth today has two distinct kinds of crust.

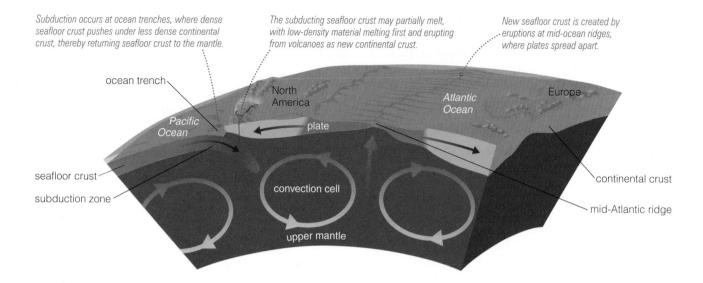

Subduction occurs at ocean trenches, where dense seafloor crust pushes under less dense continental crust, thereby returning seafloor crust to the mantle.

The subducting seafloor crust may partially melt, with low-density material melting first and erupting from volcanoes as new continental crust.

New seafloor crust is created by eruptions at mid-ocean ridges, where plates spread apart.

Figure 4.18 Interactive Figure

Plate tectonics acts like a giant conveyor belt for Earth's lithosphere.

(Figure 4.18). This movement explains many of the major features of Earth's geology.

Seafloor spreading occurs at mid-ocean ridges because they are places where mantle material rises upward toward the surface. As it gets close to the surface, the lower pressure allows it to partly melt; the molten material then erupts to the surface, cooling and contracting as it spreads sideways. This explains both the formation of the basaltic seafloor crust and the characteristic shapes of the ridges. Worldwide along the mid-ocean ridges, new crust covers an area of about 2 square kilometers every year, enough to replace the entire seafloor within about 200 million years—and thus explaining the geologically young age of seafloor rocks.

Over tens of millions of years, any piece of seafloor crust gradually makes its way across the ocean bottom, then finally gets recycled into the mantle in the process we call **subduction**. Subduction occurs where a seafloor plate meets a continental plate, which is generally somewhat offshore at the edge of a sloping *continental shelf*. The continental rocks are less dense than those on the seafloor; thus, as the dense seafloor crust of one plate pushes under the less dense continental crust of another plate, it can pull the entire surface downward to form a deep *ocean trench*. At some trenches, the ocean depth is more than 8 kilometers.

Beneath a subduction zone, the descending seafloor crust heats up and may begin to melt as it moves deeper into the mantle. If enough melting occurs, the molten rock may erupt upward. As you can see in Figure 4.18, the process of subduction tends to make the melting occur under the edges of the continents, which is why so many active volcanoes tend to be found along those edges. This volcanic activity explains the presence of many coastal mountain ranges (as well as many older mountain ranges that are no longer located along coasts). Moreover, the lowest-density material tends to melt first, which is why the continental crust emerging from these landlocked volcanoes is lower in density than seafloor crust.

Spreading and subduction are not the only ways in which plates interact. Two continental plates crashing into each other can push each other upward, providing a second way of creating mountain ranges. The Himalayas are still slowly growing in this way as the plate carrying India pushes into the plate carrying most of the rest of Eurasia (Figure 4.19). In places where continental plates are pulling apart, the crust thins and

Figure 4.19

This satellite photo shows the Himalayas, which are still slowly growing as the plate carrying India pushes into the Eurasian plate. Arrows indicate the directions of plate motion. (See the plate boundaries in Figure 4.15.)

can create a large *rift valley*. The East African rift zone is an example (see Figure 4.15). This rift is slowly growing and will eventually tear the African continent apart. At that point, rock rising upward with mantle convection will begin to erupt from the valley floor, creating a new zone of seafloor spreading. A similar process tore the Arabian Peninsula from Africa, creating the Red Sea (Figure 4.20).

Places where plates slip sideways relative to each other are marked by what we call plate boundary **faults**. For example, the San Andreas Fault in California marks a line where the Pacific plate is moving northward relative to the continental plate of North America (Figure 4.21). At its current rate, this motion will bring Los Angeles (on the Pacific plate) and San Francisco Bay (on the North American plate) together in about 20 million years. The two plates do not slip smoothly against each other. Instead, their rough surfaces catch against each other. Tension builds up until it is so great that it forces a rapid and violent shift, causing an earthquake. In contrast to the usual motion of plates, which proceeds at a few centimeters per year, an earthquake can move plates by several *meters* in a few seconds. The movement can raise mountains, level cities, set off destructive tsunamis, and make the whole planet vibrate with seismic waves (much like a ringing bell).

Figure 4.20

When continental plates pull apart, the crust thins and deep rift valleys form. This process tore the Arabian Peninsula from Africa, forming the Red Sea. Arrows indicate the directions of plate motion.

> **THINK ABOUT IT** By studying the plate boundaries in Figure 4.15, explain why the west coast states of California, Oregon, and Washington are prone to more earthquakes and volcanoes than most other parts of the United States. Find the locations of recent earthquakes and volcanic eruptions worldwide. Do the locations fit the pattern you expect?

Not all earthquakes and volcanoes occur near plate boundaries. Earthquakes sometimes occur along old or buried faults that are now far from plate boundaries; some of the biggest earthquakes in U.S. history occurred in 1811 and 1812 along the New Madrid fault zone, which runs through parts of Illinois, Kentucky, Missouri, Arkansas, and Tennessee. Volcanic activity may occur any place where a plume of hot mantle material rises up to make what we call a **hot spot**. The Hawaiian Islands are the result of a hot spot that has been erupting basaltic lava for tens of millions of years. Plate tectonics gradually carries the Pacific plate over the hot spot, forming a chain of volcanic islands as different parts of the plate lie directly above the hot spot at different times (Figure 4.22). Today, most of the lava erupts on or near the Big Island of Hawaii, giving much of this island a young, rocky surface. About a million years ago, the Pacific plate lay farther to the southeast (relative to its current location), and the hot spot built the island of Maui. Before that, the hot spot created other islands, including Oahu (3 million years ago), Kauai (5 million years ago), and Midway (27 million years ago). The older islands are more heavily eroded. Midway has been eroded so much that it barely rises above sea level. The movement of the plate over the hot spot continues today, building underwater volcanoes that eventually will rise above sea level to become new Hawaiian Islands. The growth of a future island, named Loihi, is already well under way—prime beach real estate should be available there in about a million years or so. Hot spots can also occur beneath continental crust. For example, a continental hot spot is responsible for the volcanism that supplies the heat for the geysers and hot springs of Yellowstone National Park.

Figure 4.21

California's San Andreas Fault marks a boundary where plates are sliding sideways as shown by the white arrows; asterisks indicate sites and years of major earthquakes. The inset photo shows a place along the San Andreas Fault where the painted lines in a road allow us to see how far the two sides of the fault moved in an earthquake.

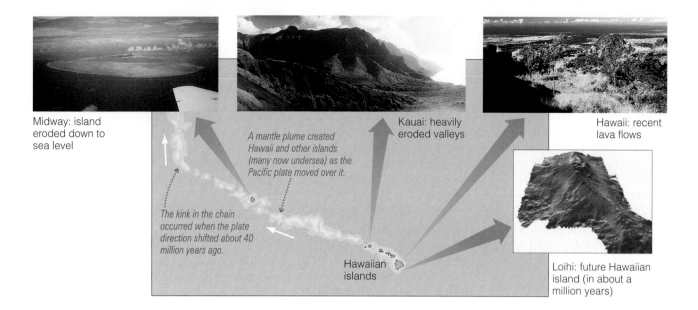

Midway: island eroded down to sea level

A mantle plume created Hawaii and other islands (many now undersea) as the Pacific plate moved over it.

Kauai: heavily eroded valleys

Hawaii: recent lava flows

The kink in the chain occurred when the plate direction shifted about 40 million years ago.

Hawaiian islands

Loihi: future Hawaiian island (in about a million years)

Figure 4.22

The Hawaiian Islands are just the most recent of a long string of volcanic islands made by a mantle hot spot. The black-and-white image of Loihi (lower right) was obtained by sonar, as it is still entirely underwater. The long chain records the past 60 million years of history of the oceanic crust in the region.

Plate Tectonics Through Time We can use the current motions of the plates to project the arrangement of continents millions of years into the past or the future. For example, at a speed of 2 centimeters per year, a plate will travel 2,000 kilometers in 100 million years. Figure 4.23 shows several past arrangements of the continents, along with one future arrangement. Note that the present-day continents were once all stuck together in a single "supercontinent," sometimes called *Pangāea* (meaning "all lands"), which began to break up about 225 million years ago.

Mapping the sizes and locations of continents at even earlier times is more difficult. However, studies of magnetized rocks (which can record the orientation of ancient magnetic fields) and comparisons of fossils found in different places around the world have allowed geologists to map the movement of the continents much further into the past. It seems that, over at least the past billion years or more, the continents have slammed together, pulled apart, spun around, and changed places on the globe. Central Africa once lay at Earth's South Pole, and Antarctica once was near the equator. The continents continue to move, and their current arrangement is no more permanent than any past arrangement.

Has Earth had plate tectonics throughout its history? We are not yet certain, but we have some evidence for plate tectonics going back quite far

Figure 4.23 Interactive Figure

Selected past, present, and future arrangements of Earth's continents.

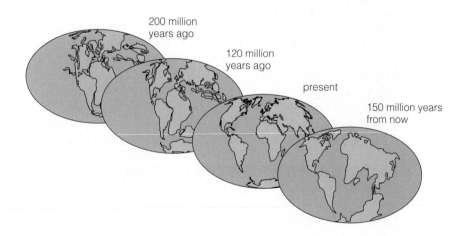

200 million years ago

120 million years ago

present

150 million years from now

in time. The oldest fairly definitive evidence comes from seismic studies that suggest the presence of an ancient subduction zone (found in Canada) that formed some 2.7 billion years ago; subduction would be a sure sign of the conveyorlike action of plate tectonics. More controversial is the recent zircon evidence suggesting that continental crust had already begun to form nearly 4.5 billion years ago. If the evidence is being properly interpreted, it suggests that plate tectonics began just shortly after the birth of Earth, since continental crust is generally formed as a direct result of the separation of rock by density along subduction zones.

Cause and Effects of Plate Tectonics   We have a fairly good understanding of how plate tectonics works, and how it is driven by Earth's internal heat and mantle convection. However, we still face at least one significant mystery: Why does plate tectonics operate only on Earth among the terrestrial worlds?

The answer is probably simple for the Moon, Mercury, and Mars: Because their small size has allowed their interiors to cool much more than Earth's interior, their lithospheres have thickened. If they have any remaining internal convection at all, it is too weak to break their thick lithospheres into plates. Venus poses a greater mystery, since it is almost the same size as Earth and therefore should have retained a similar amount of internal heat.

We still do not know why Venus lacks plate tectonics today, or whether it had plate tectonics in the past [Section 10.2]. However, we have at least one plausible hypothesis: As we'll discuss further in Chapter 10, Venus's high surface temperature has probably baked out water from its crust and upper mantle. This drying of the rock may have strengthened and thickened Venus's lithosphere so that it has resisted the fracturing that occurred on Earth; the high temperature may also make Venus's lithosphere less brittle than Earth's colder crust. If this hypothesis is correct, then we can ultimately trace the cause of plate tectonics to two factors: heat-driven mantle convection and a lithosphere thin and brittle enough to be fractured by the movement of the underlying mantle.

Whatever its cause, the effects of plate tectonics are profound. We've seen that plate tectonics is the most important geological mechanism on Earth, and it plays a key role in explaining nearly all of Earth's geological features. But its deeper significance to life lies in the fact that its recycling of rock turns out to play a crucial role in climate regulation, a topic we'll address in Section 4.5.

• Why does Earth have a protective magnetic field?

Planetary atmospheres do not necessarily last forever. All the terrestrial worlds had at least some gas released from their interiors by outgassing, but some of the gas eventually escaped to space. Atmospheric gas can be lost to space in at least three ways.

First, gas molecules move fast enough that they exceed their world's escape velocity and can simply "take off" into space. This process is often called **thermal escape**, because gas particles tend to move at higher speeds when their temperature is higher, and hence are more likely to escape when temperatures are high than low. Moreover, at any particular temperature, lightweight molecules such as hydrogen tend to move faster than heavier molecules such as oxygen; thus, light gases escape more easily. That is why none of the terrestrial worlds has any significant amounts of hydrogen gas; the large, jovian planets retain hydrogen because their large masses give them much higher escape velocities. More generally, smaller worlds have lower escape velocities, so gas escapes from them much more readily. That is why the Moon and Mercury, the two smallest of the terrestrial worlds, have become essentially airless.

Second, impacts can also blast atmospheric gas into space. Again, smaller worlds are more prone to this type of loss because of their lower escape velocities: Equivalent impacts are more likely to blast material upward with escape velocity on a smaller world than a larger world.

Third, gas can be lost through a mechanism known as **solar wind stripping**, which occurs when particles from the solar wind in effect sweep atmospheric gas particles into space. This mechanism acts slowly, but calculations suggest that without the protection of Earth's global magnetic field, over billions of years the solar wind would have swept away much of our planet's atmosphere.

Requirements for a Global Magnetic Field   A **magnetic field** can affect charged particles or magnetized objects in its vicinity. For example, if you've ever used a compass, you know that Earth has a magnetic field that determines the direction in which the compass needle points. The global extent of Earth's magnetic field gives us a strong clue that the field is generated *inside* our planet.

You are probably familiar with the general pattern of the magnetic field created by an iron bar magnet (Figure 4.24a). Earth's magnetic field is generated by a process more similar to that of an *electromagnet,* in which the magnetic field arises as a battery forces charged particles (electrons) to move along a coiled wire (Figure 4.24b). Earth does not contain a battery, of course, but charged particles move with the molten metals in its liquid outer core (Figure 4.24c). Internal heat causes the liquid metals to rise and fall (convection), while Earth's rotation twists and distorts the convection pattern of these molten metals. The result is that electrons in the molten metals move within Earth's outer core in much the same way that they move in an electromagnet, generating Earth's magnetic field.

We can generalize what we know about Earth's magnetic field to other worlds. There are three basic requirements for a global magnetic field:

Figure 4.24

Sources of magnetic fields.

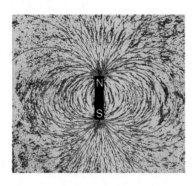

a This photo shows how a bar magnet influences iron filings (small black specks) around it. The *magnetic field lines* (red) represent this influence graphically.

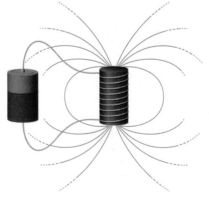

b A similar magnetic field is created by an electromagnet, which is essentially a wire wrapped around a bar and attached to a battery. The field is created by the battery-forced motion of charged particles (electrons) along the wire.

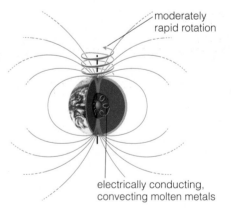

moderately rapid rotation

electrically conducting, convecting molten metals

c Earth's magnetic field also arises from the motion of charged particles. The charged particles move within Earth's liquid outer core, which is made of electrically conducting, convecting molten metals.

1. An interior region of electrically conducting fluid (liquid or gas), such as molten metal.
2. Convection in that layer of fluid.
3. At least moderately rapid rotation of the planet.

Earth is unique among the terrestrial worlds in meeting all three requirements, which is why it is the only terrestrial world in our solar system with a strong magnetic field. The Moon has no magnetic field, either because it lacks a metallic core altogether or because its core has long since solidified and ceased convecting. Mars likewise has virtually no magnetic field today, probably because of similar core solidification. Venus probably has a molten metal layer much like that of Earth, but either its convection or its 243-day rotation period is too slow to generate a magnetic field. Mercury poses a slight enigma: It possesses a measurable magnetic field despite its small size and slow, 59-day rotation. The reason may be tied to the fact that Mercury has a very large metal core, which may still be partly molten and convecting.

The Magnetosphere and the Solar Wind The magnetic field protects Earth's surface and atmosphere from most of the energetic particles of the solar wind because it creates a **magnetosphere** that acts like a protective bubble surrounding our planet. The magnetosphere deflects most of the solar wind particles while channeling a few toward the poles, where they can cause auroras (Figure 4.25). The magnetosphere itself is invisible, but we can map its presence with devices that work much like compass needles and we can detect particles that become trapped within it (in the charged particle belts, also known as *Van Allen belts*).

The magnetosphere generally deflects particles while they are still high above our atmosphere, and it therefore prevents the solar wind from stripping Earth's atmospheric gas away. Indeed, evidence for this protective function comes from studying our neighboring planets. As we'll discuss in Chapter 8, Mars today apparently has much less atmospheric gas than it did in the distant past, and we suspect that Mars lost much of its gas when its interior cooled to the point that it no longer generated a strong

Figure 4.25   Interactive Figure

Earth's magnetosphere acts like a protective bubble, shielding our planet from the charged particles of the solar wind.

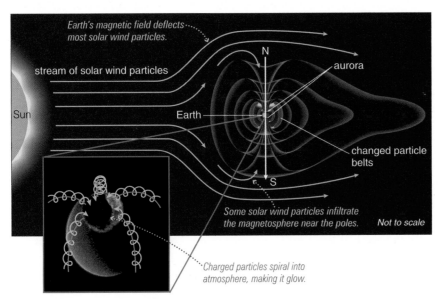

a This diagram shows how Earth's invisible magnetosphere (represented in purple) deflects solar wind particles. Some particles accumulate in charged particle belts encircling our planet. The inset is a photo of a ring of auroras around the North Pole; the bright crescent at its left is part of the day side of Earth.

b This photograph shows an aurora visible over the coast of Norway. In a video, you would see these lights dancing about in the sky.

magnetic field and protective magnetosphere. (Mars probably also lost gas due to impacts.) Careful studies of the isotopic composition of Venus's atmosphere suggest that it, too, has lost gas to solar wind stripping, as we would expect given its lack of magnetic field. However, Venus has such a thick atmosphere that its overall gas loss has been proportionally small.

4.5   CLIMATE REGULATION AND CHANGE

At the beginning of this chapter, we stated that three crucial ingredients in Earth's long-term habitability have been volcanism, plate tectonics, and the magnetic field. We have seen that volcanism's most important role has been in releasing the gases that formed our oceans and atmosphere, while the magnetic field has helped prevent atmospheric gas from being lost to space. For plate tectonics, we have seen the way it is responsible for shaping our planet's surface; in fact, plate tectonics plays an even more important role, because it helps regulate the climate.

You are probably aware that Earth's climate has not been perfectly stable through time: Even during human history our planet has experienced what we call *ice ages,* and as we look back through geological time, we find other periods of far more severe cold or warmth. Nevertheless, the climate has been sufficiently stable for life to exist continually for nearly 4 billion years. Because life on Earth needs liquid water [Section 5.3], we infer that the oceans have remained at least partially liquid throughout this long period of time. Although the temperature range in which water can be liquid may seem wide to us humans, when we compare it to temperatures found on other worlds, we realize that Earth's climate has been remarkably stable through time.

At first, Earth's stable climate might not seem surprising. After all, the primary source of heat for the atmosphere and oceans is the Sun, and Earth's orbit about the Sun should not have changed since its formation. However, theoretical models of the Sun and observations of other Sun-like stars reveal an important fact: Stars gradually brighten with age. Models suggest that the Sun today may be as much as 30% brighter than it was when Earth formed, which means the young Earth received a lot less solar warmth and light than it does today. How, then, could our planet have been warm enough for liquid water in the distant past, and why hasn't our planet overheated as the Sun brightened? To answer these and other questions about Earth's long-term climate, we must begin by investigating why Earth is warm enough for liquid water in the first place.

- ### How does the greenhouse effect make Earth habitable?

Most people assume that Earth is warm enough for liquid water simply because we lie at the "right" distance from the Sun, but this clearly is not the whole story. The Moon lies at the same distance, but its daytime temperatures rise to 125°C (257°F)—well above the normal boiling point of water—while its nighttime temperatures plummet to a frigid −175°C (−283°F). Moreover, the fact that the Sun has brightened with time means that even if we were at the "right" distance when Earth was young, that same distance would be too close to the now-brighter Sun.

In fact, it's fairly easy to calculate the temperature Earth would have, based solely on its distance from the Sun and the amount of incoming sunlight absorbed by its surface. Such a calculation shows that the **global average temperature**—that is, the average temperature for the entire planet—would be −1°C (+30°F), well below the freezing point of water.

But Earth is not frozen. The actual global average temperature today is about 15°C (59°F), and geological evidence shows that it has been warmer at various times in the past. Something must be making our planet much warmer than we would expect based on its distance from the Sun alone, and that something is what we call the **greenhouse effect**.

Figure 4.26 shows the basic mechanism of the greenhouse effect. Sunlight consists mostly of visible light, which passes easily through most atmospheric gases and reaches Earth's surface. Some of this visible light gets absorbed by the ground, while the rest is reflected back to space. The ground must return the energy it absorbs back to space, because if it didn't the energy would make the ground heat up rapidly. However, the fact that the ground doesn't glow in the dark tells us that the ground does not return the energy in the same visible-light form that it absorbs it. Instead, the ground returns the energy in the form of infrared light.

The greenhouse effect works by "trapping" some of the infrared light, thereby slowing its return to space. This trapping occurs because some atmospheric gases can absorb the infrared light. Gases that are particularly good at absorbing infrared light are called **greenhouse gases**. The most important greenhouse gases are water vapor (H_2O), carbon dioxide (CO_2), and methane (CH_4). These gases absorb infrared light effectively because their molecular structures make them prone to begin rotating or vibrating when struck by an infrared photon (an individual "piece" of light); diatomic molecules such as nitrogen (N_2) and oxygen (O_2) generally cannot rotate or vibrate in these ways and hence do not absorb infrared light.

After a greenhouse gas molecule absorbs the energy of an infrared photon, it quickly releases the energy by emitting a new infrared photon. However, the new photon will be emitted in some random direction that is unlikely to be the same direction from which the original photon came. The new photon is then usually absorbed by another greenhouse molecule, which does the same thing. The net result is that greenhouse gases tend to slow the escape of infrared radiation from the lower atmosphere, while their molecular motions heat the surrounding air. In this way, the greenhouse effect makes the surface and the lower atmosphere warmer than they would be from sunlight alone. A blanket offers a good analogy: You stay warmer under a blanket not because the blanket itself provides any heat, but rather because it slows the escape of your body heat into the cold outside air. The more greenhouse gases that are present, the greater the degree of surface warming. On Earth, the naturally occurring greenhouse effect is strong enough to raise the global average temperature by about 31°C from what it would be without greenhouse gases. Without this warming, our planet would be frozen over.

Incidentally, you are probably aware that the greenhouse effect is often in the news, usually portrayed in a negative light as part of an environmental problem. But as we have just seen, the greenhouse effect is not a bad thing in and of itself, since life on Earth would not exist without it. Why, then, is the greenhouse effect discussed as an environmental problem? The reason is that human activity is adding more greenhouse gases to the atmosphere, thereby strengthening the greenhouse effect and further warming our planet ("global warming"). While the precise effects of this human-induced warming are not known and are somewhat controversial [Section 10.5], we need only look to our hot neighbor Venus to see that changes to the greenhouse effect should not be taken lightly. While the greenhouse effect makes Earth livable, it is also responsible for the searing 470°C (878°F) temperature of Venus—proving that it's possible to have too much of a good thing.

Visible light passes through the atmosphere.

Some visible light is reflected by clouds, haze, and the surface.

The surface absorbs visible light and emits thermal radiation in infrared.

Greenhouse gases absorb and reemit infrared radiation, thereby heating the lower atmosphere.

Figure 4.26 Interactive Figure

The greenhouse effect makes the surface and lower atmosphere much warmer than they would be without greenhouse gases such as water vapor, carbon dioxide, and methane.

THINK ABOUT IT Carbon dioxide makes up less than 1% of our atmosphere today, while nitrogen and oxygen together make up some 98% of our atmosphere. Why, then, do we focus on carbon dioxide when we talk about Earth's climate?

• What regulates Earth's climate?

The case of Venus leads us to a crucial question about Earth's hospitable climate. As we'll discuss in Chapter 10, Venus's extreme greenhouse effect occurs because its atmosphere contains almost 200,000 times as much carbon dioxide as Earth's atmosphere. But Venus and Earth are nearly the same size and both were made from similar materials, so volcanic outgassing should have released similar amounts of carbon dioxide on both worlds. Moreover, outgassing from modern-day volcanoes on Earth shows that they do indeed release plenty of carbon dioxide, and over time, volcanoes must have outgassed nearly as much carbon dioxide into Earth's atmosphere as we find in the atmosphere of Venus. Where, then, did all of Earth's carbon dioxide end up?

Geological studies reveal the answer: Most of Earth's carbon dioxide is locked up in **carbonate rocks**—sedimentary rocks such as limestone that are rich in carbon and oxygen. By estimating the total amount of carbonate rock on Earth, we find that these rocks contain about 170,000 times as much carbon dioxide as our atmosphere, which means that Earth does indeed have almost as much total carbon dioxide as Venus. Venus lacks carbonate rock (for reasons we'll discuss in Chapter 10), so all of its carbon dioxide remains in its atmosphere. Keep in mind that this difference between the two planets in carbon dioxide location makes all the difference in the world: If Earth's carbon dioxide were in our atmosphere rather than in carbonate rocks, our planet would be nearly as hot as Venus and certainly uninhabitable.

Of course, the fact that Earth's carbon dioxide is locked up in rocks leads to a deeper question: How did it get there? The answer lies with a mechanism closely tied to plate tectonics.

The Carbon Dioxide Cycle The mechanism by which carbon dioxide has been removed from Earth's atmosphere and by which the current small amount of atmospheric carbon dioxide remains stable is called the inorganic **carbon dioxide cycle**, or the **CO_2 cycle** for short. Let's follow the cycle as illustrated in Figure 4.27, starting with the carbon dioxide in the atmosphere:

- Atmospheric carbon dioxide dissolves in rainwater, creating a mild acid.

- The mildly acidic rainfall erodes rocks on Earth's continents, and rivers carry the broken-down minerals to the oceans.

- In the oceans, calcium from the broken-down minerals combines with dissolved carbon dioxide and falls to the ocean floor, making carbonate minerals.* These can accumulate at the bottom of the oceans to build large quantities of rock such as limestone over time.

---

*During the past half-billion years or so, the carbonate minerals have been made by shell-forming sea animals, falling to the bottom in the seashells left after the animals die. Without the presence of animals, chemical reactions would do the same thing—and apparently did for most of Earth's history.

Volcanoes outgas CO_2.

Atmospheric CO_2 dissolves in rainwater.

Rainfall erodes rock on land; rivers carry broken-down minerals to the sea.

Broken-down minerals react with dissolved CO_2 to form carbonate rocks.

Carbonate rocks subduct and melt, releasing CO_2.

Figure 4.27

This diagram shows how the CO_2 cycle continually moves carbon dioxide from the atmosphere to the ocean to rock and back to the atmosphere. Note that plate tectonics (subduction in particular) plays a crucial role in the cycle.

- Over millions of years, the conveyor belt of plate tectonics carries the carbonate rocks to subduction zones, and subduction carries them down into the mantle.

- As they are pushed deeper into the mantle, some of the subducted carbonate rock heats up and releases its carbon dioxide, which then outgasses back into the atmosphere through volcanoes.

In summary, the reason that Earth has so little carbon dioxide in its atmosphere is that most of the carbon dioxide was dissolved in the oceans, where chemical reactions converted it to carbonate minerals. In fact, about 60 times as much carbon dioxide is dissolved in the oceans as is present in our atmosphere, though this amount still pales in comparison to the 170,000 times as much that the CO_2 cycle has locked up in rock.

The CO_2 Cycle as a Thermostat The CO_2 cycle acts as a thermostat for Earth because of the way that changes in temperature feed back into the cycle. You are probably familiar with what we generally call **feedback processes**—processes in which a change in one property amplifies (positive feedback) or counteracts (negative feedback) the behavior of the rest of the system. For example, if someone brings a microphone too close to a loudspeaker, it picks up and amplifies small sounds from the speaker. These amplified sounds are again picked up by the microphone and further amplified, causing a loud screech. This sound feedback is an example of *positive feedback,* because it automatically amplifies itself. The screech usually leads to a form of *negative feedback:* The embarrassed person holding the microphone moves away from the loudspeaker, thereby stopping the positive sound feedback.

The CO_2 cycle has a built-in form of negative feedback that returns Earth's temperature toward "normal" whenever it warms up or cools down. This negative feedback occurs because the overall rate at which carbon dioxide is pulled from the atmosphere is extremely sensitive to temperature: the higher the temperature, the higher the rate at which carbon dioxide is removed. Figure 4.28 shows how it works. Consider first what happens if Earth warms up a bit. The warmer temperature means more evaporation and rainfall, pulling more CO_2 out of the atmosphere. The reduced atmospheric CO_2 concentration leads to a weakened greenhouse effect that counteracts the initial warming and cools the planet

Figure 4.28

The carbon dioxide cycle acts as a thermostat for Earth through negative feedback processes. Cool temperatures cause atmospheric CO_2 to increase, and warm temperatures cause atmospheric CO_2 to decline.

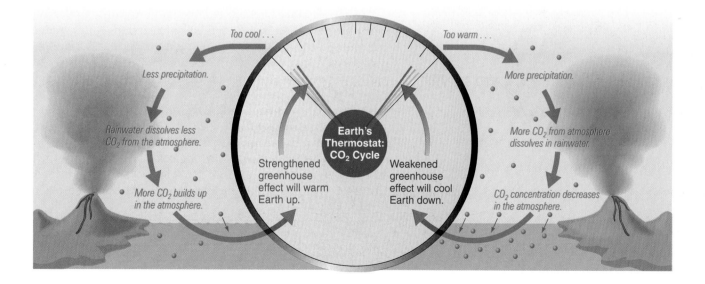

back down. Similarly, if Earth cools a bit, precipitation decreases and less CO_2 is dissolved in rainwater, allowing the CO_2 released by volcanism to build back up in the atmosphere. The increased CO_2 concentration strengthens the greenhouse effect and warms the planet back up.

Overall, the natural thermostat of the CO_2 cycle has allowed the greenhouse effect to strengthen or weaken just enough to keep Earth's climate in a range that has allowed for liquid water, regardless of what other changes have occurred on our planet. And, because subduction plays a critical role in the CO_2 cycle, we now see the importance of plate tectonics to Earth's climate: Without plate tectonics, there would be no CO_2 cycle to regulate our planet's surface temperature.

• How does Earth's climate change over long periods of time?

While Earth's climate has remained stable enough for the oceans to remain at least partly liquid throughout history, the climate has not been perfectly steady. Numerous warmer periods and numerous ice ages have occurred. Such variations are possible because the CO_2 cycle does not act instantly. When something begins to change the climate, it takes time for the feedback mechanisms of the CO_2 cycle to come into play, because these mechanisms depend on the gradual action of mineral formation in the oceans and of plate tectonics. Calculations show that the time to stabilize atmospheric CO_2 through the CO_2 cycle is about 400,000 years. That is, if the amount of CO_2 in the atmosphere were to rise due to, say, increased volcanism, it would take some 400,000 years for the CO_2 cycle to restore temperatures to their current values.*

Ice Ages   **Ice ages** occur when the global average temperature drops by a few degrees. The slightly lower temperatures lead to increased snowfall, which may cover continents with ice down to fairly low latitudes. For example, the northern United States was entirely covered with glaciers during the peak of the most recent ice age, which ended only about 10,000 years ago. In fact, relative to temperatures over at least the past 200 million years, we are still in an ice age. This ice age has persisted for the past 35 million years or so, with periods of deeper cold interspersed with periods of warmer temperatures, such as the present. Remarkably, recent evidence indicates that we can enter or leave a cold period very rapidly, within as short a time as a few decades.

The causes of ice ages are complex and not fully understood. Over periods of tens or hundreds of millions of years, the Sun's gradual brightening and the changing arrangement of the continents around the globe have at least in part influenced the climate. During the past few million years—a period too short for solar changes or continental motion to have a significant effect—the ice ages appear to have been strongly influenced by small, cyclical changes in Earth's rotation and orbit. These cyclical changes are often called *Milankovitch cycles*, after the Serbian scientist who suggested their role in climate change.

For example, while Earth's current axis tilt is about $23\frac{1}{2}°$, the tilt varies over time between about 22° and 25° (Figure 4.29). These small changes affect the climate by making seasons more or less extreme. Greater tilt

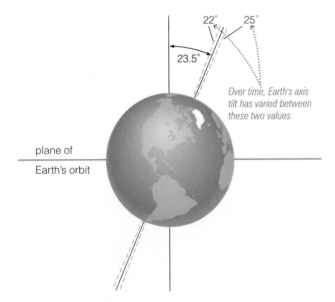

22°   25°

23.5°

Over time, Earth's axis tilt has varied between these two values.

plane of
Earth's orbit

Figure 4.29

Small changes in Earth's axis tilt affect the climate: Greater tilt tends to mean a warmer climate.

*This time scale applies to ocean/atmosphere equilibrium only. The time scale for crust recycling is much longer, while shorter-term climate variations in atmospheric CO_2 concentration can occur through factors besides the inorganic CO_2 cycle, such as cycling of carbon dioxide by life or the addition of CO_2 to the atmosphere through human activity.

means more extreme seasons, with warmer summers and colder winters. The extra summer warmth tends to prevent ice from building up, making the whole planet warmer on average. In contrast, smaller tilt means less extreme seasons that tend to keep polar regions colder and darker on average, allowing ice to build up. Earth's past periods of smaller axis tilt correlate well with colder climate and ice ages, especially when considered along with other cyclical changes in Earth's rotation and orbit.

Snowball Earth   Geologists have recently discovered evidence of several particularly long and deep ice ages between about 750 and 580 million years ago, and another similar set between about 2.4 and 2.2 billion years ago. During these periods, glaciers appear to have advanced all the way to the equator. If so, the global temperature must have dropped low enough for the oceans to begin to freeze worldwide. This ocean freezing would have set up a positive feedback process that would have cooled Earth even further, because ice reflects much more sunlight than water. (About 90% of incoming sunlight is reflected by ice, compared to about 5% by water.) With less sunlight being absorbed, the surface would cool, which in turn would allow the oceans to freeze further. Geologists suspect that in this way, our planet may have entered the periods we now call **Snowball Earth**. We do not know why these episodes occurred or precisely how extreme the cold became. However, some models suggest that at the peak of a Snowball Earth period, the global average temperature may have been as low as $-50°C$ ($-58°F$) and the oceans may have been frozen to a depth of 1 kilometer or more.

How did Earth recover from a snowball phase? Figure 4.30 shows the current model. Even if Earth's surface got cold enough for the ocean surface to freeze completely, the interior still remained hot. As a result, volcanism would have continued to add CO_2 to the atmosphere. Oceans covered by ice would have been unable to absorb this CO_2 gas, and the CO_2 content of the atmosphere would have gradually built up and strengthened the greenhouse effect. As long as the ocean surfaces remained frozen, the CO_2 buildup would have continued. Models suggest that, over a period of 10 million years or so, the CO_2 content of our atmosphere may have increased 1,000-fold.

Eventually, the strengthening greenhouse effect would have warmed Earth enough to start melting the ocean surface ice. The feedback processes that started the Snowball Earth episode then moved quickly in reverse. As the ocean surface melted, more sunlight would have been

Figure 4.30

The CO_2 cycle rescues Earth from a snowball phase.

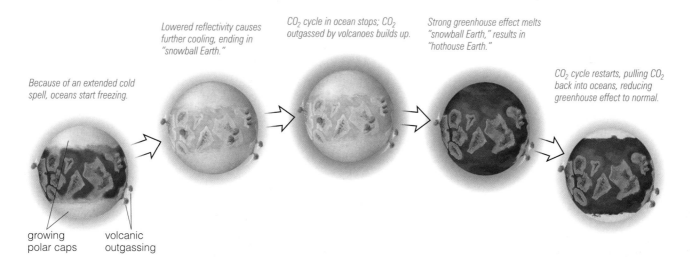

Because of an extended cold spell, oceans start freezing.

Lowered reflectivity causes further cooling, ending in "snowball Earth."

CO_2 cycle in ocean stops; CO_2 outgassed by volcanoes builds up.

Strong greenhouse effect melts "snowball Earth," results in "hothouse Earth."

CO_2 cycle restarts, pulling CO_2 back into oceans, reducing greenhouse effect to normal.

growing polar caps   volcanic outgassing

absorbed (again, because liquid water absorbs more and reflects less sunlight than ice), warming the planet further. Moreover, because the CO_2 concentration was so high, the warming would have continued well past current temperatures—perhaps taking the global average temperature to higher than 50°C (122°F). Thus, in just a few centuries, Earth would have emerged from a snowball phase into what we might call a hothouse phase. Geological evidence supports the occurrence of dramatic increases in temperature at the end of each Snowball Earth episode. Earth then slowly recovered from the hothouse phase as the CO_2 cycle removed carbon dioxide from the atmosphere.

 THINK ABOUT IT Suppose Earth did not have plate tectonics. Could the planet ever recover from a snowball phase? Explain.

The Snowball Earth episodes must have had severe consequences for any life on Earth at the time. Indeed, the end of the Snowball Earth episodes roughly coincides with a dramatic increase in the diversity of life on Earth (the increase called the *Cambrian explosion* [Section 6.3]). Some scientists suspect that the environmental pressures caused by the Snowball Earth periods may have led to a burst of evolution. If so, we might not be here today if not for the dramatic climate changes of the Snowball Earth episodes.

Earth's Long-Term Habitability We have covered a lot of ground in this chapter, but it has given us a clear picture of the major factors that have kept our planet habitable for the past 4 billion or more years. Let's briefly review a few of the key points:

- Volcanic outgassing released most of the gases that made the atmosphere and the water vapor that condensed to form the oceans.

- Earth has kept its atmosphere at least in part because the magnetic field has protected atmospheric gases from being stripped away by the solar wind.

- The greenhouse effect warms our planet enough for water to be liquid, but not so much that the water would boil away.

- This moderate greenhouse effect is maintained by the self-regulating mechanism of the CO_2 cycle, which depends on plate tectonics.

- Even with the regulation provided by the CO_2 cycle, the climate still goes through changes influenced by variations in Earth's axis tilt and other properties of its rotation and orbit.

- The regulatory mechanism sometimes breaks down, leading to periods such as the Snowball Earth episodes, but the CO_2 cycle ultimately brings the climate back into balance.

These ideas should help us understand the prospects for finding other habitable worlds, especially those that might have the long-term habitability that could allow for the evolution of intelligent species and civilizations. Of course, these ideas also leave several questions unanswered. For example, should we expect to find plate tectonics and a CO_2 cycle on other worlds similar to Earth, or was some rare "luck" involved in Earth's getting these regulatory mechanisms? We might also wonder how long

Earth's climate can continue to regulate itself as the Sun brightens with time, a topic we'll discuss in Chapter 10. Finally, and perhaps of the most immediate importance, we might wonder whether we humans could alter the regulatory mechanisms enough to cause serious consequences for current life on Earth or for our own civilization. Unfortunately, the answer appears to be yes, but we'll save this discussion for Section 10.5.

4.6 THE PROCESS OF SCIENCE IN ACTION: FORMATION OF THE MOON

Earlier in this chapter, we stated without evidence the idea that the Moon formed when a "giant impact" blasted away much of the material in the young Earth's outer layers. If you think about it, this is a rather astonishing idea, since it postulates a single event at a time so far in the distant past that we have little hope of finding rocks that survived the event and could tell its tale. For this chapter's case study in the process of science in action, we'll explore how and why this remarkable idea has gained widespread acceptance in the scientific community.

- ### How did the giant impact model win out over competing models?

The existence of the Moon has long been puzzling. As we discussed in Chapter 3, the Moon counts as one of the "exceptions to the rules" when we consider the overall formation of the solar system, because it is far larger in comparison to its planet (Earth) than any other moon in the solar system, except for Pluto's moon Charon. So how did the Moon form?

Three Models, All Flawed During the mid-twentieth century, three competing models were advanced to explain the Moon's existence. The first held that the Moon formed along with Earth through the same process of accretion; in essence, this idea suggested that the two worlds were born together. The second model suggested that the Moon had been an independent "planet" orbiting the Sun that was somehow captured into Earth's orbit. The third model suggested that a young, molten Earth had been spinning so rapidly that it split into two pieces, casting off the piece that became the Moon.

All three models had difficulties right from the start. The joint formation model just didn't seem to work when scientists tried to calculate exactly what might have happened. If you try to build a planet and such a large moon in close proximity, gravitational interactions between them disrupt the process. Moreover, the Moon's average density is much lower than Earth's, which doesn't make sense if both worlds accreted from the same material as Earth. If the Moon had formed in this way, scientists were missing some key ingredient in their models.

The capture model seemed too improbable. It is difficult for a planet to gravitationally capture a passing object under any circumstances, because the passing object has its own orbital energy carrying it around the Sun. This energy cannot simply disappear, so an object can be captured only if it somehow loses some of its orbital energy. Captures therefore are most likely for small objects that lose orbital energy to friction with gas surrounding a planet. Mars probably captured its two small moons in this way, back at a time when it had a more extended atmosphere. Jupiter and the other jovian planets probably also captured many of their small moons this way, back when they were still surrounded by gas from the solar nebula. But Earth never had an atmosphere thick enough to have slowed an object the size of the Moon. The only way that Earth could have captured the Moon would have been if another, similarly sized object was passing by at precisely the same time as the Moon, and if the Moon and this other object had exchanged just enough energy for the Moon to end up in orbit of Earth. It's possible in principle, but highly unlikely.*

The splitting model also suffered from improbability. For example, it seemed unlikely that Earth could ever have been spinning fast enough to spin off the Moon. Still, like the other two models, it could not be ruled out completely at the time.

The *Apollo* missions to the Moon ended this debate, because study of rocks brought back by the astronauts ruled out all three models. The key finding was that the Moon rocks differed significantly in composition from that of Earth rocks. This immediately ruled out the joint formation model in which both worlds would have accreted from essentially identical material. The capture and splitting models were ruled out by chemical processing that had apparently occurred in the Moon rocks: The Moon rocks contained virtually no **volatile**, or easily vaporized, ingredients. In this context, volatile ingredients include not only things such as water but also elements such as lead and gold that vaporize at lower temperatures than other metals and rocks. Because these volatile elements should have been mixed in with other elements in any accreting object, the Moon could not have accreted first and been captured later. And because these volatiles are present in Earth, the Moon should also have had them if it split off from a spinning, molten planet.

The Giant Impact Model With all three models fatally flawed, it was back to the drawing board, which meant taking a closer look at the clues. Two key pieces of evidence soon began to stand out:

1. The Moon's average density is much smaller than Earth's, and in fact is about the same as the density of Earth's mantle. This suggests that the Moon lacks a large iron core like that of Earth and the other terrestrial planets and instead is made almost entirely from material like that of Earth's mantle.
2. The overall composition of the Moon rocks looked quite similar to the composition of Earth's mantle material, except for the lack of volatile elements. Since heating could cause volatile elements to vaporize and escape into space, the rock composition suggested that the Moon was built from mantlelike material that had been strongly heated before it collected to form the Moon.

Taken together, the evidentiary implication seemed almost obvious: The Moon was made from material that accreted in Earth orbit after first being violently blasted out of Earth's mantle. The idea that it was made from mantle material would explain the Moon's general resemblance to Earth's mantle, and the violence of being blasted out would explain the heat necessary to have allowed volatiles to vaporize and escape. But what could have blasted out a large portion of Earth's mantle?

*There is one case in the solar system in which a fairly large moon apparently *was* captured: The orbital characteristics of Neptune's moon Triton make it almost certain to be a captured orbit [Section 9.3]. We do not know how such a large moon (it has about one third the mass of our Moon) was captured; however, because Neptune is much more massive than Earth (about 17 times Earth's mass), a capture event would have had a higher probability.

The Moon's scarred surface suggests an answer. As we discussed in Section 4.3, the early history of the solar system was marked by the heavy bombardment, during which leftover planetesimals smashed into the young planets. Earlier in time, the remaining planetesimals were likely to be more numerous and larger. In fact, models of planetary formation suggest that the late stages of accretion must have been extraordinarily violent [Section 3.3]. Rather than the four terrestrial planets that exist in the inner solar system today, there may have been a dozen or more planet-size bodies. The current planets are the survivors of the shattering collisions that must have occurred. Thus, impacts of planet-size bodies were not only possible but even likely in the early history of the solar system. Such a "giant impact" could easily have had enough energy to blast much of Earth's mantle into space.

Today, sophisticated computer models are used to test the giant impact idea, and it seems to work. The outcome of a giant impact depends on many factors, including the mass and speed of the incoming object, and the precise place and angle at which it strikes Earth. By testing many scenarios, scientists have developed the model of the Moon's formation summarized in Figure 4.31. According to this model, a Mars-size object blasted into the young Earth. The impact must have occurred after Earth had differentiated but before the age of the oldest Moon rocks; radiometric dating and other isotopic evidence tells us that the impact occurred within about 10 to 20 million years after Earth's iron sunk to the core, which occurred more than 4.50 billion years ago. The blast shattered and melted our planet, splashing out molten debris from the mantle. Much of this material fell back to Earth, but some remained in orbit. There,

Figure 4.31

Artist's conception of the Moon's formation by a giant impact. As shown, the Moon formed quite close to a rapidly rotating Earth, but over billions of years tidal forces have slowed Earth's rotation and moved the Moon's orbit outward.

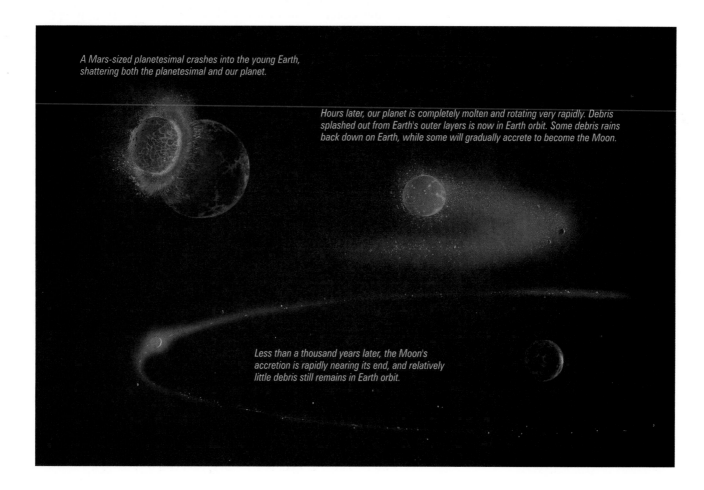

A Mars-sized planetesimal crashes into the young Earth, shattering both the planetesimal and our planet.

Hours later, our planet is completely molten and rotating very rapidly. Debris splashed out from Earth's outer layers is now in Earth orbit. Some debris rains back down on Earth, while some will gradually accrete to become the Moon.

Less than a thousand years later, the Moon's accretion is rapidly nearing its end, and relatively little debris still remains in Earth orbit.

with its volatile content having vaporized and escaped, the material accreted to make the Moon.

• Does the giant impact model count as science?

The giant impact model works so well and so successfully explains the compositional differences between Earth and the Moon that it is widely accepted among planetary scientists. But is it really "scientific" to invoke a single event for which we may never have more than indirect evidence?

One way to decide whether the giant impact model should count as science is to see how it stacks up against the hallmarks of science presented in Chapter 2. Looked at this way, the giant impact model certainly qualifies as science. It invokes a natural explanation for the origin of the Moon, one that even seems likely given what we know about the collisions that must have occurred in the solar system's early history. It also makes testable predictions. The most important of these predictions are the ones about composition — the idea that the Moon formed from mantle material leads to specific predictions about the composition of Moon rocks, and these predictions match the evidence. Note that the fact that the evidence was discovered before the model was proposed is not important here: Just as Kepler came up with his model of planetary motion to explain data that Tycho had already collected, the key point is that the model has been worked out in detail and it successfully matches the observations. The model will also be subject to ongoing tests in the future. For example, the *Apollo* astronauts collected Moon rocks from only a handful of sites on the Moon. In the future, we will presumably collect rocks from many other parts of the Moon. If these rocks were to turn up compositional surprises, it would cause us to reconsider and perhaps even discard the giant impact model. Thus, the giant impact model exhibits all the hallmarks of science: It is natural and testable, and we can imagine future discoveries that would cause us to call it into question.

THINK ABOUT IT   Considering the evidence for the giant impact model, do you think it qualifies as a hypothesis or a theory, or something in between? Note that even scientists disagree on this question, so be sure to defend your opinion.

The giant impact model has important consequences both to our understanding of the solar system and to the search for life in the universe. For our solar system, it may help explain other "exceptions to the rules." If a giant impact really was as likely as we have presumed, then Earth should not have been the only planet to suffer one. Indeed, we have reason to think that at least three other planets also suffered giant impacts. Mercury's surprisingly large iron core is easily explained if a giant impact also blasted away much of its mantle, but without leaving a moon behind. Pluto's moon Charon also shows characteristics we'd expect if it formed in a giant impact. A giant impact might also account for the huge axis tilt of Uranus, which rotates essentially on its side (although an alternative idea has recently been proposed). We do not yet have the direct evidence we would need to test any of these other giant impact ideas, but they seem reasonable in light of what we have learned about our own Moon.

In terms of life, the consequences of the giant impact model lie with roles the Moon has played in shaping Earth's biological history. The Moon is the primary cause of Earth's tides (the Sun contributes only about one third as much to Earth's tides as the Moon), and many living organisms have biological cycles tied to the tides. If there had been no giant impact and no Moon, these types of cycles might not have arisen. The Moon also plays a role in Earth's long-term climate stability. Recall that small changes in Earth's axis tilt can significantly change the climate, bringing on or ending ice ages. Models show that Earth's axis tilt would vary much more if we did not have the Moon; the Moon's gravity exerts a stabilizing influence on axis tilt. Indeed, as we'll see in Chapter 8, evidence suggests that Mars undergoes much more extreme changes in axis tilt, changes that are possible because it lacks a large moon. Some scientists therefore wonder if the existence of the Moon, along with the axis tilt stability that it brings, was necessary to our own evolution on Earth. If it was—and if the Moon really is the result of a random giant impact— then the possibility of finding intelligent life elsewhere may depend on the likelihood of giant impacts that result in forming large moons around terrestrial planets.

THE BIG PICTURE
Putting Chapter 4 into Context

In this chapter, we've explored the interconnections between geology and habitability, learning about the conditions that have made life on Earth possible. As you continue your studies, keep in mind the following "big picture" ideas:

- We can read Earth's geological history by studying the geological record. This history is not mere speculation. It is recorded in ways that we can verify independently in rocks and fossils from many places around the world and through the well-verified technique of radiometric dating for determining ages. While we probably won't ever know every detail of Earth's history, we already have a complete enough picture to understand the major processes and events that have shaped our planet.

- Earth's surface has been shaped by active geology driven by internal heat. This geology made life possible by causing the outgassing of the material that made our oceans and atmosphere, while also creating a magnetic field that may have helped preserve the atmosphere. Moreover, through the action of plate tectonics, geology has kept our planet's climate stable enough for water to remain liquid for the past 4 billion years or more.

- Geology plays a role in making life possible, but life, once it takes hold, can also change the conditions on a planet. The oxygen in our atmosphere is a direct consequence of life. Today life affects our planet in another way: Our advanced civilization is capable of changing the way our climate functions, with consequences that we cannot fully predict.

- We do not yet know how common habitable planets may be in the universe, but geology clearly plays a crucial role in habitability. We'll return to geological considerations many more times throughout this book.

4.1 GEOLOGY AND LIFE

• How is geology crucial to our existence?
Geology appears to be crucial to our existence in at least three ways: **Volcanism** released most of the gas that made the atmosphere and the water vapor that condensed to form the oceans; **plate tectonics** is crucial to the climate regulation that has kept Earth habitable over the long term; and Earth's **magnetic field** has probably helped preserve the atmosphere from being stripped by the solar wind.

4.2 RECONSTRUCTING THE HISTORY OF EARTH AND LIFE

• What can we learn from rocks and fossils?
Rocks and fossils preserve a record of the conditions under which they formed. By studying **mineral** structure and chemical and isotopic composition, we can learn such things as when the rock or fossil formed, how it formed, and what kinds of conditions prevailed on Earth when it formed.

• How do we learn the age of a rock or fossil?

Radiometric dating is based on carefully measuring the proportions of radioactive isotopes and their decay products within rocks. The ratio of the two changes with time and provides a reliable measure of a rock's or fossil's age.

• What does the geological record show?

The geological record allows us to reconstruct Earth's history, which we summarize with the **geological time scale**. We divide this history into four **eons** (the Hadean, Archean, Proterozoic, and Phanerozoic), subdividing the last one into three **eras** and shorter **periods**. The time scale extends back to Earth's birth a little over 4.5 billion years ago.

4.3 THE HADEAN EARTH AND THE DAWN OF LIFE

• How did Earth get an atmosphere and oceans?
The water and gases that made our atmosphere were originally trapped inside our planet. They were released by volcanic outgassing. Water vapor condensed to form the oceans and the other gases made Earth's early atmosphere. Life has since transformed the atmosphere, most importantly by adding molecular oxygen.

• Could life have existed during Earth's early history?
Earth may have had oceans and continents early in its history, but the first half-billion years or so were the time of the **heavy bombardment**. Large impacts must have been common, and a few could have been **sterilizing impacts** that would have killed any life on Earth, or slightly smaller impacts that would have allowed only deep-sea or underground life to survive.

4.4 GEOLOGY AND HABITABILITY

• What is Earth like on the inside?

In order of decreasing density and depth, the interior structure consists of the **core**, **mantle**, and **crust**. The crust and part of the mantle together make up the rigid **lithosphere**. Internal heat allows the mantle rock to convect slowly.

• How does plate tectonics shape Earth's surface?

Plate tectonics has led to many unique features of our geology, especially **sea-floor spreading** zones and the building of continents along **subduction zones**. The shifting of plates has completely changed Earth's geological appearance many times in the past billion years.

• Why does Earth have a protective magnetic field?

Earth's magnetic field is generated by the combination of its molten outer core, convection in that outer core, and a moderately rapid rotation rate. The magnetic field creates a **magnetosphere** that acts like a protective bubble surrounding our planet.

4.5 CLIMATE REGULATION AND CHANGE

• How does the greenhouse effect make Earth habitable?
Greenhouse gases such as carbon dioxide, methane, and water vapor absorb infrared light emitted from a planet's surface. The absorbed photons are quickly reemitted, but in random directions. The result acts much like a blanket, warming the planet's surface. Without the warming due to the greenhouse effect, Earth would be frozen.

• **What regulates Earth's climate?**

Earth's long-term climate is remarkably stable due to feedback processes that tend to counter any warming or cooling that occurs. The most important feedback process is the **carbon dioxide cycle**, which naturally regulates the strength of Earth's greenhouse effect.

• **How does Earth's climate change over long periods of time?**
Earth's temperature has remained in a range allowing liquid water at least since the end of the heavy bombardment, but there have been periods of unusual warmth or cold. Recent **ice ages** are tied to small changes in Earth's rotation and orbit. Geological evidence also shows more extreme variations in the past, including the **Snowball Earth** episodes that ended more than 500 million years ago.

4.6 THE PROCESS OF SCIENCE IN ACTION: FORMATION OF THE MOON

• **How did the giant impact model win out over competing models?**

The giant impact model is the only model of the Moon's formation that successfully explains the differences in composition between Earth and the Moon. According to this model, the Moon formed from mantle material splashed out of Earth by the impact of a Mars-size object.

• **Does the giant impact model count as science?**
The giant impact model shows all the hallmarks of science: It is natural and testable, and we can imagine future discoveries that would cause us to call it into question.

EXERCISES AND PROBLEMS

REVIEW QUESTIONS
Short-Answer Questions Based on the Reading

1. Briefly describe why the study of geology is important to the study of life in the universe.

2. What do we mean by the *geological record?* Why is it important?

3. Describe the three basic types of rock. How does recycling occur among these rock types? What is a *mineral?* How do we study rocks in the laboratory?

4. How are sedimentary *strata* made, and how do they enable us to determine the relative ages of rocks and fossils?

5. Describe the technique of *radiometric dating,* and explain how we know it is reliable. Be sure to explain what we mean by a *radioactive isotope,* parent and daughter isotopes, and a *half-life.*

6. How do fossils form? Do most living organisms leave fossils? Do most fossils contain organic matter? Explain.

7. Summarize the *geological time scale.* What are *eons, eras,* and *periods?*

8. How old is Earth? How do we know?

9. Briefly describe how *outgassing* led to the origin of our oceans and atmosphere. How did Earth's early atmosphere differ from Earth's current atmosphere?

10. What was the *heavy bombardment?* When did it occur? What were *sterilizing impacts,* and when did the last one occur?

11. Briefly describe Earth's *core-mantle-crust* structure, and how it obtained this structure. What is the lithosphere? What is mantle convection?

12. Briefly describe the conveyorlike action of *plate tectonics* and the evidence for this action. How does plate tectonics account for the observed differences in the *seafloor crust* and the *continental crust?*

13. Describe how plate tectonics shapes important geological features of Earth, including mid-ocean ridges, continents, mountain ranges, rift valleys, and earthquakes. How did Hawaii form?

14. What evidence do we have for the operation of plate tectonics in Earth's distant past? Why do we think Earth has plate tectonics?

15. What are the three requirements for a planetary magnetic field, and how does Earth meet them? How does the magnetic field protect our atmosphere?

16. Briefly describe the mechanism by which the *greenhouse effect* warms a planet. What are the most common *greenhouse gases?*

17. What has happened to most of the carbon dioxide outgassed through Earth's history? Describe the *carbon dioxide cycle* and how it helps regulate Earth's climate.

18. What are *ice ages,* and what may cause them? What do we mean by *Snowball Earth* periods, and how does Earth recover from them?

19. Briefly summarize the key ways in which geology is important to Earth's long-term habitability.

20. How do we think the Moon formed, and what evidence supports this model? Why were other models ruled out? Should we be surprised that a giant impact could have affected our planet?

Does It Make Sense?

Decide whether each statement makes sense (or is clearly true) or does not make sense (or is clearly false). Explain your reasoning clearly; because not all of these have definitive answers, your explanation is more important than your chosen answer.

21. We can expect that if there are paleontologists a few million years from now, they will find the fossil remains of almost every human who ever lived.

22. Nearly all the rocks I found in the lava fields of Hawaii are igneous.

23. The most common rock type in the strata of the Grand Canyon is sedimentary rock.

24. Although Earth contains its densest material in its core, it's quite likely that terrestrial planets in other star systems would contain their lowest-density rock in their cores and their highest-density rock in their crusts.

25. If you had a time machine that dropped you off on Earth during the Hadean eon, you'd be quickly killed by a sterilizing impact.

26. If there were no plate tectonics on Earth, our planet would be far too hot to have liquid oceans.

27. Without the greenhouse effect, there probably would be no life on Earth.

28. If nitrogen were a greenhouse gas, our planet would be far hotter than it is.

29. We can learn a lot about Earth's early history by studying the Moon.

30. Science can never determine with confidence the times or sequence of events that occurred millions or billions of years ago.

Quick Quiz

Choose the best answer to each of the following. Explain your reasoning with one or more complete sentences.

31. A rock's type (igneous, metamorphic, or sedimentary) tells us: (a) its age; (b) its chemical composition; (c) how it was made.

32. To learn a rock's age, we must: (a) determine its chemical composition; (b) identify its mineral structure; (c) measure the ratios of different isotopes within it.

33. Radiometric dating now allows us to determine Earth's age to an accuracy of about: (a) a billion years; (b) 20 million years; (c) a few thousand years.

34. The timing of the last sterilizing impact is important because it probably tells us: (a) when life could have taken long-term hold on Earth; (b) when Earth's iron sank to its core; (c) when the dinosaurs became extinct.

35. We learn about the heavy bombardment by studying: (a) craters and rocks from the Moon; (b) zircon mineral grains; (c) Earth's oldest igneous rocks.

36. Earth has retained a lot of internal heat primarily because of its: (a) distance from the Sun; (b) large iron core; (c) relatively large size.

37. Plate tectonics is best described as a process that: (a) recycles rock between Earth's surface and upper mantle; (b) brings metal from Earth's core to the surface; (c) allows continents to plow through the crust.

38. Earth has far less atmospheric carbon dioxide than Venus because: (a) Earth was born with less of this gas; (b) Earth's carbon dioxide was lost in the giant impact that formed the Moon; (c) Earth's carbon dioxide is locked up in carbonate rocks.

39. If Earth had more greenhouse gases in its atmosphere, it would: (a) heat up; (b) cool off; (c) accelerate plate tectonics.

40. *Snowball Earth* refers to: (a) one of a series of very deep ice ages that occurred more than 500 million years ago; (b) the idea that Earth would be frozen without the greenhouse effect; (c) any of the ice ages that have occurred in the past few million years.

INVESTIGATE FURTHER
In-Depth Questions to Increase Your Understanding

Short-Answer/Essay Questions

41. *Understanding Radiometric Dating.* Imagine you had the good fortune to find a meteorite in your backyard that appears to be a piece of material from the early history of the solar system. Qualitatively, how would you expect its ratio of potassium-40 and argon-40 to be different from that of other rocks in your yard? Explain why, in a few sentences.

42. *Dating Planetary Surfaces.* We have discussed two basic techniques for determining the age of a planetary surface: studying the abundance of impact craters and radiometric dating of surface rocks. Describe each technique briefly. Which technique seems more reliable? Which technique is more practical? Explain.

43. *Earth Without Differentiation.* Suppose Earth had never undergone differentiation. How would Earth be different? Write two or three paragraphs discussing likely differences. Explain your reasoning carefully.

44. *Earth Without Plate Tectonics.* Suppose plate tectonics had never begun on Earth. How would Earth be different? Write two or three paragraphs discussing likely differences. Explain your reasoning carefully.

45. *Earth Without the Moon.* Suppose the giant impact that formed the Moon had never occurred. How would you expect Earth to be different? Explain your reasoning carefully.

46. *Feedback Processes in the Atmosphere.* As the Sun gradually brightens in the future, how can the CO_2 cycle respond to reduce the warming effect? Which parts of the cycle will be affected? Is this an example of positive or negative feedback?

47. *Experiment: Geological Properties of Silly Putty.* Roll room-temperature Silly Putty into a ball and measure its diameter. Place the ball on a table and gently place one end of a heavy book on it. After 5 seconds, measure the height of the squashed ball. Repeat the experiment two more times, the first time warming the Silly Putty in hot water before you start and the second time cooling it in ice water before you start. How do the different temperatures affect the rate of "squashing"? How does the experiment relate to planetary geology? Explain.

48. *Experiment: Planetary Cooling in a Freezer.* To simulate the cooling of planetary bodies of different sizes, use a freezer and two small plastic containers of similar shape but different size. Fill each container with cold water and put both into the freezer at the same time. Checking every hour or so, record the time and your estimate of the thickness of the "lithosphere" (the frozen layer) in the two tubs. How long does it take the water in each tub to freeze completely? Describe in a few sentences the relevance

of your experiment to planetary geology. Extra credit: Plot your results on a graph with time on the x-axis and lithospheric thickness on the y-axis. What is the ratio of the two freezing times?

Quantitative Problems

Be sure to show all calculations clearly and state your final answers in complete sentences.

49. *Geological Time.* Geological time scales are often written in ways that can mask their significance. For each of the following pairs of times, state which one is larger and by how much:
    a. 25,000 years, 0.1 million years
    b. 4 million years, 0.05 billion years
    c. 0.1 billion years, 1 million years

50. *Dating Lunar Rocks.* You are analyzing Moon rocks that contain small amounts of uranium-238, which decays into lead with a half-life of about 4.5 billion years.
    a. In one rock from the lunar highlands, you determine that 55% of the original uranium-238 remains; the other 45% decayed into lead. How old is the rock?
    b. In a rock from the lunar maria, you find that 63% of the original uranium-238 remains; the other 37% decayed into lead. Is this rock older or younger than the highlands rock? By how much?

51. *Carbon-14 Dating.* The half-life of carbon-14 is about 5,700 years.
    a. You find a piece of cloth painted with organic dyes. By analyzing the dye in the cloth, you find that only 77% of the carbon-14 originally in the dye remains. When was the cloth painted?
    b. A well-preserved piece of wood found at an archaeological site has 6.2% of the carbon-14 that it must have had when it was living. Estimate when the wood was cut.
    c. Suppose a fossil is 570,000 years old, which is 100 half-lives of carbon-14. What fraction of its original carbon-14 would remain? Use your answer to explain why carbon-14 generally is not useful for dating fossils of this age.

52. *Martian Meteorite.* Some unusual meteorites thought to be chips from Mars contain small amounts of radioactive thorium-232 and its decay product, lead-208. The half-life for this decay process is 14 billion years. Analysis of one such meteorite shows that 94% of the original thorium remains. How old is this meteorite?

53. *Big Moons.* Use the data in Appendix E to compare Earth's Moon, Jupiter's moon Ganymede, Saturn's moon Titan, and Pluto's moon Charon. Rank them in order of their actual size (radius, in kilometers). Then rank them in order of the ratio of their size to the size of the planet they orbit. What can you conclude from the differences in the two rankings? Which moons seem most "unusual" compared to other moons in the solar system?

54. *Internal vs. External Heating.* In daylight, Earth's surface absorbs about 400 watts per square meter. All of Earth's internal radioactivity produces a total of 3 trillion watts, which leak out through the surface. Calculate the internal heat flow (watts per square meter) averaged over Earth's surface. Compare quantitatively to solar heating, and comment on why internal heating drives geological activity.

55. *Plate Tectonics.* Typical motions of one plate relative to another are 1 centimeter per year. At this rate, how long would it take for two continents 3,000 kilometers apart to collide? What are the global consequences of motions like this?

56. *More Plate Tectonics.* Consider a seafloor spreading zone creating 1 centimeter of new crust over its entire 2,000-kilometer length every year. How many square kilometers of surface will this create in 100 million years? What fraction of Earth's surface does this constitute?

Discussion Questions

57. *The Age of Earth.* Some people still question whether we have a reasonable knowledge of the age of Earth or the ability to date events in Earth's history. Based on what you have learned about both relative and absolute ages on the geological time scale, do you think it is reasonable for scientists to be confident of ages found by radiometric dating? Is there any scientific reason to doubt the reliability of our chronology of Earth? Explain.

58. *Plate Tectonics and Us.* Based on what you learned in this chapter, can you imagine cases in which civilizations might arise on planets without plate tectonics? Defend your opinion.

59. *Implications for Other Worlds.* Overall, do you think that Earth's geological features are likely to be rare or common on other worlds? How does your answer affect your opinion of the prospects of discovering life or civilizations on other worlds?

60. *Evidence of Our Civilization.* Discuss how the geological processes will affect the evidence of our current civilization in the distant future. For example, what evidence of our current civilization will survive in 100,000 years? in 100 million years? Do you think that future archaeologists or alien visitors will be able to know that we existed here on Earth?

WEB PROJECTS

61. *Local Geology.* Learn as much as you can about how geological features in or near your hometown were formed. Write a one- to three-page summary of your local geology.

62. *Volcanoes and Earthquakes.* Learn about one major earthquake or volcanic eruption that occurred during the past decade. Report on the geological conditions that led to the event, as well as on its geological and biological consequences.

63. *Formation of the Moon.* Scientists continue to model and study the giant impact thought to have formed the Moon. Look for and report on one recent discovery that may shed more light on how or when the giant impact occurred.

5
The Nature of Life on Earth

LEARNING GOALS

5.1 DEFINING LIFE

- What are the general properties of life on Earth?
- What is the role of evolution in defining life?
- What is life?

5.2 CELLS: THE BASIC UNITS OF LIFE

- What are living cells?
- What are the molecular components of cells?
- What are the major groupings of life on Earth?

5.3 METABOLISM: THE CHEMISTRY OF LIFE

- What are the basic metabolic needs of life?
- How do we classify life by its metabolic sources?

5.4 DNA AND HEREDITY

- How does the structure of DNA allow for its replication?
- How is heredity encoded in DNA?
- How does life evolve?

5.5 LIFE AT THE EXTREME

- What kinds of conditions can life survive?
- Are extremophiles really extreme?

5.6 THE PROCESS OF SCIENCE IN ACTION: EVOLUTION AS SCIENCE

- Is evolution a fact or a theory?
- Are there scientific alternatives to evolution?

There is grandeur in this view of life, with its several powers, having been originally breathed by the Creator into a few forms or into one; and that, whilst this planet has gone cycling on according to the fixed law of gravity, from so simple a beginning endless forms most beautiful and most wonderful have been, and are being, evolved.

Charles Darwin, *The Origin of Species*, 1859

To this point in the book, we've talked only generally about the search for life in the universe and about the conditions that make life possible on Earth. Now, we are ready to begin talking about life itself and how we will search for it. But what exactly are we searching for, and where should we look? To answer these questions, we need to know something about the nature of life and how life interacts with its planetary environment.

The prime difficulty we face in discussing the nature of life is that we have only one example to study: life on Earth. Life on Earth is remarkably diverse, with organisms ranging in size from the microscopic to huge plants and animals, and life of some kind is found in almost every conceivable place on and near the planet's surface. Nevertheless, all living organisms on Earth share common biochemical characteristics that tell us that all species are related to one another. Although no one knows for sure, it seems likely that other types of biochemistry should be possible, in which case life on other worlds might be quite different from life on Earth.

Still, if we want to know what life might be like on other worlds, the best place to start is to learn as much as possible about life here on this world. In this chapter, we'll explore the nature of life on Earth. Along the way, we'll see that life elsewhere would almost certainly share at least a few characteristics with life on Earth, and we'll gain a few clues about where we might most profitably focus our search for life in the universe.

5.1 DEFINING LIFE

What is life? This seemingly simple question lies at the heart of research into life in the universe. After all, if we are interested in the possibility of life elsewhere, we must know what it is that we are looking for. Unfortunately, defining life has proved to be surprisingly difficult, even when we consider only life on Earth. It is all the more difficult when we consider life elsewhere, because we cannot be sure that life on other worlds would resemble life on Earth physically or chemically. Given the difficulty of defining life, the only sensible way to proceed is by studying the one example of life that we know, hoping it will yield fundamental insights into how life operates and into the environmental conditions required to support life. In this first section, we will explore general characteristics of life on Earth and attempt to come up with at least some reasonably useful definition of life.

• What are the general properties of life on Earth?

A cat and a car have much in common. Both require energy to function—the cat gets energy from food, and the car gets energy from gasoline. Both can move at varying speeds and can turn corners. Both expel waste products. But a cat clearly is alive, while a car clearly is not. What's the difference?

In the case of a cat and a car, we can find many important differences without looking too far. For example, cats reproduce themselves, while cars must be built in factories. But as we look deeper into the nature of life, it becomes increasingly difficult to decide what characteristics separate living organisms from rocks and other nonliving materials. Indeed, the question can be so difficult to answer that we may be tempted to fall back on the famous words of U.S. Supreme Court Justice Potter Stewart, who, in avoiding the difficulty of defining pornography, wrote: "I shall not today attempt further to define [it].... But I know it when I see it."* If living organisms on other worlds turn out to be much like those on Earth, it may prove true that we'll know them when we see them. But if the organisms are fairly different from those on Earth, we'll need clearer guidelines to decide whether or not they are truly "living."

One way to seek distinguishing features of life is to study living organisms, looking for common characteristics. Given the difficulty of defining life, you probably won't be surprised to learn that there are exceptions to almost any "rule" we think of. Nevertheless, biologists have identified at least six key properties that appear to be shared by most or all living organisms on Earth, all of which are summarized in Figure 5.1. Let's briefly investigate each property.

Order The materials in living organisms always exhibit some type of order. For example, the molecules in living cells are not scattered randomly about but instead are arranged in patterns that make cell structures. These structures, in turn, make possible all the other properties of life that we will discuss. Note that order alone does not make something living: A book also has order, because words are not scattered randomly on the pages, but it is not alive. The same is true for rock crystals, which have their atoms arranged in an orderly way, and even for the individual molecules of life such as proteins or DNA; these molecules clearly have order, but we consider them only to be building blocks of life, not life itself.

Nevertheless, it seems reasonable to expect that all living things will show order. In logical terms, we say that order is a *necessary condition* for life, because something cannot be alive without order. However, order is not a *sufficient condition* for life, because order alone does not make something alive.

/THINK ABOUT IT The idea of necessary and sufficient conditions is important in science. To make sure you understand it, decide whether each of the following conditions is necessary or sufficient (or neither or both) for the given effect: (1) condition: breathing; effect: human survival while sleeping; (2) condition: living in New York City; effect: living in the United States; (3) condition: meeting all requirements for a college degree; effect: receiving a college degree.

---

*From Potter Stewart's concurring opinion in *Jacobellis v. Ohio*, 378 U.S. 184, 198 (1964).

Figure 5.1

Six key properties of life.

a Order: Living organisms exhibit order in their internal structure, as is apparent in this microscopic view of spiral patterns in two single-celled organisms.

b Reproduction: Organisms reproduce their own kind. Here, a single-celled organism (an amoeba) has already copied its genetic material (DNA) and is now dividing into two cells.

c Growth and development: Living organisms grow and develop in patterns determined at least in part by heredity. Here, we see a Nile crocodile emerging from its shell.

d Energy utilization: Living organisms use energy to fuel their many activities. These tube worms which live near a deep-sea volcanic vent, obtain energy from chemical reactions made possible in part by the heat released from the volcanic vent.

e Response to the enviornment: Life actively responds to changes in its surroundings. Here, we see a jackrabbit's ears flush with blood; the blood flow adjusts automatically to help the animal maintain a constant internal temperature by adjusting the heat loss from the ears.

f Evolutionary adaptation: Life evolves in a way that leads to organisms that are adapted to their enviornments. Here, a pygmy seahorse is camouflaged in its coral surroundings.

Reproduction Living organisms reproduce or are products of reproduction. Simple life-forms, such as bacteria, reproduce by dividing to make nearly exact copies of themselves. More complex organisms may reproduce in more sophisticated ways—including sexual reproduction, in which offspring inherit genetic material from two parents.

Reproduction seems necessary to a definition of life; after all, without reproduction, there would be no way for life as a whole to survive the deaths of individuals. Nevertheless, it's difficult to simply say that all life is capable of reproduction, because there are some exceptions. For example, a mule is sterile and cannot reproduce, but it is the product of reproduction between two closely related animals (a horse and a donkey) and is clearly alive.

Any attempt to include reproduction in the definition of life also exposes borderline cases about which even scientists disagree. For example, *viruses* are generally much smaller than bacteria and are incapable of reproducing on their own: If you place a virus by itself in a dish full of nutrients, nothing will happen. However, when a virus infects a living organism, it can reproduce by commandeering the organism's reproductive machinery for its own purposes. The fact that viruses can reproduce when they infect other organisms but not when they are on their own seems to put them somewhere between the nonliving and the living. Another borderline case concerns the infectious proteins known as *prions*, which are thought to be the agents of *mad cow disease*. Prions are still fairly mysterious, but they appear to be abnormal forms of protein molecules that somehow cause normal protein molecules to change into the abnormal prion form. In other words, they make copies of themselves by causing other molecules to change rather than by actually replicating themselves. Most biologists therefore put them on the nonliving side of the gray region between nonlife and life, though they present at least some ambiguity.

Growth and Development   Living organisms grow and develop in patterns directed at least in part by **heredity**—traits passed to an organism from its parent(s). The property of growth and development appears necessary to life in that all organisms grow or develop during at least some periods in their life cycles, but it is not sufficient to constitute life. For example, fire grows and develops as it spreads through a forest, but a fire is not alive. As we will discuss later in this chapter, all life on Earth passes on its heredity through the molecules known as DNA. (Some viruses use a related molecule called RNA, but we will leave viruses and prions out as we discuss "life" in the rest of this chapter.)

KEY BIOLOGICAL DEFINITIONS

This box summarizes the definitions of a few of the most important biological terms used in this chapter and throughout the book.

TERMS RELATED TO EVOLUTION:

evolution (biological): The gradual change in populations of living organisms that is responsible for transforming life on Earth from its primitive origins to the great diversity of life today.

evolutionary adaptation: An inherited trait that enhances an organism's ability to survive and reproduce in a particular environment.

theory of evolution: The theory, first advanced by Charles Darwin, that explains how and why living organisms evolve through time.

natural selection: The primary mechanism by which evolution proceeds. More specifically, natural selection refers to the process by which, over time, advantageous genetic traits naturally win out (are "selected") over less advantageous traits because they are more likely to be passed down through succeeding generations.

species: Precise definitions vary, but for our purposes we can consider a species to be a population of organisms that is genetically distinct from other groups of organisms. Species are given two-part scientific names consisting of a genus followed by a more specific category; for example, humans belong to the genus *Homo* and are fully classified as *Homo sapiens.*

TERMS RELATED TO HEREDITY:

heredity: The characteristics of an organism passed to it by its parent(s), which it can pass on to its offspring. The term can also apply to the transmission of these characteristics from one generation to the next. Hereditary information is encoded in DNA.

gene: The basic functional unit of an organism's heredity. A single gene consists of a sequence of DNA bases (or RNA bases, in some viruses) that provide the instructions for a single cell function (such as building a protein).

genome: The complete sequence of DNA bases in an organism, encompassing all of the organism's genes along with noncoding DNA in between.

genetic code: The specific set of rules by which the sequence of bases in DNA is "read" to provide the instructions that make up genes.

DNA (deoxyribonucleic acid): The basic hereditary molecule of life on Earth. A DNA molecule consists of two strands, twisted in the shape of a double helix, along each of which lies a long sequence of *DNA bases.* The four DNA bases are adenine (A), cytosine (C), guanine (G), and thymine (T), and they can be paired across the two DNA strands only so that A pairs with T and C pairs with G.

Energy Utilization   Living organisms use energy to create and maintain patterns of order within their cells, to reproduce, and to grow. Life without energy utilization is simply not possible (though some organisms can survive temporarily in dormant states). Of course, energy utilization is not sufficient to constitute life; any electrical or gas-powered appliance uses energy to function.

We can gain further insight into the importance of energy utilization by considering what is sometimes called the *thermodynamics* of life. Thermodynamics is a branch of science that deals with energy and the rules by which it operates. Recall the *law of conservation of energy* [Section 3.4], which tells us that energy can be neither created nor destroyed but only transformed from one form of energy to another; this law is sometimes referred to as the "first law of thermodynamics." The **second law of thermodynamics** states that, when left alone, the energy in a system undergoes conversions that lead to increasing disorder. Living organisms are a perfect example of this law's importance: If you place a living organism into a sealed box, it will eventually use up the available energy and therefore no longer be able to build new molecules or fuel any of the molecular processes needed for life. Its molecules will therefore become more disordered with time—for example, the molecules may decay or may lose the orderly relationships they maintain with other molecules when the organism is alive—causing the organism to die. Thus, to maintain order and survive, a living organism must have a continual source of energy that it can use to counter the tendency for disorder to take over. Living organisms get this energy from the environment, either through food or through chemical interactions with the environment. The environment, in turn, gets its energy either from an internal source such as the heat of the planet itself, or from an external source such as sunlight.

KEY BIOLOGICAL DEFINITIONS *continued*

RNA (ribonucleic acid): A molecule closely related to DNA, but with only a single strand and a slightly different backbone and set of bases, that plays critical roles in carrying out the instructions encoded in DNA.

TERMS RELATED TO THE MODERN CLASSIFICATION OF LIFE:

cell: The basic structure of all life on Earth, in which the living matter inside is separated from the outside world by a barrier called a *membrane.* Cells come in two basic types: *eukaryotes,* which contain a distinct *cell nucleus* with its own membrane, and *prokaryotes,* which lack cell nuclei.

domains of life: All known species of life fall into one of three broad domains: *Bacteria, Archaea,* and *Eukarya;* the last includes all plants and animals, as well as fungi and many microbes.

tree of life: A representation of biochemical and genetic relationships between species; the three major branches of the tree are the three domains (Bacteria, Archaea, and Eukarya).

TERMS RELATED TO CELLULAR CHEMISTRY:

organic molecule: Generally, any molecule containing carbon, but with some exceptions among molecules that are most commonly found independent of life. For our purposes in this book, the important exceptions are carbon dioxide (CO_2) and carbonate minerals, which we do not consider to be organic.

organic chemistry: The chemistry of organic molecules.

biochemistry: The chemistry of life.

amino acids: The molecules that form the building blocks of proteins. Most organisms construct proteins from a particular set of 20 amino acids, although several dozen other amino acids can be found in nature. More technically, an amino acid is a molecule containing both an *amino group* (NH or NH_2) and a *carboxyl group* (COOH).

protein: A large molecule assembled from amino acids according to instructions encoded in DNA. Proteins play many roles in cells; a special category of proteins, called **enzymes,** catalyzes nearly all the important biochemical reactions that occur within cells.

catalysis: The process of causing or accelerating a chemical reaction by involving a substance or molecule that is not permanently changed by the reaction. The unchanged substance or molecule involved in catalysis is called a **catalyst.** In living cells, the most important catalysts are the proteins known as *enzymes.*

metabolism: The many chemical reactions that occur in living organisms to provide cellular energy and nutrients.

Thus, life probably is not possible on a world that lacks a long-term source of energy input to the environment.

Response to the Environment   All living organisms interact with their surroundings and actively respond in at least some ways to environmental changes. For example, some simple organisms may move to a region where the temperature is more suited to their growth, and warm-blooded mammals may sweat, pant, or adjust blood flow to maintain a constant internal temperature. Like all the other properties on our list, response to the environment is a necessary but not sufficient condition for life. Many human-made devices also respond to changes in the environment; for example, a thermostat can respond to changes in temperature by turning on heating or cooling systems.

Evolutionary Adaptation   Life has changed dramatically over time as the organisms that lived billions of years ago have gradually evolved into the great variety of organisms found on Earth today. Life evolves as a result of the interactions between organisms and their environments, leading over time to evolutionary adaptations that make species better suited to their environments. When the adaptations are significant enough, organisms carrying the adaptations may be so different from their ancestors that they constitute an entirely new species.

Before we continue, it's worth noting that, like life itself, the familiar term **species** is not so easy to define in a precise way. Traditionally, a species was defined as a group of organisms that share some set of common characteristics and are capable of interbreeding with one another to produce fertile offspring. Thus, for example, horses and donkeys represent different species because the result of their interbreeding is an infertile mule. However, while this definition of species works fairly well for animals and most plants, it does not work for organisms that reproduce asexually, including microorganisms that reproduce through cell division. As a result, biologists today recognize species as groups of organisms that are genetically distinct from other groups, though the precise border between one species and another is not always clear, especially with microorganisms. Once a species is identified, it is given a scientific name that consists of two parts. The first part is the **genus**, which describes the "generic" category to which the organism belongs, while the second part distinguishes multiple species within the same genus. (You may recognize that the term *genus* is related to "generic" and that the term *species* is related to "specific.") The full name is always written in italics, with the genus capitalized. For example, humans are scientifically classified as *Homo sapiens,* meaning that we are one specific species that has been identified within the genus *Homo* (the others are all extinct). Horses and donkeys are, respectively, *Equus caballus* and *Equus asinus,* names that show that both belong to the same genus (*Equus*).

• What is the role of evolution in defining life?

All six properties of life that we have discussed are important, but biologists today regard evolutionary adaptation as the most fundamental and unifying of all these properties. It is the only property that can explain the great diversity of life on Earth, and an understanding of it allows us to understand how all the other properties of life came to be. Modern understanding of the capacity for evolutionary adaptation is described by the **theory of evolution**. Because this theory is so central to modern biology, let's briefly investigate the origin of the theory and the evidence that supports it.

An Ancient Idea   The word *evolution* simply means, "change with time," and the idea that life might evolve through time goes back more than 2,500 years. The Greek scientist Anaximander (c. 610–547 B.C.) promoted the idea that life originally arose in water and gradually evolved from simpler to more complex forms. A century later, Empedocles (c. 492–432 B.C.) suggested that creatures poorly adapted to their environments would perish, foreshadowing the modern idea of evolutionary adaptation. Many of the early Greek atomists [Section 2.1] probably held similar beliefs, though the evidence is sparse. Aristotle, however, maintained that species are fixed and independent of one another and do not evolve. This Aristotelian view eventually became entrenched within the theology of Christianity, and evolution was not taken seriously again for some 2,000 years. In the mid-1700s, scientists began to suspect that many fossils represented extinct ancestors of living species. Then, in the early 1800s, French naturalist Jean Baptiste Lamarck suggested that the best explanation for the relationship between fossils and living organisms is that life-forms evolve by gradually adapting to perform successfully in their environments.

Lamarck's idea of evolution by adaptation represented the first clear attempt to explain what we now consider the "observed facts" of evolution. That is, observations of how fossils differ in different layers of the geological record and of relationships between living species make it quite clear that life has changed over time. However, Lamarck was unable to come up with a successful theory to explain *how* evolution occurs. His hypothesis concerning the mechanism of evolution, called "inheritance of acquired characteristics," suggested that organisms develop new characteristics during their lives and then pass these characteristics on to their offspring. For example, Lamarck would have imagined that weight lifting would enable a person to create an adaptation of great strength that could be genetically passed to his or her children. While this hypothesis may have seemed quite reasonable at the time, it has not stood up to scientific scrutiny and therefore has been discarded as a model of how evolution occurs. It has been replaced by a different model, proposed by the British naturalist Charles Darwin.

The Mechanism of Evolution   Charles Darwin described his theory of evolution in his book *The Origin of Species*, first published in 1859. In this book, Darwin laid out the case for evolution in two fundamental ways. First, he described his observations of living organisms (made during his voyages on the HMS *Beagle*) and showed how they supported the idea that evolutionary change really does occur. Second, he put forth a new model of *how* evolution occurs, backing up his model with a wealth of evidence. In essence, the geological record and the observed relationships between species together provide strong evidence that evolution *has* occurred, while Darwin's theory of evolution explains *how* it occurs.

As is the case with most scientific theories, the underlying logic of Darwin's model is really quite simple. As described by biologist Stephen Jay Gould (1941–2002), Darwin built his model from "two undeniable facts and an inescapable conclusion":

- *Fact 1: overproduction and competition for survival.* Any localized population of a species has the potential to produce far more offspring than the local environment can support with resources such as food and shelter. This overproduction leads to a competition for survival among the individuals of the population.

- *Fact 2: individual variation.* Individuals in a population of any species vary in many heritable traits (traits passed from parents to offspring). No two individuals are exactly alike, and some individuals possess traits that make them better able to compete for food and other vital resources.

- *The inescapable conclusion: unequal reproductive success.* In the struggle for survival, those individuals whose traits best enable them to survive and reproduce will, on average, leave the largest number of offspring that in turn survive to reproduce. Therefore, in any local environment, heritable traits that enhance survival and successful reproduction will become progressively more common in succeeding generations.

It is this unequal reproductive success that Darwin called **natural selection**: Over time, advantageous genetic traits will naturally win out (be "selected") over less advantageous traits because they are more likely to be passed down through many generations. This process explains how species can change in response to their environment—by favoring traits that improve adaptation—and thus is the primary mechanism of evolution.

Evidence for Evolution by Natural Selection   Darwin backed up his logical claim that evolution proceeds through natural selection by documenting cases in which related organisms are adapted to different environments or lifestyles. He found a particularly striking example among the finches of the Galápagos Islands (Figure 5.2). The 13 finch species on

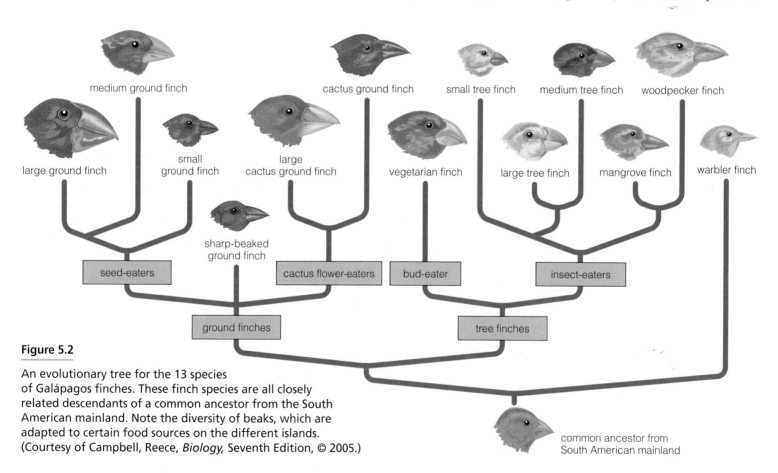

Figure 5.2

An evolutionary tree for the 13 species of Galápagos finches. These finch species are all closely related descendants of a common ancestor from the South American mainland. Note the diversity of beaks, which are adapted to certain food sources on the different islands. (Courtesy of Campbell, Reece, *Biology,* Seventh Edition, © 2005.)

these young volcanic islands closely resemble one another and mainland South American finches, but these particular species are found nowhere else in the world. Moreover, different islands have different species, and each species is adapted to particular microenvironments and lifestyles. For example, some live on the ground while some live in cacti or brush, and some have beaks adapted for cracking seeds while others have insect-catching beaks. Darwin realized that natural selection could explain this situation. He presumed that an ancestral pair of finches reached the Galápagos from the mainland (perhaps by being blown off course by winds). Over time, local populations of island finches gradually adapted to different environments and lifestyles, ultimately becoming the distinct species that he observed.

Darwin recognized similar patterns among many other species in the Galápagos and elsewhere in his round-the-world voyage. He also discovered fossils of extinct organisms that were clearly related to modern organisms, yet different in key respects. For example, in Brazil he found fossils of giant armadillos that he realized must be the ancestors of the modern armadillos found in the same region. The pieces of the puzzle gradually came together in Darwin's mind: He realized that natural selection not only explained the differences between closely related modern species like the finches, but that it also explained the fact that larger changes can occur over longer periods of time, with the result that entire species can become extinct and new ones can take their places.

Darwin also found strong support for his theory of evolution by looking at examples of *artificial selection*—the selective breeding of domesticated plants or animals by humans. Over the past few thousand years, humans have gradually bred many plants and animals into forms that bear little resemblance to their wild ancestors. For example, Figure 5.3 shows how artificial selection has created a variety of vegetables from a single common ancestor. Dogs also are examples of the power of artificial selection: Dog breeds as different as Rottweilers and Chihuahuas were bred from a common ancestor within just a few thousand years. Darwin recognized that if artificial selection could cause such profound changes in just a few

Cabbage

Terminal bud

Lateral buds

Brussels sprouts

Cauliflower

Flower clusters

Leaves

Kale

Broccoli

Flowers and stems

Wild mustard

Stem

Kohlrabi

Figure 5.3

The six vegetables shown all look and taste quite different, but they were all bred by humans from the same wild ancestor (wild mustard). This is an example of artificial selection, which is much like natural selection except that humans (rather than nature) decide which individuals in each generation survive and breed offspring. (Courtesy of Campbell, Reece, *Biology,* Seventh Edition, © 2005.)

thousand years, natural selection could do far more over the many millions of years during which it has operated.

Today, we can observe natural selection occurring right before our eyes. In many places on Earth, species have changed in time spans as short as a few decades in response to human-induced environmental changes. On a microbial level, natural selection is what allows a population of bacteria to become resistant to specific antibiotics; those few bacteria that acquire a genetic trait of resistance are the only ones that survive in the presence of the antibiotic. Indeed, bacterial cases of natural selection pose a difficult problem for modern medicine, because bacteria can quickly develop resistance to almost any new drug we produce. As a result, pharmaceutical companies are constantly working to develop new antibiotics as bacteria become resistant to existing ones. Viruses can evolve even faster, which is one reason it has proved so difficult to fight viral diseases such as influenza and AIDS.

The Molecular Basis of Evolution   Darwin's theory of evolution by natural selection tells us that species adapt and change by passing hereditary traits from one generation to the next. However, Darwin did not know precisely how these traits were communicated across generations, nor did he know why there is always variation among individuals or how new traits can appear in a population. Today, thanks to discoveries in molecular biology made since Darwin's time, we know the answers to all these questions. In particular, we now know that organisms are built from instructions contained in a molecule called **DNA** (short for "*d*eoxyribo*n*ucleic *a*cid"), and biologists can now trace how evolutionary adaptations are related to changes that occur through time in DNA. We'll discuss how DNA makes evolutionary adaptations possible when we discuss

SPECIAL TOPIC: CHARLES DARWIN AND THE THEORY OF EVOLUTION

Charles Robert Darwin was born into a wealthy and educated family in England on February 12, 1809. His father was a physician, and he had two famous grandfathers. His paternal grandfather, Erasmus Darwin (1731–1802), was a renowned physician and scientist who was a strong proponent of the idea that life evolved gradually. His maternal grandfather, Josiah Wedgwood, started the famous Wedgwood Pottery and China company that still bears his name. Darwin's mother died when he was just 8 years old, but his father and his extended family provided him with a generally happy childhood.

At his father's urging, Darwin enrolled in medical school at age 16. However, he was so horrified by the sight of operations, then done without anesthesia, that he left after just 2 years. He next enrolled in Christ College at Cambridge University, intending to become a minister. While there, he began to indulge a childhood love for the study of nature. Shortly after graduating in 1831, Darwin was offered the opportunity to serve as the naturalist aboard a ship of exploration—the HMS *Beagle.* Darwin was 22 years old when the *Beagle* set sail on December 27, 1831. The voyage lasted nearly 5 years, and proved to be the defining event of Darwin's life.

The *Beagle* spent much of its voyage exploring the coasts of South America and nearby islands. While the crew conducted surveys, Darwin went ashore to observe the geology and life and collected numerous specimens to take back to England. He also read extensively dur-

ing the voyage, and one book that had recently been published proved particularly influential: Charles Lyell's *Principles of Geology,* which presented the case for an ancient Earth sculpted by gradual geological processes. Darwin was given the book by a friend who expected Darwin to disagree with its conclusions; instead, Darwin found that his own observations of geology gave only further credence to Lyell's theory.

Meanwhile, Darwin became intrigued by the many adaptations he observed for species in varied environments. He was particularly impressed by the animal life he observed during a 5-week stay on the Galápagos Islands, which lie approximately 1,000 kilometers (600 miles) due west of the South American coast of Ecuador. He focused special attention on the Galápagos finches, concluding that the different bird species must have evolved from a common mainland ancestor (see Figure 5.2). However, at the time he returned to England, he still did not understand how the evolutionary changes occurred.

In 1838, Darwin read Thomas Malthus's *An Essay on the Principle of Population,* in which Malthus famously argued that populations are capable of growing faster than food supplies can support. The essay helped Darwin crystallize the idea of natural selection by making clear that individuals within a population must compete for survival (the idea embodied in Fact 1 on p. 155). He then began intensive study of how humans bred domestic plants and animals, which helped him understand the variation in populations (Fact 2 on p. 156). By 1842, Darits structure and function in Section 5.4, but for now the key point is that our understanding of DNA means that we understand the specific mechanisms by which natural selection occurs.

Our detailed understanding of how evolution proceeds on a molecular level, coupled with all the other evidence for evolution collected by Darwin and others, puts the theory of evolution by natural selection on a solid foundation. That is, it is a *scientific theory* [Section 2.3] that has withstood countless tests and challenges, each of which has only strengthened its foundation. Like any scientific theory, the theory of evolution can never be proved beyond all doubt. However, as we'll discuss further in Section 5.6, no credible scientific alternative to the theory of evolution has been proposed, and the evidence in the theory's favor is overwhelming. Indeed, it is difficult to imagine any aspect of biological science that can be understood without being examined in the context of the theory of evolution. That is why evolution has become the unifying theme of all modern biology.

THINK ABOUT IT The idea that life changes through time is quite ancient, and it was already well supported by observations of fossils before Darwin was even born. Moreover, Lamarck recognized that evolution occurs as a result of adaptations about a half-century before Darwin advanced the idea of natural selection. Given these facts, explain why we credit Darwin with the theory of evolution.

• What is life?

Now that we have examined the fundamental properties of life on Earth, let's return to our original goal in this section: Can we come up with a definition of life?

SPECIAL TOPIC: CHARLES DARWIN AND THE THEORY OF EVOLUTION *continued*

win was convinced that natural selection held the key to evolution, and he began to draft the text that would eventually be published as *The Origin of Species.*

Darwin is said to have been a pleasant man. He was an ardent opponent of slavery, and while he was not a feminist by contemporary standards, he believed that women should be treated with dignity and respect. He was deeply concerned with the impact his theory would have on those who believed in the biblical story of creation. That is probably why he did not publish his theory immediately—he wanted to take time building his case, in hopes that his theory would be so strong that it would be accepted by all without anyone taking offense. Indeed, Darwin might have delayed publication indefinitely if not for a manuscript he received from another scientist, Alfred Russel Wallace, on June 18, 1858.

Wallace had been observing geology and life in Indonesia and had independently come to the same conclusion as Darwin: that life evolves through natural selection. After reading Wallace's draft paper, Darwin worried that "all my originality will be smashed." Fortunately, both Darwin and Wallace were willing to share credit. Their first papers on the theory of evolution were read back-to-back at a scientific meeting in London on July 1, 1858. A little over a year later, Darwin finally published *The Origin of Species.* All 1,250 copies in the first printing sold out on the first day. Within a decade, Darwin's theory was accepted by

the vast majority of biologists, an acceptance that has grown stronger ever since.

Darwin never had a taste for arguing about evolution, leaving that to other scientists (especially Thomas Huxley, who called himself "Darwin's bulldog"). He continued his scientific work, publishing several more books on evolution and related topics. In his personal life, Darwin married a cousin, Emma. He and Emma had ten children, but two died in infancy and a third died at age 10. Darwin died on April 19, 1882, at the age of 73. A parliamentary petition won him burial in London's Westminster Abbey, where he lies next to Sir Isaac Newton.

Charles Darwin and his son William, photographed in 1842.

Based on the central role of evolution, our simplest definition might be that *life is something that can reproduce and evolve through natural selection.* This definition is probably sufficient for most practical purposes, but some cases may still challenge this definition. For example, scientists can now write computer programs (lines of computer code) that can reproduce themselves (create additional sets of identical lines of code). By adding programming instructions that allow random changes, so that the programs can compete and change through a computer analog of natural selection, scientists can even make "artificial life" that evolves on a computer.

> **THINK ABOUT IT** Do you think computer programs that can reproduce and evolve are alive? Why or why not? Would your opinion change if these programs evolved to the point where they could write their own computer code or exchange e-mail with us? What if they wrote other programs that operated machinery to build other computers? Do you think it is possible to create true life on a computer?

Another issue with this definition concerns the origin of life. Darwin's theory of evolution does not tell us how the first life got started. For that, as we'll discuss in Chapter 6, we presume there must have been some type of molecular or *chemical evolution* (as opposed to biological evolution) that went on until the first living organism arose. The idea of chemical evolution is no more surprising than that of natural selection—certain chemical processes are energetically favored under certain circumstances, and laboratory experiments show that under the right conditions, chemicals can evolve in complexity much like life. However, it begs the question of whether we would recognize a clear distinction between, for example, the last case of chemical evolution and the first living organism capable of biological evolution. No one knows the answer; some scientists think there must have been a clear "first" living organism, while others think that the emergence of life may have been marked by a more gradual transition.

The fact that we have such difficulty distinguishing the living from the nonliving on Earth suggests that we should be cautious about constraining our search for life elsewhere. No matter what definition of life we choose, the possibility always remains that we'll someday encounter something that challenges our definition. Nevertheless, the ability to reproduce and evolve through natural selection seems likely to be shared by most, if not all, life in the universe. Thus, this idea provides a useful starting point as we consider how to explore the possibility of extraterrestrial life.

5.2 CELLS: THE BASIC UNITS OF LIFE

Now that we have discussed general properties of life, we are ready to look more specifically at the nature of life on Earth. In this section, we will explore cells, the basic units of life. We will then be prepared to consider in Section 5.3 how cells make use of energy and in Section 5.4 how cells reproduce and evolve.

• What are living cells?

All living organisms are made of **cells**—microscopic units in which the living matter inside is separated from the outside world by a barrier called

a Bacteria

b Amoebas

c Plant cells

d Animal cells

a **membrane*** (Figure 5.4). Thus, cells are the basic structures of life on Earth. Some organisms consist only of a single cell. Other organisms, like oak trees and people, are complex structures in which trillions of cells work cooperatively, dividing various tasks among specialized cells of different types.

Given the great diversity of life on Earth, you may be surprised to learn that all living cells share a great many similarities. For example, all pass on their hereditary information in the same basic way with DNA, and many other chemical processes are nearly the same in all cells. These similarities, which we'll discuss in more detail later, are profoundly important to our understanding of the origin of life. As far as we know, there is no reason why all living cells must share these characteristics. That is, while life elsewhere might also be composed of cells, we should not expect those cells to have the same biochemistry as cells on Earth. Thus, the many similarities among all cells on our planet suggest a startling conclusion: *All life on Earth shares a common ancestor.* In other words, every living organism on Earth is related to every other one because all evolved over billions of years from the same origin of life.[†]

Earth Life Is Carbon-Based

Life on Earth is made from more than 20 different chemical elements. However, just four of these elements—oxygen, carbon, hydrogen, and nitrogen—make up about 96% of the mass of typical living cells. Most of the remaining mass consists of just a few other elements, notably calcium, phosphorus, potassium, and sulfur (Figure 5.5).

Given that oxygen dominates Figure 5.5, you might be tempted to say that life on Earth is "oxygen-based." However, most of the oxygen in living cells is actually part of water molecules (H_2O). The molecules that account for a cell's structure and function owe their remarkable qualities to a different element: carbon. For this reason, we say that life on Earth is **carbon-based**.

Why is carbon so important to life on Earth? The primary answer to this question lies in how carbon can combine with other elements to

Figure 5.4

Microscopic views of four types of living cells. In each cell, a membrane separates the living matter inside the cell from the outside world.

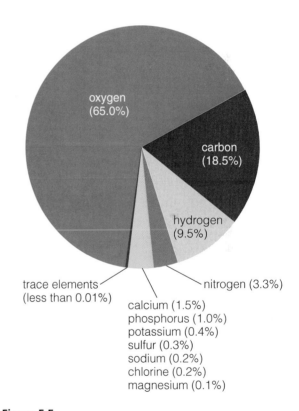

Figure 5.5

This pie chart shows the chemical composition of the human body by weight; this composition is fairly typical of all living matter on Earth. The trace elements include (in alphabetical order) boron, chromium, cobalt, copper, fluorine, iodine, iron, manganese, molybdenum, selenium, silicon, tin, vanadium, and zinc. (Adapted from Campbell, Reece, Simon, *Essential Biology,* Second Edition, © 2004.)

---

*Some organisms, such as some slime molds, do not perfectly fit this picture of discrete cells, because they consist of a large mass of protoplasm containing thousands of nuclei. Nevertheless, the basic idea that living tissue is contained in a package separated from the external environment still holds.

[†]We have not yet identified every type of living organism on Earth, so it is still conceivable that we'll someday discover organisms right here on Earth (perhaps deep underground or in other isolated ecosystems) that use a different biochemistry and hence seem to have come from a separate origin of life. If so, it would greatly expand our understanding of biology, since we'd have more than one form of life to study up close in our laboratories.

Figure 5.6

A water molecule has a single oxygen atom bonded to two hydrogen atoms. (Adapted from Campbell, Reece, Simon, *Essential Biology,* Second Edition, © 2004.)

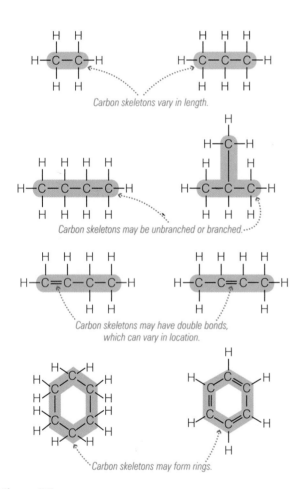

Carbon skeletons vary in length.

Carbon skeletons may be unbranched or branched.

Carbon skeletons may have double bonds, which can vary in location.

Carbon skeletons may form rings.

Figure 5.7

These diagrams represent several relatively simple hydro-carbons—organic molecules consisting of a carbon skeleton attached to hydrogen atoms. The carbon skeletons are highlighted in green. Each single line represents a single chemical bond; a double line represents a double bond. Note that every carbon atom has a total of four bonds (a double bond counts as two single bonds). (Adapted from Campbell, Reece, Simon, *Essential Biology,* Second Edition, © 2004.)

make complex molecules. The atoms in any molecule are linked together by what we call **chemical bonds**, which essentially involve sharing of electrons between the individual atoms of a molecule. Different elements are capable of making chemical bonds in different ways. For example, hydrogen atoms generally can bond with only one other atom at a time, while oxygen atoms generally can bond with at most two other atoms at a time. We can see these properties in a water molecule (Figure 5.6): The oxygen atom has two chemical bonds, one to each of the two hydrogen atoms, while each hydrogen atom has only a single chemical bond to the oxygen atom.

Carbon is a particularly versatile chemical element because it can bond to as many as four atoms at a time. This allows carbon atoms to link together in an endless variety of carbon "skeletons" varying in size and branching patterns (Figure 5.7). The carbon atom sometimes uses two of its bonds to link with the same atom, forming what we call a *double bond;* notice the double bonds in the lower set of molecules of Figure 5.7.

We refer to carbon molecules generically as **organic molecules**. The simplest organic molecules consist of carbon skeletons bonded only to hydrogen atoms; these simple organic molecules are often called *hydrocarbons* to reflect the fact that they contain only hydrogen and carbon. In more complex organic molecules, one or more carbon atoms are bonded to something besides hydrogen and other carbon atoms (Figure 5.8).

Silicon-Based Life (and Other Similar Ideas)   When we consider the possibility of extraterrestrial life, it's natural to wonder whether it might be based on an element besides carbon. In truth, we cannot say for sure whether other elements would work. However, given the importance to life on Earth of carbon's ability to form four bonds at once, we might expect that any other elemental basis for life would have to have the same bonding capability. Among the elements common on Earth's surface—and likely to be common on other planets—silicon is the only element besides carbon that can have four bonds at once. As a result, science fiction writers have often speculated about finding silicon-based life on other worlds.

Unfortunately for science fiction, silicon has at least three strikes against it as a basis for life. First and most important, the bonds formed by silicon are significantly weaker than equivalent bonds formed by carbon. As a result, complex molecules based on silicon are more fragile than those based on carbon—probably too fragile to form the structural components of living cells. In particular, complex silicon-based molecules cannot exist long in water, which is also generally thought to be necessary to life [Section 7.1]. Second, unlike carbon, silicon does not normally form double bonds; instead, it forms only single bonds. This limits the range of chemical reactions that silicon-based molecules can engage in as well as the variety of molecular structures that can form. Third, carbon can be mobile in the environment in the form of gaseous carbon dioxide, but silicon dioxide is a solid (for example, quartz is made from silicon dioxide) that offers no similar mobility. Given the three strikes against silicon, most scientists consider it unlikely that life can be silicon-based. Moreover, observational evidence on Earth also argues against silicon: Silicon is about 1,000 times as abundant as carbon in Earth's crust, so the fact that life here is carbon-based despite the greater abundance of silicon suggests that carbon will always win out over silicon as a basis for life.

A few other elements have also been suggested as possibilities for replacing carbon on other worlds, but most scientists believe carbon's natural advantages will still win out. We have found carbon-based (organic) molecules even in space (as identified in meteorites and interstellar clouds), suggesting that carbon chemistry is so easy and so common that even if life with another basis were possible, carbon-based life probably would arise first and then reproduce so successfully that it would crowd out the possibility of any other type of life.

• What are the molecular components of cells?

All the major components of cells are made from complex organic molecules. Today, biologists know the precise chemical structure of a great many of these molecules, and this knowledge has enabled them to gain a deep understanding of the biochemistry of life. If you take a course in biology, you will learn about much of this biochemistry; here we focus only on generalities about the molecules of life. The large molecular components of cells fall into four main classes: *carbohydrates, lipids, proteins,* and *nucleic acids.* Let's briefly investigate the properties of each class that are most important to life.

Carbohydrates   You're probably familiar with **carbohydrates** as a source of food energy—the sugars and starches known to athletes and dieters as "carbs." In addition to providing energy to cells, carbohydrates make important cellular structures. For example, a carbohydrate called *cellulose* forms the fibers of cotton and linen and is the main constituent of wood. Life on other worlds would presumably need molecules to play these same energy-storing and structural roles, but we do not yet know whether such molecules would have to resemble carbohydrates on Earth or if they could be very different in their chemistry. As a result, we will have little more to say about carbohydrates in this book.

Lipids   Like carbohydrates, **lipids** can store energy for cells. The types of lipids that store energy are more commonly known as *fats.* Thus, despite the bad reputation of fat, it is actually critical to living cells. Lipids also play a variety of other roles in cells on Earth, but from the standpoint of life in the universe, perhaps their most important role is as the major ingredients of cell membranes. That is, lipids form the barriers that make it possible for cells to exist. Moreover, as we'll discuss in more detail in Chapter 6, the membrane-forming role of lipids is thought to have played a critical role in the origin of life: Lipids can spontaneously form membranes in water and probably did so on the early Earth. Other organic molecules would have been trapped inside the space formed by the membranes, making what were in essence tiny chemical factories. These tiny chemical factories may have facilitated the chemical reactions that ultimately led to the origin of life.

Proteins: Key Evidence for a Common Ancestor of Life   The molecules called **proteins** are often described as the workhorses of cells, because they participate in such a vast array of functions. Some proteins serve as structural elements in cells. Others, called **enzymes**, are crucial to nearly all the important biochemical reactions that occur within cells—including the copying of genetic material (DNA)—because they serve as *catalysts* for these reactions. A **catalyst** is any substance (not necessarily a single molecule) that either facilitates a chemical reaction that would not otherwise occur or accelerates a reaction that would otherwise occur

Figure 5.8

In a more complex organic molecule, at least one bond links a carbon atom to something besides hydrogen or another carbon atom. Here, one of the carbon atom's four bonds links it to an *amino group* (which consists of a nitrogen atom and two hydrogen atoms), highlighted in green. (Adapted from Campbell, Reece, Simon, *Essential Biology,* Second Edition, © 2004.)

much more slowly; the catalyst itself is not changed by the process. Enzymes are catalysts because they greatly accelerate the reactions in which they are involved, even though they enter and leave the reactions essentially unchanged. Moreover, because an enzyme is left unchanged after it catalyzes a reaction, a single enzyme can catalyze a specific chemical reaction many times without needing to be rebuilt.

All proteins, whether they serve as enzymes or in other roles, are large molecules built from long chains of smaller molecules called **amino acids**. The "amino" in amino acids refers to the *amino group* that they all share—a nitrogen atom bonded to two hydrogen atoms and a carbon atom (see Figure 5.8); amino acids also always contain what is called a *carboxyl group* (COOH). Different types of amino acids are distinguished by the different sets of atoms also bonded to the central carbon.

The nature of the amino acid chains that make proteins in living organisms provides important evidence supporting the idea that all life on Earth shares a common ancestor. Biochemists have identified more than 70 different amino acids, but most life on Earth builds proteins from only 20 of them. (Two additional amino acids are known to be used in rare cases by particular microorganisms, and scientists suspect that other cases of rare amino acids may yet be discovered.) If life on Earth had more than one common ancestor, we might expect that different organisms would use different sets of amino acids, but they don't. Moreover, naturally occurring amino acids come in two slightly different forms, distinguished by their **handedness** (or *chirality*): The "left-handed" and "right-handed" versions are mirror images of each other (Figure 5.9). Amino acids found in nonbiological circumstances generally consist of a mix of the left- and right-handed versions, but living cells use only the left-handed versions of amino acids to build proteins. Again, the fact that all life on Earth makes use of the same versions of amino acids suggests a common ancestor. Carbohydrates provide some similar evidence, as life on Earth uses mainly the right-handed versions of sugars.

Figure 5.9

Any particular amino acid comes in two forms, distinguished by their *handedness.* These diagrams show the left- and right-handed versions of the amino acid *alanine.* Notice that the two versions are mirror images of each other.

THINK ABOUT IT Large impacts can blast meteorites into space, allowing rocks from one world to travel to another. As we'll discuss in Chapter 6, some scientists hypothesize that microscopic life might survive such impacts and thereby might have migrated between the inner worlds of our solar system. Suppose we discover life on Mars and, while it also has proteins, we find that it builds them from a different set of amino acids than does life on Earth, and they are all the right-handed versions. Would this support or contradict the hypothesis that life migrated between Earth and Mars?

Nucleic Acids Perhaps no cellular molecule is more famous than DNA, which is the basic hereditary material of all life on Earth. A second important nucleic acid, RNA (short for "ribo*nucleic acid*"), helps carry out the instructions contained in DNA. Thus, the nucleic acids DNA and RNA are responsible for allowing cells to function according to precise, heritable instructions. Changing a cell's DNA changes the inherent nature of an organism; indeed, it is changes to DNA that allow species to evolve. We do not know whether other types of molecules could replace nucleic acids in life elsewhere, but it is difficult to imagine life existing in any form without a molecule or molecules to serve the hereditary functions of DNA and RNA. These molecules are so important that we'll devote Section 5.4 to discussing them in more detail.

• What are the major groupings of life on Earth?

Up until just a few decades ago, we were generally able to classify life only by its outward appearance. If you look around, appearances suggest what most people believed for thousands of years—that life exists in the two basic forms that we call plants and animals. The first evidence of a different reality surfaced around the same time as the Copernican revolution. While Galileo turned his telescopes to the heavens, other scientists began to employ similar lens technology to study the microscopic world. The precise origin of the microscope is not known, but the first practical microscopes used for scientific study were built by the Dutch scientist Anton Van Leeuwenhoek (1632–1723; last name pronounced "*LAY-ven-hook*"). During decades of observations beginning around 1674, Leeuwenhoek discovered the world of microscopic life. He was the first to realize that drops of pond water are teeming with microorganisms—a discovery now repeated by almost every elementary school student. He also discovered bacteria and studied the microscopic structure of many plant and animal cells.

With hindsight, it may seem surprising that anyone could have thought that all microorganisms might just be tiny plants or animals, since we now know that microorganisms are far more genetically diverse than larger organisms. But that is exactly what happened. If you look at an old-enough biology textbook, you'll see that life was classified into two "kingdoms": the plants and the animals, and microbes were generally just stuck into one of those two. In the 1960s, biologists expanded the list to five kingdoms, with two (*protista* and *monera*) reserved for microorganisms; the third new addition was *fungi*, by then recognized to be different from plants. However, as our understanding of biochemistry improved during the ensuing decades, biologists began to consider whether life could be classified by its cell structure or biochemistry (including genetics), rather than by its outward appearance. Today, we know that classification by the biochemistry of cells gives us much deeper insights into the relationships between different living species than does classification by appearances alone.

Microscopic Life   Plant and animal cells look different under a microscope; for example, plant cells have what we call a *cell wall* that helps protect the cell membrane, while animal cells do not. However, if you examine many different types of cells under a microscope, you'll find that plant and animal cells overall don't seem that different from one another, at least in comparison to the variety of other types of cells that you'll see. From their microscopic structural appearances, living cells on Earth appear to come in two basic types distinguished by the presence or absence of a **cell nucleus**—an internal membrane that effectively walls off the genetic material (DNA) from the rest of the cell. Organisms that lack cell nuclei are called **prokaryotes** and organisms that have cell nuclei are called **eukaryotes** (Figure 5.10).

All prokaryotes are single-celled organisms (though some of these organisms form colonies in which they live or work together), but they nonetheless represent an incredibly wide range of different species. Examples of prokaryotes include the well-known bacteria *E. coli* and *Salmonella*, both of which are common causes of food poisoning. Eukaryotes may be either single-celled—amoebas are examples—or multicellular. All multicellular organisms are eukaryotes, which means the eukaryotes encompass all plants and animals.

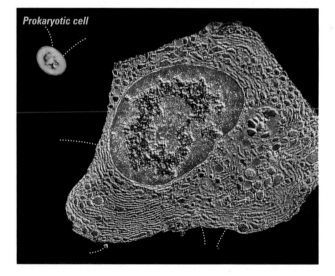

Figure 5.10

These microscopic photographs contrast a typical prokaryotic cell and a typical eukaryotic cell. The eukaryotic cell (an amoeba) is about ten times larger in diameter than the prokaryotic cell (a bacterium); this size difference is typical, though prokaryotic cells come in a fairly wide range of sizes. The eukaryotic cell is also much more complex, with numerous specialized structures and a *cell nucleus* containing its DNA. The prokaryotic cell lacks a distinct nucleus, though its DNA is still concentrated near the center.

Because we are more familiar with plants and animals than with microbes (meaning any single-celled organism), most people assume that microscopic life is a "minor" part of life on Earth. Also, because we tend to associate bacteria with disease, many people assume that microbes are generally harmful. In fact, both assumptions are wrong. Although we humans like to think of ourselves as the dominant form of life on Earth, this is not the case. Much as the Copernican revolution proved that we are not the center of the universe, recent inventories of life show that we are not even the center of Earth's "biological universe." We do not yet know the total amount of microscopic life on Earth, because we do not know how far life extends underground and we are only beginning to get accurate measurements of the amount of life in seawater and soil, but even these limited measurements make clear that tiny microbes are far more dominant in terms of mass and volume than we are (see Cosmic Calculations 5.1). Moreover, most microbes (including most bacteria) are harmless to humans, and many are crucial to our survival. For example, bacteria in our intestines provide us with important vitamins, and bacteria living in our mouths prevent harmful fungi from growing there. Other microbes play crucial roles in cycling carbon and other vital chemical elements between organic matter and the soil and atmosphere; for example, microbes are responsible for decomposing dead plants and animals. Indeed, plant and animal life would be doomed if microbes somehow disappeared from Earth. In contrast, microbes could survive just fine without plants and animals, as they did during most of the history of life on Earth [Section 6.3].

The Three Domains of Life on Earth Because of the visible distinction between cells with and without nuclei, biologists long assumed that prokaryotes and eukaryotes must represent two fundamentally different

COSMIC CALCULATIONS 5.1

The Dominant Form of Life on Earth

We can use estimation to show that microbes far outweigh human beings on Earth. We first estimate the total mass of the approximately 6 billion human beings on Earth. If an average person is 50 kilograms (110 pounds), the total human mass is about:

$$6 \times 10^9 \text{ persons} \times 50\frac{\text{kg}}{\text{person}} = 3 \times 10^{11} \text{ kg}$$

We next estimate the mass of microbes in the oceans. The density of microbes varies significantly with location and depth, but a rough average is *1 billion* (10^9) microbes per liter of water. Multiplying this value by the total volume of ocean water (a Web search reveals this to be about 1.4×10^9 km^3) gives us an estimate of the total number of microbes in the ocean:

$$\text{Total microbes} \approx \underbrace{10^9 \tfrac{\text{microbes}}{\text{liter}}}_{\substack{\text{density of microbes} \\ \text{in ocean water}}} \times \underbrace{(1.4 \times 10^9 \text{ km}^3)}_{\text{volume of ocean water}} \times \underbrace{\left(10^3 \tfrac{\text{m}}{\text{km}}\right)^3 \times \left(10^3 \tfrac{\text{liter}}{\text{m}^3}\right)}_{\text{convert from km}^3 \text{ to liters}}$$

$$= 10^9 \tfrac{\text{microbes}}{\text{liter}} \times (1.4 \times 10^9 \text{ km}^3) \times \left(10^9 \tfrac{\text{m}^3}{\text{km}^3}\right) \times \left(10^3 \tfrac{\text{liter}}{\text{m}^3}\right)$$

$$= 1.4 \times 10^{30} \text{ microbes}$$

To find the total mass of microbes, we next need to know the typical microbe mass. A typical bacterium measures about 1 micrometer on a

side, which means it has a volume of about 1 cubic micrometer. There are 1 million (10^6) micrometers per meter, so the volume of a bacterium is:

$$1 \text{ μm}^3 \times \left(10^{-6}\tfrac{\text{m}}{\text{μm}}\right)^3 = 10^{-18} \text{ m}^3$$

Because life is made mostly of water, we can use the density of water (1,000 kg/m^3) as the density of a microbe. Multiplying the microbe volume by the density, we estimate that the typical microbe mass is:

$$10^{-18} \text{ m}^3 \times 10^3\tfrac{\text{kg}}{\text{m}} = 10^{-15} \text{ kg}$$

We combine our results to find the total mass of microbes in the oceans:

$$\text{total mass of microbes} \approx (\text{number of microbes}) \times (\text{mass per microbe})$$

$$= 1.4 \times 10^{30} \text{ microbes} \times 10^{-15}\tfrac{\text{kg}}{\text{microbe}}$$

$$= 1.4 \times 10^{15} \text{ kg}$$

We can compare to the mass of human beings by dividing:

$$\frac{\text{total mass of microbes}}{\text{total mass of humans}} \approx \frac{1.4 \times 10^{15} \text{ kg}}{3 \times 10^{11} \text{ kg}} \approx 5,000$$

The total mass of microbes in the oceans is roughly 5,000 times that of all humans combined.

lines of life. However, analysis of cellular biochemistry has shown that this is *not* the case. Instead, based on biochemical and genetic similarities and differences, biologists now classify all life into three broad **domains** known as the **Bacteria**, the **Archaea**, and the **Eukarya**. The biochemical and genetic relationships between these domains are illustrated in what biologists usually call the **tree of life**. Figure 5.11 represents the basic structure of the tree of life as it is known today.

The tree of life is proving to be an extremely valuable organizational tool for biologists, and there are at least three key ideas that emerge from it that you should remember as you continue your study of astrobiology:

1. The idea of classifying life into three domains by biochemistry and genetics gives us an organizational scheme that is fundamentally different from the older idea of classifying life into "kingdoms" based solely on structural and physiological differences among species. Notice, for example, that all plants and animals are now seen to represent just two small branches of domain Eukarya. (If it helps, you can think of the domains as "superkingdoms" that represent a higher level of structure than the old kingdoms.)

2. The tree gives us a new and fundamentally different view of relationships among microbes. Bacteria and Archaea used to be grouped together as prokaryotes because they all lack cell nuclei, but we now place them in different domains because they differ in basic features of their biochemistry. For example, Bacteria and Archaea have different types of lipid structures in their cell membranes and they make their cellular proteins in different ways. Indeed, Archaea appear to be somewhat more closely related to Eukarya than to Bacteria.

3. The true diversity of life on Earth is found almost entirely within the microscopic realm. Biochemically and genetically, we humans (and all other animals) are much more closely related to mushrooms than most microbes are related to one another. Moreover, because the vast majority of microscopic species probably remain undiscovered, we are likely to find much more diversity as we continue to study biology. We will certainly discover many more branches within the three domains, and it is even possible that entirely new domains will be discovered in the future.

In addition, the branch lengths shown in the tree of life represent the amounts of genetic difference between different species. Thus, as we trace the branches back toward the "root" shown in Figure 5.11, we are presumably looking back to species that split from common ancestors at earlier times. The closer the we get to the root, the closer we must be to finding an organism that resembles a common ancestor of all life on Earth. As we'll discuss in Chapter 6, this idea is being used to probe possible scenarios for the origin of life on Earth.

5.3 METABOLISM: THE CHEMISTRY OF LIFE

Why are cells so important to life on Earth? More to the point, is it possible that life elsewhere might exist without having a fundamental organizational unit like the cell? To answer these questions, we must understand the processes that take place inside living cells. These processes, which are all chemical in nature, make up what we call **metabolism**. More specifically, metabolism is a blanket term that refers to the many

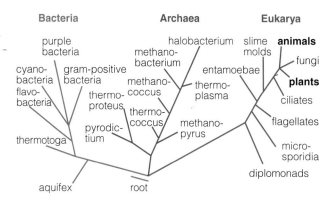

Figure 5.11

The tree of life has three major domains: Bacteria, Archaea, and Eukarya; note that all plants and animals represent just two small branches of the domain Eukarya. (Only a few of the many known branchings within each domain are shown in this diagram; it also remains possible that additional domains will be discovered.)

chemical reactions that occur in living organisms and are involved in providing energy or nutrients to cells.

• What are the basic metabolic needs of life?

Most of the important chemical reactions that occur in cells share a common characteristic: Without the help provided by the cell itself, the reactions would occur too slowly to be useful for life. In this sense, a cell's primary purpose is to serve as a tiny chemical factory in which desired chemical reactions occur much more rapidly than they could otherwise, thereby making it possible to turn simple molecules into the great variety of complex organic molecules needed by living organisms. As is also the case in many factories, cellular work sometimes involves breaking down molecules as well as building them. Like any manufacturing process, this biochemical manufacturing process requires two basic things:

1. *A source of raw materials* with which to build new products. In the case of living cells, the key raw materials are molecules that provide the cell with carbon and other basic elements of life.
2. *A source of energy* to fuel the metabolic processes that break down old molecules and manufacture new ones.

Given the large variety of molecules involved in metabolic processes, you might think that cells would need an equally large variety of sources of raw materials and energy to survive. However, cells have the ability to build incredible variety from a limited set of starting materials. Part of this ability comes from the remarkable variety of enzymes in living cells. Each enzyme is specialized to catalyze particular chemical reactions needed in cellular manufacturing. The remarkable diversity of enzymes in living organisms today is a testament to the power of evolution. The instructions for enzyme creation are encoded in DNA and hence have been evolving for billions of years.

a The molecular structure of ATP. To understand the key parts of the structure, notice that the right side of the molecule shows three identical "phosphate groups," with the third one highlighted in pink. The portion of the molecule shown in yellow is ADP (adenosine *di*phosphate, because it has two of the phosphate groups), and the entire molecule, including the pink portion, is ATP (adenosine *tri*phosphate, because it has three phosphate groups.)

b Cells recycle ATP. The ATP molecule gives up energy when it splits into ADP (yellow) and a phosphate group (pink). Energy input puts the ATP molecule back together.

Figure 5.12

(Adapted from Campbell, Reece, Simon, *Essential Biology,* Second Edition, © 2004.)

The Role of ATP   Another reason why cells can produce so much variety from so little input is that, regardless of where they get their energy, all cells put the energy to work in the same basic way. Every living cell uses the same molecule, called **ATP** (short for "*a*denosine *tri*phosphate"), to store and release energy for nearly all its chemical manufacturing (Figure 5.12a). This usage of ATP vastly simplifies the manufacturing process, because it means that a cell needs an outside energy source only for the purpose of producing ATP, rather than for producing the full variety of organic molecules in cells. Once ATP is produced, it can be used to provide energy for any cellular reaction. Moreover, the nature of ATP makes it completely recyclable. Each time a cell draws energy from a molecule of ATP, it leaves a closely related by-product, called ADP (short for "*a*denosine *di*phosphate"), that can be easily turned back into ATP (Figure 5.12b).

The fact that all life on Earth uses the same molecule (ATP) for energy storage offers further evidence for a common origin of life. There's no known reason why other molecules could not fill the role of ATP. Thus, the fact that all living cells use ATP suggests that they all evolved from a common ancestor that made use of this remarkable molecule.

Carbon Sources and Energy Sources   Because carbon compounds are the primary raw materials needed for life, the needs of metabolism essentially come down to the need for a carbon source and an energy source.

We humans, like all animals, meet both of these needs with food. The food we eat gets digested and carried in molecular form by our bloodstreams to individual cells. There, the cells make use of the molecules from our food sources. Some of the molecules are used as the carbon source for cellular manufacturing, while others undergo chemical reactions that release energy the cell can use to fuel its ATP cycle. Of course, not all organisms get their carbon and their energy by eating other organisms. Plants, for example, get energy from sunlight, and some microorganisms get energy from chemical reactions that take place inside rocks or in hot springs. Let's next turn our attention to the variety of carbon and energy sources used by different forms of life on Earth.

• How do we classify life by its metabolic sources?

Living cells on Earth get their carbon and energy from a surprisingly wide variety of sources. Indeed, as we'll see, this wide variety makes life elsewhere seem more likely. For example, we already know of worlds within our own solar system, such as Mars and Europa, that may well have the necessary materials and energy source for metabolism, suggesting that life could at least in principle exist on those worlds.

Today, astrobiologists classify life into four major categories by its metabolic sources, summarized in Table 5.1. We can understand their rather long and technical names by looking first at their carbon sources and then at their energy sources.

Carbon Sources: Autotrophs and Heterotrophs
Cells need a source of carbon from which to build the skeletons of their organic molecules. In the broadest sense, cells can get their carbon in either of two ways:

1. Some cells get carbon by consuming preexisting organic compounds, that is, by eating. For example, we humans acquire carbon by eating plants or other animals. Any organism that gets its carbon by eating is called a **heterotroph;** the word comes from *hetero* meaning "others" and *troph* meaning "to feed." All animals are heterotrophs, as are many microscopic organisms.
2. Some cells get carbon directly from the environment by taking in carbon dioxide from the atmosphere or carbon dioxide dissolved in water. An organism that gets its carbon directly from the environment is called an **autotroph,** meaning "self-feeding." For example, trees and most other plants are autotrophs.

If you look at the first column of Table 5.1, you'll see that the first two entries are both autotrophs and the second two are both heterotrophs. However, in each case the entry carries a prefix of either *photo* (meaning "light") or *chemo* (meaning "chemicals"). These prefixes describe the energy sources that go with the carbon sources of either eating (heterotroph) or environmental carbon dioxide (autotroph).

Energy Sources: Light or Chemicals
Broadly speaking, the energy source that a living cell uses to make ATP can be one of three things (see Section 9.4 for more details about chemical energy for life):

1. Some cells get energy directly from sunlight, using the process we call *photosynthesis.* For example, plants acquire their energy from sunlight. Organisms that get energy from sunlight are given the prefix *photo.*

TABLE 5.1 *Metabolic Classifications of Living Organisms*

Metabolic Classification	Carbon Source	Energy Source	Examples
photoautotroph	carbon dioxide	sunlight	plants, photosynthetic bacteria
chemoautotroph	carbon dioxide	inorganic chemicals (e.g., iron, sulfur, ammonia)	some bacteria and archaea, especially in extreme environments
photoheterotroph	organic compounds	sunlight	some bacteria and archaea
chemoheterotroph	organic compounds	organic compounds	animals, many microbes

2. Some cells get energy from food; that is, they take chemical energy from organic compounds they've eaten and use it to make their own ATP. Since the energy comes from chemical reactions, these organisms get the prefix *chemo*.

3. Some cells get energy from *inorganic* chemicals—chemicals that do *not* contain carbon—in the environment. This type of energy source is different in character from organic food, and cells that get energy directly from the environment require neither sunlight nor other organisms to survive. However, because the energy still comes from chemical reactions, these organisms also get the prefix *chemo*.

The Four Metabolic Classifications   We can now put the carbon and energy sources together to understand the four metabolic classifications in Table 5.1. The first row of the table shows the **photoautotrophs**, which get their energy from sunlight (*photo*) and their carbon from carbon dioxide in the environment (making them autotrophs). This category therefore includes plants, as well as microorganisms that obtain their energy through photosynthesis.

The second row shows the **chemoautotrophs**, which obtain energy from chemical reactions (*chemo*) involving inorganic chemicals and carbon from environmental carbon dioxide (making them autotrophs). These are in some ways the most amazing organisms, because they need neither organic food nor sunlight to survive. For example, the archaea known as *Sulfolobus* (a genus that includes many distinct species) live in volcanic hot springs and obtain energy from chemical reactions involving sulfur compounds. As is the case with *Sulfolobus,* chemoautotrophs are often found in environments where most other organisms could not survive. For much the same reason, they may also be the organisms most likely to be found on other worlds, since a wider range of conditions seems suitable to them than to other forms of life.

The third row shows the **photoheterotrophs**, which get energy from sunlight (*photo*) but get their carbon by consuming other organisms or the remains of such organisms (making them heterotrophs). This category is much rarer, but some prokaryotes do indeed get their carbon by eating organic compounds while making ATP with energy from sunlight. Examples include bacteria known as *Chloroflexus*, which obtain their carbon from other bacteria but their energy from photosynthesis. These organisms live in lakes, rivers, hot springs, and some aquatic environments very high in salt content.

The fourth row shows **chemoheterotrophs**, which get both their energy and their carbon from food. This category therefore includes us and all other animals, as well as many microorganisms. That is, we are chemoheterotrophs because we extract chemical energy (*chemo*) from food and carbon from eating (making us heterotrophs).

> **THINK ABOUT IT** Classify each of the following into one of the four metabolic categories listed in Table 5.1: (a) an organism that gets its energy from chemicals near an undersea volcano and gets its carbon from carbon dioxide dissolved in water; (b) a tomato plant; (c) a fly.

Metabolism, Water, and the Search for Life The four metabolic classifications are quite general, so they ought to apply equally well to life elsewhere as to life on Earth. Moreover, any type of complex metabolism requires the existence of some kind of structure that allows carbon and energy to come together to manufacture (and break down) the molecules needed for life. Thus, unless we are failing to imagine an entirely different potential mode of operation, it seems likely that all living organisms must have a fundamental structure that functions much like cells on Earth. This crucial observation means that we can search for life in the universe by searching for cells rather than having to search for a much broader variety of possible structures.

This leaves us with one final ingredient to consider in metabolism: liquid water. On Earth, water plays three key roles in metabolism. First, metabolism requires that organic chemicals be readily available for reactions. Liquid water makes this possible by allowing organic chemicals essentially to float within the cell (because the chemicals dissolve in water). Second, metabolism requires a means of transporting chemicals to and within cells, and of transporting waste products away; water is the medium of this transport. Third, water plays a role in many of the metabolic reactions within cells; for example, water molecules are necessary for the reactions that store and release energy in ATP.

All living cells on Earth depend on liquid water to play these three roles, and this dependence limits the conditions under which we find life on our planet: We find life only in places where it is neither too cold nor too hot for liquid water to exist. Indeed, while we've seen that life on Earth can use a variety of different carbon and energy sources, liquid water is one thing that no organism on Earth can survive without. (Some organisms can become dormant and grow or metabolize in the absence of liquid water, but they cannot survive permanently in such conditions.) Does this need for liquid water also apply to life on other worlds? Certainly, some kind of liquid seems necessary, but we'll save discussion of possibilities other than water for Chapter 7.

5.4 DNA AND HEREDITY

In the previous two sections, we studied two key features of life on Earth that are likely to be crucial to life anywhere else as well: the structural units of cells and the metabolic processes that keep cells alive. A third feature that seems generally needed for all life is some means of storing information; that is, a set of "operating instructions" for the cell and a way of passing these instructions down through the generations. This information is what we generally call an organism's *heredity*.

All living things on Earth encode their hereditary information in the molecule known as DNA (although some viruses use RNA). That is, DNA

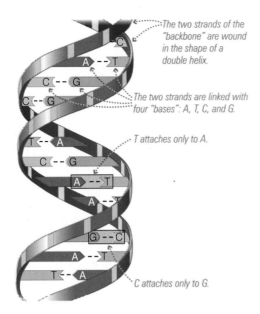

The two strands of the "backbone" are wound in the shape of a double helix.

The two strands are linked with four "bases": A, T, C, and G.

T attaches only to A.

C attaches only to G.

Figure 5.13

This diagram represents a DNA molecule, which looks much like a zipper twisted into a spiral. The important hereditary information is contained in the "teeth" linking the strands. These "teeth" are the DNA bases. Only four DNA bases are used, and they can link up between the two strands only in specific ways: T attaches only to A, and C attaches only to G. (The color coding is arbitrary and is used only to represent different types of chemical groups; in the backbone, blue and yellow represent sugar and phosphate groups, respectively.) (Adapted from Campbell, Reece, Simon, *Essential Biology*, Second Edition, © 2004.)

holds the "operating instructions" for living organisms on Earth. DNA also allows organisms to reproduce, because it can be accurately copied. In this section, we will explore how DNA determines the nature of an organism and allows reproduction. We will also discuss how rare errors in the copying of DNA can lead to evolutionary adaptations, thus giving us the molecular-level understanding of natural selection that we first discussed in Section 5.1.

• How does the structure of DNA allow for its replication?

The molecular structure of DNA, a *double helix,* is one of the most familiar scientific icons of our time (Figure 5.13). A helix is a three-dimensional spiral, such as you would make by extending a Slinky toy; a double helix has two intertwined strands, each in the shape of a helix. The structure looks much like a zipper twisted into a spiral. The fabric edges of the zipper represent the "backbone" of the DNA molecule, while the zipper teeth that link the two strands represent molecular components called **DNA bases**. The chemical structure of the backbone is interesting and important in its own right, but it is the DNA bases that hold the key to heredity. Life on Earth makes use of only four DNA bases: adenine (abbreviated A), guanine (G), thymine (T), and cytosine (C).

The key to DNA's ability to be duplicated by cellular machinery lies in the way the four DNA bases pair up to link the two strands: T can pair up only with A, while C can pair up only with G. Figure 5.13 shows this pairing by representing the different bases with different shapes. For example, the shape of A, which is depicted as ending with an open triangle, fits only into the notch in T. Similarly, what is shown as the curved end of G fits only into the curved notch in C. These diagrams are only schematic representations—there aren't literally notches and curves of these shapes—but the real chemical bases work much the same way: Their actual shapes and sizes determine how they pair up.

The process by which DNA is copied, called **DNA replication,** is illustrated in Figure 5.14. Step 1 begins with the complete double helix. In Step 2, the two strands separate, "unzipping" the links between the paired bases. Step 3 shows how, once the strands have been "unzipped," each strand can serve as a template for making a new strand. Because the "teeth" of each new strand must link to the existing strands according to the base pairing rules—T goes only with A, and C goes only with G— each new strand will be *complementary* to an existing one. (By saying that two strands are complementary, we mean that, while they are not identical, they contain the same information because knowing the base sequence on one strand automatically tells us the base sequence on the other strand.) The end result, shown in Step 4, is two identical copies of the original DNA molecule. When a cell divides, one copy goes to each daughter cell. Because cell division is the key to passing down genetic material from one generation to the next, DNA replication explains the basis of heredity.

Although the DNA copying process is easy in principle, the actual mechanics are fairly complex. More than a dozen special enzymes are involved in the various steps, performing tasks such as unzipping the double helix, making sure the correct bases pair up, checking for and correcting any errors in the copying process, and rezipping the new DNA molecules. This complexity is one reason why errors sometimes occur in DNA replication; as we'll discuss shortly, these errors are crucial to evolution. The com-

1 DNA double helix

2 Strands separate.

3 Each strand serves as a template for a new complementary strand.

Other bases, floating in the cell, come in and attach according to the pair rules.

4 Two identical "daughter" DNA molecules

plexity of replication also makes it extremely unlikely that DNA could have been the original hereditary molecule for life on Earth. Instead, most biologists believe that DNA evolved from a simpler self-replicating molecule—probably a form of RNA—that carried hereditary information in the earliest living organisms [Section 6.2]. However DNA evolved, it has proved remarkably successful—it is now the hereditary material for every known organism on Earth. The basic copying process shown in Figure 5.14 explains how all life on our planet, from the smallest bacteria to humans, passes its genetic information from one generation to the next.

Figure 5.14

DNA replication. DNA copies itself by "unzipping" its two strands, each of which then serves as a template for making a new, complementary strand built in accord with the base pairing rules: A goes only with T, and C goes only with G. The end result is two identical copies of the original DNA molecule. (Adapted from Campbell, Reece, Simon, *Essential Biology,* Second Edition, © 2004.)

• How is heredity encoded in DNA?

Besides having the ability to be replicated, DNA also determines the structure and function of the cells within any living organism. In essence, the "operating instructions" for a living organism are contained in the precise arrangement of chemical bases (A, T, C, and G) in the organism's DNA. Today, biologists have technology that allows them to rapidly determine the sequence of bases in almost any strand of DNA. This technology has been used to determine the DNA sequences that code for many cell functions, as well as to determine the complete DNA sequences of many living organisms. For example, the Human Genome Project, completed in 2003, was a 13-year effort in which scientists ultimately determined the order of all three billion bases that make up the DNA of a human being. (In humans, this DNA is spread among the 46 *chromosomes* found in normal human cells.)

Genes and Genomes   Within a large DNA molecule, isolated sequences of DNA bases represent the instructions for a variety of cell functions. For example, a particular sequence of bases may contain the instructions for building a protein, for building a piece of RNA, or for carrying out or regulating one of these building processes. The instructions representing any individual function—such as the instructions for building a single protein—make up what we call a **gene.** A gene is the basic functional unit of an organism's heredity—a single gene consists of a sequence of DNA bases (or RNA bases, in some viruses) that provides the instructions for a single cell function.

Interestingly, among plants, animals, and other Eukarya, most of the DNA is *not* part of any gene; that is, much of the DNA does not appear to carry the instructions for any particular cell function. For example, this so-called **noncoding DNA** (sometimes called "junk DNA") makes up more than 95% of the total DNA in human beings, and similarly large fractions of the DNA of many other eukaryotes. Biologists suspect that most of this noncoding DNA represents evolutionary artifacts—pieces of DNA that may once have had functions in ancestral cells but that no longer are important, much like the appendix is an organ that no longer plays an important role in our bodies. However, recent discoveries suggest that at least some of the noncoding DNA may play important roles that are not yet fully understood.

The complete sequence of DNA bases in an organism, encompassing all of the organism's genes as well as all its noncoding DNA, is called the organism's **genome.** Different organisms have genomes that vary significantly both in total length (number of bases) and in their numbers of genes. For example, some simple microbes have DNA that extends only a few hundred thousand bases and contains only a few hundred genes.* We humans have a genome that contains an estimated 20,000 to 25,000 genes among its sequence of some three billion DNA bases. Note that, genetically speaking, we are by no means the most complex organisms on Earth. Rice, for example, has significantly more genes than we do—recent sequencing of the rice genome shows that it has about 37,000 genes, though it has a shorter total DNA sequence. Other organisms have far more DNA than people. For example, the simple plant known as the "whisk fern" (*Psilotum nudum*) has more than 70 times as many bases in its genome as humans, though most of this extra DNA is probably noncoding.

Every member of a particular species has the same basic genome. However, there is always some variation among individuals. For example, while all human beings have the same set of genes, the genes of different individuals differ slightly; for example, they may vary here or there in their precise sequence of DNA bases. These differences in the genes of individuals explain why we are not all identical, and they are also the source of the individual variation that underlies the theory of evolution (see Fact 2 on p. 156). Moreover, with a few exceptions, every cell in a living organism contains the same set of genes. Different cell types, such as muscle cells or brain cells, differ only because they *express*, or actually use, different portions of their full set of genes. Thus, the DNA found in almost any cell in any organism contains the complete instructions for building an organism of that species. This fact underlies the science of *cloning*, in which a single cell from a living organism is used to grow an entirely new organism with an identical set of genes.

The Genetic Code   A strand of DNA contains a long, unbroken sequence of DNA bases; for example, a particular sequence might contain the bases ACTCAGCTTCAACGG.... For a sequence like this to be useful as the instructions for a cell function, there must be a set of rules for how to "read" the sequence. That is, the rules must specify how to break the long sequence into individual "words," as well as where to start reading and where to stop reading the words that represent the instructions for a single gene. The set of rules for reading DNA is called the **genetic code** (Figure 5.15). More specifically, genetic "words" consist of three DNA bases in a row. For the purpose of protein building, each word represents either a particular amino acid or a "start reading" or "stop reading" instruction.

---

*Many viruses are far simpler, with just a few thousand bases and a handful of genes. Mitochondria within eukaryotic cells, which are thought to have had free-living ancestors, are also much simpler than the simplest bacteria sequenced to date. For example, human mitochondria have fewer than 17,000 DNA base pairs, representing fewer than 40 genes.

second base

first base	T	C	A	G	third base
T	TTT, TTC — Phenylalanine; TTA, TTG — Leucine	TCT, TCC, TCA, TCG — Serine	TAT, TAC — Tyrosine; TAA — stop; TAG — stop	TGT, TGC — Cysteine; TGA — stop; TGG — Tryptophan	T C A G
C	CTT, CTC, CTA, CTG — Leucine	CCT, CCC, CCA, CCG — Proline	CAT, CAC — Histidine; CAA, CAG — Glutamine	CGT, CGC, CGA, CGG — Arginine	T C A G
A	ATT, ATC, ATA — Isoleucine; ATG — Met or start	ACT, ACC, ACA, ACG — Threonine	AAT, AAC — Asparagine; AAA, AAG — Lysine	AGT, AGC — Serine; AGA, AGG — Arginine	T C A G
G	GTT, GTC, GTA, GTG — Valine	GCT, GCC, GCA, GCG — Alanine	GAT, GAC — Aspartic acid; GAA, GAG — Glutamic acid	GGT, GGC, GGA, GGG — Glycine	T C A G

Figure 5.15

The genetic code. This table shows how three-base-pair "words" of DNA code for particular amino acids or a start or stop instruction. For example, you can find the "word" CAG by looking along the left for C as the first base, along the top for A as the second base, and along the right for G as the third base; you'll then see that CAG codes for the amino acid glutamine. Notice that in most cases, the first two letters alone determine the amino acid; for example, if the first two letters are CT, the amino acid is always leucine regardless of the third letter. This suggests that the current genetic code evolved from an earlier version that used only two-letter "words" rather than three-letter "words." (Adapted from Campbell, Reece, Simon, *Essential Biology*, Second Edition, © 2004.)

Because the genetic words consist of three DNA bases in a row and there are four DNA bases to choose from (A, C, T, G), the total number of words in the genetic code is $4^3 = 4 \times 4 \times 4 = 64$; you can see all 64 words spelled out in Figure 5.15. Notice that this is significantly more than the number of amino acids used to make proteins, which is 20 for most organisms. Thus, the genetic code contains a fair amount of redundancy. For example, the genetic words ACC and ACA both represent the same amino acid. Moreover, a close examination of the genetic code also offers a hint about the likely evolution of DNA: The codes for most amino acids really depend on just the first two bases in the three-base genetic words. For example, all four of the three-base words starting with AC—ACC, ACA, ACT, and ACG—code for the same amino acid (threonine). This suggests that the genetic code once depended only on two-base words rather than three-base words. Most biologists now believe that early life-forms used only a two-base language, which later evolved into the current three-base language of the genetic code.

THINK ABOUT IT Note that a two-base language would allow only $4 \times 4 = 16$ possible words—not enough for all the amino acids used by living organisms today. What does this imply about proteins in early life-forms? Does it seem likely that even earlier organisms might have used only a one-base language? Explain.

Another important feature of the genetic code is that it is the same in nearly all living organisms on Earth. Only a few organisms show any variations at all on this code, and the variations that do exist are minor. (Variations in the genetic code are also found in mitochondria, structures within eukaryotic cells that contain their own DNA.) Nevertheless, the fact that some variations occur tells us that not all the specifics of the genetic code were inevitable. If we think of the genetic code as a language, the fact that nearly all organisms use the same genetic code is as if everyone on Earth spoke the same language, even though other languages are possible. This common language of the genetic code is further evidence for a common ancestor of all life on Earth.

The Role of RNA   While the sequence of bases in a gene holds the instructions for its function, the actual implementation of these instructions is quite complex. As with DNA replication, many enzymes are involved in carrying out genetic instructions. In addition, the molecule RNA plays a particularly important role in these functions. A molecule of RNA is quite similar in structure to a *single* strand of DNA, except that it has a slightly different backbone and one of its four bases is different from one of the DNA bases (RNA uses a base called *uracil* (U) in place of DNA's thymine (T)).

RNA participates in carrying out genetic instructions in several different ways. For example, in the process of building a protein, a molecule of RNA is first assembled along one strand of DNA, essentially transcribing the DNA instructions for use in another part of the cell. (This process is called *transcription*.) This molecule of RNA then goes to a site in the cell where amino acids are actually assembled into proteins. Other molecules of RNA collect amino acids from within the cell and bring them to this site, where yet another form of RNA helps attach the amino acids into the chains that make proteins. (This second process is called *translation*, because it effectively *translates* the genetic instructions into an actual protein.) If you take a biology course, you will learn much more about the many roles of RNA.* However, for our purposes in this book, it is not necessary to know much more about RNA except the fact that it is vital for carrying out the instructions encoded in DNA.

• How does life evolve?

One of the most remarkable aspects of our current knowledge of DNA is that it has allowed us to further confirm Darwin's theory of evolution by natural selection. In particular, while Darwin had to base his theory on the variation in populations that he could directly observe, we now know precisely how such variation occurs at the molecular level. The key to this knowledge lies in understanding how DNA molecules gradually change through time. Based on what we've already said about DNA replication and protein building, we can see how and why changes in DNA occur.

Mutations: The Molecular Basis of Evolution   Despite its complexity, the process of DNA replication proceeds with remarkable speed and accuracy. Some microbes can copy their complete genomes in a matter of minutes, and copying the complete three-billion-base sequence in human DNA takes a human cell only a few hours. In terms of accuracy, the copying process generally occurs with less than one error *per billion* bases copied. Nevertheless, errors do sometimes occur. For example, the wrong base may occasionally get attached in a base pair, such as linking C to A rather than to G. In other cases, an extra base may be accidentally inserted into a gene, a base may be deleted, or an entire sequence of bases might be duplicated or eliminated. Absorption of ultraviolet light or nuclear radiation or the action of certain chemicals (carcinogens) can also cause mistakes to occur. Any change in the base sequence of an organism's DNA is called a **mutation.**

Mutations can affect proteins in a variety of ways. Some mutations have no effect at all. For example, suppose a mutation causes the genetic

---

*Readers familiar with the different types of RNA will recognize that messenger RNA (mRNA) is involved in transcription, transfer RNA (tRNA) is responsible for collecting amino acids, and both of these meet up to assemble proteins on structures called ribosomes, which are built from proteins and ribosomal RNA (rRNA).

word ACC to change to ACA in a gene that makes a protein. Because both of these words code for the same amino acid (threonine; see Figure 5.15), this mutation will not change the protein made by the gene. Other single-base mutations—such as changing ACC to CCC—will change a single amino acid in a protein. In some cases, such a change will alter a protein only slightly, hardly affecting its functionality. But in other cases, the change can be much more dramatic. For example, the cause of sickle-cell disease (Figure 5.16), which kills some 100,000 people each year worldwide, can be traced to a single mutation in the gene that makes hemoglobin in which the base A changed to the base T in just one place within the gene.* Mutations that add or delete a base within a gene tend to have the most dramatic effects on protein structure. The reason is that the genetic code has no "punctuation" or spacing between words; instead of saying something like "the fat cat ate the rat," for example, it says "thefatcatatetherat." Thus, if a letter (base) is added to or deleted from such a sequence, the result will be nonsense from that point on. For example, inserting an "a" so that the sequence becomes "the*a*fatcatatetherat" would cause it to be read as "the *a*fa tca tat eth era t."

Mutations that change proteins are often lethal, because the cell may not be able to survive without the correctly structured protein. However, if the cell survives, the mutation will be copied every time its DNA is replicated. In that case, the mutation represents a permanent change in the cell's hereditary information. If the cell happens to be one that gets passed to the organism's offspring—as is always the case for single-celled organisms and can be the case for animals if the mutation occurs in an egg or sperm cell—the offspring will have a gene that differs from that of the parent. It is this process of mutation that leads to variation among individuals in a species. Each of us differs slightly from all other humans because we each possess a unique set of genes with slightly different base sequences.

THINK ABOUT IT   Ultraviolet radiation from the Sun can cause mutations in the DNA of skin cells. Based on what you've learned, explain why this is potentially dangerous (and, indeed, is the cause of skin cancer). How would sunscreen help prevent such mutations?

Mutations therefore provide the basis for evolution. Given that each individual of a species possesses slightly different genes, it is inevitable that some genes will provide advantageous adaptations to the environment. As we discussed in Section 5.1, the combination of individual variation and population pressure leads to natural selection, in which the advantageous adaptations will preferentially be passed down through the generations. Thus, what was once a random mutation in a single individual can eventually become the "normal" version of the gene for an entire species. In this way, species evolve through time. Notice that, while we often associate the word *mutation* with harm, evolution actually proceeds through the occasional beneficial mutation. Although such beneficial mutations may be relatively rare compared to other mutations, natural selection allows these beneficial mutations to propagate preferentially, so tremendous changes can accrue over time.

Normal red blood cell   **Sickled red blood cell**

Figure 5.16

These microscopic views contrast a normal human red blood cell with a blood cell found in patients with sickle-cell disease. The sickle shape makes it easier for the blood cells to clog tiny blood vessels, which can lead to debilitating disease. Sickle-cell disease occurs in people whose gene for hemoglobin differs from the "normal" gene in just a single DNA base. (Adapted from Campbell, Reece, Simon, *Essential Biology*, Second Edition, © 2004.)

---

*Humans have two copies of each gene, and sickle-cell disease generally occurs only in people who have the sickle-cell mutation in both copies of the gene. From an evolutionary standpoint, this mutation remains prevalent in the population because it actually confers an advantage—malaria resistance—to people with only one copy of the mutated gene.

Evolution sometimes occurs in an even more dramatic way: In some cases, organisms can transfer entire genes to other organisms, a process called *lateral gene transfer.* This process is one of the primary ways that bacteria gain resistance to antibiotics. We humans have also learned to use this process for our benefit through what we call *genetic engineering,* in which we take a gene from one organism and insert it into another. For example, genetic engineering has allowed us to produce human insulin for diabetic patients: The human gene for insulin is inserted into bacteria, and these bacteria produce insulin that can be extracted and used as medicine. Lateral gene transfer can change a species more rapidly than individual mutations, but mutations are still the underlying basis, since they created the genes in the first place.

DNA and Life on Other Worlds   It is difficult to imagine life that does not have heredity, because it seems crucial for any form of life to have some means of storing its operating instructions and passing them on to its offspring. We've seen that DNA is the carrier of heredity for all life on Earth, though as we'll discuss in Chapter 6, we have good reason to believe that very early life on Earth used RNA for this role. Should we expect DNA, or RNA, to also be the heredity molecule for life elsewhere? We do not yet know whether other, quite different molecules might be able to carry hereditary information in the same way as DNA. However, it seems a near certainty that any form of life anywhere else will have some molecule that plays the same functional role that DNA plays on Earth.

5.5   LIFE AT THE EXTREME

We've discussed all the fundamental characteristics of life on Earth: the basic structure of cells, the metabolism of cells, and the means by which cells store and pass on their heredity. We've also discussed why these characteristics seem likely to be shared, at least in a general sense, by any life we find elsewhere. In essence, our bottom-line conclusion is that life

MOVIE MADNESS: THE ANDROMEDA STRAIN

Most movie aliens are complex critters—"bugeyed monsters" that come to Earth in high-speed, hi-tech spacecraft to undertake a campaign of havoc and destruction. But a different threat is offered by *The Andromeda Strain* (1971) as the U.S. military sends a satellite into space to scoop up some simple alien life—green goo attached to small meteors—hoping that microbe-sized aliens might be useful for biological warfare.

You might argue that, as a scheme for developing new weaponry, this approach is a long shot. But not in the movies. The satellite returns to Earth, and some of the alien microbes escape into the air to promptly wipe out a small New Mexico town. (Aliens nearly always prefer landing in the American Southwest.) Unaware of the military's game plan, a team of cranky civilian scientists is brought together in a top-secret Nevada bio lab to figure out what's afoot and stop the infection from space.

Most of the movie is taken up with the step-by-step analysis performed by the scientists as they methodically try to isolate and understand the meteor-borne microbes. They find that the space critters have no nucleic acids, no amino acids, and none of the usual features

of biology on Earth. In fact, these extraterrestrial entities are crystals—the entire organism is all made of the *same* molecule (like salt, except with a much more complicated molecule). This is a bit hard to swallow, but there's more: each time the alien bugs reproduce (roughly every few seconds), they also mutate. And unlike for earthly life, the mutations are not random. Each cell mutates simultaneously and identically to every other.

By the end of the film, this peculiar property makes all the weighty work of the scientists irrelevant. The alien microbes mutate into a harmless form and drift out to sea. But think about it: any species in which all individuals mutate the same way at the same time will be unable to adapt to a changing environment. Every microbe is, and remains, identical. You should feel sorry for the Andromeda strain, because it has no Darwinian evolution and, consequently, no chance of long-term survival. As a strategy for life, it strains credulity.

In real life, NASA has a program to ensure that materials brought to Earth by future spacecraft will be carefully handled to forestall any chance of contamination by alien microbes. This news should be a relief to residents of the Southwest.

elsewhere ought to share a lot of common features with life on Earth. But don't be tempted to think this means that life elsewhere will look like us—that is, like humans or any other animals. In fact, most life on Earth does not look much like "us." We've already noted that microbes are far more common than multicellular Eukarya. Perhaps even more startling, in recent decades biologists have discovered that life can survive in an astonishing variety of environments that would be lethal to humans.

• What kinds of conditions can life survive?

Deep on the ocean floor are places where volcanic activity releases hot water and rock into the surrounding ocean. Minerals and other dissolved chemicals can cause the water to turn black, which is why some of the vents and their plumes of hot water are called *black smokers* (Figure 5.17). The water coming out of such vents can be heated to temperatures above 350°C (660°F), far above the normal boiling point of 100°C (212°F). However, the ocean pressure at these depths is so great that the water remains liquid despite its high temperature.

If you took any "ordinary" organism and placed it in the extremely hot water near a volcanic vent, it would die quickly because the high temperature would cause many of its critical cell structures to fall apart. Yet in recent decades scientists have discovered life—mostly microbes of the domain Archaea*—thriving in the extremely hot water around black smokers and other vents. For example, an organism called *Pyrolobus fumarii,* which was actually discovered *in* the walls of a black smoker (its name means "fire lobe of the chimney"), can grow in water heated to as high as 113°C (235°F). And in 2003, researchers discovered another species of Archaea living near volcanic vents that can grow in even hotter water. This species does not yet have an official name but is being called "Strain 121" because it can grow in water as hot as 121°C (250°F); in addition, it can survive in the lab for up to 2 hours at temperatures of 130°C (266°F). Both *P. fumarii* and Strain 121 are chemoautotrophs that get their carbon from dissolved carbon dioxide and their energy from inorganic chemical reactions that occur in the hot water. Similar organisms thrive in hot springs on Earth's surface, such as in the springs around Yellowstone National Park (Figure 5.18).

Organisms that survive in extremely hot water are sometimes called *thermophiles,* meaning "lovers of heat" (the suffix *phile* means "lover") or, for those living at the highest temperatures, *hyperthermophiles.* More generally, organisms that survive in extreme environments of any kind are called **extremophiles**, or "lovers of the extreme." Extremophiles are quite varied, though most of them are members of the domains Bacteria or Archaea. Some can live in "normal" as well as extreme conditions, while others can survive only in the extreme conditions. For example, many hyperthermophiles die when brought to "normal" temperatures because their enzymes have evolved to function only at the high temperatures in which they live. Indeed, many extremophiles are anaerobic (meaning they live without oxygen), and they are poisoned by the oxygen on which our own lives depend.

Hot environments are not the only extreme conditions favored by some organisms. The dry valleys of Antarctica receive so little rain or

---

*Many known species of Archaea live in extreme environments, but extreme conditions are *not* a general feature of domain Archaea. Indeed, some of the most common organisms in "ordinary" environments on Earth are Archaea. For example, some 20–50% of the living cells in cool ocean water typically represent various species of Archaea.

Figure 5.17

This photograph shows a black smoker—a volcanic vent on the ocean floor that spews out extremely hot, mineral-rich water. Organisms like *P. fumarii* and Strain 121 survive here in water above the normal boiling temperature.

Figure 5.18

A hot spring in Yellowstone National Park; to judge its size, notice the walkway winding along the lower right. The different colors in the water are from different bacteria that survive in water of different temperatures.

Figure 5.19

The main photo shows a cold, dry valley in Antarctica. These valleys are among the driest deserts on Earth (the ice is runoff from surrounding regions), yet they are still home to life. The inset shows a slice of rock (about 1.8 centimeters across) from a dry valley. The colored zones contain microbes that live inside the rock in the airspaces between tiny mineral grains. The organisms are dormant and frozen for most of the year, but can grow during the approximately 500 hours per year when sunlight warms the rock above the freezing point.

snowfall that they are among the driest deserts on Earth, and temperatures in these valleys rarely rise above freezing (Figure 5.19). Nevertheless, there is life in the dry valleys living *inside* rocks. We often think of rocks as solid, but most rocks are composed of individual mineral grains packed together, leaving small spaces between the grains. Even in the dry valleys, these spaces within the rocks occasionally contain water from the rare rain- or snowfall. Sunlight can penetrate up to a few millimeters into the rock before being completely absorbed, so the layers just below the rock's surface can have temperatures slightly above freezing despite the freezing temperatures around them. Amazingly, there are microbes that survive in these tiny pockets of liquid water inside rocks in freezing cold valleys.

Other extremophiles live in conditions far too cold, acidic, alkaline, or salty for "ordinary" life to survive, and some may offer examples of the types of organisms we might find on other worlds. For example, there are some species that can survive at temperatures as low as −20°C, as long as even a thin film of liquid water is available. (These cold-loving organisms are called *psychrophiles*, essentially the opposite of *thermophiles*.) Some microbes can even survive high doses of radiation. A bacterial species known as *Deinococcus radiodurans* can survive radiation more than 1,000 times that which would be lethal to humans and other animals. These remarkable organisms actually thrive in radioactive waste dumps! They could survive the radiation exposure on a world without ozone and even in space, and they can survive extremely dry conditions as well.

One particular group of extremophiles is of special interest for its possible relevance to life on Mars. Microbes called *endoliths* (meaning "within rocks") can live several kilometers below the surface of Earth in water that fills the pores within rock. One community of endoliths, discovered in 1995, consists of bacteria living deep beneath the surface of Oregon and Washington in a rock formation known as the Columbia River Basalt. These organisms are chemoautotrophs that get their energy for metabolism from chemical reactions between the water and the surrounding rock, and they get their nutrients from chemicals within the rock itself and from carbon dioxide that has filtered down from the surface. Although these microbes clearly share the same common ancestors as the rest of terrestrial life, they tell us something significant about life's ability to survive in remarkably diverse environments. In particular, the subsurface environment in which they live almost certainly exists in similar form on other planets, such as Mars, even when the surface may be too cold for liquid water. Moreover, endoliths may be quite common. No one knows exactly how many of them exist here on Earth, but some estimates suggest that the total mass of subsurface organisms living in rock may exceed that of all the life on Earth's surface.

Another amazing adaptation to extreme conditions is found in **endospores**—special "resting" cells produced by some bacteria (Figure 5.20). Endospores allow the organisms that create them to become dormant, neither growing nor dying in extremely inhospitable conditions. (The nondormant organisms are not necessarily extremophiles; some live under more "normal" conditions.) For example, endospores of the bacterium *Bacillus anthracis*, which causes the deadly disease anthrax, can survive a complete lack of water, extreme heat or cold, and most poisons. Some endospores can survive even in the vacuum of space, which is why planetary scientists worry that our interplanetary spacecraft could potentially contaminate other worlds with life from Earth. Moreover, some endospores can remain dormant for centuries—and perhaps even longer—raising the possibility that life could survive journeys aboard meteorites that are blasted off one planet and land on another [Section 8.4].

• Are extremophiles really extreme?

From our human point of view, the environments in which extremophiles survive truly are extreme. But if an extremophile could think, it would probably claim that its environment is quite normal and that ours is the one that is extreme. So who's right, humans or the extremophiles?

In some sense, it's all just relative. Any species would naturally consider its own environment to be normal and others to be extreme. A more important question we might ask is which environment is more common. Surprisingly, if we look at the history of Earth, so-called extreme environments have been much more common than an environment suitable for humans. Earth's atmosphere may have contained oxygen at a level suitable for human life for only the past few hundred million years, or about 10% of Earth's history [Section 6.3]. Indeed, for the first couple of billion years after life first arose on Earth, extremophiles were the only organisms that could survive. Even today, it's an open question whether extremophiles are more or less common than organisms that live in conditions favorable to humans. All in all, extreme life appears to be much more the norm than is life that lives in an environment suitable to humans.

The study of extremophiles has several important implications for the search for extraterrestrial life, but two are particularly important. First, the fact that extremophiles apparently evolved earlier than other forms of life [Section 6.1] suggests that we should begin the search for life elsewhere by searching for similar extreme organisms. Second, the fact that extremophiles can survive such a broad range of conditions suggests that life may be possible in many more places than we would have guessed only a few decades ago: Any world containing an environment in which some type of extremophile might survive becomes a good candidate for the search for life.

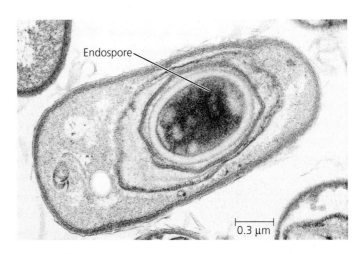

Figure 5.20

This microscopic photo shows an endospore created by the bacterium *Bacillus anthracis*. There are actually two cells here, one inside the other. The outer cell produced the specialized inner cell, which is the endospore. The endospore has a thick, protective coat. Its interior is dehydrated, and no metabolism occurs. Under harsh conditions, the outer cell may disintegrate, but the endospore can survive. When the environment becomes more hospitable, the endospore absorbs water and resumes growth.

5.6 THE PROCESS OF SCIENCE IN ACTION: EVOLUTION AS SCIENCE

In this chapter and throughout this book, we treat evolution as an established fact, a position consistent with official statements by virtually every scientific society, including the National Academy of Sciences, the American Association for the Advancement of Science, the National Association of Biology Teachers, and the American Astronomical Society. However, if you live in the United States, you've almost certainly heard about public battles in which the scientific idea of evolution is portrayed as being controversial. How can an idea so well accepted in science be considered so differently by many among the general public? The answer lies in differences between the way science works and the way that people often seek knowledge in their daily lives, and especially in the difference between the evidence-based approach of science and the faith-based approach of religion. Once this difference is understood, much of the supposed conflict between science and religion is not an issue, which explains why the vast majority of religious denominations see no inherent conflict between their faiths and the science of evolution. For this chapter's case study in the process of science, we'll explore why most scientists and most theologians agree that evolution and faith can coexist without difficulty.

• Is evolution a fact or a theory?

By this point in the book, you should already recognize that the question of whether evolution is a fact or a theory offers a false choice, much like asking whether gravity is a fact or a theory [Section 2.4]. Gravity is a fact in that objects really do fall down and planets really do orbit the Sun, but we use the *theory* of gravity to explain exactly how and why these things occur. The theory of gravity is not presumed to be perfect and indeed has at least one known flaw (its inconsistency with quantum mechanics on very small scales). Moreover, Newton's original theory of gravity is now considered only an approximation to Einstein's improved theory of gravity, which itself will presumably be found to be an approximation to a more complete theory that has not yet been discovered.

The same idea holds for evolution. Nearly all scientists consider evolution to be a fact, because both the geological record and observations of modern species make clear that living organisms really do change with time. We use the *theory* of evolution to explain how and why these changes occur. For example, we use the theory of evolution to understand the changes in species recorded in the geological record, the genetic relationships between modern species, and the way that bacteria can rapidly acquire resistance to antibiotics.

The theory of evolution clearly explains the major features of life on Earth, but scientists still debate the details of the theory. For example, there is considerable debate about the rate at which evolution proceeds: Some scientists suspect that evolution is "punctuated" with periods of rapid change followed by long periods in which species remain quite stable, while others suspect that evolution proceeds at a steadier pace. This debate can be quite heated between individual scientists, but it does not change the overall idea that life evolves and it will eventually be settled by additional evidence. Similarly, scientists often debate the precise relationships between species, especially those that are extinct. For example, we do not yet have enough evidence to put together a complete evolutionary tree for relationships among all dinosaur species, and the relationship between extinct dinosaurs and modern birds is not yet fully understood.

Indeed, we can draw an analogy between Darwin's original theory of evolution by natural selection and Newton's original theory of gravity: Just as Newton's theory captured the main features of gravity but proved to be incomplete, Darwin's theory clearly captures the main features of evolution but is not complete. Moreover, like Newton's theory, Darwin's also has been modified and improved with time. Just as Einstein's general theory of relativity allowed us to understand gravity in more realms than Newton's original theory (by refining it, not refuting it), so does our modern understanding of DNA and mutations allow us to understand biological changes and relationships beyond those that Darwin was able to understand or was even aware of. And like the theory of gravity, the theory of evolution remains a work in progress. We expect to learn more about evolution as we continue to study relationships between species, how DNA works (especially the noncoding regions), and the biochemistry of life. Perhaps someday we'll even be able to broaden the theory through the study of *comparative evolution,* in which we'll explore the similarities and differences among living organisms on multiple worlds.

Does all this mean that you need to *believe* that evolution really occurred? From the standpoint of learning science, what counts is *understanding* evolution; belief is up to you. Remember, all the evidence in the world can never prove any scientific theory true beyond all doubt [Section 2.3]. Even if we had a complete geological record, with precise dating of every species that ever lived, you could still choose to believe, for example, that the fossils had been placed there intact at some single moment in the past, rather than having been deposited as the evidence suggests. All we can say from a scientific viewpoint is that a tremendous wealth of evidence points to the idea that life on Earth has evolved through time, that these evolutionary changes have been driven by natural selection, and that natural selection occurs on a molecular level as genes are modified through mutations of DNA.

• Are there scientific alternatives to evolution?

In recent years, the public controversy over evolution has centered largely around the question of whether it should be the only idea about our origins taught in science classes or whether it should be taught alongside other, competing ideas. Other ideas certainly offer different visions of how we came to exist; for example, the idea that God created Earth and the universe a mere 6,000 years ago is obviously quite different from the idea that life has evolved gradually over the past 4 billion years. The question is whether this idea or other competing ideas qualify as science.

The best way to determine whether any alternatives to evolution qualify as science is to consider them against the hallmarks of science that we discussed in Chapter 2. Let's start by showing that the theory of evolution *does* satisfy the standards of science. The first hallmark states that science seeks explanations for observed phenomena that rely solely on natural causes. The theory of evolution clearly does this, as it explains the geological record and observed relationships between species through the mechanism of natural selection and other natural causes.

The second hallmark states that science progresses through the creation and testing of models. Our understanding of evolution has indeed progressed in this way. The idea of evolution won out over Aristotle's competing idea of species that never changed. As the fact of evolutionary change gained acceptance, the first model proposed to explain these changes came from Lamarck (see Section 5.1); his model was later discarded because Darwin's alternative model explained the observations so much more successfully. Our current, molecular model of evolution is a refinement of Darwin's original model, and we can expect further refinements to the theory in the future as continued study turns up new evidence.

The theory of evolution also satisfies the third hallmark, which states that a scientific model must make testable predictions that would lead us to revise or abandon the model if the predictions do not agree with observations. Our modern, molecular theory of evolution clearly qualifies. For example, it predicts that diseases can and will evolve in response to medicines designed to combat them, a prediction borne out in the rapid way that many diseases acquire drug resistance. It also predicts that genetically similar species should respond to medicines in similar ways, a prediction confirmed by the fact that we can test many medicines in other primates and they do indeed have effects similar to those they have in humans. The theory of evolution also provides a road map that we can use to modify organisms through genetic engineering; in this sense, every genetically engineered grain of rice or kernel of corn represents a success of the predictive abilities of the theory of evolution.

In fact, even Darwin's original theory made testable predictions. For natural selection to be possible, Darwin had to assume that living organisms have some way of passing on their heritable traits from parent to offspring. So although he did not predict the existence of DNA per se, his theory clearly predicted that some type of mechanism had to exist to carry the hereditary information. Similarly, now that we know about DNA and the genetic code, the theory of evolution predicts that closely related species should be genetically similar, a prediction that has been confirmed in just the past few years by genome sequencing. For example, in the ordering of their base sequences, the DNA of humans and chimpanzees is 98.5% identical. If we were not closely related to chimpanzees, we would not expect such similar genomes.

Now that we have established that evolution qualifies as science, we next turn to the question of whether any of the alternative models that have been suggested for inclusion in the classroom might also qualify as science. Since the time that Darwin first published his theory, the main alternatives have been religious ideas about creation. Here, we run into an immediate problem: There are so many different religious ideas about creation that we can't even define the potential alternatives clearly. For example, many Native American religious beliefs speak of creation in terms that bear little resemblance to the Judeo-Christian tradition found in Genesis. Even among people who claim a literal belief in the Bible, there are differences in interpretation about creation. Some biblical literalists argue that the creation must have occurred in just 6 days, as the first chapter of Genesis seems to say, while others suggest that the term "day" in Genesis does not necessarily mean 24 hours and therefore that the story in Genesis is compatible with a much older Earth and with evolution.

Nevertheless, a few groups have tried to claim that scientific evidence supports some alternative to evolution. In the 1980s, an idea called "creation science" emerged and its proponents tried to find scientific evidence to support the idea that Earth was created a mere 6,000 years ago. However, to support this "young Earth" view, they not only had to reject evolution, but they also had to reject the tremendous weight of evidence that supports an old Earth and an old universe—evidence based on such things as radiometric dating of rocks, astronomical measurements of distances to other stars and galaxies (since their light has obviously had time to reach us), and even tree ring data that go back more than 6,000 years. For all this evidence to be wrong, we would have to have fundamental errors in our basic understanding of the laws of nature, an idea that seems implausible, given the many successes of modern physics and chemistry.

More recently, some people have begun to advance an idea called "intelligent design," or ID; this idea holds that living organisms are too complex to be explained by natural selection, and so must have been designed by some transcendent entity or power. For example, proponents of this idea point to features of the human eye as suggesting design rather than natural processes, and some believe they see evidence of "digital code" in the arrangement of the bases in the genomes of living organisms.

For the vast majority of scientists, the primary problem with these claims of intelligent design is that they do not seem to stand up to scientific scrutiny. The features that the ID proponents cite as evidence for design are to most scientists well explained by natural selection. Nevertheless, good scientists will always allow the possibility that evidence of design might someday be found. Moreover, even if no such evidence is found, absence of evidence would not preclude a role for a Designer.

The greater problems with intelligent design from a scientific perspective show up when we test it against the hallmarks of science. In particular, ID clearly fails to meet the first hallmark—that science seeks explanations for observed phenomena that rely solely on natural causes. The very idea of a transcendent Designer implies something that natural processes cannot explain, no matter whether the Designer is or isn't explicitly named as God. As a result, some ID proponents have sought to redefine science to allow nonnatural explanations. The problem with such a redefinition is that it would render science impotent. As a simple analogy, consider the collapse of a bridge. If you choose to believe that the collapse was an act of God, you might well be right—but this belief won't help you design a better bridge. We learn to build better bridges only by assuming that collapses happen through natural causes that we can understand and learn from. In precisely the same way, it is the scientific quest for a natural understanding of life that has led to the discovery of relationships between species, genetics, DNA, and much of modern agriculture and medicine. Many of the scientists who made these discoveries, including Charles Darwin himself (see the quotation at the beginning of this chapter), believed deeply that they could see God's hand in creation. But if they had let their belief stop them from seeking natural explanations, they would have discovered nothing.

Intelligent design also fails to meet the second and third hallmarks of science, because it does not offer a predictive model that can be tested. The assumption of a Designer might or might not be correct, but it does not tell us how life would be different from what we'd see if there were no Designer. Moreover, as we've discussed, scientists continually modify the theory of evolution as new evidence requires. In the unlikely event that we found evidence that strongly contradicted the current theory—for example, fossil evidence proving that people and dinosaurs existed at the same time—scientists would willingly discard the theory and go back to the drawing board. In contrast, because most proponents of intelligent design are motivated by their religious faith, their belief in ID is unshakable.

The bottom line is that science and faith are different things, and the relative worth of one does not override the worth of the other. Whether you choose to believe the theory of evolution is up to you, and if you do believe it, you can choose whether to believe that it occurred through random chance or with the help of a guiding hand. But whatever your beliefs, the theory of evolution is a clear and crucial part of modern science, and it is integral to an understanding and appreciation of modern biology. And more important for the discussion in this book, the theory of evolution frames our ideas about how to search for life beyond Earth. No competing model offers any similar scientific benefits.

THE BIG PICTURE
Putting Chapter 5 into Context

In this chapter, we have surveyed the nature of life on Earth and explored some of the implications of this survey for the search for life elsewhere. As you continue in your studies, keep in mind the following "big picture" ideas:

- If we are going to search for life, it's useful to think about just what it is we are searching for. Defining life turns out to be surprisingly difficult, but at a minimum it seems that life must be capable of reproducing and evolving. Thus, evolution plays a central role in the definition of life as well as in our understanding of life on Earth.

- Life on Earth has at least three key features that are likely to be shared by any life we find elsewhere: (1) Life has a fundamental structural unit, which we call the cell; (2) living cells undergo metabolism, by which we mean chemical reactions that keep the cells alive; and (3) living cells have a heredity molecule, which is DNA for life on Earth, that allows them to store their operating instructions and to pass these instructions to their offspring.

- Life on Earth survives under a much wider range of conditions than we would have guessed a few decades ago, suggesting that life elsewhere might similarly be found in a fairly broad range of environments. This fact greatly increases the number of worlds on which we might hope to find life.

SUMMARY OF KEY CONCEPTS

5.1 DEFINING LIFE

• What are the general properties of life on Earth?

Six key properties of life on Earth are order, reproduction, growth and development, energy utilization, response to the environment, and evolutionary adaptation.

• What is the role of evolution in defining life?

The **theory of evolution**, which holds that life changes over time through the mechanism of **natural selection**, is the unifying principle of modern biology. It holds this central role because it successfully explains all the other properties of life, the observations we make in the geological record, and the observations we make of relationships between living organisms.

• What is life?

No known definition of life works in all circumstances, but for most purposes the following definition will suffice: Life is something that can reproduce and evolve through natural selection.

5.2 CELLS: THE BASIC UNITS OF LIFE

• What are living cells?

Cells are the basic units of life on Earth, as they serve to separate the living matter inside them from the outside world. The barrier that marks this separation is called the **cell membrane**.

• What are the molecular components of cells?

The major molecular components of cells fall into four main classes: **Carbohydrates** are sugars and starches that provide energy and build many cellular structures; **lipids**, which include fats, are the main ingredients of structures including cell membranes and also store cellular energy; **proteins** play a vast number of roles in cells, and the proteins known as **enzymes** act as **catalysts** to facilitate biochemical reactions; **nucleic acids**, which include **DNA** and **RNA**, are most important for the roles they play in heredity. Commonalities between the molecules used in different organisms, such as the fact that all life on Earth builds proteins from only left-handed versions of **amino acids**, provide strong evidence for the idea that all life evolved from a common ancestor.

• What are the major groupings of life on Earth?

Modern biologists classify life into three **domains**: **Bacteria**, **Archaea**, and **Eukarya**. The **tree of life** shows relationships among species within the three domains; note that all plants and animals are just two small branches of the Eukarya.

5.3 METABOLISM: THE CHEMISTRY OF LIFE

• What are the basic metabolic needs of life?

(1) A source of raw materials to build cellular structures, with carbon as the most important of these materials. (2) A source of energy to fuel metabolic processes.

• How do we classify life by its metabolic sources?

We classify life by its carbon source as either a **heterotroph** that gets its carbon by eating or an **autotroph** that takes carbon directly from the environment in the form of atmospheric or dissolved carbon dioxide. We then subclassify these categories by energy source, using the prefix *photo* for life that gets energy from sunlight and the prefix *chemo* from life that gets energy either from eating or from inorganic chemical reactions.

5.4 DNA AND HEREDITY

• How does the structure of DNA allow for its replication?

The double helix of DNA consists of two strands connected by the DNA bases. The bases connect according to precise pairing rules (T attaches only to A, and C attaches only to G), so that when the strands separate, each can serve as a template for making a new DNA molecule that is identical to the original one.

• How is heredity encoded in DNA?

The precise sequence of the bases in a DNA molecule contains the instructions for assembling proteins and other cell functions. A segment of DNA that codes for a single cell function or protein is called a **gene**, and the complete base sequence of an organism represents its **genome**. The "language" used to translate from the base sequence into proteins is called the **genetic code**.

• How does life evolve?

Life evolves because the copying of DNA is not perfect, although the error rate is quite small. The occasional, random copying errors, called **mutations**, can change the instructions in DNA. Most mutations either are lethal or have no effect at all, but a few carry benefits that can then be transmitted to offspring when the DNA replicates.

5.5 LIFE AT THE EXTREME

• What kinds of conditions can life survive?

Many living organisms can survive in a surprisingly wide range of conditions. These **extremophiles** include microbes that can survive in temperatures above the normal boiling point of water, in the dry deserts of Antarctica, deep underground in the tiny pores of rocks, and even under exposure to high levels of radiation.

• Are extremophiles really extreme?

The conditions that we consider extreme are, overall, probably more typical of the conditions found on Earth during most of its history than the conditions we enjoy on the surface today. Many other worlds may have similar conditions, suggesting that extremophiles may in fact be more common in the universe than life similar to plants and animals.

5.6 THE PROCESS OF SCIENCE IN ACTION: EVOLUTION AS SCIENCE

• Is evolution a fact or a theory?

This question offers a false choice, because fact and theory are not considered to be opposites in science: Evolution is a well-confirmed theory based on a wide variety of observational and experimental evidence.

• Are there scientific alternatives to evolution?

While there are many alternative explanations for our existence, including ideas such as creation science or intelligent design, none of these ideas qualifies as a *scientific* alternative to evolution.

EXERCISES AND PROBLEMS

REVIEW QUESTIONS
Short-Answer Questions Based on the Reading

1. Briefly describe the six key properties that appear to be shared by most living organisms on Earth.

2. What is *natural selection?* Summarize the logic by which Darwin came to the "inescapable conclusion" that evolution occurs by natural selection. Describe some of the evidence that supports Darwin's *theory of evolution.*

3. Briefly describe the evidence that points to a single common ancestor for all life on Earth.

4. Why do we say that living cells are *carbon-based?* Briefly discuss whether life elsewhere could be based on something besides carbon.

5. Briefly describe each of the four main classes of cellular molecules: *carbohydrates, lipids, proteins, and nucleic acids.* What are *enzymes,* and where do they fit into this picture?

6. What are *amino acids?* What do we mean by their *handedness?* How do amino acids offer further evidence for a common ancestor for all life on Earth?

7. What are the three *domains* of life? Which domain do we belong to?

8. What do we mean by the *tree of life?* List three important ideas that we learn from the tree and that differ from older ideas about biology.

9. What is *metabolism,* and what are the two basic metabolic needs of any organism? Explain the four metabolic classifications listed in Table 5.1.

10. Why is water so important to life on Earth? List the three major roles that water plays in metabolism.

11. Describe the double helix structure of *DNA.* How does a DNA molecule replicate?

12. What is a *gene?* A *genome?* The *genetic code?*

13. What are *mutations,* and what effects can they have? Briefly explain why mutations represent the molecular mechanism of natural selection.

14. What are *extremophiles?* Give several examples of organisms that live in extreme environments. What are the implications of the existence of extremophiles for the search for extraterrestrial life?

15. Describe several ways in which the theory of evolution is analogous to the theory of gravity.

16. Explain how evolution exhibits each of the three hallmarks of science, and discuss why alternatives such as creationism and intelligent design do not show these hallmarks.

TEST YOUR UNDERSTANDING

Surprising Discoveries?

Suppose we found an organism on Earth with the characteristics described. In light of our current understanding of life on Earth, should we be surprised to find such an organism existing? Why or why not? Explain clearly; because not all of these have definitive answers, your explanation is more important than your chosen answer.

17. A single-celled organism that builds proteins using 45 different amino acids.

18. A single-celled organism that lives deep in peat bogs, where no oxygen is available.

19. A multicellular organism that reproduces without passing copies of its DNA to its offspring.

20. A single-celled organism that can survive in a dormant state even in the complete absence of any liquid water.

21. A multicellular organism that can grow and reproduce even in the absence of water.

22. A bacterium with cells that lack the molecule ATP.

23. A species of Archaea that lives in the 1,000°C molten rock of a volcano.

24. A species of Archaea that lives in the walls of a nuclear reactor.

25. Two different animal species whose genomes are more than 99% identical.

26. A species of Bacteria that has a genome 99% identical to that of humans.

Quick Quiz

Choose the best answer to each of the following. Explain your reasoning with one or more complete sentences.

27. Which of the following is *not* a key property of life? (a) the maintenance of order in living cells; (b) the ability to evolve over time; (c) the ability to violate the second law of thermodynamics.

28. *Natural selection* is the name given to: (a) the occasional mutations that occur in DNA; (b) the mechanism by which advantageous traits are preferentially passed on from parents to offspring; (c) the idea that organisms can develop new characteristics during their lives and then pass these to their offspring.

29. Which of the following is *not* considered a key piece of evidence supporting a common ancestor for all life on Earth? (a) the fact that all life on Earth is carbon-based; (b) the fact that all life on Earth uses the molecule ATP to store and release energy; (c) the fact that all life on Earth builds proteins from the same set of left-handed amino acids.

30. An organism's heredity is encoded in: (a) DNA; (b) ATP; (c) lipids.

31. An enzyme consists of a chain of: (a) carbohydrates; (b) amino acids; (c) nucleic acids.

32. Which of the following is not a source of energy for at least some forms of life on Earth? (a) inorganic chemical reactions; (b) energy release from plutonium; (c) consumption of pre-existing organic compounds.

33. People belong to domain: (a) Eukarya; (b) Archaea; (c) Bacteria.

34. Which of the following mutations would you expect to have the greatest effect on a living cell? (a) a mutation that changes a single base in a region of noncoding DNA; (b) a mutation that changes the third letter of one of the three-base "words" in a particular gene; (c) a mutation that deletes one base in the middle of a gene.

35. Generally speaking, an *extremophile* is an organism that: (a) thrives in conditions that would be lethal to humans and other animals; (b) could potentially survive in space; (c) is extremely small compared to most life on Earth.

36. Based on what you have learned in this chapter, it seems reasonable to think that life could survive in each of the following habitats *except:* (a) rock beneath the martian surface; (b) a liquid ocean beneath the icy crust of Jupiter's moon Europa; (c) within ice that is perpetually frozen in a crater near the Moon's south pole.

INVESTIGATE FURTHER
In-Depth Questions to Increase Your Understanding

Short-Answer/Essay Questions

37. *Rock Life?* How do you know that a rock is not alive? In terms of the properties of life discussed in this chapter, clearly describe why a rock does not meet the criteria for being alive.

38. *The History of Evolution.* Many people assume that Charles Darwin was the first person to recognize that life evolves, but this is not true. Write a few paragraphs summarizing the history of ideas about evolution and explaining why we give Darwin credit for the theory of evolution even though he was not the first person to realize that evolution occurs.

39. *Genetic Variation.* One of the underlying facts (Fact 2 on p. 156) that explains natural selection is that individuals in a population of any species vary in many heritable traits. Based on what you have learned about the molecular basis of evolution, explain why individuals of the same species are *not* expected to be genetically identical.

40. *Artificial Selection.* Suppose you lived hundreds of years ago (before we knew about genetic engineering) and wanted to breed a herd of cows that provided more milk than cows in your current herd. How would you have gone about it? Explain, and describe how your breeding would have worked in terms of the idea of artificial selection. How does this breeding offer evidence in favor of the idea of natural selection?

41. *Ingredients of Life.* Study the ingredients of life as shown in Figure 5.5, and consider them in light of what you've learned about the overall chemical composition of the universe. Would you expect the ingredients to be rare or common on other worlds? Explain.

42. *A Separate Origin?* Suppose that we someday discover life on Mars. How might we be able to determine whether it shares a

common origin with life on Earth (perhaps suggesting that life traveled on meteorites between the two planets) or whether it is from a completely separate origin of life? Explain clearly.

43. *Dominant Life.* While most of us tend to think of ourselves as the dominant form of life on Earth, biologists generally argue that the dominant life consists of microbes of the domains Archaea and Bacteria. In two to three paragraphs, explain why microbes seem more dominant than us.

44. *The Human Power to Destroy.* We clearly have the ability to destroy ourselves today, perhaps as the result of nuclear war, or perhaps through some type of environmental catastrophe. But is there anything we could do with our current abilities that would allow us to wipe out *all* life on Earth? Explain why or why not.

45. *The Search for Life.* Based on what you have learned about life on Earth, what are we searching for when we search for life elsewhere? For example, are we searching only for worlds with surface oceans and oxygen-rich atmospheres like Earth, or for something else? Write one to three paragraphs describing the types of worlds that we can consider as potential homes for life.

46. *Evolution and God.* Does the theory of evolution preclude the existence of God? Clearly explain your answer.

Quantitative Problems

Be sure to show all calculations clearly and state your final answers in complete sentences.

47. *Atomic Numbers in Life.* A typical bacterium has a volume of about 1 cubic micrometer. A typical atom has a diameter of about 0.1 nanometer. Approximately how many atoms are in a bacterium?

48. *Oxygen Atoms in People.* Figure 5.5 shows that oxygen makes up about 65% of the mass of a human being. A single oxygen atom has a mass of 2.66×10^{-26} kg. (a) Use this fact to estimate the number of oxygen atoms in *your* body. (*Hint:* If you know your weight in pounds, you can convert to kilograms by dividing by 2.2.) (b) Compare your answer to the number of stars in the observable universe (which is roughly 10^{22}).

49. *Cellular Energy.* A typical eukaryotic cell, such as a cell in the human body, uses about 2×10^{-17} joule of energy each second. The breakdown of a single molecule of ATP (in which a phosphate separates from ATP to make ADP; see Figure 5.12) releases about 5×10^{-20} joule of energy. (a) About how many molecules of ATP must be broken down and reassembled each second to keep a eukaryotic cell alive? (b) How many times does this ATP recycling occur each *day* in a typical cell? (c) The human body has roughly 10^{14} cells. Approximately how many cycles of the ATP reaction occur each day in your body?

50. *The Genetic Code.* Suppose that, as evidence suggests, very early life on Earth used a genetic code that consisted of only two-base "words" rather than three-base "words." Could such life have made use of the same set of 20 amino acids that life uses today? Explain, using quantitative arguments.

Discussion Questions

51. *Science and Religion.* Science and religion are often claimed to be in conflict. Do you believe this conflict is real and hence irreconcilable, or is it a result of misunderstanding the differing natures of science and religion? Defend your opinion.

52. *Computer Life.* Although scientists have already developed computer programs capable of reproducing themselves and evolving, few people consider such programs to be alive. But consider future developments in computing and robotic technology. Do you think we'll ever make something out of electronics that is truly alive? Could it also be intelligent? If so, what civil rights should we give to such "artificial" intelligent life?

53. *Genetic Engineering and Future Evolution.* For billions of years, evolution has proceeded through mutations and natural selection. Today, however, we have the ability to deliberately alter DNA in what we call "genetic engineering." How do you think this ability will affect the future evolution of life? How will it affect future human evolution on Earth? Based on your answers, should we expect extraterrestrial civilizations to be naturally evolved or to be products of their own genetic engineering? Discuss and defend your opinions.

54. *Gene Transfer and GMOs.* In some cases, organisms can transfer entire genes to other organisms. This fact causes some people to worry that organisms that we have genetically engineered—commonly referred to as GMOs, for "*g*enetically *m*odified *o*rganisms"—may transfer their genes to other organisms in unexpected ways. For example, a crop engineered with a gene that gives it resistance to some pest may transfer its gene to weeds, giving them the same resistance. Discuss how GMOs might affect other organisms. Overall, what, if any, controls do you think the government should put on the use of GMOs?

WEB PROJECTS

55. *The Dover Opinion.* In December 2005, a U.S. District Court issued its opinion on a case concerning the teaching of intelligent design (ID), deciding that ID does not belong alongside evolution in science classes. The full text of the opinion, commonly called the Dover opinion, is available online. Read the opinion, and discuss its implications to the ongoing public controversy about what belongs in science classes.

56. *Darwin on Evolution.* You can find online the entire text of Charles Darwin's *The Origin of Species*. Read the final chapter, in which Darwin addresses potential criticisms of his theory. Evaluate how well he has presented his case. How much stronger does the theory seem today than at the time Darwin first described it in 1859? Summarize your conclusions in a one-page essay.

57. *Extreme Life.* Look for information about a recent discovery of a previously unknown type of extremophile. Describe the organism and the environment in which it lives, and discuss any implications of the finding for the search for life beyond Earth. Summarize your findings in a one-page report.

6
The Origin and Evolution of Life on Earth

LEARNING GOALS

6.1 SEARCHING FOR LIFE'S ORIGINS
- When did life begin?
- What did early life look like?
- Where did life begin?

6.2 THE ORIGIN OF LIFE
- How did life begin?
- Could life have migrated to Earth?

6.3 THE EVOLUTION OF LIFE
- What major events have marked evolutionary history?
- Why was the rise of oxygen so important to evolution?

6.4 IMPACTS AND EXTINCTIONS
- Did an impact kill the dinosaurs?
- Did impacts cause other mass extinctions?
- Is there a continuing impact threat?

6.5 HUMAN EVOLUTION
- How did we evolve?
- Are we still evolving?

6.6 THE PROCESS OF SCIENCE IN ACTION: ARTIFICIAL LIFE
- How might we create artificial life?
- Should we create artificial life?

Small as is our whole system compared with the infinitude of creation, Brief as is our life compared with the cycles of time, we are so tethered to all by the beautiful dependencies of law, that not only the sparrow's fall is felt to the uttermost bound but the vibrations set in motion by the words that we utter reach through all space and the tremor is felt through all time.

Maria Mitchell (1818–1889), first woman elected to the American Academy of Arts and Sciences

We explored the habitability of our planet in Chapter 4, learning how and why Earth has remained a suitable home for life during the past 4 billion or more years. Then, in Chapter 5, we saw how evolution allowed life to take advantage of this habitability, as natural selection led to a great diversity of life that now exists in almost every place on our planet that has even the tiniest amounts of liquid water. But we have yet to discuss the deepest questions of all: How did life arise in the first place, and what events shaped life's evolution into its current diversity of species?

The complete answers to these questions still elude us, but scientists have put together a fairly detailed outline of what is most likely to have occurred. We know that the early Earth was a natural laboratory for organic chemistry, and we have at least some ideas about how this chemistry might have led to life. The geological record shows us how life evolved subsequently, and it is full of important surprises. For example, plants and animals appeared only relatively recently, and we've discovered that external forces, such as the impacts of asteroids or comets, can dramatically alter the course of evolution. In this chapter, we'll survey current ideas about the origin and subsequent evolution of life on Earth, in effect studying how we ourselves came to be.

6.1 SEARCHING FOR LIFE'S ORIGINS

The geological record details much of the history of life on Earth, and the theory of evolution tells us how life has changed from the forms that existed long ago to those found on our planet today. However, neither the geological record nor ideas of biological evolution are likely to tell us precisely how life first arose on Earth. The geological record becomes increasingly incomplete as we look further back in time, and no known rocks survive from Earth's first half-billion years. The theory of evolution explains how one species can evolve into others, but does not tell us how the first living organisms came to be. Thus, any attempt to understand the origin of life must be based on careful study of limited clues about what existed in Earth's early history, along with laboratory experiments that can help us reconstruct the processes that may have occurred on the young Earth. In this section, we'll explore the clues that tell us about early life on Earth, and in the next section we'll explore how laboratory experiments shed light on a possible mechanism for the origin of the first life.

a These large mats at Shark Bay, Western Australia, are colonies of microbes known as "living stromatolites"; they stand about knee-high. Microbes near the top generate energy through photosynthesis.

b The bands visible in this section of a modern-day mat are formed by layers of sediment adhering to different types of microbes.

c This section of a 3.5-billion-year-old stromatolite shows a structure nearly identical to that of a living mat.

Figure 6.1

Rocks called stromatolites offer evidence of photosynthetic life as early as 3.5 billion years ago.

• When did life begin?

Perhaps the first thing we'd like to know about the origin of life on Earth is when it occurred. The only way to approach this question is through the study of fossils. If we find a fossil of a particular age, then we know life already existed at that time. For example, a 3.5-billion-year-old fossil tells us that life on Earth arose *before* 3.5 billion years ago. Because the geological record is incomplete and because we may not yet have discovered the oldest intact fossils, we do not know exactly how long life has existed on Earth. Nevertheless, three lines of fossil evidence all point to the idea that, geologically speaking, life arose quite early in Earth's history.

Stromatolites The first line of evidence comes from **stromatolites** (from the Greek for "rock beds"), rocks that are characterized by a distinctive, layered structure. In size, shape, and interior structure, ancient stromatolites look virtually identical to sections of mats formed today by colonies of microbes sometimes called "living stromatolites" (Figure 6.1). Living stromatolites contain layers of sediment intermixed with different types of microbes. Microbes near the top generate energy through photosynthesis, and those beneath use organic compounds left as waste products by the photosynthetic microbes. The living stromatolites grow in size as sediments are deposited over them, forcing the microbes to migrate upward in order to remain at the depths to which they are adapted. This gradual migration creates the layered structures.

The similarity of structure between the ancient stromatolites and the modern-day mats suggests a similar origin, implying that stromatolites are fossil remnants of early life. Chemical analysis offers further support for this conclusion, as stromatolites contain chemical and isotopic evidence of biological origin. Thus, while it is possible for geological processes of sedimentation to mimic the layering that we see in stromatolites, we are fairly confident that they offer evidence for the existence of microbial colonies as far back as 3.5 billion years ago. Moreover, if the microbes that made the stromatolites are like the microbes in the living mats today, then the implication is that at least some of these ancient microbes produced energy by photosynthesis. Because photosynthesis is a fairly sophisticated metabolic process, we presume that it must have taken at least a moderately long time for this process to evolve in living organisms. In other words, if we are correct in concluding that stromatolites tell us that photosynthetic life already existed some 3.5 billion years ago, then we can infer that more primitive life must have existed even earlier, and that the origin of life itself substantially predates this.

Microfossils The second line of evidence comes from individual fossilized cells. Finding ancient microscopic fossils, or microfossils, is quite challenging, both because rocks become increasingly rare with age and because the oldest rocks have been altered by geological processes in ways that tend to destroy microfossils within them. Moreover, unlike dinosaur bones that are fairly "obvious" fossils, it can be quite difficult to determine whether an interesting-looking microscopic structure is biological or mineral in origin. As a result, claims of ancient microfossils often generate significant scientific controversy, with competing hypotheses attempting to explain their origin.

Not surprisingly, the greatest controversy surrounds the oldest claimed microfossils, which come from a 3.5-billion-year-old rock formation in northwestern Australia. Microscopic photos of sections of this rock show structures that look like individual cells (Figure 6.2). This structure, along with chemical analysis showing the presence of organic carbon and an assumption that the structures resided in sedimentary rock from a shallow sea, originally led the discoverers of these structures (a team led by William Schopf of UCLA) to propose that they represented fossils of early photosynthetic organisms. However, subsequent analysis has shown that the rock is not from a shallow sea as originally assumed, but instead must have formed near a deep-sea volcanic vent similar to a black smoker (see Figure 5.17). The structures therefore cannot be fossils of photosynthetic microbes (since sunlight does not penetrate to the deep sea), and some scientists have used this fact to argue that the structures are not fossils at all, but rather were formed by some nonbiological process. There is also a possibility that, while these structures were found associated with rock known to be 3.5 billion years old, they may have come to this location more recently via flowing water from younger rock. Nevertheless, their shape and organic content still offer a reasonable argument for biological origin, so they may yet prove to represent ancient microfossils of deep-sea microbes. Only further study, and perhaps the discovery of similarly ancient structures in other places, will settle the controversy.

Figure 6.2

Microfossils of ancient living cells? This microscopic photograph shows structures that some researchers believe to be ancient fossil cells dating to 3.5 billion years ago. Other researchers, however, argue that the structures were formed by nonbiological processes.

THINK ABOUT IT   Consider the controversy over these microfossils in light of what you have learned about the process of science in this book. Do you think the controversy indicates strength or weakness in our methods of scientific inquiry? Defend your opinion.

Other microfossils date the origin of life almost as far back with less controversy. At two locations in southern Africa, rocks that date to between 3.2 and 3.5 billion years old also show what appear to be fossilized cells. Although some researchers argue that these might also have nonbiological origin, so far they seem more likely to be true fossil cells. Moreover, these microfossils may lend further support to the hypothesis of biological origin for the microfossils in the 3.5-billion-year-old Australian rock. Despite the wide separation between Australia and Africa today, these particular African and Australian rocks appear to share similar origins, suggesting that they were geologically linked in the past. Thus, if one set of these rocks holds microfossils, we would expect the same to be true of the other set.

Microfossils in rocks dating to between 2.7 and 3.0 billion years old show more conclusive evidence for life: They contain particular molecules (such as a variety of hydrocarbons) that almost certainly indicate

Figure 6.3

This ancient rock formation on the island of Akilia (off the coast of southern West Greenland) may hold the oldest known evidence for life on Earth.

biological origin. Overall, we conclude that microfossil evidence clearly points to the existence of life before about 3.0 billion years ago and may well tell us that life existed before 3.5 billion years ago.

Isotopic Evidence   The third line of evidence for an early origin of life comes from isotopic analysis of some of the most ancient rocks on Earth. Living organisms can change the ratios of isotopes from their background, nonliving values. For example, carbon has two stable isotopes: carbon-12, with six protons and six neutrons in its nucleus, and carbon-13, which has one extra neutron. Carbon-12 is far more common, but any inorganic carbon sample always contains a small proportion of carbon-13 atoms mixed in with the more numerous carbon-12 atoms. (Typically, carbon-13 accounts for about 1 out of every 89 carbon atoms.) When living organisms metabolize carbon, they incorporate carbon-12 atoms into cellular molecules slightly more easily than they do carbon-13 atoms. As a result, living organisms—and fossils of living organisms—always show a slightly lower fraction of carbon-13 atoms than that found in inorganic material.

On an island off the coast of Greenland, this lower carbon-13 ratio has been found in rocks that are more than 3.85 billion years old,* suggesting that the rocks contain remnants of life from that time (Figure 6.3). However, the claim of biological origin (made by a team led by the University of Colorado's Stephen Mojzsis) has been challenged by a few other scientists. The rocks are metamorphic, meaning they have been transformed substantially by high pressure or heat, which would explain why no intact microfossils remain within them. For these rocks to have contained life, we must presume that they were sedimentary before they underwent the metamorphic transformation; if they were volcanic (igneous) rocks, then it is much more difficult to see how they could have been home to or preserved evidence for living microbes. Scientists are still debating the nature of these rocks and hence the question of whether their carbon isotope ratio shows evidence of life. While the controversy is by no means settled, recent evidence seems to be swaying the debate in favor of the hypothesis that they really do contain evidence that life already existed by 3.85 billion years ago.

The new evidence falls into two basic categories. First, if the Greenland rocks hold evidence of life, we would expect to find similar evidence in other rocks dating to the same general time. This is indeed the case, as other rocks dating to some 3.8 billion years ago have been found to show similar carbon isotope ratios. Second, life can also alter the isotopic ratios of other elements, such as iron, nitrogen, and sulfur. Recent studies show that these isotopes also are present in characteristic ratios for life within the ancient rocks, just as we would expect if the carbon isotope data are a result of biological origin. Further study should eventually resolve the issue, one way or the other.

Implications of the Early Origin of Life   We have seen that the stromatolite evidence suggests the presence of fairly advanced life by nearly 3.5 billion years ago. The microfossil evidence, while more controversial, is consistent with this dating. Carbon isotope evidence, if it stands up to further scrutiny, pushes the existence of life back to at least 3.85 billion years ago. Given the fact that the geological record is so sparse for such

---

*We say that the rocks are *older* than 3.85 billion years because the rocks containing the isotopic evidence are sediments that cannot be dated. However, they are cut through by igneous rocks that date to 3.85 billion years, which means the sediments must be even older.

early times, we would expect to find evidence of life in such ancient rocks only if life had already been widespread on Earth. Thus, if we are interpreting all the data correctly, it is likely that life first arose considerably earlier than 3.85 billion years ago. Geologically speaking, this would mean life arose quite early in Earth's history (see Figure 4.10).

In addition, remember from Chapter 4 that the young Earth probably experienced numerous large impacts during the heavy bombardment, some of which may have been large enough to have extinguished any life on Earth at the time, or at least to have extinguished life near the surface. We cannot know exactly when the last such impact might have occurred, but studies of cratering rates on the Moon suggest it was most likely between about 4.2 and 3.9 billion years ago.

Overall, we conclude that life arose and colonized our planet in a period of no more than a few hundred million years, and perhaps in as little as 100 million years or less. By itself, this early origin of life proves nothing about life elsewhere, since it is always possible that Earth was the lucky beneficiary of a highly improbable event. However, if we assume that what happened here would be typical of what might happen elsewhere, then the early origin of life is profoundly important: It suggests that we could expect life to arise similarly rapidly on any other world with similar conditions. Because we expect many other worlds to have conditions similar to those that prevailed on the young Earth, this idea gives us reason to think that life might be quite common in the universe.

• What did early life look like?

The earliest living organisms presumably went extinct long ago, replaced by others with evolutionary adaptations that allowed them to outcompete their ancestors. Nevertheless, just as we know that sharks and alligators are evolutionarily much older than primates (because the geological record shows them coexisting with dinosaurs while primates emerged much later), it seems reasonable to think that some modern-day microbes are more closely related than others to the earliest living organisms. Study of these more primitive species should help us understand what early life looked like, which in turn should help us investigate the question of how life arose.

Mapping Evolutionary Relationships   The geological record tells us a great deal about how life evolved, but it is not the only way to study evolution. Because living species have evolved from common ancestors, the base sequence in the DNA of living organisms provides a sort of map of the genetic changes that have occurred through time. Thus, by comparing the genomes of different organisms, we should be able to reconstruct the evolutionary history of much of life on Earth.

To understand how the technique works, consider the DNA of an organism that long ago became the common ancestor of all life today. Mutations created variations on this DNA, and each new species therefore had slightly different DNA sequences than did the older species from which it evolved. Lateral gene transfer [Section 5.4] may also have been common among early living organisms, changing genomes even more rapidly. Over millions and billions of years, continuing evolution led to new species with DNA molecules increasingly different from the DNA of the common ancestor. But, always, the new molecules were built by changes to the older ones so that, in principle, the changes are traceable in the precise base sequences of living organisms.

Determining the sequence of bases in an organism's DNA is a difficult and time-consuming task, and to date only a small fraction of known species have had their complete genome sequences determined. In many more cases, biologists have compared smaller pieces of the DNA of many species.* By comparing the DNA sequences in similar genes among many different species, biologists can map the evolutionary history of the genes. For example, two species with very similar DNA sequences probably diverged relatively recently in evolutionary history, while two species with very different DNA sequences probably diverged much longer ago.

These types of DNA sequence comparisons are what has enabled biologists to map out the relationships shown in the tree of life (see Figure 5.11). That is, DNA studies tell us that life can be divided into the three domains that we discussed in Chapter 5—Bacteria, Archaea, and Eukarya—and also tell us about the branching patterns within each domain. Despite the many uncertainties that remain in the tree of life, the branching patterns still tell us a lot about evolutionary history. For example, the fact that animals and plants represent two branches that split off in about the same place from other Eukarya tells us that all animals and plants are quite similar genetically, at least in comparison to organisms on most other branches. Moreover, organisms on branches located closer to the "root" of the tree must contain DNA that is evolutionarily older, suggesting that they more closely resemble the organisms that lived early in Earth's history.

Conclusions About Early Life   The genetic studies that have led to mapping the tree of life tell us that species near the root of the tree must be more similar than other species to the common ancestor of life on Earth. Although it is not shown explicitly in the version of the tree in Figure 5.11, it appears that the organisms closest to the root are extremophiles such as those living near deep-sea volcanic vents or in hot springs like those in Yellowstone [Section 5.5]. Recall that these organisms are adapted to life in hot water (making them *hyperthermophiles*) and use chemical energy rather than photosynthesis to fuel their metabolism. In other words, it seems likely that the earliest life on Earth resembled the hot-water extremophiles that still survive today.[†]

• Where did life begin?

We can combine what we've learned about the probable nature of early organisms with geological considerations to come up with ideas about *where* life first arose on Earth. Let's begin by considering locales that we can probably rule out.

It seems unlikely that life could have arisen on the land surface. The early atmosphere contained practically no molecular oxygen, so our planet could not have had a protective layer of ozone. Ozone (O_3) is a form of oxygen produced in the upper atmosphere by interactions between ordinary oxygen (O_2) and ultraviolet light from the Sun. Today, ozone shields Earth's surface from the Sun's dangerous ultraviolet radiation. Before the ozone layer existed, any surface life would have been exposed to high lev-

---

*A particularly common technique relies on determining the sequence of bases in molecules of ribosomal RNA (rRNA), which tells us the sequence of the DNA that coded for it; rRNA is thought to evolve more slowly than most genes, so it provides a good proxy for overall evolutionary changes.

[†]One important caveat: A not-quite-sterilizing impact could have extinguished all life except that living near deep-sea vents. In that case, we'd still find that present-day life evolved from a hyperthermophile ancestor, even though earlier life might have had greater variety.

els of this radiation. While we can't rule out the possibility that life might have arisen in such an environment—indeed, some organisms today (such as *D. radiodurans* [Section 5.5]) can survive high-radiation conditions—the environment would have been much more hospitable underwater (because water also absorbs ultraviolet light) or in rocks beneath the surface.

The next idea, popular a few decades ago, suggested shallow ponds as the first home for life. If the ponds were full of organic compounds (an idea we'll discuss in the next section), these compounds would have become more concentrated as the ponds evaporated, making it easier for them to combine into more complex molecules that could lead to life. However, more recent studies suggest that shallow ponds would have lacked a source of chemical energy sufficient to support the origin of life, and any life in these ponds would probably also have suffered lethal effects from ultraviolet radiation and/or high salt concentrations.

Given the DNA evidence suggesting that early organisms resembled modern-day extremophiles that live near deep-sea vents and in hot springs, it now seems reasonable to favor these locations as the first home of life. They offer plenty of chemical energy to fuel chemical reactions that might have led to life, and they are protected from the Sun's ultraviolet radiation by the water that lies above them. Of course, we cannot rule out the possibility that life originated elsewhere, such as within rocks, and later adapted to these hot-water environments.

We have at least some reason to favor deep-sea vents over surface hot springs as the earlier home for life. Smaller impacts were more common than larger ones during the heavy bombardment, so partial vaporization of the oceans should have been more common than complete vaporization. No place on Earth was completely safe, but the deep ocean was clearly safer than springs at or near the surface. We may never know for certain where life first arose on Earth, but deep-sea vents look like a good bet.

6.2   THE ORIGIN OF LIFE

Even the simplest living organisms today seem remarkably advanced. Metabolic processes involve many intricate molecules and enzymes working together. The complex chemistry of DNA and RNA is deeply intertwined with the proteins and enzymes that help in making them [Section 5.4]. Indeed, every cellular component and process depends on many other components and processes. Given the complexity and interdependency of these processes, it might at first seem difficult to conceive of ways in which they might have come to be. However, over the past few decades, laboratory experiments have given us insights into the chemical processes that likely occurred on the early Earth. While these experiments have not yet told us precisely how life first arose—and it's possible that they never will—we'll see in this section that they give us good reason to think that life may have started through natural, chemical processes.

——Before we begin, it's worth noting two important caveats. First, the laboratory experiments generally try to re-create the chemical conditions that should have prevailed on the early Earth, an assumption that makes sense if life originated here. However, it is conceivable that life migrated to Earth from another world—a possibility we will also discuss in this section—in which case it might have arisen in a somewhat different chemical environment. Second, we have not said anything about the possibility that life arose through any kind of divine intervention; as we have discussed in prior chapters, that possibility falls outside the realm of science and instead is a matter of personal faith.

Figure 6.4

Stanley Miller poses with a re-creation of the original Miller–Urey experiment. The flask to the left contains liquid water that represents the sea; the gases, representing the atmosphere, consist of water vapor, methane, ammonia, and hydrogen. The central flask was supplied with energy in the form of electrical discharges. Below this flask, the gas was cooled so that it could condense and flow back into the flask with the liquid water. Chemical reactions in the experiment produced a wide variety of organic molecules.

• How did life begin?

Life today is based on the chemistry of a wide variety of organic molecules, making it logical to assume that the first life was somehow assembled from organic molecules produced by chemical reactions on the early Earth. Such reactions do not occur naturally today, because Earth's oxygen-rich atmosphere prevents complex organic molecules from forming readily outside living cells: Oxygen is such a highly reactive gas that it tends to attack chemical bonds, removing electrons and destroying organic molecules. However, the oxygen in our atmosphere is a product of life, produced by photosynthesis, which means it could not have been present before life arose. It therefore seems reasonable to think that the young Earth might have been much like a giant laboratory for organic chemistry.

The Miller–Urey Experiment As early as the 1920s, some scientists recognized that Earth's early atmosphere should have been oxygen-free, and they hypothesized that sunlight-fueled chemical reactions could have led to the spontaneous creation of organic molecules. (This idea was proposed independently by Russian biochemist A. I. Oparin and British biologist J. B. S. Haldane.) This hypothesis was put to the test in the 1950s in a famous experiment credited to Stanley Miller and Harold Urey, now known as the **Miller–Urey experiment** (Figure 6.4).

The original Miller–Urey experiment used small glass flasks to simulate chemical conditions that they thought represented those on the early Earth. One flask was partially filled with water to represent the sea and heated to produce water vapor. Gaseous methane and ammonia were added and mixed with the water vapor to represent the atmosphere. These gases flowed into a second flask, where electric sparks provided energy for chemical reactions. Below this flask, the gas was cooled so that it could condense to represent rain and then was cycled back into the water flask. The water soon began to turn a murky brown, and a chemical analysis (performed after letting the experiment run for a week) showed that it contained many amino acids and other organic molecules. Thus, the experiment appeared to support the idea that organic compounds would have formed spontaneously and in vast quantities on the early Earth, covering our planet and filling at least the top layers of the oceans with a rich "organic soup" containing all the building blocks of life.

However, other scientists soon realized that the methane and ammonia gas mixture used by Miller and Urey probably was *not* representative of Earth's early atmosphere, so the significance of their results was called into question. That the relevance of the Miller–Urey experiment is an even hotter scientific controversy today demonstrates the way that science often progresses in fits and starts.

The major issue concerns exactly what the early atmosphere was made of and, in particular, how much hydrogen it contained. When analyzed in more detail, the Miller–Urey experiment shows that hydrogen is a key ingredient needed to form large quantities of organic compounds; oxygen in any form, even when tied up in carbon dioxide, tends to suppress the formation of organic compounds. Today, we assume that Earth's early atmosphere was produced by volcanic outgassing [Section 4.2]. Outgassing may have occurred at a somewhat higher rate in the distant past than it does today, but we expect the gases to have been similar: largely water (H_2O), carbon dioxide (CO_2), nitrogen (N_2), and methane (CH_4), along with sulfur-bearing gases and hydrogen (H_2). The water condensed to form the oceans, leaving the other gases as potential ingredients for the early atmosphere. The key question is how much hydrogen remained relative to carbon dioxide. If there was a lot of hydrogen, then the Miller–Urey results might still be valid, suggesting that our young planet did indeed have a rich organic soup. If there was little hydrogen, then the oxygen in carbon dioxide would have kept the amount of organic material far lower than the Miller–Urey experiment suggests.

Prior to 2005, it was generally assumed that hydrogen would have escaped from the early atmosphere at the same rate that it escapes today, which is rapid enough that little hydrogen would have been present at any time. However, new calculations, still subject to verification, suggest that conditions in Earth's early upper atmosphere could have slowed the escape considerably, allowing hydrogen to make up as much as 30% of Earth's early atmosphere. Until this issue is settled, we will not know whether spontaneous chemical reactions would have turned Earth's surface and oceans into a rich organic laboratory or left them with only small amounts of locally produced organic compounds.

Other Sources of Organic Molecules   Although the jury is still out on precisely how much organic material might have been made through processes like those in the Miller–Urey experiment, we know of at least two other potential sources of organic molecules on the early Earth.

The first of these potential "other" sources is chemical reactions near deep-sea vents. As these undersea volcanoes heat the surrounding water, a variety of chemical reactions can occur between the water and the minerals. These chemical reactions would have occurred spontaneously in the conditions thought to have prevailed in the early oceans, and they should have resulted in the production of the same types of organic molecules thought to have been necessary for the origin of life.

The other additional source of organic molecules may have been material from space. Analysis of meteorites shows that they often contain organic molecules, including complex molecules such as amino acids. Similarly, telescopic and spacecraft studies of comets suggest that they, too, contain organic molecules. Apparently, organic molecules can form under the conditions present in interplanetary space and can survive the plunge to Earth. Moreover, recent research shows that the building blocks of life should also have been produced as ultraviolet light from the young Sun caused chemical reactions on the surfaces of dust grains in the spinning disk of the solar nebula. This dust, laden with organic molecules, could have "rained down" on the young Earth. The many impacts of the heavy bombardment could have brought additional organic material from asteroids and comets. The heat and pressure generated by the impacts may have further facilitated the production of organic molecules in Earth's atmosphere and oceans. All in all, a lot of organic material could have come to Earth from space.

It's likely that all three sources of organic molecules—chemical reactions in the atmosphere, chemical reactions near deep-sea vents, and material from space—played a role in shaping the chemistry of the early Earth. More important, given three different ways of obtaining organic molecules, it seems likely that at least parts of the early Earth would have contained substantial amounts of the organic molecules needed for life. The likelihood of organic molecules forming near deep-sea vents is particularly encouraging, given the evidence suggesting that life may have originated in such locales.

The Transition from Chemistry to Biology   The existence of at least three potential sources of organic molecules suggests that at least some locations on the young Earth had all the building blocks needed to make life. The next question, then, is how these building blocks might have assembled themselves to make a living cell.

Variations on the original Miller–Urey experiment have since produced all the essential building blocks of life, but, to paraphrase the late Carl Sagan, these building blocks represent only the notes of the music of life, not the music itself. Viewed in terms of simple probability, the likelihood of a set of simple building blocks ramming themselves together to form a complete living organism is at least as small as that of letting monkeys loose in a roomful of musical instruments and hearing Beethoven's Ninth Symphony. It simply wouldn't happen, even if the experiment was repeated over and over again for billions of years. There must have been at least a few intermediate steps—each involving a chemical pathway with a relatively high probability of occurring—that eased the transition from chemistry to biology.

One way to explore the transition is to work backward from organisms living today. Heredity today is shaped by DNA, which serves this function primarily because of its ability to replicate. Thus, we presume that early life must also have had a self-replicating molecule. However, this molecule probably was not DNA: Double-stranded DNA seems far too complex, and its replication far too intertwined with RNA and proteins, to have been the genetic material of the first living organisms. We are therefore looking for a molecule that is simpler than DNA but still capable of making fairly accurate copies of itself. The most obvious candidate is RNA.

RNA World

RNA is much simpler than DNA because it has only one strand rather than two and its backbone structure requires fewer steps in its manufacture. But it still possesses hereditary information in the ordering of its bases, and in principle it can serve as a template for making copies of itself. However, for a while there seemed to be a problem with this idea. In modern organisms, neither DNA nor RNA can replicate itself. Both require the help of enzymes. These enzymes are proteins that are made from genetic instructions contained in DNA and carried out with the help of RNA. This fact seemed to present a "chicken and egg" dilemma: RNA cannot replicate without enzymes, and the enzymes cannot be made without RNA.

A way around this dilemma was discovered in the early 1980s by Thomas Cech and his colleagues at the University of Colorado, Boulder. They found that RNA can catalyze biochemical reactions in much the same way as enzymes (work for which Cech shared the Nobel Prize in 1989). We now know that RNA molecules play this type of catalytic role in many cellular functions, and we call such RNA catalysts *ribozymes* (by analogy to enzymes). Follow-up work has shown that some RNA molecules can at least partially catalyze their own replication. These discoveries have led biologists to envision that modern, DNA-based life may have arisen from an earlier **RNA world**, in which RNA molecules served both as genes and as chemical catalysts for copying and expressing those genes.

How might an RNA world have gotten started? The first requirement would have been the spontaneous production of self-replicating strands of RNA. Even under the most optimistic assumptions, the concentration of organic molecules on the early Earth would have been far too low to allow those building blocks to assemble spontaneously into full-fledged RNA molecules. RNA assembly almost certainly would have required some sort of catalytic reaction to facilitate it. Here, again, laboratory experiments offer evidence for such a process.

Experiments show that several types of inorganic minerals can facilitate the self-assembly of complex, organic molecules. Minerals of the type that geologists call *clay** may have been especially important to the origin of life. Clay is extremely common on Earth and in the oceans, where it forms through simple weathering of silicate minerals; indeed, the oldest zircon grains [Section 4.2] suggest the widespread abundance of clays more than 4.4 billion years ago, so we expect clay to have been common at the time of the origin of life. Moreover, clay minerals contain layers of molecules to which other molecules, including organic molecules, can adhere. When organic molecules stick to the clay in this way, the mineral surface structure can force them into such close proximity that they react with one another to form longer chains. Laboratory experiments with this process have already produced strands of RNA up to nearly 100 bases in length. Although these strands are still quite primitive compared to what might be needed to start any sort of RNA life, the experiments suggest that clays in Earth's early oceans could have served as natural chemical "factories" that could potentially have produced self-replicating (or partially self-replicating) molecules of RNA. Indeed, given the millions of years and the countless grains of clay that could have facilitated chemical reactions, it seems reasonable to *expect* that self-replicating RNA molecules would at some point have been produced.

Assuming that self-replicating RNA did indeed come to exist, it would have been rapidly modified by a molecular analog to natural selection: The RNA molecules that replicated faster and more accurately soon came to dominate the population. Copying errors introduced mutations, ensuring the production of many variations of successful molecules. This allowed the molecular evolution to continue, as those RNA molecules best suited to the particular environmental conditions were more likely to continue replicating. Molecular evolution in the RNA world would have progressed much faster if the RNA and other complex molecules could have been confined within structures similar to living cells. The advantages of confining organic molecules within such "pre-cells" (also often called *vesicles*) are twofold. First, keeping the molecules concentrated and close together should have increased the rate of reactions among them, making it more likely that cooperative relationships like those in modern cells would evolve. Second, an enclosure would have isolated its contents from the outside in a way that should have facilitated natural selection among RNA molecules. For example, suppose a particular self-replicating RNA molecule assembled amino acids into a primitive enzyme that sped up replication (Figure 6.5a). If the enzyme floated freely within the ocean water, it might just as easily have helped the replication of other RNA molecules as of the one that made it. But inside a pre-cell, the enzyme would help only the RNA that made it, giving this RNA an advantage over less capable RNA molecules in other pre-cells (Figure 6.5b).

How might such pre-cell enclosures have formed? Once again, laboratory experiments suggest that they may have appeared easily and naturally in at least two ways. First, if we cool a warm-water solution of amino acids, they can form bonds among themselves to make an enclosed, spherical structure (Figure 6.6a). Although they are not alive, these structures exhibit many lifelike properties. For example, they can grow in size by absorbing more short chains of amino acids until they reach an unstable size, at which point they split to form "daughter" spheres. They can also

---

*In this context, clay refers to silicate minerals with a particular physical structure; this mineralogical definition is somewhat different from what you may think of as clay in the context of pottery or sculpture.

A strand of RNA serves as a template for its own replication.   *Amino acids can also attach to the RNA, which links them into small proteins.*   *The proteins then act as simple enzymes to speed up the RNA replication.*

a This diagram shows a self-replicating RNA molecule that has evolved the capability to produce a primitive enzyme that helps its own replication.

b If the RNA and the enzyme are isolated from the outside environment inside a pre-cell, then only the molecules in this particular pre-cell will benefit from the new enzyme, a fact that can speed up the molecular evolution.

Figure 6.5

Self-replicating RNA could have rapidly evolved through a molecular analog to natural selection. (Adapted from Campbell, Reece, Simon, *Essential Biology,* Second Edition, © 2004.)

a These microscopic spheres were made by cooling a warm-water solution of amino acids. They are not alive, but they exhibit many lifelike properties.   **b** These microscopic membranes are made from lipids that, when mixed with water, spontaneously form enclosed droplets.

Figure 6.6

Laboratory versions of pre-cells. (Adapted from Campbell, Reece, Simon, *Essential Biology,* Second Edition, © 2004.)

Figure 6.7

This microscopic photo (made with the aid of fluorescent dyes) shows short strands of RNA (red) contained within a lipid pre-cell (green circle), both of which formed with the aid of catalysis by clay minerals beneath them.

selectively allow some types of molecules to cross into or out of the enclosure. Some of these spheres even store energy in the form of an electrical voltage across their surfaces, which can be discharged in a way that facilitates reactions inside them. The second type of membrane forms spontaneously when we mix lipids with water (Figure 6.6b). (Recall that lipids are one of the four major types of cellular components [Section 5.2].)

Perhaps even more significantly, recent experiments show that lipid pre-cells can form on the surface of the same clay minerals that help assemble RNA molecules, sometimes with RNA inside them (Figure 6.7). This work, which we'll discuss in more detail in Section 6.6, gives further credence to the idea that an RNA world may have been born on the early Earth with the catalytic assistance of clay minerals.

If and when the RNA world got started, the molecular natural selection would have inevitably led RNA molecules to gain complexity and evolve more efficient replication pathways. At some point, the RNA pre-cells would have become sufficiently good at reproducing and evolving that we would have considered them to be "alive," although the process was probably gradual: There might never have been a particular moment when we would have been able to say that the "first living cell" had appeared on the scene. At this point, biological natural selection could take over, and it seems easy to understand why the RNA world would have eventually given way to the present DNA world: DNA is a more flexible hereditary material and is less prone to copying errors than RNA. Thus, because DNA is structurally so similar to RNA, it seems natural that the RNA molecules serving hereditary functions would have eventually evolved DNA, while other RNA molecules would have continued the cellular roles that they still play today.

Putting It All Together   Let's review the sequence we've discussed here as the possible explanation for the origin of life (Figure 6.8).

1. Through some combination of atmospheric chemistry, chemistry near deep-sea vents, and molecules brought to Earth from space, the early Earth had at least localized areas with significant amounts of organic molecules that could serve as building blocks for more complex organic molecules.
2. More complex molecules, including short strands of RNA, grew from the organic building blocks, probably with the aid of reactions catalyzed by clay minerals. Some of the RNA molecules became capable of self-replication.
3. Membranes formed spontaneously, perhaps also with the aid of clay minerals, and enclosed some of these complex molecules in "pre-cells" that facilitated the development of cooperative molecular interactions.

Figure 6.8

A summary of the steps by which chemistry on the early Earth may have led to the origin of life. (Adapted from Campbell, Reece, Simon, *Essential Biology*, Second Edition, © 2004.)

1. Organic precursor molecules appear.

2. RNA molecules become self-replicating.

RNA

3. Membrane-enclosed pre-cells arise.

4. True cells with RNA genome appear.

RNA genome

primitive cell

5. Modern cells with DNA genome evolve.

DNA genome

4. Natural selection among the RNA molecules in pre-cells gradually led to an increase in complexity, until eventually some of these structures became true living organisms.

5. Natural selection then rapidly improved and diversified life. DNA became the favored hereditary molecule, and life has continued to evolve ever since.

We may never know for certain whether life actually originated in this way, in some similar way, or in some completely different way. Nevertheless, this scenario seems quite reasonable and perhaps even "easy," given geological time scales. It seems especially reasonable given that a number of different components of the scenario have been demonstrated in laboratory experiments. Even if life did not originate in this way, it seems that it could have—which suggests that the actual path to life must have been equally easy, or else it would have followed the path we've described. In summary, we have good reason to believe that the origin of life was a *likely* consequence of conditions on the early Earth, in which case it might be equally likely that life arose on many other worlds.

THINK ABOUT IT We've noted that the probability of life's arising by randomly mixing simple organic building blocks is so small as to seem impossible. Yet, in the scenario we've described, the likelihood of getting life seems quite good. In your own words, describe why these two probabilities are so different.

• Could life have migrated to Earth?

Although our scenario suggests that life could have arisen easily and naturally here on Earth, it is also possible that life arose somewhere else first—for example, on Venus or Mars—and then migrated to Earth within meteorites.

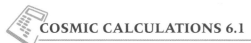

COSMIC CALCULATIONS 6.1

Bacteria in a Bottle I: Lessons for Early Life

Once the first organisms took hold, how quickly could they have spread and evolved? A thought experiment* offers insight into this question. Suppose you place a single bacterium in a nutrient-filled bottle at noon, and that this species is capable of replicating by cell division every minute. The original bacterium grows until it divides into two bacteria at 12:01. These two bacteria divide into 4 bacteria at 12:02, which divide into 8 bacteria at 12:03, and so on. Then the number of bacteria in the bottle at any time t minutes after noon is:

$$\text{\# bacteria at } t \text{ minutes after } 12:00 = 2^t$$

We'll explore general characteristics of this **exponential growth** (t is in the exponent) in Cosmic Calculations 6.2. Here, to understand how rapidly early life could have spread, let's consider the volume of a bacterial colony. A typical bacterium is 10^{-7} m (0.1 micrometer) across, which means it has a volume of about $(10^{-7} \text{ m})^3 = 10^{-21} \text{ m}^3$. So the volume of bacteria at any time t minutes after noon is

$$\text{bacterial volume} = 2^t \times 10^{-21} \text{ m}^3$$

Our two formulas tell us that after 60 minutes the number of bacteria is $2^{60} \approx 1 \times 10^{18}$, or a *million trillion*; and their volume is about $2^{60} \times 10^{-21} \text{ m}^3 \approx 0.001 \text{ m}^3$, or 1 cubic centimeter. By this point in a small bottle, the bacteria will be close to exhausting their nutrients. But let's imagine they could somehow continue to multiply. By the end of the second hour, they would number an astonishing $2^{120} \approx 1.3 \times 10^{36}$, and their volume would be $2^{120} \times 10^{-21} \text{ m}^3 \approx 1.3 \times 10^{15} \text{ m}^3$—large enough to cover the surface of the Earth to a depth of about 2 meters (see Problem 50 at end of chapter). Continuing the calculations, you'd find that the bacteria would exceed the total volume of the world's oceans (about $1.3 \times 10^{18} \text{ m}^3$) at $t = 130$ minutes. Note that changing the doubling time from one minute to a year hardly matters; a time of $t = 130$ years rather than 130 minutes is still geologically insignificant.

Although the bacteria could not really continue this hypothetical growth, the implication should be clear: The first self-replicating organisms would have spread rapidly as far as conditions allowed, leaving the door wide open for biological evolution through natural selection.

*This thought experiment is adapted from one created by Professor of Physics Albert A. Bartlett of the University of Colorado.

The idea that life could travel through space to land on Earth, sometimes called *panspermia,* once seemed outlandish. After all, it's hard to imagine a more forbidding environment than that of space, where there's no air, no water, and constant bombardment by dangerous radiation from the Sun and stars. However, the presence of organic molecules in meteorites and comets tells us that the building blocks of life can survive in the space environment, and we've already discussed some Earth microbes that are capable of surviving at least moderate periods of time in space [Section 5.5]. It therefore seems possible that life could migrate from one planet to another, if it can hitch a suitable ride.

Figure 6.9

Chemical analysis of this meteorite, known as ALH84001, indicates that it came from Mars. The small block shown for scale to the lower right is 1 cubic centimeter, about the size of a typical sugar cube. (We will discuss this particular meteorite further in Section 8.5.)

The Possibility of Migration   We know that meteorites can and do travel from one world to another. Among the more than 20,000 meteorites that scientists have identified and cataloged, careful chemical analysis has so far revealed about three dozen with compositions that clearly suggest that they came from Mars (Figure 6.9); even more have been found that come from the Moon. Apparently, these meteorites were blasted from their home worlds by large impacts, then followed orbital trajectories that eventually caused them to land on Earth. Observations of these meteorites, along with theoretical calculations based on the amount of material blasted into space by impacts, suggest that over time the inner planets have exchanged many tons of rock. In a sense, Earth, Venus, and Mars have been "sneezing" on each other for billions of years, offering the possibility of microscopic life hitchhiking between worlds on one of the meteorites.

For a living microbe to arrive intact on Earth after such a journey, it would have to survive at least three potentially lethal events: the impact that blasts it off the surface of its home world, the time it spends in the harsh environment of interplanetary space, and the fiery plunge through our atmosphere. Examination of martian meteorites suggests that neither the first nor last of these events poses insurmountable obstacles. The interiors of martian meteorites (such as ALH84001) show only minimal disruption, suggesting that microbes inside these rocks could survive both the initial impact and the later fall to Earth. The larger question is whether they could survive their time in space.

The chance of surviving the trip between planets probably depends on how long the meteorite spends in space. Once a rock is launched into space, it orbits the Sun until its orbit carries it directly into the path of another planet. Most meteorites will orbit for many millions of years before reaching Earth, even if they come from a world as nearby as Venus or Mars. It seems highly unlikely that living organisms could survive in space for millions of years. However, a few meteorites are likely to be launched into orbits that cause them to crash to Earth during one of their first few trips around the Sun. For example, calculations suggest that about 1 in 10,000 meteorites may travel from Mars to Earth in a decade or less. Because experiments in Earth orbit have already shown that some terrestrial microbes can survive at least 6 years in space, it seems quite reasonable to imagine microbes from Mars arriving safely on Earth.

While migration between planets seems possible, similar considerations almost certainly rule out the possibility of migration from other star systems. Under the best of circumstances, meteorites from planets around other stars would spend millions or billions of years in space before reaching Earth; any living organisms would almost surely be killed by exposure to cosmic rays during this time. Moreover, calculations suggest that the probability of a rock from another star system hitting Earth is extremely low, which also explains why we have never yet found a meteorite from beyond our own solar system.

Reasons to Consider Migration   Given the reality that the inner planets exchange rocks in the form of meteorites, the key question probably is not whether life *could* migrate through space but whether we have any reason to suppose it originated elsewhere rather than right here on Earth. Many scientists have debated this question, with the debate taking many different twists. Today, most ideas about migrating life fall into one of two broad categories.

The first broad idea suggests that life does not form as easily as we have imagined, at least under the conditions present on the early Earth. In this view, the only explanation for life on Earth (other than invoking the supernatural) would be migration from elsewhere. Although this idea in some sense only moves the problem of life's origin to another place, it at least allows for the possibility that another world had conditions that were more conducive to rapid development of life, or that life arose on a world that offered more time. The primary drawback to this idea is that other worlds in our solar system should have had no more time available than Earth, and we know of no compelling reason why any of these worlds would have offered better conditions for an origin of life. The only way to get significantly more time for an origin of life is to suppose that life migrated from another star system, but we have already explained why that possibility seems highly unlikely.

The second broad idea suggests that life forms so easily that we should expect to find life originating on any planet with suitable conditions. In that case, the origin of life in our solar system would have occurred on whichever planet got those conditions *first;* for example, if the very early Venus or very early Mars had suitable conditions for life before Earth, life from one of those worlds might have migrated to Earth and taken hold on our planet as soon as conditions allowed. In essence, this idea suggests that life never got the chance to originate indigenously on Earth because life from another planet got here first.

Implications of Migration to the Search for Life Beyond Earth

While ideas about microbes migrating *to* Earth are speculative, it seems a near-certainty that microbes *from* Earth have many times made the journey to Mercury, the Moon, Venus, and Mars. After all, there have been many large (but not sterilizing) impacts on Earth in the almost 4 billion years during which we know life existed here, offering plenty of opportunity for hitchhiking microbes. Thus, if it were possible for Earth life to survive on any of these other worlds, we should actually *expect* to find it there. As we'll discuss in Chapter 7, we can almost certainly rule out the possibility of survival on the Moon and Mercury, and probably on Venus as well. Mars, however, may well have habitats that could provide at least temporary refuge to terrestrial microbes, and Mars may have been globally habitable in the distant past.

The likelihood of such interplanetary migration raises at least two important issues in astrobiology. First, if we someday find life on Mars, we will have to wonder if it is native or if it arrived there from Earth. The only way we may ever be confident that Mars life is not transplanted Earth life will be if its biochemistry is too different from that of terrestrial life to allow for a common ancestor. Even in that case, we might wonder whether the martian life could have descended from a species that arose on Earth very early but was extinguished by a sterilizing impact before the end of the heavy bombardment. After all, if life arose and was extinguished multiple times on the early Earth, there is no reason to expect the biochemistry to have been the same with each separate origin of life.

Second, the possibility of life migrating among the planets raises the question of whether we could ever distinguish between an indigenous origin of life on Earth and an origin based on migration from elsewhere. To date, no one knows how we might distinguish between these possibilities. It's conceivable that a fossil record from Mars might suggest an earlier origin of life there, though even then we might not be certain that this life came to Earth. Venus poses a more intractable problem: As we'll discuss in Chapter 10, it is possible that Venus once had oceans and a habitable climate in which life might have arisen. However, Venus now is so hot that any fossil record would almost certainly have been destroyed by the heat and subsequent geological activity.

Despite these potential uncertainties, the major lessons of our study of life's origins still hold: One way or another, life arose on Earth quite soon after conditions first allowed it, and even if life migrated here from another world, we have good reason to think that it evolved naturally, through chemical processes that favor the creation of complex, organic molecules and the subsequent molecular evolution of self-replicating molecules.

6.3 THE EVOLUTION OF LIFE

Regardless of how or where it originated, life on Earth has been evolving throughout the 4 billion or so years during which it has existed on Earth. Careful studies of the geological record provide the key data with which we attempt to re-create the evolutionary time scale, while genome comparisons offer data that help us map relationships between species. In this section, we'll briefly retrace the history of life on Earth as it is now understood, which should in turn help us understand the possibilities for finding similarly complex life on other worlds.

- ## What major events have marked evolutionary history?

Reconstructing 4 billion years of history from limited clues is an obviously difficult task. Nevertheless, we have identified at least a few of the key events that have marked the evolution of early life into its current diversity, and that help us understand our own origins. As you read here, you may wish to look back at the geological time scale in Figure 4.10, which indicates the major events.

Early Microbial Evolution The earliest organisms must have been quite simple, but they undoubtedly had at least a few enzymes and a rudimentary metabolism. Their cells must have resembled modern prokaryotic cells, lacking cell nuclei and other complex structures that we now find only in eukaryotes. Moreover, because the atmosphere at that time was essentially oxygen-free, all early life must have been **anaerobic**, meaning that it did not require molecular oxygen; by contrast, we are **aerobic** organisms, because we cannot survive without molecular oxygen. Based on the hypothesis that life originated near deep-sea vents or hot springs, we expect that the first microorganisms were *chemoautotrophs* (see Table 5.1)—organisms that obtained their carbon from carbon dioxide dissolved in the oceans and their energy from chemical reactions involving inorganic chemicals. Some modern Archaea that appear to be fairly close to the root of the tree of life, such as those thriving in hot sulfur springs, obtain their energy in this way through chemical reactions involving hydrogen, sulfur, and iron compounds. Because similar compounds were abundant on the early Earth, perhaps especially so in hot springs and near deep-sea vents, it seems reasonable to assume that early life used the available inorganic chemical energy. (See Section 9.4 for discussion of possible chemical pathways.)

Natural selection probably caused rapid diversification among the early life-forms. Modern DNA replication involves a variety of enzymes that help keep the mutation rate low. Early organisms, with a much more limited set of enzymes, probably experienced many more errors in DNA copying. Because more errors mean a higher mutation rate, evolution would have been rapid among early microbes. As life diversified, many new metabolic processes evolved, making some of the new organisms biochemically quite different from their ancestors. Because of the rapid pace of early evolution, it may not have taken long to establish many of the major branches in the tree of life (see Figure 5.11).

Fossil evidence supports the idea of rapid diversification. Recall that stromatolites suggest the presence of organisms that obtained energy by photosynthesis some 3.5 billion years ago, and some of the oldest microfossils also resemble modern photosynthetic organisms. Because photosynthesis is a complex metabolic pathway, its early emergence indicates that early evolution was rapid.

Photosynthesis probably evolved through multiple steps. At first, some organisms may have developed light-absorbing pigments that absorbed excess light energy—especially ultraviolet—that was harmful to life near the ocean surface. Over time, some of these pigments evolved to enable the cell to make use of the absorbed solar energy. Modern organisms known as purple sulfur bacteria and green sulfur bacteria may be much like the early photosynthetic microbes. These organisms use hydrogen sulfide (H_2S) rather than water (H_2O) in photosynthesis and therefore do not produce any oxygen. Photosynthesis using water, which produces oxygen as a by-product, probably came later and ultimately caused the buildup of oxygen in Earth's atmosphere. The timing of the oxygen buildup is still uncertain, but it probably did not get underway much before about 2.4 billion years ago.

The rise of oxygen created a crisis for life, because oxygen attacks the bonds of organic molecules. Many species of microbes probably went extinct, and those that survived had to somehow avoid the detrimental effects of oxygen. Some avoided these effects because they lived in (or migrated to) underground locations where the oxygen did not reach them. We still find many anaerobic microbes in such locales today, living in soil or deeper underground in rocks. Others survived because the oxygen content of the atmosphere rose gradually, allowing them time to evolve new metabolic processes and protetctive mechanisms that enabled them to thrive rather than die in the presence of oxygen. Plants and animals, including us, still use the metabolic processes that evolved in response to the "oxygen crisis" faced by living organisms some 2 billion or more years ago.

The Evolution of Eukarya The evolution of Eukarya was the crucial first step in our own eventual evolution; we are, after all, members of this domain. Moreover, although the "complexity" of cellular processes is sometimes a subjective judgment, overall the structure of eukaryotic cells seems better suited to the evolution of advanced organisms than that of prokaryotic cells. Even single-celled eukaryotes exhibit much more diversity in structure than exists among all prokaryotes, and multi-celled eukaryotes enjoy diversity far beyond that. Because more variations are possible on complex structures than on simple ones, the complexity of eukaryotic cells allowed for the selection of many more adaptations than were possible in prokaryotic cells. Indeed, multicellularity appears to have evolved independently in several different branches of Eukarya, suggesting that the complex structure of eukaryotic cells opened the door for the evolution of more advanced organisms. If we ever find life elsewhere that is more like us than like simple microbes, it's likely that it will have cell structures more like those of modern Eukarya than of Bacteria or Archaea.

When did Eukarya arise? Despite the fact that modern Eukarya have more complex cellular structures than Bacteria and Archaea, genome studies do not suggest any substantial differences in the evolutionary ages of the three domains. That is, it is quite likely that members of all three domains—Bacteria, Archaea, and Eukarya—split from a common ancestor early in Earth's history. Early Eukarya could not yet have had cell nuclei and other complex intracellular structures. These must have come later, and the oldest known fossils that clearly show cell nuclei date to about 2.1 billion years ago. However, because cell nuclei do not fossilize well, it's possible that Eukarya began to have such structures much earlier but we are unable to recognize them in the fossil record.

Modern, complex Eukarya probably evolved through a combination of at least two major adaptations that arose in their simpler ancestors. First, some early species of Eukarya may have developed specialized infoldings of their membranes that compartmentalized certain cell functions, ultimately leading to the creation of a cell nucleus (Figure 6.10a). Second, some relatively large ancestral host cell absorbed smaller prokaryotes with specialized functions. The absorbed prokaryotes developed a **symbiotic relationship** with the host cell, that is, a relationship in which both the invading organism and the host organism benefit from living together (Figure 6.10b).

The key evidence favoring symbiosis (the development of a symbiotic relationship) comes from the existence of structures within eukaryotic cells that look much like tiny prokaryotes trapped inside. The most important of these "cells within cells" are **mitochondria**, the cellular organs in which oxygen helps produce energy (by making molecules of ATP), and **chloroplasts**, structures in plant cells that produce energy by photosynthesis. Mitochondria and chloroplasts even have their own DNA and reproduce themselves within their eukaryotic homes. Moreover, DNA sequences in mitochondria and chloroplasts clearly group them with domain Bacteria rather than with Eukarya. Thus, it seems a near-certainty that mitochondria and chloroplasts were originally free-living bacteria. Assuming that these bacteria had already evolved their ability to make efficient use of oxygen (in the case of mitochondria) or to carry out photosynthesis (in the case of chloroplasts), a symbiotic relationship might have developed quite easily. The host cell would benefit from the energy produced by the incorporated bacteria, while the bacteria would benefit from the protection offered by the host cell. Further support for this hypothesis comes from the many cases in which living organisms today form symbiotic relationships. Other evidence comes from the fact that some present-day bacteria show striking similarities to mitochondria and chloroplasts, suggesting that they share a common ancestry.

The Cambrian Explosion   We have seen that life on Earth existed at least 3.5 billion years ago (and perhaps hundreds of millions of years be-

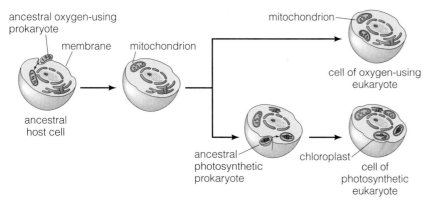

Figure 6.10

Hypotheses concerning the origin of eukaryotes. (Adapted from Campbell, Reece, Taylor, Simon, *Biology Concepts & Connections*, Fifth Edition, © 2006.)

a Some large prokaryotes may have developed specialized membrane infoldings that compartmentalized certain cell functions, ultimately leading to the creation of a cell nucleus.

b Mitochondria and chloroplasts may have evolved as small prokaryotes invaded a larger host cell, forming a symbiotic relationship.

fore that), and all three domains of life were well established by at least 2.1 billion years ago. However, the fossil record tells us that all this life remained microscopic (aside from microbes organized into colonies) until much later. The earliest fossil evidence for complex, multicellular organisms—all of which are Eukarya—dates to only about 1.2 billion years ago. Thus, microbes had our planet to themselves for more than the 2 billion years after the origin of life. Even today, the total biomass of microbes far exceeds that of multicellular organisms like fungi, plants, and animals [Section 5.2]. But mass isn't everything, and we have a special interest in understanding the evolution of multicellular life, even if it is comparatively rare on our world.

In particular, we have a special interest in animal evolution: Animals may represent only one small branch on the tree of life, but it's *our* branch. Moreover, we generally assume that extraterrestrial intelligence, if it exists, will belong to animal-like beings from other worlds. The fossil record suggests that animal evolution progressed slowly at first, with relatively little change seen between fossils from 1.2 billion years ago and those from a half-billion years later. But then something quite dramatic happened.

In the broadest sense, biologists classify animals according to their basic "body plans." For example, the basic body plan shared by mammals and reptiles is fundamentally different from that of insects. Animals are grouped by body plan into what biologists call **phyla** (singular, *phylum*), the next level of classification below kingdoms (such as the plant kingdom and the animal kingdom [Section 5.2]). Mammals and reptiles both belong to the phylum *Chordata,* which represents animals with internal skeletons. Insects, crabs, and spiders belong to the phylum *Arthropoda,* which represents animals with body features such as jointed legs, an external skeleton, and segmented body parts. Classifying animals into phyla is an ongoing effort by biologists, but modern animals appear to comprise about 30 different phyla, each representing a different body plan.

Remarkably, nearly all of these different body plans, plus a few others that have gone extinct, make their first known appearance in the geological record during a period spanning only about 40 million years—less than about 1% of Earth's history. This remarkable flowering of animal diversity appears to have begun about 545 million years ago, which corresponds to the start of the Cambrian period (see Figure 4.10).* Hence, it is called the **Cambrian explosion**.

 THINK ABOUT IT One early trend in the evolution of multicellular organisms was a trend toward larger size. Briefly discuss how larger size might have conferred an evolutionary advantage.

The fact that the Cambrian explosion marks the only major diversification of body plans in the geological record presents us with two important and related questions: Why did the Cambrian explosion occur so suddenly, at least in geological terms, yet so long after the origin of eukaryotes, and why hasn't any similar diversification happened since?

No one knows the answers to these questions, but we can identify at least four possible contributing factors. First, the oxygen level in our atmosphere may have remained well below its present level until about the time of the Cambrian explosion. Thus, the dramatic change in animal life may have occurred at least in part because oxygen reached a critical level for the survival of larger and more energy-intensive life-forms.

A second factor, which some scientists think may have been much more important, was the evolution of genetic complexity. As eukaryotes evolved, they developed more and more genetic variation in their DNA, which opened up ever more possibilities for further variation. Perhaps the Cambrian explosion marks a point in time when organisms had become sufficiently complex that a great diversity of forms could evolve over a short time.

The third factor may have been climate change. Recall that geological evidence points to a series of snowball Earth episodes [Section 4.5] ending around the time of the Cambrian explosion. The extreme climate conditions that marked these episodes may have exerted evolutionary pressure that aided the diversification of life and then fueled the Cambrian explosion when the environmental conditions eased.

The fourth factor may have been the absence of efficient predators. Early predatory animals were probably not very sophisticated, so some adaptations that later might have been snuffed out by predation were given a chance to survive if they arose early enough. Thus, the beginning of the Cambrian period may have marked a window of opportunity for many different adaptations to gain a foothold in the environment.

This last idea may partly explain why no similar explosion of diversity has taken place since the Cambrian. Once predators were efficient and widespread, it may have been much more difficult for entirely new body forms to find an available environmental niche. In addition, the fact that certain body forms were already selected during the Cambrian explosion may have limited other options. That is, while more body plans may have been possible than actually arose, once these were in existence there may not have been clear evolutionary pathways to others. Alterna-

---

*Although the Cambrian explosion is among the most vivid events in the geological record, evidence indicates that the lineages we first see in the Cambrian explosion actually began evolving earlier. The Cambrian explosion didn't involve only animals; other groups, such as algae, also diversified.

tively, perhaps the various body forms that arose during the Cambrian explosion represent the full range of forms possible, at least within the constraints of the genetic variability available on Earth. In any case, we and nearly all other animals living today can trace our ancestry to species that arose during the Cambrian explosion.

The Colonization of Land   Because most early microfossils are found in sediments that were originally deposited in the oceans, it's difficult to know exactly when life first migrated onto land. However, given the wide variety of environments in which microbes survive today and the fact that many different genetic lineages seem to have appeared quite early in evolutionary history, it's likely that microbial life quickly established itself wherever it could find liquid water and protection from ultraviolet radiation. Plenty of such locations are available on land—including underground and anyplace where water can pool under a shelter of overhanging rock—so it is hard to imagine reasons why microbial life would not have taken hold on land quite early. However, the situation is different for multicellular organisms. While microbes may have thrived on land, larger organisms, including all animals, remained confined to the oceans (and other bodies of water) even after the Cambrian explosion.

For larger organisms, surviving on land was more difficult than surviving in the oceans, primarily because it required evolving a means of obtaining water and mineral nutrients without simply absorbing them from surroundings. The timing of the development of the ozone layer may also have played a role in the late colonization of land. Recall that ozone is a molecule (O_3) made from oxygen atoms, so a protective ozone layer could not exist until the atmosphere contained some threshold level of oxygen. Uncertainties in both the oxygen levels through time and the level needed for a substantial ozone layer make it difficult to know when the ozone layer first appeared. But until it did, life on any exposed land surface would have been difficult or impossible.

Fossil evidence shows that plants (and perhaps fungi as well) were the first large organisms to develop the means to live on the land. The colonization of land by plants appears to have begun about 475 million years ago. DNA evidence suggests that plants evolved from a type of alga. Some ancient algae might have survived in salty shallow-water ponds or along lake edges. Because such locales occasionally dry up, natural selection would have favored adaptations, such as thick cell walls, that allowed the algae to survive during periods of dryness. Cell walls would have given the organisms structure that would have helped them survive on land. The first fully land-based organisms would have had even more advantages, because there were no land animals around to eat them. Large plants gradually developed complex bodies with some parts specialized for energy collection above ground (where sunlight is available) and other parts specialized for collecting water and nutrients from the soil.

Once plants moved onto the land, it was only a matter of time until animals followed them out of the water. Within about 75 million years, amphibians and insects were eating land plants. By the beginning of the Carboniferous period, about 360 million years ago, vast forests and abundant insects thrived around the world (Figure 6.11). These Carboniferous forests were important not only as a major step in evolution, but also because they became an important part of our modern economy. Much of the land area of the continents was flooded by shallow seas during the Carboniferous period, hindering the decay of dead plants. Thick layers of dead organic matter piled up in the stagnant waters. Over millions of years, as these layers were buried, pressure and heat gradually converted the

Figure 6.11

This painting, based on fossil evidence, shows a forest of the Carboniferous period. The large trees with straight trunks are seedless plants called lycophytes. The tree on the left with feathery branches is a horsetail. The plants near the forest floor are ferns. The dragonfly was about the size of a modern raven.

organic matter to coal. Nearly all the coal that has helped fuel our industrial age is, in fact, the remains of these forests of the Carboniferous period.

> **THINK ABOUT IT** Based on the preceding discussion, explain why coal is called a "fossil fuel." What other fossil fuels do we use to generate energy?

• Why was the rise of oxygen so important to evolution?

We are oxygen-breathing animals, so there's no question that the rise of oxygen was critical to our eventual emergence. More generally, the rise of oxygen was important to evolution because oxygen can react so strongly with organic molecules. While these reactions can kill organisms that are not adapted to oxygen's presence, they also offer the possibility of much more efficient cellular energy production (that is, making molecules of ATP [Section 5.3]) than is possible through anaerobic processes. Thus, as aerobic organisms evolved, they were able to develop adaptations that demanded much more energy than would have been available to their anaerobic ancestors. The rise of oxygen ignited an explosion of eukaryotic diversification and, as we've discussed, may have helped fuel the Cambrian explosion.

The Origin of Atmospheric Oxygen Molecular oxygen is a highly reactive gas that would disappear from the atmosphere in just a few million years if it were not continually resupplied by life. Fire, rust, and the discoloration of freshly cut fruits and vegetables are everyday examples of **oxidation reactions**—chemical reactions that remove oxygen from the atmosphere. Many elements and molecules can participate in oxidation reactions. Today, most reactions that remove oxygen from the atmosphere occur in living organisms that use oxygen, including ourselves. Before oxygen-breathing organisms evolved, oxidation reactions primarily involved volcanic gases, dissolved iron in the oceans, and surface minerals (especially those containing iron) that could react with oxygen. Such reactions essentially "rust" the minerals, causing them to turn reddish in color. In the oceans, oxidation reactions with dissolved iron create minerals that precipitate to the bottom, forming "red beds" on the

Figure 6.12

Cyanobacteria split water and release oxygen in photosynthesis and are thought to have been responsible for the rise of oxygen in Earth's atmosphere.

a The blue-green color of this lake (in Cape Cod, Massachusetts) is the result of a population explosion, or "bloom" of cyanobacteria.

b This micrograph shows individual cyanobacteria from the lake.

ocean floor. On land, the reddish color of much of Earth's rock and clay is a direct result of oxidation reactions involving surface minerals.

The fact that free oxygen would not last long without life tells us that our atmosphere must have been essentially oxygen-free before life existed. Moreover, while today we recognize plants as a major source of oxygen, we know that plants arrived relatively recently on the evolutionary scene. Where, then, did the oxygen come from? Remarkably, it seems that we owe our oxygen atmosphere to microscopic bacteria sometimes called "blue-green algae" but more technically known as **cyanobacteria** (Figure 6.12). Fossil evidence shows that cyanobacteria were producing oxygen by at least 2.7 billion years ago, and perhaps for hundreds of millions of years before that. It took at least 2 billion years for oxygen to build up in the atmosphere to its present levels, but in the end the oxygen we breathe originally entered the air through the action of microscopic cyanobacteria.

Timing of the Oxygen Rise

The precise timing of the rise of oxygen is difficult to study, because we have no direct way to sample air from hundreds of millions or billions of years ago. However, we can learn about oxygen content from a variety of other clues. For example, fossils of oxygen-breathing organisms can tell us a minimum amount of oxygen that must have been in the atmosphere for them to survive. Careful study of rock chemistry offers even more clues.

Studies of rocks that are between about 2 and 3 billion years old, especially rocks of a type called *banded iron formations* (Figure 6.13), show that the atmosphere during that time contained less than 1% of the amount of oxygen it contains today. The banded iron formations were made from iron-containing minerals dissolved in the oceans, and such iron minerals cannot dissolve if there is substantial oxygen in the atmosphere and ocean. More recent studies, based on sulfur isotope ratios in ancient rocks, constrain the timing of the rise of oxygen more tightly. Atmospheric oxygen alters the atmospheric chemistry of sulfur compounds in the atmosphere (SO_2 and H_2S from volcanic outgassing) in ways that change the ratios of sulfur isotopes that end up in surface rock. The oldest rocks showing sulfur isotopes in a ratio that indicates the presence of atmospheric oxygen are about 2.35 billion years old. If these data are

Figure 6.13

This rock is an example of a banded iron formation (BIF) formed more than 2 billion years ago. Such rocks could have formed only before the atmosphere contained significant amounts of oxygen. The pen is included for scale.

being properly interpreted, it means that the abundance of atmospheric oxygen must have been less than 20 parts per million (0.002%) up until 2.35 billion years ago. At that point, sometimes called the "great oxidation event," oxygen began to build up in the atmosphere.

The timing of the great oxidation event poses a mystery, however: Given that cyanobacteria began to produce oxygen at least 2.7 billion years ago—or at least 350 million years before the great oxidation event at 2.35 billion years ago—what took so long? We do not yet know the details, but presumably the oxygen was being removed from the atmosphere by non-biological processes, such as oxidation of surface rock and ocean minerals, as rapidly as the cyanobacteria could make it.

An even greater mystery concerns what happened once the buildup began. Other isotopic evidence suggests that the "great oxidation event" wasn't really that great, and that oxygen levels remained far below modern levels for at least the next billion years, and perhaps all the way up to about the time of the Cambrian explosion. Although a variety of hypotheses have been proposed to explain this slow buildup of oxygen, we do not yet have sufficient data to decide which ones might work and which ones don't.

Regardless of the reason, it now seems likely that oxygen levels remained far too low for complex animals until somewhere near the time of the Cambrian explosion, which occurred when Earth was already nearly 4 billion years old. The oxygen-breathing animals that evolved at that time probably needed oxygen levels of at least 10% of the modern value. It's possible that the oxygen level was higher than that, but the first clear evidence of an oxygen level near its current value appears in the geological record only about 200 million years ago. That is when we first find charcoal in the geological record, implying that enough oxygen was present in the atmosphere for fires to burn. Thus, if you had a time machine and could randomly spin the dial to take you back to any point in Earth's history, you'd have less than about a 1 in 10 chance—and perhaps only about a 1 in 20 chance—of appearing at a time recent enough so that you could step out and breathe the air.

THINK ABOUT IT   What does the late appearance of substantial atmospheric oxygen tell you about the difference between our planet's being habitable *in general* and being habitable *for us?*

Implications for Life Elsewhere   The importance of oxygen to advanced life on Earth and the timing of its rise could potentially have important implications for life on other worlds. Our study of the origin of life gives us reason to think that life might be common on worlds with conditions like those of the early Earth. But the fact that it took so long for oxygen to build up in the atmosphere on Earth should make us wonder about the likelihood of getting oxygen-breathing life on other worlds. Could Earth have been "lucky" to ever get conditions that allowed the buildup of oxygen? If so, perhaps life on most other worlds would never evolve past microscopic forms; life might then be common, but advanced or intelligent life quite rare. Alternatively, maybe Earth was "unlucky" in having conditions that prevented the oxygen buildup for so long. In that case, other worlds might have complex plants and animals by the time they are just 1 to 2 billion years old, instead of having to wait until they are 4 billion years old. For the time being, we have no way to distinguish between these and other, intermediate possibilities. We will therefore continue our study with the assumption that Earth has been "typical," until and unless we learn otherwise.

6.4 IMPACTS AND EXTINCTIONS

Once animals colonized the land, the evolutionary path that led to humans becomes much clearer. Reptiles evolved from amphibians; early dinosaurs and mammals followed by about 245 million years ago. But the fossil record shows that the path was not smooth. In particular, the fossil record shows a number of striking transitions in the nature of living organisms. The most famous transition in the fossil record defines the boundary between the Cretaceous and Tertiary periods, which dates to about 65 million years ago. Dinosaur fossils exist below this boundary, but not above it. Somehow, after some 180 million years as Earth's dominant animals, the dinosaurs went extinct in a geological blink of the eye. Significant evidence now points to the idea that this extinction, and perhaps others, may have been caused by the impact of asteroids or comets crashing into Earth.

• Did an impact kill the dinosaurs?

There's no doubt that major impacts have occurred on Earth in the past. Meteor Crater in Arizona (Figure 6.14) formed about 50,000 years ago when a metallic asteroid roughly 50 meters across crashed to Earth with the explosive power of a 20-megaton nuclear bomb. Although the crater is only a bit more than 1 kilometer across, the blast and ejecta probably battered an area covering hundreds of square kilometers. Meteor Crater is relatively small and recent, and it is a popular tourist stop because it is so obvious. But it is not alone: Geologists have identified more than 150 impact craters on our planet, and many others have presumably been destroyed by erosion and other geological processes. Could larger impacts have altered the entire course of evolution?

The K–T Boundary Layer In 1978, while analyzing geological samples collected in Italy, a scientific team led by Luis and Walter Alvarez (father and son) made a startling discovery. They found that the thin layer of sediments that marks the Cretaceous–Tertiary boundary, called the **K–T boundary** for short (the *K* comes from the German word for "Cretaceous," *Kreide*), is unusually rich in iridium—an element that is rare on Earth's surface (because Earth's iridium sank to the core when our planet underwent differentiation [Section 4.4]) but more common in meteorites. Subsequent studies found the same iridium-rich sediment marking the K–T boundary at many other sites around the world (Figure 6.15). The Alvarez team proposed a stunning hypothesis: The extinction of the dinosaurs was caused by the impact of an asteroid or comet. They calculated that it would take an asteroid about 10–15 kilometers in diameter to produce as much iridium as is apparently distributed worldwide in the K–T boundary layer.

In fact, the death of the dinosaurs was only a small part of the biological devastation that seems to have occurred 65 million years ago. The geological record suggests that up to 99% of all living plants and animals died around that time and that up to 75% of all existing plant and animal *species* were driven to extinction. This makes the event a clear example of a **mass extinction**—the rapid extinction of a large percentage of all living species. Could it really have been caused by an impact?

Figure 6.14

Meteor Crater in Arizona was created about 50,000 years ago by the impact of an asteroid about 50 meters across. Because the asteroid hit Earth at high speed, it left a crater more than 1 kilometer across and almost 200 meters deep (compare to the size of the parking lot and buildings at the crater's left). The K–T impact was about 200 times larger.

A layer rich in iridium and soot tells us a huge impact occurred at this point in geological (and biological) history.

Figure 6.15

Around the world, sedimentary rock layers that mark the 65-million-year-old K–T boundary share evidence of the impact of a comet or asteroid. Fossils of dinosaurs and many other species appear only in rocks below the iridium-rich layer.

Figure 6.16

This computer-generated image, based on measurements of small local variations in the strength of gravity, reveals a buried impact crater (dashed circle) in the northwest corner of Mexico's Yucatán Peninsula. The crater measures about 200 kilometers across, and about half of it lies underwater off the coast. (The solid white lines show the coastline and borders of Mexican states.)

Figure 6.17

This painting shows an asteroid or comet moments before its impact on Earth, some 65 million years ago. The impact, known as the K–T impact, probably caused the extinction of the dinosaurs, and if it hadn't occurred, the dinosaurs might still rule Earth today.

More Evidence for the Impact   There's still some scientific controversy about whether the impact was the sole cause of the mass extinction or just one of many causes, but there's little doubt that a major impact coincided with the death of the dinosaurs. Key evidence comes from further analysis of the K–T sediment layer. Besides being unusually rich in iridium, this layer contains four other unusual features: (1) high abundances of several other metals, including osmium, gold, and platinum; (2) grains of "shocked quartz," quartz crystals with a distinctive structure that indicates they experienced the high-temperature and high-pressure conditions of an impact; (3) spherical rock "droplets" of a type known to form when drops of molten rock cool and solidify in the air; and (4) soot (at some sites) that appears to have been produced by widespread forest fires.

All these features point to an impact. The metal abundances look much like what we commonly find in meteorites rather than what we find elsewhere on Earth's surface. Shocked quartz is also found at other known impact sites, such as Meteor Crater in Arizona. The rock "droplets" presumably were made from molten rock splashed into the air by the force and heat of the impact. The soot probably came from vast forest fires ignited by impact debris. Some debris would have been blasted so high that it rose above the atmosphere, spreading worldwide before falling back to Earth. On their downward plunge, friction would have heated the debris particles until they became a hot, glowing rain of rock. Like falling hot coals, the particles ignited fires as they reached the ground.

In addition to the evidence within the sediments, scientists have identified a large, buried impact crater that appears to match the age of the sediment layer. The crater, about 200 kilometers across, is located on the coast of Mexico's Yucatán Peninsula, about half on land and half underwater (Figure 6.16). Its size indicates that it was created by the impact of an asteroid or a comet measuring about 10 kilometers across, large enough to account for the iridium and other metals. (It is named the *Chicxulub crater*, after a nearby fishing village.)

The Mass Extinction   If the impact was indeed the cause of the mass extinction, here's how it probably happened: On that fateful day some 65 million years ago, the asteroid or comet slammed into Mexico with the force of a hundred million hydrogen bombs (Figure 6.17). It apparently hit at an angle, sending a shower of red-hot debris across the continent of North America. A huge tsunami sloshed more than 1,000 kilometers inland. Much of North American life may have been wiped out almost immediately. Not long after, the hot debris raining around the rest of the world ignited fires that killed many other living organisms. Indeed, the entire sky may have been bright enough to roast most life on land.

Dust and smoke remained in the atmosphere for weeks or months, blocking sunlight and causing temperatures to fall as if Earth were experiencing a global and extremely harsh winter. The reduced sunlight would have stopped photosynthesis for up to a year, killing large numbers of species throughout the food chain. This period of cold may have been followed by a period of unusual warmth: Some evidence suggests that the impact site was rich in carbonate rocks, so the impact may have released large amounts of carbon dioxide into the atmosphere. The added carbon dioxide would have strengthened the greenhouse effect, so that the months of global winter immediately after the impact might have been followed by decades or longer of global summer.

The impact probably also caused chemical reactions in the atmosphere that produced large quantities of harmful compounds, such as nitrous oxides. These compounds dissolved in the oceans, where they probably were responsible for killing vast numbers of marine organisms. Acid rain may have been another by-product, killing vegetation and acidifying lakes around the world.

Perhaps the most astonishing fact is not that up to 75% of all plant and animal species died but that some 25% survived. Among the survivors were a few small mammals. These mammals may have survived in part because they lived in underground burrows and managed to store enough food to outlast the global winter that immediately followed the impact.

The evolutionary impact of the extinctions was profound. For 180 million years, dinosaurs had diversified into a great many species large and small, while most mammals (which had arisen at almost the same time as the dinosaurs) had remained small and rodentlike. With the dinosaurs gone, mammals became the new animal kings of the planet. Over the next 65 million years, the small mammals rapidly evolved into an assortment of much larger mammals—ultimately including us. So had it not been for the K–T impact, dinosaurs might still rule Earth.

Note that, while this general picture seems to make sense, the jury is still out on whether it tells the entire story. The geological record can be difficult to read when we are trying to understand events that happened over just a few years rather than a few million years. Some scientists suspect that the dinosaurs were already in decline and the impact was only the last straw. Others suggest that major volcanic eruptions also may have played a role. A few have deeper doubts about the role of an impact. Only additional data will settle the remaining questions.

• Did impacts cause other mass extinctions?

The K–T extinction seems quite clear, but it's generally difficult to measure past extinction rates precisely. The primary problem is that identifying the extinction of a species requires finding its *last* occurrence in the fossil record, which means we can be misled if we've yet to find a more recent occurrence of the species. Nevertheless, we have enough data to be sure that extinction rates vary considerably with time.

Figure 6.18 shows current data on the extinction rate for plants and animals over the past 500 million years. The data reveal at least five major

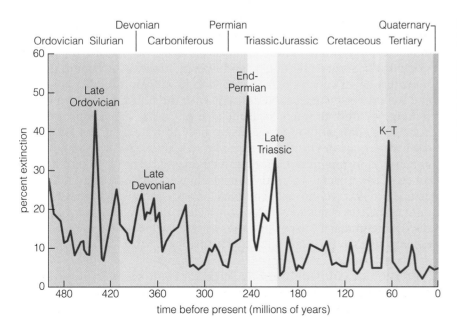

Figure 6.18

This graph shows data concerning the approximate percentage of plants and animals to go extinct with time over the past 500 million years. Peaks represent mass extinctions, with names shown for five major events. (The data actually are for families, a higher level of classification than genus and species.) (Adapted from Campbell, Reece, *Biology*, Seventh Edition, © 2005.)

mass extinctions, including the K–T extinction, and numerous smaller extinction events. Could some of these extinctions also have been caused by impacts?

Impacts and Other Hypotheses Simple probability makes impacts a reasonable hypothesis. On average, impacts the size of the K–T event should happen about every 100 million years or so. And, indeed, impacts appear to have coincided with at least some of the other mass extinctions. For example, the sedimentary strata marking the most severe mass extinction—the Permian extinction about 245 million years ago—show many of the same features as the strata marking the K–T boundary. In 2004, a research team (led by Luann Becker of UC Santa Barbara) claimed to find evidence off the northwest coast of Australia for a crater dating to about this time. Other geologists are still working to verify this claim. In any event, the K-T example suggests that we should consider impacts as possible causes for the Permian and other extinctions, even if we never identify craters left by the impacts. Remember that most impacts occur in the oceans (because oceans cover nearly $\frac{3}{4}$ of Earth's surface), and seafloor crust is almost completely recycled in about 200 million years [Section 4.4]; this recycling would destroy any evidence of a crater on the seafloor.

Besides impacts, scientists have come up with a number of other hypotheses about possible causes of mass extinctions. For example, some geologists hypothesize that episodes of unusually active volcanism may have led to climate change and high extinction rates. Other types of climate change may also have caused mass extinctions. For example, if there have been more recent periods of cooling similar to the known snowball Earth episodes (which ended more than 500 million years ago [Section 4.5]), the cold climate may have driven many species to extinction.

Another set of hypotheses envisions extinctions tied to changes in the mutation rate. While many mutations occur simply as the result of copying "errors" within cells, others are caused by external influences. For example, ultraviolet light can cause mutations, which is why sun exposure can lead to skin cancer. If the concentration of Earth's ozone layer varies with time, then the amount of ultraviolet light reaching the surface would also vary. Perhaps some of the extinctions occurred when the ozone layer thinned, allowing solar ultraviolet light to cause many more mutations.

Mutations can also be caused by high-energy particles that stream continuously from the Sun (making up the *solar wind*). Recall that Earth's magnetosphere deflects most of these particles, preventing them from reaching the surface [Section 4.4]. However, studies of magnetized rocks show that Earth's magnetic field varies significantly in strength with time, and sometimes reverses itself entirely, with the north magnetic pole becoming the south magnetic pole and vice versa. These magnetic reversals occur every few million years on average, and the magnetic field may disappear altogether for thousands of years while a reversal is in progress. The mutation rate might spike upward during this time due to the absence of the normal protection from high-energy particles. Although magnetic field reversals happen much more frequently than mass extinctions, some reversals may have occurred at times when life was more susceptible to major change and thus might have played a role in extinction events.

Supernovae and Gamma-Ray Bursts Some scientists hypothesize that more distant events could trigger mass extinctions on Earth, including *supernovae,* the explosions of massive stars [Section 3.2]. Supernovae are rare events. Out of the more than 100 billion stars in the Milky Way Galaxy, we expect only about one star per century to explode in a supernova. Most of these supernovae occur far from our solar system. Nevertheless, because the Sun orbits the center of the galaxy independently of other stars, different sets of stars make up our galactic neighborhood at different times. Simple probability calculations suggest that our planet must occasionally be located within a few tens of light-years of an exploding star. Supernovae can generate prodigious numbers of very-high-energy particles called *cosmic rays.* Thus, when a supernova occurs near Earth, we might expect a big upward spike in the number of cosmic rays reaching Earth and causing mutations.

A related idea suggests that mass extinctions could be caused by *gamma-ray bursts*—bursts of gamma rays [Section 3.4] from space that last just minutes or less—most of which are produced by unusually powerful supernovae. Atmospheric models suggest that a gamma-ray burst occurring within a few thousand light-years of Earth could generate enough gamma rays to destroy half of Earth's ozone layer, thereby leading to a massive die-off through exposure to solar ultraviolet light. Probability arguments suggest that Earth should have been exposed to such a nearby gamma-ray burst at least once in the past billion years, and a few scientists have attempted to link a gamma-ray burst to the Ordovician mass extinction some 450 million years ago.

Lessons from Mass Extinctions   Whether or not we ultimately learn the causes of past mass extinctions, they clearly have had tremendous effects on the evolution of life. With each mass extinction, many of the dominant species on the planet have disappeared, creating changes in environmental conditions and predator–prey relationships. These changes allow new species to evolve over the millions of years that follow. Just as the K–T event apparently paved the way for the rise of mammals, other extinctions may have caused similarly critical junctures in the evolutionary path that made our present existence possible.

The topic of mass extinctions also holds a cautionary lesson for our species today. Human activity is driving numerous species toward extinction. The best-known cases involve relatively large and wide-ranging animals, such as the passenger pigeon (extinct since the early 1900s) and the Siberian tiger (nearing extinction). But most of the estimated 10 million or more plant and animal species on our planet live in localized habitats, and most of these species have not even been cataloged. The destruction of just a few square kilometers of forest may mean the extinction of species that live only in that area. According to some estimates, human activity is driving species to extinction so rapidly that half of today's species could be gone within a few centuries or less. On the scale of geological time, the disappearance of half the world's species in just a few hundred years would certainly qualify as another of Earth's mass extinctions, potentially changing the global environment in ways that we are unable to predict.

THINK ABOUT IT   The geological record suggests that the dominant animal species are nearly always victims in a mass extinction. If we are causing a mass extinction, do you think we will be victims of it? Or will we be able to adapt to the changes so that we survive even while many other species go extinct? Defend your opinion.

Figure 6.19

Damage from the 1908 impact over Tunguska, Siberia, shown in a photo taken many years later.

• Is there a continuing impact threat?

The discovery that at least one mass extinction is tied to an impact has spurred scientific concern over whether our civilization might be vulnerable to future impacts. How serious is this threat?

Small particles hit Earth almost continuously, burning up in our atmosphere as **meteors** (sometimes called "shooting stars," a misleading name that arose long before we knew their true source). Most meteors are caused by particles no bigger than a pea. The particle itself is too small for us to see, but it enters the atmosphere at such high speed (typically between about 45,000 and 250,000 km/hr) that it burns up and heats the surrounding air, producing the meteor flash. An estimated 25 million particles enter our atmosphere and burn up as meteors each day. Interestingly, these particles add a total of about 20,000 to 40,000 tons to Earth's mass each year. This sounds like a lot in human terms, but it is negligible compared to Earth's total mass of 6 *billion trillion* (6×10^{21}) tons.

Somewhat larger objects entering our atmosphere may be heated to the point where they explode, producing an extraordinarily bright flash called a *fireball*. (Some so-called UFO sightings are actually fireballs.) If debris from the explosion hits the ground, we may find some of it as the rocks we call *meteorites*. An impacting object more than a few meters across may survive the plunge through our atmosphere intact, carrying so much energy that it excavates a visible crater, such as Meteor Crater (see Figure 6.14).

No large impacts have left craters since the dawn of human civilization, but we know of at least one smaller impact that devastated a fairly large area. In 1908, a tremendous explosion occurred over Tunguska, Siberia, flattening and setting fire to the surrounding forest (Figure 6.19). Seismic disturbances were recorded up to 1,000 kilometers away, and atmospheric pressure fluctuations were detected at distances of almost 4,000 kilometers. The explosion, now estimated to have released energy

MOVIE MADNESS: ARMAGEDDON

In 1994, a lot of people who believed the dinosaurs were the victims of a unique accident changed their minds. That was the year Comet Shoemaker–Levy 9 smacked into Jupiter, leaving entrance wounds the size of planet Earth.

Folks realized that death by rock isn't all that improbable. If it happened to the dinos, it could happen to us. An errant asteroid a dozen miles across might someday careen into our planet and raise enough dust and burn enough forests to darken the world for years. We'd all slowly starve.

Alerted to this possibility for havoc and destruction, Hollywood lost little time in showing how ingenuity and some gutsy guys (with the emphasis on the latter) could save us even if Nature hurls a large space rock our way. TV specials and theatrical films soon appeared, showing Earth under mortal threat from ballistic boulders.

In the film *Armageddon*, the incoming object is as big as Texas, and a mere few weeks away. That's a real slap in the face for astronomers. Picture this: An asteroid as big as Ceres (the largest rock in the asteroid belt) is headed our way, and the astronomers find the darn thing only when it's as close as Mars?

The end of the world is nigh, but not to worry. NASA is in high gear to divert this king-size clod and decides the best thing to do is to blow

it into two large pieces with a nuclear bomb. The two pieces will diverge slightly and sail harmlessly by opposite sides of Earth. Needless to say, geeky NASA personnel aren't up to this kind of macho mission, so the space agency recruits a bunch of oil-rig roughnecks to plant and detonate the bomb. The NASA folk refer to this group of gritty misfits as "the wrong stuff."

In fact, it's a bad idea to try to blow up an incoming asteroid. The chances are that you'd only turn a single shell into buckshot. More practical schemes envision fastening some sort of rocket engine to the side of the rock and slowly nudging it out of the way. Another approach, at least for asteroids that we find when still far from Earth, is to paint the asteroid white and let the gentle pressure of sunlight do the job. Neither scheme involves roughnecks (or human pilots at all).

The threat, of course, is real. Astronomers estimate that rocks comparable to the one that did in the dinosaurs (together with countless other species) will slam into Earth roughly every 50 million years. But by carefully keeping tabs on those asteroids that cross Earth's orbit, we can see disaster coming. We'll have years to mount a defense.

The bottom line is that this is one kind of disaster we can probably avoid. After all, unlike the dinos, we've got a space program.

equivalent to that of several atomic bombs, is thought to have been caused by a small asteroid no more than about 40 meters across. Atmospheric friction caused it to explode completely before it hit the ground, so it left no impact crater. We have also witnessed one far larger impact on another world: In 1994, astronomers recorded the impact of Comet Shoemaker–Levy 9 (SL9) as it slammed into Jupiter; the comet had broken into pieces before the impact, and each piece crashed into Jupiter with energy equivalent to that of a million hydrogen bombs (Figure 6.20).

Objects of similar size to that of the Tunguska event probably strike our planet every century or so. They have gone unnoticed, presumably because they have always hit remote areas, with most striking the ocean. Nevertheless, the death toll would be enormous if such an object struck a densely populated area. Larger impacts are correspondingly rarer. Based on statistics of past impacts, Figure 6.21 shows how often, on average, we expect Earth to be hit by objects of different sizes.

Note that while the chance that our civilization is in any imminent danger of a major impact seems relatively low, it is not negligible. Moreover, the question is not *whether* a future impact will occur, but *when*. Many known asteroids have orbits around the Sun that cross Earth's orbit or pass near enough to Earth's orbit that they may someday be gravitationally perturbed into an Earth-crossing orbit. These objects have a real chance of striking Earth at some future time. Some of these objects are in the 1-kilometer size range, and a few are as big as the 10-kilometer K–T impactor. NASA has set a goal of detecting 90% of all large, near-Earth asteroids by 2008. Unfortunately, we do not yet have a similar way of studying the threat from comets, because they reside so far from the Sun. By the time we saw a comet plunging toward us from the outer solar system, we'd have at best a few years to prepare for the impact.

If we were to find an asteroid or a comet on a collision course with Earth, could we do anything about it? Many people have proposed schemes to save Earth by using nuclear weapons or other means to demolish or divert an incoming asteroid, but no one knows whether current technology is really up to the task. We can only hope that the threat doesn't become a reality before we're ready.

THINK ABOUT IT Study Figure 6.21. Based on the frequency of impacts large enough to cause serious damage, do you think we should be spending time and resources to counter the impact threat? If so, how much should we invest relative to what we invest to combat other threats? Defend your opinion.

6.5   HUMAN EVOLUTION

We've traced the course of evolution from the origin of life through the extinction of the dinosaurs. We've seen that evolution took many surprising twists and turns to that point. The subsequent evolution of mammals and humans was just as interesting. In this section, we'll briefly investigate the pathway that led to our emergence as the first species on Earth capable of learning about its own origins.

• How did we evolve?

We are primates, as are all the great apes, monkeys, and prosimians (such as lemurs). The ancestor of all of today's primates lived in the trees, and many of the traits that make us so successful evolved as adaptations to

Figure 6.20

This infrared photo shows the brilliant glow of a rising fireball created when one of the pieces of Comet Shoemaker–Levy 9 (SL9) crashed into Jupiter (the round disk in the background) in 1994. Although overexposure exaggerates the size of the fireball, you can get a sense of scale by remembering that about ten Earths could fit side by side across Jupiter's diameter.

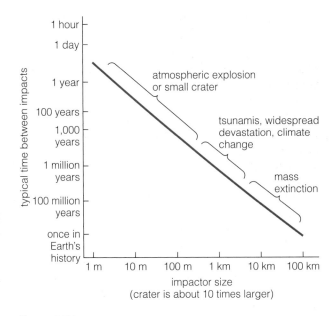

Figure 6.21

This graph shows that larger objects (asteroids or comets) hit Earth less frequently than smaller ones. The labels describe the effects of impacts of different sizes.

Figure 6.22

The evolutionary history of the major primate branches. Notice, for example, that the common ancestor of modern humans, chimpanzees, and gorillas lived between about 6 and 8 million years ago. (Adapted from Campbell, Reece, Simon, *Essential Biology,* Second Edition, © 2004.)

Figure 6.23

This famous type of illustration suggests that humans evolved along a simple pathway from apes. However, it is almost completely wrong.

tree life. For example, the limber arms that allow us to throw balls and work with tools evolved so that our ancestors could swing through trees, and our dexterous hands evolved to hang from branches and manipulate food. The eyes of primates are close together on the front of the face, providing overlapping fields of view that enhance depth perception—an obvious advantage when swinging from branch to branch. For the same reason, primates developed excellent eye–hand coordination.

Parental care is essential for young animals in trees, and primates evolved close parent–child bonds. These bonds, in turn, made it possible for primates to be born in a much more helpless state than the babies of most other types of animals. Although many primate species, including us, eventually moved down from the trees, most primates continue to nurture their young for a long time. This trait reaches its extreme in humans. Human babies are nearly helpless at birth and require parental care for more years than the offspring of any other species.

Contrary to a common myth, humans did not evolve *from* gorillas or other modern apes. Rather, modern apes and humans share a common ancestor that is now extinct. Figure 6.22 shows the evolutionary history of major primate branches. Our closest living relatives, chimpanzees and gorillas, shared a common ancestor with us just a few million years ago.

The Emergence of Humankind Even after hominids (human ancestors) diverged from the ancestors of chimpanzees and gorillas, human evolution followed a remarkably complex path. Indeed, one of the most pervasive but incorrect myths about human evolution is that it followed a simple pathway from stooped apes to upright humans (Figure 6.23).

The reality is that there have been numerous hominid species, some of which may be part of the lineage of modern humans and others that may have come to evolutionary dead ends. The exact timing of the first hominids is not known, but a fossil skull recently excavated in Chad and likely to be hominid dates to between about 6 and 7 million years ago. The skull of *Sahelanthropus tchadensis,* nicknamed Toumaï (Figure 6.24), shows features intermediate between those of apes and humans, along with features suggesting that Toumaï walked upright. Many researchers now suspect that Toumaï was the first true hominid, but more confirming fossil evidence will be needed to seal the case.

The earliest fossil skulls that appear to be identical to those of modern humans are about 100,000 years old. However, even then our ancestors shared the planet with at least one other hominid species, usually called *Neandertals,* that was quite similar and may have had a slightly larger brain on average than *Homo sapiens.* The Neandertals disappeared about 30,000 years ago, and no one yet knows why. In 2004, scientists unearthed evidence of yet another recent hominid species, *Homo floresiensis,* that lived on an Indonesian island as recently as about 12,000 years ago. These people apparently stood no more than about a meter tall, and some scientists question whether they were truly a separate species or whether the fossils represent aberrant *Homo sapiens.* Figure 6.25 summarizes hominid lineages from Toumaï to modern *Homo sapiens.* (*H. floresiensis* is not shown, since their status is still not well accepted.)

Deciphering the rest of the details of human ancestry is a rich field of research, and much remains subject to scientific debate. We will not discuss such details in this book, but before we leave the topic, it's worth dispelling two additional common myths. First, there is no longer a "missing link" in human evolution. While a few mysteries may always remain, we now know enough from the geological record and genome comparisons to see a clear path from the earliest microbes to ourselves. Second,

Figure 6.24

This computer reconstruction, based on the actual fossils, shows the skull of *S. tchadensis,* or Toumaï, first excavated in Chad in 2001.

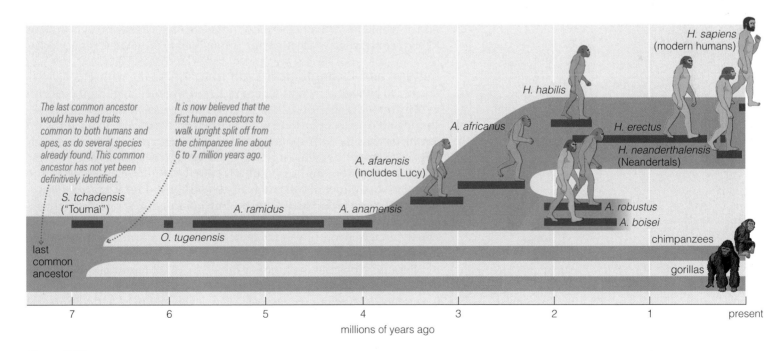

Figure 6.25

Human evolution over about the past 7 million years since our lineage separated from that of modern gorillas and chimpanzees.

despite the many species of hominids that have come and gone, all modern humans are members of the same species. That is, while people often focus on outward differences between races such as skin color or hair texture, all human genomes are virtually identical. Moreover, any small racial differences that might once have arisen have since been spread across races by the widespread interbreeding of our ancestors. Any remaining genetic differences between human races are generally much smaller than the genetic variation among the individuals in each race.

Implications for Extraterrestrial Intelligence   The fact that modern gorillas, chimps, and humans all evolved from the same ancestor has at least two important implications for understanding the possibility of extraterrestrial intelligence. First, it shows that relatively small genetic differences can make a big difference in species success. More than 98% of the DNA sequences that make up the human genome are identical to the sequences that make up the chimpanzee genome. Thus, a relatively small genetic difference is all that separates our success on this planet from the current predicament of chimpanzees, which survive naturally in only a few isolated locations in Africa. Second, it suggests that the evolution of intelligence is a complex process. Gorillas and chimpanzees have been evolving from our common ancestor just as long as we have, but we are the only species building cities and radio telescopes. This fact raises the question of whether advanced intelligence is an inevitable outcome of evolution. We'll discuss this question further in Chapter 12.

• Are we still evolving?

Given the very recent arrival of *Homo sapiens* on the evolutionary time scale, it's natural to wonder whether we are still evolving. There is some scientific debate on this issue, but overall it seems unlikely that we have undergone any significant evolutionary changes in the past 40,000 years or so. That is, if we could sequence the genome of a human from that period, it would be virtually identical to the genome of anyone living today. Nevertheless, we have clearly gone through dramatic changes as a species.

Cultural and Technological Evolution   These dramatic changes are not due to biological evolution, but rather to what we might call **cultural evolution**—changes that arise from the transmission of knowledge accumulated over generations. In other words, we humans can transmit our history, using both spoken and written language, which thereby allows us to learn from what has been done before. The know-how to build tractors, computers, and spaceships is stored not in our genes but in the cumulative product of hundreds of generations of human experience. Although some other species, including chimpanzees, demonstrate aspects of culture, humans are probably unique in having reached the point where cultural evolution is far more important to our changing nature than is biological evolution.

Because biological evolution is driven by random mutations, it tends to proceed at a slow and relatively steady rate. Cultural evolution, in contrast, tends to accelerate over time. The development of agriculture and written language took tens of thousands of years, while less than two centuries separate the beginning of the industrial revolution from the first walk on the Moon. More recently, we have begun to develop another new type of evolution that is accelerating even more rapidly—**technological evolution**.

You are probably familiar with the way new computers get faster each year. These speed increases come from an interesting coupling of technology and science: Increased computing power enables scientists to make new discoveries, which in turn lead to further increases in computing power, and so on. Similar couplings of technology and science affect almost every field of human knowledge. One of the most striking cases involves the way technology is helping us understand biological evolution. As we've discussed, much of our current knowledge about the origin and evolution of life comes from studies of genomes that are made possible by technological advances such as machines that can read DNA sequences.

Evolution Among Humans and Other Intelligent Civilizations

The same type of technology that allows us to read genomes also now allows us to reengineer living organisms (genetic engineering) in such a way that we may soon outpace nature in developing new species. It seems inevitable that we will also develop the capability to change human DNA, thus opening the door to attempts to "improve" our species. The moral and ethical dimensions of this power are already profound, and you will undoubtedly have to confront these issues many times during your life. For our purposes in this book, perhaps the primary lesson—whether good or bad—is that advanced civilizations can alter the course of evolution through their choosing, rather than remaining subject to the random processes of natural selection. Indeed, modern medicine has already taken us out of the realm of Darwinian evolution, because we routinely save individuals who would have died earlier in generations past. The course of our future evolution is in our own hands. If other advanced civilizations exist, they must similarly control their own destinies.

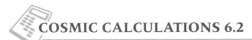

COSMIC CALCULATIONS 6.2

Bacteria in a Bottle II: Lessons for the Human Race

Recall the thought experiment of the bacteria in a bottle in Cosmic Calculations 6.1, in which we start with a single bacterium at 12:00 that replicates each minute. Suppose that bacterial growth completely fills the bottle at 1:00, exhausting all nutrients so all the bacteria die. Let's explore this issue with a series of simple questions.

Question 1: The tragedy occurs when the bottle is full at 1:00, just 60 minutes after the first bacterium started the colony at 12:00. When was the bottle *half*-full?

Answer: Most people guess 12:30, halfway through the hour of growth. But this is incorrect: The bacterial population doubled every minute, so the bottle went from half-full at 12:59 to full at 1:00.

Question 2: You are a mathematically sophisticated bacterium, and at 12:56 you recognize the impending disaster. You warn your fellow bacteria that the end is just four minutes away unless they slow their growth dramatically. Will anyone believe you?

Answer: Because the bottle was full at 1:00, it was $\frac{1}{2}$-full at 12:59, $\frac{1}{4}$-full at 12:58, $\frac{1}{8}$-full at 12:57, and $\frac{1}{16}$-full at 12:56—which means there's still 15 times as much unused bottle as used when you give your warning. Your warning may go unheeded.

Question 3: Just before the disaster strikes, a bacterial space program discovers 3 more bottles in the lab. With an immediate and massive population redistribution program among the original and 3 new bottles, how much more time will the bacteria buy?

Answer: You may be tempted to think that 3 more bottles should give the colony 3 more hours, but in fact it gives them only 2 more minutes: Because the growth occurs through doubling, the colony will fill 2 bottles at 1:01 and 4 bottles at 1:02. In fact, *nothing* could allow the growth to continue much longer, because the doublings would soon lead the bacteria to impossible volumes (see Problem 51 at the end of the chapter).

We can draw several general conclusions. First, because populations of living organisms tend to grow exponentially, numbers can rise very rapidly. This explains the inevitable population pressure that helped Darwin realize the role of natural selection (see Fact 1 on p. 155). Second, exponential growth must always be a short-term, temporary phenomenon; for living organisms, the growth typically stops due to predation or a lack of sufficient nutrients or energy. Third, these laws about growth apply to all species—our intelligence cannot make us immune to simple mathematical laws. This is a critical lesson, because human population has been growing exponentially for the past few centuries (see Figure 12.14). Of course, our intelligence gives us one option not available to bacteria. Exponential growth can stop only through some combination of an increase in the death rate or a decrease in the birth rate. Unlike bacteria, we can *choose* to stop our exponential growth with changes to our birth rate before we "fill" our planet.

6.6 THE PROCESS OF SCIENCE IN ACTION: ARTIFICIAL LIFE

In this chapter, we have discussed some of the laboratory experiments through which scientists seek to understand the origin of life. Some researchers are attempting to go even further, by trying to actually create life in the lab. Their efforts are rather different from those of the fictional Dr. Frankenstein, who sewed together dead body parts and jolted a living, humanlike creature into existence with high voltage. Today's researchers are trying to put together novel organisms that can reproduce and grow, but on a microbial scale.

Success could have a variety of implications. For example, it would help us understand how life likely got started on Earth, as well as help us decide how best to look for life in extraterrestrial environments such as subsurface aquifers on Mars or in a subterranean ocean of Europa. We might be able to evolve organisms that could provide us with a variety of benefits, medical and otherwise. Of course, it might also be possible to create dangerous organisms, forcing us to confront the ethical dilemmas of our work. For this chapter's case study in the process of science in action, we turn our attention to current work in artificial life.

• How might we create artificial life?

It's still far beyond our means to create from elementary chemicals a bacterium that's comparable to those found in nature. Nonetheless, hundreds of researchers are working to spawn A-life ("artificial life") by either rearranging bits and pieces of existing organisms or trying to build an extremely simple living cell in the lab. The former scheme is referred to as a "top-down" approach, and is similar to hot-rodding, where a car is stripped down and rebuilt with different components, producing a vehicle that is completely unlike any existing auto.

Engineering New Species from Existing Organisms   Craig Venter, the man behind the private-sector effort that first sequenced the human genome, has embarked on a top-down program to make designer organisms. His goal, shared by many others, is to create microbes that do useful things, such as fighting malaria or converting atmospheric carbon dioxide into methane, a trick that could be valuable for reducing our dependence on fossil fuels. As his basic building block, he starts with *Mycoplasma genitalium,* a familiar bacterium often responsible for urinary tract infections. *M. genitalium* has a tiny genome even for a microbe (about 580,000 base pairs, compared with about three billion base pairs for a human). It has only 486 proteins, and research has suggested that it could live with only about half that number. So Venter's team is trying to remove as much of the organism's genes as they can, while still leaving it viable and able to reproduce. Once they have stripped down the *Mycoplasma's* genome to the fewest number required to survive, they will insert this minimalist DNA into the hollow shell of another *M. genitalium* cell from which the original DNA has been removed. By this method, the team hopes to produce a bare-bones microbe—the simplest thing (other than viruses) that is still alive. They can then customize this elemental organism for specialized tasks by fitting it out with some additional genes. In 2002, A-life researchers managed to create a polio virus by lashing together a short, synthetic strand of DNA. Venter's bare-bones *M. genitalium,* if he can make it work, will have a strand many dozens of times longer.

Note that this is not really creating life from scratch, but merely stream-lining and modifying an existing microbe. It's a more efficient method for producing desirable organisms than simply breeding them and selecting those that have the desired properties. That scheme depends on muta-tion to produce a range of characteristics, and you simply cull those that are, by chance, closer to what you want. But Venter's approach deliber-ately introduces genes that will result in the desired behavior. Top-down engineering promises to give us entirely new species of great practical value, and it will also undoubtedly teach us much about how cells work. However, it's less likely to make clear how life got started on Earth, and thereby help us understand whether terrestrial biology is a fortunate ac-cident, or a virtually inevitable development. The bottom-up approach to A-life more directly addresses that question.

Making Life from Raw Ingredients   As we discussed in Section 6.2, it seems likely that the young Earth had plenty of organic building blocks for life, either produced by chemical reactions on Earth or brought here from space. The many possible ways in which the building blocks could have formed or arrived has led some people to believe that the creation of life in the laboratory should be just around the corner.

Few researchers think it's that easy. But several have dared to take a bottom-up approach to making A-life in the lab. One of these is Jack Szostak, a Harvard University geneticist who was working on the genes of yeast. He decided to switch gears when he read of the work of Thomas Cech, who discovered that RNA could serve as both a blueprint for re-production and a catalyst for making proteins. Cech suggested that RNA may have been the basis of life before DNA arrived on the scene, an idea we discussed earlier as "RNA world." This vision of an RNA world encour-aged Szostak to leave his yeast cells behind, and try building synthetic or-ganisms out of strands of this older nucleotide.

There were already research hints as to how RNA life might have arisen, since prior lab experiments had shown that short strands of RNA can form when a dilute organic solution washes over the surface of clay or rock. If, by chance, one of those RNA strands could reproduce, it would have staying power. Those that could do this most quickly, and with the fewest number of copying errors, would soon dominate their environ-ment. In this way, a robust form of RNA life could have evolved. But did it really happen that way?

Szostak seeks to gain insight into this question by attempting to pro-duce simple RNA-based cells that can replicate. If he succeeds, he will have created a new life-form after beginning with only nonliving materi-als—a rather different accomplishment than the top-down A-life efforts described above. He begins with a sample of RNA fragments: thousands of trillions of short nucleotide (base) sequences. These are taken from some simple RNA strands that are known to have special talents. They can, for example, grab onto a particular type of molecule, or duplicate parts of it. He then puts his fragment collection into a test tube, and uses a tech-nique called "in vitro selection" to fast-forward Darwinian evolution. He gives his collection the opportunity to chemically react, and then screens the samples to find, for example, those that have managed to replicate a short sequence of RNA. Keep in mind that these are merely organic mol-ecules: They're not life. Szostak then filters out the winners in this test, and makes trillions of copies of those (some copy errors, or mutations, are allowed in this process). He runs the experiment again with the im-perfectly cloned winners. Those that are even better, or at least faster, are retained for the next round—and so forth through dozens of cycles. In essence, Szostak is trying to emulate what might have happened on the early Earth by compressing time with laboratory evolution. It is somewhat analogous to producing large kernels of corn by repeatedly using seeds from only the largest ears of the crop.

So far, Szostak's lab has made strands that can replicate other RNA sequences, although they are short and the replication isn't always very accurate. But the big prize is an RNA strand that can really sift through the material of an organic solution and build a copy of itself. If Szostak succeeds, it will be possible to watch this self-replicating molecule evolve in the laboratory, because the sample will soon be dominated by the type of RNA that makes copies of itself, rather than of other RNA.

We've noted that freely floating organic molecules aren't cells. The molecules Szostak hopes to build need to be confined by a wall to keep aggressive compounds in the cruel outside world from dismantling them, while at the same time allowing the building blocks necessary for reproduction to enter their mini-habitat. Szostak therefore also seeks natural processes that could make pre-cells to enclose the RNA. He has already had some success: The clay-catalyzed lipid membranes that we discussed earlier, with RNA inside them (see Figure 6.7), were produced in his lab. When these blobs pass through porous rock-like materials, they split into smaller bubbles, with much of their RNA content still present. This is an obvious candidate for a mechanism that might have produced the first living cells, and it's remarkable that clay can both help assemble complex RNA and facilitate the development of cell membranes.

Overall, while Szostak doesn't yet have RNA that will make a complete protein or reproduce itself, he reckons that the creation of a replicating strand is well within reach. If so, we may soon know whether the origin of life could have occurred as easily as we have imagined under conditions that existed on the young Earth.

• Should we create artificial life?

There's reason to think that the top-down approach to A-life is likely to work sooner, as it starts from living organisms such as *Mycoplasma genitalium*, and modifies it with genes from other life. This is at least somewhat analogous to the Frankenstein scenario, of building something new out of existing parts, albeit at the level of single-cell life. The payoff in producing nature's own nanobots for specific tasks is obvious. We've mentioned the possibility of generating hydrocarbons for fuel, but A-life could also be used to target cancer cells or clean up toxic waste. Still, there's the danger that someone might eventually use this technology to engineer deadly organisms that have no natural enemies, a potent agent for biowarfare. Although Szostak's work is less intent on creating useful life and more focused on learning how biology began, it too poses a potential risk for misuse.

These dangers are recognized by the researchers themselves. While they're not overly worried about the threat A-life might pose (they point out that it would be extremely fragile and would have a difficult time living outside the laboratory environment), the scientists have occasionally convened panels to consider the ethical implications of their work. Do humans have the moral right to create new types of life? We don't seem to mind the development of hybrid corn or improved cattle, but the production of an entirely new species is likely to be more controversial.

There's an alternative to the A-life scenario that avoids most of these ethical dilemmas, and that is to build virtual life by modeling the functions of living cells with computer software. This might not seem particularly interesting, but in fact it would allow us to do biological experiments at the keyboard, without the difficulty (and potential danger) of using real microbes as lab rats. Indeed, it's somewhat analogous to testing aircraft with computer simulations of the air flow over their wings and fuselage, as opposed to building scale models and putting them in wind tunnels.

Software life could be useful for determining the effects of new drugs, or even for predicting the consequences of removing or adding genes (genetic engineering). *Escherichia coli* (*E. coli*) is a bacterium that resides in your gut, and is one of the best studied of all organisms. But it has more than 4,000 genes, of which approximately 1,000 have functions that are still not understood. Consequently, creating a computer model of a real *E. coli* is still beyond us. However, borrowing an idea from the top-down researchers, computer scientists are now programming a stripped-down model of the bacterium, with only about a thousand genes. This effort might bear fruit in the near future, and even though the model is only an approximation to the actual microbe, it might still be useful in understanding how life works. Building a complete computer model of *E. coli*, while not on the immediate horizon, would surely encourage scientists to modify it to more closely resemble human cells. Suppose that could be done. Then, even aside from the insight it would give into how biology functions, it would permit us to, for example, quickly evaluate drugs for fighting cancer without the necessity of testing them on laboratory animals or people.

Of course, even if the life is only programmed on a computer, once we know the necessary DNA sequences, it would be possible to put it together for real. Thus, one way or other, we are likely to be forced to confront the ethical dilemmas of creating artificial life. Indeed, with all the research that is already under way, it may be too late to put the genie back in the bottle, even if we wanted to. The future of biological science—and perhaps of our species as a result—will largely depend on the ethical choices that we make as modern biotechnology continues its technological evolution.

◯ THE BIG PICTURE
Putting Chapter 6 into Context

In this chapter, we have completed our overview of life on Earth. We have built on our understanding of Earth's habitability (Chapter 4) and of the nature of life (Chapter 5) to develop a modern picture of the origin and evolution of life. As you continue in your studies, keep in mind the following "big picture" ideas:

- We may never know precisely how life arose on Earth. However, we have found *plausible* scenarios for the origin of life based on natural, chemical processes. These scenarios are based on solid evidence found in the geological record, in comparisons of genomes between species, and in laboratory experiments.

- Earth has supported life for most or all of the past 4 billion or more years, but life remained relatively simple for most of this period. Animal life exploded in diversity only about 545 million years ago (with the Cambrian explosion), and plants colonized the land only about

475 million years ago. If aliens had observed Earth during about the first 90% of its history, they would have found a planet that was home to nothing more than microscopic species.

- The course of evolution has been drastically changed at least several times by mass extinctions, and at least one of these mass extinctions is clearly linked to an asteroid or comet impact. Thus, as we'll discuss in more detail later in the book, the likelihood of impacts must be an important consideration in considering the habitability of other planets.

- From the time of the first living organism to today, the evolution of life on our planet has been shaped by natural selection. However, we have developed or are on the verge of developing the capability to engineer existing species, including ourselves, and perhaps to create entirely new species. This power must be available to any advanced civilization and therefore is an important consideration in the search for extraterrestrial intelligence.

SUMMARY OF KEY CONCEPTS

6.1 SEARCHING FOR LIFE'S ORIGINS

• When did life begin?

Stromatolite and sulfur isotope evidence tells us that life existed by about 3.5 billion years ago; microfossil evidence is consistent with this view. Carbon and other isotopic evidence may push the time back to 3.85 billion years ago. Life must have existed even earlier than the oldest fossil evidence of it, though we do not know exactly how much earlier.

• What did early life look like?

Although early life is long gone, genetic comparisons allow us to determine which modern organisms are evolutionarily oldest, suggesting that they are most similar to early life. These studies suggest that early life resembled modern extremophiles living in hot water near deep-sea vents or in hot springs.

• Where did life begin?

Based on the genetic studies and availability of chemical energy, deep-sea vents or hot springs are good possibilities. Because deep-sea vents are protected from impacts that might vaporize surface hot springs, they seem the most likely place for the origin of life.

6.2 THE ORIGIN OF LIFE

• How did life begin?

According to laboratory studies, the most likely scenario holds that organic molecules, either produced chemically or brought here from space, were found in ocean locations where clay and other minerals were common. Clay helped catalyze the building of RNA strands that became enclosed in lipid pre-cells. Some RNA molecules were able to partially or completely self-replicate, allowing natural selection among them to improve their replication until true life emerged.

• Could life have migrated to Earth?

If life originated first on Venus or Mars, it may have migrated to Earth when impacts blasted rocks from one world to another. Meteorites from Mars show that life could in principle survive the journey in some cases, and then take hold near deep-sea vents or elsewhere. Longer migrations, such as from planets around other stars, are highly unlikely.

6.3 THE EVOLUTION OF LIFE

• What major events have marked evolutionary history?
Life probably diversified rapidly after its origin, but remained microscopic for more than 2 billion years. Keys to the eventual transition included the origin of oxygen-producing photosynthesis, which released the oxygen now in our atmosphere, and the evolution of cell nuclei and other complex structures in eukaryotes. Multicellular animals diversified in the **Cambrian explosion**, starting about 545 million years ago. Plants and animals migrated onto land not long after.

• Why was the rise of oxygen so important to evolution?
Aerobic processes, using oxygen, offer the possibility of much more efficient cellular energy production than anaerobic processes, and thus can lead to much greater evolutionary diversification. The precise timing of the rise of oxygen is not well known, but began before about 2.35 billion years ago and probably did not reach levels near those of the present until the time of the Cambrian explosion or later.

6.4 IMPACTS AND EXTINCTIONS

• Did an impact kill the dinosaurs?

It may not have been the sole cause, but a major impact clearly coincided with the **mass extinction** that killed the dinosaurs about 65 million years ago. Sediments from the time contain iridium and other evidence of an impact, and a crater of the right age lies buried beneath the Yucatán coast of Mexico.

• Did impacts cause other mass extinctions?
At least five mass extinctions have occurred in the past 500 million years. Although only the K-T extinction is clearly linked to an impact, some evidence suggests that impacts have played a role in other extinctions. Other possible causes include periods of active volcanism, severe climate change, and changes to mutation rates, possibly influenced by changes in Earth's magnetic field or by distant supernovae or gamma-ray bursts.

• Is there a continuing impact threat?
Impacts certainly pose a threat, though the probability of a major impact in our lifetimes is small. Impacts like the Tunguska event occur more frequently (about once a century), and would be catastrophic if they occurred over cities.

6.5 HUMAN EVOLUTION

• How did we evolve?

Humans share a common ancestor with modern gorillas and chimpanzees, an ancestor that lived between about 6 to 8 million years ago. The earliest *Homo sapiens* emerged about 100,000 years ago.

• Are we still evolving?
Genetically, humans have probably changed little in at least the past 40,000 years. However, we are now changing in new ways, through **cultural evolution** and **technological evolution**.

6.6 THE PROCESS OF SCIENCE IN ACTION: ARTIFICIAL LIFE

• How might we create artificial life?
Scientists are seeking to make artificial life in two basic ways. A "top-down" approach starts with existing organisms and genetically strips them down, then tries to rebuild them into new species. A "bottom-up" approach starts in the laboratory with the raw ingredients of life, and seeks to reproduce life in much the same way that it presumably originated billions of years ago.

• Should we create artificial life?
Researchers trying to make A-life in the lab are confident they will succeed within a few years. There has already been discussion about whether this is ethically desirable, even though there could be important practical benefits. In addition to these moral concerns, some people worry that A-life might be the source of new diseases or toxins for which terrestrial life would have little resistance.

EXERCISES AND PROBLEMS

REVIEW QUESTIONS
Short-Answer Questions Based on the Reading

1. What are the three lines of fossil evidence that point to an early origin of life on Earth? Discuss each line and what it tells us about when life arose. What are the implications of an early origin to the possibility of life elsewhere?

2. How do studies of DNA sequences allow us to reconstruct the evolutionary history of life? What living organisms appear to be most closely related to the common ancestor of all present life?

3. Based on current evidence, what locations on Earth seem likely for the origin of life? What locations can we rule out?

4. What was the *Miller–Urey experiment,* and how did it work? Why is its relevance now subject to scientific debate? How else might Earth have obtained the organic building blocks of life?

5. What do we mean by an "RNA world," and why do scientists suggest that such a world preceded the current "DNA world"?

6. Briefly summarize current ideas about the sequence of events through which life may have originated on Earth. What role(s) might clay or other inorganic materials have played?

7. Briefly discuss the possibility that life migrated to Earth. Also discuss the possibility that Earth life might have migrated to other worlds, and the implications of possible migration to the search for life elsewhere.

8. Why do we think that evolution would have proceeded rapidly at first, and what fossil evidence supports this conclusion?

9. Briefly discuss the early evolution of life, from the first organisms to the development of photosynthesis and oxygen production.

10. How do we think that eukaryotes evolved? What time constraints can we place on when eukaryotes first got cell nuclei?

11. What was the *Cambrian explosion?* Briefly discuss ideas about what might have caused it and why no similar event has happened since.

12. How and when did life colonize land? Why did it take so long after the origin of life in the oceans?

13. How do we know that the early Earth could not have had an oxygen atmosphere? Where did the oxygen in our atmosphere come from? How did the introduction of oxygen affect early life?

14. Summarize the history of the oxygen buildup as it is understood today, and describe key mysteries that still remain. When did oxygen reach current levels?

15. What was the *K–T impact,* and how is it thought to have led to the demise of the dinosaurs? What evidence supports this scenario? How did this event pave the way for our existence?

16. Briefly discuss the evidence for other mass extinctions, and list a few of their possible causes.

17. Discuss the threat that future impacts may pose to us and our planet, and how we know that the threat is real.

18. Describe several adaptations that evolved so primates could live in trees and that have proved useful to us as humans.

19. When did hominids arise, and when did modern humans arise? What is the significance of the Toumaï skull, assuming it is a hominid?

20. Briefly describe and clarify a few common misconceptions about human evolution.

21. What do we mean by cultural and technological evolution? What implications do they have for extraterrestrial intelligence?

22. Briefly describe two main approaches to creating artificial life. Then describe the possibility and potential benefit of constructing computer programs that can mimic the biological functions of cells, and the ethical aspects of making artificial life.

TEST YOUR UNDERSTANDING

Would You Believe It?

Each of the following statements describes a hypothetical future discovery. In light of our current understanding of Earth and evolution, briefly discuss whether each discovery seems plausible or surprising.

Explain clearly; because not all of these have definitive answers, your explanation is more important than your chosen answer.

23. We discover evidence of life, in the form of a particular ratio of carbon-12 to carbon-13, in rock that was originally formed in sediments and is 3.9 billion years old.

24. We discover an intact fossil of a eukaryotic cell, with a cell nucleus, that is 3.0 billion years old.

25. We discover a preserved, 3.5-billion-year-old microfossil that apparently had a genome genetically just like that of many modern animals.

26. We discover clear evidence that life arose on a high mountaintop, not in the oceans.

27. We discover a fossil of a large dinosaur that lived approximately 750 million years ago.

28. We discover that, contrary to present belief, oxygen was abundant in Earth's atmosphere at the time when life arose.

29. We discover a crater from the impact of a 10-kilometer asteroid that dates to about 2,500 years ago.

30. We discover an asteroid about 3 kilometers across that is on a collision course with Earth.

31. We find fossil remains of an early primate that lived about 50 million years ago and was, from all appearances, identical to a modern gorilla.

32. The first life created in the laboratory has an RNA genome, rather than a DNA genome.

Quick Quiz

Choose the best answer to each of the following. Explain your reasoning with one or more complete sentences.

33. The origin of life on Earth most likely occurred: (a) before 4.5 billion years ago; (b) between about 4.5 and 3.5 billion years ago; (c) between about 3.0 and 2.5 billion years ago.

34. The first living organisms probably were: (a) cells without nuclei that used RNA as their genetic material; (b) cells with nuclei that used RNA as their genetic material; (c) cells with nuclei that used DNA as their genetic material.

35. The importance of the Miller-Urey experiment is that: (a) It proved beyond doubt that life could have arisen naturally on the young Earth. (b) It showed that natural chemical reactions can produce building blocks of life. (c) It showed that clay can catalyze the production of RNA.

36. "RNA world" refers to: (a) the possibility that life migrated from Mars; (b) the idea that RNA was life's genetic material before DNA; (c) the idea that early life was made exclusively from RNA, needing no other organic chemicals.

37. Early life arose in an oxygen-free environment, but if any of these microbes had somehow come in contact with oxygen, the most likely effect would have been: (a) nothing at all; (b) to increase their metabolic rates; (c) to kill them.

38. The oxygen in Earth's atmosphere was originally released by: (a) outgassing from volcanoes; (b) plants; (c) cyanobacteria.

39. The *Cambrian explosion* refers to: (a) a dramatic increase in animal diversity beginning about 545 million years ago; (b) the impact that killed the dinosaurs; (c) the sudden emergence of eukaryotic life in the geological record dating to about 2.1 billion years ago.

40. Which statement about Earth's ozone layer is *not* true? (a) It protects us from dangerous solar radiation. (b) It did not exist when life first arose on Earth. (c) It first formed a few hundred million years after life colonized the land.

41. The hypothesis that an impact killed the dinosaurs seems: (a) well supported by geological evidence; (b) an idea that once made sense but now can be ruled out; (c) just one of dozens of clear examples of impacts causing mass extinctions.

42. Which of the following is a likely benefit of creating artificial life? (a) We'll gain insight into the origin of life on Earth. (b) The new life could eliminate our dependence on fossil fuels. (c) The technique could allow us to bring vanished species back to life.

INVESTIGATE FURTHER
In-Depth Questions to Increase Your Understanding

Short-Answer/Essay Questions

43. *A Brief History of Life on Earth.* Take all the ideas about the origin and evolution of life on Earth and try to condense them into a one- to three-page essay on the history of life on Earth. Or, if you prefer, try to capture the ideas in a poem.

44. *Geology and Life.* In Chapter 4, we discussed the role of plate tectonics and the CO_2 cycle in climate regulation on Earth. Suppose neither of these processes had ever operated. Could life still have arisen on Earth as discussed in this chapter? If so, how far could evolution have progressed before the lack of climate regulation would have blocked further major developments? Write a one-page essay summarizing and explaining your answers.

45. *Keys to Our Existence.* Identify and describe four crucial events in evolutionary history without which our current existence would have been highly unlikely. Explain your reasoning clearly.

46. *Extinction and Oxygen.* Suppose we somehow kill off a large fraction of the photosynthetic life on Earth. What consequences would this have for the oxygen content of our atmosphere? Explain your reasoning.

47. *Impact Movie Review.* View one of the Hollywood movies concerning the threat of an impact to our civilization, such as *Deep Impact* or *Armageddon*. Based on what you have learned in this chapter, write a one- to two-page critical review in which you include discussion of whether the impact scenario is realistic.

48. *Artificial Life Review.* Numerous science fiction stories and movies involve the creation of artificial life. Review one such story or movie, and identify at least three ideas in it that either do or do not meet the standards of being testable by science. Describe each in detail.

Quantitative Problems

Be sure to show all calculations clearly and state your final answers in complete sentences.

49. *Bacterial Evolution.* Suppose that a mutation occurs in about 1 out of every 1 million bacterial cells, and suppose that you have a bacterial colony in a bottle like that described in Cosmic Calculations 6.1 (in which the bacteria divide each minute). Given the number of bacteria in the bottle after 1 hour, approximately how many bacteria would have some type of mutation? What does this tell you about why bacteria often evolve resistance to new drugs?

50. *Deep in Bacteria.* In Cosmic Calculations 6.1, we calculated that the volume of bacteria after 120 doublings would be $1.3 \times 10^{15} \text{ m}^3$.

The total surface area of Earth is about $5.1 \times 10^{14} \text{ m}^2$. Use these facts to calculate the average depth of the bacteria at that time, if we spread them evenly over Earth's entire surface.

51. *Bacterial Universe.* Suppose the bacteria described in Cosmic Calculations 6.1 and 6.2 could continue to multiply and spread out.
    a. Recall that the observable universe extends about 14 billion light-years in all directions from us. Calculate its volume in cubic light-years, then convert your answer to cubic meters.
    b. How long would it take for the bacteria to reach this volume? (*Hint:* You can proceed by trial and error, testing different values of t in the formula.)
    c. Even assuming that nutrients and energy were available, why couldn't the bacteria really grow this fast?

52. *Human Population Growth.* During the twentieth century, human population grew with a doubling time of about 40 years, reaching about six billion in 2000. Suppose this growth rate continued. What would human population be in 2200? in 2600? Do these populations seem possible on Earth? Explain.

53. *Impact Energy.* Consider a comet about 2 kilometers across with a mass of 4×10^{12} kg. Further assume that it crashes into Earth at a speed of 30,000 meters per second (about 67,000 miles per hour).
    a. What is the total energy of the impact, in joules? (*Hint:* The "kinetic energy" formula tells us that the impact energy in joules will be $\frac{1}{2} \times m \times v^2$, where m is the comet's mass in kilograms and v is its speed in meters per second.)
    b. A 1-megaton nuclear explosion releases about 4×10^{15} joules of energy. How many such nuclear bombs would it take to release as much energy as the comet impact?
    c. Based on your answers, comment on the degree of devastation the comet might cause.

54. *Impact Probability.* Impacts the size of the Tunguska event occur about once every century or two on average. Estimate the probability that the next impact will occur over a major city, thereby killing hundreds of thousands or millions of people. Be sure to explain all the numbers in your estimate clearly.

Discussion Questions

55. *Our Bacterial Ancestry.* Some of Darwin's early detractors complained that evolution implied we were descended from monkeys or apes. In fact, as we saw in this chapter, our evolution is built on far more primitive organisms. The oxygen we breathe was produced by bacteria and is processed in our cells by mitochondria that probably represent bacteria living symbiotically within us. Does our relationship to bacteria affect the way we should view ourselves as a species? Defend your opinion.

56. *The Missing Link.* As we discussed in this chapter, there is no longer a critical "missing link" in human evolution, and the theory of evolution has never suggested that humans evolved *from* present-day apes. Nevertheless, huge numbers of Americans profess belief in both of these myths about evolution. Why do you think these erroneous claims continue to be popular? What can or should be done to better educate the public?

57. *Evolution by Choice.* Consider the technology we are likely to have in the near future that would enable us to genetically engineer our own species, allowing us to choose the path of our future evolution. How do you think society can or should regulate the use of this awesome power? Do you think its potential benefits outweigh its risks, or vice versa? Overall, do you think

it likely that advanced civilizations, if they exist, have engineered their own evolution? Defend your opinions.

58. *The Origin of Life.* NASA's Astrobiology home page frequently covers new discoveries about the origin and evolution of life. Learn about one recent important discovery, and write a short essay summarizing the discovery and how it affects our understanding of how life might have evolved on Earth.

59. *Impact Programs.* The discovery that impacts could pose a threat to our civilization has led to calls for new programs to help alleviate the threat. In a few cases, legislation has even been pro-posed to implement such programs. Learn about one proposal for dealing with the impact threat. Write a short essay explaining the proposal and discussing your opinion of its merits.

60. *Extinction.* Learn more about how biologists estimate the rate at which human activity is driving species to extinction. What conclusions can we draw about the present rate of extinction? Based on this rate, are we in danger of causing a mass extinction comparable to past mass extinctions on Earth? Summarize your findings in a one-page essay.

61. *Artificial Life.* Find the Web sites of three research groups who are trying to produce artificial life, and write a one-page summary of each of their goals and methods.

7
Searching for Life in Our Solar System

LEARNING GOALS

7.1 ENVIRONMENTAL REQUIREMENTS FOR LIFE
- Where can we expect to find building blocks of life?
- Where can we expect to find energy for life?
- Does life need liquid water?
- What are the environmental requirements for habitability?

7.2 A BIOLOGICAL TOUR OF THE SOLAR SYSTEM: THE INNER SOLAR SYSTEM
- Does life seem plausible on the Moon or Mercury?
- Could life exist on Venus or Mars?

7.3 A BIOLOGICAL TOUR OF THE SOLAR SYSTEM: THE OUTER SOLAR SYSTEM
- What are the prospects for life on jovian planets?
- Could there be life on moons or other small bodies?

7.4 THE PROCESS OF SCIENCE IN ACTION: SPACECRAFT EXPLORATION OF THE SOLAR SYSTEM
- How do robotic spacecraft work?
- What are some key spacecraft missions for astrobiology?

... for what can more concern us than to know how this world which we inhabit is made; and whether there be any other worlds like it, which are also inhabited as this is?

Bernard Le Bovier De Fontenelle, *Conversations on the Plurality of Worlds,* **1686**

Having studied Earth's habitability and life, we now turn to the search for life elsewhere in our solar system. Because there are many places to look—including other planets and their moons, and thousands of known asteroids and comets—we need a strategy to help focus our efforts on the worlds most likely to be habitable.

The first step in such a strategy is to determine where it makes sense to look, so we begin this chapter by discussing the environmental requirements that we expect to be necessary for life on any world. With those requirements in mind, we'll then take a biological tour of the solar system, seeking to determine where the requirements might be met. This will enable us to decide which worlds deserve the greatest attention, both in research and in our studies in this book.

Keep in mind that, for our solar system at least, we are looking primarily for microbes or other simple life. We have already learned enough to be confident that no other advanced civilization has ever arisen in our solar system. Still, the discovery of life of any kind would be profound, both to our understanding of biology and to philosophical considerations of our place in the universe. Even if we don't find life in our own solar system, the search itself will teach us much about the characteristics that can make a planet habitable and will thereby help us when we extend the search to other planetary systems.

7.1 ENVIRONMENTAL REQUIREMENTS FOR LIFE

There's no place like home—at least, not within our own solar system. No world besides Earth has an atmosphere that we could breathe or abundant surface water that we could drink. No other world has a combination of surface temperature and pressure under which we could survive outside without a space suit. Few worlds have atmospheres that offer any protection from dangerous ultraviolet radiation from the Sun or from high-energy particles from space. Indeed, without undertaking major engineering projects to build self-contained environments (or greatly altering the basic conditions of a planet through "terraforming" [Section 8.4]), we have no hope of long-term survival on any other world in our solar system.

However, when we discuss habitability, we generally mean an environment in which life of *some* kind might survive, not necessarily human life. This greatly broadens the possibilities. After all, we could not have survived even on our own planet for much of its history, yet life flourished just the same. Past and present life on Earth has managed to thrive in a far greater variety of environments than we ourselves can endure [Section 5.5]. If we are going to identify potentially habitable worlds in our solar system, we must specify the range of environments that we can consider acceptable for life. We've touched on some of these ideas in previous chapters. Here, we'll try to tie them all together into a clear list of environmental requirements for life.

• Where can we expect to find building blocks of life?

Perhaps the most obvious requirement for life is a set of chemical elements with which to make the components of cells. Life on Earth uses about 25 of the 92 naturally occurring chemical elements, although just four of these elements—oxygen, carbon, hydrogen, and nitrogen—make up about 96% of the mass of living organisms (see Figure 5.5). Thus, a first requirement for life might be the presence of most or all of the elements used by life.

This requirement probably can be met by almost any world. Recall that essentially all chemical elements besides hydrogen and helium were produced by stars [Section 3.2]. Although all of these "heavy elements" are quite rare compared to hydrogen and helium, they are found just about everywhere. Moreover, the elements oxygen, carbon, and nitrogen—arguably the most crucial elements for life—are the third-, fourth-, and sixth-most-abundant elements in the universe, respectively. The proportions of heavy elements varies: While they make up about 2% of the chemical content (by mass) of our solar system, they make up less than 0.1% of the mass in some very old star systems. Nevertheless, every star system we've studied has at least some amount of all the elements used by life.

The nature of solar system formation gives us additional reason to expect the elements of life to be common on other worlds. Recall that, according to the nebular theory of solar system formation, the planets were built when solid particles condensed from gas in the solar nebula, and these particles then accreted into planetesimals and ultimately into planets, moons, asteroids, and comets. The first step in this process—condensation—affects only the heavier elements or hydrogen compounds containing heavy elements, because pure hydrogen and helium always remains gaseous [Section 3.4]. As long as condensation and accretion can occur,* we expect the resulting worlds to contain the elements needed for life.

Note that this basic argument doesn't change even if we allow for life quite different from life on Earth. Life on Earth is carbon-based and, as we discussed in Chapter 5, we have good reason to think that life elsewhere would also be carbon-based. However, we can't absolutely rule out the possibility of life with another chemical basis. The set of elements (or their relative proportions) used by life based on some other element might be somewhat different from that used by carbon-based life on Earth. But the elements are still products of stars that should be found everywhere. No matter what kind of life we are looking for, we are likely to find the necessary elements on almost every planet, moon, asteroid, and comet in the universe.

---

*Observations demonstrate that condensation and accretion can occur in star systems like ours, in which the proportion of heavy elements is about 2% by mass. We do not yet know if these processes also occur in systems with much smaller abundance of heavy elements.

A somewhat stricter requirement is the presence of these elements in molecules that can be used as ready-made building blocks for life, just as the early Earth probably had at least moderate abundances of amino acids and other complex molecules [Section 6.2]. Recall that Earth's organic molecules likely came from some combination of three sources: chemical reactions in the atmosphere, chemical reactions near deep-sea vents in the oceans, and molecules brought to Earth from space. The first two sources can occur only on worlds with atmospheres or oceans, respectively. But the third source should have brought similar molecules to nearly all worlds in our solar system.

Studies of meteorites and comets suggest that organic molecules are widespread among both asteroids and comets. Because every world was pelted by asteroids and comets during its early history [Section 4.3], every world should have received at least some organic molecules; interplanetary dust may also have contained organic molecules that rained down on young worlds. However, organic molecules tend to be destroyed by solar radiation on surfaces unprotected by atmospheres. Moreover, while these molecules might stay intact beneath the surface—as they evidently do on asteroids and comets—they probably cannot react with each other unless some kind of liquid or gas is available to move them about. Thus, given that it makes sense to start our search with worlds on which organic molecules are likely to be involved in chemical reactions, we should concentrate on worlds that have either an atmosphere, or a surface or subsurface liquid medium such as water, or both.

• Where can we expect to find energy for life?

In addition to a source of molecular building blocks, life requires an energy source to fuel metabolism [Section 5.3]. Recall that life on Earth uses a wide variety of energy sources. Some organisms get energy directly from sunlight through photosynthesis. Others get energy by consuming organic molecules (for example, by eating photosynthetic organisms) or through chemical reactions with inorganic compounds of iron, sulfur, or hydrogen.

Sunlight is available everywhere in our solar system, though it becomes much weaker with increasing distance from the Sun. The energy available in sunlight decreases with the *square* of the distance from the Sun (Figure 7.1). For example, if we could put a leaf of a particular size on a world twice as far from the Sun as Earth, the leaf would receive only one-fourth as much energy as the same leaf on Earth (over the same period of time). At 10 times Earth's distance from the Sun—roughly the distance of Saturn—it would receive only $\frac{1}{10^2} = \frac{1}{100}$ of the energy it would receive on Earth. Photosynthetic life on such a world would have to be either much larger than life on Earth (giving it a larger surface area for collecting light), much more efficient at collecting solar energy, or much slower in its metabolism and reproduction. For the far outer solar system, sunlight almost certainly is too weak to support life.

Chemical energy sources also place constraints on life. Chemical reactions can occur under a wide variety of circumstances, but only if the potential reactants are brought into contact with each other. This means that the ongoing reactions needed to provide energy for life can occur only on worlds where materials are being continually mixed. On a practical level, this probably requires either an atmosphere to mix gases or a liquid medium to mix materials on or below a world's surface—the same requirements we found for obtaining the building blocks of life.

Figure 7.1

Any given amount of sunlight is spread over a larger area with increasing distance from the Sun. As shown in this diagram, the area over which the sunlight is spread increases with the square of the distance: At 2 AU the sunlight is spread over an area $2^2 = 4$ times as large as at 1 AU, and at 3 AU the sunlight is spread over an area $3^2 = 9$ times as large as at 1 AU. (Recall that 1 AU is the average Earth–Sun distance, or about 150 million kilometers.) Thus, the energy contained in sunlight (per unit area) *decreases* with the square of the distance from the Sun.

• Does life need liquid water?

In addition to a source of building blocks and a source of energy, there is one more ingredient that is essential to all life on Earth: liquid water. Recall that water plays at least three vital roles for life on Earth [Section 5.3]: It dissolves organic molecules, making them available for chemical reactions within cells; it allows for the transport of chemicals into and out of cells; and it is involved directly in many of the metabolic reactions that occur in cells. It is difficult to imagine life in the absence of a liquid substance to play these roles. But could these roles be fulfilled by some liquid other than water?

Potential Liquids for Life   No one knows whether other liquids could support life in the absence of liquid water, but there are a number of constraints to consider. For example, a substance that might fulfill the roles of water must, like water, be fairly common. On Earth, the only liquid besides water commonly found is molten rock, which is so hot that it's difficult to imagine life surviving within it.* However, several other common substances might take liquid form on colder worlds.

Table 7.1 lists the temperature range over which water and three other potential candidates—ammonia, methane, and ethane—remain liquid. The given temperature ranges are those that apply under the atmospheric pressure on Earth. At different pressures, or with the presence of dissolved minerals, the ranges can be different. For example, salt water can remain liquid at temperatures slightly below 0°C—sea water freezes at about −2°C and water fully saturated with salt freezes at −21°C—and under sufficient pressure water can remain liquid at temperatures well above 100°C. (Near deep-sea vents, the pressure keeps water liquid at temperatures as high as about 375°C.) Despite such variation in melting and boiling temperatures, the ranges in Table 7.1 provide a useful comparison.

 THINK ABOUT IT   Oil (petroleum) is also found in liquid form on Earth. Why doesn't oil seem likely as a liquid medium for the origin of life? (*Hint:* Remember that oil is a fossil fuel.)

Advantages of Water   Although we cannot rule out the other liquids in Table 7.1, water has at least three advantages that make it seem far more suitable as a liquid medium of life. First, as you can see in the table, water remains liquid over a wider and higher range of temperatures. (The liquid range for ethane is almost as wide, but at much lower temperatures.) A wider range makes it more likely that the substance can stay liquid through changes in the weather or climate. A higher temperature range facilitates chemical reactions. As a general rule, chemical reactions proceed more rapidly at higher temperatures because the molecules themselves move more rapidly. Typically, the rate of a given chemical reaction doubles with each 10°C increase in temperature. Thus, chemical reactions in water should generally proceed much more rapidly than similar reactions in liquid ammonia, methane, or ethane. Any life using these other liquids would therefore need to have a much slower metabolism

TABLE 7.1 *Potential Liquids for Life*

Freezing and boiling points (under 1 atmosphere of pressure) for common substances that may be found in liquid form in our solar system. The last column gives the width of the liquid range, found by subtracting the freezing point from the boiling point.

Substance	Freezing Temperature	Boiling Temperature	Width of Liquid Range
Water (H_2O)	0°C	100°C	100°C
Ammonia (NH_3)	−78°C	−33°C	45°C
Methane (CH_4)	−182°C	−164°C	18°C
Ethane (C_2H_6)	−183°C	−89°C	94°C

---

*Rock becomes molten only at temperatures above about 700°C. Given that the most extreme high-temperature extremophiles yet known can't survive in water heated above 130°C, it seems difficult to imagine life at 700°C.

Figure 7.2

Most substances are denser as solids than as liquids, so when solid and liquid forms exist together, the solid form sinks. Water is an exception; solid water in the form of ice is less dense than liquid water, and so ice floats in liquid water.

Figure 7.3

Within individual water molecules, the electrons tend to be distributed in a way that makes one side have a net positive charge and the other side have a net negative charge.

than life on Earth. The slower rate of reactions may also make it less likely that life would arise in the first place in these other liquids, because the origin of life probably requires many complex chemical reactions.

The second advantage of liquid water involves an oddity in the way water freezes. Most substances are denser as solids than as liquids, but water is a rare exception (Figure 7.2): Ice is *less* dense than liquid water, which is why ice floats. This property helps life survive on Earth. In the winter, when surface temperatures are low enough for water to freeze, floating ice forms a layer on the tops of lakes and seas. This layer of ice insulates the water beneath, allowing it to remain liquid—which allows life to survive within it. No other liquid in Table 7.1 shares this property with water. That is, all the others sink when frozen.

The flotation of solid ice may be even more important to long-term climate stability. During periods when Earth cools a bit, such as the ice ages [Section 4.5], lower temperatures allow more ice to form on the surface. If this ice sank, the surface would still be covered with liquid water, which in turn would freeze and sink. This process would continue until no liquid water was left at the surface—that is, until lakes and oceans were completely frozen. Instead, because ice floats, a cool period thickens the insulating layer of ice on the surfaces of lakes and oceans, making it less likely that they will freeze completely.

The third advantage of liquid water over the other liquids in Table 7.1 comes from the way that electrical charge is distributed within water molecules (Figure 7.3). Within individual water molecules, the electrons tend to be distributed in a way that makes one side have a net positive charge and the other side have a net negative charge. (Molecules with such charge separation are called *polar* molecules; the term *polar* comes from the positive and negative charges being concentrated at opposite ends, or "poles," of the molecule's axis and not from anything related to temperature.) This charge separation affects the way in which water dissolves other substances. Molecules and salts that also have charge separations dissolve in water easily. Molecules that do not have any charge separation—such as molecules of oil—do not dissolve in water.

On Earth, the charge separation property of water is critical to life. Living cells have membranes that do not dissolve in water, so the membranes effectively protect the interior contents of cells. If we place living cells in liquid ethane, methane, or ammonia—molecules with less charge separation than water—their membranes may come apart. Charge separation also makes possible the formation of a special type of chemical bond, called a *hydrogen bond,* that is important to the biochemistry of life on Earth. (The formation of hydrogen bonds as water freezes also explains why ice is less dense, and hence why it floats, because these bonds force the molecules into a slightly expanded structure.) Thus, among the liquids listed in Table 7.1, Earth-like living cells can survive only in liquid water.

The Bottom Line on Water   We've identified three advantages of water over other liquid candidates for life: (1) a wider and higher range of temperatures over which it remains liquid; (2) the fact that solid water floats; and (3) the fact that the charge separation of water molecules allows for types of chemical bonds that are not possible with the other liquids. These advantages make a strong case for the need for liquid water as a basis for life, but the case is not definitive. For example, it's possible to imagine circumstances under which the third advantage might be turned around. If life elsewhere had cell membranes made of molecules with different charge separation properties, they might dissolve in water but not in ammonia, methane, or ethane. We do not know if this is possible, but if it is, then life might exist in one of those other liquids while finding water to be poisonous or too hot.

The bottom line is that, while water clearly is an excellent medium for supporting life, we do not know if other liquids could also support life. It's difficult to tell whether the advantages we've cited for water are truly universal or are specific only to terrestrial life. Given this uncertainty, along with the fact that other liquids probably exist on (or within) some of the worlds in our solar system, we should not be too quick to rule out the potential for those worlds to support life. In general, however, water is much more abundant than other liquids both in our solar system and elsewhere (largely due to the fact that oxygen is the third-most-abundant element in the universe, after hydrogen and helium), so it remains our prime candidate for a liquid medium.

What are the environmental requirements for habitability?

We began this section with the goal of making a list of environmental requirements for life that can help us decide which worlds to focus on as we begin the search for possible abodes of life in our solar system. In the broadest terms, we've found that the environment must satisfy three major requirements:

1. It must have a source of molecules from which to build living cells.
2. It must have a source of energy to fuel metabolism.
3. It must have a liquid medium—most likely liquid water—for transporting the molecules of life.

The first requirement is probably met by most if not all worlds. The second requirement is somewhat more limiting, but there are still plenty of worlds that should have sufficient sunlight or chemical energy for life. The third requirement—the need for a liquid—is probably the most stringent. Moreover, any world that meets this third requirement stands a good chance of meeting the first two as well. A liquid like water can facilitate chemical reactions with inorganic planetary materials, which offers at least a potential source of energy for life—and, indeed, a source tapped by many microbes on Earth. Thus, based on our current understanding of life, we can consolidate the requirements into a single "litmus test" for habitability: *A world can be habitable only if it has a liquid medium, probably meaning liquid water but possibly including one of the other liquids listed in Table 6.1.*

This requirement for a liquid certainly narrows the possibilities for life in our solar system, but not as much as you might at first guess. The wide variety of habitats in which we find both liquid water and life on Earth, including the deep ocean and in rocks buried deep underground, tells us that habitability requires only the presence of a liquid *somewhere*, not necessarily on the surface. As we'll discuss in the rest of this chapter, a large fraction of the worlds in our solar system have probably met this condition at least at some point in the past, and many still do.

There are at least two important caveats to our discussion. First, we may need to distinguish between the ability of a world to support an origin of life and its ability to support life that might have arrived from elsewhere. For example, while most scientists have concluded that life could not have arisen indigenously in Jupiter's atmosphere, some scientists have imagined organisms that could survive there if they somehow arrived from somewhere else (see Section 7.3). Second, we must remember that everything we know about life comes from the study of life on only a single world—our own. We've already noted the difficulty of knowing whether the advantages of water as a liquid medium are truly universal or specific to life on Earth. Other apparent attributes of life may similarly be less crucial elsewhere than they are on Earth. It is therefore possible that our discussions of habitability are based on too narrow a view of life, in which case our litmus test may be too narrow as well. Indeed, science fiction writers have imagined all sorts of bizarre life-forms existing under conditions far outside those we've examined here. Nature might be even more inventive. Nevertheless, a search for life must start somewhere, and it makes sense to begin by looking for the conditions that seem most likely to support life. If it turns out that our initial search is too narrow, we can always expand it in the future.

7.2 A BIOLOGICAL TOUR OF THE SOLAR SYSTEM: THE INNER SOLAR SYSTEM

Imagine living a century from now, when we have sent orbiters and landers to every major world in our solar system and when humans may even be living and working on other worlds. At that time, a "life in the universe" course might begin with a true biological tour of the solar system. One world at a time, we would discuss its habitability, the characteristics of its life or why it does not have its own indigenous life, and the prospects of survival (or the results of actual experiments) for any life from Earth that we might transport there.

We cannot undertake such a complete biological tour today. However, we have already learned enough to say which worlds seem most likely to be habitable—that is, to meet the environmental requirements for life—and which worlds seem least likely to be capable of harboring any indigenous life. This information will help guide the search for life in our solar system, because it will help us determine where to focus our efforts. And, as a practical matter, it will help us decide where to devote our attention in this book. In this section, we'll focus on our neighbors in the inner solar system: our Moon, and the terrestrial planets Mercury, Venus, and Mars.

• Does life seem plausible on the Moon or Mercury?

Although Mercury is a planet and the Moon orbits planet Earth, the two worlds share many characteristics. Both are pockmarked with craters (Figure 7.4). Both are much smaller than Earth (see Figure 4.1)—so small that by now they have lost most of their internal heat, leaving them with no ongoing volcanism and without significant tectonic activity. The lack of volcanism means no outgassing to release gases into an atmosphere, and their small sizes mean weak gravity that has long since allowed any past atmospheric gases to escape to space. That is why both worlds are essentially airless.* Mercury and the Moon also share the distinction of being among the places least likely to be habitable in the solar system, primarily because neither world is likely to have any liquids anywhere.

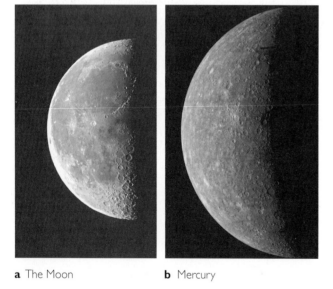

a The Moon   **b** Mercury

Figure 7.4

Similar views of the Moon and Mercury, shown to scale. See Figure 4.1 for a size comparison to Earth.

---

*Technically, the Moon and Mercury have extremely thin atmospheres created by diffusion of atoms and molecules from the interior and by release of gas molecules by impacts (mostly microscopic impacts of high-energy particles and photons), but these atmospheres are far thinner than even what we would consider to be an excellent vacuum in laboratories on Earth.

The Moon As we discussed in Section 4.6, the Moon has a far smaller proportion of water and other volatiles than Earth, probably as a result of vaporization that occurred in the giant impact thought to have formed the Moon. The only place the Moon may have any significant amount of water is in craters near its poles. The bottoms of these polar craters are perpetually shielded from sunlight, making them extremely cold and dark. As a result, ices deposited over the eons by impacts (or possibly by outgassing from the lunar interior early in the Moon's history) may lie frozen at the crater bottoms (Figure 7.5). However, even if such ices are present in lunar craters, they exist only because the crater bottoms have never warmed enough to melt them into liquid form. Thus, ice in lunar craters probably isn't relevant to the Moon's habitability, though it may be useful to *us* as a source of water for future colonies on the Moon.

Mercury We are less certain about the presence of water on Mercury. Because it formed so close to the Sun, Mercury must have accreted primarily from planetesimals that lacked any ices. But, as was the case with Earth, a few planetesimals from farther out in the solar system may have brought water and other ices to Mercury during its formation. If so, some of these materials may be chemically bound in subsurface rock on Mercury. However, Mercury's high average density suggests that, like Earth, it was rocked by a giant impact some time shortly after it underwent differentiation. The giant impact is thought to have ejected much of Mercury's original mantle into space, leaving behind a planet consisting predominantly of an iron-rich core. If it really occured, this impact likely would have ejected any water that accreted during the planet's formation.

Even if Mercury has water in its surface rock, it's unlikely that any of this water could ever take liquid form, because Mercury's surface temperatures are nearly always far above the boiling point or far below the freezing point. The combination of a 58.6-day rotation period and an 87.9-day orbital period gives Mercury days and nights that last about three Earth months each. As the closest planet to the Sun, the dayside temperatures get quite hot—about 425°C. But with no atmosphere to retain heat, temperatures plummet to a frigid −175°C during the long nights. Some observational evidence suggests that Mercury, like the Moon, may have ices brought by impacts in crater bottoms near its poles. (We should learn more from the MESSENGER mission to Mercury; see Section 7.4.) As is the case for the Moon, some of Mercury's polar craters have bottoms that probably have not been exposed to sunlight for billions of years, and hence where temperatures are perpetually far below freezing.

The lack of liquid water may make the Moon and Mercury lost causes when it comes to hope for finding life, but they are still important to our general understanding of planetary systems. In particular, studies of the Moon and Mercury teach us a great deal about the origin and history of our solar system and help us learn why some worlds are habitable and others are not. Thus, while the Moon and Mercury are almost certainly lifeless, they are still important targets of study for planetary scientists.

• Could life exist on Venus or Mars?

Prospects for past or present life look far better when we turn to our nearest planetary neighbors, Venus and Mars. Venus has been called our "sister planet," because it is the nearest planet to Earth in distance and is nearly identical to Earth in size. Mars is considerably smaller than Earth (see Figure 4.1), but spacecraft photos of its surface reveal an eerily Earth-like

North Pole
X

Figure 7.5

This color-coded map shows evidence for water ice in craters near the Moon's north pole. The portion of the Moon shown includes all latitudes north of 70°N, and the colors represent the concentration of hydrogen atoms measured by NASA's Lunar Prospector mission in the late 1990s. (The measurements were actually made by neutron detection techniques.) The dark blue and purple regions have the highest hydrogen concentrations, which are probably due to the presence of water ice on or just below (within a half-meter of) the surface. Note that these regions are very near the pole, where some crater bottoms are permanently shielded from the Sun. Similar results were found for the Moon's south pole.

Figure 7.6

Clouds are all that can be seen in this ultraviolet image of Venus from the *Pioneer Venus Orbiter;* no surface features can be seen at all.

Figure 7.7

This photograph shows a small patch of the surface of Venus; part of the lander that took the picture is in the foreground. The Soviet Union sent several landers to Venus during the 1970s and early 1980s. Because of the high temperatures and corrosive atmosphere, none of them functioned for more than about 2 hours after landing.

landscape (see Figure 1.5). Both planets have at least some potential for supporting past and present life, although Mars looks far more promising.

Venus   Venus is completely enshrouded in thick clouds (Figure 7.6), and past generations of scientists could only guess about the surface conditions. Some imagined a pleasantly warm planet. Although Venus is closer to the Sun, it is not *that* much closer—about two thirds as far from the Sun as is Earth. Thus, if Venus had an Earth-like atmosphere, we might expect it to be only a little hotter than Earth. In fact, calculations show that with an Earth-like atmosphere, Venus's global average temperature would be about 35°C (95°F). This type of thinking led past generations of science fiction writers to speculate that Venus might be a lush, tropical paradise.

THINK ABOUT IT   A good way to think about the expected temperature difference between Venus and Earth is to consider the 1-to-10-billion scale model of our solar system from Chapter 3. Recall that at this scale, the Sun is about the size of a grapefruit, and Venus and Earth are located about 10 and 15 meters away, respectively. Imagine that you have a grapefruit that is extremely hot. If the heat feels comfortable at a distance of 15 meters, how would it feel at 10 meters? Discuss the difference qualitatively, and relate it to the expected temperature difference between Venus and Earth.

Such dreams evaporated in 1962, when the *Mariner 2* spacecraft flew past Venus and discovered that its surface is scorching hot. (Some earlier radio observations from Earth had hinted at the high temperature, but other explanations for the data seemed possible prior to the spacecraft observations.) The surface temperature is an incredible 470°C (about 880°F)—easily hot enough to melt lead. Moreover, this temperature persists planetwide, both day and night. All the while, a thick atmosphere bears down on the surface with a pressure 90 times that on Earth's surface—equivalent to the pressure at a depth of nearly 1 kilometer (0.6 mile) in the oceans. Besides Venus's crushing pressure and searing temperature, the atmosphere of Venus contains sulfuric acid and other chemicals that are toxic to us. Indeed, landers sent to Venus by the Soviet Union provided data for only a brief period before being disabled by the high temperatures (Figure 7.7). Far from being a beautiful sister planet to Earth, Venus resembles a traditional view of hell.

What causes such extreme conditions on Venus? The answer is the *greenhouse effect*—the same effect that makes our own planet so comfortable [Section 4.5], but operating much more strongly. Recall that, in the absence of the greenhouse effect, Earth would be frozen over. Our planet is habitable because a moderate greenhouse effect traps enough heat to raise temperatures above the freezing point of water. The greenhouse effect operates in precisely the same way on Venus, but it raises temperatures much more.

The strong greenhouse effect on Venus (often called a *runaway greenhouse effect,* for reasons we will discuss in Chapter 10) comes primarily from carbon dioxide in its atmosphere. Earth has a modest greenhouse effect because carbon dioxide makes up only a fraction of 1% of our atmosphere. In contrast, carbon dioxide makes up more than 96% of Venus's far thicker atmosphere. Interestingly, we can explain this difference in atmospheric carbon dioxide by contrasting the fate of carbon dioxide gas on each planet. Both planets have outgassed similar total amounts of carbon dioxide over the course of their histories. However, while almost all of the carbon dioxide outgassed on Earth is now locked up in carbonate rocks or dissolved in the oceans (170,000 times as much as is in our atmosphere [Section 4.5]), all of Venus's outgassed carbon dioxide remains in its atmosphere. Thus, Venus's high temperature results from the extremely strong greenhouse effect produced by its atmospheric carbon dioxide, and the high pressure results from the sheer amount of this gas.*

The high surface temperature all but rules out the possibility of life on the surface of Venus today. It is far too hot for liquid water, let alone liquid ammonia, methane, or ethane. However, the surface may have been habitable in the past. Recall that carbon dioxide gets locked up in Earth's carbonate rocks through the mechanism of the carbon dioxide cycle (see Figure 4.27). This cycle depends on both plate tectonics and the presence of oceans: Carbon dioxide dissolves in the oceans, where it reacts chemically to form carbonate minerals. Venus has no similar cycle because it lacks oceans (and apparently lacks plate tectonics as well) and therefore has no mechanism for removing carbon dioxide from the atmosphere. But the situation may have been quite different in the distant past. As we'll discuss further in Chapter 10, the fact that the Sun should have been much dimmer earlier in its history may have allowed Venus to have oceans prior to about 4 billion years ago. In fact, if you consider "Earth-like" to mean conditions found on Earth today, it's conceivable that ancient Venus may have been more "Earth-like" than Earth itself was at that time.

If Venus once had oceans—a big "if"—it's possible that life arose at that time. And if so, there's one place on Venus where microbes might still survive: in the clouds. At altitudes of about 50 kilometers above the surface, the greenhouse effect is far weaker and droplets containing liquid water can and do exist. The clouds are extremely acidic, but their sulfur content could allow chemical reactions that might provide sufficient energy for extremophiles adapted to survive in this environment. The European Space Agency's *Venus Express* mission, which arrived in Venus orbit in April 2006, may be able to search for evidence of such life, if it exists.

If life does not exist in the Venusian clouds, we may never know whether Venus had life in the distant past. Crater counts suggest that Venus's entire surface is less than about a billion years old, meaning that volcanism or tectonic processes have reshaped or paved over rocks from earlier times. Thus, if life arose and survived until the runaway greenhouse effect made Venus uninhabitable, any fossil evidence would almost certainly have been destroyed long ago.

Mars   While Venus overheated early on, Mars went in the other direction. There's little doubt that Mars once had flowing water with a thicker and warmer atmosphere, but some 3 billion years ago, climate change caused the planet in effect to freeze over. Any remaining surface water froze, and Mars has since lost so much atmospheric gas that the surface pressure is too low for liquid water. Nevertheless, the evidence of past water makes Mars a prime candidate for past life. Moreover, Mars retains enough internal heat that liquid water may be possible underground, in which case life might still survive there.

---

*Calculations show that carbon dioxide alone accounts for most but not all of Venus's high temperature. Other greenhouse gases—notably sulfur dioxide (SO_2), carbon monoxide (CO), and hydrochloric acid (HCl)—also make significant contributions.

Indeed, Mars seems such a good candidate for past or present life that, if it turns out never to have had it, we might have to revisit our assumptions about the likelihood of life arising under the "right" conditions. Because Mars is such a good candidate for habitability and life, we'll defer discussion of it for now, and devote all of Chapter 8 to its possibilities.

7.3   A BIOLOGICAL TOUR OF THE SOLAR SYSTEM: THE OUTER SOLAR SYSTEM

Beyond Mars, you might expect that the weakening strength of sunlight would make life increasingly less likely. But there are still plenty of worlds with internal heat that could keep some water liquid and provide energy for life. In this section, we'll continue our biological tour in the outer solar system, starting with the jovian planets, then examining a few of their large moons, and finally considering the many small bodies of our solar system, including Pluto and other Kuiper belt objects, comets, and asteroids.

• What are the prospects for life on jovian planets?

The four jovian planets—Jupiter, Saturn, Uranus, and Neptune—are very different from the terrestrial worlds. They are far more massive, lower in density, and composed largely of hydrogen, helium, and hydrogen compounds like water, methane, and ammonia. Figure 7.8 shows the bulk properties of the jovian planets, with Earth shown for scale.

The jovian planets are also different from the terrestrial planets in their interior structures, which scientists have deduced from observa-

---

MOVIE MADNESS: *2001—A SPACE ODYSSEY*

It's the year 2001, and five clean-cut astronauts and a conniving computer named HAL are on their way to Jupiter. Yes, we know, this didn't really happen, and 2001 is now ancient history. But this Stanley Kubrick epic is a cinema classic, with music and scenes that remain a part of our popular culture, and special effects that were ahead of their time when the movie came out in 1968. If you haven't seen it, you should.

So why are astronauts and a wayward computer going to this behemoth world? Is it to study Jupiter's complex, churning atmosphere? Umm, no, that's not it. Are they on a mission to examine the giant planet's imposing magnetic field? Er, negative on that. Perhaps they're searching for signs of simple life on the moons Europa, Callisto, and Ganymede, all of which could have unseen oceans beneath their crusty exteriors? Actually, no, that's not it either.

No, these jovian rocket jockeys are headed to the king of the planets because some unknown race of aliens has been messing with our solar system. Four million years ago, some kindly extraterrestrials took a look at Earth and decided our planet's simian population had potential. They planted a dark gray monolith that, when touched by the local apes, converted them from a bunch of howling half-wits into tool-using primates.

This is an amazing tale (and one that, if it were true, would confound all paleontologists), but the aliens weren't content merely to have introduced intelligence on Earth. They also buried a second mono-

lith on the Moon, figuring that a brainy species might eventually develop enough technology to leave its planet and find this weird artifact. The lunar monolith, in turn, directed us to Jupiter.

So that's why the interplanetary spacecraft *Discovery One* is on its way to the biggest world of the solar system. The hope, it seems, is that by journeying to this massive planet, *Homo sapiens* can finally learn whatever profound secrets these altruistic aliens wish to share.

2001 is as much a space oddity as a space odyssey. Forget the fact that the message found at Jupiter is mostly a puzzling psychedelic light show. Ignore the fact that the ship's smooth-talking, onboard computer is more scheming than Machiavelli. No, in the end the most distressing thing about *2001* (other than the dismaying truth that in 2001 the real space program wasn't even close to sending manned missions to Jupiter) is that our past and our destiny are simply some extraterrestrials' science-fair experiment. We're no more than a bunch of choreographed puppets.

Someday it's quite likely that we will travel to many other worlds of the solar system, not just in our mind's eye or in the virtual worlds of our theaters and planetaria, but in honest-to-goodness spacecraft. However, we will make these voyages of discovery based on our own curiosity and quest for knowledge—and not because some control-freak aliens pointed to the gas-giant worlds of the outer solar system with their monolith fingers and said, "Go there."

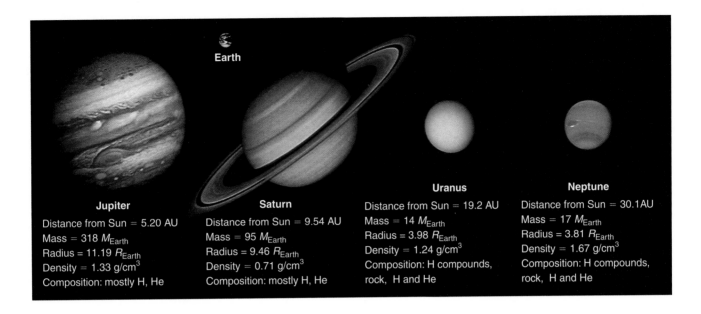

Earth

Jupiter
Distance from Sun = 5.20 AU
Mass = 318 M_{Earth}
Radius = 11.19 R_{Earth}
Density = 1.33 g/cm^3
Composition: mostly H, He

Saturn
Distance from Sun = 9.54 AU
Mass = 95 M_{Earth}
Radius = 9.46 R_{Earth}
Density = 0.71 g/cm^3
Composition: mostly H, He

Uranus
Distance from Sun = 19.2 AU
Mass = 14 M_{Earth}
Radius = 3.98 R_{Earth}
Density = 1.24 g/cm^3
Composition: H compounds, rock, H and He

Neptune
Distance from Sun = 30.1AU
Mass = 17 M_{Earth}
Radius = 3.81 R_{Earth}
Density = 1.67 g/cm^3
Composition: H compounds, rock, H and He

tions and theoretical models of how their materials behave, given the planetary masses and sizes. These planets lack anything resembling Earth's solid surface. Instead, as shown in Figure 7.9, their outer layers contain visible clouds surrounding extended layers of gaseous hydrogen, mixed with helium and hydrogen compounds. Deeper in their interiors, the high pressure compresses the hydrogen into liquid form, and within Jupiter and Saturn the pressure becomes so high that the hydrogen takes on a metallic form that was only recently synthesized in laboratories on Earth. Near their centers, each has a core of rock, metal, and hydrogen compounds, but the pressure is so great that these materials would be in a phase different from anything we ever see on Earth. If you plunged into any one of these worlds, you would just continue downward until you were crushed by the increasing pressure. Ultimately, your remains would sink into a hot sea of liquid hydrogen.

Because these planets are so far from the Sun, temperatures in their upper atmospheres are extremely cold. However, observations show that all must be quite hot in their deep interiors. Because these worlds lack solid surfaces like those of the terrestrial worlds, this heat cannot fuel any geological activity. However, the heat ensures that at some altitudes their atmospheres are warm enough for liquid water, and chemical reactions—some powered by frequent lightning that has been observed in

Figure 7.8

The four jovian planets of our solar system, shown to scale. Note that the tables list distance, mass, and radius in terms of Earth units; that is, M_{Earth} is Earth's mass and R_{Earth} is Earth's radius; 1 AU is Earth's distance from the Sun.

Figure 7.9

These diagrams compare the interior structures of the jovian planets, shown approximately to scale. All four have cores of rock, metal, and hydrogen compounds; the cores of Uranus and Neptune are differentiated into separate layers of rock/metal and hydrogen compounds. All four cores have about the same mass (ten Earth masses), so the planets differ primarily in the amount of material and the depth of the layers around the cores.

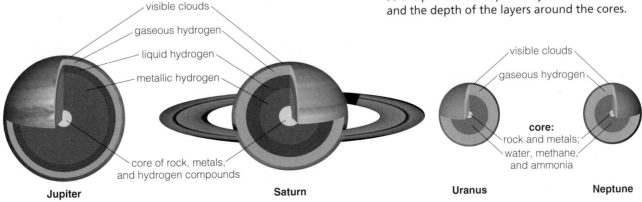

visible clouds
gaseous hydrogen
liquid hydrogen
metallic hydrogen
core of rock, metals, and hydrogen compounds

Jupiter   Saturn

visible clouds
gaseous hydrogen
core:
rock and metals;
water, methane, and ammonia

Uranus   Neptune

jovian planet atmospheres—could conceivably provide energy for life. So is it reasonable to imagine life in the jovian atmospheres?

Jupiter and Saturn   Let's start by considering Jupiter, the largest and nearest of the four jovian planets. As Figure 7.9 shows, Saturn is so similar in its interior structure that the same considerations should apply to it as well.

Jupiter's temperature is far below freezing at its cloud tops, but the temperature rises rapidly with depth. Indeed, Jupiter has several cloud layers, each formed as different types of gases condense. Clouds containing droplets of liquid water can form at a depth of a little over 100 kilometers into Jupiter's atmosphere, which is just a little over 1% of the way from the highest clouds to the center. Given the comfortable temperature and the presence of liquid water, we might therefore wonder if Jupiter could be habitable and host life at this depth in its atmosphere.

Unfortunately, Jupiter's atmosphere appears to present a fatal problem when we consider the possibility of life originating there: Jupiter has strong, vertical winds with speeds that would make a hurricane seem like a gentle breeze in comparison. Any complex organic molecules that might form would quickly be carried to depths at which the heat would destroy them, making it difficult to see how any life could arise. We might imagine microbes reaching Jupiter on meteorites from elsewhere, but again the vertical winds make their survival seem impossible. Such microbes would be thrown quickly onto a nonstop elevator ride between cloud layers that are unbearably cold and others that are insufferably hot.

The only way that anyone has imagined life surviving in Jupiter's atmosphere is by supposing that it might have some sort of buoyancy that allowed it to stay at the right altitude while the vertical winds rushed by it. However, such buoyancy would require large gas-filled sacs, making the organisms themselves enormous. Given that we cannot envision a way for microbes to survive, there seems to be no way for large, buoyant organisms to evolve in the first place. They might survive if they arrived on Jupiter from elsewhere, but we know of no way that such large organisms could survive a journey through space, even if they existed elsewhere. As a result, most scientists do not consider Jupiter to be habitable. For essentially the same reasons, Saturn is also unlikely to be habitable.

Uranus and Neptune   Uranus and Neptune are probably just as unlikely as Jupiter and Saturn as candidates for habitability. Their atmospheres are much colder than those of Jupiter and Saturn, due mainly to their greater distance from the Sun. If they have clouds in which liquid water droplets can form, we have not yet been able to detect them. Even if they do have such clouds deep in their atmospheres, vertical winds similar (though slower) to those on Jupiter and Saturn would also make life unlikely.

Uranus and Neptune have at least one possible zone of habitability that Jupiter and Saturn lack: their outer cores of water, methane, and ammonia. Theoretical models suggest that these materials may be in liquid form, making for very odd "oceans" in the deep interiors of Uranus and Neptune. The high pressures, strange mix of liquids, and lack of any obvious way to extract energy from these "oceans" makes life seem highly unlikely, but we cannot rule it out. However, at present we lack any technology that could allow us to search for life in the cores of these giant planets; even if life might exist there, the search for it will have to await new technological developments.

• Could there be life on moons or other small bodies?

Each of the four jovian planets reigns over a set of moons that make their systems look much like solar systems in miniature. Together, the jovian planets have at least 150 moons, and additional small moons undoubtedly still await discovery. The outer solar system also contains a myriad of other small bodies [Section 3.3], starting with the asteroids of the asteroid belt, moving out to the Kuiper belt with its "dwarf planets" like Pluto and Eris as well as many smaller comets, and on to the Oort cloud that may contain as many as a trillion icy comets. Could any of these myriad worlds be habitable?

Large Moons   Perhaps surprisingly, the answer is yes, at least for a few of the largest moons, such as Jupiter's moons Europa, Ganymede, and Callisto, and Saturn's moon Titan. This answer is surprising because even these large moons are fairly small by planetary standards. None are as large as Mars, and if they were similar to the terrestrial planets in composition and in their heat sources, they would be too small to have retained internal heat that could support enough geological energy to fuel life. However, the moons of the jovian planets differ from the terrestrial planets in an important way: They tend to contain a great deal of ice, because they formed in parts of the solar system where it was cool enough for ices to condense from the solar nebula (see Figure 3.22). Because ices melt at

COSMIC CALCULATIONS 7.1

Newton's Version of Kepler's Third Law

How do we know the masses of distant objects? In many cases, we can use a modified version of Kepler's third law ($p^2 = a^3$). Recall that this law applies only to objects orbiting the Sun (see Cosmic Calculations 2.1). However, Newton found that Kepler's original law was just a specific case of a more general law, usually called *Newton's version of Kepler's third law*:

$$p^2 = \frac{4\pi^2}{G(M_1 + M_2)}a^3$$

where M_1 and M_2 are the masses of the two objects, p is their orbital period, and a is the distance between their centers. The term $4\pi^2$ is simply a number ($4\pi^2 \approx 4 \times 3.14^2 \approx 39.44$), and $G = 6.67 \times 10^{-11}\frac{m^3}{kg \times s^2}$ is the gravitational constant.

This law gives us the power to measure the masses of distant objects. Any time we measure an orbiting object's period (p) and orbital distance (a), Newton's equation allows us to calculate the sum $M_1 + M_2$ of the two objects involved in the orbit. If one object is much more massive than the other, we essentially learn the mass of the massive object, as the following example shows.

Example: Use the fact that Earth orbits the Sun in 1 year at an average distance of 150 million kilometers (1 AU) to calculate the mass of the Sun.

Solution: Newton's version of Kepler's third law becomes:

$$p_{Earth}^2 = \frac{4\pi^2}{G(M_{Sun} + M_{Earth})}a_{Earth}^3$$

Because the Sun is much more massive than Earth, the sum of their masses is nearly the mass of the Sun alone: $M_{Sun} + M_{Earth} \approx M_{Sun}$. Using this approximation, we find:

$$p_{Earth}^2 \approx \frac{4\pi^2}{GM_{Sun}}a_{Earth}^3$$

To find an expression for the mass of the Sun we multiply both sides by M_{Sun} and divide both sides by $(p_{Earth})^2$:

$$M_{Sun} \approx \frac{4\pi^2 a_{Earth}^3}{Gp_{Earth}^2}$$

Because G is given above with units of seconds and meters, we must use the same units for Earth's orbital period ($p_{Earth} = 1$ year $\approx 3.15 \times 10^7$ seconds) and average orbital distance ($a_{Earth} = 1$ year $\approx 1.5 \times 10^{11}$ m). We find:

$$M_{Sun} \approx \frac{4\pi^2(1.5 \times 10^{11}\ m)^3}{\left(6.67 \times 10^{-11}\frac{m^3}{kg \times s^2}\right)(3.15 \times 10^7\ s)^2} = 2.0 \times 10^{30}\ kg$$

Simply by knowing Earth's orbital period and distance from the Sun and the gravitational constant G, we have used Newton's version of Kepler's third law to find that the Sun's mass is about 2×10^{30} kilograms.

much lower temperatures than do metal or rock, these moons can have internally driven "ice geology"—such as internal convection involving gradual movement of icy material—with much less internal heat than is needed for the "rock geology" of the terrestrial planets.

In addition, at least some of the jovian moons—most notably Jupiter's moons Io and Europa—have an ongoing source of internal heat quite different from any heat source on the terrestrial planets. (We'll discuss this source, *tidal heating,* in Chapter 9.) The available heat can in principle melt ice, allowing the possibility of liquid water in their interiors. In at least one case, Europa, current evidence strongly suggests a deep, subsurface ocean of liquid water. Numerous other moderate- to large-size moons in the outer solar system also show evidence of past or present geological activity that could have allowed (or still allow) for some liquid medium.

The prospects for habitability are so good for some of these large moons that we'll defer their discussion, in order to give them our full attention in Chapter 9. The prospects are much poorer for the more numerous but much smaller moons, so we can consider them along with other small bodies of the solar system.

Small Bodies   Our brief biological tour now brings us to the numerous smaller bodies in our solar system: small and medium-size moons, asteroids, comets, and the large Kuiper belt objects [Section 3.3]. We'll also consider in this group the two small moons of Mars, Phobos and Deimos (Figure 7.10), since they probably were once asteroids orbiting the Sun independently. We suspect they were captured by Mars early in its history, when it had an extended atmosphere that created the friction necessary to slow the passing asteroids enough for them to end up as moons.

The small sizes of all these objects probably rule out present life, because they almost certainly lack any liquid medium today. These bodies are too small to have any leftover internal heat that might melt any ices contained within them. Moreover, the comets and Kuiper belt objects generally orbit so far from the Sun that they are in a perpetual state of deep freeze. The lone exceptions are when impacts occur, which might melt small amounts of ice into water for short periods of time, and those rare comets that have had their original orbits perturbed enough to send them plunging into the inner solar system. As these comets come sunward, the same processes that cause their ices to vaporize into gas could conceivably melt small amounts of ice into liquid water. However, given that such periods would last only a short time on each orbit, and that such comets spend only small amounts of their multibillion-year histories on orbits that take them close to the Sun, present life does not seem likely.

The prospects for life on small bodies may have been better in the past, though it still seems unlikely. Recall that we have substantial evidence of complex organic molecules in both asteroids and comets, and studies of meteorites show that many of them must have contained liquid water during the earliest history of the solar system. The liquid water may have persisted over time periods as long as a few tens of millions of years. Could life have arisen during these time periods? We cannot rule it out, but we have found no evidence of past life in meteorites from the asteroid belt, and most scientists consider it unlikely that life could have originated on any of the countless small bodies in the solar system.

Tour Recap   We have completed our brief biological tour of the solar system. Although every world, large and small, has the raw chemical ele-

a Phobos   b Deimos

Figure 7.10

Mars has two small moons, Phobos (13 kilometers across) and Deimos (8 kilometers across), that are probably captured asteroids and much like many of the small moons of the outer solar system. Their small sizes make them unlikely to have any liquid water or life.

ments needed for life, the possibilities for life are much more limited when we focus on the environmental requirements we have found from the study of life on Earth. In particular, the need for liquid water, or possibly some other liquid medium, means we can effectively rule out finding life on the numerous small worlds of our solar system. We have similarly ruled out life on Mercury and the Moon, and found life to be unlikely on the jovian planets. That leaves the slim chances for life in the atmosphere of Venus, and the much better chances for life on Mars and a few of the moons of the jovian planets. We will discuss these cases further in the coming chapters.

It's important to remember that just because a world is an unlikely candidate for harboring life does not mean it's uninteresting or unimportant to our understanding of life in the universe. In particular, every world has at least something to teach us about how planetary systems form and why some worlds are habitable and others are not. This in turn helps us understand the prospects for life not only in our own solar system, but among other star systems as well. We still have much to learn about these issues, which is why every world in our solar system will remain a target of scientific interest for decades or even centuries to come.

7.4 THE PROCESS OF SCIENCE IN ACTION: SPACECRAFT EXPLORATION OF THE SOLAR SYSTEM

We have discussed a lot of details about the different worlds in our solar system during our biological tour, and we'll cover even more details in the next three chapters. How have we learned so much about all these objects?

Much of our knowledge comes from telescopic observations, using both ground-based telescopes and telescopes in Earth orbit such as the Hubble Space Telescope. In one case—our Moon—we have learned a lot by sending astronauts to explore the terrain and bring back rocks for laboratory study. In a few other cases, we have samples of distant worlds that have come to us as meteorites; we also have samples of comet dust returned by the *Stardust* mission. But most of the data fueling the recent revolution in our understanding of the solar system have come from robotic spacecraft. To date, we have sent robotic spacecraft to all of the eight official planets as well as to many moons, asteroids, and comets; the spacecraft called *New Horizons* is currently en route to Pluto. For this chapter's case study in the process of science in action, we'll briefly explore how these robotic spacecraft work.

• How do robotic spacecraft work?

The spacecraft we send to explore the planets are robots suited for long space journeys and jam-packed with specialized equipment for scientific study (Figure 7.11). All spacecraft have power sources such as solar cells, propulsion systems, devices to point cameras and other instruments precisely at their targets, and computers that control their major components. Robotic spacecraft operate primarily with preprogrammed instructions. They carry radios for communication, which allow them to receive additional instructions from Earth and to send home the data they collect. Most robotic spacecraft make one-way trips from Earth, never physically returning but sending their data back from space in the same way we send radio and television signals around the world.

Figure 7.11

The *Cassini* spacecraft before launch. It is now nearly a billion miles away as it orbits Saturn. Notice major components, such as rocket thrusters at the bottom, the communications dish at the top, and various scientific instruments arrayed all around the main skeleton of the spacecraft. The *Huygens* probe (which landed on Titan) was not yet attached when this photo was taken. (For details on what you are seeing, go to the *Cassini* Web site.)

Broadly speaking, the robotic missions we send to explore other worlds fall into four major categories:

- **Flyby:** A spacecraft on a flyby goes past a world just once and then continues on its way.

- **Orbiter:** An orbiter is a spacecraft that orbits the world it is studying, allowing longer-term observation during its repeated orbits.

- **Lander** or **probe:** These spacecraft are designed to land on a planet's surface or probe a planet's atmosphere by flying through it. Some landers have carried rovers to explore wider regions.

- **Sample return mission:** A sample return mission requires a spacecraft designed to return to Earth carrying a sample of the world it has studied.

The choice of spacecraft type depends on both scientific objectives and cost. In general, a flyby is the lowest-cost way to visit another planet, and some flybys gain more "bang for the buck" by visiting multiple planets. For example, *Voyager 2* flew past Jupiter, Saturn, Uranus, and Neptune before continuing on its way out of our solar system (Figure 7.12).

Flybys Flybys tend to be cheaper than other missions because they are generally less expensive to launch into space. Launch costs depend largely on weight, and onboard fuel is a significant part of the weight of a spacecraft heading to another planet. Once a spacecraft is on its way, the lack of friction or air drag in space means that it can maintain its orbital trajectory through the solar system without using any fuel at all. Fuel is needed only when the spacecraft needs to change from one trajectory (orbit) to another. Moreover, with careful planning, some trajectory changes can be made by taking advantage of the gravity of the planets.

For example, if you look closely at the *Voyager 2* trajectory in Figure 7.12, you'll see that the spacecraft made significant changes in path as it passed by Jupiter and Saturn. In effect, it made these changes for free by using gravity to bend its trajectory rather than by burning fuel. (This technique is known as a "gravitational assist," and it essentially speeds up the spacecraft significantly while slowing the planet by an unnoticeable amount.) The boost in speed can be quite significant: During its February 2007 flyby of Jupiter, the speed of the *New Horizons* spacecraft should increase by about 20%, shaving more than 3 years off the time it would otherwise take to reach Pluto.

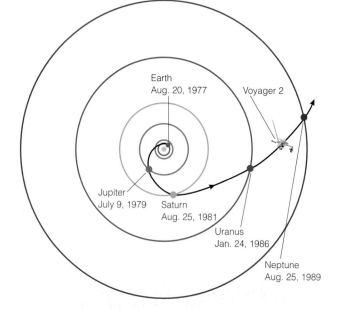

Figure 7.12

The trajectory of *Voyager 2*, which made flybys of each of the four jovian planets in our solar system.

THINK ABOUT IT Study the *Voyager 2* trajectory in Figure 7.12. Given that Saturn orbits the Sun every 29 years, Uranus orbits the Sun every 84 years, and Neptune orbits the Sun every 165 years, would it be possible to send another flyby mission to all four jovian planets if we launched it now? Explain.

Although a flyby offers only a relatively short period of close-up study, it can provide valuable scientific information. Flybys generally carry small telescopes, cameras, and spectrographs. Because these instruments are brought within a few tens of thousands of kilometers (or closer) of other worlds, they can obtain much higher-resolution images and spectra than even the largest current telescopes viewing these worlds from Earth. In addition, flybys sometimes give us information that would be difficult to obtain from Earth. For example, *Voyager 2* helped us discover Jupiter's rings and learn about the rings of Saturn, Uranus, and Neptune through views in which the rings were backlit by the Sun. Such views are possible only from beyond each planet's orbit.

Flybys may also carry instruments to measure local magnetic field strength or to sample interplanetary dust. The gravitational effects of the planets and their moons on the spacecraft itself provide information about object masses and densities. Like the backlit views of the rings, these types of data cannot be gathered from Earth. Indeed, most of what we know about the masses and compositions of moons comes from data obtained by spacecraft that have flown past them.

Orbiters   An orbiter can study another world for a much longer period of time than a flyby. Like the spacecraft used for flybys, orbiters often carry cameras, spectrographs, and instruments for measuring the strength of magnetic fields. Missions to Venus, Mars, and Saturn's moon Titan have also used radar to make precise altitude measurements of surface features. Radar works by sending radio waves from the spacecraft to bounce off the surface: The time it takes for the bounced signals to return to the spacecraft tells how far they traveled, allowing measurements of surface altitude. The precise way in which the signal is reflected by the surface can tell us about the terrain. In the cases of Venus and Titan, radar observations give us our only detailed surface views, because we cannot otherwise see through the thick cloud cover on these worlds.

An orbiter is generally more expensive than a flyby for an equivalent weight of scientific instruments, primarily because it must carry added fuel to change from an interplanetary trajectory to a path that puts it into orbit around another world. Careful planning can minimize the added expense. For example, recent Mars orbiters have saved on fuel costs by carrying only enough fuel to enter highly elliptical orbits around Mars. The spacecraft then settled into the smaller, more circular orbits needed for scientific observations by skimming the martian atmosphere at the low point of every elliptical orbit. Atmospheric drag slowed the spacecraft with each orbit and, over several months, circularized the spacecraft orbit. (This technique of using the atmosphere to slow the spacecraft and change its orbit is called *aerobraking*.)

To date, we have sent orbiters to the Moon and to the planets Venus, Mars, Jupiter, and Saturn, as well as to an asteroid called Eros (Figure 7.13). At present, orbiters are studying Venus, Mars, and Saturn. Another orbiter, called MESSENGER (which stands for "MErcury Surface, Space ENvironment, GEochemistry, and Ranging"), is on its way to Mercury. Launched in 2004, MESSENGER won't reach Mercury orbit until 2011. The reason for the long trip is that MESSENGER needs to slow down in order to enter an orbit around Mercury, and it would have been prohibitively expensive to launch it with enough fuel to slow itself down with rocket thrusts. Therefore, scientists planned a trajectory that allows it to use the gravity of planets to slow down, essentially doing the reverse of the type of speed boost picked up by the *New Horizons* spacecraft as it goes past Jupiter. By the time it reaches Mercury orbit, MESSENGER will have circled the Sun 15 times while passing once by Earth, twice by Venus, and three times past Mercury.

Landers and Probes   The most "up close and personal" study of other worlds comes from spacecraft that send probes into the atmospheres or landers to the surfaces. For example, in 1995, the *Galileo* spacecraft—which orbited Jupiter for more than 5 years—dropped a probe into Jupiter's atmosphere. The probe collected temperature, pressure, composition, and radiation measurements for about an hour as it descended, teaching us a

Figure 7.13

The asteroid Eros, photographed by the NEAR spacecraft. NEAR orbited Eros for a year before ending its mission with a soft landing on the surface. The inset photo was taken just before the spacecraft landed.

The aeroshell protects the rover from fiery temperatures as it enters the Martian atmosphere. Six minutes to landing.

With the parachute deployed, three retrorockets fire their engines, suspending the lander 30–50 feet above the Martian surface.

Protected by large airbags, the lander falls away from the parachute, landing safely on Mars. The lander bounces for several minutes, traveling hundreds of meters.

After bouncing to a stop, the airbags are retracted and the lander's petals deploy, creating a ramp for the rover to descend.

The rover deploys the solar arrays, wheels, cameras, and other instruments and begins its exploration of the Martian surface.

Figure 7.14

This artist's conception shows how the *Spirit* and *Opportunity* rovers used parachutes and air bags to land safely on Mars.

great deal about Jupiter's winds and atmospheric conditions before it was destroyed by Jupiter's high interior pressures and temperatures. The *Cassini* spacecraft, currently in orbit of Saturn, also carried a probe, called *Huygens*, that descended to the surface of Saturn's moon Titan while studying the atmosphere on its way down. We'll discuss the findings from *Huygens* in Chapter 9, when we cover Titan in more detail.

On planets with solid surfaces, a lander can offer close-up surface views, local weather monitoring, and the ability to carry out automated experiments. Some landers carry robotic rovers able to venture across the surface, such as the *Spirit* and *Opportunity* rovers that landed on Mars in 2004 and operated long beyond their planned lifetimes. Landers typically require fuel to slow their descent to a planetary surface, but clever techniques can reduce cost. For example, the recent Mars landers hit the surface of Mars at crash-landing speed, but were protected by cocoons of air bags deployed on the way down. These air bags allowed the landers to bounce along the surface of Mars before finally coming to rest (Figure 7.14).

Sample Return Missions   While probes and landers can carry out experiments on surface rock or atmospheric samples, the experiments must be designed in advance and the instrumentation must fit inside the spacecraft. These limitations make scientists long for missions that will scoop up samples from other worlds and return them to Earth for more detailed study. To date, sample return missions have been sent only to the Moon. Astronauts collected rock samples during the *Apollo* missions, and the then–Soviet Union sent robotic spacecraft to collect rocks from the Moon in the early 1970s. Many scientists are working toward a sample return mission to Mars, and they hope to launch such a mission within the next decade or so. A slight variation on the theme of a sample return mission is the *Stardust* mission, which collected comet dust on a flyby and then returned it to Earth in early 2006.

• What are some key spacecraft missions for astrobiology?

Every space mission provides at least some insight into issues important to astrobiology and, today, most missions are planned with at least some astrobiology considerations in mind. Table 7.2 lists some of the most significant robotic missions of the past and present.

A quick study of the table should make clear that we are living during an exciting time. Many missions that are already operating should provide

TABLE 7.2 *Selected Robotic Missions to Other Worlds*

Name	Type	Destination	Lead Space Agency	Arrival Year	Mission Highlights
New Horizons	Flyby	Pluto	NASA	2015[†]	First mission to Pluto, launched January 2006
Rosetta	Flyby and lander	Comet 67P/ Churyumov-Gerasimenko	ESA[*]	2014[†]	Launched in 2004, carries lander and orbiter for comet
MESSENGER	Orbiter	Mercury	NASA	2011[†]	Studies Mercury's surface, atmosphere, and interior
Mars Reconnaissance Orbiter	Orbiter	Mars	NASA	2006	High-resolution mapper will identify future landing sites
Venus Express	Orbiter	Venus	ESA[*]	2006	Studying Venus from orbit
Stardust	Flyby and sample return	Comet Wild 2	NASA	2006	Collected dust from comet in 2004, returned to Earth in 2006
Hayabusa	Lander and sample return	Asteroid Itokawa	JAXA[**]	2005	Collected asteroid sample in 2005, to return to Earth in 2007 (return may not be possible due to spacecraft problems)
Deep Impact	Flyby and "lander"	Comet Tempel 1	NASA	2005	"Lander" hit at 10 km/sec, flyby spacecraft studied the impact
Cassini	Orbiter	Saturn	NASA	2004	Included probe (*Huygens*) that landed on Titan
Mars Exploration Rovers	Lander	Mars	NASA	2004	Rovers *Spirit* and *Opportunity* explored surface
Mars Express	Orbiter	Mars	ESA[*]	2004	Climate studies of Mars
Mars Odyssey 2001	Orbiter	Mars	NASA	2001	Detailed study of Martian surface features and composition
Near Earth Asteroid Rendezvous (NEAR)	Orbiter	Eros (asteroid)	NASA	2000	First spacecraft dedicated to in-depth study of an asteroid
Mars Global Surveyor	Orbiter	Mars	NASA	1997	Detailed imaging of surface from orbit
Mars Pathfinder	Lander	Mars	NASA	1997	Carried *Sojourner*, the first robotic rover on Mars
Galileo	Orbiter	Jupiter	NASA	1995	Dropped probe into Jupiter; close-up study of moons
Magellan	Orbiter	Venus	NASA	1990	Detailed radar mapping of surface of Venus
Voyager 1	Flyby	Jupiter, Saturn	NASA	1979, 1980	Unprecedented views of Jupiter and Saturn; continuing out of solar system
Voyager 2	Flyby	Jupiter, Saturn, Uranus, Neptune	NASA	1979, 1981, 1986, 1989	Only mission to Uranus and Neptune; continuing out of solar system
Vikings 1 and 2	Orbiter and lander	Mars	NASA	1976	First landers on Mars

[*] European Space Agency.   [**] Japan Aerospace Exploration Agency.   [†] Scheduled arrival year.

new insights into the prospects for finding life in the solar system. For example, numerous missions are currently studying Mars, which is a prime target because it seems so likely to have been habitable in the past and perhaps to be habitable underground even today. *Venus Express* may shed light on the possibility of life in Venus's atmosphere. Studies of the dust returned by the *Stardust* mission should teach us both about comets and about the origin of our solar system, while studies of the aftermath of the *Deep Impact* blast into Comet Tempel 1 should tell us about comet structure.

Over the longer term, all the world's major space agencies have high hopes of launching numerous and diverse future missions to study our solar system. Of course, these hopes are subject to the vagaries of politics and budgets for scientific exploration. If they come to fruition, however, the next couple of decades will see spacecraft that answer many fundamental questions about the habitability of worlds in our solar system, and perhaps even tell us whether any of them have supported—or still support—life.

○ THE BIG PICTURE
Putting Chapter 7 into Context

In this chapter, we have discussed the general requirements for habitability, and taken a biological tour of the solar system as we know it today. As you continue in your studies, keep in mind the following "big picture" ideas:

- The general environmental requirements for life are much broader than the requirements for complex beings such as humans. The search for life beyond Earth is not necessarily a search for complex animals or intelligence, but in most cases a search for simpler life-forms and microbes.

- Our solar system contains a vast number of worlds, but most of them are unlikely to have life because they lack a liquid medium of any kind. Nevertheless, a few worlds may meet the criteria for habitability in at least some regions of their surfaces, subsurfaces, or atmospheres. More worlds may have had liquid water or other liquids in the distant past.

- Much of our current knowledge about the solar system and the potential for life comes from studies conducted by robotic spacecraft. We are living during a time when many spacecraft are simultaneously exploring different worlds in our solar system, and more missions are being planned for the future.

SUMMARY OF KEY CONCEPTS

7.1 ENVIRONMENTAL REQUIREMENTS FOR LIFE

- **Where can we expect to find building blocks of life?**
The chemical elements needed for life should be present on almost any world, but a smaller number of worlds will contain more complex organic molecules that can serve as building blocks for life. Still, the fact that these building blocks are present in asteroids and comets suggests that we'll find them in many places.

- **Where can we expect to find energy for life?**

Energy for life can come from sunlight or chemical reactions. Sunlight weakens with distance from the Sun, and is unlikely to be sufficient at large distances. Chemical energy is probably available in many more places, and is likely on any world with a substantial atmosphere or a liquid medium that can mix and support chemical reactions.

- **Does life need liquid water?**

Life almost certainly requires some liquid, and water has at least three advantages over other liquids: a wider and higher range of temperatures in which it is liquid; the fact that ice floats; and the type of chemical bonding made possible by charge separation within water molecules. Nevertheless, we can't completely rule out other liquids, such as liquid ammonia, methane, or ethane.

- **What are the environmental requirements for habitability?**
Life requires a source of molecules from which to build living cells, a source of energy for metabolism, and a liquid medium for transporting chemicals. In practice, these requirements probably come down to a need for liquid water, so the possibility of liquid water is the main requirement we search for in looking for habitable worlds.

7.2 A BIOLOGICAL TOUR OF THE SOLAR SYSTEM: THE INNER SOLAR SYSTEM

- **Does life seem plausible on the Moon or Mercury?**
The Moon and Mercury are probably not habitable, since neither has liquid water nor any other liquid medium for life.

- **Could life exist on Venus or Mars?**

It is far too hot for liquid water to exist on or under the surface of Venus, making life seem unlikely. However, life might be possible high in Venus's atmosphere, where clouds contain droplets of water. Mars almost certainly had habitable conditions in the distant past, and might still have habitable regions underground.

7.3 A BIOLOGICAL TOUR OF THE SOLAR SYSTEM: THE OUTER SOLAR SYSTEM

- **What are the prospects for life on jovian planets?**

The jovian planets have depths at which liquid water can exist in their atmospheres, but strong vertical winds make life seem unlikely. Uranus and Neptune may have "oceans" of water and other liquids in their deep interiors, but at present we have no way to search such depths for life.

- **Could there be life on moons or other small bodies?**
A few large moons may contain liquid water or other liquids, and thus seem like potential candidates for life. Smaller moons and other small bodies probably do not have any liquids at present, though many may have had liquid water in the distant past.

7.4 THE PROCESS OF SCIENCE IN ACTION: SPACECRAFT EXPLORATION OF THE SOLAR SYSTEM

- **How do robotic spacecraft work?**

Spacecraft can be categorized as flyby, orbiter, lander/probe, or sample return missions. In all cases, robotic spacecraft carry their own propulsion, power, and communication systems, and can operate under preprogrammed control or with updated instructions from ground controllers.

- **What are some key spacecraft missions for astrobiology?**
All spacecraft missions have at least some connection to astrobiology. Among the most important missions at present are those studying Mars, Titan, and Venus.

EXERCISES AND PROBLEMS

REVIEW QUESTIONS
Short-Answer Questions Based on the Reading

1. Why do we expect the elements of life to be widely available on other worlds? How does the requirement of organic building blocks further constrain the prospects of habitability?

2. How does the strength of sunlight vary with distance from the Sun? Discuss the implications for photosynthetic life.

3. Under what conditions does it seem reasonable to imagine a chemical energy source for life?

4. Why is a liquid medium important for life? Why does water seem the most likely liquid medium for life? Briefly discuss a few other potential liquids for life.

5. Summarize the three major environmental requirements for life. Overall, what "litmus test" seems appropriate for constraining our search for habitable worlds, and why?

6. Why do the Moon and Mercury seem unlikely to be habitable? Does evidence for ice in lunar craters affect the answer? Explain.

7. Why is Venus so much hotter than Earth? How does this heat affect the possibility of life on Venus? Explain why Venus may

nonetheless have been habitable in the past and might still be habitable in some of its clouds.

8. Why does Mars seem such a good candidate for life?

9. Briefly discuss the possibility of life on Jupiter and Saturn.

10. In the context of habitability, how do the cases of Uranus and Neptune differ from those of Jupiter and Saturn? Explain.

11. What characteristics make some of the large moons of jovian planets seem like potential candidates for habitability?

12. What are the prospects for habitability of the many small bodies in the solar system, and why?

13. Describe and distinguish between space missions that are *flybys, orbiters, landers* or *probes,* and *sample return missions.* What are the advantages and disadvantages of each mission type?

14. For a few of the most important past, present, or future robotic missions to the solar system, describe their targets, types, and mission highlights.

TEST YOUR UNDERSTANDING

Would You Believe It?

Each of the following gives a statement that a future explorer might someday make. In each case, decide whether the claim seems plausible in light of current knowledge. Explain clearly; because not all of these have definitive answers, your explanation is more important than your chosen answer.

15. On the smallest moon of Uranus, my team discovered a vast, subsurface ocean of liquid water.

16. New spacecraft images show lakes of liquid water on both Pluto and its moon Charon.

17. By drilling about a kilometer down into the Moon's surface, we are now able to pump up liquid water for the new Moon colony.

18. I was part of the first group of people to land on Venus, where we found huge, ancient cities that had been hidden from view by cloud cover.

19. We sent a robotic airplane into the atmosphere of Jupiter, but we could not keep it at a steady altitude and it was quickly ripped apart.

20. On a moon of Neptune, we discovered photosynthetic life with a metabolism that operates nearly a hundred times faster than that of any photosynthetic organism on Earth.

21. We deposited bacteria that get energy from chemical reactions with sulfur compounds into the upper clouds of Venus, and they are surviving.

22. The drilled sample showed no signs of life on asteroid B612, but we found many complex organic molecules.

23. We cut holes in the frozen methane surface of the lake on Titan, so that we could search for swimming organisms in the liquid methane underneath it.

24. The drilled sample from Mars brought up rock that contained microscopic droplets of liquid water.

Quick Quiz

Choose the best answer to each of the following. Explain your reasoning with one or more complete sentences.

25. Oxygen and carbon are: (a) rarer than almost all other elements; (b) found only on worlds close to a star; (c) the third- and fourth-most-abundant elements in the universe.

26. On an asteroid that is twice as far as Earth from the Sun, the strength of sunlight would be: (a) twice as great as on Earth; (b) $\frac{1}{2}$ as great as on Earth; (c) $\frac{1}{4}$ as great as on Earth.

27. Compared to liquid water, liquid methane is: (a) colder; (b) hotter; (c) denser.

28. Frozen lakes often have liquid water beneath their icy surfaces primarily because: (a) Earth's internal heat keeps the water liquid; (b) ice floats and provides insulation to the water below; (c) sunlight penetrates the ice and warms the water below.

29. Temperatures on Mercury are: (a) always very hot; (b) very hot in the day and very cold at night; (c) about the same as those on Venus.

30. On Venus, liquid water: (a) does not exist anywhere; (b) exists only deep underground; (c) exists only high in the atmosphere.

31. The reason that Venus is so much hotter than Earth is: (a) it has many more volcanoes; (b) its closer distance to the Sun makes sunlight dozens of times stronger; (c) its thick, carbon dioxide atmosphere creates a far stronger greenhouse effect.

32. Life is probably not possible in Jupiter's atmosphere because: (a) it is too cold there; (b) there is no liquid water at all; (c) winds are too strong.

33. Which ingredient are you most likely to find if you randomly choose a small moon of one of the jovian planets to examine? (a) water ice; (b) organic molecules; (c) an abundance of heavy metals, such as gold.

34. The *Cassini* spacecraft: (a) flew past Pluto; (b) landed on Mars; (c) is orbiting Saturn.

INVESTIGATE FURTHER

In-Depth Questions to Increase Your Understanding

Short-Answer/Essay Questions

35. *Solar System Tour.* Based on the brief tour in this chapter, which world in our solar system (besides Earth) do you think is most likely to have life? Explain why.

36. *Galileo Spacecraft.* In 2003, scientists deliberately ended the *Galileo* mission to Jupiter by causing the spacecraft to plunge into Jupiter's atmosphere. They did this to avoid any possibility that the spacecraft might someday crash into Europa, which could potentially have "contaminated" this moon with microbes from Earth. Do you think that the scientists should also have been worried about contaminating Jupiter itself? Why or why not?

37. *Greenhouse Effect.* The text (in Chapter 4) makes the statement that the greenhouse effect on Venus proves "that it is possible to have too much of a good thing." Explain this statement in two or three paragraphs.

38. *Transplanting Life.* Suppose you could genetically engineer organisms on Earth in any way that you chose. What, if any, features could you give them that would enable them to survive on: (a) the Moon or Mercury; (b) Venus; (c) Jupiter; (d) a comet?

39. *Mission Plan.* Suppose you could send a robotic mission of any type (flyby, orbiter, probe/lander) to any one of the places dis-

cussed in this chapter as being *unlikely* to be habitable. Which place and type of mission would you choose, and why? Defend your choice clearly in a one- to two-page essay.

40. *Science Fiction Life.* Choose a science fiction book or movie that describes some form of alien life that falls outside the bounds of the type of life we have considered in this chapter. Write a one- to two-page critical review in which you discuss the plausibility of the life-form.

Quantitative Problems

Be sure to show all calculations clearly and state your final answers in complete sentences.

41. *Understanding Newton's Version of Kepler's Third Law I.* Imagine another solar system, with a star of the same mass as the Sun. Suppose there is a planet in that solar system with a mass twice that of Earth orbiting at a distance of 1 AU from the star. What is the orbital period of this planet? Explain. (*Hint:* The calculations for this problem are so simple that you will not need a calculator.)

42. *Understanding Newton's Version of Kepler's Third Law II.* Suppose a solar system has a star that is four times as massive as our Sun. If that solar system has a planet the same size as Earth orbiting at a distance of 1 AU, what is the orbital period of the planet? Explain. (*Hint:* The calculations for this problem are so simple that you will not need a calculator.)

43. *Earth Mass.* The Moon orbits Earth in an average time of 27.3 days at an average distance of 384,000 kilometers. Use these facts to determine the mass of Earth. (*Hint:* You may neglect the mass of the Moon, since its mass is only about $\frac{1}{80}$ of Earth's.)

44. *Jupiter Mass.* Jupiter's moon Io orbits Jupiter every 42.5 hours at an average distance of 422,000 kilometers from the center of Jupiter. Calculate the mass of Jupiter. (*Hint:* Io's mass is very small compared to Jupiter's.)

45. *Pluto/Charon Mass.* Pluto's moon Charon orbits Pluto every 6.4 days with a semimajor axis of 19,700 kilometers. Calculate the *combined* mass of Pluto and Charon. Compare this combined mass to the mass of Earth, which is about 6×10^{24} kg. Can you determine the individual masses of Pluto and Charon from the given data? Explain.

46. *Mission to Pluto.* The *New Horizons* spacecraft will take about 9 years to travel from Earth orbit to Pluto. About how fast will it be traveling on average? Assume that its trajectory is close to a straight line. Give your answer in AU/year and in km/hr. (*Hint:* You can find needed data in Appendix D.)

Discussion Questions

47. *Bizarre Forms of Life.* Discuss some potential forms of life that have appeared in science fiction and fall outside the general types of life that we've discussed in this chapter. Which forms seem more plausible, and why? Overall, do you think the definition of life we've used in this chapter is too constraining? Defend your opinion.

48. *Artificial Life.* Imagine that future humans decide to breed new organisms tailored to as many different environments as possible. Discuss some of the places in our solar system where we could potentially plant such artificially created species, even if life probably would not arise naturally in those places. Do you think it likely that we will someday develop life-forms for other worlds? What are the philosophical ramifications of being able to custom-tailor life for worlds that don't have any natural life?

49. *Future Astrobiology Course.* Imagine that you are living a century from now and are taking a course about life in our solar system. Based on the current rate of exploration and reasonable rates for the future, how much more do you think we will know then about life in our solar system than we know now? Speculate about some of the discoveries you think may occur in the next century.

WEB PROJECTS

50. *Project Apollo.* Learn more about NASA's *Apollo* project, the only set of missions that has ever sent humans to another world. Describe the goals and objectives of each of the *Apollo* missions. Which were successful, and which were not? What lessons does *Apollo* offer for future attempts to send humans to the Moon and beyond? Summarize your findings and your opinions about lessons for the future in a one- to two-page essay.

51. *Planetary Missions.* Visit the Web page for one of the missions listed in Table 7.2. Write a one- to two-page summary of the mission's basic design, goals, and current status.

8
Mars

LEARNING GOALS

8.1 FANTASIES OF MARTIAN CIVILIZATION

- How did Mars invade popular culture?

8.2 A MODERN PORTRAIT OF MARS

- What is Mars like today?
- What are the major geological features of Mars?
- What geological evidence tells us that water once flowed on Mars?

8.3 THE CLIMATE HISTORY OF MARS

- Why was Mars warmer and wetter in the past?
- Why did Mars change?
- Is Mars habitable?

8.4 SEARCHING FOR LIFE ON MARS

- Is there any evidence of life on Mars?
- How do we plan to search for life on Mars?
- Should we send humans to Mars?

8.5 THE PROCESS OF SCIENCE IN ACTION: MARTIAN METEORITES

- Is there evidence of life in martian meteorites?

I remember being transfixed by the first lander image to show the horizon of Mars. This was not an alien world, I thought. I knew places like it in Colorado and Arizona and Nevada. There were rocks and sand drifts and a distant eminence, as natural and unselfconscious as any landscape on Earth. Mars was a place. I would, of course, have been surprised to see a grizzled prospector emerge from behind a dune leading his mule, but at the same time the idea seemed appropriate.

Carl Sagan, in *Cosmos*

The idea of a civilization on Mars was once taken so seriously that the term *Martians* became nearly synonymous with alien life. Spacecraft sent to Mars in the 1960s shattered this fictional image of a world of cities and sophisticated beings, revealing a cold, dry, cratered landscape. But while we can rule out a martian civilization, the existence of past or present microbial life on Mars remains a subject of intense scientific investigation and debate.

Substantial evidence suggests that water once flowed on Mars, and it seems likely that Mars once had surface or subsurface environments similar to those in which life thrived on the early Earth. If life arose on Mars (or was transplanted there on meteorites from the early Earth), we may be able to find its fossil remains. It's even possible that life still survives somewhere on Mars, perhaps underground where volcanic heat can keep some water liquid.

Whether or not we someday discover martian life, by studying Mars we are sure to learn a great deal about planetary habitability. Based on current hypotheses concerning the origin of life on Earth, Mars appears to have met all of the requirements for life to have arisen there as well. If we find life, it will help us understand more about biology in general. If we don't find life, Mars will still teach us much about the range of conditions under which life can arise. In this chapter, we'll discuss our past and present understanding of the habitability of Mars and the nature of our search for life on the red planet.

8.1 FANTASIES OF MARTIAN CIVILIZATION

Shining brightly and noticeably red in the nighttime sky, Mars has long captured the human imagination. It figures prominently in many ancient myths and legends. But the idea that Mars might harbor an advanced civilization arose only a little over 200 years ago. Before we discuss what we know about Mars today, let's go back in time and look at how Mars came to be so closely associated with the possibility of extraterrestrial life. For modern-day context, look at Figure 8.1, which shows a high-resolution image of one side of Mars as revealed by recent spacecraft observations, and look back at Figure 1.5, which shows the surface of Mars from the *Spirit* rover.

Figure 8.1

The full disk of Mars, as photographed by orbiting spacecraft. The horizontal "gash" across the center is the giant canyon Valles Marineris.

• How did Mars invade popular culture?

The story begins with the noted English astronomer William Herschel (1738–1822). Though best known for discovering the planet Uranus, Herschel made numerous other astronomical discoveries, usually with help from his sister and fellow astronomer Caroline Herschel (1750–1848). The Herschels often observed Mars through their telescopes, noting its polar ice caps and discovering that the length of its day (24 hours 37 minutes) is similar to an Earth day. In a talk presented to Britain's Royal Society in 1784, William Herschel also claimed that Mars possessed an atmosphere and that consequently, "its inhabitants probably enjoy a situation in many respects similar to our own." With the mention of "inhabitants," the possibility of living beings on the red planet had been broached by a respected scientist in an academic setting, and Martians were assumed to exist. (It should be noted that Herschel was not overly particular when it came to populating the cosmos. As far as he was concerned, everything in the solar system was inhabited, including the Moon and the Sun.)

During the following century, Mars rose to the top of the astronomical charts. Herschel had spoken of permanent spots on the planet, and in the 1860s Angelo Secchi in Rome noted some peculiar, streaky features as well. But it was the work of Giovanni Schiaparelli, director of the Milan Observatory, that transformed the perception of Mars for the next hundred years. In 1877, Schiaparelli claimed to see a network of 79 linear features that he called *canali,* by which he meant the Italian word for "channels." However, it was often translated incorrectly as "canals." Coming amid the excitement that followed the 1869 opening of the Suez Canal, Schiaparelli's discovery soon inspired visions of artificial waterways built by a martian civilization. Schiaparelli himself remained skeptical of such claims, and it's not clear whether he even thought the *canali* contained water. But his work caught the imagination of a young Harvard graduate, Percival Lowell (1855–1916).

Lowell, whose degree was in mathematics, came from a wealthy and distinguished New England family. His brother Abbott became famous as a president of Harvard, and his sister Amy gained fame as a poet. After spending a few years as a businessman and as a traveler in the Far East, Percival Lowell turned to astronomy. Impassioned by his belief in the martian canals and enabled by his wealth, Lowell commissioned the building of an observatory. He was aware that not all astronomers could make out Schiaparelli's *canali,* but rather than questioning the existence of the canals, he instead guessed that they were difficult to observe because most astronomical observatories were affected by poor "seeing" (a term used to describe the clarity and stability of the air above an observatory). Seeing varies from place to place and from night to night and tends to be best in places with dry air and high altitude. After considerable research, Lowell chose Flagstaff, Arizona, for his new observatory, because he believed it had the best overall seeing conditions in the United States. The Lowell Observatory opened in 1894 and is still operating today. Over a period of two decades, Lowell meticulously mapped nearly 200 canals that he claimed to see on Mars, identifying each of them with a Latin name.

In 1895, Lowell published his first detailed maps of the canals, as well as the first of three books in which he explained not only his belief that the canals were constructed by a martian civilization but also his conclusions about the nature of that civilization. Lowell correctly deduced that Mars was generally arid, and he guessed that the polar caps were made of water ice (in reality, they also contain carbon dioxide ice). He therefore imagined that the canals were built to carry water from the poles to thirsty cities nearer the equator. From there it was a short leap to imagine the Martians as an old civilization on a dying planet. The global network of canals convinced Lowell that the Martians were citizens of a single, global nation. Such ideas provided the "scientific" basis for H. G. Wells' *The War of the Worlds*, published in 1898.

> **THINK ABOUT IT** Think of as many popular references to a civilization of "Martians" as you can. For example, consider science fiction stories or movies, television shows, advertisements, or music that refers to intelligent beings on Mars. What do these references tell us about the influence of science on the public imagination?

Lowell was an effective advocate for the canals, but they do not really exist. Even in his own time, other scientists shot holes through most of Lowell's claims. Numerous astronomers, including some who had initially accepted the existence of canals, soon began to question Lowell's observations. These other astronomers could not see the canals when they put their eyes to the telescope—not even when using telescopes larger than Lowell's—and the canals failed to show up in photos made through telescopes as well. Moreover, while some canal-like features may have seemed to be present when seeing was poor, they were absent when seeing was best—the opposite of what we would expect if they were really there.

Other scientists questioned Lowell's basic assumptions and interpretations. Writing in 1907, Alfred Russel Wallace—the codiscoverer with Darwin of evolution by natural selection [Section 5.1]—used physical arguments to suggest that Mars must be too cold for liquid water to flow and that the polar caps must be dry ice (frozen carbon dioxide) rather than water ice. He also pointed out a major flaw in Lowell's interpretation of the canals: Lowell's canals all followed straight-line paths for hundreds or thousands of miles, but real canals would be built to follow natural contours of topography (for example, to go around mountains). In summarizing this argument, Wallace wrote that such a network of canals, "as Mr. Lowell describes, would be the work of a body of madmen rather than of intelligent beings."*

What was Lowell seeing? In only a few cases do his canals correspond to real features on Mars. For example, the canal he claimed to see most often (which he called *Agathodaemon*) coincides with the location of the huge canyon network now known as Valles Marineris (see Figure 8.1). A few other canals also roughly follow the contours of real features on Mars, but most of the canals were pure fantasy. Figure 8.2 compares a telescopic photo of Mars with one of Lowell's maps of the same regions. You can probably see how the dark and light regions match up in the photo and the drawing, but seeing any canals requires a vivid imagination.

Lowell's story illustrates both the pitfalls and the triumphs of modern science. The pitfall is that individual scientists, no matter how upstanding and dedicated, may still bring personal biases to bear on their scientific work. In Lowell's case, he was so convinced of the existence of canals and Martians that he simply ignored all evidence to the contrary. But the story's ending shows why modern science ultimately is so successful. Despite Lowell's stature, other scientists did not accept his claims on faith. Instead, they sought to confirm his observations and to test his

*Excerpted in K. Zahnle, "Decline and Fall of the Martian Empire," *Nature,* vol. 412, July 12, 2001.

Figure 8.2

Can you see how the markings on Mars in the telescopic photo on the left might have resembled the geometrical features in the drawing by Percival Lowell on the right? Try squinting your eyes.

Radio Listeners in Panic,
Taking War Drama as Fact

Many Flee Homes to Escape 'Gas Raid From Mars'—Phone Calls Swamp Police at Broadcast of Wells Fantasy

Figure 8.3

This front-page story from the *New York Times* described the panic caused by Orson Welles's 1938 broadcast of *The War of the Worlds.*

Figure 8.4

The surface of Mars photographed by the *Viking Lander 2* in 1979, showing a thin coating of ice on the rocks and soil. The inset shows a working model (actually, a spare spacecraft) of the *Viking* landers, identical to those that landed on Mars, on display at the National Air and Space Museum in Washington, D.C.

underlying assumptions. They found that Lowell's claims fell short on all counts. As a result, Lowell became an increasingly isolated voice as he continued to advocate a viewpoint that was clearly wrong.

The story also illustrates the divide that often persists between scientists and the general public. Although nearly all scientists had abandoned belief in Martians within the first few decades of the twentieth century, among the public the canal myth persisted much longer. Public belief in Martians remained widespread enough to create a famous panic during Orson Welles's 1938 radio broadcast of *The War of the Worlds*—many people thought an invasion was actually under way (Figure 8.3). Books and movies about martian civilization remained popular throughout the twentieth century. Even today, a few die-hard proponents claim to see evidence of a past martian civilization, but recent high-resolution photographs make their claims increasingly implausible.

8.2   A MODERN PORTRAIT OF MARS

The public debate about martian canals and cities was not entirely put to rest until NASA began sending spacecraft to Mars. In 1965, NASA's *Mariner 4* spacecraft flew to within 6,000 miles of the martian surface, transmitting a few dozen television-quality images of the landscape below. When the photos were given to President Lyndon Johnson, he looked at them with surprise. Mars's surface was littered with craters, not canals. The red planet looked as lifeless as the Moon. Measurements of the atmospheric pressure and temperature made from the spacecraft indicated a cold, dry planet seemingly incapable of supporting life.

Nevertheless, all was not lost when it came to the potential for life on Mars. There was no evidence of any intelligent beings, but the thin atmosphere and the polar caps left open the possibility of the existence of microbes or perhaps even some primitive plants or animals. On July 20, 1976, seven years to the day after Neil Armstrong's history-making walk on the Moon and nearly a century since Schiaparelli's description of *canali,* the thin skies above Mars were pierced by a NASA space probe. *Viking Lander 1* touched down without difficulty on the Chryse Planitia, a sprawling, rock-strewn plain about 1,300 kilometers north of the martian equator. It was followed nearly two months later by its twin, *Viking Lander 2,* which landed almost halfway around the planet. Meanwhile, two *Viking* orbiters studied the planet from above.

When the *Viking* landers' cameras opened their eyes in the frigid martian air, they found a bleak landscape with red dust and scattered rocks. No creatures stared back at the cameras, and no plants were huddled in the weak sunlight. For months the images continued to come in, but the view scarcely changed. Nothing grew other than some occasional patches of frost, and nothing moved other than windblown dust (Figure 8.4). Though neither lander could move from the spot at which it had settled, each had a robotic arm with which it collected soil for some onboard experiments designed to look for microbes (see Section 8.4).

The *Viking* orbiters and landers provided a wealth of scientific data about Mars. But they also left many questions unanswered, and the scientific community was itching for follow-up missions. Unfortunately, budgetary and political considerations, along with the failure of two Russian missions to Mars (*Phobos 1* and *2*) and one American mission (*Mars Observer*), all conspired to stop spacecraft exploration of Mars for some 20 years. The long mission drought did not end until July 4, 1997, with

Figure 8.5

The view from the *Pathfinder* lander (partially visible in the foreground); the scattered rocks were probably carried to the site by an ancient flood. The little rover, *Sojourner*, is at the upper right, studying a rock that scientists named Yogi.

the landing of *Pathfinder* and its little rover, *Sojourner* (Figure 8.5), which was named for Sojourner Truth, an African American heroine of the Civil War era. Although *Sojourner* could travel only a few tens of meters from *Pathfinder*, this was enough to check the chemical composition of many nearby rocks.

The *Pathfinder* mission marked the beginning of a new era of intensive scientific study of Mars. By early 2006, four orbiters and two roving landers were all studying Mars (see Table 7.2). By combining data from these new missions with earlier *Viking* and *Pathfinder* data and telescopic studies, we are beginning to put together a realistic portrait of the past and present habitability of Mars.

• What is Mars like today?

The present-day surface of Mars may look much like some deserts or volcanic plains on Earth, but its thin atmosphere makes it unlike any place on Earth. Table 8.1 summarizes some basic Mars data. Note that the surface temperature is usually well below freezing, while the atmospheric pressure is less than 1% that on the surface of Earth—making the air so thin that no human could survive outside for more than a few minutes without a pressurized space suit. The air contains only trace amounts of oxygen, so we could not breathe it. The lack of oxygen also means that Mars lacks an ozone layer, so much of the Sun's damaging ultraviolet radiation passes unhindered to the surface.

Nevertheless, martian conditions are much less extreme than those on the Moon (mainly because of the moderating effects of the atmosphere), and it's easy to imagine future astronauts living and working in airtight research stations while occasionally donning space suits for outdoor excursions. Martian surface gravity is about 40% that on Earth, so everyone and everything would weigh about 40% of Earth weight. Astronauts could walk around easily even while wearing space suits with heavy backpacks. It would also be easy to adapt to patterns of day and night, since the martian day is only about 40 minutes longer than an Earth day.

The Lack of Surface Liquid Water The low atmospheric pressure explains one of the key facts relevant to the search for life on Mars: No liquid water exists anywhere on the surface of Mars today. We know this

TABLE 8.1 *Basic Mars Data*

Average Distance from Sun	1.52 AU = 227.9 million km
perihelion distance	206 million km
aphelion distance	249 million km
Orbital Period	1.881 Earth yr
Equatorial Radius	3,397 km
Mass (Earth = 1)	0.107
Rotation Period	24 hr 37 min
Surface Gravity (Earth = 1)	0.38
Atmospheric Composition	95% CO_2, 2.7% N_2, 1.6% argon
Average Surface Temperature	−50°C
Average Surface Pressure	0.007 bar[*]

[*] 1 bar ≈ sea level pressure on Earth

not only because we've studied most of the surface in reasonable detail, but also because the surface conditions would not allow liquid water to be present as lakes, rivers, or even puddles. In most places and at most times, Mars is so cold that any liquid water would immediately freeze into ice. Even when the temperature rises above freezing, as it often does at midday near the equator, the air pressure is so low that liquid water would quickly evaporate. In other words, liquid water is *unstable* on Mars today: If you donned a space suit and took a cup of water outside your pressurized spaceship, the water would almost immediately either freeze or evaporate away (or some combination of both). However, as we'll discuss shortly, Mars almost certainly had liquid water in the past, and still has ample water ice and some water vapor, and perhaps even pockets of liquid water underground.

SPECIAL TOPIC: THE FACE ON MARS

Among the tens of thousands of photographs snapped as part of the *Viking* missions to Mars in the late 1970s, one achieved special fame. The *Viking 1* orbiter snapped a picture showing a feature on Mars that looked remarkably like a human face (Figure 1). The "face on Mars" soon became a cottage industry, with proponents arguing that it must be the work of a lost Martian civilization or of alien beings visiting our solar system.

To scientists, it seemed a near-certainty that the human likeness was a coincidence of geology and camera angle. However, the people promoting an artificial origin for the "face" claimed that NASA was conspiring to hide the truth—despite the fact that it was first described as a face by a NASA scientist in a NASA press release. A few proponents even claimed that a failed NASA mission to Mars (*Mars Observer* in 1992) had actually achieved orbit successfully and was imaging the face while NASA hid the results from the public.

Today, the truth is clear for all to see: Figure 2 compares the original *Viking* photo with a more recent and higher-resolution photo from *Mars Global Surveyor*. The "face" was a product of shadows and the human tendency to see patterns where none exist—just as we see faces in clouds, a man in the Moon, and canals on Mars.

Figure 2

Two photos of the "face on Mars." The photo on the left is another image from *Viking 1* in 1976 (a blow-up view of the face in Figure 1). The much higher-resolution photo on the right is from the *Mars Global Surveyor* mission.

A few die-hard proponents still claim either that it was a face that has been heavily altered by natural erosion or that there is still a NASA conspiracy to hide the truth (though Mars is also now being studied by a European mission, so the conspiracy would have to be global). It's hard to blame them for their desire to see evidence of an extraterrestrial civilization. After all, there's a certain appeal to all the adventures that past science fiction writers imagined taking place among the inhabitants of Mars. But if you look hard enough, you'll always find coincidental patterns: For example, in searching through hundreds of other photos from Mars missions, NASA scientists came upon a crater that seems to show a "happy face" on Mars (Figure 3).

Figure 1

This *Viking 1* image shows the "face on Mars" and the surrounding area. The face is on a tall mesa about 1 kilometer across. Some of the other features were claimed to represent a "fortress" and "pyramids." The black dots are instrument artifacts.

Figure 3

This *Viking* image shows a crater on Mars that appears to show a "happy face."

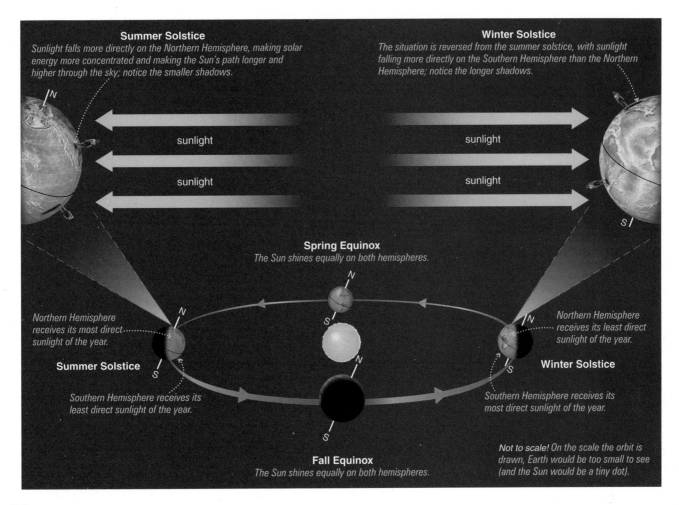

Summer Solstice
Sunlight falls more directly on the Northern Hemisphere, making solar energy more concentrated and making the Sun's path longer and higher through the sky; notice the smaller shadows.

Winter Solstice
The situation is reversed from the summer solstice, with sunlight falling more directly on the Southern Hemisphere than the Northern Hemisphere; notice the longer shadows.

sunlight

sunlight

sunlight

sunlight

Spring Equinox
The Sun shines equally on both hemispheres.

Northern Hemisphere receives its most direct sunlight of the year.

Summer Solstice

Southern Hemisphere receives its least direct sunlight of the year.

Northern Hemisphere receives its least direct sunlight of the year.

Winter Solstice

Southern Hemisphere receives its most direct sunlight of the year.

Fall Equinox
The Sun shines equally on both hemispheres.

Not to scale! On the scale the orbit is drawn, Earth would be too small to see (and the Sun would be a tiny dot).

Figure 8.6 | Interactive Figure |

Earth's seasons are caused by the tilt of the axis. Notice that the axis points in the same direction (toward Polaris) throughout the year, which means the Northern Hemisphere is tipped toward the Sun on one side of the orbit and away from the Sun on the other side. The same is true for the Southern Hemisphere, but on opposite sides of the orbit.

Martian Seasons and Winds The tilt of the martian axis is about 25°—only slightly greater than Earth's 23.5° tilt—so Mars has seasons much like those on Earth. However, because the martian year is nearly twice as long as an Earth year, each season lasts nearly twice as long on Mars.

The martian seasons also differ from seasons on Earth in another important way that is due to the nature of Mars's orbit. Seasons on Earth are caused by the tilt of the axis. Earth's axis remains pointed in the same direction (toward the north star, Polaris) throughout the year, so on one side of the orbit, the Northern Hemisphere is tilted toward the Sun and on the other side, it is tilted away from the Sun. Therefore, as shown in Figure 8.6, the Northern Hemisphere receives more direct sunlight when the Southern Hemisphere receives less direct sunlight, and vice versa. That is why the two hemispheres experience opposite seasons.

Earth's orbit is so close to being circular that axis tilt is the only significant influence on the strength of sunlight reaching the surface at different times of year. On Mars, however, the ellipticity of the orbit introduces a second important effect: Mars is significantly closer to the Sun during its southern hemisphere summer and farther from the Sun during its southern hemisphere winter (Figure 8.7). Mars therefore has more

Seasons on Mars

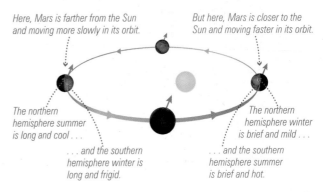

Here, Mars is farther from the Sun and moving more slowly in its orbit.

But here, Mars is closer to the Sun and moving faster in its orbit.

The northern hemisphere summer is long and cool . . .

The northern hemisphere winter is brief and mild . . .

. . . and the southern hemisphere winter is long and frigid.

. . . and the southern hemisphere summer is brief and hot.

Figure 8.7

The ellipticity of Mars's orbit makes seasons more extreme (hotter summers and colder winters) in the southern hemisphere than in the northern hemisphere.

October 1996: Late winter, with polar cap near maximum size.

January 1997: Midspring, notice shrinkage of polar cap.

March 1997: Early Summer, with polar cap near minimum size.

Figure 8.8

These images from the Hubble Space Telescope show Mars in polar projection so that the north pole is in the center of the images. Notice how the polar cap shrinks as summer approaches, due to sublimation of carbon dioxide ice. This sublimation at the summer pole occurs at the same time that carbon dioxide condenses at the winter pole. These changes drive winds from the summer pole to the winter pole.

extreme seasons in its southern hemisphere—that is, shorter, hotter summers and longer, colder winters—than in its northern hemisphere. (The shorter duration of the southern summer arises because, in accord with Kepler's second law [Section 2.2], Mars moves faster in the portion of its orbit closer to the Sun.)

The seasonal changes create the major feature of Mars's weather: winds blowing from the summer pole to the winter pole. Polar temperatures at the winter pole drop so low (about −130°C) that carbon dioxide condenses into "dry ice," forming a polar cap of carbon dioxide frost that can be as much as a meter thick at the pole and can extend as far south as latitude 40°. At the same time, frozen carbon dioxide at the summer pole sublimates (goes directly from solid to gas phase) into carbon dioxide gas (Figure 8.8). During the peak of summer, nearly all the carbon dioxide may sublimate from the summer pole, leaving only a residual polar cap of water ice (Figure 8.9). The atmospheric pressure therefore increases at the summer pole and decreases at the winter pole, driving strong pole-to-pole winds. Overall, as much as one third of the total carbon dioxide of the martian atmosphere moves seasonally between the north and south polar caps.

200 km

Figure 8.9

This image, from *Mars Global Surveyor*, shows the residual north polar cap during northern hemisphere summer. The white material is primarily water ice, with dust mixed in.

THINK ABOUT IT   Understanding the difference between sublimation and evaporation is important to visualizing the behavior of ices and liquids on Mars. It's easy to watch evaporation on Earth, because any puddle of water soon evaporates. Sublimation, though less obvious, is also common. An easy way to see sublimation is with frozen carbon dioxide, also known as "dry ice." How can you tell that dry ice is sublimating? (If you've never seen dry ice, it is readily available; try looking up local sources of it on the Web.)

The direction of the pole-to-pole winds on Mars changes with the alternating seasons. Sometimes these winds initiate huge dust storms, particularly when the more extreme summer approaches in the southern hemisphere (Figure 8.10). At times, airborne dust shrouds the surface so much that no surface markings can be distinguished and large dunes form and shift with the strong winds.

The martian winds often spawn *dust devils,* swirling winds that you may have seen over desert sands or dry dirt on Earth. Dust devils look much like miniature tornadoes, but they rise up from the ground rather than coming down from the sky. The air in dust devils is heated from below by the sunlight-warmed ground, and dust devils swirl because of the way the rising air interacts with prevailing winds. Dust devils on Mars are especially common during summer in either hemisphere and can be far larger than their counterparts on Earth. The Mars rovers have even photographed dust devils (Figure 8.11), and some have apparently passed directly over the rovers without causing damage. In one case, a dust devil apparently "cleaned" dust off the solar panels of the rover *Spirit,* restoring power that had been lost as dust accumulated.

Winds generated by the temperature differences near the edges of the polar caps sometimes initiate local dust storms. Occasionally, local dust storms in low southern latitudes expand during summer into huge dust storms that enshroud the entire planet (see Figure 8.10). At times the martian surface becomes almost completely obscured by airborne dust. As the dust settles out, it can change the surface appearance over vast areas (for example, by covering dark regions with brighter dust); such changes fooled past astronomers into thinking they were seeing seasonal changes in vegetation.

Color of the Martian Sky   Martian winds and dust storms also leave Mars with perpetually dusty air, which helps explain the colors of the martian sky. The air on Mars scatters visible sunlight much like the air on Earth; that is, it scatters blue light more than red light (which is why our sky is blue). However, because there is much less air on Mars than Earth, it scatters less light. Thus, by itself this scattering would make the martian sky a deep blue. The dust in the atmosphere has an almost opposite effect. Dust tends to absorb blue light, so by itself it would leave the sky with a brownish-pink color. The actual color of the martian sky comes from the combination of the effects of air and dust. The result depends on the time of day, the season, and how much dust is in the air. Most of the time, the daytime martian sky ends up yellow-brown. However, it sometimes ends up with other colors, such as blue or even green, particularly in the mornings and evenings.

• What are the major geological features of Mars?

The surface of Mars may be desolate and barren today, but it was not always so. Many surface features appear to have been shaped by liquid water, leading scientists to conclude that Mars must once have had a much more hospitable climate. But before we discuss the evidence for surface water and ideas about the climate history of Mars, it's useful to get our bearings by looking at the large-scale geographic features of the planet.

A Map of Mars   Figure 8.12 shows the full surface of Mars, with the poles at the top and bottom and the equator running horizontally across the middle (in much the same way that an atlas shows the full globe of Earth). Mars is about half as large in diameter as Earth, so its surface area

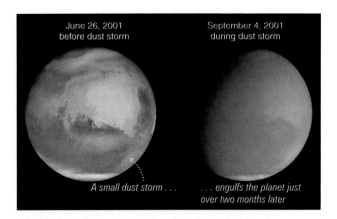

Figure 8.10

These two Hubble Space Telescope photos contrast the appearance of the same face of Mars in the absence (left) and presence (right) of a global dust storm.

Figure 8.11

This photograph shows a dust devil on Mars, photographed by the *Spirit* rover in early 2005.

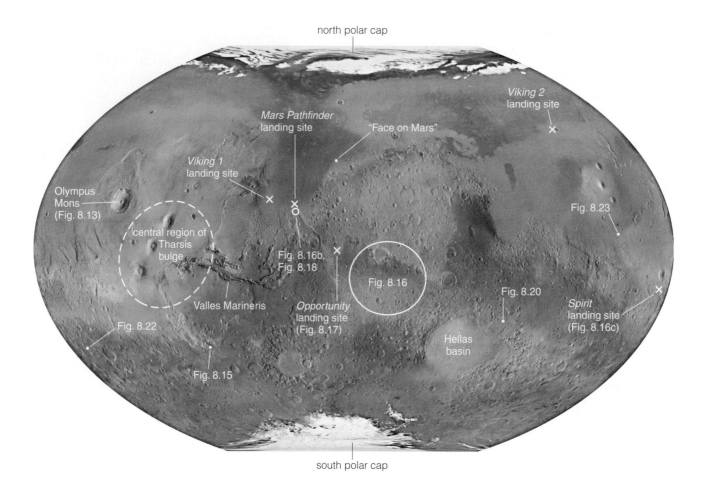

north polar cap

Viking 2
landing site

Mars Pathfinder
landing site

"Face on Mars"

Viking 1
landing site

Olympus
Mons
(Fig. 8.13)

Fig. 8.23

central region of
Tharsis
bulge

Fig. 8.16b,
Fig. 8.18

Fig. 8.16

Valles Marineris

Opportunity
landing site
(Fig. 8.17)

Fig. 8.20

Spirit
landing site
(Fig. 8.16c)

Fig. 8.22

Hellas
basin

Fig. 8.15

south polar cap

Figure 8.12   Interactive Figure

This image showing the full surface of Mars is a composite made by combining more than 1,000 images with more than 200 million altitude measurements from the *Mars Global Surveyor* mission. Several key geological features are labeled, and the locations of features shown in close-up photos elsewhere in this book are marked.

is about one fourth that of Earth (because surface area is proportional to the square of the radius). Thus, if you placed this map next to a map of Earth on the same scale, the Earth map would be four times as large in area. Moreover, because water covers about three fourths of Earth's surface, the total land area of Mars is about the same as the total land area of Earth.

THINK ABOUT IT   Try to find a map of Earth that has about four times the area (twice as long and wide) as Figure 8.12. (If you cannot find a map of the right size, use any Earth map and try to photocopy it at the appropriately scaled size.) Compare the size of various Earth features, such as continents and oceans, to that of various Mars features.

Many key geographical features are clearly visible in Figure 8.12. After the polar caps, the most striking feature is probably the dramatic difference in terrain around different parts of Mars. Much of the southern hemisphere has relatively high elevation and is scarred by numerous large impact craters, including the very large crater known as the Hellas Basin. In contrast, the northern plains show few impact craters and tend to be below the average martian surface level. This difference in crater crowding tells us that the southern highlands are a much older surface than the northern plains: Recall that all worlds should have been heavily scarred with craters during the heavy bombardment, so surface regions that still show these craters must be quite old (3.8 billion years or more) while surface regions that have few craters must have had their early craters erased by subsequent geological processes [Section 4.3].

Volcanism was probably the most important of the processes responsible for erasing ancient craters on Mars, although tectonics and erosion also played a part. The northern plains show features that are characteristic of lava flows, suggesting that eruptions of an extremely fluid lava covered up the older impact craters. Interestingly, we can see faint "ghost" craters in some of these regions, suggesting that the lava flows were not thick enough to completely erase the underlying features and confirming that the entire planet was once densely cratered. Plenty of mysteries still remain, however. For example, no one knows why volcanism should have affected the northern plains so much more than the southern highlands or why the two regions differ so much in elevation.

Other evidence of volcanism on Mars comes from its towering volcanoes. You can probably recognize the volcanoes in Figure 8.12 by their dome shapes and central calderas (the "craters" in the centers of volcanoes). Many are concentrated on or near the continent-size Tharsis Bulge. Tharsis, as it is usually called, is some 4,000 kilometers across, and most of it rises several kilometers above the average martian surface level. It was probably created by a long-lived plume of rising mantle material that bulged the surface upward and provided the molten rock for the eruptions that built the giant volcanoes. The Tharsis volcanoes dwarf any found on Earth. Olympus Mons (Figure 8.13), located near the northwest edge of Tharsis, is probably the largest mountain in the solar system. Its base is some 600 kilometers across, large enough to cover an area the size of Arizona, and is rimmed by a cliff that in places is 6 kilometers high. Its peak stands about 26 kilometers above the average martian surface level, or some three times as high as Mount Everest stands above sea level on Earth.

East of Tharsis and just south of the equator is the long, deep system of valleys called *Valles Marineris*. Named for the *Mariner 9* spacecraft that first imaged it, Valles Marineris is as long as the United States is wide and almost four times as deep as the Grand Canyon. No one knows exactly how Valles Marineris formed. However, if you look closely at Valles Marineris in Figure 8.12, you'll see that its west end shows extensive cracks where it runs up against the Tharsis Bulge, suggesting a connection between the two features. Valles Marineris may have formed through tectonic stresses accompanying the uplift of material that created the Tharsis Bulge, cracking the surface and leaving the tall cliff walls of the valleys. A few features of the valleys may have been shaped by flowing water, and in some places the canyon walls show evidence of layering that may have been caused either by numerous lava flows or by deposits of sediments. Moreover, the canyon is so deep in places that some of its walls probably were once several kilometers underground, where they may have been exposed to subsurface liquid water. For this reason, Valles Marineris may be one of the best places to look for fossil evidence of past martian life.

Martian Geological History The varying terrain and geographical features of Mars provide important clues to its geological history. In particular, detailed comparisons of crater counts on different parts of Mars allow us to divide the surface broadly into regions of three different ages (Figure 8.14). The oldest and most heavily cratered regions—which include the heavily cratered southern highlands—must still be essentially unchanged from the way they looked at the end of the heavy bombardment,

20 km

Figure 8.13

Olympus Mons, photographed from orbit. For perspective on its size compared to the entire planet, you should also look at this giant volcano toward the far left of Figure 8.12.

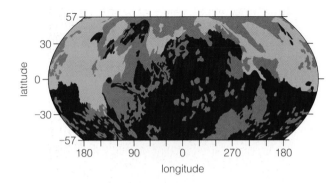

Figure 8.14

This simplified map of Mars shows how different surface regions have different ages based on crater counts; the color coding is summarized in Table 8.2 below. (Compare this simplified map to the more detailed view in Figure 8.12.)

TABLE 8.2 *Geological Eras on Mars*

Era	Color Code in Figure 8.14	Time
Early (Noachian)		4.6–3.8 billion years ago
Middle (Hesperian)		3.8–1.0 billion years ago
Recent (Amazonian)		1.0 billion years ago to the present

some 3.8 billion years ago. The youngest regions are those with the fewest craters, such as the regions around the volcanoes of Tharsis. Crater counts suggest that these young regions are less than about 1 billion years old. In between, Mars has large patches of "middle-aged" surface where crater counts suggest an age from about 1 billion to 3.8 billion years old. For simplicity, we will refer to these different age surfaces as dating to the early, middle, and recent eras of Mars, as summarized in Table 8.2.* Keep in mind that ages based on crater counts are quite rough; we will get more precise ages only when we are able to collect rocks from the different regions and determine their ages through radiometric dating [Section 4.2].

THINK ABOUT IT Does Earth's surface have regions of different ages similar to what we find on Mars? Explain.

By examining the types of geological features that appear on surfaces of different ages on Mars, we can get an idea both of what processes helped shape the surface and of when they operated. For example, we can look at features that indicate volcanic eruptions, such as lava flows or volcanoes, and deduce the history of volcanism. Such studies show that frequency of volcanic eruptions on Mars has decreased steadily since at least about 3.5 billion years ago, just as we would expect for a planet small enough to have lost much of its internal heat by now.

Nevertheless, Mars may not yet be geologically "dead," a fact that has important implications to the possibility of martian life. Although martian volcanoes show enough impact craters on their slopes to suggest that they have been inactive for at least tens of millions of years, analysis of martian meteorites (meteorites that appear to have come from Mars [Section 6.2]) offers a different perspective. Radiometric dating of these meteorites shows some of them to be made of volcanic rock that solidified from molten lava as little as 180 million years ago—quite recent in the 4.5-billion-year history of the solar system. Given this evidence of geologically recent volcanic eruptions, it is likely that martian volcanoes will erupt again someday, though not necessarily in our lifetimes. In that case, Mars must still retain at least some internal heat, and almost certainly enough to melt some underground ice into liquid water.

• What geological evidence tells us that water once flowed on Mars?

Large-scale features of Mars are important to our overall understanding of the planet, but features that might have been carved by liquid water are of particular interest for our investigation of Mars's habitability. The martian surface shows ample evidence of liquid water in its past, which is the prime reason we consider Mars a candidate for past or present life.

Evidence from Orbit The first evidence of past water came from photos taken by *Mariner 9* and the *Viking* orbiters, some of which showed features that that look much like dry riverbeds on Earth seen from above

---

*Planetary scientists sometimes refer to the early period of Mars as the *Noachian* (pronounced "*no-AH-ki-an*") era, the middle period as the *Hesperian* era, and the recent period as the *Amazonian* era.

(Figure 8.15). These channels were almost certainly carved by running water, though no one knows whether the water came from runoff after rainfall, from erosion by water-rich debris flows, or from an underground source. From counts of the craters in and near the channels, it appears that they are at least 2–3 billion years old, meaning that water has not flowed through them since that time. Nevertheless, they tell us an important story about the martian past: The branching nature of the channels suggests that they were carved gradually, so liquid water must have been stable at or just below the surface at that time. Because the low temperature and low atmospheric pressure make liquid water unstable at the surface today, we conclude that Mars must have had a much warmer and thicker atmosphere during at least some times in its distant past.

Other orbital evidence also argues that Mars had rain and surface water long ago. Figure 8.16a shows a broad region of the ancient, heavily cratered southern highlands. Notice the indistinct rims of many large craters and the relative lack of small craters. Both facts argue for ancient rainfall, which would have eroded crater rims and erased small craters altogether. Figure 8.16b shows a close-up of a crater floor with sculpted patterns that suggest it was once the site of a lake. Ancient rains may have filled the crater, allowing sediments to build up from material that settled to the bottom. The sculpted patterns in the crater bottom may have been created as erosion by wind exposed layer upon layer of sedimentary rock, much as erosion by the water of the Colorado River exposed the layers visible in the walls of the Grand Canyon on Earth. Figure 8.16c shows a three-dimensional perspective of a surface region where a channel between two craters suggests that they were once lakes with water flowing between them.

Surface Evidence   Surface studies have provided additional evidence for past water on Mars. The *Pathfinder* landing site was in the midst of

Figure 8.15

This photo, taken by a *Viking* orbiter, shows what appear to be dried-up riverbeds. Toward the top of the image we see many individual tributaries, which merge into the larger "river" near the lower right. Counts of craters near the channels indicate that they formed at least 2–3 billion years ago.

Figure 8.16

More orbital evidence of past water on Mars.

a This photo shows a broad region of the southern highlands. The eroded rims of large craters and the relative lack of small craters suggest erosion by rainfall.

b The close-up view of this crater floor shows sculpted patterns that probably represent layers of sedimentary rock, presumably laid down at a time when water filled the crater.

c This computer-generated perspective shows how a martian valley forms a natural passage between two possible ancient lakes (shaded blue); the lower crater is Gusev, where the *Spirit* rover landed. Vertical relief is exaggerated 14 times to reveal the topography.

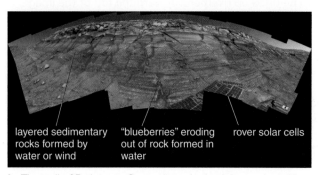

a Zooming in on a knee-high rock outcrop near the rover's landing site. The close-up shows a piece of the rock about 3 cm across. The layered structure, odd indentations, and small sphere all support the idea that the rock formed from sediments in standing water.

layered sedimentary rocks formed by water or wind

"blueberries" eroding out of rock formed in water

rover solar cells

b The wall of Endurance Crater shows bedrock layers altered by water. Close-ups show hematite "blueberries" eroding out of the rock. The changing tilts of the rock layers hint at changing wind or waves during deposition.

Figure 8.17

Photos from the *Opportunity* rover on Mars.

Figure 8.18

This image from Europe's *Mars Express* orbiter shows outflow channels likely carved by floodwaters. The inset is a perspective view of the central region.

what appears from orbit to be a great floodplain. The surface view from *Pathfinder* (see Figure 8.5) supports the idea that a flood passed through the region, because rocks are scattered and stacked against each other in the same way that we see them after floods on Earth.

Even stronger evidence comes from the recent rover missions to Mars. The *Spirit* rover has explored Gusev crater, suspected site of an ancient lake (see Figure 8.16c). *Opportunity* landed in the Meridiani Plains, where orbital spacecraft had detected spectroscopic hints of minerals that form in water. The twin rovers carried dozens of cameras, several instruments to identify rock composition, and even a grinder to expose fresh rock for analysis.

Both rovers found strong evidence that liquid water was once plentiful on Mars. Rocks at the *Opportunity* landing site contain tiny spheres—nicknamed "blueberries" although they're neither blue nor as large as the berries we find in stores—and odd indentations suggesting that they formed in standing water, or possibly by groundwater percolating through rocks (Figure 8.17). Compositional analysis shows that the abundant "blueberries" contain the iron-rich mineral hematite, and other rocks contain the sulfur-rich mineral jarosite. Both of these minerals form in water, and chemical analysis seems to support the case for formation in an acidic, salty environment such as a sea or ocean. In addition, the layering of the sedimentary rocks suggests a changing environment of waves and/or wind, furthering the case for a past sea or ocean in the Meridiani Plains.

The specific case for water in the Meridiani Plains is not fully closed, because some scientists have proposed volcanic processes, or even asteroid impacts, that might mimic the features more commonly attributed to water. Nevertheless, taken in combination with all the other evidence we have discussed, it seems a near-certainty that Mars had abundant flowing water in its distant past.

Possible Floods and Oceans Craters in younger regions of the martian surface show far less erosion than do the ancient craters of the southern highlands, suggesting that little or no rain has fallen during the past 2–3 billion years or more. Nevertheless, many surface features show evidence of more recent water flows. This is not as far-fetched as it may sound. Although liquid water is unstable on the surface of Mars today, a large quantity of water that suddenly erupts onto the landscape may take a short time to freeze or evaporate completely. Thus, a catastrophic release of floodwaters—from beneath the surface, for example—could survive long enough to carve channels and other surface features.

Some features on surfaces dating to middle or recent eras on Mars do indeed appear to have been shaped by catastrophic floods. As noted earlier, the *Pathfinder* landing site was in the midst of a region (called Ares Vallis) that shows many features suggesting past flooding, though it is also possible that glaciers were responsible (Figure 8.18). Assuming that the channels were carved by liquid water, we can trace the flow upstream to its source region. Because no surface reservoir is visible there, the water must have emerged from underground. Moreover, the floods appear to have occurred at least sporadically throughout the past few billion years, suggesting that liquid water has been present in the martian crust for most or all of the planet's history.

Some large volcanoes show water-carved channels on their flanks, suggesting a link between volcanic heating and some of the floods. Figure 8.19 shows a volcano with numerous downhill channels flowing outward in

Figure 8.19

These photos show the volcano Hadriaca Patera (center), with its caldera just to the right of the border of the overlapping photos. Note the many channels flowing downhill from the caldera and the wide flood channel (called Dao Vallis) toward the lower right.

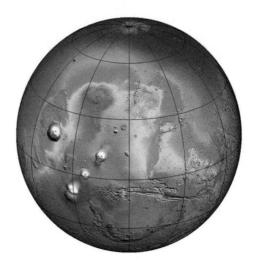

Figure 8.20

This map shows Mars color-coded by elevation. Blue areas are the farthest below the average surface level, and red and brown areas are the highest above it. Note that the entire north polar region is quite low in elevation, suggesting the possibility that this low-lying region once held an ocean. The evidence for an ocean is still weak, however.

all directions from its central caldera. Toward the lower right, we see a much wider channel that was probably carved by floodwaters released during one or more eruptions. These features suggest the past existence of underground pockets of liquid water near volcanoes—a potential habitat for life. If martian volcanoes remain active today, such pockets of water might still exist.

Most of the large flood channels appear to have drained into the northern plains, where the water may have collected before freezing, evaporating, or being absorbed into the surface. Could this mean that the northern lowlands once contained an ocean of liquid water? The topography suggests the possibility (Figure 8.20), but scientists are actively debating whether the geological features seen from orbit support the idea. Some features along the edges of the low regions look like shorelines, but we'll probably need surface studies from future landers to decide whether an ocean really existed.

Martian Water Today   Mars may lack surface liquid water today, but a lot of water still exists in the form of ice. We see water ice in the polar caps, and enough ice sublimates into gas that we find water vapor and ice crystals in the martian atmosphere, which sometimes form clouds in the martian sky. Moreover, orbital studies strongly suggest that a significant amount of water ice is present in the martian subsoil around much of the planet (Figure 8.21).

Could any of this ice still melt into liquid water and flow on Mars today? Evidence for this intriguing possibility comes from orbital photos

Figure 8.21

This map, made with observations from *Mars Odyssey*, represents the hydrogen content of the martian surface soil. The blue areas contain the most hydrogen, probably because they represent regions in which the top meter or so of the surface soil contains water ice.

Figure 8.22

This orbital photograph from *Mars Global Surveyor* shows gullies on a crater wall. The gullies may have been formed by water melting under the protective cover of snowpack.

Figure 8.23

The plate-like features in this orbital photograph from *Mars Express* resemble ice patterns seen in Antarctica, leading some scientists to suggest that they represent a "frozen sea" covered by a thin coating of ash or dust.

showing gullies on crater and channel walls that look strikingly similar to the gullies we see on almost any eroded slope on Earth (Figure 8.22). One hypothesis suggests that the gullies form when snow accumulates on the crater walls in winter and melts away in spring. Because the gullies are relatively small (note the scale bar), they should be gradually covered over by blowing sand during martian dust storms. Thus, gullies that are still clearly visible must be no more than a few million years old. Geologically speaking, this time is short enough to make it quite likely that water flows are still forming gullies today. In addition, recent images from *Mars Express* show features that some scientists interpret as "frozen seas" of ice just beneath the surface (Figure 8.23). The lack of craters on the ice plates themselves tells us that they are geologically young, opening the possibility that they might occasionally be exposed to the surface where they could potentially melt.

Even if water does still flow on occasion, Mars clearly was much warmer and wetter at times in the past than it is today. Ironically, Percival Lowell's supposition that Mars was drying up has turned out to be basically correct, although in a different way than he imagined. Thus, it is time to turn our attention to the question of why Mars changed so much, and what this change might mean for the possibility of life.

8.3   THE CLIMATE HISTORY OF MARS

From the time that *Mariner 9* and *Viking* first revealed evidence of past water on Mars, scientists have actively debated the strength of that evidence. At first, some scientists argued that channels might have been carved by lava flows or liquid carbon dioxide, but more detailed studies made this idea seem implausible. Today, scientists debate the various roles of liquid water, mud, ice, and volcanism in shaping particular features, but overall the case seems all but closed: One way or other, liquid water must have been stable or nearly stable on Mars during at least some periods of time prior to about 2–3 billion years ago. For that to have been possible, both the atmospheric pressure and temperature must have been significantly higher than they are today. Mars in the past offered a much more hospitable climate than it does today, and perhaps one in which life could have arisen and taken hold.

•   Why was Mars warmer and wetter in the past?

It's easy to conclude that Mars must have been warmer and wetter in the past, but more challenging to explain why. The basic answer presumably lies with the greenhouse effect. Recall that the greenhouse effect can make a planet's surface much warmer than it would be otherwise. A moderate greenhouse effect keeps our own planet Earth from freezing over [Section 4.5], while an extremely strong greenhouse effect is responsible for the blistering temperatures on Venus [Section 7.2].

Today, Mars has such a thin atmosphere that it has only a weak greenhouse effect, despite the fact that 95% of its atmosphere is composed of the greenhouse gas carbon dioxide (see Table 8.1). But if Mars had a thicker atmosphere in the past—as it must have, since that is the only way to explain the past stability of liquid water—the greenhouse effect would have been much stronger. In fact, computer simulations show that Mars today could have liquid water if it had a carbon dioxide atmosphere about 400 times as dense as the current martian atmosphere; this atmosphere would give Mars a surface pressure about three times that on Earth today and a temperature above freezing.

Could a much thicker carbon dioxide atmosphere be the explanation for the warmer climate of the martian past? On one level, the idea seems reasonable: Mars has plenty of ancient volcanoes, and calculations show that these volcanoes could have easily supplied the atmosphere with 400 times as much carbon dioxide as it has today. Moreover, if martian volcanoes outgassed carbon dioxide and water in the same proportions as do volcanoes on Earth, Mars would have also had enough water to fill oceans tens or even hundreds of meters deep. However, there's one problem with this explanation: While a 400-times-denser atmosphere could allow for liquid water on Mars today, remember that the Sun was dimmer in the distant past [Section 4.5]. Thus, even more greenhouse warming would have been needed to allow for liquid water when Mars was young.

Current climate models are unable to account for the necessary additional warming with carbon dioxide gas alone, so scientists are seeking other ways to explain the past climate. At least three general categories of explanation have been proposed. One set of hypotheses suggests that the greenhouse effect may have been further enhanced by clouds or by other gases, such as methane. Another hypothesis suggests that the young Sun might not have been quite as dim as is generally assumed: The brightness of the young Sun is calculated from stellar models that use the Sun's current mass. If the Sun somehow lost a small fraction of its original mass over time—which might have happened if the solar wind had been much stronger and carried more gas away—then it would not have been quite so dim at early times. A third hypothesis suggests that Mars never had an extended period of warmth, but instead had only intermittent periods of rainfall and warmth, perhaps triggered by the heat of large impacts.

We do not yet know which, if any, of these hypotheses is correct. However, none of them work unless Mars had a substantially thicker atmosphere and stronger greenhouse effect to begin with. In other words, we cannot be sure if the climate was ever quite "Earth-like," but it was certainly once much warmer and wetter than it is today.

• Why did Mars change?

Given that Mars must once have had a much denser atmosphere with a much stronger greenhouse effect, we can explain the current extremely different conditions only if Mars somehow lost a vast quantity of carbon dioxide gas. This loss would have weakened the greenhouse effect until the planet essentially froze over. Where did all this gas go? Some of the carbon dioxide condensed to make the polar caps, some may be chemically bound to surface rock, and some still makes up the martian atmosphere today. However, the bulk of the gas was probably lost to space.

Loss of Carbon Dioxide and Water The precise way in which Mars lost its carbon dioxide gas is not clear, although some gas was almost certainly lost by being blasted away by large impacts. However, recent data suggest that an even more important loss mechanism was tied to a change in Mars's magnetic field (Figure 8.24). Early in its history, Mars probably had molten metals in its core, much like Earth today [Section 4.4]. Circulating with convection and Mars's rotation, these metals should have produced a magnetic field and a protective magnetosphere around Mars. However, the magnetic field weakened as the small planet cooled and the core solidified, leaving the atmosphere vulnerable to solar wind particles.

COSMIC CALCULATIONS 8.1

The Surface Area–to–Volume Ratio

The total amount of heat contained in Mars or any other planet depends on the planet's *volume*, but this heat can escape to space only from the planet's *surface*. As heat escapes, more heat flows upward from the interior to replace it, until the interior is no hotter than the surface. Thus, the time it takes for a planet to lose its internal heat is related to the ratio of the *surface area* through which it loses heat to the *volume* that contains heat:

$$\text{surface area–to–volume ratio} = \frac{\text{surface area}}{\text{volume}}$$

A spherical planet (radius r) has surface area $4\pi r^2$ and volume $\frac{4}{3}\pi r^3$, so the ratio becomes:

$$\underset{\text{(for a sphere)}}{\text{surface area–to–volume ratio}} = \frac{4\pi r^2}{\frac{4}{3}\pi r^3} = \frac{3}{r}$$

Because r appears in the denominator, we conclude that *larger objects have smaller surface area–to–volume ratios.* (Although we've considered a sphere, this idea holds for objects of any shape.)

Example: Compare the surface area–to–volume ratios of the Moon and Earth.

Solution: Dividing the surface area–to–volume ratios for the Moon and Earth, we find:

$$\frac{\text{surface area–to–volume ratio (Moon)}}{\text{surface area–to–volume ratio (Earth)}} = \frac{3/r_{\text{Moon}}}{3/r_{\text{Earth}}} = \frac{r_{\text{Earth}}}{r_{\text{Moon}}}$$

The radii of the Moon and Earth are $r_{\text{Moon}} = 1,738$ km and $r_{\text{Earth}} = 6,378$ km:

$$\frac{\text{surface area–to–volume ratio (Moon)}}{\text{surface area–to–volume ratio (Earth)}} = \frac{6,378 \text{ km}}{1,738 \text{ km}} = 3.7$$

The Moon's surface area–to–volume ratio is nearly four times as large as Earth's, which means the Moon would cool four times faster if all else were equal. In fact, Earth has retained heat much longer, because its larger size gave it more heat to begin with and because Earth has a higher proportion of radioactive elements. (See Problem 51 for a similar analysis of Mars.)

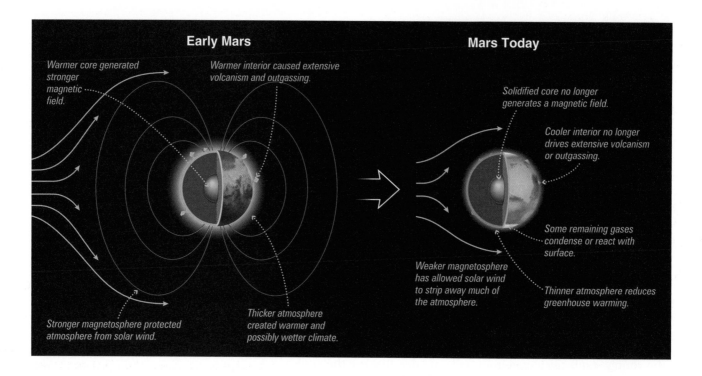

Figure 8.24

Some 3 billion years ago, Mars underwent dramatic climate change, probably because it lost its global magnetic field, leaving its atmosphere vulnerable to the solar wind.

These solar wind particles could have stripped gas out of the martian atmosphere and into space. Had Mars kept its magnetic field, it might have retained much of its atmosphere and might have a more Earth-like climate today.

Much of the water once present on Mars is also probably gone for good. Like the carbon dioxide, some water vapor may have been stripped away by the solar wind. However, Mars also lost water in another way. Because Mars lacks an ultraviolet-absorbing stratosphere, atmospheric water molecules would have been easily broken apart by ultraviolet light from the Sun. The hydrogen atoms that broke away from the water molecules would have been lost rapidly to space through thermal escape—the process in which low-mass gas atoms can reach escape velocity and escape into space [Section 4.4]. With these hydrogen atoms gone, the water molecules could not be made whole again. Initially, oxygen from the water molecules would have remained in the atmosphere, but over time this oxygen was lost too. The solar wind probably stripped some of the oxygen away from the atmosphere, and the rest probably was drawn out of the atmosphere through chemical reactions with surface rock. This oxygen literally rusted the martian rocks, giving the "red planet" its distinctive tint.

In summary, Mars changed primarily because of its relatively small size. It was big enough for volcanism and outgassing to release plenty of water and atmospheric gas early in its history, but too small to maintain the internal heat needed to prevent this loss of water and gas. As its interior cooled, its volcanoes quieted and stopped releasing gas into the atmosphere, while its relatively weak gravity and the loss of its magnetic field allowed existing gas to be stripped away to space. If Mars had been as large as Earth, so that it could still have outgassing and a global magnetic field, it might still have a pleasant climate today. Mars's distance from the Sun also helped seal its fate: Even with its small size, Mars might still have some flowing water if it were significantly closer to the Sun, where the extra warmth could melt the water that remains frozen underground and at the polar caps.

Mars Climate and Axis Tilt

With the gas that once warmed the planet now gone, there is little hope that Mars will ever again have the warm, wet climate it had before about 2–3 billion years ago. However, recent studies suggest that Mars still undergoes significant climate change over geologically short periods of hundreds of thousands of years. This climate change arises from changes in the martian axis tilt, and it may have significant implications to the potential habitability of Mars.

Recall that Earth experiences long-term climate cycles, such as ice ages, due to small changes in its rotation and orbit, including small changes in axis tilt. Earth's axis tilt doesn't change much—varying only between about 22° and 25°—because our large Moon exerts a gravitational pull that stabilizes it. Mars lacks a large moon, and its two tiny moons (Phobos and Deimos) are far too small to offer any stabilizing influence on its axis. In addition, because Mars is closer than Earth to Jupiter, Jupiter's gravity more strongly perturbs Mars as it orbits the Sun. Together, calculations suggest that the lack of a stabilizing moon and the effects of Jupiter should cause Mars to experience wild swings in its axis tilt over periods of between 100,000 and 1 million years.

Models suggest that the martian axis tilt varies over time from 0° to as much as 80°, which means that the current 25° is significantly smaller than the average. These changes in tilt would have dramatic effects on the climate (Figure 8.25). When Mars's axis tilt is small, the poles may stay in a perpetual deep freeze for tens of thousands of years. With more carbon dioxide frozen at the poles, the atmosphere becomes thinner, lowering the pressure and weakening the greenhouse effect. When the axis is highly tilted, the summer pole becomes warm enough to allow substantial amounts of water ice to sublimate, along with carbon dioxide, into the atmosphere. The pressure therefore increases, and Mars becomes warmer as the greenhouse effect strengthens. The martian polar regions show layering that probably reflects changes in climate due to the changing axis tilt (Figure 8.26).

Even at the greatest tilts, the atmospheric pressure probably does not become high enough to allow liquid water to pool in surface lakes or ponds. Nevertheless, models suggest that liquid water might form just beneath the surface or at rock/ice boundaries on the surface whenever the tilt is greater than about 40°—and because a 40° value of tilt is probably about the average over time, Mars could have such zones of liquid water during most epochs.

• Is Mars habitable?

Mars clearly has the elements needed for life, and energy is available for life in the form of sunlight (on the surface) and chemical energy (underground). Thus, the question of whether Mars is habitable hinges ultimately on the availability of liquid water.

The geological evidence strongly suggests that Mars once had abundant liquid water at its surface, meaning that Mars *was* habitable some time before about 2–3 billion years ago. The only question is whether it was habitable for a long period of time—making an indigenous origin of life seem plausible—or only for shorter, intermittent periods. If water was

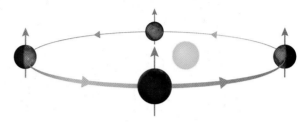

a When the axis is highly tilted, the summer pole receives fairly direct sunlight and becomes quite warm.

b When the axis tilt is small, the poles receive little sunlight at any time of year.

Figure 8.25

Mars's axis tilt probably varies dramatically, causing climate change because of the effect on the seasons.

Alternating layers of icy dust and dusty ice.

5 km

Figure 8.26

This photo from Mars orbit shows a region about 40 kilometers wide at the edge of the north polar cap. Note the alternating layers of ice and dust, which are probably the result of climate changes caused by the changing axis tilt of Mars.

present only intermittently, such as after impacts, then Mars may only intermittently have been habitable. But if the water was present for millions or tens of millions of years, then young Mars may have been quite similar to the young Earth. In that case, Mars may have been habitable during the same period of time in which life arose on Earth [Section 6.1].

The lack of liquid water means the *surface* of Mars is not habitable at the current time. However, we've seen evidence suggesting that pockets of liquid water could still exist underground, kept warm by remaining volcanic heat. In that case, Mars may currently have underground zones of habitability. Moreover, the climate changes tied to the changing martian axis tilt imply that, averaged over geological time, Mars may still have small amounts of surface water at rock/ice boundaries about half the time. On Earth, we find microbes that live in thin films of liquid water at such boundaries, which opens the intriguing possibility that Mars could have surface habitable zones during most epochs.

Unless we are drastically misinterpreting the evidence, the conclusion seems clear. The surface of Mars was habitable during some periods of its early history, and it might still sometimes be habitable when the axis tilt is greater, while the subsurface probably contains habitable zones even today. Given the apparent habitability of Mars, it is time for us to turn our attention to the search for actual life.

8.4 SEARCHING FOR LIFE ON MARS

While we have some confidence in the past and present habitability of Mars, we do not yet know whether Mars has ever actually had life. The only way to learn whether life existed in the past is to search for fossil

---

▶ MOVIE MADNESS: MISSIONS TO MARS

There was a time, not so long ago, when the term *Martian* was just about synonymous with "space alien." You could frequently meet the Martians at the local cinema, where they were busy invading Earth and ruining everyone's whole day.

The classic example of this type of smooth move was in *War of the Worlds,* which has so far spawned one scary radio play and two moderately scary films. In H. G. Wells's story, sophisticated Martians abandon their turf to grab ours. Mars, you see, was drying up and dying. Earth, on the other hand, was a world with abundant water: a sanctuary for our desperate neighbors.

But space probes of the last four decades have shown that the martian surface is apparently lifeless, and there are no obvious indications of technologically sophisticated inhabitants, either dead or alive. So Hollywood, ever flexible, has switched gears. Now it is Earthlings who go to Mars—often to find hidden signs of habitation that would startle astrobiologists.

In the film *Red Planet,* our descendants try to rebuild Mars into a kinder, gentler world in order to escape environmental disaster on Earth. Robotic craft are sent to melt the polar caps and sow the planet with blue-green algae in a barely plausible bid to produce a warm, breathable atmosphere. In the course of this terraforming project, visiting humans stumble across some complicated life-forms—indigenous

Martians—which look like economy-size lice. The lice eat everything from space suits to spacemen, and frankly it's a puzzle why they haven't eaten Mars itself.

As implausible as this may be, astronauts in another space opera, *Mission to Mars,* find something even less reasonable. While checking out an odd mountain in Mars's Cydonia region, they discover that it's really a massive alien "face," disguised by dust and rock. Venturing inside, the astronauts eventually learn that the ancient martian civilization that built the face was wiped out by an asteroid a half-billion years ago. Just before abandoning their planet, the Martians had launched a rocket to seed Earth with their DNA. This molecular emigration supposedly produced the Cambrian explosion that began the reign of multicellular life on Earth. One has to wonder how the Martians could be confident that the jellyfish and trilobites that resulted from their seeding would eventually evolve to humans that would look pretty much like … the long-gone Martians!

Life on the Red Planet may, indeed, have once existed, and perhaps still does. But our current understanding of conditions on Mars strongly suggests that this life would never have resembled either voracious lice or us. It's probable that any real Martians would be visible only in a microscope.

evidence in martian rocks, and the only way to learn whether life exists today is to find it. To date, only very limited searches for life have been carried out on Mars, but much more is planned for the future.

• Is there any evidence of life on Mars?

The discovery of life on Mars would forever alter our view of life in the universe, so it should be no surprise that many scientists are working hard in hopes to be the first to find evidence for such an important discovery, if life on Mars actually exists. As we'll discuss, some scientists already claim to have found such evidence. But are they interpreting data correctly, or are they engaged in the same type of wishful thinking that led Percival Lowell astray?

The answer is the subject of heated scientific debate, but at the moment the vast majority of scientists are skeptical of claimed evidence for martian life. Nevertheless, it is worth examining the claims, both to illustrate why there is scientific controversy and because they may point us toward ways of resolving the question in the future.

The claims of evidence of life fall into three main categories: claims based on results from the *Viking* landers, recent claims based on evidence of methane in the martian atmosphere, and claims based on studies of martian meteorites found on Earth. We'll examine the first two categories here; we'll save discussion of the martian meteorites for Section 8.5.

The Viking Experiments
One obvious way to search for life on Mars is to study the soil to see whether it contains living microbes. This type of search was first carried out by the two *Viking* landers in 1976.

Each of the landers was equipped with several onboard, robotically controlled experiments, along with a robotic arm for scooping up soil samples (Figure 8.27) to test in the experiments; the arm could even push aside rocks to get at shaded soil that was less likely to have been sterilized by ultraviolet light from the Sun. Three experiments were designed expressly to look for signs of life. A fourth was designed to analyze the general content of martian soil.

The first of the three biology experiments (called the *carbon assimilation experiment*) mixed a sample of martian soil with carbon dioxide (CO_2) and carbon monoxide (CO) gas brought from Earth. In some runs of the experiment, the soil was also mixed with water. The aim was to see if any of the carbon dioxide or carbon monoxide would become incorporated into the soil, as would be the case if living organisms were using either gas as a source of carbon in their metabolism. The carbon dioxide and carbon monoxide from Earth could be distinguished from the same gases in the martian atmosphere because they had been "tagged" with radioactive carbon-14 (rather than the far more common, stable isotope carbon-12). Sure enough, the carbon assimilation experiment found that the carbon-14 was incorporated into the soil, a result that seemed to suggest that metabolism was occurring. As a further test, however, the experiment was repeated with soil first heated for 3 hours to 175°C (347°F)—hot enough to break the chemical bonds between carbon and other atoms and presumably kill any carbon-based organisms that might have been present. The results showed that the tagged carbon was still incorporated into the soil, making it seem more likely that the incorporation was due to some sort of geochemical process rather than to biological activity.

The second biology experiment (the *gas exchange experiment*) mixed martian soil with a "broth" containing organic nutrients from Earth. The

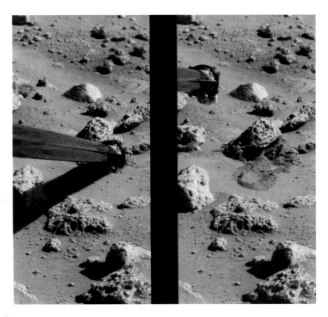

Figure 8.27

This pair of before (left) and after (right) photos from the *Viking 2* lander show how the robotic arm pushed away a small rock on the martian surface. (You can see the entire robotic arm in the inset of Figure 8.4.)

experiment looked for gases that might be released by the respiration of martian microbes, including hydrogen (H_2), nitrogen (N_2), oxygen (O_2), methane (CH_4), and carbon dioxide (CO_2). Again, the initial results were promising for life: As soon as the soil was exposed to the nutrients, oxygen was released into the chamber. However, further analysis suggested that the oxygen was coming from chemical rather than biological reactions. The oxygen release was somewhat characteristic of photosynthesis, but it occurred in the dark rather than in sunlight, making photosynthesis seem implausible. Moreover, the oxygen was present even when the soil was exposed only to water vapor, rather than to the nutrients, a result inconsistent with metabolism. Finally, in a result similar to that of the first biology experiment, the reactions continued even when the soil was heated to temperatures that should have easily killed any organisms present.

The third biology experiment (the *labeled release experiment*) also mixed martian soil with organic nutrients. The nutrients were tagged with radioactive carbon-14 and sulfur-35 so that, if they were consumed by martian microbes, by-products of metabolism and respiration would be released as gases and be detectable by virtue of their radioactivity. Thus, if life were present, we would expect the level of radioactivity to rise as the organisms consumed the nutrients and released the radioactive gases, and then level off as the nutrients were used up. This is precisely what happened. Moreover, unlike the cases for the first two experiments, heating of the soil in this experiment produced results consistent with life: Heating the soil to 50°C (122°F) substantially reduced the amount of radioactivity, and heating it to 160°C (320°F) eliminated any sign of the tagged isotopes in the chamber gas. This is just what we would expect if the heat were killing living organisms that caused the reactions.

It is this third experiment that leads a small fraction of scientists to think that *Viking* found evidence of life on Mars. However, most scientists are skeptical because of the results of a fourth experiment (the *gas chromatograph/mass spectrometer*) that measured the abundance of organic molecules in the martian soil. This experiment started by heating the soil to temperatures as high as 500°C (930°F)—high enough to kill any organisms, break apart any organic molecules they contained, and release the molecules as gases. The gases were passed through a *gas chromatograph*, a device that separates different gases as they pass through it, and then analyzed with a *mass spectrometer*, an instrument that measures the masses of molecules. The results showed no measurable level of organic molecules in the martian soil at all. (The only organic molecules detected were solvents left over from the cleaning of the instrument before it was launched from Earth.) Moreover, based on the sensitivity of the experiment, these results implied that organic molecules could not be present in the tested martian soil at a concentration of more than a few parts per billion—some ten million times lower than the concentration in typical Earth soil. This apparent lack of organic material seemed to rule out a biological explanation for any of the results in the other three experiments, suggesting that the seemingly positive result of the third experiment was more likely produced by some chemical process mimicking a biological result.

Methane on Mars   In 2004, scientists working with the European Space Agency's *Mars Express* mission announced an unexpected and potentially important result: Instruments on the spacecraft had detected methane gas in the martian atmosphere. This news came as a surprise to most scientists, because methane gas cannot last more than a few centuries in the martian atmosphere before chemical reactions transform it into other gases.* Thus, if the methane detection is real, it implies that Mars has a source of ongoing methane production somewhere on or beneath its surface.

Some scientists still doubt that the data are being properly interpreted, but a few ground-based observations made to check the result seem to confirm a methane signature in spectra of Mars. Where could the methane be coming from? We know of at least three possibilities: comet impacts, volcanic activity, or life. The first possibility is highly unlikely, since impacts are such rare events and an impact would have to have occurred quite recently for methane to remain in the atmosphere. That leaves us with volcanism or life.

A few scientists have looked at correlations between the location of methane and water vapor in the martian atmosphere and used the results to argue that we might actually be seeing the signature of microscopic life on Mars. However, volcanism also seems a reasonable explanation, especially since other evidence suggests that Mars has at least some low level of ongoing volcanic activity. Nevertheless, even this result could have implications for the existence of life: If there is enough volcanic activity to produce a detectable level of methane, it may also mean there is enough volcanic heat to make pockets of underground liquid water, raising the possibility of subsurface life.

• How do we plan to search for life on Mars?

The world scientific community has ambitious plans for continuing the search for life on Mars through new space missions. The orbits of Earth and Mars bring the two planets to closest approach about every 26 months (Figure 8.28), and scientists hope to take advantage of every upcoming alignment to send new and ever more sophisticated spacecraft to Mars.

Upcoming Mission Plans   The most recent alignment, which made Mars bright in our skies during the fall of 2005, was used to time the launch of the *Mars Reconnaissance Orbiter* (*MRO*). *MRO* reached Mars orbit in March 2006 and is capable of much higher-resolution images of the surface than prior orbiters. It is also equipped with instruments to study the mineral content of the surface from orbit and to search for subsurface water ice. Its findings should help scientists choose sites for future landing missions.

The next planned lander, a NASA mission called *Phoenix*, is scheduled to be launched in August 2007, with arrival at Mars in May 2008 (Figure 8.29). *Phoenix* will not be a rover, but will have a robotic arm for scooping soil and onboard experiments for analyzing it. It will land in the north polar region, where it should be able to dig into surface ice and search for evidence of buried organic material. For the following launch window to Mars, in fall 2009, NASA is building the *Mars Science Laboratory*, a rover that will be considerably larger and more sophisticated than the *Spirit* and *Opportunity* rovers. If all goes well, the *Mars Science Laboratory* will arrive on the martian surface in May 2010. It will combine the capa-

---

*The methane is oxidized to form water and carbon dioxide; the oxidation occurs because there is always some amount of free oxygen in the martian atmosphere that is made by the breakup of atmospheric carbon dioxide molecules.

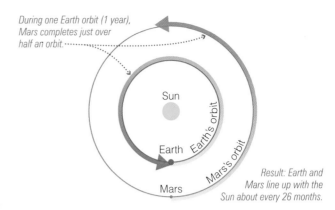

During one Earth orbit (1 year), Mars completes just over half an orbit.

Sun

Earth   Earth's orbit

Mars   Mars's orbit

Result: Earth and Mars line up with the Sun about every 26 months.

Figure 8.28

Mars takes almost 2 years to orbit the Sun compared to Earth's 1 year. Thus, if you start from an alignment as shown here, after 1 year Earth will be back in the same place, but Mars will be only halfway around. It therefore takes a second year before the two planets line up again. Using Mars's more precise orbital period of 1.88 years, we find the alignments occur every 780 days, or about 26 months.

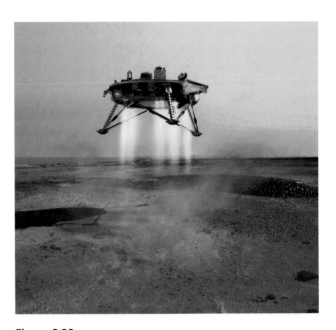

Figure 8.29

This painting shows the *Phoenix* lander, using rocket thrusters to land on Mars. The landing is planned for May 2008.

bilities of the earlier rovers with onboard experiments for analyzing soil and rock in much more detail.

Although any of these missions could in principle find evidence of life, they are designed primarily to make further assessments of the habitability of Mars. A more thorough robotic search for life will require more sophisticated and expensive instruments, so it seems reasonable to start by deciding where and how best to look before actually conducting the search. Nevertheless, scientists are already working on plans for robotic missions that could conduct an actual search for life in the next decade, perhaps by drilling deeper into the surface, or by exploring larger regions of the terrain. An even better way to search for life would be to bring martian rock back to laboratories on Earth, and scientists hope that a sample return mission could occur as early as a decade from now.

Preventing Contamination—In Both Directions   Given the likelihood that some Earth organisms could survive in at least a few locations on Mars, it's important to make sure that our robotic missions don't accidentally contaminate Mars with life from Earth. Otherwise, microbes that hitched a ride from Earth aboard a spacecraft might fool us into thinking we'd found evidence for martian life. The possibility of contamination also poses an ethical issue: It's at least conceivable that terrestrial life could outcompete any indigenous martian life, driving the martian life to extinction. Do we have a right to do something that could endanger native life on another planet? Clearly, the best way to avoid these problems is to prevent contamination in the first place. An international treaty, signed in 1967, requires that any spacecraft sent to Mars must have a less than 1 in 1,000 chance of causing contamination. Today, scientists strive for even lower contamination probabilities by sterilizing spacecraft before they are launched.

Similarly, but in the other direction, the prospect of a sample return mission has caused some people to fear we might unleash dangerous martian microbes on Earth. Could martian microbes cause disease for which we are unprepared or outcompete terrestrial organisms on our own turf? We cannot completely rule out any danger, but it is quite unlikely, because disease-causing microbes are highly adapted to the species they infect. For example, diseases that infect plants generally do not infect animals. Indeed, "species jumping" by diseases is quite rare and generally occurs only between species that are evolutionarily close. HIV (the virus that causes AIDS), for example, is thought to have jumped from chimpanzees to humans, but on an evolutionary level this is a fairly small jump between different species of primates. Thus, even if martian microbes were accidentally released and subsequently survived on Earth, it's unlikely that they would cause disease. In addition, because martian meteorites must frequently land on Earth, life that hides in martian rocks has almost certainly reached Earth already. The fact that we do not see any harmful effects from this "natural contamination" makes it unlikely that any martian life can harm Earth life.

Nevertheless, it pays to be cautious, given the high stakes involved, and samples brought back from Mars will surely be transported in sealed containers that would not break open even if they were to crash on Earth. Once here, they will be quarantined, and subject to biological tests such as exposing terrestrial microbes to them. Biologists already know how to deal with dangerous terrestrial microbes, such as the Ebola virus, and scientists are developing protocols to ensure safe handling of any harmful martian organisms.

THINK ABOUT IT   Should we allow samples from Mars to be brought to Earth, or should they be studied only in space, such as on the Space Station or at a Moon colony? Defend your opinion.

• Should we send humans to Mars?

A long-term dream of many people, and a part of NASA's vision for coming decades, is to send humans to Mars. Sending people is far more difficult than sending robots. Even with the most advanced rockets that anyone now has on the drawing board for the next couple of decades, the trip to Mars would take at least 4 months. That means a human mission would have to carry not only the weight of the astronauts and their living quarters, but also enough food, air, and water to last the trip. Shielding against dangerous radiation would also be necessary, which means having an onboard "storm cellar" in case a violent flare erupts on the Sun. Once on Mars, astronauts might conceivably get water from the subsurface ice and chemically extract oxygen from martian water or rock, but they would still need food, which would have to be either taken along with them or sent separately aboard other spacecraft. Astronauts would also need to come home, which means they'd need fuel for the return journey. That would add far more weight to the human mission, unless we can manufacture the fuel for the return mission on Mars rather than bring it from Earth (an idea that is being actively explored). Still, no matter how you look at it, the enormous amount of stuff required for a human mission ensures it would cost at least as much as dozens of robotic missions, and of course it would pose many dangers to the crew.

Scientific Pros and Cons   While the cost and inherent danger of sending humans to Mars would be very high, the scientific payoff could potentially be even higher. We humans are far more capable than any robot, and a team of scientists with vehicles for traveling around the planet and equipment for drilling into the crust might well answer our questions about martian life long before they would be answered by robotic explorers. However, sending humans to Mars also has at least one significant scientific drawback: It vastly complicates the issue of avoiding contamination by terrestrial organisms. People are veritable warehouses of microbes: The number of bacteria in the average person's mouth, for example, is far greater than the number of people who have ever lived. We harbor microbes on our skin, in our breath, in our food, and in our excrement. Preventing all these microbes from escaping into the martian environment during an extended stay on the planet would be nearly impossible.

The scientific pros and cons of sending humans to Mars are fairly clear, but the history of the space program shows that human exploration has rarely been driven by science. The manned space program began for political reasons, largely as part of a "race" between the United States and the Soviet Union. For example, while the *Apollo* program provided valuable scientific data about the Moon, its primary purpose was to prove to the world that the Americans could get there before the Soviets. If we decide to send humans to Mars, the decision will also probably be based more on social and political considerations than scientific ones.

Terraforming Mars   Some people dream of eventually establishing permanent colonies on Mars. For the near future, any such colonies would have to be self-contained environments, and no one would dare venture outside without a space suit. But for the more distant future, some people wonder if we might be able to alter the martian environment in ways that would make it more hospitable to us. Such changes go by the name **terraforming,** because they would tend to make the planet more Earth-like.

Proposals to terraform Mars envision changing the environment so that the atmospheric pressure and temperature become greater. Raising the temperature might be accomplished by adding a greenhouse gas to the atmosphere, while raising the pressure would require adding a large amount of gas to the atmosphere. One suggestion involves manufacturing chlorofluorocarbons (CFCs), which are strong greenhouse gases, and releasing them into the martian atmosphere. If we could strengthen the greenhouse effect enough, the warmer temperatures might begin to release frozen carbon dioxide from the martian polar caps and elsewhere beneath the surface, which would further increase the atmospheric pressure and strengthen the greenhouse effect. There still wouldn't be oxygen to breathe, but if the pressure rose enough, we might be able to walk around on Mars carrying only an oxygen tank (and some protection from ultraviolet radiation) rather than having to wear a full, pressurized space suit. Such conditions might also allow plants to survive outdoors, making it much easier to grow food, and eventually increasing the concentration of atmospheric oxygen.

The idea might just work, but putting it into practice wouldn't be easy. Because CFCs tend to be broken apart by sunlight, we would have to manufacture them continually and in great abundance in order to start the greenhouse warming. Calculations suggest that we would need a manufacturing capability about a million times greater than our recent CFC-manufacturing capability on Earth and would need to keep it up for a *few hundred thousand years* before the surface warmed enough to drive substantial quantities of carbon dioxide into the atmosphere. Thus, if it is possible at all, we have plenty of time to consider the ethical issues of terraforming, which could be quite significant if Mars turns out to have life: Do we have a right to alter a planet in a way that could harm its native life?

THINK ABOUT IT A similar ethical issue surrounds endangered species on Earth. Some people say that we have no right to drive any species to extinction—an idea that was embodied in the U.S. Endangered Species Act. Others say that potential extinctions must be weighed against the human and economic costs of preventing them. Where do you stand on this issue? Does your answer affect your opinion of whether it would be ethical to terraform Mars? Explain.

Interestingly, some of the ethical issues involved in Mars colonization were explored by science fiction writers well before the idea of terraforming ever arose. In particular, back in the days when people believed in canals and a dead or dying martian civilization, many stories dealt with the conditions under which humans might colonize Mars. So for our last word on the topic of human colonization, we turn to a science fiction story called "The Million-Year Picnic," written in 1946 by Ray Bradbury. It tells the story of a human family who escapes to Mars just as people on Earth are finishing off our civilization through hatred and war. On Mars, the family finds plenty of water and the vacant cities left by extinct Martians. The story ends with the family on the banks of a canal, where one of the children asks his father about a promise made earlier:

"I've always wanted to see a Martian," said Michael. "Where are they, Dad? You promised."

"There they are," said Dad, and he shifted Michael on his shoulder and pointed straight down.

The Martians were there. Timothy began to shiver.

The Martians were there—in the canal—reflected in the water. Timothy and Michael and Robert and Mom and Dad.

The Martians stared back up at them for a long, long silent time from the rippling water....

8.5 THE PROCESS OF SCIENCE IN ACTION: MARTIAN METEORITES

As we briefly noted earlier, one claim of evidence for life on Mars comes from study of rocks from Mars that have fallen to Earth—the so-called *martian meteorites* [Section 6.2]. The story begins in 1984, when a team of American scientists scooped up a 1.9-kilogram meteorite (see Figure 6.10) from the Allan Hills region of Antarctica. It was cataloged as "ALH84001": "ALH" for Allan Hills, "84" for the year in which it was found, and "001" to indicate that it was the first meteorite found on the expedition. It did not immediately draw special attention, but an analysis a decade later showed that it was one of the rare martian meteorites. It then proved itself special even among this small group of rocks, and was subject to intense study. In 1996, a team of researchers (led by David McKay at NASA) made an astonishing claim: They said that ALH84001 holds fossil evidence of past life on Mars. Because this claim would be so important if true, and because it has proved so controversial, we use it as this chapter's case study of the process of science in action.

• Is there evidence of life in martian meteorites?

To evaluate the claims about ALH84001, we must begin by understanding the rock. Scientists are fairly confident (though with some doubts) that ALH84001 really is a meteorite from Mars. It is definitely not an Earth rock, because its relative abundance of the isotopes oxygen-16, oxygen-17, and oxygen-18 are significantly different from that found in terrestrial rocks. But neither does it match what we'd expect from a piece of an asteroid or a rock from the Moon. Most important, gas trapped within ALH84001 appears very similar to that of the martian atmosphere in its chemical and isotopic composition—and distinctly different from any other known source of gas in our solar system—leading to the suspicion that it came from Mars. As discussed in Chapter 6, about three dozen such martian meteorites have been found to date, each presumably launched into space as a consequence of a large impact on Mars.

ALH84001 was singled out for more intense study than other martian meteorites for a simple reason: While other known martian meteorites are geologically young, radiometric dating showed ALH84001 to be a piece of igneous rock that solidified about 4.5 billion years ago. Thus, it formed quite shortly after Mars was born, and must have resided on Mars throughout the times when liquid water flowed on the surface. Scientists

TABLE 8.3 *The History of Meteorite ALH84001*

Time	Event
4.5 billion years ago	Solidifies from molten rock in the southern highlands of Mars
4.0–4.5 billion years ago	Affected by nearby impacts, but not launched into space
3.9 billion years ago	Infiltrated by water, leading to the formation of carbonate grains within the rock
16 million years ago	Blasted into space by an impact on Mars
13,000 years ago	Falls to Earth in Antarctica
December 27, 1984	Found by scientists
October 1993	Recognized as a martian meteorite
August 1996	Announcement that ALH84001 contains possible evidence of martian life

Figure 8.30

Top: Microscopic chains of magnetite crystals produced by bacteria on Earth. Bottom: Similar chains of magnetite crystals found in the carbonate grains of ALH84001. Does the similar appearance of the martian crystals suggest a similar biological origin?

therefore wondered if it might tell us something about the past habitability of Mars.

History of the Meteorite   Careful study tells us quite a lot about the history of ALH84001. Radiometric dating tells us its age, while study of its structure reveals evidence of later shocks, probably due to the effects of impacts that occurred long before the one that ultimately launched it into space. The meteorite also contains carbonate grains (about 0.1 to 0.2 millimeter in diameter) that date to about 3.9 billion years ago and tell us that the rock must have been infiltrated by liquid water from which the carbonate minerals precipitated out—evidence that is at least consistent with the idea that Mars once had flowing water.

We can determine the timing of the impact that blasted ALH84001 into space by looking for effects of exposure to *cosmic rays*, high-energy particles that leave telltale chemical signatures on anything unprotected by an atmosphere. The results tell us that ALH84001 spent about 16 million years in space, which means the impact that started its journey occurred on Mars about 16 million years ago. By studying decay products from radioactive isotopes produced by the cosmic rays, we learn when cosmic rays stopped disturbing the meteorite—which must be when it fell to Earth and gained the protection of Earth's atmosphere. Such analysis shows that ALH84001 landed in Antarctica about 13,000 years ago. Table 8.3 summarizes the history of ALH84001.

Evidence of Life   The claimed evidence of life in ALH84001 comes from detailed studies of its carbonate grains and the surrounding rock. In brief, four types of evidence have been cited as pointing to the existence of biology on Mars:

- The carbonate grains have a layered structure, with alternating layers of magnesium-rich, iron-rich, and calcium-rich carbonates. On Earth, this type of layering generally occurs only as a result of biological activity.

- The carbonate grains contain complex organic molecules known as *polycyclic aromatic hydrocarbons,* or *PAHs.* These molecules can be produced by both biological and nonbiological processes, and they have indeed been found in many meteorites that are not from Mars. However, their abundance in ALH84001 is much higher than that in other meteorites, and on Earth these molecules are most commonly produced by the decay of dead organisms or by reactions between such decay products and the environment (for example, in the burning of fossil fuels).

- Under a microscope, we see crystals of the mineral magnetite within the iron-rich layers of the carbonate grains. The sizes, shapes, and arrangements of these crystals appear to match those of magnetite grains that on Earth occur only when made by bacteria (Figure 8.30).

- Most intriguingly, highly magnified images of the carbonate grains reveal rod-shaped structures that look much like recently discovered *nanobacteria* on Earth (Figure 8.31). Terrestrial nanobacteria are about $\frac{1}{100}$ the size of ordinary bacteria, and some biologists question whether they truly are living organisms. However, they appear to contain DNA, which suggests that they are indeed a form of life. If so, could the similar-looking structures in the martian meteorite be fossil life from Mars?

While none of these lines of evidence alone would prove biological activity, the original investigators argued that, on the whole, biology seemed a much more likely explanation than nonbiological processes. They felt that it would be a "simpler and thus better" scientific explanation (in effect invoking Occam's razor [Section 2.3]) if only a single process—biology—could account for each observation than if a different process was required to explain each result.

Alternative Explanations   The four lines of evidence for fossil life in ALH84001 might seem to make a strong case for past life on Mars. However, other scientists have proposed alternative, nonbiological mechanisms that might have produced each observed phenomenon. Let's look at these alternatives in the same order that we presented the evidence:

- There may be nonbiological ways to get layered carbonate. For example, several pulses of hot water with different dissolved elements might have passed through the rock and laid down the different mineral layers.

- Other meteorites prove that PAHs can be produced by chemical rather than biological processes, and their high abundance might also be explained by terrestrial contamination during the time the rock resided in Antarctica.

- The resemblance between the magnetite crystals in the meteorite and those made by bacteria on Earth may be coincidental, and some scientists have proposed nonbiological ways in which the crystals and chains might have been formed.

- Similarly, the rod-shaped structures don't tell us anything other than that they *look* a lot like nanobacteria. Looks alone do not prove anything, and most scientists suspect that structures with this appearance could be produced nonbiologically; they might even be artifacts of the way the meteorite slices were processed in the laboratory.

One of the most important issues in evaluating the evidence is whether some of it could be the result of terrestrial contamination. Scientists have developed control tests that check for contamination during handling and laboratory analysis of the meteorite, and no such contamination appears to have occurred. However, the rock spent 13,000 years sitting in Antarctica before scientists found it; could Earth life have contaminated it during that time?

At first, scientists were fairly confident that no contamination had occurred on the Antarctic ice, because analysis showed that ALH84001 contains a higher abundance of organic molecules deep within the rock than nearer the surface. If the organic molecules represented terrestrial contamination, we would expect just the opposite. Nevertheless, subsequent analysis found modern, living terrestrial bacteria inside the rock! This clear contamination probably explains the presence of PAHs in the rock. Though it may not affect the other lines of evidence, it certainly weakens any case to be made about martian life in ALH84001.

THINK ABOUT IT   Given the fact that ALH84001 has apparently been contaminated by terrestrial bacteria, do you think we could ever be sure that a martian meteorite holds evidence of life on Mars? Defend your opinion.

1,000 nm

Figure 8.31

Top: Terrestrial nanobacteria in a sample of volcanic rock from Sicily. Bottom: Rod-shaped structures, each about 100 nanometers in length, found in ALH84001. Are they fossils of nanobacteria from Mars?

Summary of the Controversy The extraordinary debate over whether we already have discovered evidence of life beyond Earth will undoubtedly continue as scientists subject ALH84001 and other martian meteorites to further scrutiny. The techniques that have been applied in the analysis to date are the best available, but new technologies will allow the martian meteorites to be studied in even more detail over the coming years. Still, unless a surprising and major breakthrough occurs, it is unlikely that the meteorites by themselves can provide definitive evidence of life.

It is also possible that further study will prove that ALH84001 does not contain fossil life. But even in that case, we still won't know whether there has ever been life on Mars. Most Earth rocks do not contain fossils. Thus, whether or not life existed on Mars, it wouldn't be surprising if the first ancient Mars rock we examine lacks fossils. In all likelihood, a final verdict on the question of life on Mars will have to wait at least until future space missions scoop up and return to Earth samples of rock from a variety of locations on Mars.

THE BIG PICTURE
Putting Chapter 8 into Context

In this chapter, we have discussed past fantasies about martian civilization, our current understanding of the habitability of Mars, and the search for life on the Red Planet. As you continue your studies, keep in mind the following "big picture" ideas:

- Mars holds a special allure not only because of legitimate scientific questions, but also because past fantasies led many people to imagine a martian civilization. Mars and Martians became deeply embedded in modern popular culture, helping generate great public interest in Mars exploration.

- Different regions of the martian surface appear to be almost frozen in time, representing different eras in the planet's history. As a result, we can piece together at least a partial story of Mars from its earliest times to the present. We find a planet that has gone through dramatic change. Its surface, once warm and wet, is now dry and frozen.

- According to present understanding, Mars almost certainly was a habitable planet in the past and may still have habitable zones underground. This makes Mars a prime target in the search for life beyond Earth.

8.1 FANTASIES OF MARTIAN CIVILIZATION

• How did Mars invade popular culture?

Superficial similarities between Mars and Earth led to speculation about martian civilization. Astronomer Percival Lowell thought he saw canals built by an advanced society, but the canals do not really exist.

8.2 A MODERN PORTRAIT OF MARS

• What is Mars like today?

Mars is cold and dry, with an atmospheric pressure so low that water is unstable. Martian weather is driven largely by seasonal changes that cause carbon dioxide alternately to condense and sublime at the poles, creating pole-to-pole winds and sometimes leading to huge dust storms.

• What are the major geological features of Mars?

Mars has regions that are densely cratered and must be very old, and other regions with fewer craters that must be much younger. Giant volcanoes dot certain regions of Mars, and we also see evidence of past tectonics, which probably created Valles Marineris.

• What geological evidence tells us that water once flowed on Mars?

Orbital images of eroded craters, dry river channels, and floodplains all point to past water flows, and supporting evidence is found in chemical analysis of Martian rocks. Any periods of rainfall seem to have ended at least 2–3 billion years ago, but some flooding may have occurred later. Mars today still has water ice underground and in its polar caps, and could possibly have pockets of underground liquid water.

8.3 THE CLIMATE HISTORY OF MARS

• Why was Mars warmer and wetter in the past?

Mars's atmosphere must once have been much thicker with a much stronger greenhouse effect, though we do not yet know whether this made Mars warm and wet for a long, extended period of time or only intermittently.

• Why did Mars change?

Change must have occurred due to loss of atmospheric gas, which weakened the greenhouse effect. Some gas was probably blasted away by impacts, but more probably was stripped away by the solar wind, which was able to reach the atmosphere as Mars cooled and lost its magnetic field and protective magnetosphere. Water was probably also lost because ultraviolet light could break apart water molecules in the atmosphere, and the lightweight hydrogen then escaped to space.

• Is Mars habitable?

Mars almost certainly had a habitable surface during its wet periods more than 2–3 billion years ago. Its surface or near-surface might still sometimes be habitable at times when its axis tilt is greater than it is now, and the subsurface may still have habitable regions today.

8.4 SEARCHING FOR LIFE ON MARS

• Is there any evidence of life on Mars?

The *Viking* experiments produced results that some scientists think may be evidence of life, but nonbiological explanations seem more likely. Recent observations of methane in the atmosphere may be due to volcanism. Overall, there are some possible hints of life on Mars, but no definitive evidence.

• How do we plan to search for life on Mars?

Space scientists plan an ongoing series of Mars missions, timed for the close approaches of Mars to Earth that occur about every 26 months.

• Should we send humans to Mars?

Scientifically, humans could probably answer questions about life much more quickly than robotic missions, but humans also pose a risk of contamination. Ultimately, the question will probably be decided by considerations beyond science alone.

8.5 THE PROCESS OF SCIENCE IN ACTION: MARTIAN METEORITES

• Is there evidence of life in martian meteorites?

Four lines of evidence have been presented as suggesting the presence of past life in a martian meteorite, but each also has a potential nonbiological explanation. As a result, we do not have a definitive answer to the question, and more study and more evidence are needed.

REVIEW QUESTIONS

Short-Answer Questions Based on the Reading

1. Briefly summarize the evidence from little over a century ago, both real and imagined, that led to widespread belief in a martian civilization.

2. What would it be like to walk on Mars today? Briefly discuss the conditions you would experience.

3. Why isn't liquid water stable at the martian surface today? What happens to water ice that melts on Mars?

4. How do martian seasons differ from Earth seasons? Briefly explain why seasonal changes on Mars are accompanied by pole-to-pole winds.

5. Give a brief overview of the geography of Mars, explaining the nature of key features such as the Tharsis Bulge, Olympus Mons, Valles Marineris, the southern highlands, and the northern lowlands.

6. How do we know that different regions of the martian surface date to different eras in the past? What have we learned about changes in martian volcanism during the past eras?

7. Summarize the evidence suggesting that Mars must have been warm and wet, possibly with rainfall, in its distant past.

8. What evidence suggests that water might still flow at or beneath the martian surface today? Why do we think that Mars might still have subsurface liquid water today?

9. How must the martian atmosphere have been different when the surface was warmer and wetter? What uncertainties remain in our understanding of the cause of the higher temperatures in the past?

10. What happened to the gas that made up the much thicker atmosphere of the distant past, and what happened to the abundant water? What role did Mars' small size play in these changes?

11. How does the tilt of the martian axis change with time, and why does it change so much? What implications do these changes have for the martian climate?

12. Based on all the geographic and geological evidence, summarize the current view about the past and present habitability of Mars.

13. Briefly summarize the *Viking* experiments and their results. Do the results constitute evidence of life? Explain.

14. What is the potential significance of atmospheric methane to the search for life on Mars? Are there other explanations for the methane?

15. Briefly summarize plans for Mars exploration over the next few years. Why do we send missions to Mars only about every 26 months?

16. Discuss the issue of biological contamination in either direction between Earth and Mars. How serious is each issue? What steps can we take to prevent contamination in each direction?

17. Summarize the scientific pros and cons of sending humans to Mars. What other considerations are likely to play a role in decisions about such missions?

18. What do we mean by *terraforming* Mars? Why might it be tempting for future human colonists? Is it something we could do within our lifetimes?

19. How do we know that ALH84001 really came from Mars, and how have we learned its history?

20. Briefly summarize the possible evidence of past life discovered in studies of ALH84001 and why this evidence generates controversy.

TEST YOUR UNDERSTANDING

Surprising Discoveries?

Suppose we were to make the following discoveries. (These are *not* real discoveries.) In light of your understanding of Mars, decide whether the discovery would be considered plausible or surprising. Explain clearly; because not all of these have definitive answers, your explanation is more important than your chosen answer.

21. The first human explorers on Mars discover that the surface is littered with the ruins of an ancient civilization, including remnants of tall buildings and temples.

22. We discover a string of active volcanoes in the heavily cratered southern highlands.

23. We find underground pools of water on the slopes of one of the Tharsis volcanoes.

24. We discover that Mars was subjected to global, heavy rainfall about 1 billion years ago.

25. Photos from future orbiters show that new gullies have formed alongside some of the ones already seen in crater walls from orbiting spacecraft.

26. We find a lake of liquid water filling a small crater close to one of the dry river channels.

27. The first fossils discovered on Mars come from the canyon walls of Valles Marineris.

28. A sample return mission finds fossil evidence not only of martian microbes, but also of larger photosynthetic plants that live on the exposed surfaces of martian rocks.

29. We discover evidence that the martian polar caps have in the past extended more than twice as far toward the equator as they do now.

30. We find rocks on Mars showing clearly that the planet once had a global magnetic field nearly as strong as Earth's magnetic field.

Quick Quiz

Choose the best answer to each of the following. Explain your reasoning with one or more complete sentences.

31. When we say that liquid water is *unstable* on Mars, we mean that: (a) a cup of water would shake uncontrollably; (b) it is impossible for liquid water to exist on the surface; (c) any liquid water on the surface would quickly either freeze or evaporate.

32. Mars's seasonal winds are driven primarily by: (a) dust; (b) sublimation of carbon dioxide ice; (c) sublimation of water ice.

33. Olympus Mons is: (a) a giant volcano; (b) a huge canyon network; (c) a continent-size plateau.

34. We can recognize the oldest surface regions of Mars by the fact that they have: (a) the most impact craters; (b) the most volcanoes; (c) the most evidence of past water flows.

35. Minerals in surface rock studied by the martian rovers seem to tell us that: (a) they probably formed in water; (b) they probably were formed by impacts; (c) they hold fossil evidence of life.

36. Rivers on Mars: (a) have never existed; (b) existed in the past but are dry today; (c) continue to have flowing water today.

37. Which must be true if Mars was warmer and wetter in the past? (a) Mars was once closer to the Sun. (b) Mars once had a much thicker atmosphere. (c) Mars must somehow have avoided the effects of the heavy bombardment.

38. Which of the following fundamental properties of Mars could explain why it once had a global magnetic field but later lost it? (a) its small size; (b) its larger distance than Earth from the Sun; (c) a rotation rate that is slightly slower than Earth's.

39. Under the leading scenario, if Mars once had much more carbon dioxide in its atmosphere, most of this carbon dioxide is now: (a) gone, because it escaped to space; (b) frozen at the polar caps; (c) locked up in the form of carbonate rocks, just like on Earth.

40. The *Viking* experiments found: (a) no evidence of life on Mars; (b) clear evidence of life on Mars; (c) some results consistent with life, but others that were inconsistent with life.

INVESTIGATE FURTHER
In-Depth Questions to Increase Your Understanding

Short-Answer/Essay Questions

41. *Hold Your Breath.* If you held your breath, would it be safe to walk outside on Mars? Why or why not?

42. *Miniature Mars.* Suppose Mars was significantly smaller than its current size—say, the size of our Moon. How would this have affected its potential habitability? Explain.

43. *Larger Mars.* Suppose Mars was significantly larger than its current size—say, the same size as Earth. How would this have affected its potential habitability? Explain.

44. *Civilization on Mars.* Based on what we can see on the surface of Mars, does it seem possible that Mars once had a civilization with cities on the surface but that the evidence has now been erased or buried underground? Explain.

45. *Martian Fossil Hunting.* On Earth, we cannot find fossil evidence of life dating to times prior to about 3.8 billion years ago. If life ever existed on Mars, is it possible that we would find older fossils than we find on Earth? Explain.

46. *Future Landing Site.* Suppose you were in charge of a mission designed to land on Mars. Assume the mission carries a rover that can venture up to about 50 kilometers from the landing site. What landing site would you choose? Write a one-page summary of why you think your site is a good target for a future mission.

47. *Human Mission Requirements.* Assume that a mission will carry humans to Mars on a journey that takes a few months in each direction and allows the explorers to spend about 2 years on the martian surface. Make a list of key provisions that would be needed for the mission, explaining the purpose of each item. In addition, briefly discuss whether you think any of these provisions could be found or manufactured on Mars rather than having to be brought from Earth.

48. *Terraforming Mars.* Make a list of the pros and cons of terraforming Mars, assuming that it is possible. Overall, do you think it would be a good idea? Write a short defense of your opinion.

49. *Mars Movie Review.* Watch one of the many science fiction movies that concern trips to Mars (such as *Total Recall* or *Red Planet*). In light of what you now know about Mars, does the movie give a realistic view of the planet? Are the plot lines that concern Mars plausible? Write a critical review of the movie, focusing on these issues.

50. *Martian Literature.* Read a book of science fiction about Mars, such as H. G. Wells's *The War of the Worlds,* Ray Bradbury's *The Martian Chronicles,* or any of the Edgar Rice Burroughs books about Martians. Write a critical review of the book, being sure to consider whether it still merits interest in light of current scientific understanding of Mars.

Quantitative Problems

Be sure to show all calculations clearly and state your final answers in complete sentences.

51. *Interior Heat.* Compare the surface area–to–volume ratios (that is, total surface area divided by total volume) of the Moon, Earth, and Mars. What does your answer tell you about how quickly each world should have cooled with time? What does your answer tell you about the implications of planetary size to habitability?

52. *Mars's Elliptical Orbit.* Mars's distance from the Sun varies from 1.38 AU to 1.66 AU. How much does this change the globally averaged strength of sunlight over the course of the martian year? Give your answer as a percentage by which sunlight at perihelion (the orbital point closest to the Sun) is stronger than that at aphelion (the farthest orbital point). Comment on how this affects the martian seasons. (*Hint:* Remember that light follows an inverse square law; see Figure 7.1.)

53. *Atmospheric Mass of Earth.* What is the total mass of Earth's atmosphere? Use the fact that, under Earth's gravity, the sea level pressure of 1 bar is equivalent to 10,000 kilograms pushing down on each square meter of the surface. Also remember that the surface area of a sphere of radius r is $4\pi r^2$.

54. *Atmospheric Mass of Mars.* The weaker gravity of Mars means that 1 bar of pressure on Mars would be that exerted by about 25,000 kilograms pushing down on each square meter of the martian surface. Based on this approximation, the atmospheric pressure on Mars (see Table 8.1), and the size of Mars, estimate the total mass of Mars's atmosphere. Compare to Earth's atmospheric mass from Problem 53.

55. *Past Gas on Mars.* Models suggest that Mars today could have liquid water on its surface if the atmosphere were about 400 times as dense as it actually is. What would the atmospheric mass be in that case? How does this compare to the present mass of Earth's atmosphere? Does it seem plausible that Mars might once have had this much gas? Explain why or why not.

Discussion Questions

56. *The Role of the Martians.* Percival Lowell may have been sadly mistaken in his beliefs about Martians, but he succeeded in generating intense public interest in Mars. If he had never made

his wild claims about canals and civilization, do you think we would be exploring Mars with the same fervor today? Defend your opinion.

57. *Lessons from Mars.* Discuss the nature of the climate change that occurred on Mars some 3 billion years ago. Do you think this climate change holds any important lessons for us as we consider potential climate changes that humans are causing on Earth? Explain.

58. *Human Exploration of Mars.* Should we send humans to Mars? If so, when? How much would you be willing to see spent on such a mission? Would you volunteer to go yourself? Discuss these questions with your classmates, and try to form a class consensus regarding the desirability and nature of a human mission to Mars.

WEB PROJECTS

59. *Martian Photo Journal.* By now, we have many thousands of photos of Mars taken both on the surface and from orbit, and virtually all of them can be found on the Web. Make your own photo journal of "Mars's Greatest Photo Hits" by choosing ten of your favorite photos. For each one, write a short descriptive caption and explain why you chose it.

60. *Current Mars Missions.* Pick one of the Mars missions that is currently operating and visit its Web site. Write a short report about the mission's history, goals, and accomplishments to date.

61. *Future Mars Missions.* Pick one of the Mars missions that is being planned or considered for the future, and visit its Web site. Write a short report about the purpose of the mission and its current status.

9
Life on Jovian Moons

The atmosphere of Titan.

LEARNING GOALS

I was drawn by the sirens of Titan
Carried along by their call
Seeking for a way to enlighten
Searching for the sense of it all
Like a kiss on the wind I was thrown to the stars
. . .
I was drawn by the sirens of Titan (as are we all)
As are we all

Al Stewart, from his song "The Sirens of Titan,"
based on the Kurt Vonnegut novel

Jupiter orbits the Sun at more than five times Earth's distance, and the other jovian planets (Saturn, Uranus, and Neptune) lie much farther from the Sun. Sunlight in this distant realm is faint—too weak to provide much warmth. Nevertheless, several of the moons in these frigid outer reaches of our solar system are now considered to be possible places to find life beyond Earth. In this chapter, we will investigate these distant worlds.

We'll begin by examining the general characteristics of these moons and how and why it might be possible for some of them to have liquid water (or other liquids). We'll then turn our attention to the most promising potential abodes of life. Looking first at the moons of Jupiter, we'll discuss the evidence suggesting that Europa has a subsurface ocean that might contain more water than the oceans of Earth, making it a prime target in the search for life. We'll also investigate the likelihood that conditions similar to those on Europa might also be found on two other moons of Jupiter: Ganymede and Callisto. We'll then move further outward into the solar system, discussing the latest developments concerning the possible habitability of moons around Saturn, Uranus, and Neptune, including Saturn's remarkable moon Titan, which is blanketed by an atmosphere thicker and denser than Earth's.

Aside from addressing our general curiosity about the habitability of jovian moons, our discussion in this chapter will lead to an intriguing possibility: It's conceivable that these cold and distant moons could be the most numerous homes to life in our solar system. If any of these moons do indeed prove to harbor living things, the possibilities for finding biology elsewhere in the universe will be greatly broadened.

9.1 THE MOONS OF THE OUTER SOLAR SYSTEM

In previous chapters, we discussed life on Earth in some detail and investigated the prospects for finding life on Mars. We now turn our attention to more distant reaches of the solar system, notably the realm of the four jovian planets—Jupiter, Saturn, Uranus, and Neptune. As we discussed in Chapter 7, the jovian planets themselves seem unlikely to be habitable.

However, these planets are orbited by many moons, which we call **jovian moons** because they orbit jovian planets.

The idea of finding life in the cold outer solar system once seemed far-fetched, but several of the jovian moons now seem potentially habitable. In this section, we'll introduce the major moons and explore the mechanisms thought to make it possible for at least some of these moons to have liquid water (or other liquids) on or within them, thus creating possible habitats for life.

• What are the general characteristics of the jovian moons?

As of early 2006, the four jovian planets were known to have at least 150 moons between them, more than 20 times that of all other planets combined. While some of the larger of these moons have been known to astronomers for nearly four centuries, many of the smaller ones have been found only with the improved telescopes and space probes of the last few decades. Probes in particular have given us the chance to study these worlds from close by. While previous generations knew jovian moons only as points of light, today we have photographs of many of them that show enough detail that we can name individual craters and other features. One of the biggest surprises to emerge from our modern reconnaissance of these moons is learning that some of them might be homes to life.

Discovering the Moons
Among Galileo's many discoveries [Section 2.2], one of the most notable was finding the four large moons of Jupiter. Having heard of the telescope's recent invention, Galileo built his own homemade versions beginning in 1609. At the time, the telescope was thought to be primarily a toy, or possibly useful for defense. Galileo, however, did something different with his telescopes: He turned them toward the sky. On January 7, 1610, while gazing at Jupiter, he saw what at first seemed to be three small stars in its vicinity. What intrigued Galileo was that the three were close to one another and in a line. The following night, he looked again and was surprised to note that the three stars had moved relative to Jupiter, but not in the direction or to the extent expected of stars. A few days later he noted a fourth point of light, and within a week he realized that these four "stars" always stayed close to Jupiter and were clearly in orbit around it (Figure 9.1). In March 1610, Galileo published his results in a pamphlet he called *The Starry Messenger*, claiming to have found four bodies moving around the giant planet "as Venus and Mercury around the Sun." These four bodies are what we now call the **Galilean moons** of Jupiter; proceeding outward from that planet, we know them individually as Io, Europa, Ganymede, and Callisto (Figure 9.2).

Other scientists soon discovered additional moons in the outer solar system. The accomplished Dutch scientist Christiaan Huygens (1629–1695) found the largest of Saturn's moons—Titan—in 1656. Before the close of the seventeenth century, Giovanni Domenico Cassini (1625–1712), an Italian astronomer who became director of the Paris Observatory, had discovered four more moons around the ringed planet.*

Figure 9.1

A page from Galileo's notebook written in 1610. His sketches show four "stars" near Jupiter (the circle), but in different positions at different times (and sometimes hidden from view). Galileo soon realized that the "stars" were moons orbiting Jupiter.

---

*Saturn's rings were first sighted by Galileo, but the resolution of his telescope was too low for him to make out what they were. Huygens was the first to realize that the rings do not touch Saturn's surface, and Cassini showed that the rings were not solid but instead were marked by a dark division, which we still call the *Cassini division*.

Figure 9.2

This set of photos, taken by the *Galileo* spacecraft, shows global views of the four Galilean moons as we know them today. Sizes are shown to scale.

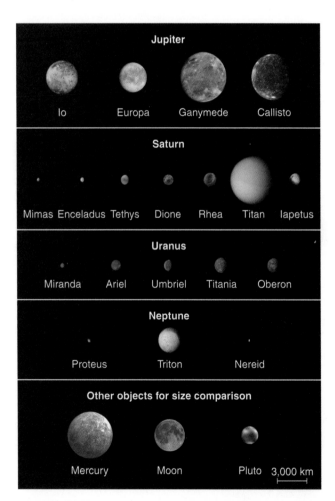

Figure 9.3

The larger moons of the jovian planets, with sizes (but not distances) shown to scale. Mercury, our own Moon, and Pluto are included for comparison.

Even today, astronomers continue to make new discoveries of moons orbiting the jovian planets, though all of the larger moons have surely been discovered by now. The new discoveries involve small moons—usually no more than a few kilometers across—detected with the aid of new telescopes or spacecraft. For example, nearly two dozen small moons were discovered in photographs taken by the *Voyager* probes in the 1970s and 1980s, and as recently as 2004 and 2005 astronomers used a telescope in Hawaii, as well as the *Cassini* spacecraft and the Hubble Space Telescope, to find 16 new moons of Saturn and two new moons of Pluto.

Sizes and Orbits of the Moons The jovian moons come in a wide range of sizes. While many small ones are not much bigger than a single mountain on Earth, others are much larger. The two largest—Jupiter's moon Ganymede and Saturn's moon Titan—are bigger than the planet Mercury. The three other Galilean moons (Io, Europa, and Callisto) and Neptune's moon Triton are bigger than Pluto. Figure 9.3 shows a montage of jovian moons larger than about 300 kilometers in diameter.

Almost all the moderate- and large-size moons orbit their planet in much the same way that planets orbit the Sun: They orbit nearly in the equatorial plane of their host world, moving in the same direction as their planet's spin. In this sense they resemble miniature solar systems, which suggests that they were formed in a smaller-scale version of the same processes that gave birth to the planets [Section 3.3]. As shown in Figure 9.4, the gravity of the jovian planets drew in gas and dust from the surrounding solar nebula. Like the solar nebula as a whole, this gas and dust formed a swirling disk-shaped cloud around each jovian planet. Condensation and accretion then built moons that shared the orbital properties of the original disks of gas and dust.

The small jovian moons differ from their larger brethren both in appearance and in the properties of their orbits. Most have an irregular shape and often resemble peanuts, potatoes, or other snack foods (Figure 9.5). This is hardly surprising: The lesser gravity of these small objects is too weak to force the rigid material of which they're composed into spheres.

Some small moons may have formed "in place" like most of the larger moons, and others come in groups that share common orbital characteristics, suggesting that they may be fragments of larger moons that broke apart. But many of the smaller moons also have orbits that are highly elliptical or inclined to the equator of their host planet; in some cases, the moons orbit backward relative to their host planet's spin. These orbital characteristics are telltale signs of moons that are captured asteroids or comets, which probably originally orbited the Sun independently. As we discussed briefly in Section 4.6, it's not easy for a planet's gravity to grab a wayward asteroid or comet: Because energy must be conserved, the passing object must somehow lose some of its orbital energy or else it would simply fly on by and continue its orbit around the Sun. The leading hypothesis holds that captures occurred when the jovian planets were young, at a time when they had large, extended atmospheres that could have served to slow down small bodies as they passed nearby, thereby removing orbital energy and allowing the objects to be captured.

Surprisingly, one large moon also has orbital characteristics suggesting it is a captured object: Neptune's moon Triton, which orbits backward relative to its planet's rotation. This orbit makes it a near-certainty that Triton once orbited the Sun as an independent object, although capturing such a large object poses a trickier problem than capturing smaller moons. Recent research suggests that Triton could have been captured if it had once been orbited by its own satellite. In that case, it could have been captured while passing close to Neptune, because the excess orbital energy that Triton had to lose could have been carried off by the satellite as it was ejected from its orbit. Regardless of the specific mechanism of the capture, for several decades scientists have assumed that Triton was once one of the *Kuiper belt objects* (KBOs) of which we now consider Pluto to be a member [Section 3.3]. Indeed, the fact that Triton is larger than Pluto is one reason why some astronomers suspected that Pluto might not be the largest such object in the outer solar system, a suspicion confirmed with the 2005 discovery of Eris—an ice-rich world that is larger than Pluto and nearly a hundred times as far from the Sun as Earth.

Composition of the Moons

As we discussed in Chapter 3, the outer solar system was cold enough to allow ices to condense along with metal and rock. We therefore would expect the jovian moons to be made of a mixture of ice and rock, and that is indeed the case for most of them. The average densities of most jovian moons are significantly lower than that of Earth, reflecting the fact that they contain substantial quantities of ice, which is low in density. Within individual moon systems, we see variations in composition reflecting the fact that the moons formed at different distances from a hotter, central planet. For example, the Galilean moons show a decrease in density with distance from Jupiter, just as we would expect if they formed from a swirling cloud of gas that was hotter in the center than in its outer regions: Io's density indicates that it has virtually no water in any phase, Europa is mostly rock with water ice (and perhaps a liquid ocean) only in its outer layers, and Ganymede and Callisto have more significant amounts of water ice relative to their amounts of rock.

In addition to the compositional variation within individual moon systems, there is also variation in the composition of moons as we move from one planet to the next; these differences came about because of temperature differences in the overall solar nebula. Water ice condensed easily at the temperatures near Jupiter, but methane and other ices condensed only at the colder temperatures found at significantly greater distances from the Sun. As a result, Jupiter's moons contain significant quantities of water ice but no other ices. Moons of the more distant planets contain higher overall proportions of ice than of rock, and contain not only water ice but also methane and other ices.

Synchronous Rotation of the Moons

Nearly all jovian moons share an uncanny trait: Like our own Moon, they always keep the same face turned toward their planet. This behavior, called **synchronous rotation**,

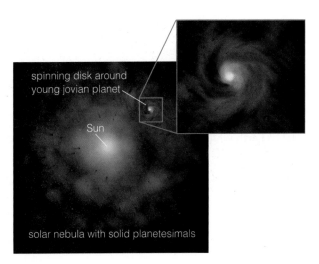

Figure 9.4

The young jovian planets are thought to have been surrounded by disks of gas and dust much like the solar nebula as a whole but smaller in size.

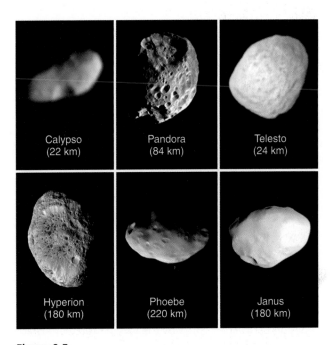

Figure 9.5

A montage of six of Saturn's smaller moons (not to scale). Because they are not spherical, the sizes in parentheses represent approximate lengths along their longest axes.

Figure 9.6

The fact that we always see the same face of the Moon means that the Moon must rotate once in the same amount of time that it takes to orbit Earth once, an idea you can understand by walking around a model of Earth while imagining that you are the Moon. The same idea applies to the synchronous rotation of jovian moons.

a If you do not rotate while walking around the model, you will not always face it.

b You will face the model at all times only if you rotate exactly once during each orbit.

means that each moon completes exactly one rotation around its axis while it makes one orbit around the planet. You can see how this works with a simple demonstration (Figure 9.6). Place a ball on a table to represent a planet like Earth, and walk around the ball so that your head represents an orbiting moon. If you do not rotate as you walk around the ball, you'll be facing away from it by the time you are halfway around your orbit. The only way you can face the ball at all times is by completing exactly one rotation while you complete one orbit.

Synchronous rotation is not a coincidence: Rather, it develops naturally for any moon that orbits relatively close-in to its planetary parent, and it arises as a consequence of the same gravitational effects that lead to tides on Earth. Recall that the strength of gravity declines with distance, following an inverse square law [Section 2.4]. As a result, the Moon's gravitational pull on the near side of Earth is slightly stronger than that on the far side. This *difference* in the strength of gravity is the cause of the tides (Figure 9.7). Note that there are two tidal bulges, one facing the Moon and one opposite, which is why there are two high tides per day as Earth rotates through the bulges. Simply put, the reason why there are two bulges is that the oceans on the side nearest to the Moon are being pulled out from Earth, while the oceans on the opposite side bulge because Earth is being pulled away from under them. A better way to look at tides is to recognize that the gravitational attraction toward the Moon gets progressively weaker with distance throughout Earth and that this varying attraction tends to stretch the entire Earth—land and ocean—along the Earth–Moon line. The tidal bulges are more noticeable for the ocean than for the land only because liquids flow more readily than solids. (Recall that solids *can* flow, though slowly, which is why Earth's solid mantle can undergo convection [Section 4.4].)

The gravitational attraction to the Moon is weakest here...   *...and strongest here.*

The difference in gravitational attraction tries to pull Earth apart, raising tidal bulges both toward and away from the Moon.   *Not to scale!*

Figure 9.7

Tides on Earth are created by the varying force of attraction between different parts of Earth and the Moon. There are two daily high tides as any location on Earth rotates through the two tidal bulges. The diagram highly exaggerates the tidal bulges, which actually raise the oceans only about 2 meters and the land only about a centimeter.

THINK ABOUT IT The Sun exerts a stronger gravitational force on Earth than does the Moon—after all, Earth orbits the Sun, not the Moon. So why is the Moon rather than the Sun primarily responsible for Earth's tides? (The Sun's tidal effect on Earth is only about one third as strong as the Moon's.) Do you think other planets can have any significant effect on tides on Earth? Explain.

Because tidal forces stretch Earth itself, the process necessarily creates some friction, called **tidal friction**. As shown in Figure 9.8, this tidal friction has important consequences for Earth and the Moon, because it

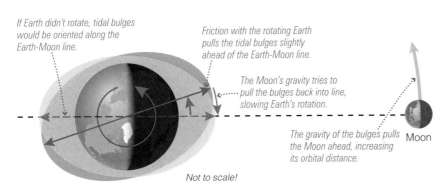

If Earth didn't rotate, tidal bulges would be oriented along the Earth-Moon line.

Friction with the rotating Earth pulls the tidal bulges slightly ahead of the Earth-Moon line.

The Moon's gravity tries to pull the bulges back into line, slowing Earth's rotation.

The gravity of the bulges pulls the Moon ahead, increasing its orbital distance.

Moon

Not to scale!

Figure 9.8

Earth's rotation pulls its tidal bulges slightly ahead of the Earth–Moon line, leading to gravitational effects that very gradually slow Earth's rotation and increase the Moon's orbital distance.

allows Earth's rotation to pull the bulges slightly ahead of the Earth–Moon line. This slight misalignment of the tidal bulges with the Earth–Moon line means the Moon's gravity is always pulling back on the bulges, causing Earth's rotation to slow gradually. At the same time, the gravity of the bulges pulls the Moon slightly ahead in its orbit, causing the Moon to move gradually farther from Earth. These effects are barely noticeable on human time scales; for example, tidal friction increases the length of Earth's day by only about 1 second every 50,000 years. (On short time scales, this effect is overwhelmed by other effects on Earth's rotation.) But the effects add up over billions of years. Early in Earth's history, a day may have been only 5 or 6 hours long and the Moon may have been one tenth or less its current distance from Earth.

We can understand the Moon's synchronous rotation by turning the situation around. Just as the Moon raises tides on Earth, Earth must raise tides on the Moon—in fact, much stronger tides, because Earth's gravity is much stronger than the Moon's. If the Moon rotated rapidly, these tides would generate substantial tidal friction that would tend to slow the Moon's rotation. The tidal friction would disappear only when the Moon's rotation slowed so much that its tidal bulges always stayed on the Earth–Moon line—which would mean that the Moon was always keeping the same face to Earth in synchronous rotation. In other words, no matter how fast the Moon may have been rotating at its birth, it was inevitable that tidal friction would slow this rotation until it was synchronous. At that point, the synchronous rotation was permanently "locked in," because there was no more tidal friction to slow the Moon's rotation further.

Tidal friction similarly explains the synchronous rotation of the jovian moons.* If at its birth a moon was rotating faster than it orbited, tidal friction slowed its rotation until the rotation and the orbit were synchronized. Similarly, if a moon was once spinning slower than it orbited, tidal friction accelerated its rotation until it was synchronized with its orbit. Given the large sizes of the jovian planets compared with their moons, synchronous rotation likely set in for the close-in moons within no more than a few million years.

• Why do we think that some moons could harbor life?

Most of the jovian moons are almost certainly lifeless. As we discussed in Chapter 7, they are too small and too far from the Sun to have any reasonable likelihood of having liquid water or other liquids. However, the

---

*Europa shows evidence of a very slight deviation from sychronous rotation, which may be an effect of a subsurface ocean.

situation may be different for a few of the larger moons. One large moon, Titan (which we'll discuss in Section 9.3), appears to have a temperature and atmospheric pressure that at least sometimes allow liquid methane to form lakes or rivers on its surface. In addition, several moons may have just enough internal heat to keep water liquid in their interiors—making their interiors potentially habitable.

Tidal Heating   Based on what we've learned about internal heat on the terrestrial planets, it might seem surprising to find much internal heat in any of the jovian moons. After all, only Ganymede and Titan are even as large as Mercury, and we know that Mercury's interior has cooled enough over time that it no longer supports any active volcanism. But when the *Voyager 1* spacecraft passed Jupiter in 1979, we got definitive proof that at least some distant moons have plenty of internal heat. Confirming a theoretical prediction made just weeks before *Voyager*'s arrival,

COSMIC CALCULATIONS 9.1

The Strength of the Tidal Force

Recall from Section 2.4 that the force of gravity acting between two objects is:

$$F_g = G\frac{M_1 M_2}{d^2}$$

The tidal force that a planet exerts on a satellite is the *difference* between the gravitational force on the near and far sides of that satellite. One way to get a sense of this difference is to consider a "test mass"—say, a 1-kg rock—on the satellite's surface. If d is the distance between the center of the planet and the center of a satellite of radius r_{sat}, then on the near side of the satellite the distance from the planet's center and the 1-kg rock is $d - r_{sat}$ and on the far side it is $d + r_{sat}$. We then calculate the gravitational force in both positions and subtract to get the tidal force acting on the satellite per kilogram of mass.

Example: Calculate the tidal force that Earth exerts on the Moon (per kilogram) and compare to the effect of the Moon's own gravity. Useful data: $M_{Earth} = 5.97 \times 10^{24}$ kilograms, $r_{Moon} = 1.74 \times 10^6$ meters, the Earth–Moon distance is $d = 3.84 \times 10^8$ meters, and $M_{Moon} = 7.35 \times 10^{22}$ kilograms. As long as you use G in standard units and the masses and distances in kilograms and meters, your answer will come out in *newtons* (1 N ≈ 0.225 lb).

Solution: We find the tidal force F_{tidal} that Earth exerts on the rock by subtracting the gravitational force F_g when the rock is on the far side of the Moon (where the rock's gravitational attraction to Earth is weaker) from the gravitational force when it is on the near side (where the gravitational attraction is stronger):

$$F_{tidal} = F_g(\text{on near side}) - F_g(\text{on far side})$$

Let's first consider the near-side term:

$$F_g(\text{near side}) = G\frac{M_{Earth} M_{rock}}{(d_{Earth\text{-}Moon} - r_{Moon})^2}$$

$$= \left(6.67 \times 10^{-11}\frac{m^3}{kg \times s^2}\right)\frac{(5.97 \times 10^{24}\text{ kg})(1\text{ kg})}{(3.84 \times 10^8\text{ m} - 1.74 \times 10^6\text{ m})^2}$$

$$= 0.00273\text{ N}$$

For the far-side term, we find:

$$F_g(\text{far side}) = G\frac{M_{Earth} M_{rock}}{(d_{Earth\text{-}Moon} + r_{Moon})^2}$$

$$= \left(6.67 \times 10^{-11}\frac{m^3}{kg \times s^2}\right)\frac{(5.97 \times 10^{24}\text{ kg})(1\text{ kg})}{(3.84 \times 10^8\text{ m} + 1.74 \times 10^6\text{ m})^2}$$

$$= 0.00267\text{ N}$$

The difference gives us the tidal force (per kilogram):

$$F_{tidal} = F_g(\text{on near side}) - F_g(\text{on far side})$$

$$= 0.00273\text{ N} - 0.00267\text{ N} = 0.00006\text{ N}$$

For comparison, the Moon's gravitational force acting on the rock is:

$$F_g = G\frac{M_{Moon} M_{rock}}{r_{Moon}^2}$$

$$= \left(6.67 \times 10^{-11}\frac{m^3}{kg \times s^2}\right)\frac{(7.35 \times 10^{22}\text{ kg})(1\text{ kg})}{(1.74 \times 10^6\text{ m})^2}$$

$$= 1.6\text{ N}$$

Note that this 1.6 N gravitational force (which is the weight of the rock on the Moon and is equivalent to 1.6 N × 0.225 lb/N = 0.36 pounds) is more than 15,000 times greater than the 0.00006 N tidal force that Earth exerts on the rock. Clearly, the tidal force is quite small in comparison to the gravitational force, though over time it has been large enough to bring the Moon into synchronous rotation. (The tidal force was stronger when the Moon was closer to Earth.)

the spacecraft photographed active volcanic eruptions on Io, the innermost of the four Galilean moons. The *Galileo* orbiter then provided much more detailed views (Figure 9.9). How did scientists predict this surprising volcanism? They realized that jovian moons can have a type of internal heating different from that found on any of the terrestrial worlds, and that Io would be the extreme example of this heating.

Recall that Earth and the other terrestrial worlds were all hot inside at their births, with heat from accretion, differentiation, and radioactive decay [Section 4.4]. Over time, heat has escaped, and only radioactive decay continues to supply new heat. In small worlds, like the Moon and Mercury, the supply of new heat has not been enough to overcome the loss of escaped heat. If the same sources were all that could heat the jovian moons, they probably would be cool by now too. To explain the internal heating of Io, we must consider a different process. Scientists call this **tidal heating**, because it arises from effects of tidal forces.

Io's tidal heating arises from a combination of two factors: (1) its proximity to Jupiter means it experiences a strong tidal force from the massive planet; and (2) Io has a slightly elliptical orbit, which causes the strength and direction of the tidal force to change slightly as Io moves through each orbit. Figure 9.10 shows the result: The constantly changing orientation of Io's tidal bulges means that Io is continuously being flexed in different directions, which generates friction inside it in much the same way that flexing warms Silly Putty* Tidal heating generates tremendous heat on Io—more than 200 times as much heat (per gram of mass) as the radioactive heat driving much of Earth's geology. This heat makes Io the most volcanically active place in the solar system. Material from Io's volcanic vents reaches temperatures of more than 2,000°C (3,600°F), and eruptions often spew plumes of sulfur and other gases to heights of hundreds of kilometers.

Although the combination of the elliptical orbit and Jupiter's tidal force seem to explain tidal heating, there is still another question to ask: Why is Io's orbit slightly elliptical, when most large satellites have nearly circular orbits? The answer lies with an "orbital resonance" that occurs between Io, Europa, and Ganymede: The three moons periodically line up because, during the time Ganymede takes to complete one orbit of Jupiter (about 7 days), Europa completes two orbits and Io completes four orbits (Figure 9.11). The effect is much like that of pushing a child on a swing. If timed properly, a series of small pushes can add up to a *resonance* that causes the child to swing quite high. For the three moons, the periodic alignments mean they experience repeated gravitational tugs that act much like the repeated pushing of the child on the swing (hence the term *orbital resonance*). In this case, these gravitational tugs cause the orbits to be more elliptical than they would be otherwise.

Like synchronous rotation, we expect orbital resonances to arise quite naturally. The orbital resonances of Io, Europa, and Ganymede probably came about because of a feedback with tides that the moons raise on Jupiter. Just as our Moon raises tides on Earth that are gradually causing the Moon to move farther from Earth (and slowing Earth's rotation), the Galilean moons also raise tides in Jupiter's atmosphere that cause the moons to move farther from Jupiter. The effect is greatest on the closest-in moon, Io, so its orbit should have tended to move outward until it

---

*If you are looking for the ultimate source of the energy that drives tidal heating, it comes from Jupiter's rotation, which is gradually slowing much like Earth's rotation, although at a rate too small to be observed.

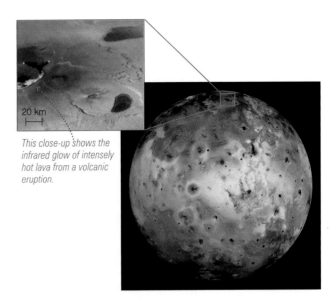

This close-up shows the infrared glow of intensely hot lava from a volcanic eruption.

Figure 9.9 Interactive Photo

Io is the most volcanically active body in the solar system. Most of the black, brown, and red spots on the surface are related to recently active volcanoes.

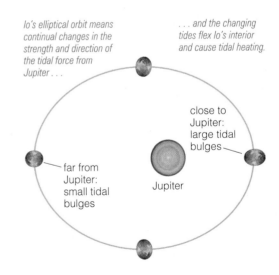

Io's elliptical orbit means continual changes in the strength and direction of the tidal force from Jupiter . . .

. . . and the changing tides flex Io's interior and cause tidal heating.

close to Jupiter: large tidal bulges

far from Jupiter: small tidal bulges

Jupiter

Figure 9.10

Tidal heating arises on Io from the combination of its elliptical orbit and the strong tidal force exerted on it by Jupiter. The bulges and orbital eccentricity are exaggerated.

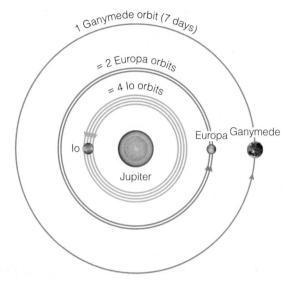

Figure 9.11

Io, Europa, and Ganymede share an orbital resonance that returns them to the positions shown every 7 days, and the recurring gravitational tugs explain why all three orbits are slightly elliptical. The effect on Io is greatest, since it is closest to Jupiter.

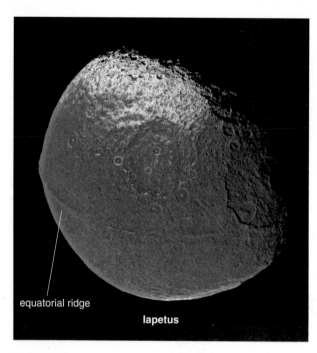

Iapetus

equatorial ridge

Figure 9.12

Saturn's moon Iapetus (about 1,440 kilometers in diameter) has a steep ridge running along its equator, suggesting that it must once have been surprisingly warm inside.

achieved a resonance with the next moon out, Europa. Then Io and Europa continued to move outward in lockstep until they achieved a resonance with the third moon, Ganymede. This is presumably how Io, Europa, and Ganymede came to have orbital resonances among themselves. Indeed, these moons are now moving slowly outward in lockstep toward Callisto, though it will be many billions of years before they reach a resonance with this fourth moon.

Other Heat Sources   Tidal heating certainly explains Io's volcanic activity and probably explains Europa's internal heat. As we'll see in the next two sections, however, it may not be enough to explain the heating that apparently occurs or has occurred in the past on a few other jovian moons. Saturn's moon Iapetus, for example, has a striking equatorial bulge that indicates it must once have been warm enough to be fluid inside (Figure 9.12), but tidal heating does not seem to explain how it became hot enough for its interior to melt (or become sufficiently "soft" to flow). One possible explanation lies with radioactive decay, although there is great scientific debate over whether or how Iapetus and other moons could have incorporated enough radioactive material during their formation to explain the level of heating that they have apparently experienced.

9.2   LIFE ON JUPITER'S GALILEAN MOONS

Tidal heating has turned Io into a veritable hell, and its lack of water and extreme volcanic activity essentially rule it out as a home for life. But tidal heating has somewhat lesser effects on Europa and Ganymede (and no effect on Callisto, since it does not participate in the orbital resonance). In the case of Europa, in particular, it is possible that the level of tidal heating is "just right" for life. Without tidal heating, Europa would be wrapped in solid ice. However, its appearance offers hints that this moon has a subsurface ocean that makes it a possible home to life. In this section, we'll discuss the prospects of habitability and life for Europa, as well as the prospects for the outer two Galilean moons, Ganymede and Callisto.

• Does Europa have an ocean?

Even before the *Voyager* spacecraft snapped the first detailed photos of Europa in 1979, scientists already knew that Europa is covered with an ice shell. Spectroscopic observations made from Earth indicated the presence of water ice on the surface, although the amount was unknown. Theoretical studies, based on our ideas about the formation of moons, suggested that Europa was likely built of a mixture of rock and water. If heat had caused the moon to differentiate sometime in the past, then the less dense water would have migrated to the surface. The *Voyager* spacecraft confirmed at least part of this speculation: Europa's exterior is bright white and entirely ice-covered. Also, its surface is remarkably smooth, with few features rising higher than about a kilometer. Faint ridge lines—giant cracks in the ice—crisscross the surface, looking almost like a spider web of highways.

The *Galileo* orbiter revealed Europa in much more detail (Figure 9.13). Notice the small number of recognizable impact craters, which tells us that Europa's surface has been repaved in recent geological times. However, it is extremely difficult to tell whether the resurfacing has been caused by the action of liquid water or of ice that is relatively warm and therefore soft enough to flow (as glaciers flow on Earth).

By using the *Galileo* spacecraft to measure subtle variations in Europa's gravitational field, scientists were able to determine its internal layering. Crudely speaking, Europa seems to consist of a central metallic (probably iron) core, overlaid with a thick sheath of silicate rock and an 80–170-kilometer-thick outer skin of water or water ice. From the gravity measurements alone, the water layer could contain just about any combination of solid ice, liquid water, or slush (partially melted ice), because all of these have about the same density (1 g/cm^3). As a result, over the past couple of decades scientists have debated which form of water is most likely. The debate is not yet fully settled, but as we'll discuss shortly, careful analysis of data from the *Galileo* spacecraft suggests that the most likely answer is the model shown in Figure 9.14: Europa probably has a brittle, icy crust underlaid by a layer of warmer ice that can flow easily and undergo convection, and a liquid water ocean beneath that. Surface temperatures on Europa are brutally cold (typically −150°C), so its icy crust must be as stiff as granite.

Evidence for a Liquid Ocean

The spacecraft pictures provide strong evidence suggesting that liquid water, not just more ice, underlies Europa's icy skin. One key piece of evidence comes from the surprising lack of impact craters. Europa has only about a few dozen large craters (10 kilometers or more across), a number that should accumulate in only a few tens of millions of years, not in the billions of years since the solar system's formation. Clearly, something has erased the craters on Europa. An obvious candidate is resurfacing by an occasional breakthrough of subsurface water or slushy ice.

Another suggestive piece of evidence for an underground ocean is the appearance of Europa's so-called *chaotic terrain,* which resembles photos of an arctic ice pack (Figure 9.15). The surface is clogged with iceberg-like blocks—some as small as football fields and others as large as a city—crisscrossed by ridge lines and suspended in what appears to have been a slushy ocean that froze. These blocks are often separated, as if they have rafted away from their original positions. Imagine putting your palms down on an assembled jigsaw puzzle and then spreading them slightly. Gaps will form between the pieces, and the picture printed on them will become disjointed. Europa's ridge lines similarly form a telltale picture on the ice, a picture that in places has fractured into small, jostled pieces. Individual ridges can be traced by the eye from one block to another, clearly showing the motion that must have occurred. Further evidence for an ocean comes from features suggesting that liquid water has gushed from below and frozen in spaces between some of the ridges (Figure 9.16).

While the photographic evidence suggests that Europa's ice-pack exterior has been churned by an underlying ocean, it still leaves a shadow of a doubt. Could it be that the ocean that produced the tortured and broken surface existed millions of years ago and is now frozen, which would mean that Europa's intriguing face is merely the frozen remnant of an earlier time? Perhaps the icy crust sits atop relatively warm soft or slushy ice, and the surface features are the result of this warm ice convecting upward from below. In that case, Europa's frigid skin might extend all

Figure 9.13

A global view of Europa, as seen from the *Galileo* spacecraft. Colors are enhanced to bring out subtle details.

Figure 9.14

These artist's drawings show the leading model for Europa's interior layering. Notice the liquid water ocean under the outer layers of ice.

Figure 9.15

Chaotic terrain on Europa. This landscape suggests that liquid water (or slushy ice) has welled up from below, breaking apart the surface and then freezing in place. The pieces can be mentally put back into place by matching up the details of the ice blocks. However, such a re-assembly indicates that some of the pieces are missing; they presumably sank, melted, or crumbled during the reshuffle.

Figure 9.16

The smooth, dark terrain just to the right of the center of this photo mosaic could represent an area where liquid water broke through the surface and covered over some of the ridged landscape.

the way to the moon's rocky interior. But there's another line of evidence, arguably even more convincing than the photos, that makes the case for a liquid ocean.

In 1996, a magnetometer aboard the *Galileo* spacecraft detected a magnetic field near Europa. Because moons seldom have a magnetic field (though, as we'll discuss shortly, Ganymede does), researchers asked themselves what was causing magnetism on Europa. Jupiter has a strong magnetic field (which causes it to be a relatively powerful source of natural radio static), and Europa orbits within this field. Because Jupiter's magnetic equator is tilted with respect to Europa's orbit, its field at Europa's position is constantly changing as the giant planet spins on its axis. Just as a moving magnet produces an electric current in a coil of wire, so too could Jupiter's changing field induce currents in an electrical conductor within Europa. Of course, there's no giant coil of wire in Europa, but a salty ocean would conduct electricity in much the same way. The currents in such an electrically conducting ocean would act to set up a magnetic field that opposes Jupiter's—and thus would change as Jupiter's field changes. Additional measurements with *Galileo*'s magnetometer showed that Europa's magnetic field does indeed change as Jupiter spins. Most researchers consider this the best evidence yet that a liquid, salty ocean exists under Europa's icy surface. The magnetometer data require that Europa's subsurface water be global in extent, not limited to just a few isolated liquid pockets. They also imply that this ocean could be as salty as Earth's seas. Some of *Galileo*'s instruments also found evidence for what appear to be salts on Europa's surface—possible seepage from a briny deep.

If we tally up the evidence for an ocean on Europa, we can list the following:

- The relatively small number of craters implies that the moon's surface is young, perhaps only a few tens of millions of years old, indicating that it has been recently repaved.

- Various features on the surface (chaotic and flooded terrain) suggest that liquid water sometimes wells up from below.

- Europa has a magnetic field that is likely caused by currents produced in something that conducts electricity—like a salty ocean—as Jupiter's magnetic field changes.

- Calculations show that tidal heating can supply enough heat to keep most of Europa's ice melted beneath a solid ice crust.

While no single piece of evidence would make an overwhelmingly strong case by itself, together the evidence gives us good reason to suspect that the model in Figure 9.14 is correct, and that a liquid ocean really does exist beneath the surface of Europa. Still, the case is not definitive. For that, we'll need new space missions.

Proving the Case for an Ocean NASA scientists are hoping to send an orbiter to Europa, equipped with instruments that would include a laser altimeter and long-wavelength radar. The altimeter would be used to seek proof that an ocean lurks under the ice. The gravitational tug-of-war that Europa undergoes every 3.6 days in its elliptical path around Jupiter causes distortions in Europa's shape. If a deep, liquid ocean exists under the ice, Europa's surface would regularly bulge in and out by up to

30 meters. If there is only solid ice beneath the surface, however, then Europa won't stretch so easily and the altitude change in the bulge would be only about 1 meter. The difference can be reliably measured by the altimeter, and should settle the case regarding the existence of the ocean. The radar can also probe the surface, looking for radio reflections from a subsurface water interface. This same technique was used in Antarctica to discover a lake (Lake Vostok) that lies hidden beneath a 5-kilometer-thick deck of ice.

If there is an ocean on Europa, how big might it be? The current model (shown in Figure 9.14) suggests that the frozen crust (including the region of warmer, convecting ice) is about 20–25 kilometers thick. Given that our models of Europa's interior indicate that the total thickness of the water/ice layer is between about 80 and 170 kilometers, a 20–25-kilometer-thick crust leaves plenty of room for a deep ocean. In fact, if a global, subsurface ocean really exists, it could easily be 100 kilometers or more deep—some ten times as deep as the deepest ocean trenches on Earth. A 100-kilometer-deep europan ocean would contain roughly twice as much water as all of Earth's oceans combined.

• Could Europa have life?

The oceans on Earth teem with life, and in Chapter 6 we discussed reasons why many scientists suspect that life on Earth first arose in the oceans, and most likely near deep-sea volcanic vents. If Europa really does have a deep water ocean, tidal heating has probably kept it liquid for billions of years. Might life have arisen in the europan ocean as it arose on Earth, and could Europa be home to life today?

In Section 7.1, we identified the three key environmental requirements for life: (1) a source of elements and molecules from which to build living organisms, (2) a source of energy for metabolism and growth, and (3) a liquid medium for transporting the molecules of life. If Europa has an ocean, then it clearly meets the third requirement. As we discussed in Chapter 7, nearly all worlds probably contain at least some quantity of the 25 elements that make up life on Earth. Europa probably is no exception; its rock/water composition likely has all these elements, thereby satisfying the first requirement. Moreover, these elements would certainly be present at the rock/water boundary of the ocean floor, which is where we might expect any life to have arisen. That leaves open the question of whether Europa satisfies the second requirement: Does it have an energy source for life?

You might at first guess that the existence of a liquid water ocean would automatically mean there is energy for life. After all, if there is liquid water, there must be enough energy from tidal heating (and radioactive decay) to keep the water temperature above freezing. However, the existence of energy in the form of heated water is not by itself enough to allow the energy to be put to use. That is why fish can't live off ocean heat alone.

In general, extracting energy from a reservoir of warm water is possible only if there is also an adjacent "sink" of much colder water and a substantial difference in temperature between the two. And while we can build machines to take direct advantage of such differences in temperature, life has an additional requirement: There must be chemicals present that can react to generate energy for biological use. In practice, this additional requirement is almost certainly met when we have temperature differences at a rock/water boundary, because molecules in the rock and water can react together in a variety of useful ways. (We'll discuss some of the specific types of reactions that can supply energy to life in Section 9.4.)

Thus, to decide whether there might be life in a europan ocean, we must ask whether there is enough energy in a useful form to support biology. We can separate this question into two parts. First, is there enough energy to support an origin of life? Second, if so, is there enough energy to support a reasonable total biomass of ongoing life? Let's begin by considering the question of energy for an origin of life.

Energy for an Origin of Life   From an energy standpoint, the idea that life on Earth might have begun near deep-sea vents makes sense. The volcanic vents heat the water near them to very high temperatures, creating a large temperature difference with the surrounding cooler water. This temperature difference would have facilitated chemical reactions between the water and the rock erupted from the vents, leading to the formation of complex organic molecules and perhaps to the origin of life. We might therefore wonder whether the same type of vents exist on Europa, thereby providing the energy needed for an origin of life.

We cannot see through Europa's icy crust, which is why we don't even know for certain that a liquid ocean lies beneath it. Clearly, then, we have no way at present to know whether Europa has volcanic activity on an ocean floor. Nevertheless, the possibility of volcanic vents seems reasonable. Our models of Europa's interior suggest that it should have a rocky ocean floor, and tidal heating and the decay of radioactive elements may provide enough energy to melt interior rock that could erupt into the ocean. Europa might even have large undersea volcanoes. In that case, it certainly seems plausible to imagine an independent origin of life on Europa.

Energy to Support Ongoing Life   If life did arise in a europan ocean, how widespread could it be today? While deep-sea vents might lead to enough energy for an origin of life, they could not by themselves support more than a small total biomass because they simply don't make enough energy available to organisms. This fact might surprise you if you think about the great communities of life that live near deep-sea vents on Earth today (Figure 9.17), but most of this life actually gets its energy from materials that filter down from above, such as dead organisms and oxygen produced by photosynthetic life near the surface. Only a small fraction of the life near Earth's deep-sea vents lives solely off energy from the vents themselves. Thus, for life to be similarly abundant or widespread on Europa, it would need some other energy source in addition to the chemical reactions near deep-sea vents.

On Earth, photosynthesis is the best-known source of energy for life, and the one that ultimately powers humans via the food chain. Photosynthesis could not directly provide energy for life in a europan ocean, because sunlight cannot penetrate through more than a few tens of meters of ice. However, if pockets of liquid water exist near the top of the ice crust, then photosynthesis might provide energy to organisms living in these pockets, and this energy might then filter downward as organisms (dead or alive) cycle between the crust and the ocean. Even in this case, however, remember that because Europa is about five times as far from the Sun as Earth, sunlight is about $5^2 = 25$ times weaker at Europa than

Figure 9.17

On Earth, we find abundant life near deep-sea volcanic vents. However, while the vent supplies energy used by some microbes, most of the life around it—including the tube worms visible in this photo—actually gets energy from materials that filter down from above.

at Earth. Overall, it seems unlikely that photosynthesis would play any role on Europa, and in the best case it could support only a much smaller abundance of life than it supports on Earth.

Before the origin of photosynthesis on Earth, biochemical reactions may have been facilitated by energy sources such as lightning, ultraviolet radiation from the Sun, and heat released by impacts of asteroids and comets. Unfortunately, none of these energy sources is likely to be useful to life in europan seas. Europa has no atmosphere for lightning, ultraviolet light does not penetrate the ice, and the time during which impacts were frequent enough to provide significant energy ended some 4 billion years ago.

There is, however, another possible scheme for producing energy on Europa's icy surface. High-energy particles accelerated and trapped in Jupiter's magnetic field, as well as ultraviolet light from the Sun, regularly slam into the surface ice. These particles and photons hit the surface with enough energy to break up molecules in the ice, leading to the production of small quantities of other molecules such as hydrogen peroxide (H_2O_2), molecular oxygen (O_2), and hydrogen (which quickly escapes); this process explains why Europa has an extremely thin atmosphere (not noticeable to the eye, but detected by instruments). These molecules can facilitate energy-producing reactions, and they should be mixed into the top meter of the europan surface by the frequent churning caused by small meteorites. If all these molecules were ultimately to end up in the ocean below, they could provide energy to support life there. Unfortunately, we don't know how much of the outer ice actually gets cycled into the water below, or how often. In addition, the total amount of energy that might be available in this way is at least ten thousand times less than the amount of energy that photosynthesis generates on Earth, so an ocean on Europa could not support anywhere near as much life as do the oceans on our planet.

One other known process might yield some energy for life on Europa. Some of the potassium contained in the rocky material that makes up this moon would dissolve in the ocean (and be frozen in the ice above). The energy from the natural decay of one of potassium's radioactive isotopes would produce both hydrogen and oxygen molecules. Rough estimates suggest that these molecules could then facilitate chemical reactions that might support a small biomass.

The Case for Possible Life on Europa   Summarizing the case for life on Europa, we can say:

- There is good, indirect evidence that a liquid-water ocean exists.

- We expect the elements needed for life to be present in that ocean and on its floor.

- There are possible energy sources to support life, but the total available energy is small compared to the energy available for life on Earth.

Taken together, the evidence makes Europa seem like a good candidate for the possibility of life. That said, we should caution that it seems unlikely that sufficient sources of energy could exist in the europan ocean to support macro-fauna—the complex sea creatures of our aquariums, for example. If life exists in Europa's oceans, it is probably quite simple and small, perhaps analogous to the most primitive single-celled organisms that have existed on Earth.

THINK
ABOUT IT

Given the various uncertainties about a liquid ocean and available energy on Europa, do you consider it likely or unlikely that we will find life there? Why?

The only way to find out for sure whether life exists on Europa will be to go there. Although the next mission to Europa, still in its planning phases (and currently delayed due to NASA's overall budge woes), will almost certainly be an orbiter designed to determine whether the ocean really exists, subsequent missions could land on the surface. If a lander found any water that might have been brought up from below, we could analyze it for various organic molecules such as amino acids. The lander might also melt and filter a sample of surface ice—preferably a large sample since we expect the abundance of any life on Europa to be low—in search of evidence of life. Note that we'd need to be careful to sterilize the spacecraft so that we don't accidentally transport biology from Earth to Europa and then "discover" it there. Indeed, in 2003, NASA deliberately ended the *Galileo* mission by causing the spacecraft to plunge into Jupiter's atmosphere, expressly to prevent any possibility that the spacecraft might someday crash into and contaminate Europa with hitchhiking microbes from Earth. If the results from a lander are encouraging—for example, if they show the presence of organic molecules in the ice—we could then dream and scheme about probes that would manage to work their way through the ice and explore the eternally dark ocean depths where alien life might swim.

• Could other moons of Jupiter have life?

Europa is the most likely of the Galilean moons to be habitable. However, both Ganymede and Callisto are also composed of significant amounts of water ice. Could they, too, have underground oceans, and hence the possibility of life?

MOVIE MADNESS: *2010: THE YEAR WE MAKE CONTACT*

The monoliths are back, and they've brought plenty of buddies.

In the film *2010,* a sequel to the classic *2001,* humans are once again prompted by some enigmatic black slabs to head for the outer solar system. It seems that an unseen race of aliens is trying to make an important point about Jupiter and its moons. So we oblige them by sending yet another batch of confused astronauts on a billion-kilometer joy ride.

However, unlike the mission in *2001,* this crotchety crew actually returns to Earth, bringing with them a few sage words of alien advice: "All these worlds are yours except Europa. Attempt no landing there."

What's the deal? After all, this is *our* solar system. So why are some unknown, unseen entities marking Europa off-limits ? That's like Mom forbidding you to go into a basement closet. Of course that's the one place you'll find most interesting, and Arthur C. Clarke, the author of the novel *2010,* knew that there was, indeed, an interesting closet in the jovian system. The *Voyager* spacecraft had shown Europa to be an ice-covered world that could have a huge liquid ocean, and maybe even life.

In *2010,* this possibility is subtly exploited. The astronauts learn that there are chlorophyll-equipped critters somewhere below the europan ice. Alas, life in such a deep, dark habitat is a bit of a drag. So just as the humans arrive, the sophisticated aliens who built the monoliths decide to reengineer Jupiter, turning it into a mini-sun. They do this for the benefit of the primitive life-forms on Europa. The new star eventually warms their moon and converts it to something that looks like Earth during the Mesozoic era. The Europans, we presume, are destined to crawl out of their formerly ice-capped seas and find a monolith that will promote them to intelligent beings, the way our simian ancestors were improved in *2001.*

We lose Jupiter, but we gain a second, dimmer sun in our skies (no doubt a headache for astronomers and a source of confusion for migratory birds). But one wonders why these unseen extraterrestrials are so keen to meddle in the biological evolution of other worlds. Perhaps they just want our distant descendants to have some intelligent company right here in the solar system. It's a nice thought, but in the meantime someone needs to tell the space agencies that it's "hands off" Europa for the next few hundred millennia!

Ganymede   Ganymede, the third from Jupiter of the four Galilean moons, is the largest moon in the solar system. Like Europa, Ganymede has a surface of hard, brittle ice. However, while Europa's impact craters have been mostly erased by other geological processes, Ganymede appears to have both young and old surface regions, sometimes separated by remarkably sharp boundaries (Figure 9.18). Some regions are dark and densely cratered, suggesting that they look much the same today as they did billions of years ago. Other regions are light-colored with few craters, suggesting that they are geologically younger and have had their ancient craters erased; the young regions also exhibit strange grooves. The most likely explanation for the younger regions of ice is that they were created by "water eruptions" that occurred when internal heat caused ice below the surface to melt, erupt as watery "lava," and then freeze. The grooves were probably made by tectonic stresses that stretched the icy crust.

The idea that water sometimes erupts onto Ganymede's surface implies that partial melting must occasionally occur in the underlying ice, but could Ganymede have a full-fledged ocean like that thought to exist on Europa? The *Galileo* spacecraft turned its magnetometer to Ganymede just as it did to Europa, and made two intriguing discoveries. First, Ganymede apparently has its own intrinsic magnetic field—one generated within the moon—which may indicate that it has a molten, convecting core somewhat like Earth's outer core. Second, the magnetometer data showed that a small part of Ganymede's magnetic field varies with Jupiter's 10-hour rotation, just like Europa's magnetic field. Again, this is interpreted to indicate a field that is produced in electrically conducting material under the surface—most likely a salty ocean. Another bit of evidence for an underground ocean is the presence of salts on Ganymede's surface, which could conceivably be brine brought up from below. Overall, a liquid water ocean seems fairly likely to exist beneath Ganymede's frozen surface, although the case for its existence is less strong than the case for an ocean within Europa.

What source of heat could be keeping water melted under Ganymede's surface? Because of its greater distance from Jupiter, Ganymede has much less tidal heating than Europa. However, Ganymede's larger size also means it should retain heat better than Europa. Perhaps the heat produced over time by the combination of tidal heating and radioactive decay has been just enough to sustain a layer of subsurface liquid water.

While the possibility of a subsurface ocean is encouraging from the standpoint of habitability, the lesser heating on Ganymede means that the ice cover would be much thicker than on Europa—probably at least 150 kilometers thick. This would make finding life in a subsurface ocean far more difficult and the transport of possible nutrients from the surface considerably less efficient. In addition, the pressure in Ganymede's interior is high enough to create high-density forms of ice that likely lie beneath any liquid water ocean. As a result, it is also unlikely that Ganymede has a rock/water boundary at its ocean bottom like that on either Earth or Europa, a fact that would further reduce the available energy for life. Life, if it exists on Ganymede at all, would probably be less abundant and less evolved than seems possible on Europa. It might also be so deep below the surface that we'd have little hope of gaining access to it.

Callisto   Callisto is the farthest out of the four Galilean moons. Figure 9.19 shows that its entire surface is densely pockmarked by craters (the bright, circular patches) that must date back to the heavy bombardment. Other surface features are more difficult to interpret. For example, the close-up photo shows a dark, powdery substance that is concentrated

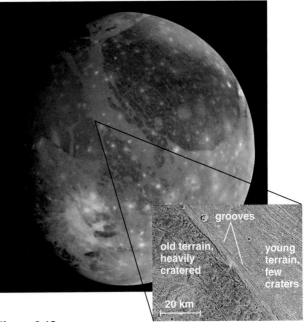

Figure 9.18

Ganymede, the largest moon in the solar system, has both old and young regions on its surface of water ice. The dark regions are heavily cratered and must be billions of years old, while the light regions are younger landscapes where tectonic faulting and eruptions of liquid water or slush have presumably erased ancient craters; the long grooves in the light regions were probably formed by tectonic stresses. Notice that the boundary between the two types of terrain can be quite sharp.

Figure 9.19

Callisto is heavily cratered, indicating an old surface that nonetheless may hide a deeply buried ocean. The inset shows how a dark powder appears to cover low-lying areas of the surface.

in low-lying areas, leaving ridges and crests bright white. The dark powder may be debris left behind when ice sublimates into gas from Callisto's surface.

Gravity measurements suggest that Callisto is mostly a ball of mixed ice and rock, overlaid by several hundred kilometers of water ice. Because its interior doesn't seem to be fully differentiated, we conclude that Callisto was never very warm inside and that neither radioactive decay nor tidal heating ever warmed this moon enough to melt it through. Surprisingly, however, the Galileo spacecraft found an induced magnetic field for Callisto, too, suggesting—as for Europa and Ganymede—the presence of a salty, subsurface ocean.

If a subsurface ocean exists on Callisto, what heat source could keep it liquid? Unlike the other three Galilean moons, Callisto doesn't participate in the orbital resonances that cause tidal heating, meaning that the warmth required to maintain a liquid ocean would almost surely have to come from radioactive decay. This meager heat source might be sufficient due to the insulating properties of a thick, icy skin. In addition, the water might contain salts and ammonia that act like antifreeze to help keep it liquid at low temperatures.

If Callisto really does have a deep, unseen ocean, we arrive at the astonishing possibility that three of Jupiter's large moons could have liquid water oceans and, perhaps, life. From the standpoint of intelligent aliens studying our solar system, they might have more examples of differing biology to explore by spending time near Jupiter than on Earth. Nevertheless, energy considerations make it unlikely that any of these moons has life in the abundance or diversity that we find for life on Earth. Callisto probably had even less energy available to life than Europa or Ganymede, and would therefore be the least likely to have life in its ocean.

 THINK ABOUT IT Suppose we had the technology to send landers that could somehow reach subsurface oceans on any of Jupiter's moons. Which moon would you choose to explore first? Why?

9.3 LIFE AROUND SATURN, AND BEYOND

Prospects for habitability and life dim considerably as we go beyond Jupiter. Although a few outer solar system moons have tidal heating, we do not find more examples quite as extreme as those of, say, Io and Europa. The greater distance from the Sun also dampens the possibility of obtaining energy from sunlight, and cold temperatures would presumably slow any metabolic reactions. Nevertheless, the outer solar system offers some intriguing prospects for life—or at least for interesting organic chemistry that could be a precursor to life.

• Could Titan have life?

The best-studied candidate for habitability beyond Jupiter is Saturn's moon Titan. The second-largest moon in the solar system after Ganymede, Titan is the only solar system moon to have a substantial atmosphere. In fact, Titan's atmosphere is even thicker than Earth's. The surface pressure is about 1.5 times that on Earth, which means that if you could visit Titan, the pressure would feel fairly comfortable even without a space suit. The temperature, however, would not. Here, where sunlight is nearly 100 times as weak as on Earth, the surface temperature is a frigid −180°C (−290°F).

It is not, however, completely hopeless. The ultraviolet light that hits Titan's atmosphere produces a wide range of organic molecules (the main contributors to the observed smog). Over billions of years, some of these compounds should have accumulated as a deep layer of organic sediment on the surface—these sediments may be the material in the dunes (see Figure 9.25). Occasional impacts by comets or asteroids would provide enough heat locally to melt any water ice and create pockets of warm water that might persist for a thousand years or so. While it is not clear that life could form in such a short time period, some interesting chemistry would certainly have occurred along the way. Moreover, if Titan really has icy volcanism, it might hint at the existence of liquid, subsurface aquifers of a water/ammonia mixture; in that case, it's conceivable that there could be volcanic "hot springs" where temperatures might rise slightly above 0°C. Perhaps life could have originated in such places and gained a foothold.

Another suggested possibility for powering the metabolism of simple life comes from chemical reactions in the upper atmosphere. Among other things, these reactions produce acetylene—an energy-rich, heavy compound that accumulates on the surface (the *Huygens* probe detected it). When acetylene reacts with oxygen, it releases a lot of energy (think of welding torches). On bitterly cold Titan, acetylene would react more slowly with hydrogen, releasing energy and producing methane as a waste product. Whether there is life that actually feeds on acetylene in this manner is, of course, only speculation, but at least a possible food source is available. One further possibility for life on Titan is deep beneath the surface. Models of Titan's interior suggest that, much like Ganymede or Callisto, it could have a liquid ocean far beneath its icy crust. However, this ocean would probably consist of a very cold ammonia/water mixture, rather than just liquid water, and the cold temperature would make life seem less likely.

Even if Titan lacks biology, studying this moon will give us valuable insights into chemical processes that might have been important to the beginnings of life on the early Earth. For example, scientists would like to know whether amino acids on Titan form equally in both right- and left-handed forms or if they exist mostly in left-handed forms like those used by life on Earth (see Figure 5.9). Moreover, we often give lip service to the fact that our ideas about the conditions that would produce life might be too conservative, but we don't often discuss alternatives. If future missions find living things on this small and hostile world—a place where the liquid environment and energy sources are quite different from those on Earth today—it will most assuredly be life as we *don't* know it.

• Could other moons of Saturn have life?

No other moon of Saturn is even close in size to Titan or to any of Jupiter's Galilean moons. Rhea, the next largest of Saturn's moons after Titan, is less than one third of Titan's size in diameter and less than one half the diameter of Europa (smallest of the Galilean moons). We expect such relatively small moons to retain far less radioactive heat than larger moons, and any tidal heating also should be at least slightly less effective. Nevertheless, several of Saturn's moons show evidence of past geological activity and one—Enceladus—has surprised scientists by showing that it is still geologically active.

Cassini photos of Enceladus show a moon with some very young surface regions. The bluish "tiger stripes" in Figure 9.26 (which are not really

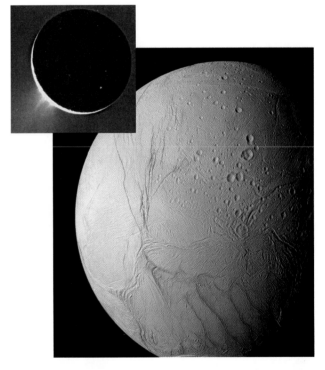

Figure 9.26

Active Enceladus. The blue "tiger stripes" in the main photo are regions of fresh ice that must have recently emerged from below. The colors are exaggerated; the image is a composite made at ultraviolet, visible, and infrared wavelengths. The inset shows Enceladus backlit by the Sun, with fountains of ice particles (and water vapor) clearly visible as they spray out to the lower left.

blue in color, since the image was made with ultraviolet and infrared as well as visible light) show regions near the moon's south pole that are measurably warmer than the surrounding terrain, and close-up examination suggests that they are cracks or grooves through which material can well up from below. These regions appear to be covered by "fresh" ice—ice that has solidified on the surface within no more than the past few thousand years, and possibly within just the past few decades or less. Moreover, images taken by *Cassini* as it looked at Enceladus backlit by the Sun show that fountains of ice particles and water vapor are spraying out from the tiger stripe regions (inset in Figure 9.26). This is clear evidence of icy volcanism on Enceladus. Spectral analysis of the tiger stripes also shows some simple organic molecules.

When you put it all together, Enceladus seems to have a subsurface liquid—probably an ammonia/water mixture—that drives its icy volcanism, at least some simple organic molecules, and enough heat to power all this activity. The astonishing conclusion: Enceladus may have subsurface habitable zones. The low temperature expected for a liquid ammonia/water mixture may make life seem unlikely, but we cannot rule it out. This unexpected finding has at least two important implications for the search for life in our solar system and the universe. First, while we know that Enceladus has some tidal heating, we are not yet certain how to explain all the geological activity that we see. Apparently, we don't yet fully understand the way that moons can be heated, and as a result conditions for liquid water (or lower-temperature ammonia/water mixtures) might be more widespread than we had imagined. Second, the fact that we have been so surprised by Enceladus should tell us that other surprises are likely to await us. Our basic ideas about where to look for life are probably still valid as starting points, but it might be wise to keep an open mind about other places as well.

• Could moons of Uranus or Neptune have life?

As we have progressed through this chapter, we have gradually moved down the scale of potential habitability. Europa seems reasonably likely to be habitable; Ganymede and Callisto seem somewhat less likely; and Titan and Enceladus, with their cold temperatures, seem to stretch to the limit the prospects for habitability. Nevertheless, our surprising findings make us ask whether there could be still other habitable places in our solar system.

After the Galilean moons and Titan, the next largest moon in our solar system is Neptune's moon Triton. As we discussed earlier, Triton orbits Neptune "backward," suggesting that it was somehow captured from what was once an independent orbit around the Sun. The question of how such a capture might have occurred is quite interesting, but here we are more concerned with the issue of potential habitability. Triton was photographed close-up only once, by *Voyager 2* in 1989. Its icy surface is an enigmatic mix of terrain that in some places is smooth and in others is crinkled into patterns resembling the skin of a cantaloupe (Figure 9.27). There are few impact craters. On the basis of the crater count on Triton today, its last resurfacing must have been no more than 10–100 million years ago. Clearly, Triton has had internal heat, possibly left over from tidal heating that would have occurred as it was being captured into its present orbit, and probably with an additional contribution from radioactive decay. Some researchers think this heat may be sufficient to occasionally cause ice volcanoes to erupt from a liquid ocean beneath the surface. The liquid would be much colder than ordinary liquid water, and

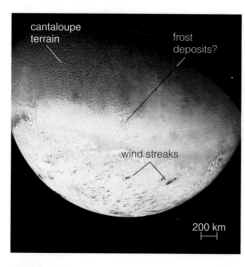

Figure 9.27

The southern hemisphere of Neptune's moon Triton, photographed by *Voyager 2* in 1989.

probably would consist of water mixed with ammonia, methane, or other melted ices.

Perhaps the most important idea emerging from our inspection of Triton is that even a moon in Neptune's distant realms, where surface temperatures are horrifically cold (colder than $-230°C$, or $-382°F$), might be geologically active and could possibly hide a liquid ocean and a refuge for life. And, given the surprise of finding the possibility of a subsurface liquid mixture on a moon as small as Enceladus, we might just find similar surprises if and when we send future spacecraft to orbit and study the satellite systems of Uranus and Neptune.

Overall, while we do not yet know if any of the moons of the jovian planets in our solar system has life, we've found at least six that seem at least potentially habitable. The lesson is clear: If similar moons are also numerous around the planets of other stars, such moons might be the most common habitats for life in the universe.

9.4 THE PROCESS OF SCIENCE IN ACTION: CHEMICAL ENERGY FOR LIFE

In our discussions of Europa, we touched on the fact that simply having a heat source is not by itself enough to make energy available for life. To support an origin of life, there must also be a chemical pathway by which complex molecules can be made, and to support ongoing life there must be chemical reactions that life can tap to fuel metabolism.

Not so long ago, the only known chemical pathways for metabolism were those used by plants and animals. However, as scientists have studied microbes living in "extreme" environments on Earth [Section 5.5], they have discovered many other metabolic pathways. These chemical pathways typically occur at interfaces between rock and water (such as in water-infiltrated rock underground), or through reactions with minerals in the hot water near deep-sea vents. If there is life on Europa or other jovian moons, it is likely to get its energy from the same types of chemical reactions. We therefore focus this chapter's case study in the process of science in action to the ways that life can extract chemical energy. As you'll see, this topic illustrates the interdisciplinary nature of astrobiology, in which biologists discover new forms of life on Earth, chemists and biochemists figure out how these organisms obtain energy, and planetary scientists then seek other worlds that might offer the same types of energy sources.

• What is the role of disequilibrium in life?

From a chemical energy standpoint, the basic requirement for life is a situation in which chemicals naturally exist in a state that is "unbalanced," which we describe as a state of **disequilibrium**. The idea is similar to that of a scale on which you place objects on both sides to see which is heavier. If the two sides weigh the same, the scale is balanced. But if you then add a little extra to one side, the scale quickly tips because it is no longer balanced. In a similar way, if there is disequilibrium among chemicals, they will start to react just like the scale starts to tip. With a scale, the movement quickly stops, because it can move only so far. But with chemicals, the reactions can continue as long as the disequilibrium remains. The idea will be clearer if we go into a bit of chemistry.

Any mixture of atoms and molecules can naturally undergo chemical reactions that may rearrange the atoms in such a way as to form or break chemical bonds. Left to themselves, however, chemical reactions will ultimately come to an **equilibrium** that represents a balance between the reacting atoms and molecules and the product atoms and molecules. For example, molecular hydrogen and oxygen can react together to make water. We can write this chemical reaction as:

$$H_2 + \tfrac{1}{2}O_2 \longleftrightarrow H_2O$$

The double arrow indicates that the reaction can proceed in both directions, and the $\tfrac{1}{2}$ in front of the O_2 indicates that the reaction requires only half as many oxygen molecules as hydrogen molecules.

If we begin by mixing hydrogen and oxygen, at first the reaction will proceed only toward making water molecules, since there is no water present initially. But eventually the reaction will proceed at equal rates in both directions, at which time we will have chemical equilibrium. In some cases, nearly all the hydrogen and oxygen will be converted to water, while in other cases little of it may be converted to water. The relative amounts of hydrogen, oxygen, and water at equilibrium depend on the external circumstances (such as pressure, temperature, and the presence of other chemicals). For example, the reaction between H_2 and O_2 needs a "push" to get it started. In a room filled with these two gases, this push might come from lighting a match. The energy from the match gets the reaction started; as H_2 and O_2 combine to form H_2O, additional energy is released that can then trigger more molecules to combine. This sequence can occur extremely rapidly, and the amounts of energy released can cause an explosion.

Now, imagine that the reaction between hydrogen and oxygen is occurring in a small flask that is inside a large room and that the reaction has come to equilibrium at the room's temperature. Suppose we do something suddenly that disturbs the equilibrium, such as adding excess hydrogen and oxygen molecules. Because the excess hydrogen and oxygen means that the reaction is no longer in equilibrium, we will have created a state of *disequilibrium*. This disequilibrium will cause the rate of water formation to speed up until the equilibrium is restored. Note that, once we force the chemicals into their disequilibrium, the rest of the process is completely natural: The reaction rate changes automatically in such a way as to bring the relative amounts of hydrogen, oxygen, and water back to their equilibrium values. Under the right circumstances, these reactions moving back toward equilibrium can release chemical energy that might be used by life to fuel metabolism.

No one is adding chemicals to the mixture of life, so the key to making chemical energy available for life lies in having some *natural* set of circumstances that can create and maintain a state of disequilibrium. In that case, the ongoing disequilibrium means that the reactions are always trying to move back toward equilibrium, thereby offering a continuous source of chemical energy. Natural processes that maintain chemical disequilibrium turn out to be quite common on Earth, and probably exist to some extent on any geologically active world.

For example, chemical disequilibrium inevitably exists near deep-sea volcanic vents, because mixing between the high-temperature vent water and the surrounding low-temperature ocean water creates conditions in which minerals and water will undergo chemical reactions. Because the vents continually release hot water, the disequilibrium can be maintained for long periods of time. Another place where we find ongoing disequilibrium is at interfaces between rock and water. The rock and water will naturally undergo chemical reactions, and as long as there is a supply of water that continuously circulates and comes in contact with the rock, the reactions will remain out of equilibrium. Thus, both deep-sea vents and rock/water interfaces offer places where chemical disequilibrium can provide ongoing energy that could be utilized to create complex molecules. This energy can also string complex molecules together into complicated structures—a process that may have been important to the origin of life—or to support the metabolism of living organisms.

• What types of chemical reactions supply energy for life?

Chemical disequilibrium offers the *potential* of providing chemical energy for life, but realizing this potential requires chemical reactions that life can actually use. A particular class of chemical reactions turns out to be especially important for life on Earth—reactions called **redox reactions**. Redox reactions involve an exchange or reshuffling of electric charge (which occurs through movement of electrons) between the reacting atoms or molecules.

Let's consider again what happens when hydrogen and oxygen combine to make water. Viewed on a molecular level, the reaction occurs in two steps. First, a hydrogen molecule decomposes into two protons (hydrogen nuclei) and two electrons:

$$H_2 \longrightarrow 2\,H^+ + 2\,e^-$$

(H^+ represents a single proton, which is positively charged, and e^- represents a single negatively charged electron.) Next, the two protons and two electrons combine with an oxygen atom (half of an oxygen molecule) to make a water molecule:

$$\tfrac{1}{2}O_2 + 2\,H^+ + 2\,e^- \longrightarrow H_2O$$

Viewed in this way, the production of water is a redox reaction, because electrons are effectively transferred from hydrogen to oxygen. Because hydrogen gives up the electrons, we say that it is the *electron donor* for the overall reaction. Because the oxygen takes on the electrons, we say that it acts as the *electron acceptor.*

In accepting electrons, the electrical charge of the oxygen is *reduced* (because electrons are negatively charged); hence the first three letters in *red*ox refer to this process of **reduction** of electrical charge. The charge of the hydrogen is increased, but because this increase occurs as the result of action by oxygen, we say that the hydrogen has become *oxidized*. In fact, oxygen is so efficient at grabbing electrons from other chemicals (atoms or molecules) that chemists have come to use the term **oxidation** to describe the process of losing electrons in general, even when the electrons are lost to something besides oxygen. Thus, the overall process of making water from hydrogen and oxygen, $H_2 + \tfrac{1}{2}O_2 \longrightarrow H_2O$, is called a redox reaction because the oxygen gets *reduced* while the hydrogen gets *oxidized*.

A redox reaction always involves the transfer of one or more electrons from an electron donor (which becomes oxidized) to an electron acceptor (which becomes reduced). The transfer of electrons gives off energy that can then drive other chemical reactions, including the biochemical reactions of life.

Redox Reactions on Earth   Most of the key energy-generating chemical reactions used by life on Earth are redox reactions. For example, the basic process of aerobic respiration in animals involves combining a sugar acquired by eating, such as glucose ($C_6H_{12}O_6$), with oxygen acquired by breathing. The reaction makes carbon dioxide and water, and releases energy in the process:

$$C_6H_{12}O_6 + 6\,O_2 \longrightarrow 6\,CO_2 + 6\,H_2O + energy$$

This is a redox reaction because the glucose donates electrons (it is oxidized) while the oxygen accepts electrons (it is reduced).*

Many cellular energy-generating processes proceed through chains of redox reactions, sometimes called *electron transport chains* because a series of redox reactions means a series of electron transfers. In photosynthesis, for example, the chain begins when chlorophyll absorbs sunlight. The energy from the sunlight creates disequilibrium in the cell, and this disequilibrium then offers energy that the cell utilizes through a chain of redox reactions.

Possible Redox Reactions for Other Worlds

Redox reactions are especially important when we consider the prospects for life in extreme environments, either on Earth or on other worlds. Many Earth organisms use fairly simple redox reactions as their primary source of energy. For example, bacteria known as *Thiobacillus ferrooxidans,* which can thrive in highly acidic conditions such as in mine tailings, obtain energy by oxidizing iron:

$$2\,Fe^{+2} + \tfrac{1}{2}O_2 + 2\,H^+ \longrightarrow 2\,Fe^{+3} + H_2O$$

(Fe^{+2} and Fe^{+3} represent iron atoms missing two and three electrons, respectively.) In this case, the iron is the electron donor (it is oxidized) and the oxygen is the electron acceptor (it is reduced). Note that neither pre-existing organic molecules nor sunlight is needed for this reaction, which means reactions like this one could have been used by early life-forms on Earth—and might be used by life living underground on other worlds.

Many other redox reactions produce energy for various microbes on Earth, including reactions involving molecular hydrogen and sulfur. These are especially important when we consider possible energy sources for an origin of life. For example, both iron and sulfur are common in the disequilibrium environments of hot springs and deep-sea vents, and thus could be involved in redox reactions that might ultimately lead to life. At rock/water interfaces, which exist any place there is liquid water underground, chemical reactions between water and iron in rock can produce molecular hydrogen, which can then be used in redox reactions for biochemistry.

As a result of the understanding of this chemistry gained by studying life on Earth, we now have reason to think that life could exist on many other worlds. In particular, any geologically active world with liquid water may have places where chemical disequilibrium persists for long periods of time, such as near underwater volcanic vents or anywhere where rock and water come into contact. At these places, redox reactions can provide energy that could power biochemical reactions that might ultimately

---

*You can often recognize what is being oxidized or reduced in redox reactions, even without knowing precisely how electrons are rearranged. In the aerobic respiration reaction, for example, the C in glucose is being oxidized because it ends up being combined with a greater number of O atoms. Glucose has equal numbers of C and O atoms, but CO_2 has twice as many O as C atoms and so is more oxidized. The oxygen is being reduced because it ends up being combined with more hydrogen; it has no H atoms on the left side of the reaction, but has two H atoms per O atom on the right. Thus, for example, the C in CH_4 is more reduced than the C in CO_2, which is more oxidized.

lead to life, and that could support life once it arises. That is why places such as underground pockets of liquid water on Mars or the subsurface ocean of Europa seem so promising as potential abodes of life.

THE BIG PICTURE
Putting Chapter 9 into Context

In this chapter, we have considered the moons of the outer solar system and found that several of them might offer conditions suitable for life. As you continue in your studies, keep in mind the following "big picture" ideas:

- Our own Moon, a now-dead relic of an early collision, may have misled us into thinking that only planets can harbor life. In fact, moons exhibit enormous variety. Some of the moons of the outer solar system are as large as small planets, a few might have liquid oceans, and one even has a substantial atmosphere. Life might well be possible in such places.

- The solar system moon most likely to be habitable, Europa, is kept warm inside by a mechanism quite different from that which warms Earth's interior. Tidal heating, the result of orbital resonances that occur among three large moons of Jupiter, can provide a continuous source of heat for billions of years. Because orbital resonances can arise quite naturally, tidal heating may be common among moons of jovian planets throughout the universe.

- The icy moons of the outer solar system force us to rethink our basic concept of "habitability." If any of these moons are indeed homes to life, then the range of habitability is much broader than we might have guessed from studies of terrestrial worlds.

- From a chemical energy standpoint, life requires conditions in which there is a natural and ongoing source of chemical disequilibrium. Such conditions probably exist on almost any geologically active world that has liquid water, either at underwater volcanic vents or simply at the interfaces of rock and even tiny amounts of liquid water.

SUMMARY OF KEY CONCEPTS

9.1 THE MOONS OF THE OUTER SOLAR SYSTEM

- **What are the general characteristics of the jovian moons?**
The moons of jovian planets range greatly in size, from a few kilometers across to somewhat larger than Mercury. They tend to have ice mixed in with their rock: water ice for all the jovian moons; plus ammonia, methane, and other ices for moons of the more distant jovian planets. Nearly all are in **synchronous rotation**, keeping one side perpetually turned toward their host planet.

- **Why do we think that some moons could harbor life?**

Some moons apparently retain substantial internal heat as a result of **tidal heating**, along with radioactive decay. Tidal heating explains the tremendous volcanic activity on Io and the heating thought to melt subsurface ice on Europa. A few other moons may also have liquid water or other liquids, and thus would seem to meet the minimum requirements for life.

9.2 LIFE ON JUPITER'S GALILEAN MOONS

- **Does Europa have an ocean?**

Photos of Europa's surface show numerous features, including a relative lack of impact craters, suggesting that liquid or slush from below has occasionally gushed through and repaved the surface. In addition, Europa has a magnetic field that makes sense only if we assume it has a salty ocean. These observational data, combined with the known tidal heating of Europa, make it likely that it has a subsurface ocean that may contain twice as much water as the oceans of Earth.

- **Could Europa have life?**

While it's probable that Europa has both a liquid water environment and the elements necessary for life, possible energy sources for life are much more limited than on Earth. Volcanic vents on the ocean floor could provide some energy, perhaps enough for life to have arisen but probably not enough to support life in great abundance. A few other energy sources may contribute additional energy, but overall we would expect any life on Europa to be simple and small.

- **Could other moons of Jupiter have life?**

Ganymede, the solar system's largest moon, is somewhat akin to Europa, in that it has an icy crust, at least part of which seems to undergo "repaving" by liquid or slush from below. A change in this moon's magnetic field as it orbits Jupiter also suggests a subsurface, salty ocean. Callisto, the farthest out of the Galilean moons, undergoes no tidal heating and appears to have an ancient surface. Nonetheless, magnetic measurements suggest a possible hidden ocean. Thus, both Ganymede and Callisto could conceivably offer conditions for life, although energy sources are even more limited on these moons than on Europa.

9.3 LIFE AROUND SATURN, AND BEYOND

- **Could Titan have life?**

Titan has an atmosphere even thicker than Earth's, but it's largely composed of nitrogen, argon, ethane, and methane. The *Huygens* probe showed that fluid hydrocarbons have coursed over Titan's landscape, but the bitterly cold temperatures on this moon would greatly slow chemical reactions, making metabolism difficult and decreasing the chances for life. However, it is possible that Titan sometimes has surface or near-surface pockets of liquid water, and a subsurface ocean of a cold ammonia/water mixture. Some energy sources for life might also be available.

- **Could other moons of Saturn have life?**

Much to scientists' surprise, *Cassini* observations show that the relatively small moon Enceladus must have a subsurface liquid that drives the fountains of ice and water vapor observed to be emerging from the moon. Thus, it is possible that Enceladus has zones of habitability. While we have no direct evidence for similar possibilities on other moons of Saturn, the case of Enceladus tells us not to rule them out too quickly.

- **Could moons of Uranus or Neptune have life?**

Life seems less likely on such distant moons, but it is still possible that some could have habitable zones beneath their surfaces. Neptune's moon Triton shows evidence of icy volcanism, suggesting it might have liquid beneath its surface. Other moons seem like much longer shots, but the lesson of Enceladus tells us we should study them further before concluding that they lack liquids that might sustain life.

9.4 THE PROCESS OF SCIENCE IN ACTION: CHEMICAL ENERGY FOR LIFE

- **What is the role of disequilibrium in life?**

Life as we know it can exist only in places where natural conditions maintain a state of chemical **disequilibrium**. This disequilibrium can cause chemical reactions that may be used to create complex molecules, to string complex molecules together into complicated structures, or to support metabolism of living organisms.

- **What types of chemical reactions supply energy for life?**

Life on Earth gains energy from **redox reactions** in which one molecule gains electrons and another loses them. The redox reactions used by terrestrial life include those that occur near deep-sea vents and at underground rock/water interfaces, suggesting that the same types of reactions might be used by life on worlds that have similar conditions, such as Mars (underground rock/water interfaces) and Europa (deep-sea vents).

REVIEW QUESTIONS
Short-Answer Questions Based on the Reading

1. Briefly explain how the larger *jovian moons* tend to differ in general from the smaller ones. How does the formation process of the moons explain these differences?

2. Briefly describe the cause of the tides on Earth, why they lead to *tidal friction,* and how tidal friction affects Earth's rotation and the orbit of the Moon.

3. What is *synchronous rotation,* and why is it so common among the jovian moons?

4. What is *tidal heating?* Briefly explain how it can arise and persist as a result of *orbital resonances.* How does tidal heating affect Io?

5. Describe the evidence suggesting that Europa has a liquid water ocean beneath its icy crust. How might future observations confirm this idea?

6. What energy sources might be available to life on Europa? Overall, what can we say about the likelihood and abundance of life on Europa?

7. Describe the evidence for subsurface oceans on Ganymede and Callisto. What are the prospects for life on these worlds?

8. Why was Titan chosen for such intense study by the *Cassini/Huygens* mission? Why is it surprising to find methane in Titan's atmosphere?

9. Based on recent data, describe the general nature of Titan and discuss its prospects for life.

10. What makes recent discoveries about Enceladus so surprising? Could Enceladus be habitable? What lessons does Enceladus hold for our more general search for life in the universe?

11. Could Triton be habitable? Briefly discuss the possibility of finding habitable moons around Uranus or Neptune.

12. What do we mean by chemical *equilibrium* and chemical *disequilibrium?* Why is disequilibrium necessary for life?

13. What are *redox reactions?* Give a couple of examples.

14. Based on our understanding of the chemistry of life, where should we expect such chemistry to be possible? What are the implications of this idea to the search for life beyond Earth?

TEST YOUR UNDERSTANDING

Evaluate the Claims

Each of the following statements makes some claim. Evaluate the claim, writing a few sentences describing why you think it is valid or invalid (or clearly true or false). Explain clearly; because not all of these have definitive answers, your explanation is more important than your chosen answer.

15. Io is riddled with volcanoes because of its proximity to Jupiter's strong magnetic field.

16. Europa is likely to have fishlike organisms the size of whales swimming in its ocean.

17. While Europa, Ganymede, and Callisto are all candidate locations for life, we expect that the most abundant and diverse life would be found within Callisto.

18. The fact that our Moon keeps one side always facing Earth is an astonishing coincidence.

19. Titan is simply too cold to have any life.

20. Triton might have life that uses liquid ammonia, rather than liquid water, as its transport medium.

21. Io doesn't have a significant atmosphere because it lacks a source of outgassing.

22. Orbital resonances like that between Io, Europa, and Ganymede are the results of extremely rare accidents, so we would not expect tidal heating to be important in other planetary systems.

23. If there is life on Enceladus, it probably gets its energy from sunlight.

24. If our solar system is typical, then other star systems might have an average of five to ten worlds on which liquid water (or a mixture of water and some other liquid) exists in at least some places.

Quick Quiz

Choose the best answer to each of the following. Explain your reasoning with one or more complete sentences.

25. The moons of Saturn may have large amounts of ammonia and methane ice, while those of Jupiter do not because: (a) Methane and ammonia come only from comets that exist in the Oort cloud. (b) Jupiter's strong magnetic field encourages water ice to form. (c) The greater cold at Saturn's distance from the Sun means that ices of ammonia and methane could condense there but not at Jupiter.

26. Which statement about synchronous rotation is true? (a) It can develop only on moons that are born with slow rotation. (b) It occurs commonly as a result of tidal forces exerted on moons by their parent planet. (c) It can develop only on moons with liquid oceans.

27. Io is covered in volcanoes while Europa is covered in ice because: (a) Io is larger in size than Europa. (b) Io receives much more sunlight than Europa. (c) Io is subject to stronger tidal heating than Europa.

28. Which of the following is *not* an indication of liquid water beneath Europa's icy surface? (a) the moon's changing magnetic field; (b) the moon's average density; (c) surface features that look like jumbled icebergs.

29. Photosynthesis is an unlikely source of energy for life in europan seas primarily because: (a) The moon's ice cover is too thick. (b) Sunlight at the distance of Jupiter is too weak. (c) There is no soil on which plants could grow.

30. It's assumed that, even if Europa has life, the total amount of that life will be small. That's because: (a) Europa is only about as big as our own Moon, and consequently there's not much room for life. (b) The ocean will be cold, slowing down metabolism. (c) There are likely to be only limited sources of energy for life.

31. The chances for life on Titan's surface are considered slim, mainly because: (a) There's little oxygen in the atmosphere. (b) The liquid methane and ethane rain would be lethal. (c) The surface temperature is far below the freezing point of water.

32. Where might we find liquid water on Titan? (a) in lakes and rivers that appear to exist on the surface; (b) beneath the surface near sources of "icy volcanism"; (c) as droplets in high-altitude clouds.

33. Why were scientists so surprised to find active geology on Enceladus? (a) because it is so small in size; (b) because it lacks any possibility of tidal heating; (c) because it is so far from the Sun.

34. Chemical disequilibrium is likely to be present at all the following places *except:* (a) volcanic vents on ocean floors; (b) in solid ice exposed to the extreme cold of space; (c) underground aquifers where thin films of liquid water move over rock.

INVESTIGATE FURTHER
In-Depth Questions to Increase Your Understanding

Short-Answer/Essay Questions

35. *Lessons for Life.* Considering everything we've learned about the possible habitability of jovian moons, make a list of what you think are the three most important lessons that apply to the search for life in *other* solar systems. Describe the importance of each lesson clearly, and conclude by summarizing how the study of jovian moons has changed our perspective about life in the universe.

36. *Exploring Europa 1.* Although Europa is a promising place to look for life, penetrating its thick, icy crust will be difficult. Suggest a possible way of making a spacecraft that could enter the europan ocean. If it is technically feasible, do you think we should do it soon, or wait until we have further evidence of life? Defend your opinion.

37. *Exploring Europa 2.* One suggestion for determining whether Europa has life is to send an orbiter that passes close to the surface of the planet, drops a "bowling ball" that makes a crater on the surface, and then catches the ice thrown up by the impact. This sample ice would be brought back to Earth for analysis. Briefly describe what we would be looking for in such an experiment, and what it might teach us. How do you think such an experiment would affect the cost of a mission to Europa compared to, say, a mission that orbits but does not return to Earth? Explain.

38. *Europan Fish.* On Earth, fish breathe oxygen that is dissolved in the ocean. Do you expect that we will find dissolved oxygen in Europa's ocean? Why or why not? Based on your answer, if we could somehow transport fish to Europa, is it possible that they could survive in the europan ocean? What other types of life from Earth might survive on Europa?

39. *Latest Cassini Results.* Find the latest press release or discovery from the *Cassini* mission to Saturn (as posted on NASA's *Cassini* Web site). What has been discovered? Does the discovery alter any of the ideas discussed in this chapter? Explain.

40. *Life on Titan.* Several possibilities have been suggested for the support of life on Titan, including acetylene that falls onto the surface, and the possible presence of subsurface, liquid water aquifers. If you could plan one new Titan lander, how would you design it to search for life?

41. *Top-Priority Mission.* Suppose that you were chosen to design one robotic mission, and one mission only, to search for life in the outer solar system. Which world would you investigate? Defend your choice.

42. *Migrating Life.* As we discussed in Chapter 6, there is a decent likelihood that life might at some point have traveled from Earth to Mars (or vice versa) on meteorites. Discuss the likelihood that life from Earth or Mars could have made its way to, and taken root on, Europa or other jovian moons.

43. *Movie Review.* A number of science fiction movies have concerned jovian moons, including *2010, Outland,* and *Gattaca.* Choose one, watch it, and write a critical review. Be sure to comment on the accuracy of any scientific content in the movie.

44. *The Sirens of Titan.* The moon Titan plays the title role in Kurt Vonnegut's novel *The Sirens of Titan.* Although the book never intended to give a realistic portrayal of Titan, it helped popularize the moon. Read the novel, and write a short critical review, focusing on how Vonnegut's portrayal of Titan differs from reality.

Quantitative Problems

Be sure to show all calculations clearly and state your final answers in complete sentences.

45. *Orbital Resonances I.* Using the data in Appendix D, identify an orbital resonance relationship between Titan and Hyperion. (*Hint:* If the orbital period of one were 1.5 times the other, we would say that they are in a 3:2 resonance.)

46. *Orbital Resonances II.* Using the data in Appendix D, identify an orbital resonance that affects Enceladus. In light of the *Cassini* mission findings about Enceladus, how might this resonance be important?

47. *Tidal Force on the Moon.* In Cosmic Calculations 9.1, we found the tidal force that Earth exerts on the Moon today. Following a similar procedure, calculate the tidal force Earth would have exerted on the Moon shortly after the Moon formed, when it was only about $\frac{1}{10}$ its current distance from Earth. How much greater was that tidal force than it is today? What does this tell you about how a factor-of-10 change in distance affects the tidal force?

48. *Tidal Force on Io.* Using the procedure from Cosmic Calculations 9.1, calculate the tidal force exerted on Io by Jupiter. Compare to the tidal force that Earth exerts on the Moon, and comment on the implications to Io's volcanism.

49. *Tidal Force on Europa and Ganymede.* Using the procedure from Cosmic Calculations 9.1, calculate the tidal force exerted on Europa and Ganymede by Jupiter. Compare to the tidal force exerted on Io (from Problem 48), and comment on the expected strength of tidal heating on each world.

50. *Astrology Explained (or Not).* Recall that astrology claims that the positions of the planets in the sky at a person's moment of birth forever influence the person's life. In an attempt to "explain" this claim, some astrologers have proposed that the source of the influence could be tidal effects from the planets. (a) Using the method of Cosmic Calculations 9.1, calculate the tidal force exerted by Jupiter on a baby being born on Earth. For the purposes of the calculation, assume the baby is 50 centimeters long, so that the distance of the baby's "near" side to Jupiter is Jupiter's distance *minus* 25 centimeters (0.25 meter), and the distance to the baby's "far" side from Jupiter is Jupiter's distance *plus* 25 centimeters; for Jupiter's distance, use 780 million kilometers (which is about the average). (b) Jupiter, of course, is not the

only gravitational influence on the baby. For example, there is also a gravitational force acting between the baby and the doctor supervising the delivery. Calculate the gravitational force that is pulling the baby toward the doctor. Assume the baby's mass is 3 kilograms and the doctor's mass is 50 kilograms, and that they are 0.5 meter apart during delivery. (c) Compare the tidal effect of Jupiter on the baby to the gravitational pull between the baby and doctor. Based on your answer, is it plausible to claim that tidal effects of the planets can influence the baby's life?

Discussion Questions

51. *Importance of Life.* In coming decades, scientists hope to devote a lot of effort to searching for life on jovian moons. How important do you think it would be if we found life on any of these moons? Answer in terms of both scientific importance and philosophical importance. Overall, do you think the possible benefits justify the expense required to undertake this search?

52. *Limited Thinking.* Throughout this book, we have generally assumed that life would be at least somewhat similar to Earth life; for example, we assume it would be carbon-based, most likely use liquid water, and gain energy from sunlight or chemistry. Are these assumptions still valid in light of what we've learned about jovian moons? Under what circumstances might we need to broaden them? Explain.

WEB PROJECTS

53. *Galileo Photo Journal.* Visit the Web site for NASA's *Galileo* mission to Jupiter, and study some of the images of Europa, Ganymede, and Callisto. For one of these moons, create your own photo journal in which you include at least ten photos along with a paragraph or two for each photo that explains how it relates to the question of habitability on that moon.

54. *Europa Orbiter.* Find out the current status of NASA's plans to send a mission to study Europa. What will the mission do? When will it be launched? What are its science goals? Write a one- to two-page summary of your findings.

55. *Prove the Book Wrong (or Right).* In this chapter, we presented a number of tentative findings from the *Cassini–Huygens* mission. Visit the mission Web site and try to find at least one new discovery that either shows an idea discussed in this book to be wrong or lends further evidence to support it. Write a one-page summary of what you learn.

10
The Nature and Evolution of Habitability

LEARNING GOALS

10.1 THE CONCEPT OF A HABITABLE ZONE

- How does a planet's location affect its prospects for life?
- Could life exist outside the habitable zone?

10.2 VENUS: AN EXAMPLE IN POTENTIAL HABITABILITY

- Why is Venus so hot?
- Could Venus have once been habitable, and could life still exist there?

10.3 SURFACE HABITABILITY FACTORS AND THE HABITABLE ZONE

- What factors influence surface habitability?
- Where are the boundaries of the Sun's habitable zone today?

10.4 THE FUTURE OF LIFE ON EARTH

- How will the Sun's habitable zone change in the future?
- How long can life survive on Earth?

10.5 THE PROCESS OF SCIENCE IN ACTION: GLOBAL WARMING

- What is the evidence for global warming?
- What are the potential consequences of global warming?

This dead of midnight is the noon of thought,
And Wisdom mounts her zenith with the stars.

Anna Letitia Barbauld (1743–1825),
A Summer's Evening Meditation

Although the other worlds of our solar system may have life, there's little doubt that none will have the diversity and abundance of life found on Earth. Why is this? After all, Mars and Venus are roughly similar in size and composition to our own planet, and both have atmospheres; yet conditions on those worlds are dramatically different from those on Earth.

On one level, this question seems easy to answer: Earth has abundant liquid water on its surface, and the other planets do not. But as we probe deeper, we find that the underlying causes for *why* Earth is teeming with life, and other worlds are not, are more subtle. For example, Mars had surface liquid water in the past, but no longer does—implying that the potential for life can change with time.

In this chapter, we'll take a broad approach to the question of habitability in our solar system, seeking to understand the precise factors—such as size and distance from the Sun—that make Earth so different from its neighboring worlds. In doing so, we'll gain deep insight into the likelihood of finding Earth-like worlds around other stars. After all, we can expect Earth-like planets to be common only if they can exist over a fairly wide range of sizes and distances. Learning what makes Earth habitable will help us understand how its habitability might change in the future, either through long-term, natural changes that will ultimately occur in our solar system or through human-induced changes that could have much more immediate and serious consequences for our own civilization.

10.1 THE CONCEPT OF A HABITABLE ZONE

We all know of urban legends—for example, that small pet alligators flushed down toilets have supposedly grown to adulthood and are roaming the sewers of New York. Indeed, urban legends occur across a wide variety of subject areas, and astrobiology is no exception. So here's another one to think about: If Earth moved a mile closer to the Sun, would we all burn up? According to an urban legend that has spread widely over the past few years, the answer is yes, and the legend also holds that we'd freeze if our planet moved just a mile farther away.

But in truth, we are no more likely to burn up (or freeze) if Earth moved a mile either closer or farther from the Sun than city sanitation employees are likely to become an alligator's meal while at work. The legend is nonsense, as you can realize just by thinking about Earth's orbit around the Sun: The orbit is an ellipse, not a circle, and Earth's distance from the Sun varies from a minimum of about 147.1 million kilometers each January to a maximum of about 152.1 million kilometers each July.

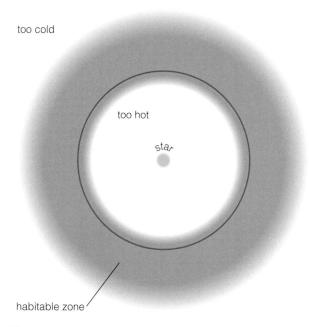

Figure 10.1

A planet with abundant liquid water on its surface can in principle exist only within some particular range of distances from a star. This range defines the star's habitable zone, and the size of the range depends on characteristics of the star.

Thus, according to the legend, our whole planet would burn up each January and freeze each July. In reality, this 5-million-kilometer variation in distance has virtually no effect on the weather at all, a fact that becomes obvious when you remember that the Northern Hemisphere has summer during the time that Earth is *farthest* from the Sun and winter when Earth is closest to the Sun. If the distance variation were important, the whole planet would have to be significantly warmer in January than in July, and it's not.

But like most urban legends, this one holds at least one deeply buried kernel of truth, because there must be some distance from the Sun that would be too hot for life to survive on Earth, and some distance at which it would be too cold. The distance isn't a mile, nor even a few million miles, but in principle we should be able to determine what it is. We'll devote much of this chapter to determining this distance.

• How does a planet's location affect its prospects for life?

If there are distances from the Sun at which our planet would become too hot or too cold for life, then there must also be some range of distances that is "just right" for a world like Earth. This range defines what we call the Sun's **habitable zone;** that is, the habitable zone is the range of distances from the Sun within which we could in principle move our planet without fundamentally changing the characteristics that make it home to abundant life (Figure 10.1). Other stars also have their own habitable zones, meaning distances at which a world similar in size to Earth, and with a similar atmosphere, would be habitable, although the sizes of habitable zones are different for different types of stars [Section 11.1].

Note that this definition captures the essence of what we mean by a habitable zone, but there are several important caveats to keep in mind:

1. The concept of a habitable zone is based on the range of distances at which *worlds similar to Earth* could exist. In other words, a habitable zone is a zone in which it is possible for a world to have abundant liquid water on its surface.
2. Simply being in a star's habitable zone is *not* sufficient to make a world habitable. The Moon presents an obvious case in point: As a companion to our planet, it is located at essentially the same distance from the Sun as Earth, but it is not habitable.
3. Habitable zones evolve with time. In particular, because stars like the Sun tend to brighten as they age, we expect the habitable zone to move outward over time.

In summary, *at any particular time, a star's habitable zone is the range of distances around it at which a planet could potentially have surface temperatures that would allow for abundant liquid water.*

We have considered a number of places besides Earth in our solar system that might either have life now or have had life in the past, including Mars, Europa, and Titan. However, none of these places seem likely to have *surface* liquid water or life at present; instead, their life would be in underground locations where liquid water might be present. This distinction may not be that important for the search for life within our own solar system, because future space missions should eventually allow us to search for life beneath the surface of worlds like Mars and Europa. However, the distinction between surface life within a habitable zone and subsurface life beyond it is important when we consider the challenges of finding life in other star systems.

The surface pressure and temperature on Mars are too low for liquid water to exist there today. Does this imply that Mars is beyond the outer boundary of the Sun's habitable zone? Why or why not?

We are unlikely to be able to travel to the stars anytime soon [Section 13.1], so telescopic images and spectra offer our only realistic hope of finding life on extrasolar planets or moons. Surface life may well create spectral signatures in a planet's atmosphere that would allow us to detect it [Section 11.2], but subsurface life is unlikely to cause enough atmospheric change to make an unmistakable spectral signature. Thus, while we may be able to find subsurface life (if it exists) within our own solar system, we have little hope of finding it on planets or moons around other stars. So the search for life beyond the realm of the Sun is by necessity a search for surface life on worlds within the habitable zones of their stars.

• Could life exist outside the habitable zone?

There are several ways that habitability might present itself outside the habitable zone. One is in small pockets of subsurface groundwater, such as that which may still exist on Mars today [Section 8.3]. Because this water would be kept liquid primarily by geological conditions rather than solar heat, a Mars-like planet could have subsurface habitability even if the planet itself lies beyond the outer edge of the habitable zone. It's not far-fetched to imagine that Mars-like planets could be more common than Earth-like planets.

The possibility of life in subsurface oceans, such as that thought to exist on Europa [Section 9.2], could make habitability even more common. We expect moons in the outer regions of any solar system to contain large amounts of water ice, because water is the most common of the ices that can condense in regions far from a forming star. Orbital resonances like those that contribute to the tidal heating [Section 9.1] of Io, Europa, Ganymede, and Enceladus could occur in any case where a large planet has multiple moons, and the heating could potentially melt subsurface ice. Heat from radioactive decay might also contribute to melting interior ice. Because tidal heating and radioactive decay can supply heat in the absence of sunlight, internally heated moons could exist around planets orbiting at almost any distance from a star (except very close-in, where all ice would melt and evaporate away).

Another intriguing possibility concerns *surface* habitability on Earth-size planets outside habitable zones. Theoretical calculations suggest that an Earth-size planet's own internal heat could keep its surface warm enough for liquid water *if* the surface were protected against heat loss by a thick hydrogen atmosphere. Such a hydrogen atmosphere is not possible on an Earth-size planet in our solar system, because solar heat would cause the hydrogen to escape into space fairly rapidly. But, as we will discuss further in Chapter 11, it is possible that Earth-size planets are sometimes ejected from solar systems in the process of forming, and sent into interstellar space. When an Earth-size planet is born, it might have a hydrogen atmosphere for a short time, particularly if it forms at a greater distance from its star or around a star cooler than our Sun. If such a planet is ejected into interstellar space before it loses this atmosphere, its thick hydrogen envelope might remain for many billions of years. Indeed, such "free-floating Earths" could conceivably be quite common in interstellar space, though their relatively small size and low brightness would make them extremely difficult to detect. If they exist, such planets could have surface oceans, as well as geothermal and chemical energy much like that sustaining life underground and around deep-sea vents on Earth. Although the total available energy would probably not sustain a huge amount of life, at least some might be possible.

We could also conceive of habitability outside the habitable zone if life can use a liquid medium other than water, such as liquid ethane, which as you'll remember from Table 7.1 has a much lower freezing point. Any liquid will evaporate rapidly when atmospheric pressure is low, so the presence of surface liquids of any type requires an atmosphere. In general, this means we can hope to find surface liquids only on worlds large enough to hold significant atmospheres, but not so large that they become giants like the jovian planets, where strong vertical winds probably preclude life. In our solar system, Titan is the only world that meets this criterion. Nevertheless, moons like Titan might be relatively common among other star systems. Thus, if life based on other liquids is possible, there could be many habitable worlds similar to Titan.

In summary, it's quite possible that the majority of habitable worlds in the universe are not located within stellar "habitable zones." From this standpoint, the concept of a habitable zone might seem obsolete because it doesn't account for the potential habitability of Europa-like or Titan-like moons, of Mars-like subsurface water, or of "free-floating Earths." Nevertheless, while life might be common on such worlds, it would be extremely difficult to detect. Also, it seems extremely unlikely that such worlds could give rise to complex life, let alone advanced civilizations. Thus, the concept of a habitable zone is still quite useful in searching for life beyond Earth, and it is critical when we are searching for intelligent life.

10.2 VENUS: AN EXAMPLE IN POTENTIAL HABITABILITY

One of our primary goals in this chapter is to define the boundaries of the habitable zone in our own solar system, so that we can then extend the definition to the search for planets within the habitable zones of other stars. A useful first step in defining these boundaries is to determine which planets currently lie within the Sun's habitable zone. We can clearly rule out Mercury, which is too close to the Sun, and all the planets beyond the asteroid belt, which are too far. That narrows the possibilities down to Venus, Earth, and Mars.

We have already explored the cases of Earth and Mars in some depth. Earth is obviously within the habitable zone. Mars must be at least near the borderline of the habitable zone, since it apparently had liquid water on its surface in the past. But what about Venus?

• Why is Venus so hot?

The surface of Venus is far too hot for liquid water [Section 7.2], but we can't directly attribute this heat to Venus's distance from the Sun. Venus orbits the Sun at a distance about 72% of Earth's distance; because the intensity of sunlight follows an inverse square law (see Figure 7.1), this makes the intensity of sunlight a little less than twice as great at Venus as at Earth (because $\frac{1}{0.72^2} \approx 1.9$). We'd expect this extra sunlight to make Venus warmer than Earth, but not nearly by enough to account for its searing, 470°C (878°F) surface temperature. Calculations show that if Venus had the same atmosphere as Earth, its average temperature would be only about 30°C hotter than Earth's—making Venus somewhat like the tropical planet envisioned in old science fiction.

As we discussed in Chapter 7, the immediate cause of Venus's high temperature is an extreme greenhouse effect produced primarily by atmospheric carbon dioxide. Venus has about 200,000 times as much carbon dioxide in its atmosphere as Earth, and carbon dioxide is a greenhouse gas that traps heat near a planet's surface [Section 4.5]. However, given their similar sizes and compositions, we expect Venus and Earth to have had similar levels of volcanic outgassing—and the released gas ought to have had about the same composition on both worlds. We therefore are left with a deeper question about Venus's extreme heat: Why is Venus's atmosphere so different from Earth's?

Our understanding of planetary geology suggests that huge amounts of water and carbon dioxide should have been outgassed into the atmospheres of both Venus and Earth. Venus's atmosphere does indeed have huge amounts of carbon dioxide, but it has virtually no water. Earth's atmosphere has only small amounts of either gas. Thus, we can divide the question of why the planets have such different atmospheres into two parts: (1) Why did Venus keep its atmospheric carbon dioxide and Earth did not? (2) Where did all the outgassed water go on each planet?

The Abundance of Carbon Dioxide on Venus   Let's start with the question about the carbon dioxide. In fact, Venus and Earth both have similar total amounts of carbon dioxide. The difference is in where it is located. On Venus, the carbon dioxide is all in its atmosphere. On Earth, nearly all the carbon dioxide is locked up in carbonate rocks or dissolved in the oceans [Section 4.5], through the action of the carbon dioxide cycle (see Figure 4.27). This cycle is possible because carbon dioxide dissolves in water, where it can undergo chemical reactions to make carbonate minerals (minerals rich in carbon and oxygen) such as limestone. Venus lacks a similar cycle because it has no liquid water in which carbon dioxide can dissolve. We conclude that the difference in carbon dioxide is a direct consequence of a difference in liquid water: Earth's carbon dioxide became incorporated into rock because Earth has oceans and a carbon dioxide cycle; Venus's carbon dioxide is all in its atmosphere because it has no oceans and hence no carbon dioxide cycle.

The Lack of Water on Venus   The differences between Venus and Earth in carbon dioxide are attributable to differences in water, so we must next ask what has happened to outgassed water on each planet. For Earth, the answer is easy: The water outgassed from volcanoes is still on Earth, but most of it is in the oceans. Venus, however, is nearly totally lacking in water in any phase. The surface is far too hot for either ice or liquid water. It is even too hot for water to be chemically bound in surface rock, and any water deeper in its crust or mantle was probably baked out long ago. The only place where Venus could conceivably have much water is in its atmosphere, but measurements show that there's little there, either. Overall, the total amount of water on Earth is about 10,000 times more than the amount on Venus.

Broadly speaking, there are two possible explanations for the lack of water on Venus. Either Venus never had much water in the first place, or Venus once had more water but somehow lost it. We cannot absolutely rule out the first explanation, but it seems unlikely. At the time the planets were forming, the temperature of the protoplanetary disk would not

These two volcanic peaks are probably much like the broad, shallow-sloped volcanoes that make up the Hawaiian Islands on Earth.

The round blobs are steep volcanoes, apparently built from a "thick" lava.

Ishtar Terra

100 km

10 km

Tectonic forces have fractured and twisted the crust.

Aphrodite Terra

10 km

Lada Terra

This central image shows the full surface of Venus, 98% of which was mapped by Magellan. Notice the three large, elevated "continents" called Ishtar Terra, Lada Terra, and Aphrodite Terra.

10 km

50 km

Impact craters, like this one, are relatively rare on Venus and are distributed uniformly over the surface.

This round "corona" was probably made by a mantle plume. It is dotted with small volcanoes (the round dots) and surrounded by tectonic stress marks.

Figure 10.2

The surface of Venus, as revealed by radar observations from the *Magellan* spacecraft. Bright regions in the radar images represent rough areas or higher altitudes.

have been very different at the orbits of Venus and Earth, so both planets should have accreted from planetesimals of similar composition. As we discussed in Chapter 3, these planetesimals probably contained little or no water, because temperatures in the inner solar system were too high for water vapor to condense into solid particles of ice. The water on Earth must have come from planetesimals that originated in more distant parts of the solar system and had their orbits perturbed in such a way that they ended up crashing into our planet. Simple physical arguments suggest that such collisions should have been equally common on Venus, so Venus should have obtained water in this same way. Thus, the more likely scenario is that Venus started out with nearly as much water locked into its crust and mantle as Earth.

As on Earth, trapped water on Venus would have been released into the atmosphere through outgassing. Radar mapping of Venus's surface shows plenty of evidence of ongoing geological activity such as tectonics and volcanism (Figure 10.2). Venus has relatively few impact craters, which tells us that the craters due to large impacts must have been erased by other geological processes. Moreover, the few impact craters are distributed fairly uniformly over the planet's surface, suggesting that the surface is about the same age everywhere. Precise crater counts indicate that the surface is about 750 million years old, implying that the entire surface was somehow "repaved" at that time. We also see numerous tectonic features on Venus, as well as volcanoes. We have not witnessed any volcanic eruptions on Venus, but the composition of the clouds suggests that the volcanoes must still be active. The clouds contain sulfuric acid, which is made from sulfur dioxide (SO_2) and water. Sulfur dioxide enters the atmosphere through volcanic outgassing, but once in the atmosphere it is steadily removed by chemical reactions with surface rocks; it would all be removed within the geologically short time of 100 million years. The existence of sulfuric acid clouds therefore implies that outgassing must continue to supply sulfur dioxide to the atmosphere. All this evidence of past and present outgassing means that lots of water vapor should have entered Venus's atmosphere in the past.

The leading hypothesis for the disappearance of Venus's water invokes some of the same processes thought to have removed water from Mars [Section 8.3]: Ultraviolet light from the Sun broke apart water molecules in Venus's atmosphere. The hydrogen atoms then escaped to space (through *thermal escape*), ensuring that the water molecules could never reform. The oxygen from the water molecules was lost to some combination of chemical reactions with surface rocks and stripping by the solar wind. Venus's upper atmosphere is vulnerable to the solar wind because Venus lacks a protective magnetic field, probably due to its slow rotation. Recall that at least moderately rapid rotation is one of the requirements for a global magnetic field [Section 4.4]; Venus, which takes 243 days to complete a single rotation on its axis, does not meet this requirement. Without a magnetic field, the solar wind can strip atmospheric gas away, and probably has stripped away at least as much or more gas from Venus as it stripped away from Mars. However, because Venus has such a thick atmosphere, this gas loss would barely be noticeable.

Acting over billions of years, the breakdown of water molecules and the escape of hydrogen can easily explain the loss of an ocean's worth of water from Venus. Indeed, because Venus would have lost any surface water it once had, outgassed water could not be recycled back into the mantle the way it is recycled on Earth. Venus should therefore have lost water from its interior continuously throughout its history, explaining why its crust and mantle must by now be extremely dry—a situation quite different from that on Earth, where significant amounts of water are chemically bound to crust and mantle rock. As we discussed in Chapter 4, this dryness may have strengthened and thickened Venus's lithosphere, explaining why Venus lacks Earth-like plate tectonics.

Proving that Venus really did lose so much water is difficult, but evidence comes from the gases that didn't escape. Recall that most hydrogen nuclei contain just a single proton, but a tiny fraction of all hydrogen atoms (about 1 in 50,000 on Earth) contain a neutron in addition to the proton, making the isotope of hydrogen that we call *deuterium* (see Figure 3.24). Water molecules that contain an atom of deuterium (called "heavy water") behave chemically just like ordinary water and can be broken apart by ultraviolet light just as easily. However, a deuterium atom is twice as heavy as an ordinary hydrogen atom and thus does not escape to space as easily when the water molecule is broken apart. If Venus lost a huge amount of hydrogen from water molecules to space, the rare deuterium atoms would have been more likely to remain behind than the ordinary hydrogen atoms. Measurements show that this is the case. The fraction of deuterium among hydrogen atoms is a hundred times higher on Venus than on Earth, suggesting that a substantial amount of water must have been broken apart and its hydrogen lost to space.

The deuterium ratio does not allow us to determine exactly how much water Venus has lost, because Venus must be subjected to occasional comet impacts. The comets bring water that enters the atmosphere, where it ultimately gets broken down and its hydrogen escapes just as with Venus's original water. Because comet water contains deuterium in a ratio slightly higher, but comparable to that found on Earth (measurements indicate that comets have about twice the deuterium-to-hydrogen ratio as does Earth), the addition of comet water tends to lower the atmospheric ratio of deuterium to hydrogen on Venus. As a result, the water loss that we can infer from the current ratio of deuterium to hydrogen in the venusian atmosphere is only a *lower limit* on the actual water loss. The measured deuterium-to-hydrogen ratio implies that Venus has lost at least the equivalent of a global layer of water several meters deep. However, because this value is a lower limit, it is likely that Venus had much more water —possibly enough to have given it oceans if the water had not been lost to space.

The Runaway Greenhouse Effect

We have explained how Venus might have lost its water, but why didn't Earth lose its water in the same way? The answer is the ocean itself. On Earth, most of the outgassed water vapor condensed into rain, forming the oceans, long before ultraviolet radiation could break apart much of the water vapor. The short-wavelength ultraviolet light that tends to break apart water molecules does not penetrate far into the atmosphere, let alone penetrate the ocean surface, so water in the oceans is protected. (Today, ultraviolet light is also absorbed by the ozone layer, but the ozone layer didn't exist when our planet was young.) To understand why Venus was unable to protect its water in a similar way, let's consider what would happen if we moved Earth to Venus's distance from the Sun.

Figure 10.3 summarizes what would probably occur. The greater intensity of sunlight would almost immediately raise the global average temperature by about 30°C, from its current 15°C (59°F) to about 45°C (113°F). Although this is still well below the boiling point of water, the higher temperature would lead to increased evaporation of water from the oceans. It would also allow the atmosphere to hold more water vapor before the vapor condensed to make rain (think of how much more humid hot days are than cold days). The combination of more evaporation and greater atmospheric capacity for water vapor would substantially increase the total amount of water vapor in Earth's atmosphere. Now, remember that water vapor, like carbon dioxide, is a greenhouse gas. The added water vapor would therefore strengthen the greenhouse effect, driving temperatures a little higher. The higher temperatures, in turn, would

Figure 10.3

This diagram shows what scientists suspect would happen if Earth were placed at Venus's distance from the Sun: A runaway greenhouse effect would cause the oceans on Earth to evaporate completely.

If Earth moved to Venus's orbit

More intense sunlight . . .

Higher temperature increases evaporation, and warmer air holds more water vapor.

Runaway greenhouse effect

. . . would raise surface temperature by about 30°C.

Additional water vapor further strengthens the greenhouse effect.

Result: Oceans evaporate and carbonate rocks decompose, releasing CO_2 further strengthening the greenhouse effect . . .

. . . making Earth hotter than Venus.

lead to even more ocean evaporation and more water vapor in the atmosphere—strengthening the greenhouse effect even further. In other words, we'd have a "positive feedback loop" in which each little bit of additional water vapor in the atmosphere would mean higher temperatures and even more water vapor. The process would careen rapidly out of control, resulting in what we call a **runaway greenhouse effect.**

The runaway process would cause our planet to heat up until the oceans were completely evaporated and the carbonate rocks had released all their carbon dioxide back into the atmosphere. By the time the process was complete, temperatures on our "moved Earth" would be even higher than they are on Venus today, thanks to the combined greenhouse effects of the released carbon dioxide and water vapor in the atmosphere. The water vapor would then gradually disappear, as ultraviolet light broke water molecules apart and the hydrogen escaped to space. In short, moving Earth to Venus's orbit would essentially turn our planet into another Venus.

If it is correct, this scenario suggests a simple explanation for the difference between Earth and Venus. Venus is about 30% closer to the Sun than Earth, and this difference was enough to be critical. On Earth, it was cool enough for water to rain down to make oceans in which carbon dioxide could dissolve and undergo chemical reactions that locked it away in carbonate rocks. As a result, our atmosphere was left with only enough greenhouse gases to make our planet pleasantly warm. On Venus, the greater intensity of sunlight made it just enough warmer that oceans either never formed or soon evaporated. Without oceans to dissolve carbon dioxide and make carbonate rock, carbon dioxide accumulated in the atmosphere, leading to a runaway greenhouse effect, which resulted in the extreme temperatures of Venus today.

 THINK ABOUT IT Moving Earth to Venus's orbit would probably cause our planet to become Venus-like. If we could somehow move Venus to Earth's orbit, would it become Earth-like? Why or why not?

Venus and the Boundary of the Habitable Zone   The fact that moving our own planet to Venus's orbit would lead it to the same fate implies that *any* planet at this distance would be doomed to a runaway greenhouse effect (assuming it had an atmosphere from outgassing). In other words, if our scenario is correct, it seems impossible for a planet to have a habitable surface with abundant liquid water at Venus's distance from the Sun. We can therefore conclude that Venus does *not* lie within the Sun's habitable zone today.

In contrast, Earth clearly *does* lie within the habitable zone, which is why our planet has not suffered a runaway greenhouse effect. Indeed, models show that a runaway greenhouse effect could not occur on the present-day Earth, at least through natural processes (as opposed to human-induced processes), an idea supported by our planet's climate history. Whenever the atmospheric concentration of greenhouse gases has increased on Earth in the past—as it has numerous times, perhaps most extremely during "hothouse" phases that follow Snowball Earth episodes [Section 4.5]—our greater distance from the Sun has helped prevent the temperature from increasing to the point at which a runaway process would occur. Putting the lessons of Earth and Venus together, we conclude that the inner boundary of the Sun's habitable zone must currently lie beyond the orbit of Venus but within the orbit of Earth.

• Could Venus have once been habitable, and could life still exist there?

Venus's closeness to the Sun may have sealed its ultimate fate, but it's possible that Venus might have had a more moderate climate in its early history. Remember that the Sun has gradually brightened with age. Thus, sunlight was less intense on all the planets when they were young. Some 4 billion years ago, the intensity of sunlight at Venus was not much greater than it is at Earth today. Rain might well have fallen on the young planet Venus, and oceans could have formed. It's even conceivable that life could have arisen on the young Venus, or been transported there on meteorites from Earth.

As the Sun gradually brightened, any liquid water or life on Venus's surface was doomed. The runaway greenhouse effect raised the temperature so high that all the water evaporated and, as we've discussed, the water is now gone forever. Moreover, any evidence of past oceans is probably also gone. Recall that Venus's entire surface appears to have been "repaved" by tectonics and volcanism; this repaving would have covered up any shorelines or other geological evidence of past oceans. Rocks from Venus probably can't tell us anything about an oceanic past either: The high temperatures would have baked out any gases that might once have been incorporated into surface rock, thereby likely eliminating the evidence that might have told us whether the rocks formed in water or ever held life.

Nevertheless, the fact that the Sun was once dimmer makes it reasonable for us to think that Venus could have had a more Earth-like early history. If it did, and if life arose, could any life still survive? The high temperatures make it seem implausible to imagine life anywhere either on Venus's surface or underground. However, as we discussed in Chapter 7, the temperature drops significantly with altitude in the atmosphere, and Venus has high-altitude clouds containing tiny amounts of liquid water (in highly acidic form). Though the prospects of finding life in these clouds are admittedly slim, it is at least within the realm of possibility to imagine that life once arose in venusian oceans and, as the runaway greenhouse effect set in, successfully adapted to life in the sky.

10.3 SURFACE HABITABILITY FACTORS AND THE HABITABLE ZONE

The case of Venus tells us that distance from the Sun is a critical factor in surface habitability, since Venus seems to have been doomed to suffer a runaway greenhouse effect no matter how habitable it may have been in its youth. But we know that distance from the Sun is not the only habitability factor, because we have seen that Mars's small size probably explains how it lost liquid water that existed on its surface in the past. In this section, we'll summarize what we have learned about surface habitability factors, and then use these lessons to consider the present-day boundaries of the Sun's habitable zone.

• What factors influence surface habitability?

Our comparative studies of Venus, Earth, and Mars have given us deep insight into the factors that make these three worlds so different. Let's build on the discussions in both this chapter and prior chapters, so that we will gain additional insight into the prospects of finding worlds with habitable surfaces around other stars.

The Role of Distance from a Central Star As we've just discussed, the first factor that affects surface habitability is a planet's distance from its home star, in our case the Sun. The effects of subtle differences in distance can result in dramatic variations in habitability, as the case of Venus shows. As we've seen, a planet at Venus's distance with an Earth-like atmosphere would be only a little warmer than Earth; the reason that Venus is not habitable is that this "a little warmer" turns out to be warm enough to start a runaway greenhouse process that heats the planet far more. Thus, when we consider the minimum distance at which surface habitability is possible, we must account not simply for solar heating but also for resultant processes that can lead to the evaporation of surface water. Similarly, when we consider the maximum distance at which surface habitability is possible, we must consider processes that might weaken greenhouse warming, causing surface water to freeze.

There's another obvious factor that will influence the range of distances that form the habitable zone: the central star's luminosity (brightness). As we will discuss in Chapter 11, stars come with a wide range of luminosities: A few stars are much more luminous than the Sun, about 10% of all stars have luminosities similar to that of the Sun, and the vast majority of stars are considerably dimmer than the Sun. Clearly, a brighter star must have a wider and more distant habitable zone than the Sun, while a dimmer star must have a narrower and closer-in habitable zone. The wider habitable zones of brighter stars would seem to increase the odds of finding planets within these zones, but another problem may make biology rare or nonexistent in these cases: The brightest stars turn out to have extremely short lifetimes—millions of years rather than billions of years—and hence may not offer enough time to nurture biology [Section 11.1]. The dimmer stars, with their narrow habitable zones, are less likely to have planets in these zones. However, they may make up for this problem in their sheer numbers, since they are so much more common than stars like the Sun (see Cosmic Calculations 10.1).

The Role of Planetary Size Clearly, distance from the Sun is not the only factor affecting surface habitability. If it were, then the Moon would be habitable because it is the same distance from the Sun as is Earth. More to the point, if distance were all that mattered, then Mars should have become more habitable as the Sun grew brighter with time; instead, it froze over. Mars's current lack of surface habitability appears to be less the result of its distance from the Sun than of its small size. As we discussed in Chapter 8, its small size allowed its interior to cool more quickly than that of a larger planet like Earth. Its core presumably cooled to the point at which it no longer had a molten layer and therefore could no longer generate a global magnetic field. The martian atmosphere was then left vulnerable to stripping by the solar wind, and through this and other mechanisms, it lost too much gas to support a strong greenhouse effect. Without a strong greenhouse effect, Mars effectively froze. The cooler interior also means that if martian volcanoes remain active at all, they do not erupt frequently enough to resupply lost atmospheric gases.

Is there a minimum or maximum size that makes it possible for a planetary surface to remain habitable over long time periods? While it's uncertain how we should set limits on maximum size, a minimum size clearly exists. We have traced Earth's long-term habitability directly to the climate regulation provided by the carbon dioxide cycle, which in turn depends on the cycle of plate tectonics [Section 4.5]. Thus, we have at least some reason to think that plate tectonics is necessary for long-term habitability on any world, in which case the size requirement for surface habitability is a size that allows plate tectonics to exist.

We do not fully understand how planetary size is related to plate tectonics, but we are fairly confident that planets that are too small will not have tectonics, and that Mars falls below the minimum. We see no evidence of ongoing plate tectonics on Mars, and this is probably because the interior cooling attributable to its small size has allowed its lithosphere to thicken so much that it cannot break into plates. But while Mars—with a radius about half that of Earth—seems too small, we don't know how much bigger a planet would need to be so that it would have both sufficient internal warmth to power tectonics and a crust thin enough to be pushed around by the convective currents below. The case of Venus illustrates the problem. Recall that, given the closeness in size of Venus and Earth, we might expect to find plate tectonics operating on both planets, but there is no evidence of Earth-like plate tectonics on Venus. Could it be that Venus's slightly smaller size was just small enough to make the difference, much as its slightly nearer distance to the Sun led to its runaway greenhouse effect? It's possible, but a more likely hypothesis ties Venus's lack of plate tectonics to its runaway greenhouse effect. According to this hypothesis, high temperatures baked water out of the rock in Venus's crust and upper mantle, making the crust too stiff to allow subduction and thereby suppressing plate tectonics. If this explanation is correct, then Venus may well have had plate tectonics if it had not been so close to the Sun, and might have had plate tectonics in its early history when the Sun was dimmer.

At this point, all we can say with confidence is that Earth is of sufficient size for long-term surface habitability and Mars is not. Beyond that, we don't know how large a planet must be to allow for the presence of liquid water over extended periods of time. Factors of which we are still unaware may well contribute to the occurrence or absence of plate tectonics and the ability to sustain an appropriate climate.

The Role of an Atmosphere   A third crucial factor in surface habitability, after size and distance, is the necessity of an atmosphere. Without sufficient atmospheric pressure, liquid water cannot be stable and abundant regardless of other factors. Moreover, an atmosphere is necessary to protect a planetary surface against harmful solar radiation. So while a world without an atmosphere might still shelter life underground, it seems implausible that any biology would creep, crawl, or fly across its landscape.

The presence or absence of a significant atmosphere sometimes depends on size. In the case of Mars, for example, we have traced its loss of surface habitability to loss of atmospheric gas, which most likely was tied to its small size and loss of magnetic field. However, there may be several other ways in which a planet either might never get or would subsequently lose an atmosphere, even if it were of the right mass and at the right distance from its star.

Remember that the early atmospheres of Venus, Earth, and Mars are all thought to have been produced through outgassing. If a planet lacked trapped gases in its interior, then no outgassing would be possible. Given that most of the gases within the terrestrial worlds are assumed to have been brought by impacts of objects from more distant reaches of the solar system, and that these objects were presumably cast onto collision courses by gravitational interactions, terrestrial planets might form without gases in a star system either that lacks gas-bearing planetesimals or in which these planetesimals for some reason never get perturbed inward. Neither situation seems likely in light of what we know about solar system formation [Section 3.3], but we can't rule them out.

For a planet that has outgassing and sufficient size to hold its gas gravitationally, we know of at least two possible ways in which it might nonetheless lose its atmosphere. First, large impacts can in principle blast significant amounts of atmospheric gas into space. This process may have played a role on Mars. If large impacts are more common in some other star systems—which may be the case in systems that lack a planet comparable in size to Jupiter at the right distance [Section 11.3]—these impacts could cause significant gas loss on Earth-size planets. Second, as we have discussed, the solar wind can strip atmospheric gas from any planet that lacks a global magnetic field, and we don't expect to find magnetic fields on slowly rotating planets. Venus has so much atmospheric gas that loss to solar wind stripping has been insignificant in comparison. Earth, however, might well have become uninhabitable (at least on its surface) if not for the protection that the magnetic field has offered against billions of years of stripping by the solar wind.

Summary of Habitability Factors   We have identified three major factors for surface habitability:

1. The planet must be neither too close nor too far from its star; that is, it must be within its star's habitable zone.
2. The planet must be large enough to retain internal heat and have plate tectonics for climate regulation. We don't know the precise minimum size required, but it is certainly larger than Mars.
3. The planet must have enough of an atmosphere for liquid water to be present on its surface. This means it must have had gases trapped in its interior, so that an atmosphere could form through outgassing, and that it has not since lost too much of this atmospheric gas to impacts or solar wind stripping. Protection against the latter may require a global magnetic field, which in turn may require at least moderately rapid rotation.

The latter two requirements depend on factors intrinsic to the planet itself. The first depends on where it forms around its star, which brings us to the topic of habitable zone boundaries.

THINK ABOUT IT   Using the three habitability factors, explain why the Moon is not habitable.

• Where are the boundaries of the Sun's habitable zone today?

The boundaries of the Sun's present-day habitable zone depend on the range of distances at which a planet of suitable size and with sufficient atmospheric pressure could have liquid water on its surface. The inner boundary marks the place where, if a planet were any closer to the Sun, a runaway greenhouse effect would be triggered. The outer boundary marks the place where, if a planet were any farther from the Sun, surface water would freeze.

The case of Venus has shown us that the inner boundary of the habitable zone does not lie as far inward as Venus's orbit. The case of Mars should tell us something about the outer boundary, though it is more ambiguous: Geological evidence makes it clear that Mars once had liquid water on its surface, in which case it must once have been within the Sun's habitable zone—and since the habitable zone moves outward with time, it would then still have to be in the habitable zone today. However, as we discussed in Chapter 8, models of the martian climate cannot yet fully account for the conditions that might have led Mars to have been quite so warm in its distant past, a fact that has led some scientists to suggest that Mars may have had liquid water only intermittently. In that case, Mars might never have been within the Sun's habitable zone, but only was close enough that special events (such as impacts) made its surface temporarily habitable.

The ambiguity in the case of Mars means we won't be able to put precise numbers on the boundaries of the habitable zone. Nevertheless, we can place some general constraints on the possible locations of the boundaries.

The Inner Boundary

To locate the inner boundary more precisely than just saying it lies between Venus and Earth, we can consider theoretical models of what would happen to Earth if we moved it to various distances closer to the Sun. These calculations suggest that a runaway greenhouse effect would occur if we placed Earth anywhere inside of 0.84 AU from the Sun, or about halfway between the orbits of Venus and Earth. (Recall that 1 AU, or astronomical unit, is Earth's average distance from the Sun, or about 150 million kilometers.)

However, another factor might cause temperatures to spin out of control even beyond 0.84 AU from the Sun. According to some models, a moderate additional warming would allow water vapor to circulate to much higher altitudes in Earth's atmosphere (into the stratosphere). At these high altitudes, water molecules would be above much of the ozone layer and hence could be broken apart by ultraviolet light from the Sun. The hydrogen would then escape to space, causing Earth to lose this water and thereby allowing more water to rise into the upper atmosphere and be lost in turn. Over time, this **moist greenhouse effect**—so-called because the upper atmosphere would become moist with water, at least until all the water was lost—might cause Earth to lose its oceans. Note that the moist greenhouse effect leads to water loss not because the temperature is outside the range in which liquid water could exist, but rather because water that evaporates from the surface can rise into the upper atmosphere, where it can be lost to space. This is a less dramatic version of the scenario we discussed when we imagined placing Earth at Venus's orbital distance, but it could be just as lethal in the long run. An Earth-like planet might suffer this moist greenhouse water loss anywhere within 0.95 AU of the Sun, but there may be other factors that would mitigate the problem: Models that describe the onset of the moist greenhouse effect have numerous recognized uncertainties (such as the effects of clouds) that might push this distance inward.

In summary, the inner boundary of the present-day habitable zone in our solar system may be at 0.84 AU if we allow for only a simple runaway greenhouse effect, but as far out as 0.95 AU if we allow for water loss by a moist greenhouse effect.

The Outer Boundary

The outer boundary of the present-day habitable zone is the distance from the Sun at which even a strong greenhouse effect could not warm a planet enough to keep liquid water from freezing. At first, we might guess that Mars is beyond the outer boundary, since the temperature is too cold for liquid water at its surface today. However, if Mars were larger and had retained a thick atmosphere, it might have enough greenhouse warming to still have a habitable surface. Thus, if it is possible in principle for a planet to keep a thick atmosphere at such distances, the outer boundary of the habitable zone could lie beyond the orbit of Mars.

Calculations suggest that this is indeed the case. If we allow for a thick atmosphere with a strong greenhouse effect, the outer boundary of the present-day habitable zone lies at about 1.7 AU, well outside Mars's average orbital distance of 1.52 AU. However, there is at least one potential problem that could bring the outer boundary in closer. If the atmosphere of a planet is too cold, the atmospheric carbon dioxide that produces greenhouse warming will condense into snowflakes and fall to the surface. This carbon dioxide snow might limit how much carbon dioxide could reside in the atmosphere, thus preventing the atmosphere from staying thick enough for a strong greenhouse effect. This scenario might occur if the atmosphere lacked dust or any other greenhouse gas to help keep the middle atmosphere warm. In that case, the outer boundary of the present habitable zone might lie at only about 1.4 AU, or just inside the orbit of Mars. Note that this case would also imply that Mars never was in the habitable zone, so it could have had surface liquid water only intermittently in the past.

The Extent of the Habitable Zone

We have found two estimates each for the distances of the inner and outer boundaries of the Sun's present habitable zone. Using the more optimistic estimates, the present-day habitable zone extends from about 0.84 to 1.7 AU. Using the more conservative estimates, it extends only from about 0.95 to 1.4 AU. Both sets of boundaries are shown in Figure 10.4. Note that, even in the more conservative case, the Sun's present-day habitable zone represents a fairly wide region in the inner solar system.

Keep in mind that we don't yet know enough about how planetary atmospheres work to know which estimate is correct—or if the truth lies somewhere in between. In addition, there might be processes that can affect the atmospheric temperature that we haven't yet discovered, and therefore have not accounted for in our calculations of the habitable zone boundaries. Thus, these estimates of the boundaries of the habitable zone should be considered just that—estimates. They might be significantly refined as we learn more in the future.

THINK ABOUT IT   About how much wider is the more optimistic view of the present-day habitable zone than the more conservative view? If planets form at random locations in the inner portions of other stars' solar systems, how would these two different views affect the likely number of planets within habitable zones around Sun-like stars in the Milky Way Galaxy?

10.4 THE FUTURE OF LIFE ON EARTH

Because the boundaries of the habitable zone are calculated under the assumption that we have a planet of suitable size, they depend only on the amount of heat and light put out by the Sun. Thus, when the Sun was dimmer in the past, the habitable zone must have been closer to our Sun.

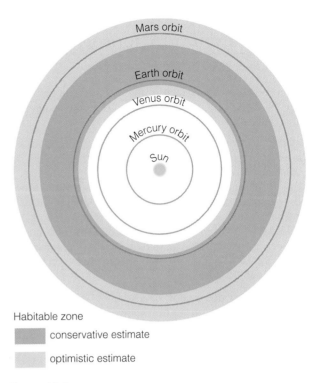

Habitable zone

■ conservative estimate

■ optimistic estimate

Figure 10.4

Boundaries of the Sun's habitable zone today. The narrower set of boundaries represents a model based on the more conservative assumptions, while the wider set represents the most optimistic scenarios.

Chances of Being in the Zone

If all other factors are equal, the likelihood of finding planets in a star's habitable zone depends on the width of the zone.

Example: The Sun's habitable zone is (optimistically) calculated to extend from about 0.84 to 1.7 AU. Consider a smaller star (of spectral type M [Section 11.1]), in which the habitable zone extends only from 0.05 to 0.1 AU. How does the probability of finding a habitable planet in this star's habitable zone compare to the probability around a Sun-like star? Given that these types of small stars outnumber Sun-like stars by approximately eight to one in our galaxy, for which class of stars should we expect more worlds in habitable zones?

Solution: The width (radius) of a habitable zone, R_{HZ}, is:

$$R_{HZ} = R_{outer} - R_{inner}$$

For a Sun-like star:

$$R_{HZ} = 1.7 - 0.84 = 0.86 \text{ AU}$$

For the smaller star:

$$R_{HZ} = 0.1 - 0.05 = 0.05 \text{ AU}$$

The probability of being in the habitable zone for the small star is only $0.05/0.86 = 0.058$ times that for a Sun-like star. However, because these stars are eight times as common, the *total number* of worlds in habitable zones around these small stars would be about $8 \times 0.058 = 0.46$, or 46%, as great as for worlds around Sun-like stars. (Of course, we have not considered factors besides the size of the habitable zone that may also be important.)

In the future, when the Sun will be brighter than it is today, the habitable zone will lie farther from the Sun. In this section, we'll discuss the way in which the Sun's habitable zone evolves with time.

• How will the Sun's habitable zone change in the future?

To determine how the habitable zone moves over time, we need to know how the Sun brightens as it ages. Fortunately, the process that causes the Sun to brighten is well understood.

The Sun shines by fusing hydrogen into helium. Each fusion reaction converts four hydrogen nuclei into one helium nucleus. Thus, the total number of *independent* particles in the solar core gradually falls with time. This gradual reduction in the number of particles causes the solar core to shrink, because there are fewer particles to generate the pressure that supports the core against the weight of overlying layers of the Sun. The slow shrinkage, in turn, gradually increases the core temperature, much as a bicycle tire pump gets warm when you compress the air in it by pushing on its piston. The gradual temperature increase causes a corresponding increase in the fusion rate, which is why the Sun slowly brightens.

Theoretical models indicate that the Sun's core temperature should have increased enough to raise its fusion rate and the solar luminosity by about 30% since the Sun was born $4\frac{1}{2}$ billion years ago. The models also allow scientists to predict the Sun's future luminosity. Observational data support the models' conclusions, as stars of particular masses do indeed vary somewhat in brightness, with older stars being brighter for any particular mass.

Figure 10.5 shows the results of calculations for the boundaries of the habitable zone from the Sun's birth until it exhausts its core hydrogen fuel some 5 billion years from now. Notice how the habitable zone gradually moves outward from the Sun, just as we would expect.

The horizontal swaths in Figure 10.5 show the distances from the Sun at which conditions have remained habitable from 4 billion years ago to the present. This zone is often called the **continuously habitable zone**, because it has been habitable at all times since the end of the heavy bombardment about 4 billion years ago. The width of the continuously habitable zone is fairly narrow if we use the more conservative assumptions, and substantially wider with the more optimistic assumptions. In fact, under the optimistic assumptions, both Earth and Mars are in the continuously habitable zone. Note also that the continuously habitable zone is defined for habitability only up to the present. If we look billions of years into the future, the habitable zone continues to move outward and the continuously habitable zone becomes narrower.

> **THINK ABOUT IT** Was Venus ever in the habitable zone? Is it in the continuously habitable zone? Under the more conservative assumptions, about when will the continuously habitable zone move beyond Earth's orbit? Explain.

• How long can life survive on Earth?

Earth has remained habitable for some 4 billion years, allowing plenty of time for life to evolve, diversify, and ultimately make our own human existence possible. However, the continuing evolution of the habitable zone means that Earth's days of habitability must eventually come to an end.

Before we begin, it's important to note that the demise of Earth's habitability is not something worth losing sleep over. Even under the most pessimistic scenarios, our planet will remain habitable for many hundreds of millions of years to come. Under more optimistic scenarios, Earth has a couple of billion years of remaining habitability. Compared to the length of time our civilization has existed so far, these are incredibly long time scales. If our species survives for this long, we will have had plenty of time to find ways to move to other planets in our solar system or in other star systems, or to otherwise prevent our perishing with Earth.

 THINK ABOUT IT Given the more immediate threats to our civilization, is it even worth thinking about what will happen to our planet millions or billions of years from now? Defend your opinion.

The End of Earth's Habitability

Thanks to the climate regulation provided by the carbon dioxide cycle, Earth has remained habitable even as our Sun has brightened by some 30% over the past 4 billion years. As the Sun continues to brighten, the carbon dioxide cycle should continue to keep the climate relatively pleasant for us for at least hundreds of millions of years to come. Eventually, however, the warming Sun will cause the cycle to break down.

If you study Figure 10.5, you'll see that under the more conservative assumptions, the inner boundary of the habitable zone will move beyond Earth's orbit in about a billion years. Thus, if these assumptions are correct, the end of habitability on Earth will come about a billion years from now. Recall that these conservative assumptions are based on the idea that a *moist greenhouse* effect will cause the oceans to evaporate away. In other words, about a billion years from now, water vapor will begin to circulate into Earth's upper atmosphere, where ultraviolet light will break apart water molecules and allow the hydrogen to escape into space, as it did with Venus. As water is lost from the upper atmosphere, more water will evaporate to take its place, until the oceans are completely gone. At that point, it seems unlikely that any life could continue to survive on Earth. Of course, no one is yet sure whether the moist greenhouse problem will really arise in about a billion years. Many feedback mechanisms in Earth's climate are not yet well understood. For example, increased cloud cover may reduce the amount of sunlight reaching Earth's surface, preventing the onset of the moist greenhouse.

Under more optimistic assumptions, Earth may remain habitable until some 3–4 billion years from now. At that time, the Sun will have grown so bright that Earth will finally suffer the fate of Venus—a runaway greenhouse effect. The rising temperature on Earth will cause increased ocean evaporation, and the buildup of water vapor in the atmosphere will increase the greenhouse effect further (see Figure 10.3). The positive feedback won't stop until the oceans have evaporated away and all the carbon dioxide has been released from carbonate rocks. Our planet will become a Venus-like hothouse, with temperatures far too high for liquid water to exist.

We know of no natural phenomena that can prevent this runaway greenhouse effect from occurring. However, if we imagine that our descendants have become sufficiently advanced in their technology, there

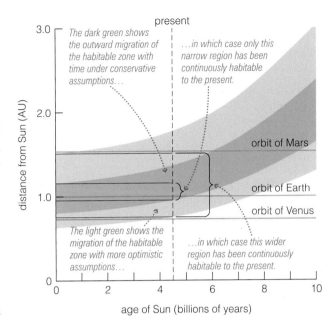

The dark green shows the outward migration of the habitable zone with time under conservative assumptions...

...in which case only this narrow region has been continuously habitable to the present.

orbit of Mars

orbit of Earth

orbit of Venus

The light green shows the migration of the habitable zone with more optimistic assumptions...

...in which case this wider region has been continuously habitable to the present.

Figure 10.5

This graph shows the Sun's habitable zone through time. The narrower region represents the habitable zone based on the more conservative assumptions, and the wider region represents the habitable zone based on the more optimistic assumptions. The horizontal swaths represent the zone that has been continuously habitable from 4 billion years ago to the present, again under the conservative (narrower) and optimistic (wider) assumptions.

are several ways they could survive in our solar system. They might protect Earth itself by building a giant sunshade in space to reduce the intensity of the light reaching Earth from the brightening Sun. Or they might simply move: When the habitable zone first moves past the orbit of Earth, Mars will be well within it, so perhaps our descendants will be able to make Mars livable with some advanced terraforming [Section 8.4]. If they are truly powerful, they might even find a way to move our planet gradually outward from the Sun to keep it within the habitable zone; by slowly moving Earth to the orbit of Mars over the next 3–4 billion years, humans could stay home and our planet would stay habitable. Another possibility is to relocate the population to large, artificial habitats constructed in space. Still, at best all these solutions can be only temporary, because the Sun itself eventually will die.

Death of the Sun   The Sun's gradual death has already begun, and in about 5 billion years it will completely run out of hydrogen in its central core. Then, over the following billion years or so, the Sun will undergo a dramatic transformation. During the first few hundred million years of this period, the Sun will gradually swell to about 100 times (or more) its present radius, and its surface temperature will drop (changing its color from yellow to red), becoming what we call a *red giant* star. Even though the Sun's surface temperature will decrease, its much greater size will pump greater energy into space than now. At its peak, the red giant will be about 1,000 times as luminous as the Sun is today, and Earth's surface temperature will rise to 700°C (1,292°F) or higher. Any remaining oceans will evaporate, and even underground life will be baked to death during this period as subsurface water boils away.

In its final death throes, the Sun will expel its outer layers into space, creating a *planetary nebula** (Figure 10.6). All that will remain of the Sun will be its hot central core; no more nuclear fusion will occur. This remaining core, known as a *white dwarf* star, will then gradually cool over time. The violence that accompanies the planetary nebula ejection will probably destroy Earth. Even if our planet somehow survives this event and continues to orbit the white dwarf Sun, the white dwarf's light will eventually become so feeble that Earth's charred surface will face a future of perpetual, frigid darkness.

Could We Still Survive?   For those undaunted by the thought of humans or other intelligent Earth beings surviving some 5 billion years into the future, it's natural to wonder whether we could also survive the death of the Sun. The obvious solution to the Sun's death is to move to another star system. Stars that are being born today might offer great homes to us in 5 billion years. When these stars die, we can move on to others born still later. As long as there are new stars with habitable planets, we could potentially survive by migrating to new homes.

However, even this type of long-term migration has its limits. The recycling of stellar material cannot continue forever, because dying stars return to space only part of the gas from which they were made. Over time, the galaxy will contain less and less interstellar gas. About 50 billion years from now, the Milky Way Galaxy will contain so little gas that new stars will no longer be born. What then?

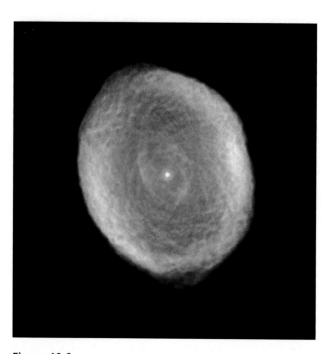

Figure 10.6

This Hubble Space Telescope photograph shows the Spirograph Nebula, an example of a planetary nebula. The gas of the nebula was expelled from a Sun-like star that had reached the end of its life. The central white dot is the white dwarf star that remains, which is essentially just the hot core of the now-dead star. Some 5 billion years from now, our Sun will suffer the same fate as the star that created this nebula.

---

*Despite the name, these structures have nothing to do with planets; the term comes from the planetlike appearance of some of these nebulae when seen through a small telescope.

Ever since Edwin Hubble first discovered that our universe is expanding, we have wondered whether the expansion will continue forever or someday stop, causing the universe to collapse back in on itself. In the past few years, astronomers have been surprised to learn that the expansion appears to be accelerating, in which case it seems that the fate of the universe is to expand forever at a greater and greater clip. However, you should keep in mind that forever is an extremely long time. It remains possible that we will someday discover other surprises that will change our view of the fate of the universe.

If the universe continues to expand after all star formation has ceased, then life will be able to continue only around those long-lived stars that still have habitable planets. But even the longest-lived stars will eventually run out of hydrogen to fuse. Within a few hundred billion years, all existing stars will burn out and die, and the now-brilliant galaxies will fade into darkness. On much longer time scales, interactions among the burned-out stars will send many of them flying into the vastness of intergalactic space, while the rest will converge toward their galactic centers, merging into gigantic black holes. At this point, the story becomes even more speculative. If the so-called "grand unified theories" of physics are correct, the stellar corpses will eventually disintegrate into swarms of subatomic particles. Meanwhile, the giant black holes will slowly evaporate into energy and particles in a process first described by the noted British physicist Stephen Hawking. At some point in the far distant future, the universe will consist of nothing but a dilute sea of subatomic particles and photons of light, each separated from others by immense distances that will grow larger as the universe endlessly expands. Our current epoch of a universe filled with stars and galaxies will have been just a fleeting moment in an eternity of darkness.

This end in darkness may seem a bit depressing, but it is, after all, inconceivably far in the future. Nevertheless, it is fair to ask whether it is truly the end or instead could be followed by something else. Some serious scientists already argue that there might be ways by which an intelli-

SPECIAL TOPIC 10.1: FIVE BILLION YEARS

The Sun's demise in about 5 billion years might at first seem worrisome, but 5 billion years is an extremely long time. It is longer than Earth has yet existed, and human time scales pale by comparison. A single human lifetime, if we take it to be about 100 years, is only 2×10^{-8}, or two hundred-millionths, of 5 billion years. Because 2×10^{-8} of a human lifetime is about 1 minute, we can say that a human lifetime compared to the life expectancy of the Sun is roughly the same as 60 heartbeats in comparison to a human lifetime.

What about human creations? The Egyptian pyramids have often been described as "eternal." But they are slowly eroding due to wind, rain, air pollution, and the impact of tourists, and all traces of them will probably have vanished within a few hundred thousand years. While a hundred thousand years may seem like a long time, the Sun's remaining lifetime is some 50,000 times longer.

On a more somber note, we can gain perspective on billions of years by considering evolutionary time scales. During the past century, our species has acquired sufficient technology and power to destroy human life totally, if we so choose. However, even if we suffer that unfortunate fate, some species (including many insects) are likely to survive.

Would another intelligent species ever emerge on Earth? There is no way to know, but we can look to the past for guidance. Many species of dinosaurs were biologically quite advanced, if not actually intelligent, when they were suddenly wiped out about 65 million years ago. Some small, rodentlike mammals survived, and here we are 65 million years later. We therefore might guess that another intelligent species could evolve some 65 million years after a human extinction. If these beings also destroyed themselves, another species could evolve 65 million years after that, and so on. But even at 65 million years per shot, Earth would have some 15 more chances for an intelligent species to evolve in 1 billion years—the length of time our planet will remain habitable under fairly conservative scenarios. Under more optimistic estimates for long-term habitability, there could be 60 or more periods—each as long as the period separating us from the dinosaurs—still to come before our planet dies. That is a lot of potential opportunities for the evolution of a species wise enough to avoid its self-destruction and to move on to other star systems by the time the Sun finally dies. Perhaps we ourselves will prove to be so wise.

gent civilization could survive even as the universe dies. For a lighthearted viewpoint, we turn to science fiction. Isaac Asimov, in his story "The Last Question," begins with a couple of people asking a supercomputer whether there is a way to reverse the decline of the universe and thereby avert a cold and dark end. The computer responds that the available data are insufficient to answer the question. Over billions of years, computers advance and humankind survives, making the question ever more important as the universe approaches its cold, dark future. By the end of the story, the computer exists solely in hyperspace, outside the time and space of our universe. The universe has reached a state of ultimate darkness, with nothing left alive. Meanwhile, the computer, which Asimov calls AC, whirs on in the timelessness of hyperspace until finally it learns how to reverse the decay of the universe:

> *For another timeless interval, AC thought how best to do this. Carefully, AC organized the program.*
>
> *The consciousness of AC encompassed all of what had once been a Universe and brooded over what was now Chaos. Step by step, it must be done.*
>
> *And AC said, "LET THERE BE LIGHT!"*
>
> *And there was light—*

10.5 THE PROCESS OF SCIENCE IN ACTION: GLOBAL WARMING

Earth may someday be doomed to a runaway greenhouse effect, but it won't be happening any time soon, even taking into account human actions. Nevertheless, considerable evidence suggests that Earth's global average temperature is currently on the rise, a trend referred to as **global warming**. You're undoubtedly aware that global warming is a hot political topic that has generated significant controversy. Most of the contro-

---

◄ MOVIE MADNESS: THE TIME MACHINE

This film, whose resemblance to H. G. Wells's original story is slightly less than marginal, features a brilliant, young New York professor from the early years of the twentieth century who travels back and forth in time, hoping to retrieve a lost girlfriend. To do so, he uses a time machine that looks as if it were kit-built from a discarded lighthouse lens, a laboratory vacuum tank, and a surplus adding machine from the Spanish-American War.

Forget the questionable physics; this story is actually about human evolution. The bulk of the film takes place in the far future, 800,000 years hence. Thanks to an industrial accident involving some retirement home developers and the Moon, all civilization on Earth has been destroyed. Humans have evolved into two distinct races: the lovable (if somewhat simple) Eloi, and the disagreeable, ugly Morlocks. The Eloi live the life of carefree forest folk, building modest structures from bamboo and hemp. The Morlocks are slightly more sophisticated (they can smelt metal) but for some reason live underground. They venture to the surface only to hunt down the Eloi, drag them to their subterranean digs, and eat them for dinner.

Does it make sense that the most technically advanced beings on the planet will be underground heavies? In our consideration of habitable zones, we've looked at surface environments, where the raw materials of a planet are exposed to the abundant energy radiated by its sun. As we know, the energy for metabolism is tough to find below the surface, which explains why the Morlocks count on the Eloi to eat the products of photosynthesis (veggies, for instance), and then they simply eat *them*. But this is a dangerously fragile lifestyle for the Morlocks, as they are dependent on a single surface species.

Aside from such dietary considerations, it's discouraging to think that, after a length of time several times as far into the future as we've come since the first *Homo sapiens*, the best our descendants can do is erect bamboo huts. What about expansion to other planets and travel to the stars? If *The Time Machine* is an accurate vision of what happens to thinking species, then we needn't bother with our SETI experiments (Chapter 12). After all, it's difficult to build radio transmitters out of bamboo.

versy concerns whether it is occurring naturally or as a result of human activity and, if the latter, what (if anything) we should do about it. Thus, for this chapter's case study in the process of science in action, we'll investigate how researchers seek to understand global warming and its potential consequences.

• What is the evidence for global warming?

The basic science behind global warming is surprisingly simple: We know that carbon dioxide is a greenhouse gas that can enhance the greenhouse effect and therefore cause a rise in a planet's surface temperature, and we know that human activity (such as the burning of fossil fuels) is adding carbon dioxide and other greenhouse gases to Earth's atmosphere. It might therefore seem natural to conclude that our activity will cause Earth's climate to warm up. However, we also know from our study of Earth's climate regulation mechanisms that there are many feedback loops that can make the reality much less straightforward than this simple analysis would suggest. As a result, the obvious starting point for the scientific study of global warming is to find out if it is indeed occurring as we might expect and, if so, whether we can tie the warming to the carbon dioxide emissions of our civilization.

Evidence of Recent Warming   You might think that determining whether our planet has been warming up during the past century or two would require nothing more than collecting temperature data from old newspapers. However, remember that global warming refers to an increase in the *average* temperature of our whole planet. We don't expect all localities to warm by the same average amount; indeed, it's possible that some places will get colder even as the planet as a whole warms up.

Today, orbiting satellites provide data that allow us to determine the global average temperature quite accurately, because they give us a view of our entire planet. We can validate these records with "ground truth" measurements recorded at more than 7,000 weather stations around the world, along with measurements of ocean temperatures generally obtained by measuring the temperature of water collected by ships' intake valves. As a result, we have reliable temperature data for the approximately three decades for which we have satellite observations of our planet.

The data become somewhat less reliable for years prior to the satellite era. Getting a good estimate of the global average temperature requires having many local measurements from around the world, and fewer such measurements are available as we look deeper into the past couple of centuries. Moreover, most past temperature records were kept in cities, which have tended to become warmer over time for reasons independent of global warming, such as through changes that occur as vegetation is paved over and as local utilities generate more heat. Scientists can often account for this "urban heat island" effect, but even then are left primarily with data for land temperatures and few records of temperatures over the oceans, which cover three fourths of Earth's surface. As a result, scientists have devoted a lot of effort in the past few years to examining past temperature data in detail. Through techniques of statistical analysis, it is possible to reconstruct a fairly reliable temperature history for most of the past two centuries, though remember that the uncertainties become larger as we look further back.

Figure 10.7 ▶

Average global temperatures from 1860 through 2005. Notice the clear global warming trend of the past few decades. The "30-year average" is the average temperature for the years 1950–1980. Data was taken from the National Center for Atmospheric Research.

a In 2002, satellite imagery showed Antarctica's Larsen B ice shelf breaking up. The melting and fragmentation of this frozen shelf of ice, as large in area as Rhode Island and more than 200 meters thick, was the most dramatic example to date of warming in the southern continent.

b These maps contrast the extent of the year-round Greenland ice sheet, shown in white, in 1979 and 2005. The brown area indicates the region that melts during the warm season. Notice that the melt region has expanded significantly, extending both further inland (to higher elevations) and further north.

Figure 10.8

Evidence of recent warming at the poles.

Figure 10.7 shows the reconstructed history of Earth's global average temperature since 1860. Despite the uncertainties, the overall conclusion is clear: Global average temperatures have risen by about 0.8°C (1.4°F) in the past century. Moreover, most of the warming (about 0.6°C) has occurred in just the last 30 years, the period for which the data are most reliable. The warming trend also seems to be accelerating. The year 2005 was the hottest since people began keeping records, and 9 of the previous 10 years also made the top-10 list.

THINK ABOUT IT Look up global temperature data for this year. How do the data from this year compare to those from 2005?

A temperature increase of 0.8°C may not sound like much, and it might not be so important if it were uniform everywhere. But it is not, and some regions are warming much more than others due to differences in topography, wind direction, and other factors. In general, polar regions are warming much more than equatorial regions, and the Northern Hemisphere is warming more rapidly than the Southern Hemisphere. (The greater proportion of land in the north is responsible for this, because water takes a great deal of energy to either heat or cool.) As a result, glaciers are on the retreat around the world, and polar ice appears to be melting. Orbiting satellites routinely observe massive ice sheets breaking off the continent of Antarctica (Figure 10.8a), and the area of the north polar ice cap has decreased by 20% from its extent little more than a decade ago, along with significant melting of Greenland's ice sheet (Figure 10.8b).

Correlation of Warming with Increased Carbon Dioxide Given the clear evidence of warming over recent decades, the next question is whether we can tie that warming to increased concentration of carbon dioxide (or other greenhouse gases) in the atmosphere. To establish causality in any case like this, we generally proceed in two steps. First, we look for a *correlation;* in this instance, evidence that temperatures rise and fall with carbon dioxide concentration. If we find a correlation, then we next ask whether the correlation implies causality; in this case, whether the temperature rise is *caused* by the carbon dioxide rise. Keep in mind that correlations do not automatically imply causality: Often, they may be co-incidental or attributable to some underlying commonality. For example, there is a correlation between the number of churchgoers in a city and the number of beer drinkers in a city, but that obviously doesn't mean that going to church causes people to drink beer; instead, it just reflects the fact that both numbers tend to go up as population increases.

We can see evidence of a correlation by looking at past data for temperatures and carbon dioxide concentration. Direct measurements of carbon dioxide concentration have been made only since 1958, but we can get longer-term data about both temperature and CO_2 from other sources, including tree rings and ice cores drilled out of the Antarctic ice sheet. Ice cores are the frozen equivalent of sedimentary rock, made up of accumulated layers of ancient, compressed snow. The thickness of these layers—like tree rings—shows the year-to-year history of snowfall. Trapped bubbles of air in the ice can be analyzed, and oxygen isotopes within them give clues to past temperatures. When temperatures are colder, the heavier isotopes aren't transported through the air as easily as when temperatures are warm. Thus, by measuring the ratio of the heavier and lighter isotopes at each layer of the ice cores, researchers can determine relative temperatures in the distant past. Similar analysis allows reconstruction of past levels of atmospheric carbon dioxide.

As of early 2006, ice core data are available for temperatures and carbon dioxide concentration going back some 650,000 years; Figure 10.9 shows the data for the past 400,000 years, along with the direct record of carbon dioxide concentration collected over the past few decades. Notice that there is indeed a correlation between temperature and carbon dioxide concentration: Periods of higher temperature tend to also be periods of higher carbon dioxide concentration. Moreover, the data show that both temperature and carbon dioxide concentration vary substantially and naturally with time. Average temperatures have risen or fallen by as much as 10°C several times during the last 650,000 years, and carbon dioxide concentration has varied naturally between less than 200 and

Figure 10.9

This diagram shows the atmospheric concentration of carbon dioxide and global average temperature over the past 400,000 years. Notice that both vary together: More carbon dioxide goes with higher temperature, and vice versa. The CO_2 data for the past half-century come from direct measurements (made at Mauna Loa; the up and down wiggles reflect annual season cycles). Most of the earlier data come from studies of air bubbles trapped in Antarctic ice (ice core samples). The concentration is measured in parts per million (ppm), which is the number of CO_2 molecules among every one million air molecules. (The ice core data have recently been extended back to 650,000 years; these newer data are not shown but still lead to the same conclusions.)

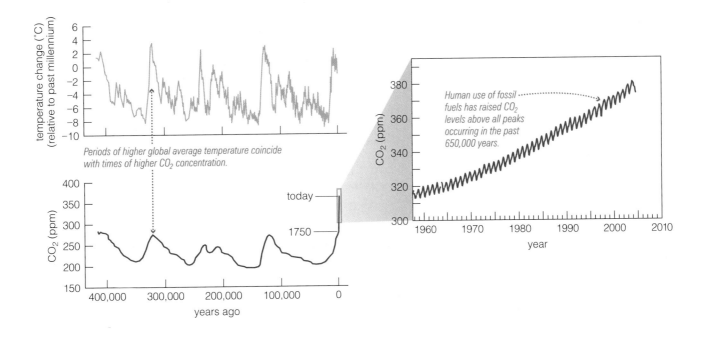

Periods of higher global average temperature coincide with times of higher CO_2 concentration.

Human use of fossil fuels has raised CO_2 levels above all peaks occurring in the past 650,000 years.

nearly 300 carbon dioxide molecules per million molecules of air (ppm). Considered against these natural changes, the temperature rise of 0.8°C over the past century may not seem all that great. But the recent rise in carbon dioxide concentration is a different story: The atmospheric CO_2 concentration is now significantly higher than it has been at any time in the past 650,000 years, and it is continuing to rise rapidly.

Global Warming Models and Uncertainties

We have seen that a rising carbon dioxide concentration correlates with a rising temperature, but does that mean it *causes* the temperature rise? More to the point, do we have any reason to think that recent global warming is a result of the increased carbon dioxide put into the atmosphere by human activity, or could it just be part of the natural variation in temperature that we know has occurred in the past?

Given our understanding of the basic mechanism of the greenhouse effect, we cannot doubt that a continually rising concentration of greenhouse gases would eventually make our planet heat up. But for geologically short time scales—anything less than tens of thousands of years—there are feedback mechanisms that could potentially alter the cause and the effect. Scientists seek to understand the mechanisms and answer our questions about global warming by creating sophisticated computer models of the climate and comparing the model predictions to actual observations. If the models can correctly mimic the real climate, then they can also be used to understand how much of the temperature increase is due to human activity rather than to natural factors, and to predict how the climate will change in the future as we continue to pump greenhouse gases into the atmosphere.

Earth's climate is incredibly complex, and many uncertainties remain in attempts to model the climate on computers. Nevertheless, today's models are the result of decades of work and refinement: Each time a model of the past failed to match real data, scientists sought to understand the missing (or incorrect) ingredients in the model and then tried again with improved models. And while even today's models are not perfect, they now match real climate data quite well, giving scientists confidence that the models do indeed have predictive value. Figure 10.10 compares model data and real data, showing good agreement. Moreover, the models suggest that global temperatures are increasing in a manner consistent with what we expect if human activity is the cause: If you include the effects of the greenhouse gases put into the atmosphere by humans, the models agree with the data, but if you leave out these effects, the models fail. Thus, while plenty of uncertainties still remain, the ability to evaluate the component of warming due to our own actions continues to improve. Today, most scientists are convinced that human activity is indeed causing global warming.

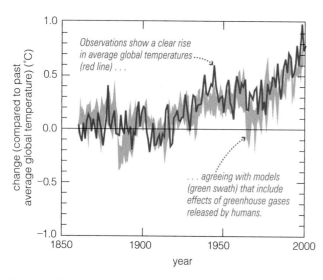

Figure 10.10

This graph compares the predictions of climate models (green swath) with observed temperature changes (red line) since about 1860. The thickness of the green swath for the models represents the range of variation between several different models. These models take into account both natural factors (such as slight changes in the Sun's brightness and the effects of volcanic eruptions) and emission of greenhouse gases by humans. The agreement is not perfect—telling us that we still have much to learn—but it is good enough to give us confidence that greenhouse gases are, indeed, causing global warming.

• What are the potential consequences of global warming?

Given that we appear to be causing global warming, the next question to ask is how it might affect us. If we can predict the future effects, then we can decide what (if anything) to do about it.

The models tell us that if current trends in the greenhouse gas concentration continue—that is, if we do nothing to slow our emissions of carbon dioxide and other greenhouse gases—the warming trend will continue to accelerate. By the end of this century, the global average temperature will be 3°C to 5°C (6°F to 10°F) higher than it is now, giving our children and grandchildren the warmest climate that any generation of *Homo sapiens* has ever experienced.

The consequences of global warming are not simply hotter weather. That alone would be important, of course, because it would shift where we could grow crops. Imagine if Northern Hemisphere wheat couldn't be grown in the American Midwest, but only in Canada, northern Europe, or Siberia. But the more serious consequences could be in changes to weather patterns brought on by the overall warming trend.

Predicting changes to local or regional weather is more difficult than predicting changes to the global average temperature, but a few consequences of global warming seem clear. Different regions of Earth will warm by different amounts; some regions will even become colder. Other regions might experience more rainfall or might become deserts. The greater overall warmth of the atmosphere would tend to mean more evaporation from the oceans, leading to more numerous and more intense storms.

Polar regions will warm the most, leading to increased melting of polar ice. This is clearly threatening to the species of these high-latitude regions (polar bears, which depend on an abundance of ice floes, are already under pressure), but it also warms the oceans everywhere, and changes their salt content as melting ice pours more fresh water into the sea. The fact that the waters of the Gulf of Mexico are at their warmest level in at least a century may be contributing to the greater strength of hurricanes that have recently blown out of the Caribbean, though it is difficult to tie specific storms to a general warming trend.

Sea level, which has already risen some 20 centimeters in the past hundred years, will continue to rise, with potentially major consequences for coastal habitats and cities. Sea level is rising primarily because water expands slightly as it warms, with an additional contribution from water added by the melting of glaciers and landlocked ice sheets. (Floating ice does not affect sea level as it melts.) By the year 2100, depending on the model used to make the projection, the oceans will have risen by between another couple dozen centimeters to a meter or more. This might be enough to flood many coastal cities, and obliterate large areas of such low-lying countries as Bangladesh. It would also make coastal storm surges far worse than they are today. Some researchers also worry that the influx of large quantities of fresh water into the (salty) seas may change the way water circulates in the oceans; changes to the flow of the Gulf Stream, for example, could have drastic climate consequences for western Europe and parts of the United States.

Other consequences could come from ecological changes, though these are even more difficult to predict. For example, it is difficult to know how forests and other ecosystems will respond to climate changes induced by global warming. As a result, we cannot easily predict the impact of such climate changes on food production, fresh water availability, or other issues critical to the well-being of human populations. Nevertheless, significant changes are well within the realm of possibility, even if we cannot be certain they will occur.

The bottom line is that we are in effect conducting a dramatic experiment on our planet in which we increase its greenhouse gas concentration far more than it would increase naturally. We do not know what the outcome of this experiment will be, or how easy or difficult it will be for us to deal with the consequences of our experiment. However, the cases of Venus and Mars show that major climate change can occur even without any human intervention. While we may not be about to do anything quite so dramatic to our own planet, we really do not know how our tampering might affect the finely balanced climate mechanisms upon which our civilization has been built.

THINK ABOUT IT If you were a political leader, how would *you* deal with the uncertain threat of global warming?

THE BIG PICTURE
Putting Chapter 10 into Context

In this chapter, we have tied together much of what we learned in other chapters to examine the concept of a habitable zone and its evolution over time. As you continue in your studies, keep in mind the following "big picture" ideas:

- The habitable zone refers to the region around a star in which a planet of suitable size (often taken to be that of Earth) could have liquid water on its surface. Despite its name, the habitable zone is not the only region around a star where habitability is possible.

- Venus, Earth, and Mars may all have been habitable in the early history of the solar system. However, only Earth has remained habitable to this day. Understanding why the climates of Venus and Mars have changed and why Earth's climate has remained comparatively stable can help us understand our climate and how we can protect it.

- The habitable zone is gradually migrating outward from the Sun. Eventually, it will lie beyond Earth's orbit, rendering our planet uninhabitable. Still, with sufficient technology, we can imagine our descendants finding a way to remain within the habitable zone until the Sun dies and then surviving by moving to other star systems.

- Given that an advanced civilization could find a way to survive as its star dies and could certainly find ways to overcome other natural threats such as asteroid impacts, it seems that nature will not impose intractable problems on us for billions of years to come. If we do not survive, it is far more likely to be the result of our own actions than the result of any natural catastrophe.

10.1 THE CONCEPT OF A HABITABLE ZONE

• **How does a planet's location affect its prospects for life?**

At any particular time, a star's **habitable zone** is the range of distances around it at which a planet could *potentially* have surface temperatures that would allow for abundant liquid water.

• **Could life exist outside the habitable zone?**

There could be many worlds that have underground or underwater life fueled by energy other than sunlight, in which case these worlds need not be (and are unlikely to be) in habitable zones. However, outside our solar system it would be difficult to detect life on such worlds, and it seems unlikely that complex, intelligent life could arise in such environments.

10.2 VENUS: AN EXAMPLE IN POTENTIAL HABITABILITY

• **Why is Venus so hot?**

Venus's distance from the Sun ultimately led to a **runaway greenhouse effect**: Venus became too hot to develop (or keep) liquid oceans like those on Earth. Without oceans to dissolve outgassed carbon dioxide and lock it away in carbonate rocks, all of Venus's carbon dioxide remained in its atmosphere, creating its intense greenhouse effect.

• **Could Venus have once been habitable, and could life still exist there?**

Early in its history, when the Sun was some 30% dimmer than it is today, Venus may have been within the Sun's habitable zone and hence could have had rain, oceans, and perhaps life. We are unlikely ever to know for sure, because geological evidence of such a period would by now be destroyed. But if Venus once had oceans and life, it is conceivable that life could still survive among liquid droplets in high-altitude clouds.

10.3 SURFACE HABITABILITY FACTORS AND THE HABITABLE ZONE

• **What factors influence surface habitability?**

According to present understanding, a planet can have a habitable surface only if it is within its star's habitable zone, is large enough to retain internal heat and have plate tectonics, and has enough of an atmosphere for liquid water to be stable on its surface.

• **Where are the boundaries of the Sun's habitable zone today?**

Using the most optimistic assumptions, the boundary currently extends from a distance of about 0.84 AU to 1.7 AU. Under more conservative assumptions, the boundary extends from about 0.95 AU to 1.4 AU.

10.4 THE FUTURE OF LIFE ON EARTH

• **How will the Sun's habitable zone change in the future?**

As the Sun ages, its luminosity gradually increases. As a result, the habitable zone gradually moves outward with time.

• **How long can life survive on Earth?**

At minimum, Earth should remain habitable for another several hundred million years. By about a billion years from now, a **moist greenhouse effect** could cause Earth's oceans to evaporate away, though natural feedback processes might prevent this from occurring so soon. In 3–4 billion years, the Sun will become bright enough so that our planet will certainly be subject to a runaway greenhouse effect, ending surface habitability.

10.5 THE PROCESS OF SCIENCE IN ACTION: GLOBAL WARMING

• **What is the evidence for global warming?**

Temperature data make clear that the global average temperature is on the rise, having risen about 0.8°C over the past hundred years and 0.6°C over the past 30 years. The rise in average global temperature has been accompanied by a rise in the atmospheric CO_2 concentration, which is due to human activity. Past climate data show that the current CO_2 concentration is significantly higher than the concentration at any time in the past 650,000 years, and climate models strongly suggest that this higher concentration is indeed the cause of global warming.

• What are the potential consequences of global warming?
Continued global warming could raise the average worldwide temperature by 3°C to 5°C during this century. Regional climate changes will be greater, and we can expect increased polar melting and a rise in sea level. Additional heat should increase ocean evaporation, which may lead to more numerous and more intense storms. Many other serious effects could also occur, but precise consequences are difficult to predict.

EXERCISES AND PROBLEMS

REVIEW QUESTIONS

Short-Answer Questions Based on the Reading

1. What is a *habitable zone*, and how is the idea useful? Is a planet in the habitable zone necessarily habitable? Explain.

2. Describe several ways in which it may be possible to have habitability outside the habitable zone. Is it possible that habitable worlds are more common outside this zone than within it? Explain.

3. Why do we think that Venus should have outgassed similar amounts of carbon dioxide and water vapor as Earth? Where is Venus's carbon dioxide today? Where is Earth's?

4. How much water is present on Venus today? How do we think that Venus lost water, and what evidence supports the idea that Venus really did lose water this way?

5. What is a *runaway greenhouse effect*, and why did it occur on Venus but not on Earth? What does this fact tell us about the inner boundary of the Sun's habitable zone?

6. Could Venus ever have had oceans and, if so, could we find geological evidence that they existed? Explain.

7. How do we expect the habitable zones of brighter stars to compare to that of the Sun?

8. Why is planetary size important to habitability? What does the case of Mars tell us about the minimum size? Can we draw any conclusions about size from the case of Venus? Explain.

9. How is an atmosphere important to surface habitability? What factors besides size and distance from the Sun might influence whether a world has an atmosphere sufficent to allow life?

10. What factors affect the location of the inner boundary of the habitable zone? Be sure to explain and consider the role of a possible *moist greenhouse effect* in such calculations.

11. What factors affect the location of the outer boundary of the habitable zone? Briefly summarize the current boundaries of our Sun's habitable zone under both the more optimistic and the more conservative scenarios.

12. Why does the Sun gradually brighten, and how does this brightening affect the location of the habitable zone over time? What do we mean by a *continuously habitable zone*?

13. How and when will Earth become uninhabitable? Why? Could humans still survive? Explain.

14. Briefly describe the eventual fates of the Sun and of the universe and what these fates might mean to our descendants (if anyone survives that long).

15. How do we determine global average temperatures over the past couple centuries? What do the data show?

16. What do ice core data tell us about the past climate and the role of greenhouse gases? How does the current concentration of atmospheric carbon dioxide compare to the concentration over past times for which we have ice core data?

17. What is the role of climate modeling in understanding global warming? How do scientists improve their models with time? Based on current models, what can we say about the cause of global warming?

18. Describe several potential future consequences of global warming, and discuss why the precise consequences are difficult to predict.

TEST YOUR UNDERSTANDING

Does It Make Sense?

Decide whether each statement makes sense or does not make sense. Explain clearly; because not all of these have definitive answers, your explanation is more important than your chosen answer.

19. If Venus were just a little bit smaller, its climate would be Earth-like.

20. Venus is not in the habitable zone now, but it may have been in the past.

21. Venus is not in the habitable zone now, but a few billion years from now it will be.

22. If we could somehow start plate tectonics on Venus, its surface would cool and it would regain the oceans it had in the past.

23. Mars will someday undergo a runaway greenhouse effect and become extremely hot.

24. We are not yet certain, but it is quite likely that Earth has suffered through a runaway greenhouse effect at least once in the past 4 billion years.

25. While the habitable zone of the Sun migrates outward over time, the habitable zones of other Sun-like stars might instead migrate inward over time.

26. If Earth someday becomes a moist greenhouse, it will have a climate that is humid but still quite comfortable.

27. The fact that additional atmospheric CO_2 must cause a greenhouse effect is all we need to know to link global warming to human activity.

28. Earth has been warmer in the past than it is today, so there is nothing to worry about with global warming.

Quick Quiz

Choose the best answer to each of the following. Explain your reasoning with one or more complete sentences.

29. The habitable zone refers to: (a) the regions of a planet where good weather allows life to exist; (b) the range of distances from a star where a planet's surface temperature is always above the freezing point of water; (c) the range of distances from a star within which water could exist in liquid form on a suitably sized planet.

30. A planet that is *not* within a habitable zone *cannot* have: (a) life; (b) subsurface oceans; (c) abundant liquid water on its surface.

31. Venus's atmosphere has much more carbon dioxide than Earth's because: (a) Venus was born in a region of the solar system where more carbon dioxide gas was present. (b) Venus lacks oceans in which carbon dioxide can be dissolved. (c) Venus has volcanoes that outgas much more carbon dioxide than those on Earth.

32. What is the likely reason for Venus's lack of water in any form? (a) The planet accreted little water during its birth. (b) The water is locked away in the crust. (c) The water was in the atmosphere, where molecules were broken apart by ultraviolet light from the Sun.

33. If Earth were to be moved to Venus's orbit, it would probably: (a) stay about the same temperature, thanks to the small amount of CO_2 in Earth's atmosphere; (b) become a tropical paradise; (c) suffer a runaway greenhouse effect and become even hotter than Venus is today.

34. The inner boundary of the Sun's habitable zone today is: (a) inside the orbit of Venus; (b) between Venus and Earth; (c) outside the orbit of Earth.

35. As the Sun ages, the habitable zone will: (a) move outward and grow wider; (b) move outward but get narrower; (c) stay about the same as it is now.

36. Which of the following could cause Earth to become uninhabitable in about 1 billion years? (a) a moist greenhouse effect; (b) a runaway greenhouse effect; (c) the death of the Sun.

37. Global warming means that: (a) Earth's average temperature is increasing; (b) every place on Earth is getting warmer; (c) Earth will soon have a greenhouse effect.

38. The current concentration of atmospheric carbon dioxide on Earth is: (a) higher than the concentration at any time during the past 650,000 years; (b) higher than in the past century, but lower than at many other times during past millennia; (c) gradually rising, but only about average for the time period during which we have ice core data.

INVESTIGATE FURTHER
In-Depth Questions to Increase Your Understanding

Short-Answer/Essay Questions

39. *Are Habitable Zone Planets Common?* Based on what you have learned so far about solar system formation and habitable zones, should we expect to find many planets in habitable zones? Explain.

40. *No Plate Tectonics.* Suppose plate tectonics magically stopped on Earth, but other geological processes (such as volcanism) continued. Would Earth's *surface* get warmer or cooler? Explain.

41. *Continuously Habitable Zone.* Is Earth in a zone that will remain continuously habitable from the Sun's birth to its death? Is any planet? Explain.

42. *Planetary Changes.* For each situation described, write two or three paragraphs explaining why the planet would or would not be habitable today.
    a. A planet the size of Mars located at the distance of Venus.
    b. A planet the size of Mars located at the distance of Earth.
    c. A planet the size of Venus located at the distance of Earth.
    d. A planet the size of Earth located at the distance of Mars.

43. *A Billion Years.* At minimum, it appears that our planet will remain habitable for at least the next billion years, give or take a couple hundred million years. How long is a billion years? Think of some ways to put this time period into perspective.

44. *Venus's History.* Many people are not surprised to learn that Venus is hotter than Earth, given that it is closer to the Sun. Explain why we cannot attribute its heat to distance from the Sun alone. How do we explain Venus's high temperature?

45. *Habitable Moons.* As we'll discuss in Chapter 11, some of the newly discovered extrasolar planets are Jupiter-like in size but are located at Earth-like distances from Sun-like stars. These planets are unlikely to be habitable themselves. Could they have moons with habitable surfaces? Explain.

46. *Greenhouse Lessons.* While it seems unlikely that human activity could cause a runaway greenhouse effect on Earth, we could still cause the climate to warm substantially. Do you think we can learn anything valuable about our potential effects on Earth's climate by studying the climate histories of Venus and Mars? If so, what? Defend your opinion.

47. *Global Warming.* Briefly summarize the evidence suggesting that global warming is occurring and is a result of human activity. Then write a short essay outlining what, if anything, we should do about it.

Quantitative Problems

Be sure to show all calculations clearly and state your final answers in complete sentences.

48. *Stellar Habitable Zone.* Consider a star slightly smaller and less luminous than the Sun, with a habitable zone that extends from 0.3 to 0.5 AU. How does the size of this star's habitable zone compare to that of the Sun, and what are the implications for the likelihood of finding planets within the zone?

49. *Massive Stellar Habitable Zone.* Consider a star that is more massive and more luminous than the Sun, with a habitable zone that extends from 2.5 to 4 AU. How does the size of this star's habitable zone compare to that of the Sun? Should we consider life to be likely around such a star? Explain.

50. *Strength of Sunlight at Earth.* The power of sunlight reaching the top of Earth's atmosphere is 1,370 watts per square meter. The amount of power flowing outward through Earth's surface caused by radioactivity within Earth is estimated to be 3 trillion watts. Which one—sunlight or radioactive heat—is providing more energy to Earth's surface, and by how much? What does your answer tell you about the relative importance of these two energy sources for life? (*Hint:* You'll need to convert the radioactive power number from a total to an amount per square meter of surface; remember that the surface area of a sphere of radius r is $4\pi r^2$.)

51. *Strength of Sunlight at Venus and Mars.* The solar energy reaching the top of Earth's atmosphere is 1,370 watts per square meter.

What is the comparable energy: (a) at the distance of Venus: (b) at the distance of Mars? (*Hint:* Remember that light follows an inverse square law [see Figure 7.1]; you'll need to look up distances in AU for Venus and Mars.)

52. *Energy Use by Cars.* There are approximately 800 million automobiles in use worldwide . If the average auto runs 1 hour a day, at 50 horsepower, what is the resultant total average power use by cars, in watts? How does this amount of power compare to the amount Earth gets from sunlight? Note that 1 horsepower is 746 watts and 1 watt = 1 joule/second. (*Hint:* You'll first need to figure out the average number of cars running at any one time, and once you find the total power for the cars, you'll need to divide by Earth's surface area in square meters so you can compare it to the solar power given in Problem 50.)

53. *Atmospheric Mass of Venus.* The atmospheric pressure on Venus is 90 bars. What is the total mass of Venus's atmosphere? You may use the fact that 1 bar is the pressure exerted by 10,000 kilograms pushing down on a square meter in Earth's gravity, and assume that Venus's gravity is essentially the same as Earth's.

54. *Melting Greenland.* Greenland is approximately 700 × 2,400 km in size, with an average ice cover that's about 1.5 kilometers thick. Suppose all the Greenland ice were to melt. By approximately how much would this raise the level of Earth's oceans? Assume that oceans cover 70% of Earth, and that water and ice are approximately the same density. (*Hint:* Dividing the volume of melted water by the surface area of the oceans will tell you how much the ocean depth will increase.)

Discussion Questions

55. *The Fate of Life in the Universe.* Consider the evidence suggesting that life is just a fleeting phase in the long-term history of the universe. Assuming this to be the case, how do you think it should influence our perspective on our own place in the universe? Why?

56. *The Politics of Global Warming.* The current scientific case for global warming seems quite strong, but the topic nonetheless generates significant political controversy. Why do you think that is the case? Do you consider global warming to be primarily a scientific issue or a political issue? Explain.

57. *Dealing with Uncertainty.* One of the difficulties in deciding what to do about global warming is the fact that the precise consequences are uncertain. In general, how do you think we as a society should deal with issues for which consequences are potentially severe but highly uncertain? How would you deal with this situation in the particular case of global warming? Explain.

WEB PROJECTS

58. *Kyoto Protocol.* What is the Kyoto protocol concerning global warming? What are its pros and cons? Do you think it is valuable and should be followed? Explain.

59. *Global Warming Scenarios.* Research data showing how the amount of global warming might differ depending on whether we decrease, keep the same, or increase our future carbon dioxide emissions. Summarize your findings.

60. *Global Warming Skeptics.* Compare the arguments from Web sites that claim that: (a) global warming is caused by fossil fuel use; and (b) there is no convincing proof that this is the case. How would you evaluate these arguments?

61. *Long-Term Survival.* Read about some exotic ideas concerning how advanced civilizations might survive as the universe grows cold and dark. Do you think any of these ideas make sense, or are they just wishful thinking? Write a short essay describing one of these ideas and your opinion of it.

11
Habitability Outside the Solar System

LEARNING GOALS

*Then felt I like some watcher of the skies, when a new
planet swims into his ken.*
John Keats (1795–1821)

So far we have focused on life on Earth and the prospect of life existing elsewhere in our solar system. We have found that while Earth is the only planet known to harbor life, several other worlds in our solar system are—or at some time in the past have been—potentially habitable. If we can translate this situation to the billions of other star systems in the Milky Way Galaxy and, by extension, to the billions of other galaxies that pepper the universe, then the total number of habitable worlds must be enormous. The existence of a vast number of such worlds would offer the prospect that among them we would find not only microbes, but also complex creatures and intelligent beings.

Unfortunately, current data are not sufficient for us to be sure that such an immense number of habitable worlds really exists. We have detected many extrasolar planets, but not yet planets that could be Earth-like. Is this simply because our technology is not yet up to the task, or could it be that such planets are rare?

We will devote this chapter primarily to exploring the question of whether habitable planets are common or rare and how we might someday observe—or even identify life on—such planets if they exist. This understanding will then help us frame the questions of whether other civilizations exist and how we might someday contact them if they do, topics we'll discuss in the chapters that follow.

11.1 DISTANT SUNS

Perhaps the first and most obvious prerequisite to finding a planet on which life might arise is selecting a "sun" that can provide sufficient light and heat to support habitable worlds. Not all stars are potential suns in this way. To understand what kinds of stars might be orbited by potentially habitable planets, we must first investigate the nature of stars in more depth.

• How do stellar life cycles affect the possibility of habitable planets?

In ancient times, almost any light in the sky was considered to be a star, and in some cases we still use this historical language. For example, we often refer to meteors as "shooting stars," even though they really are just bits of interplanetary dust entering our atmosphere. Asteroids got their name, which means "starlike," because that is how they appear when first seen through a telescope, even though they are actually chunks of rock in our own solar system. Our modern definition of a star is a large ball of gas that generates energy by nuclear fusion in its hot central core (see the definitions on p. 51). Thus, our Sun shines as a star because it fuses hydrogen into helium in its core.

Recall that a star goes through a "life cycle" that begins with its formation in a giant cloud of gas (see Figure 3.6). Before the center gets hot enough to ignite nuclear fusion, we refer to the unfinished star as a *protostar* (*proto* comes from the Greek for "earliest form of"). A star is "born" when nuclear fusion begins in its central core. A star "dies" when it finally ceases to produce energy by any kind of fusion.

All stars spend most of their lives (about 90% of the time from star birth to star death) fusing hydrogen into helium. Stars shine fairly steadily during this period, brightening gradually as they age (for the same reason the Sun is slowly brightening [Section 10.4]). Core hydrogen fusion may continue for millions or billions of years (depending on the star's mass, as we'll discuss shortly), but eventually the central core will be so depleted of hydrogen that the fusion cannot continue.

Perhaps surprisingly, a star that exhausts its core hydrogen begins to grow larger and brighter, becoming a *giant* or *supergiant* star. For example, when our own Sun becomes a red giant some 4–5 billion years from now, it will grow so large that it will engulf some of the inner planets. The reason an aging star grows so large and luminous is the immense pressure exerted on the core by the weight of the star's overlying gas. During the time that the central core burns hydrogen, the energy generation helps the core resist the pressure and maintain its size. Once the hydrogen in the central core runs out, the pressure causes the core to shrink and therefore rise in temperature. As this occurs, a layer surrounding the central core—a layer that still contains unburned hydrogen—ignites with nuclear fusion. In fact, this layer becomes so hot that the total rate of fusion is higher than it was while the central core fused hydrogen. The higher rate of fusion means a greater release of energy, which causes the star to increase in size and emit more light.

Stars can begin to fuse heavier elements during their giant or supergiant stages. Our own Sun will someday produce carbon by fusing the helium in its core, and more massive stars can ultimately create all the other elements (through a combination of core fusion reactions and reactions that occur in their final death throes). As we discussed in Chapter 3, this stellar manufacturing explains the existence of all the elements in the universe except the original hydrogen and helium that were spawned by the Big Bang, which is why we say we are made of "star stuff."

Eventually, the star will reach a point where it can fuse no other elements, and at that point the star dies. Relatively low-mass stars like our Sun end their lives comparatively gently, ejecting their outer layers into space as planetary nebulae and leaving behind the type of dead star that we call a *white dwarf* (see Figure 10.6). Higher-mass stars die in titanic explosions called *supernovae* (see Figure 3.8), in which their cores collapse to form stellar remnants that may be either *neutron stars* or *black holes.*

From the point of view of an astronomer, all of the various stages of stellar life are interesting. But in terms of the search for habitable planets, we are generally concerned only with stars that are in the long-lasting, hydrogen-fusing stage of their lives. Giants and supergiants may still have planets, but the rapidly changing nature of these stars (for example, bloating up in just a few million years or less) makes it unlikely that any planets could remain habitable for long. Dead stars may also have planets—indeed, the first confirmed discovery of extrasolar planets involved three small worlds orbiting a neutron star. But because a neutron star is no longer generating continuous sunshine of the type produced by ordinary stars, it would be an unlikely host for habitable planets.

• How do we categorize stars?

Stars vary significantly in their properties, such as mass, temperature, light output, and composition. The most luminous stellar beacons shine with a brilliance as great as a million Suns, while other stars are so dim that they are invisible to the naked eye even when relatively nearby. Once astronomers recognized the substantial differences among stars, their first task (like that of botanists and zoologists in organizing plants and animals) was to find a useful scheme for categorizing this diverse collection of objects.

Finding such a categorization was not trivial. Stars are so far away that even when viewed with large telescopes they appear as no more than points of light. The most obvious characteristic of a star is how bright it appears in the sky, and indeed a system for describing the brightness of stars visible to the naked eye was developed by the ancient Greeks (and is still used today). Unfortunately, ranking a star by how bright it appears in Earth's sky doesn't tell us how bright it *really* is, because the total amount of light it emits into space—which defines its **luminosity**—also depends on distance. An extremely luminous star that is far away may look dim, while a modestly luminous star that is nearby may shine quite brightly. For example, the brightest star in our sky, Sirius, is only a moderately luminous star that happens to be relatively close. (Sirius is about 8.6 light-years distant, making it the seventh-nearest star to Earth [other than the Sun].)

Refinement of the telescope eventually led to techniques for measuring stellar distances, which enabled astronomers to classify stars according to their true luminosities. But to learn more about their properties required yet another technological step: the invention of spectroscopy [Section 3.4]. In the late 1800s, astronomers at Harvard College Observatory, under the directorship of Edward Pickering (1846–1919), began a massive effort to study stellar spectra and thereby determine other characteristics of the stars.

The Spectral Sequence   Making a detailed stellar spectrum is a tedious process, but at the end of the nineteenth century astronomers took an important step forward when they invented a method to record the spectra of many stars at once. To do this, astronomers mounted a glass prism in front of a telescope's objective lens and then photographed a patch of sky containing a large number of stars. On the resulting photo, the image of each star was spread out into a rainbow-like streak, and the most obvious spectral lines could be seen (Figure 11.1).

Once the technique was developed, it wasn't long before thousands of stellar spectra were in hand. The next step was to study this wealth of data and try to make sense of what it was telling us. Pickering needed help with these tasks, so he began to hire assistants whom he called "computers." The job clearly required people well trained in physics and astronomy, but, in part because the task was seen as somewhat tedious, Pickering found few takers among the men graduating from what was then the all-male Harvard College. Because at that time women with equivalent educations faced enormous obstacles to securing good positions in science, a job with Pickering represented a rare chance for career advance-

Figure 11.1

This photograph shows how, by placing a prism in front of a telescope's lens, astronomers recorded the spectra of many stars at once. The individual spectra were somewhat crude, but were adequate for identifying the stronger hydrogen and helium lines that are the basis of the spectral type classification.

ment for women. So Pickering recruited women who had studied physics and astronomy at colleges such as Wellesley and Radcliffe. Although the work was indeed tedious, it was also cutting-edge research, and the women astronomers of Harvard College Observatory were to make many great discoveries (Figure 11.2).

At first, the astronomers found it difficult to make sense of the spectra. Pickering suggested a scheme in which stars were classified by the visibility of hydrogen lines in their spectra (that is, spectral lines caused by the element hydrogen), using type A to designate stars with the strongest hydrogen lines, type B for those with slightly fainter lines, and so on down the alphabet to type O for stars showing the weakest lines. Following this suggestion, one of Pickering's first "computers," Williamina Fleming (1857–1911), classified more than 10,000 stellar spectra by 1890. But the work was just beginning.

In 1896, Pickering hired Annie Jump Cannon (1863–1941), who in the course of her career would personally classify the spectra of more than 400,000 stars. Within a few years of being hired, she realized that Pickering's sequence of spectral types A to O included some redundancies and, more importantly, fell into a much more natural order than he had supposed. She concluded that there were only seven major spectral types, which could be logically ordered as **OBAFGKM**, the *spectral sequence* that legions of astronomy majors have memorized using the politically incorrect mnemonic "Oh, Be A Fine Girl, Kiss Me." Cannon also subdivided each type by number; for example, stars of spectral type G could be subclassified as G0, G1, G2, and so on to G9, with G0 being most similar to the F stars and G9 being nearly the same as a K0 star. Our Sun is now classified as spectral type G2. The astronomical community adopted Cannon's system of stellar classification in 1910.

The stellar classifications clearly were telling us something important about the nature of stars, but no one yet knew just what that was. The answer finally came in 1925, in the dissertation of another woman working at Harvard College Observatory, Cecilia Payne-Gaposchkin (1900–1979). Relying on insights from what was then the newly developing science of quantum mechanics, Payne-Gaposchkin showed that the differences in the spectral types reflected differences in the surface temperatures of the stars. A later review of twentieth-century astronomy called her work "undoubtedly the most brilliant Ph.D. thesis ever written in astronomy."

The Role of Stellar Mass

This research quickly led to another interesting discovery. One might expect that stars could have many combinations of size, temperature, mass, and composition. The truth is much simpler, a fact that was first revealed when astronomers began to make graphs of the relationship between luminosity and surface temperature. These types of graphs, called *Hertzsprung–Russell diagrams*, helped astronomers unlock the secrets of stars; we will discuss these important tools in Section 11.4.

All stars are born with basically the same composition—because they form in interstellar clouds, they are born with the same compositions as the clouds, which are almost entirely (98% or more) made of hydrogen and helium. For that reason, the physics that determines a star's characteristics is relatively straightforward. During the hydrogen-burning phase of its life, a star's surface temperature (and hence its spectral type) and total luminosity are determined almost entirely by one thing: the star's mass. Table 11.1 lists typical properties for stars of each of the seven major spectral types.

Figure 11.2

Women astronomers pose with Edward Pickering at Harvard College Observatory in 1913. Annie Jump Cannon is in the back row near the right edge of the door.

TABLE 11.1 *Typical Properties for Hydrogen-Burning Stars of the Seven Major Spectral Types*

Numbers given in solar units are values in comparison to the Sun; for example, a mass of 60 solar units means 60 times the mass of the Sun. Note that the Sun is a G star. (More specifically, the Sun's spectral type is G2.)

Spectral Type	Approximate Percentage of Stars in This Class	Surface Temperature (°C)	Luminosity (solar units)	Mass (solar units)	Lifetime (years)
O	0.001%	50,000	1,000,000	60	500 thousand
B	0.1%	15,000	1,000	6	50 million
A	1%	8,000	20	2	1 billion
F	2%	6,500	7	1.5	2 billion
G	7%	5,500	1	1	10 billion
K	15%	4,000	0.3	0.7	20 billion
M	75%	3,000	0.003	0.2	600 billion

Stellar Lifetimes Table 11.1 also shows that the lifetimes of stars vary considerably, which we can understand by studying the mass and luminosity columns. A star's mass essentially tells us how much hydrogen fuel it has available for fusion, while its luminosity tells us how brightly the star shines and hence how rapidly it is fusing its hydrogen. Note that there is an enormous range in luminosity: The most luminous stars are nearly a billion times as luminous as the dimmest. The range in mass is much smaller. Thus, for example, an O star has some 60 times as much hydrogen fuel as a G star like the Sun (from the mass column) but burns it at a rate a million times faster than the Sun (from the luminosity column). Therefore, it will go through all of its available fuel much faster than does the Sun. The rule is a general one: *The more massive the star, the shorter its lifetime.* As shown in the last column of the table, the range of lifetimes extends from just a few hundred thousand years to hundreds of billions of years.

THINK ABOUT IT   Many generations of massive O and B stars have lived and died in the history of the universe. Is it also true that many generations of G, K, and M stars have lived and died? Explain. (*Hint:* How old is the universe?)

• Which stars would make good suns?

We can use the properties listed in Table 11.1 to investigate the question of which types of stars might make suitable suns for habitable planets. We can immediately rule out stars of type O from their short lifetimes alone. Recall that, in our solar system, it took millions of years for the rocky inner planets to form by accretion [Section 3.3]. With lifetimes of less than a million years, O stars simply don't last a long enough time to allow the formation of Earth-size planets. We can probably also rule out stars of type B. Although the 50-million-year lifetime of a typical B star is probably long enough to allow planets to form, the star's death would probably occur before the process of accretion had settled down enough for life to take hold.

Spectral Types A and F Stars of types A and F, with lifetimes of 1–2 billion years, would seem to offer enough time for both the formation of planets and the beginnings of biology. After all, evidence suggests that life took hold on our own world within a few hundred million years or less of its formation [Section 6.1]. Stellar types A and F are hotter than the Sun, so their habitable zones would be wider and would lie at greater distances from their central star. By itself, the wider habitable zone could make habitable planets even more likely than they are around Sun-like stars.

One potential problem is that, because of their higher temperatures, A and F stars emit many times as much ultraviolet light as does the Sun. Biology on Earth, and perhaps biology in general, is vulnerable to high-energy ultraviolet light, which easily breaks chemical bonds in complex organic molecules. The intense radiation might well keep planetary surfaces sterile. However, there are at least two ways around this problem.

First, ultraviolet radiation does not penetrate far into the ground, oceans, or ice. If life on Earth emerged near volcanic vents in the deep oceans, as some evidence suggests, the same thing might happen on a watery planet around an A or F star. Similarly, worlds with a subsurface ocean, such as Europa may have, would offer life plenty of protection.

Second, even though our Sun emits far less ultraviolet light than A or F stars, it still emits enough to make Earth's surface sterile if not for the shielding provided by the ozone layer in the atmosphere. Planets around an A or F star might enjoy similar shielding if they had either a sufficiently thick atmosphere or an atmosphere containing sufficient oxygen. In the latter case, the additional ultraviolet light would split more atmospheric O_2 molecules, producing single oxygen atoms that would then combine to form ozone. In other words, a higher dosage of ultraviolet radiation could result in more ozone shielding.

Overall, A and F stars, despite their energetic nature, seem quite capable of hosting habitable planets. However, given the fact that complex plants and animals on Earth did not arise until our planet was some 4 billion years old, the 1- to 2-billion-year lifetime of A and F stars suggests that life on these planets would most likely be much simpler than life on Earth.

THINK ABOUT IT In light of current evidence concerning the past oxygen content of Earth's atmosphere (see Section 6.3), does it seem likely that planets around A or F stars could have ozone layers? Why or why not?

Spectral Types G, K, and M Although the possibility of habitable planets around A and F stars is intriguing, Table 11.1 suggests that statistically it's not too important: Together, these types make up only about 3% of stars. G stars, like our Sun, make up another 7%—and our own existence proves that G stars can have habitable planets. But if we want to know whether the *majority* of stars could have worlds capable of supporting life, we need to consider the smaller K and M types, which make up some 90% of all stars in the universe.

K and M stars have long lifetimes that impose no limit on their ability to have planets on which life could evolve for many billions of years. Instead, the primary issue for life around these stars is the size of their habitable zones. Because these stars are dimmer than the Sun, temperatures on orbiting planets will be lower in these systems than on planets

Figure 11.3

The approximate habitable zones around our Sun, a typical K star, and a typical M star, shown to scale. The habitable zone becomes increasingly smaller and closer-in for stars of lower luminosity. (*Note:* As discussed in Section 10.4, the habitable zone around any star moves outward with time; the zones shown here are for stars similar in age to the Sun.)

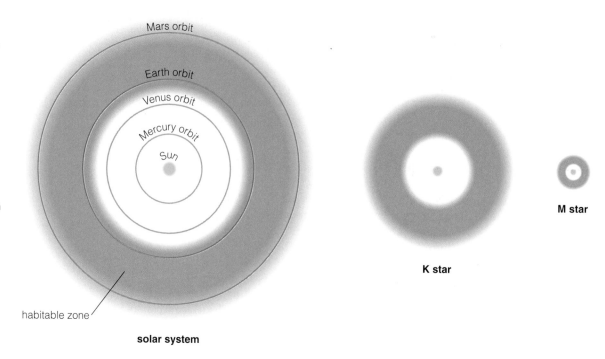

Mars orbit

Earth orbit

Venus orbit

Mercury orbit

Sun

habitable zone

solar system

K star

M star

the same distances from the Sun in our solar system. More specifically, recall that the energy in starlight falls off with the square of the distance from the star (see Figure 7.1). Thus, for a planet to receive as much radiant energy as Earth from a K star having one-fourth the Sun's luminosity, the planet would have to orbit its star at one-half Earth's distance from the Sun (because $\left(\frac{1}{2}\right)^2 = \frac{1}{4}$). For a star with a luminosity 1% of the Sun's, a similar planet would need to orbit at a distance one-tenth of Earth's distance from the Sun, which means it would follow an orbit about one-fourth the size of Mercury's.

The same reasoning tells us that the boundaries of a dim star's habitable zone would also be much closer to the star than are the boundaries of the habitable zone in our solar system.* In general, the habitable zone becomes progressively smaller and closer-in for stars of lower luminosity (Figure 11.3). Around a typical M star, for example, the habitable zone would extend only over a region roughly equivalent to from one-tenth to one-fourth Mercury's distance from the Sun in our solar system. The small size of the habitable zone certainly decreases the probability that planets would be found within this region (see Cosmic Calculations 10.1).

On the other hand, the overwhelming majority of stars are M stars, due both to their long lives—every M star ever born is still shining—and to the fact that small stars are born in larger numbers than are more massive stars. Even if their individual habitable zones are modest, there are enough M stars to host a large number of inhabited worlds. For years, this optimistic thought was moderated by two objections. First, planets orbiting close to a star get locked into synchronous rotation [Section 9.1], with one side of the planet perpetually facing the star (much as the Moon always keeps one face turned to Earth). Once this happens, the side away from the star becomes perpetually dark, and the atmosphere might be

_____

*The boundaries of the habitable zone are also affected to some extent by the star's surface temperature, which determines the wavelength at which it puts out most of its light.

expected to freeze out. Second, small stars have frequent and energetic flares*—sudden bursts of intense light and radiation—that might cook any complex life.

Recent research, however, suggests that neither of these objections is necessarily fatal. Even a modestly thick atmosphere containing carbon dioxide would circulate heat from the bright to the dark side on a synchronously locked world, keeping temperatures relatively uniform and allowing liquid water to exist over much of the planet's surface. As for flares, their dangerous ultraviolet light might actually cause the production of a protective layer of atmospheric ozone, in much the way we have described for worlds orbiting stars of types A and F. And, of course, underwater or underground life would be protected in any case. The bottom line is that M stars probably can support life. If they do, given the fact that they live so long, these stars might possibly house the galaxy's oldest biology.

Brown Dwarfs   Table 11.1 shows that an "average" (meaning middle of the range) M star has a mass about 20% that of the Sun; although it is not shown in the table, the smallest M stars have masses about 8% that of the Sun. Given that these low-mass M stars are the most plentiful of all stars, you might wonder why there aren't stars of even lower mass. The answer is that in an object with a mass less than 8% that of the Sun, gravity isn't strong enough to compress the core to high enough temperatures to sustain hydrogen fusion. As a result, such "substellar" objects, called **brown dwarfs**, are very dim and difficult to detect. Nevertheless, they have much hotter surfaces than planets, which means they emit moderate amounts of infrared radiation that can allow us to detect them if they are nearby. Thus, thanks to improved infrared telescopes, we now know that brown dwarfs are quite common (Figure 11.4). Typical brown dwarfs have masses between about 10 and 80 times that of Jupiter (the larger number is equivalent to 8% the mass of the Sun). In essence, brown dwarfs occupy a fuzzy range of masses between those of the largest planets and the smallest true stars.

Because they are so dim, brown dwarfs have no habitable zones at all. However, recent infrared observations have shown that many young brown dwarfs have disks of material surrounding them, suggesting that planets are forming in their neighborhood, and one brown dwarf appears to have a Jupiter-size planet orbiting it (Figure 11.5). Thus, while planets with habitable surfaces do not seem possible around brown dwarfs, it is possible that brown dwarfs could have planets orbited by Europa-like moons with subsurface oceans and, perhaps, life.

Multiple Star Systems   Based on our discussion so far, we have good reason to believe that the vast majority of stars are, at least in principle, capable of having habitable planets. However, there's another potential complicating factor that we have not yet considered: Many stars are not loners like the Sun, but rather are members of **multiple star systems**, in which two or more stars orbit each other closely. For example, the nearest star system to our own, Alpha Centauri, is actually a triple star

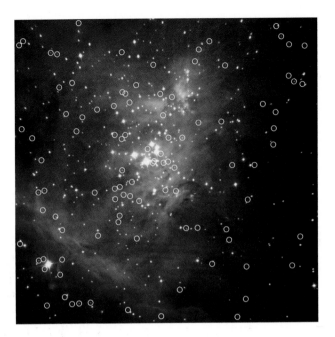

Figure 11.4

This infrared image shows many brown dwarfs (circled) in the constellation Orion.

Figure 11.5

This infrared image from the European Southern Observatory's Very Large Telescope shows a brown dwarf called 2M1207 (blue) and what is probably a Jupiter-size planet in orbit around it (red).

---

*The occurrence and size of flares, sunspots, and other types of surface "activity" on stars is related to the strength of a star's magnetic field. A star's magnetic field strength depends on its rotation rate and the depth of convection in the star's outer layers. Small stars have deeper convection, so if they rotate quickly then they can have very strong magnetic fields associated with large flares and other surface activity.

system: Its largest and brightest member ("Alpha Centauri A") is a type G star like our Sun, its second-largest ("Alpha Centauri B") is a K star that is separated from the G star by roughly the distance that separates our Sun and the planet Uranus, and its third star ("Proxima Centauri") is an M star that orbits the other two at a distance of about one-fifth of a light-year. Can multiple star systems have habitable planets? The answer to this question hinges on whether a planet in a multiple star system can have a stable orbit that keeps it within a habitable zone.

Let's consider possible planetary orbits for the most common cases, which are **binary star systems** with just two stars. (Systems with three or more stars are much less common.) Broadly speaking, there are three possible situations to consider for planetary orbits in a binary star system:

- A planet could orbit around both stars (Figure 11.6a). If the planet orbits the stellar pair at a distance considerably greater than the separation between the two stars, then gravitationally the two stars act much as one, and the planet can orbit without disruption around its distant twin masters. Computer simulations indicate that a stable orbit is generally possible for planets that orbit at a distance of more than about five times the separation of the stars. (The exact distance depends on the details of the system, such as the relative masses of the two stars and the eccentricity of their orbits.)

- A planet could orbit one star or the other (Figure 11.6b). If the two stars are themselves widely separated, then a planet near either one can orbit steadily because it will feel little disturbing effect from the second star. Computer simulations indicate that stability becomes possible when the two stars are separated by more than about five times the planet's orbital distance.

- We might imagine a planet trying to orbit between the two stars (Figure 11.6c). However, if the distance between the planet and at least one of the stars is not sufficiently different from the distance between the two stars, its orbit will not be stable. The planet will experience competing gravitational tugs from both stars that will ultimately fling it out of the system (or send it crashing into one of the stars).

As an example of the first stable case, imagine that our Sun had a companion star at Jupiter's distance (which is 5.2 times that of Earth's

Figure 11.6

Three orbital possibilities in a binary star system.

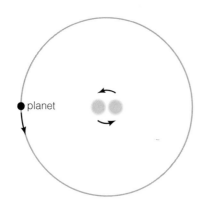

a *Potentially stable:* The planet orbits both stars; the radius of the planet's orbit is much larger than the separation between the stars.

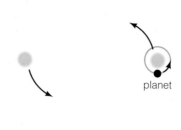

b *Potentially stable:* The planet orbits one star; the radius of the planet's orbit is much smaller than the separation between the stars.

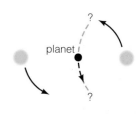

c *Not stable:* The orbit of the planet is neither much larger nor much smaller than the separation between the stars distance). In that case, Earth's orbit (as well as those of Mercury and Venus) would be stable. As an example of the second case, imagine that our Sun had a companion star at Earth's position. The orbits of the inner planets would not be stable, but Jupiter's orbit could be.

Given these orbital possibilities, what can we say about habitability in binary star systems? In fact, both stable cases seem to allow for habitable planets. In the first case, the habitable zone would be some region surrounding the two stars together. In the second case, the more distant star would probably have little influence on the planet's climate, so the habitable zone would be defined solely by the star that the planet orbits. Similar conclusions probably apply to other multiple star systems.

This is encouraging news in our search for life in the cosmos, because the majority (about 60%) of stars like the Sun are in multiple systems. Until recently, it was believed that the majority of *all* stars shared this penchant for partners. However, new research has shown that, among the galaxy's least massive (and dimmest) stars—the M stars—only about one-fourth are multiple. Since 75% of galactic stars are M stars, this means that, overall, about one-third of the galaxy's stars are multiple.

Summary of the Types of Stars That Might Have Habitable Planets

We began this section with the goal of determining what types of stars might potentially have habitable planets. Perhaps surprisingly, we have found relatively few limits. The only stars that we can rule out completely— the O and B stars—are rare. G stars like our Sun clearly can have habitable planets. Even though only 7% of the stars in our galaxy are G stars, 7% of a few hundred billion stars is still an impressive tally. The small size of the habitable zone around the most common types of stars—dim, low-mass K and M stars—suggests that habitable planets might be rare around these stars, but the great abundance of such stars might compensate. Even multiple star systems seem reasonably capable of having habitable planets.

Our studies of stars suggest that many or most stars could *potentially* have orbiting habitable worlds. Whether they *do* may depend on a number of other factors. To understand these factors, we must investigate the discoveries of extrasolar planets and what these discoveries have taught us about the formation and nature of other planetary systems.

11.2 EXTRASOLAR PLANETS: DISCOVERIES AND IMPLICATIONS

In Chapter 3, we discussed both theoretical and observational reasons why we expect that planets are common. On the theoretical side, current ideas about how stars and planets form suggest that the same general processes that gave birth to the planets of our solar system ought to occur around most other young stars. For example, it seems almost inevitable that a collapsing cloud of gas and dust will form a star surrounded by a spinning disk of debris (see Figure 3.19).

On the observational side, two types of evidence point to planets being common. First, we have seen flat disks of material orbiting many other newborn stars (see Figure 3.20). The disks of young stars often contain a few percent of the mass of their stars, which is more than sufficient for constructing planets. (The combined mass of all the worlds in our solar system amounts to less than 0.2% of the Sun's mass.) Based on the observations made to date, it appears that at least a quarter to half of all young stars are surrounded by such disks. Although this fact does not tell us directly whether planets are likely to form in such disks, it certainly lends support to our theory of how stars and planets form.

The second and more clear-cut form of observational evidence comes in actual planets discovered around other stars. As of mid-2006, nearly 200 extrasolar planets had been found around more than 160 different stars. For 19 of these stars, multiple planets (as many as three) had been found.

THINK ABOUT IT   By the time you read this book, the number of known extrasolar planets will have grown larger still. Do a quick Web search on "extrasolar planets" to find the current number. How many more extrasolar planets have been discovered since mid-2006?

These discoveries tell us that planets must be at least moderately common, but they do not yet tell us anything directly about the possibility of habitable planets, because nearly all the extrasolar planets discovered are much closer in mass to the jovian planets in our solar system than to terrestrial worlds like Earth. This fact is probably just due to technological limitations; detecting any extrasolar planet is quite difficult, and small ones are correspondingly harder to find than larger ones. Improving technology will likely make it possible for us to detect Earth-size planets soon, and in the meantime we are learning important lessons about potential habitability from the orbital characteristics of the larger planets already discovered. In this section, we will investigate planet detection techniques and the astrobiological implications of what we have learned so far.

• How do we detect planets around other stars?

The first clear-cut discovery of a planet around another Sun-like star—a star called 51 Pegasi—came in 1995. The discovery was made by Swiss astronomers Michel Mayor and Didier Queloz and was soon confirmed by a team led by Geoffrey W. Marcy and R. Paul Butler. Many more extrasolar planets have been discovered since that time, and the hunt for worlds around other stars has become an international effort involving many teams of astronomers. Several different planet-finding strategies have been used or proposed. If we strip away the details, however, there are really only two basic ways to search for extrasolar planets:

1. *Directly:* Pictures or spectra of the planets themselves constitute direct evidence of their existence.
2. *Indirectly:* Precise measurements of stellar properties (such as position, brightness, or spectra) may indirectly reveal the effects of orbiting planets, even without seeing the planets themselves.

Until technology makes direct searches routine, we can rely only on indirect searches that look at stars, seeking effects that can be attributed to orbiting planets.

Direct detection is preferable in principle, because it can tell us far more about the planet's properties. However, current telescopes are just reaching the point where direct detection is possible, and only a handful of such detections have been made to date. Let's therefore start by considering the indirect techniques responsible for most extrasolar planet discoveries, and then come back to direct detection.

Gravitational Tugs   To date, nearly all extrasolar planets have been detected by observing the gravitational tugs they exert on the stars they orbit.

This type of detection is indirect because we discover the planets by observing the stars they orbit without actually seeing the planets themselves.

Although we usually think of a star as remaining still while planets orbit around it, that is only approximately correct. In reality, all the objects in a star system, including the star itself, orbit the system's **center of mass**. To understand this concept, think of a waiter carrying a tray of drinks. To carry the tray, he places his hand under the spot at which it balances—its center of mass. If the tray has a heavy glass of water off to one side, he will place his hand a little to that side of the tray's center. Because the Sun is far more massive than all the planets combined, the center of mass of our solar system lies very close to the Sun—but not exactly at the Sun's center.

We can see how this fact allows us to discover extrasolar planets by imagining the viewpoint of extraterrestrial astronomers observing our solar system from afar. Let's start by considering only the influence of Jupiter, which exerts a much stronger gravitational tug on the Sun than the rest of the planets combined. The center of mass between the Sun and Jupiter lies just outside the Sun's visible surface (Figure 11.7). Thus, what we usually think of as Jupiter's 12-year orbit around the Sun is really a 12-year orbit around their mutual center of mass; we generally don't notice this fact because the center of mass is so close to the Sun itself. In addition, because the Sun and Jupiter are always on opposite sides of the center of mass (otherwise it wouldn't be a "center"), the Sun must orbit this point with the same 12-year period as Jupiter. The Sun's orbit traces out only a very small circle (or ellipse) with each 12-year period, because the Sun's average orbital distance is barely larger than its own radius, and with sufficiently precise measurements, extraterrestrial astronomers could detect this orbital movement of the Sun. They could thereby deduce

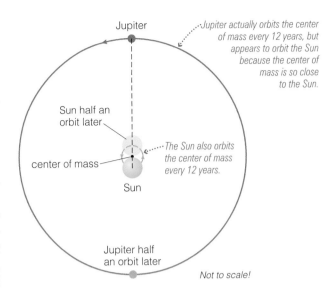

Figure 11.7

This diagram shows how both the Sun and Jupiter actually orbit around their mutual center of mass, which lies very close to the Sun. The diagram is not to scale; the sizes of the Sun and its orbit are exaggerated about 100 times compared to the size shown for Jupiter's orbit.

SPECIAL TOPIC 11.1: THE NAMES OF EXTRASOLAR PLANETS

The planets in our solar system have familiar names rooted in mythology. Unfortunately, there's not yet a well-accepted scheme for naming extrasolar planets. As a result, we generally refer to extrasolar planets by the star they orbit, such as "the planet orbiting the star named. . . ." Worse still, the stars themselves often have confusing or even multiple names, reflecting naming schemes used in star catalogs made by different people at different times in history.

A few hundred of the brightest stars in the sky carry proper names from ancient times. Many of these names are Arabic—such as Betelgeuse, Algol, and Aldebaran—because of the work of the Arabic scholars of the Middle Ages. In the early 1600s, German astronomer Johann Bayer developed a system that gave names to many more stars. In Bayer's system, each star gets a name based on its constellation and a Greek letter indicating its ranking in brightness within that constellation. For example, the brightest star in the constellation Andromeda is called Alpha Andromedae, the second brightest is Beta Andromedae, and so on. Of course, because there are only 24 letters in the Greek alphabet, this system works for only the 24 brightest stars in each constellation. About a century later, English astronomer John Flamsteed published a more extensive star catalog in which he used numbers once the Greek letters were exhausted. For example, 51 Pegasi gets its name from Flamsteed's catalog. (Flamsteed's numbers are based on position within a constellation rather than brightness, so a star may have a different designation in Bayer's system than in Flamsteed's.)

As more powerful telescopes made it possible to discover more and fainter stars, astronomers developed many new star catalogs. The names we use today usually come from one of these catalogs. For example, the star HD209458 gets its name from the Henry Draper (HD) catalog and has the number 209458 in that catalog. You may also see stars with numbers preceded by other catalog names, including Gliese, Ross, and Wolf; these catalogs are also named for the astronomers who compiled them. Moreover, because the same star is often listed in several of these catalogs, a single star can have several different names. Some of the newest planets orbit stars so faint that they have not been previously cataloged. They then carry the name of the observing program that discovered them, such as TrES-1 for the first discovery of Trans-Atlantic Exoplanet Survey, or OGLE-TR-132b for the 132nd object scrutinized by the Optical Gravitational Lensing Experiment.

Objects orbiting other stars usually carry the star name plus a letter denoting their order of discovery around that star. The host star itself is designated "a." Thus, for example, HD209458b is the first object discovered to be orbiting star number 209458 in the Henry Draper catalog; Upsilon Andromedae d is the third object discovered to be orbiting what should be the 20th brightest star (because upsilon is the 20th letter in the Greek alphabet) in the constellation Andromeda (although it isn't the 20th brightest because Bayer's measurements were not all correct). Unfortunately, this scheme doesn't tell you whether the orbiting object is thought to be a planet or another star.

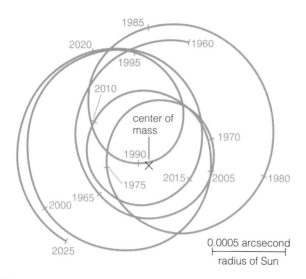

Figure 11.8

This diagram shows the orbital path of the Sun around the center of mass of our solar system as it would appear from a distance of 30 light-years away; the graph shows the motion between 1960 and 2025. Notice that the entire range of motion during this period is only about 0.0015 arcsecond, which is almost 100 times smaller than the angular resolution of the Hubble Space Telescope. Nevertheless, if alien astronomers could measure this motion, over time they could learn of the existence of planets in our solar system even without being able to detect the planets themselves.

the existence of Jupiter, even without having observed Jupiter itself. They could even determine Jupiter's mass from the Sun's orbital characteristics as it goes around the center of mass. Remember that for any particular separation between the Sun and a planet, a more massive planet pulls the center of mass farther from the Sun's center, thereby giving the Sun a larger orbit and a faster orbital speed around the center of mass.

Now let's add in the effects of Saturn, which exerts the second-greatest gravitational tug on the Sun. Saturn takes about 29 years to orbit the Sun, so by itself it would cause the Sun to orbit their mutual center of mass about every 29 years. However, because Saturn's influence is secondary to that of Jupiter, this 29-year period appears as an added complication on top of the Sun's 12-year orbit around its center of mass with Jupiter. In other words, every 12 years the Sun would return to *nearly* the same orbital position around its mutual center of mass with Jupiter, but the precise point of return would make its own even tinier orbital circle (or ellipse) with Saturn's 29-year period. By measuring this motion carefully from afar, an extraterrestrial astronomer could deduce the existence and masses of both Jupiter and Saturn after a few decades of observing.

The other planets also exert gravitational tugs on the Sun, which create further complications to the Sun's orbital motion around the solar system's center of mass (Figure 11.8). These extra effects become increasingly difficult to measure in practice, but in principle they would allow an extraterrestrial astronomer to discover all the planets in our solar system. If we turn this idea around, you'll realize that it means we can search for planets in other star systems by carefully watching for the tiny orbital motion of a star around the center of mass of its star system.

There are two techniques for observing the small orbital motion of a star caused by the gravitational tugs of planets. The first is straightforward: We make precise measurements of a star's position to see if it is, in fact, moving about in a way that tells us it is being tugged on by planets. This method is usually called the **astrometric technique**, because it requires measuring stellar positions in the sky (*astrometric* means "measurement of the stars"). The second scheme, known as the **Doppler technique**, detects orbital motion by searching for slight shifts in the positions of spectral lines in a star's spectrum. We next consider the advantages and limitations of these two techniques.

The Astrometric Technique   The astrometric technique has been used for many decades to identify binary star systems, since two orbiting stars will move periodically around their center of mass. The technique works especially well for binary systems in which the two stars are not too close together, because the stellar motions tend to be larger in those cases. In the case of planet searches, however, the expected stellar motion is much more difficult to detect.

As an example, from a distance of 10 light-years, a Jupiter-size planet orbiting 5 AU from a Sun-like star would cause its star to dance slowly over a side-to-side angular distance of only about 0.003 arcsecond—approximately the width of a human hair seen from a distance of 5 kilometers. Remarkably, with careful telescope calibration, astronomers can now measure displacements this small, and instruments currently under development will be five to ten times better yet. However, two other complications add to the difficulty of the astrometric technique.

First, the farther away the star, the smaller its side-to-side movement will appear. For example, while Jupiter causes the Sun to move by about 0.003 arcsecond as seen from 10 light-years away, the observed motion is only half as large when seen from 20 light-years away and one-tenth as large when seen from 100 light-years away. Thus, the astrometric technique works best for massive planets around relatively nearby stars.

The second complication arises from the time required to detect a star's motion. It is obviously easier to detect larger movements than smaller ones, and a planet with a larger orbit has a larger effect on its star. To understand why, consider what would happen if Jupiter moved farther from the Sun. Because the center of mass of the solar system is nearly at the balance point between the Sun and Jupiter, moving Jupiter outward would also cause the center of mass to move farther from the Sun. With the center of mass located farther from the Sun, the Sun's orbit around the center of mass—and hence its side-to-side motion as seen from a distance—would be larger. Thus, the astrometric technique works best not only for massive planets, but also in particular for massive planets that are far from their stars. However, Kepler's third law [Section 2.2] tells us that a more distant planet takes longer to complete its orbit, which means that the stellar motion it causes also takes longer. For example, while Jupiter causes the Sun to move around the center of mass with a 12-year period, Neptune's effects on the Sun show up with the 165-year period of Neptune's orbit. It would probably take a century or more of patient observation to be confident that stellar motion was occurring in a 165-year cycle.

As a result of these complications, the astrometric technique has been used to detect only two of the 170 extrasolar planets known through early 2006. However, we expect the astrometric technique to prove much more valuable in the future. Astronomers are making rapid progress in developing a technology called *interferometry* that allows two or more telescopes to work together to achieve much higher angular resolution than either telescope individually. This technology has been used with radio telescopes for decades, and it is now being applied successfully to infrared and visible-light telescopes. Several major observatories, including the twin Keck observatories on Mauna Kea, may soon be capable of using interferometry to detect at least some planets through the astrometric technique. In addition, both NASA and the European Space Agency are working on space missions that will be able to search for planets using the astrometric technique. The European mission, called *GAIA*, is slated for launch in 2010 and has the goal of performing astrometric observations of a billion stars in our galaxy with an accuracy of 10 *microarcseconds*. (A microarcsecond is one-millionth of an arcsecond.) NASA hopes to follow not long after with its *Space Interferometry Mission* (*SIM*). As currently planned, *SIM* will be capable of measuring stellar positions to a precision of just 1 microarcsecond—ten times better than *GAIA*'s precision and good enough to detect stellar motion caused by Earth-size planets around the nearest few dozen stars and motion due to Jupiter-size planets orbiting stars as far as 3,000 light-years away.

The Doppler Technique The *Doppler technique* offers another way to find the small movement of a star caused by orbiting planets, and it has proved very successful: The vast majority (more than 90%) of the extrasolar planets found to date have been found with this technique. Rather than looking for the side-to-side motion of a star against the background of the sky, as in the astrometric techniques, the Doppler technique involves studying a star's spectrum to look for telltale signs that the star is moving around a center of mass. The technique is based on searching for small shifts in the wavelengths of lines in the spectrum caused by what we call the **Doppler effect.**

train stationary

The pitch this person hears . . .

. . . is the same as the pitch this person hears.

train moving to right

Behind the train, sound waves stretch to longer wavelength (lower frequency and pitch).

In front of the train, sound waves bunch up to shorter wavelength (higher frequency and pitch).

light source moving to right

The light source is moving away from this person so the light appears redder (longer wavelength).

The light source is moving toward this person so the light appears bluer (shorter wavelength).

a The whistle sounds the same no matter where we stand near a stationary train.

b For a moving train, the sound you hear depends on whether the train is moving toward you or away from you.

c We get the same basic effect from a moving light source (although the shifts are usually too small to notice by eye).

Figure 11.9

The Doppler effect. Each circle represents the crests of sound (or light) waves going in all directions from the source. For example, the circles from the train might represent waves emitted 0.001 second apart.

You've probably noticed the Doppler effect on the sound of a train whistle near train tracks. (You can also notice the Doppler effect with emergency sirens or even just with the "buzz" of a fast car as it goes past you.) If the train is stationary, the pitch of its whistle sounds the same no matter where you stand (Figure 11.9a). But if the train is moving, the pitch will sound higher when the train is coming toward you and lower when it's moving away from you. Just as the train passes by, you can hear the dramatic change from high to low pitch—a sort of "weeeeeeee–oooooooooh" sound. To understand why, we have to think about what happens to the sound waves coming from the train (Figure 11.9b). When the train is moving toward you, each pulse of a sound wave is emitted a little closer to you. The result is that waves are bunched up between you and the train, giving them a shorter wavelength and higher frequency (pitch). After the train passes you by, each pulse comes from farther away, stretching out the wavelengths and giving the sound a lower frequency.

The Doppler effect causes similar shifts in the wavelengths of light (Figure 11.9c). If an object is moving toward us, then its entire spec-

Figure 11.10

The Doppler technique for discovering extrasolar planets.

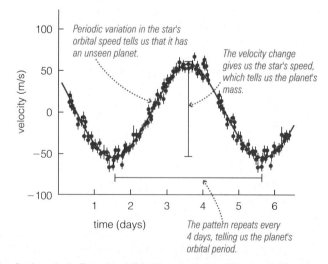

a Doppler shifts allow us to detect the slight motion of a star caused by an orbiting planet.

b A periodic Doppler shift in the spectrum of the star 51 Pegasi shows the presence of a large planet with an orbital period of about 4 days. Dots are actual data points; bars through dots represent measurement uncertainty.

trum—including the spectral lines within it—is shifted to shorter wavelengths (the spectral lines serve as reference lines for measuring the shift). Because shorter wavelengths of visible light are bluer, the Doppler shift of a star coming toward us is called a **blueshift.** If an object is moving away from us, its light is shifted to longer wavelengths; we call this a **redshift,** because longer wavelengths of visible light are redder. The faster the object is moving, the greater the amount of its blueshift or redshift.

The Doppler effect can be used to find planets in the following manner: If an orbiting planet causes its star to move alternately slightly toward and away from us (as part of its orbital motion), the starlight should shift alternately toward the blue and toward the red (Figure 11.10a). The 1995 discovery of a planet orbiting 51 Pegasi occurred when this star was found to be moving with a rhythmic wobble corresponding to an orbital speed of 57 meters per second (Figure 11.10b). The 4-day period of the star's motion tells us the orbital period of its planet.

Current techniques can measure a star's velocity to within about 1 meter per second—walking speed—which corresponds to a Doppler wavelength shift of only 1 part in 300 million. Thus, we can find planets that exert a considerably smaller gravitational tug on their stars than the tug of the planet orbiting 51 Pegasi. Moreover, by carefully analyzing Doppler shift data, we can learn of the existence of a planet and also determine its orbital properties, including its average distance from its star and the shape of its orbit (Figure 11.11). In some cases, the Doppler data are good enough to tell us whether a star has more than one planet.

The Doppler technique also tells us about planetary masses, but with an important caveat. Doppler shifts reveal only the part of a star's motion directed toward or away from us. Thus, for example, a planet whose orbit we view face-on does not cause any Doppler shift in the spectrum of its star, making it impossible to detect the planet with the Doppler technique, let alone to determine its mass (Figure 11.12a). We can observe Doppler shifts in the star's spectrum only if we view its planet orbiting at some

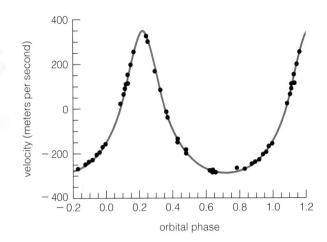

Figure 11.11

The Doppler shifts measured for the star 70 Virginis (from the work of Marcy and Butler). Note the uneven nature of the change in velocity, which tells us that the planet causing the Doppler shift is in a highly eccentric (elliptical) orbit.

Figure 11.12

The amount of Doppler shift we observe in a star's spectrum depends on the orientation of the planetary orbit as we view it from Earth.

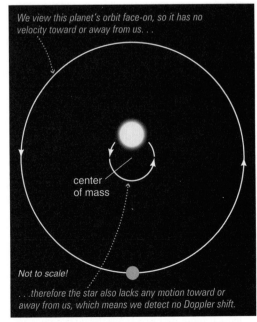

a If we view a planetary orbit face-on, we will not detect any Doppler shift at all.

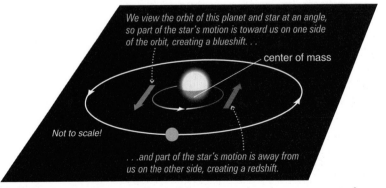

b We can detect a Doppler shift only if the planet and star have some part of their orbital velocities directed toward or away from us. The more the orbit is tilted toward edge-on, the greater the shift we'll observe.

angle other than face-on (Figure 11.12b), and the Doppler shift will tell us the star's full orbital velocity only if we are viewing it precisely edge-on. As a result, a planetary mass that we infer from its star's Doppler shift will be accurate only for a planet that we see in an edge-on orbit. In all other cases, the Doppler shift underestimates the true orbital speed of the star and therefore also causes us to underestimate the true mass of the planet. The more the orbit is tilted away from edge-on, the greater the degree of the underestimate. Thus, planetary masses inferred from the Doppler technique alone are actually *minimum* possible masses (or "lower limits") for these planets.

We can determine a precise mass only if we somehow know that we are viewing an edge-on orbit or if we have some way to measure the precise orbital angle. For example, if we also know the planet's side-to-side motion from the astrometric technique, we can combine this information with the toward and away motions from the Doppler technique to determine the orbital angle. Unfortunately, we rarely have such knowledge, which means that nearly all planetary masses found through the Doppler technique are really minimum masses rather than actual masses. Nevertheless, statistical arguments show that in two out of three cases, the planet's true mass will be no more than double this minimum mass, so the minimum mass is still a relatively good estimate.* Of course, this also means that a third of the planets discovered with the Doppler technique have masses that are more than double the minimum mass. About 6% of the time, the planet's actual mass will be more than ten times the minimum mass, which may raise the question of whether it is truly a planet or a very small star.

The Doppler technique is powerful, but it has limits. In particular, it is best suited for identifying massive planets that orbit relatively close to their star. This limitation arises because gravity weakens with distance, so a planet of a given size pulls harder on its star—making the star move faster—if it is closer. Moreover, because a close-in planet has a much shorter orbital period than a more distant planet, it takes a lot less time to observe the periodic Doppler shifts caused by close-in planets. For example, while it takes only a few weeks of observation to detect a planet with a 4-day period like the one orbiting 51 Pegasi, it would take 12 years to observe just a single orbital cycle of a planet in an orbit like that of Jupiter around our Sun. The Doppler technique also presents a practical difficulty: Current technology allows it to be applied to just one star at a time (with a particular telescope), so only a relatively small number of stars can be studied with this technique.

The limitations of the Doppler technique explain what may at first seem like surprising facts: Most of the extrasolar planets discovered to date orbit relatively close to their stars, and we have yet to discover any planets at all with Earth-like masses. Both these facts may simply be *selection effects* of the Doppler technique; that is, the technique tends to find (or "select") massive planets in close orbits much more easily than any other type of planet. Planets with masses similar to Earth would have such weak gravitational effects on their stars that we could not use the

---

*You can understand the statistical argument if you have studied trigonometry: The actual mass is the minimum mass divided by the *cosine* of the orbital inclination (where edge-on is an angle of 0° and face-on is an angle of 90°). Thus, the actual mass is more than double the minimum mass only for cosines smaller than $\frac{1}{2}$, which means for angles between 60° and 90°. By random chance, only about one-third of all stars will have such large orbital angles. Similarly, the actual mass is more than ten times the minimum mass for cosines smaller than 0.1, which means angles between about 84.3° and 90°; these angles occur randomly about 6% of the time.

Doppler technique to find them with current technology, while planets orbiting far from their stars have such long orbital periods that it might take decades of observations to detect them. Thus, the current lack of evidence for Earth-mass planets or jovian-mass planets in distant orbits does not necessarily mean that such planets are rare.

Transits A third indirect way to detect distant planets does not require the observation of the effects of gravitational tugs at all. Instead, it relies on searching for slight changes in a star's brightness caused by a planet passing in front of it. If we were to examine a large sample of stars with planets, a small number of them—typically one in several hundred—will by chance be aligned in such a way that one or more of its planets will pass directly between us and the star once each orbit. The result is a **transit**, in which the planet appears to move across the face of the star. We occasionally witness this effect in our own solar system when Mercury or Venus crosses in front of the Sun. The result, seen through a telescope, is a small black dot transiting the Sun's bright disk (Figure 11.13). This effect is impossible for us to photograph for other stars, which appear in our telescopes only as points of light. However, as the planet crosses the star's face, it blocks a little of the star's light. The resulting tiny decrease in the star's brightness may be measurable.

Figure 11.13

On June 8, 2004, a transit of Venus was visible throughout much of Europe and Asia. In this photo, Venus appears as a black dot. It blocked out approximately 0.1% of the Sun's light for about 6 hours as it slowly transited across our home star. In principle, we should be able to detect such drops in brightness when extrasolar planets cross in front of their stars. Note that the amount of sunlight blocked out is about ten times as much as would be observed from a great distance, since—being so close to us—Venus looks relatively large. (Think of how the Moon can completely block the Sun as seen from Earth, but would be a very tiny dot against the Sun if viewed from much farther away.)

 THINK ABOUT IT What kind of planet is most likely to cause a transit across its star that we could observe from Earth: (a) a large planet close to its star; (b) a large planet far from its star; (c) a small planet close to its star; or (d) a small planet far from its star? Explain.

COSMIC CALCULATIONS 11.1

Finding Masses of Extrasolar Planets

An object's momentum is defined as its mass m times its velocity v; like angular momentum [Section 3.4], momentum must be conserved. Consider a star with a single planet. Because the center of mass remains stationary between them (see Figure 11.7), the system has no momentum relative to this center of mass. The star's momentum ($m_{star} \times v_{star}$) must therefore be equal in magnitude (but opposite in direction) to the planet's momentum ($m_{planet} \times v_{planet}$):

$$m_{star}v_{star} = m_{planet}v_{planet}$$

Solving, the planet's mass is:

$$m_{planet} = \frac{m_{star}v_{star}}{v_{planet}}$$

We generally know the star's mass from its spectral type. The Doppler technique can tell us the star's velocity (v_{star}) and the planet's orbital period p; we can use the latter to find the planet's velocity (v_{planet}) *if* we already know its average orbital distance a_{planet}. Fortunately, we can generally calculate a with Newton's version of Kepler's third law (see Cosmic Calculations 7.1). Once we know a_{planet}, we find v_{planet} by realizing that each orbit of the planet represents a distance of $2\pi a$ covered in a time p (and remembering that velocity is distance divided by time):

$$v_{planet} = \frac{2\pi a_{planet}}{p_{planet}}$$

We now substitute this expression into the mass equation from above:

$$m_{planet} = \frac{m_{star}v_{star}p_{planet}}{2\pi a_{planet}}$$

If we use velocity data from the Doppler technique, this formula tells us the *minimum* mass of the planet.

Example: Estimate the mass of the planet orbiting 51 Pegasi, which has an orbital period of 4.23 days ($p = 3.65 \times 10^5$ s) and an average orbital distance $a = 7.82 \times 10^9$ m.

Solution: The data in Figure 11.10b show the star's orbital velocity to be 57 m/s. Plugging in the values to solve for the planet's minimum mass, we get:

$$m_{planet} = \frac{m_{star}v_{star}p_{planet}}{2\pi a_{planet}}$$

$$= \frac{(2.12 \times 10^{30} \text{ kg}) \times \left(57\frac{m}{s}\right) \times (3.65 \times 10^5 \text{ s})}{2\pi \times (7.82 \times 10^9 \text{ m})}$$

$$\approx 9 \times 10^{26} \text{ kg}$$

This is about half of Jupiter's mass (which is 1.9×10^{27} kg).

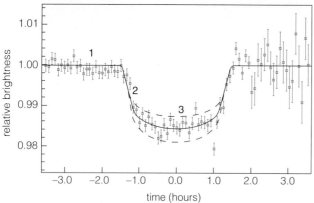

Figure 11.14

Careful measurements of the brightness of the star called HD209548 revealed that an orbiting jovian planet passes directly in front of it as seen from Earth; the numbers 1–3 on the art (top) correspond to the indicated positions on the graph.

Detecting planets through transits requires many repeated observations, because most stars exhibit intrinsic variations in luminosity. To be confident that a dip in brightness is caused by an orbiting planet rather than by variability in the star itself, we need to see the same dimming repeated multiple times. If the repeated dimming occurs with a regular period, then it is likely telling us the orbital period of a transiting planet. We can then calculate the planet's orbital distance and mass, and the amount of the drop in the star's brightness allows us to calculate the planet's radius. For example, observations show that a planet transits across the face of the star HD209548 every $3\frac{1}{2}$ days, causing a 1.7% drop in the star's brightness for about 2 hours (Figure 11.14). This tells us that the area of the planet's visible disk must be 1.7% of the area of the star's visible disk; if you remember that the area of a circle is πr^2, you'll realize that the planet's radius must be $\sqrt{0.017} = 0.13$, or 13%, of the star's radius. These data have told us that the planet has a mass, radius, and density consistent with what we expect if it really is jovian in nature. More detailed studies of how the starlight changes both during the transit and when the planet passes behind the star (called an *eclipse*) have even allowed astronomers to learn a little about the planet's atmosphere.

For an Earth-size planet, the dimming caused by a transit would be less than about 0.01% of a star's normal brightness, an amount that is difficult to measure with ground-based telescopes affected by atmospheric turbulence. Space-based telescopes, however, should be able to uncover such subtle dimming. NASA already has a mission in development, called *Kepler,* designed to look for transits caused by Earth-size (or even smaller) planets through careful monitoring of the brightness of 100,000 stars during the course of 4 years. After its launch (planned for 2008), *Kepler* should tell us whether or not solar systems in which small planets exist are plentiful. It therefore should give us our first good inventory of extra-solar planetary systems and of Earth-size planets.

Gravitational Lensing and Other Indirect Techniques

All of the planet discovery techniques we've discussed so far favor the detection of planets that orbit close to their star. Ground-based searches are also most sensitive to large planets. However, a technique known as **gravitational lensing** is able to detect even Earth-size planets in large orbits, and can find them despite the fact that they're in star systems tens of thousands of light-years away.

Gravitational lensing makes use of one of the most important predictions of Einstein's general theory of relativity: Mass distorts space. The consequence is that light rays bend slightly when passing near any large mass. The light rays are traveling in straight lines, but because matter warps space, a straight line appears bent. The effect is noticeable only in the presence of strong gravitational fields, so you'd have a hard time measuring the additional bending of a flashlight beam as it, say, grazed the rooftop of your house. But the Sun has sufficient gravity to bend starlight that passes near it, an effect first measured during a solar eclipse only a few years after Einstein first published his theory. In essence, the Sun, and for that matter, any star, acts like a lens that both magnifies and distorts the light from any object that may be behind it.

Two international teams of astronomers are using this subtle effect to search for planets in the star-choked inner regions of the Milky Way. They do so by monitoring the brightness of approximately 100 million stars located in the rich star fields near the galaxy's central regions. Because they use telescopes around the globe, they can monitor these stars continuously. About twice a day, on average, one star will, by chance, slowly pass in front of another. The two stars are not likely to be close to one another in space; they simply line up by chance for a few weeks because of the motions of stars and the rotation of the galaxy. When these alignments occur, the nearer star acts as a lens to increase the brightness (usually by a factor of two or three) of the star farther away. If the nearer star also has a planet, there's a chance that it will be suitably aligned to act as a lens as well, producing a secondary brightening of the lensed background star.

As of early 2006, three extrasolar planets had been found using this method. While that's only a tiny fraction of the known extrasolar worlds, the gravitational lensing scheme broke the record for the least-massive planet detected around a normal star—finding one that is only five times the mass of Earth (see Figure 1.2).

These successes have inspired astronomers to seek other ways of searching for planets indirectly. For example, another strategy is to search not for the planets or their effect on their host stars, but for visible disturbances of the large (and therefore, somewhat easier to see) disks of dust in which they form. The disk around the star Beta Pictoris, for example, shows ripples and gaps that are the clear fingerprints of planets within the disk. Other astronomers are searching for the thermal emission from the impacts of accreting planetesimals, and even for the kinds of emission known to come from the magnetospheres of jovian planets. As we learn more about extrasolar planets, new search methods are sure to arise.

Direct Detection   While upcoming missions like *Kepler* or the *Space Interferometry Mission* (*SIM*) may find a few Earth-*size* planets by using indirect techniques, they still won't tell us whether these planets are Earth-*like* in terms of habitability. For that we need to study the planets directly.

To see why direct detection of extrasolar planets is so difficult, recall the 1-to-10-billion scale model of the solar system that we discussed in Chapter 3 (see Figure 3.3). Remember that on this scale, the Sun is the size of a grapefruit, Earth is a pinhead orbiting 15 meters away, and Jupiter is a marble orbiting 80 meters away—while the distance to the nearest stars is equivalent to the distance across the United States. Thus, seeing an Earth-like planet orbiting the nearest star besides the Sun would be like looking from San Francisco for a pinhead orbiting just 15 meters from a grapefruit in Washington, D.C., and seeing a Jupiter-like planet wouldn't be much easier. The task is further complicated by the fact that a Sun-like star would be a *billion* times as bright as the light reflected from any of its planets. Because even the best telescopes blur the light from stars at least a little, the glare of scattered starlight tends to overwhelm the small blips of planetary light.

Because of these difficulties, only three planets so far have actually been photographed by direct imaging (Figure 11.5 shows one of them); that is, by finding them in photographs made with big telescopes. All these planets are large (many times Jupiter's mass), and all are at great distances from their home star (more than the distance from the Sun to Pluto), making it easier to spot them. But it's presently impossible to find planets similar to those of our own solar system with these straightforward, direct techniques.

One way to improve on this discouraging situation is to observe at infrared wavelengths. Planets are too cool to emit any visible light of their own, so they shine only by the light they reflect from their stars, but they emit reasonable amounts of infrared light. They are still far dimmer than

Figure 11.15

This artist's rendering shows one possible configuration for the second of NASA's planned *Terrestrial Planet Finder* (*TPF*) telescopes. It consists of five separate infrared telescopes flying in formation so that they can work together as an interferometer.

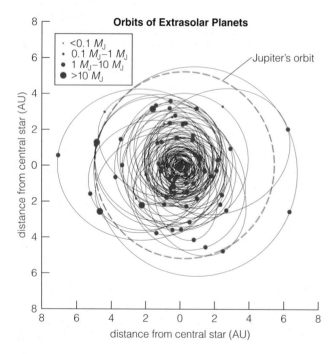

Figure 11.16

This diagram shows the orbits of the first 146 known extrasolar planets superimposed on each other, as if all the planets orbited a single star. The sizes of the dots indicate the minimum masses of the planets; the dots are located at the aphelion (farthest) point for each orbit.

their stars, but while the Sun is a billion times brighter than Jupiter in visible light, it is only about 100,000 times brighter in the (far) infrared. This substantial improvement makes direct detection much more practical with infrared light than with visible light, and the planets photographed to date have all been imaged in the infrared. In addition, scientists are developing new technologies such as *nulling interferometry*, designed to cancel (or "null") the light from a star so that we can more easily see its orbiting planets. NASA plans to test nulling technology on the *Space Interferometry Mission* (*SIM*), which may thereby be able to make crude images of a few Jupiter-size planets, though probably not of Earth-size planets.

NASA's planned follow-up mission to *SIM*, the *Terrestrial Planet Finder* (*TPF*), is tentatively envisioned as a double-barreled approach to directly imaging extrasolar planets, consisting of two, completely separate space-based telescopes. The first of these would be a more or less standard telescope, but one that is outfitted with a *coronagraph*—a small disk within the telescope that can block the light from a distant star so that we can see surrounding planets. Years after the coronagraph telescope is in orbit, *TPF*'s second mission would leave the launchpad; as currently envisioned, it would have five infrared telescopes flying in tightly controlled formation as an interferometer (Figure 11.15), using nulling interferometry to cancel out light from the star so that we can see planets. The European Space Agency (ESA) is considering a similar mission, called *Darwin*, on roughly the same time scale, and NASA and ESA are discussing possible ways of combining the two missions, perhaps with other international partners as well. In principle, either *TPF* or *Darwin* should have the power to conduct a conclusive search for Earth-size planets around 150 nearby, Sun-like stars. If the planets are there, these missions should see them. The images will be very low resolution—perhaps no more than three or four dots (pixels) for a typical planet—but the collected light should be sufficient to allow us to obtain reasonable spectra of the planets. As we'll discuss shortly, spectroscopy may allow us to determine whether a planet is habitable, and perhaps even whether it has life.

Later in this century, astronomers hope for even more powerful interferometers either in space or on the surface of the Moon. It's certainly possible that, within the lifetimes of today's college students, we'll have optical interferometers with dozens of telescopes spread across hundreds of kilometers. Such telescopes will not only be able to detect Earth-like planets around nearby stars but will also be able to obtain fairly clear images of those planets. We might see whether they have continents and oceans like Earth and perhaps be able to watch seasonal changes. Just a few centuries ago, we knew so little of the heavens that our ancestors could still imagine planets simply as "wandering stars." A mere decade ago, we had good images of the planets in our solar system but still had no conclusive evidence regarding the existence of planets elsewhere in the universe. Before the current century is out, we may have arguably better knowledge of planets around distant stars than the ancient Greeks had of our own Earth.

• What are the properties of extrasolar planets discovered so far?

Given our discussion of planet detection techniques, you probably won't be surprised by the fact that most of the planets discovered to date around hydrogen-burning stars are far larger than Earth. However, scientists have been surprised by the orbital characteristics of many of these planets.

Hot Jupiters Figure 11.16 shows the orbits of the first 146 known extrasolar planets all superimposed on each other, as if they were all orbiting a single star; the dots indicate the minimum masses of the planets found through the Doppler technique. Figure 11.17 compares the eccentricities (a technical measure of how much the orbit differs from a circle) of the planetary orbits to their distances, with the giant planets of our own solar system shown as well. If you study these figures carefully, you'll notice the following important facts:

- While most of the discovered planets have masses similar to or greater than that of Saturn or Jupiter, many orbit quite close to their central stars—some are closer to their star than Mercury is to the Sun.

- Most of the planets that are not close-in to their stars are in highly elliptical orbits, quite unlike the nearly circular paths followed by most of the planets of our own solar system. This means that the planets are much closer to their stars at some times than at others.

Recall that our theory of solar system formation suggests that massive jovian planets should form only in the cooler, more distant regions of the swirling disk around a young star (see Figure 3.22). In the warm inner regions of the disk, we expect to find rocky, terrestrial worlds. Thus, the presence of massive planets so close to their stars came as a big surprise to astronomers.

Could it be that these heavyweights are large rocky worlds? It's difficult to imagine how terrestrial planets could become so big, though we can't absolutely rule this out in all cases. However, in several cases we now have data from transits demonstrating that the planets are jovian rather than terrestrial in density, and we have every reason to suspect that the same is true for the other known massive extrasolar planets. Because the close orbits of some of the planets make them much higher in temperature than Jupiter, these planets have been nicknamed "hot Jupiters" (Figure 11.18).

Could "hot Jupiters" be habitable? It seems extremely unlikely. Their atmospheres would probably have the same strong vertical turbulence that led us to decide that Jupiter probably could not give rise to life [Section 7.3], and their proximities to their stars and elliptical orbits might only exacerbate their problems. Nevertheless, it's possible that these planets have numerous large moons, as do the jovian planets in our solar system. Some of these moons could be habitable, especially around the extrasolar planets that orbit within the habitable zones of their stars.

Implications for Solar System Formation Theory

If our theory of solar system formation tells us that jovian planets can form only in the outer regions of the disks that surround newborn stars, then how can we explain the existence of "hot Jupiters" or the eccentric orbits followed by so many of these worlds? After all, the orbits in any planetary disk should tend to become circular due to an "averaging out" of the numerous collisions involved in planetary accretion.

One possibility that scientists must always consider is that something is fundamentally wrong with our model of solar system formation. That is, perhaps jovian planets *can* form in the inner solar system. Upon discovering the "hot Jupiters," scientists indeed began to consider this possibility. However, they failed to find any fundamental flaws in our basic understanding of how planets form. So they considered other ideas and now suspect that the unusual orbits followed by most of the known extrasolar

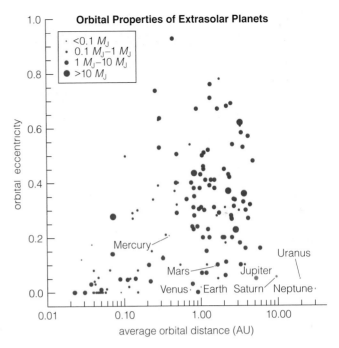

Orbital Properties of Extrasolar Planets

Figure 11.17

This graph shows the orbital radii (average distance from star) and eccentricities for the first 146 known extrasolar planets. Each red dot represents an extrasolar planet, and the green dots represent planets of our own solar system. Planets that orbit closer to their stars are farther to the left on this diagram, and the planets with more circular orbits are lower down. Notice that many extrasolar planets orbit quite close to their stars and most have larger orbital eccentricities than the planets of our solar system. (Eccentricity is a measure of how much an orbit differs from a circle; it is 0 for a perfect circle and approaches 1 for highly elliptical orbits.)

Figure 11.18

Artist's impression of the massive planet orbiting the star 51 Pegasi. At an orbital radius eight times closer to its star than Mercury is to the Sun, the temperatures on this planet are expected to be 1,000°C or greater. Planets like this are called "hot Jupiters."

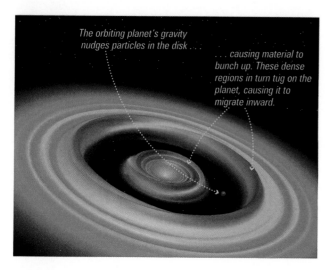

The orbiting planet's gravity nudges particles in the disk . . .

. . . causing material to bunch up. These dense regions in turn tug on the planet, causing it to migrate inward.

Figure 11.19

This figure shows a simulation of waves created by a planet embedded in a dusty disk of material surrounding its star; these waves may cause the planet to migrate inward.

planets are the result of changes that occurred in these systems after the planets first formed. That is, these planets most likely did form in the outer regions of their solar systems, and only later migrated inward or were deflected into eccentric orbits.

How might planetary migration occur? You might think that drag within the solar nebula could cause planets to migrate, much as atmospheric drag can cause satellites in low-Earth orbit to lose orbital energy and eventually plunge into the atmosphere; however, calculations show this drag effect to be negligible. A more likely scenario is that waves propagating through a gaseous disk lead to migration (Figure 11.19). A planet's gravity and motion tend to disturb the otherwise smooth disk material, generating waves that propagate through the disk. The waves cause dust to bunch up into ridges as they pass by, and these ridges exert their own gravitational pull on the planet, robbing it of energy and causing it to move inward.

Computer simulations of this rather subtle interplay of an orbiting planet and a disk of dust confirm that waves in the nebula can cause young planets to spiral slowly toward their star. In our own solar system, this migration is not thought to have played a significant role, probably because the solar wind cleared out the gas before it could have much effect. But planets may form earlier in at least some other solar systems, allowing time for jovian planets to migrate substantially inward. In some cases, the planets may form so early that they end up spiraling into their stars. Indeed, astronomers have observed some stars that have an unusual assortment of elements in their outer layers, suggesting that they may have swallowed planets.

Another possible way to account for some of the observed extrasolar planets invokes close encounters between young jovian planets in the outer regions of a disk. Such an encounter might kick one planet out of the star system entirely while sending the other inward in a highly elliptical orbit. Other possibilities include the idea that a jovian planet could migrate inward as a result of multiple close encounters with much smaller planetesimals (there is evidence that the jovian planets in our solar system have migrated a little bit by this mechanism). Orbital resonances like those that make the orbits of Jupiter's Galilean moons elliptical (see Figure 9.11) might also affect planets. Several of the multi-planet systems already discovered have orbital resonances between their planets, lending support to the idea that these resonances might play an important role in the architecture of many planetary systems.

The bottom line is that our discoveries of extrasolar planets have shown us that our prior theory of solar system formation was incomplete. It explained the formation of planets well, and also the simple layout of a solar system like ours. However, it needs modification in order to explain the differing layouts of other solar systems. While we are not yet sure of the precise mechanism, it now seems clear that planets can move from their birthplaces, making possible a much wider range of solar system arrangements than we had guessed before the discovery of extrasolar planets.

THINK ABOUT IT Look back at the discussion of the nature of science in Section 2.3, especially the definition of a scientific theory. Why does our theory of solar system formation qualify as a scientific theory even though we know that it needs modification to account for migration? Does this mean that the theory was "wrong" as we understood it before? Explain.

Implications for Habitable Worlds   Planetary migration may have important implications for the question of whether solar systems of the type we have recently found can harbor habitable planets. As we have discussed, it's unlikely that the large jovian planets could themselves support life. But could there also be smaller and potentially habitable terrestrial planets in these systems? Observationally we cannot say, because our current telescopes aren't able to detect Earth-size worlds. However, theoretical work suggests that even if terrestrial-size planets were born in these systems, they could face severe problems, because the migration of a large planet might disrupt the inner solar system.

If the migration occurs before terrestrial planets have finished forming, the material that would have accreted onto the terrestrial worlds might be swallowed instead by the larger world. Even if the formation process is essentially complete, gravitational encounters between big planets and small ones nearly always send the smaller ones scattering. As a large planet migrated inward, it would tend to fling less massive planets inward toward its star or outward to interstellar space. This suggests that terrestrial planets would not stay within the habitable zones of stars that also have "hot Jupiters" or other massive planets in highly elliptical orbits. Indeed, these same gravitational encounters might also disrupt and disperse any potentially habitable moons.

However, some recent research has called this discouraging scenario into question, and suggests that, even if giant planets spiral inward through a field of Earth-size worlds, not all of these smaller bodies would be scattered hither and yon. And in any case, the prospects look quite good for habitable planets around stars where planetary migration does not occur. Without migration, we expect rocky planets to form and to remain on stable orbits, and many of these orbits should be within the habitable zones of their stars.

In addition, even though the evidence so far might seem to point to the systems with substantial migration as being common, there's an important selection effect we must keep in mind. Remember that nearly all of the planets discovered have been found with the Doppler technique and that this technique works best for massive planets in close orbits. Therefore, it is possible that the reason we are finding so many systems with "hot Jupiters" is that they are the easiest ones to find with our current technology, not necessarily because they are the most common. Of the several hundred Sun-like stars that have been examined so far, approximately one in ten shows evidence of one planet or more—which means that we have not detected planets around the vast majority of Sun-like stars. In many cases, the reason may be that these stars have planetary systems more like our own than like the ones with "hot Jupiters," making them undetectable to our present technology. As the number of known extrasolar planets continues to grow, we will gain more insight about whether our solar system's layout is typical. But until we are able to make searches capable of finding small planets with upcoming missions like *Kepler, SIM,* and *TPF,* the question remains wide open.

• How could we detect life on extrasolar planets?

If Earth-size planets exist around other stars, we should begin to find them within the next several years and get crude images and spectra within the next several decades, thanks to improved technology and new space missions. Let's explore how these capabilities might allow us

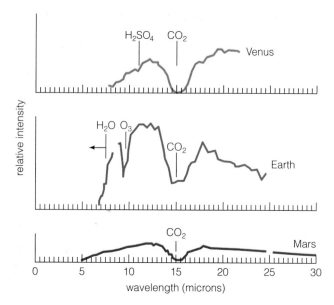

Figure 11.20

The infrared spectra of Venus, Earth, and Mars, as they might be seen from afar, showing absorption features that point to the presence of various gases in their atmospheres. While carbon dioxide is present in all three spectra, only our own planet has appreciable oxygen (and hence ozone)—a product of photosynthesis. If we could make similar spectral analyses of distant planets, we might possibly detect atmospheric gases that would indicate life.

to determine whether the planets are habitable, and perhaps whether they have life.

Suppose a telescope actually shows us a crude image—only a few pixels—of a distant world. What might we learn? To begin with, by simply watching its changing brightness, we could gain some information about the ratio of ocean to land, because seas are darker than continents. We might also find that the light from such a world changes from day to day, due to clouds, or from season to season, because of snow or ice.

A spectrum would allow us to measure many other properties. For example, even a fairly crude spectrum should allow us to gauge the surface temperature of the planet. If we can collect even more light from a distant planet, we could make a more detailed spectral analysis, one that could suggest far more convincingly that a planet is home to life. For instance, we might gain information on the types of minerals or ices on its surface. If we find that the planet has characteristics like those of Earth or Mars, we would know that it is habitable in principle though perhaps not whether it actually harbors life.

With spectra from infrared telescopes (such as *TPF* or *Darwin*), we could search for the absorption or emission features of gases in the atmosphere of a planet. Several of these gases will be easy to detect if they are present, including carbon dioxide, ozone, methane, and water vapor (Figure 11.20). While the mere presence of such gases would not necessarily point to life, their precise abundances and the combinations in which they occur could provide stronger evidence of whether life is present.

For example, Earth's atmosphere has large amounts of oxygen, the result of photosynthesis. If we found abundant oxygen in the air of another world, we would have reason to suspect the presence of life, particularly if the ratio of oxygen to the other detected gases seemed incompatible with nonbiological chemistry. To some extent, the same is true of methane, which is present in Earth's atmosphere today largely thanks to the "exhaust" gases produced by livestock and rice paddies. In early times, before photosynthesis raised the oxygen level, the atmosphere may have been altered by another metabolic process, known as *methanogenesis*, through which microbes expel methane rather than oxygen. The first billion years or so of Earth's biological history might therefore have been marked by the presence of atmospheric methane.

The bottom line is that for billions of years, life on Earth has been making its presence known to anyone with a telescope large enough to find our world and make a spectrum of its reflected light. In principle, we could identify life on other worlds in the same way. Perhaps in the next few decades we will discover abundant oxygen or methane on a distant world visible to us as no more than a dot in a telescope, providing an exciting and encouraging clue that it harbors life.

11.3 THE POSSIBILITY THAT EARTH IS RARE

We have discussed the current status of our search for extrasolar planets, along with the question of whether potentially habitable planets should be common or rare. We have found that our current knowledge is still quite incomplete. It may be that a large fraction of stars have habitable planets, or it may be that the fraction of stars with such planets is fairly low. Nevertheless, the example of our own solar system tells us that the fraction is greater than zero. Given the enormous number of stars in the galaxy, it seems highly likely that, one way or another, there are large numbers of

Earth-size planets within the habitable zones of stars. But does Earth-*size* necessarily mean Earth-*like* in terms of habitability?

• Are Earth-like planets rare or common?

Most scientists suspect that a significant percentage of Earth-size planets are likely to be Earth-like as well. However, some scientists have questioned this idea, suggesting that Earth may be the fortunate beneficiary of several kinds of planetary luck. According to this idea, sometimes called the "rare Earth hypothesis," the specific circumstances that have made it possible for evolution to progress beyond microbes to complex creatures (such as oak trees or humans) might be so rare that ours might be the only inhabited planet in a galaxy that harbors anything but the simplest life. This suggestion would have profound implications if true, particularly for the efforts to search for extraterrestrial intelligence that we will discuss in the next chapter. Let's briefly examine some of the key issues in the rare Earth hypothesis.

Galactic Constraints   Proponents of the rare Earth hypothesis suggest that Earth-like planets can form in only a relatively small region of the Milky Way Galaxy, making the number of potential homes for life far smaller than we might otherwise expect it to be. In essence, they argue that there is a relatively narrow ring at about our solar system's distance from the center of the Milky Way Galaxy that makes up a *galactic habitable zone* analogous to the habitable zone around an individual star (Figure 11.21).

According to the arguments for a galactic habitable zone, outer regions of our galaxy are unlikely to have terrestrial planets because of a low abundance of elements besides hydrogen and helium. Recall that the fraction of heavy elements varies among different stars, from less than 0.1% among the old stars in globular clusters to more than 2% among young stars in the galactic disk [Section 3.2]. Even within the galactic disk, the abundance tends to decline with distance from the center of the galaxy. Because terrestrial planets are made almost entirely of heavy elements, a lower abundance of these elements might lessen the chance that terrestrial planets could form. The inner regions of the galaxy are ruled out primarily through an argument concerning supernova rates. Supernovae are more common in the more crowded, inner regions of the galactic disk, making it more likely that a terrestrial planet would be exposed to the intense radiation from a nearby supernova. By assuming that this radiation would be detrimental to life, the proponents of a galactic habitable zone argue against finding habitable planets in the inner regions of the galaxy. Together, the constraints on finding Earth-like planets in the inner and outer regions of the galaxy leave the galactic habitable zone as a relatively narrow ring encompassing no more than about 10% of the stars in the galactic disk.

However, other scientists offer counterarguments to both sets of galactic constraints. For the heavy-element abundance, they note that Earth's mass is less than $\frac{1}{100,000}$ of the mass of the Sun. Thus, even a very small heavy-element abundance could be enough to make one or more Earth-like planets. Unless there is something about the planetary accretion process that prevents terrestrial planets from forming in systems with low heavy-element abundances, then we might find terrestrial planets around almost any star. Regarding the radiation danger from supernovae, we do not really know whether such radiation would be detrimental to

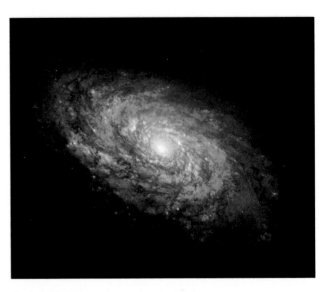

Figure 11.21

The green ring in this diagram of the Milky Way Galaxy highlights what some scientists hypothesize to be a galactic habitable zone—the only region of the galaxy in which Earth-like planets can form. However, other scientists doubt the claims that underlie this hypothesis in which case Earth-like planets could be far more widespread.

life. A planet's atmosphere might protect life against the effects of the radiation. It is even possible that the radiation could be beneficial to life by increasing the rate of mutations and thereby accelerating the pace of evolution. If these counterarguments are correct, then Earth-like planets might be found throughout much or all of the galaxy.

Impact Rates and Jupiter   Another issue raised by rare Earth proponents concerns the rates of impacts by asteroids and comets on planets in other solar systems. Recall that Earth was probably subjected to numerous large impacts—some possibly large enough to vaporize the oceans and sterilize the planet—during the first few hundred million years after our planet was born [Section 4.3]. In our solar system, the impact rate lessened dramatically after that. Might the impact rate remain high much longer in other solar systems?

Although the scars of impact craters on numerous planets and moons bear witness to huge numbers of impacts in our solar system's history, the number of *potential* impacts is far higher. Thousands of asteroids still roam the region between Mars and Jupiter. Of even greater significance, by studying the orbits of comets that enter the inner solar system, astronomers have concluded that as many as a trillion comets must orbit the Sun at distances far beyond the orbit of Pluto, making up the *Oort cloud* [Section 3.3]. Fortunately for us, these myriad objects are essentially out of reach, posing no threat to our planet. However, we have good reason to believe that their great distances can be traced directly to gravitational effects of Jupiter.

The Oort cloud comets are thought to be the survivors among an even greater number of comets that formed originally in the region of the solar system where the jovian planets were born. During the early history of our solar system, the vast majority of these comets would have experienced close encounters with a jovian planet—most of them with Jupiter.

MOVIE MADNESS: *STAR WARS*

It's "a long time ago" in someone's far-off galaxy, and the political situation is turning ugly.

The premise of the original *Star Wars,* a cinematic space opera that, like relatives, just keeps on returning (with six installments to date), is that a galaxy-wide republic has been hijacked and converted to an autocratic, evil empire by despots in gray flannel suits. This may sound vaguely reminiscent of the story of Rome, but unlike what happened to that ancient civilization, this political shift has encouraged a serious rebellion, a war among the stars. The rebels are led by Princess Leia (you can tell she's a princess because her hair is done up like twin Danish pastries), and her strategy is to take out the empire's headquarters— an enormous, spherical spacecraft known to its friends as "the Death Star." The Death Star packs weaponry that can explode a planet in seconds (calculate, if you wish, the energy required to do that!). On the other hand, the rebels have "the force" on their side—a mystical ability to change the odds of every situation based on moral merit and self-discipline.

Most of *Star Wars* is battle of the sort that's familiar to any movie fan, except that the bad guys wear brittle, white plastic suits and fly spacecraft that look like box kites. But *Star Wars* offers some interesting peeks into life as it might be elsewhere in the cosmos. The rebels have their base of operations on a large planet's moon, a not-impossible scenario since hefty moons could be habitable. Luke Skywalker, the young hero, hails from a world circling a close double star. Not a problem—research has shown that planetary orbits around double stars could be stable.

There are peculiar anachronisms in *Star Wars,* however. The Death Star is obviously extremely high-tech, and yet the principals occasionally face off using souped-up swords. Everyone jets around in spacecraft that are somehow capable of exceeding speed-of-light travel by jumping into hyperspace, and yet we often see aliens saddled up onto giant, dinosaurlike creatures—much as in *The Flintstones.*

All of that can be forgiven. But *Star Wars* takes its biggest literary license in the fact that dozens of alien races are all living contemporaneously (although it's clear that the human types are in charge). In the movie's famous cantina scene, which takes place in the wretched port city of Mos Eisley, aliens of all shapes and colors get together to do business and get drunk. In fact, the chance that any two intelligent species (let alone dozens) appear on the galactic scene within 100,000 years of one another is quite small. If there are other societies out there, they will either be far behind us or enormously beyond our level. We won't be sharing dance music and booze with them in a seedy extraterrestrial dive.

And besides that, why does a republic have a princess, anyhow?

Recall that a close encounter between a big planet and a small object like a comet will tend to send the small object flying off in a new direction. Many comets must have ultimately crashed into the Sun or into the planets of the inner solar system. The rest were flung out to the great distances at which they now reside or out of the solar system entirely.

Thus, if Jupiter did not exist, the comets might have remained in the part of the solar system where they could pose a danger to Earth. In that case, the heavy bombardment might never have ended, and huge impacts would continue to this day. From this viewpoint, our existence on Earth has been possible only because of the "luck" of having Jupiter as a planetary neighbor.

The primary question is just how "lucky" this situation might be. Our discoveries of extrasolar planets show that Jupiter-size planets are in fact quite common, though if most of them migrate inward, then giant planets in outer orbits might be rare. Until we learn much more about the layout of other planetary systems, we have no reason to conclude that we are particularly "lucky" to have Jupiter as a planetary companion.

THINK ABOUT IT The story of life on Earth is replete with disasters that served to stress terrestrial species, resulting in the rapid evolution of new, more complex organisms. For example, the K–T impact [Section 6.4] apparently led to the demise of the dinosaurs and opened the door for the rise of mammals. More recently, ice ages are thought to have played a major role in the evolution of modern humans. Do you think it's possible that a higher rate of impacts could be good rather than bad for life on another planet? Explain.

Climate Stability Another issue affecting the rarity of Earth-like planets concerns climate stability. Recall that, in comparison to Venus and Mars, Earth has had a remarkably stable climate. This climate stability has almost certainly played a major role in allowing complex life to evolve on our planet. If our planet had frozen over like Mars or overheated like Venus, we would not be here today. Advocates of the rare Earth hypothesis point to at least two pieces of "luck" with regard to Earth's stable climate.

The first piece of luck concerns the existence of plate tectonics. As we discussed in Chapter 4, plate tectonics plays a major role in the carbon dioxide cycle and hence in regulating Earth's climate. The existence of this climate-regulating mechanism is especially important given that the Sun, like all stars, brightens as it ages. The Sun is about 30% more luminous now than when Earth formed. Yet, thanks to plate tectonics and the carbon dioxide cycle, Earth has remained habitable. Plate tectonics probably was not necessary to the origin of life, but it seems to have been quite important in keeping the climate stable long enough for the evolution of plants and animals. But are we really "lucky" to have plate tectonics, or are such processes inevitable on Earth-size planets in Earth-like orbits? As far as we are aware, there is nothing particularly unusual about Earth's size, composition, or orbit. Therefore, we have no reason to believe that planets with similar characteristics should be rare and no reason to think that geological processes should be any different in other cases. On the other hand, Venus is quite similar in size to Earth but apparently lacks plate tectonics. This may be the result of Venus's runaway greenhouse effect [Section 10.2], or it may be due to factors that we do not yet understand. Nevertheless, unless some unknown, special circumstance has encouraged long-lasting plate tectonics on Earth, it seems likely that such processes will be found on other, similar planets.

The second piece of luck regarding climate stability concerns the existence of the Moon. As we discussed in Chapter 8, Mars undergoes dramatic climate changes because the tilt of its axis varies over a significant range (see Figure 8.25). Earth's tilt varies much less, contributing to climate stability. Models of Earth's rotation and orbit show that if the Moon did not exist, gravitational tugs from other planets, especially Jupiter, would cause large swings in Earth's axis tilt over periods of tens to hundreds of thousands of years. In other words, if the Moon did not exist, Earth's spin axis would be subject to the same large swings in its tilt as occur on Mars. Given that the Moon likely formed as a result of a random, giant impact [Section 4.6], it might seem that we are "lucky" to have our Moon and the climate stability it brings.

Again, however, there are other ways to look at this issue. Changes in axis tilt might warm or cool different parts of the planet dramatically, but the changes would probably occur slowly enough for life to adapt or migrate as the climate changed. In addition, our Moon's presumed formation in a random giant impact does not necessarily mean that large moons will be rare. At least a few giant impacts should be expected in any planetary system. Indeed, Earth may not be the only object in our own solar system that ended up with a large moon because of a giant impact: Pluto's largest moon (Charon) may have formed in the same way (along with its two smaller moons). Collisions are also thought to be responsible for the many moons found around other Kuiper belt objects. Thus, while luck was certainly involved in Earth's having a large moon, it might not be a rare kind of luck.

Summarizing the Arguments   The bottom line is that while the rare Earth hypothesis offers some intriguing arguments, it is too early to say whether any of them will hold up over time. For each potential argument that Earth has been lucky, we've seen counterarguments suggesting otherwise. There's no doubt that our solar system and our world have "personality"—that is, they exhibit properties that might be found only occasionally in other star systems—but we have no clear reason to think that these special properties are essential to the existence of complex or even intelligent life. Indeed, it may be that we have missed out on some helpful phenomena that could have sped evolution on Earth. We might be less lucky than we recognize, and creatures on other worlds might regard the nature of our planet with disappointment. Until we learn much more about other planets in the universe, we cannot know whether Earth-like planets and complex life are common or rare.

11.4 THE PROCESS OF SCIENCE IN ACTION: CLASSIFYING STARS

Much of this book is devoted to studying individual planets, and for obvious reasons. Planets have personality: The appearance, composition, and size of any given world depend on where and how it was formed, and what happened to it after its birth. Just because two planets are the same size hardly means that they have the same characteristics (think of Earth and Venus).

But stars are much more like clones. You might think that this is simply because all stars (other than the Sun) are quite far away. Consequently, they appear only as featureless, bright pinpoints when seen through our telescopes. But in fact, stars don't just look similar; they really are similar. Discovery of that fact has helped us puzzle out one of astronomy's fundamental mysteries: What is the life history of a star? In this section, we'll discuss a tool that was developed nearly a century ago, the **Hertzsprung–Russell (H–R) diagram**, which is used to classify the various types of stars. This diagram is as fundamental to astronomy as the periodic table is to chemistry.

• How and why did the Hertzsprung–Russell diagram develop?

Classifying objects is a familiar tactic in science. Long before biologists knew anything about DNA or even that cells exist, they were busy setting up categories for the life they could observe (this is called *taxonomy*, which comes from the Greek *tassein*, meaning "to classify," and *nomos*, meaning "a science or study"). Separating animals into reptiles, amphibians, insects, mammals, marsupials, and so on, requires no more equipment than a good eye and a pencil, but doing so eventually gives insight into the evolutionary processes that produced animal life. Even if you don't yet understand the underlying reasons for an observed diversity of objects, classifying them is a good first step.

In the late nineteenth and early twentieth centuries, astronomers were engaged in their own taxonomy projects. Even a casual observer could notice that not all stars are the same brightness. Until methods were developed for measuring colors, apparent brightness was the only known parameter that differed from star to star (other than position in the sky). A star's brightness depends on: (1) its distance, (2) its intrinsic luminosity, and (3) the dimming effects of any interstellar dust that might lie between us and the star. Even ignoring this last effect, which for nearer stars is generally not important, astronomers still have to sort out the distance of a star if they want to derive its intrinsic luminosity from its apparent brightness. They are keen to do this because, while the latter is mere happenstance, the former is a physical property of the star. This problem is akin to determining the wattage of a light bulb seen at a distance. We can easily measure its apparent brightness, but we need to know how far away it is if we want to know its intrinsic luminosity, or wattage.

In addition to luminosity, there's a second property of stars that, by the end of the nineteenth century, could be quickly determined: color. Because the cones of our retinas do not work well in low light, looking at stars with the naked eye is a poor way to judge their hue. But any color photograph of the stars of the night sky will show a range from blue to red. As discussed in Section 11.1, the spectral properties of stars simply reflect the temperatures of their outer gas layers. So knowing a star's spectral type will tell you its surface temperature (for example, the yellow Sun, a type G star, has a surface temperature around 5,500°C, as listed in Table 11.1).

In the first years of the twentieth century, Danish astronomer Ejnar Hertzsprung, who had collected data on the distances of nearby stars, noticed something peculiar: If he arranged the stars by color, from blue-white to red (spectral types B through M), the arrangement was also a sequence in intrinsic luminosity. That is, the hottest (most blue) stars were more luminous than the cooler stars. However, if he included stars that were farther away, he found that the distant red stars were, on average, intrinsically much more luminous than the nearer ones. Since he realized that the more distant sample of red stars undoubtedly included intrinsically faint objects he couldn't see, this suggested to Hertzsprung that there might be two categories of red stars: small ones (the intrinsically fainter stars) and giants (the intrinsically more luminous stars).

Hertzsprung's results were surprising because we might expect that stars could have any combination of luminosity and color. Why were the red stars separating into distinct categories? Beginning in 1911, Hertzsprung began to expand this work by plotting the colors and luminosities of stars in two clusters, the Pleiades and the Hyades. Using clusters of stars saved him from having to measure their individual distances, because all the stars in a cluster are more or less at the same distance (just as the people in Chicago are all at approximately the same distance from people in New York). He soon found that the stars were, indeed, not scattered at random across his plot, but fell into two broad groups: (1) most were in a swath running diagonally through the diagram, which was later named the **main sequence**; and (2) nearly all of the remainder were in a part of the diagram he called the *giants*.

The American astronomer Henry Norris Russell, working independently, was making a similar analysis. However, rather than examining stars in clusters, he used stars whose distances were known from trigonometric measures. Of course, this meant that his diagram was based on relatively close stars, those for which geometric calculation could be used to determine distances. But the results were quite similar to Hertzsprung's. By the 1920s, Hertzsprung's and Russell's results were codified in the H–R diagram, a graphic found in every astronomy textbook ever since (Figure 11.22).

• What can we learn from the Hertzsprung–Russell diagram?

While one might have expected to find stars all over the H–R diagram, it turned out that they were confined to only a few regions. You might compare this to characterizing humans on the basis of two easily measured parameters: height and weight. If you actually take these measures for a large number of people, and plot them on a graph with height on one axis and weight on the other, you'll find that there's a pretty narrow range within which most of the data points fall. Very short, very heavy people are rare, as are very tall, very light folk. The clustering of the data along a diagonal swath relating height and weight is a consequence of the fact that, in a general way, all people are the same. The Hertzsprung–Russell diagram suggests that stars, too, might be generally similar.

First, let's describe the parts of a modern H–R diagram where large numbers of stars are found. Note that today's diagrams are based on data of a quantity and quality unavailable a century ago. The major occupied regions are the following:

- The main sequence, which is the S-shaped line that curves from top left to bottom right. Stable stars burning hydrogen in their cores—the overwhelming majority of all stars—fall on this line. Our Sun lies about midway along the main sequence, near the center of the diagram, which is the reason the Sun is often referred to as an "average star," even though it is actually more massive than the majority of stars (the K and M stars on the lower part of the main sequence).

- At the top right are the giants and their somewhat bigger brethren, the supergiants. These stars are mostly cool and red, but very luminous. In fact, giants and supergiants are stars that used to be on the main sequence but are now at the end stages of their lives, having already exhausted their core hydrogen (see the discussion of giants and supergiants in Section 11.1).

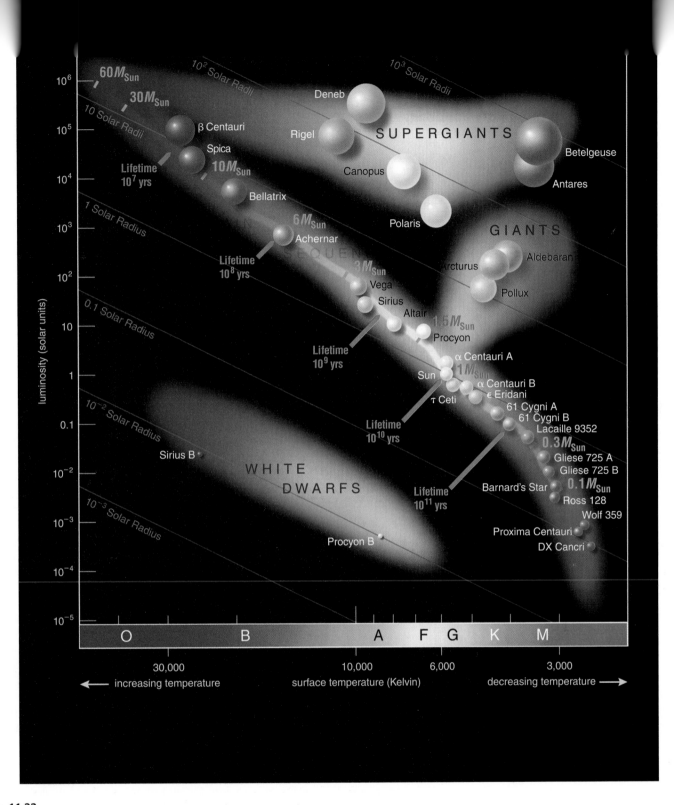

Figure 11.22

An H–R diagram, one of astronomy's most important tools, shows a star's surface temperature or spectral type along the horizontal axis and its luminosity along the vertical axis; the fact that surface temperature increases to the left (rather than to the right) is a consequence of the historical development of the spectral sequence OBAFGKM. Several of the brightest stars in the sky are plotted here, along with a few of those closest to Earth. They are not drawn to scale—the diagonal lines, labeled in solar radii, indicate how large they are compared to the Sun. The lifetime and mass labels apply only to main-sequence stars. (Star positions on this diagram are based on data from the Hipparcos satellite.)

- At the bottom left are the hot but dim white dwarfs. White dwarfs are the collapsed remains of stars like the Sun, stars that have exhausted their fuel and are slowly turning into ashen, stellar corpses.

The fact that more than nine out of ten stars are on the main sequence tells us that stars spend about 90% of their lives burning hydrogen in their cores, and their arrangement on this line tells us something about the character of stars during this long phase of their lives: The hotter they are, the more luminous they are (that is, the more energy they pump into space). This may seem obvious, and you may wonder why page space is being expended on this diagram. But it might have been true that the universe was crowded with lots of hot stars of relatively low luminosity (say, the luminosity of the Sun). A blue acetylene torch is hot, but is of low luminosity compared to a red bonfire. However, the existence of the main sequence suggests that all the stars populating it are basically built the same way, and Russell concluded that there was one variable, one parameter, that determined where on the main sequence a star would lie. That parameter turns out to be a star's mass. As we move from bottom right to top left on the main sequence, the stars become more massive (notice the mass labels in Figure 11.22). The main sequence is actually a *mass* sequence.

Think of how stars come about. Stars are formed by collapsing clouds of gas, and the composition of this gas is mostly hydrogen and helium. In this book we make a great fuss about the elements heavier than helium, because these are necessary for planets, plants, and people. But from the standpoint of making a star, which is destined to fuse hydrogen into helium in the main phase of its existence, these impurities are of small consequence. Stars form when clouds of gas collapse under their own weight, squeezing their inner cores to higher temperatures and pressures until nuclear fusion begins. This produces the heat that stops further collapse, and the resulting size and temperature are simply determined by the mass of the star itself. The greater the mass, the greater the pressure within the star. The increased pressure raises the density of matter at the core, which increases the rate at which the particles there bump into one another and fuse. So the rate at which fuel is burned—and therefore the star's luminosity—goes up as the overall mass does. In 1926, Russell and German astronomer Heinrich Vogt stated this explicitly, noting that "the evolution of stars is fixed at their birth by two inherent fundamental parameters, mass and chemical composition." This Russell–Vogt theorem is simply a reiteration of a fact that is evident from the existence of the main sequence: All hydrogen-burning stars are made from the same blueprint. They are all cousins differing only in their mass.

What about the giants and supergiants? Despite the fact that they are not on the main sequence, these luminous heavyweights do not violate the idea that mass is the most important property. They have simply changed their chemical composition by having lived long enough to burn up much of their central hydrogen fuel. They have moved on to a different "engine" for producing energy: fusing helium or even heavier elements. Paradoxically, the temporary energy boost provided by this switch in fuel causes them to swell up to the point where their surface temperatures are lowered (they become redder), even though their total luminosity has increased. From a graphical point of view, they've moved off the main sequence to take up a rather short residence in the giant or supergiant region of the H–R diagram. Eventually they exhaust these fuels and either collapse to become a white dwarf (at the bottom left of the diagram) or, if they began life as a star several times as massive as the Sun, explode in a supernova and leave behind either a neutron star or a black hole. There is no location on the H–R diagram for such pathological stellar remnants, so the heavier giants eventually, and rather suddenly, disappear from the chart.

The H–R diagram was originally just an organizational strategy, like classifying plants and animals. But we have seen that organizing leads to insights; indeed, the H–R diagram remains one of the most important tools of professional astronomers. The lesson for the process of science is clear: When you're not sure what's going on, start by organizing what you do know, and it may take you down the path to discovery.

THE BIG PICTURE
Putting Chapter 11 into Context

In this chapter, we have discussed the search for habitable planets beyond our own solar system. As you continue in your studies, keep the following "big picture" ideas in mind:

- Thanks to rapidly accumulating discoveries of planets around other stars, there's no longer any doubt that massive planets, at least, are common. However, we still do not know whether habitable planets are common or rare.

- Our discoveries of extrasolar planets have provided valuable new information about the processes involved in the birth of solar systems, forcing us to modify our existing theory to account for planetary migration, and possibly our ideas of how giant planets form.

- Although we do not yet have an answer regarding the existence of Earth-size or Earth-like planets around other stars, several planned missions will address this question. Within the next few years, we should know whether other Earth-size worlds exist. Within a few decades, we may have crude images or spectra that will tell us whether these worlds are also Earth-like.

**SUMMARY OF
KEY CONCEPTS**

11.1 DISTANT SUNS

- **How do stellar life cycles affect the possibility of habitable planets?**

The most basic requirement for a star to be able to support life on orbiting planets is that it be stable for long enough to allow that life to originate. This limits us to stars that are in the long-lived, hydrogen-burning phase of their lives. Giants and supergiants are stars that have used up the bulk of their core hydrogen fuel, and undergo substantial change on time scales of millions of years. Consequently, they are not thought to be suitable stars for hosting planets with life.

- **How do we categorize stars?**

The **spectral sequence OBAFGKM** runs from hot to cool in terms of the surface temperatures of stars. For stars in their hydrogen-burning phases, this sequence also runs in mass order. O stars are the hottest, most massive, most luminous, and shortest lived. The Sun is a G star, and the far end of the scale has the cool, dim, low-mass, and long-lived M stars.

• Which stars would make good suns?

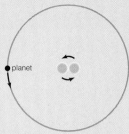

The lifetimes of O and B stars seem too short to permit life to arise on any surrounding planets. A and F stars may live long enough for life to evolve, though probably not long enough for intelligent life. G stars like the Sun clearly can have habitable planets. K and M stars have small habitable zones, but make up for that in their sheer numbers, since they are the most common types of stars. Many stars are members of multiple star systems, but they may still have habitable planets in stable orbits.

11.2 EXTRASOLAR PLANETS: DISCOVERIES AND IMPLICATIONS

• How do we detect planets around other stars?

There are two fundamental methods for detecting extrasolar planets: direct observation or measuring their indirect effects. To date, almost all discoveries have been indirect. We can look for a planet's gravitational effect on its star through the **astrometric technique**, which looks for small shifts in stellar position, or the **Doppler technique**, which looks for the back-and-forth motion of stars revealed by Doppler shifts. We can also search for **transits** in which a system becomes slightly dimmer as a planet passes in front of its star. A few planets have been found by gravitational lensing, an effect predicted by Einstein's general theory of relativity.

• What are the properties of extrasolar planets discovered so far?

The known extrasolar planets are all at least several times as massive as Earth, and most are Jupiter-like in mass. Many orbit surprisingly close to their stars—which is why we call them "hot Jupiters"—or have large orbital eccentricities. We have only limited information about sizes and compositions, but these data are consistent with the idea that the planets are jovian in nature.

• How could we detect life on extrasolar planets?

Future telescopes should allow us to obtain crude images or spectra of planets within stellar habitable zones. An image of an extrasolar planet—even if only a few pixels in size—might indicate the presence of snow or clouds, and would tell us the planet's rotation period. Spectroscopic analysis could tell us much more, and might reveal combinations of atmospheric gases, such as oxygen and methane, that would be good evidence for life.

11.3 THE POSSIBILITY THAT EARTH IS RARE

• Are Earth-like planets rare or common?

We don't know. Some of the key questions are whether our galaxy, like a star, has a relatively narrow habitable zone; whether the role Jupiter has played in lowering our solar system's impact rate is rare or critical to life; and whether Earth's relatively stable climate, due largely to plate tectonics and our large Moon, is rarely found on otherwise similar worlds. Arguments can be made on both sides of each question, and at present we lack the data to determine which side is correct.

11.4 THE PROCESS OF SCIENCE IN ACTION: CLASSIFYING STARS

• How and why did the Hertzsprung–Russell diagram develop?

By the beginning of the twentieth century, astronomers had methods for measuring both the intrinsic luminosities and colors (which are related to surface temperatures) of stars. Ejnar Hertzsprung in Denmark and Henry Norris Russell in the United States independently realized that by classifying stars by plotting these two easily observed quantities, the stars clustered into groupings. The graph based on these quantities is now called the **Hertzsprung–Russell diagram**.

• What can we learn from the Hertzsprung–Russell diagram?

In the H–R diagram, most stars fall along a continuous swath, called the **main sequence**, that runs from hot, luminous stars at the upper left to cool, dim stars at the lower right. Other stars—giants and supergiants—clump in the part of the diagram where stars are luminous but have cool surface temperatures. The stellar corpses known as white dwarfs are dim but hot, so they are found in the lower left of the diagram. Study of the H–R diagram helped astronomers realize that mass is a star's most fundamental property, and the organizational power of the diagram makes it one of astronomy's most powerful tools.

EXERCISES AND PROBLEMS

REVIEW QUESTIONS
Short-Answer Questions Based on the Reading

1. Briefly summarize the life cycles of stars from birth to death. At which stage of stellar life does it seem possible that life-bearing planets could exist? Why?

2. How was the *spectral sequence* OBAFGKM discovered? Based on Table 11.1, briefly summarize how the properties of hydrogen-burning stars change as you look down the sequence from O to M.

3. For the seven major spectral types OBAFGKM, discuss whether stars of each type are likely to have habitable planets, and explain why or why not. Could *brown dwarfs* be orbited by habitable worlds? Explain.

4. What are *multiple star systems?* What are *binary star systems?* Briefly discuss the prospects of finding habitable planets in multiple star systems.

5. Briefly explain why we are confident that planets are common, even while we do not yet know whether *habitable* planets are common.

6. Explain why a planet can cause its star to move slightly in the sky. What do we mean by the *center of mass* of a system?

7. Describe the two techniques—the *astrometric technique* and the *Doppler technique*—by which we can measure gravitational effects of planets on stars and contrast their advantages and limitations. What future technologies or missions should help us extend the power of these techniques?

8. What is a *transit?* How can we use transits to find extrasolar planets? Briefly describe the significance of transits for the upcoming *Kepler* mission.

9. Why does *gravitational lensing* occur, and how can it allow us to detect a planet?

10. Why is direct detection of extrasolar planets so difficult? Why do infrared observations help? What future technologies or missions should improve our ability to detect planets directly?

11. Briefly discuss the general properties of the known extrasolar planets. What evidence suggests they are jovian rather than terrestrial in nature? What do we mean by "hot Jupiters"?

12. Why are the "hot Jupiters" surprising? What have they taught us about the formation of planetary systems? Discuss the prospects of finding habitable worlds in systems with "hot Jupiters."

13. How might future images and spectroscopy allow us to determine whether distant planets are habitable or have life?

14. What is the *rare Earth hypothesis?* Briefly summarize the arguments used to advance it and the counterarguments against each of these.

15. What is the *Hertzsprung–Russell diagram?* How does a star in the upper left section differ from one in the lower right?

16. Briefly summarize the characteristics of stars in each of the three major regions of the H–R diagram—the *main sequence,* the giants and supergiants, and the white dwarfs.

TEST YOUR UNDERSTANDING
Would You Believe It?

Suppose that, on the date indicated, you saw the following headlines. (These are not real discoveries.) In each case, decide whether the headline is believable in light of what we currently know about extrasolar planets and our technological capabilities. Explain clearly; because not all of these have definitive answers, your explanation is more important than your chosen answer.

17. Date: February 16, 2009. Headline: Astronomers Conclude That Earth-Size Planets Don't Exist.

18. Date: January 9, 2010. Headline: Astronomers Discover Earth-Like World Orbiting Massive Star of Spectral Type B.

19. Date: June 19, 2008. Headline: Spectrum Reveals Unmistakable Evidence of Life on a "Hot Jupiter."

20. Date: November 7, 2010. Headline: New Images Show Oceans on Extrasolar Planet.

21. Date: November 7, 2050. Headline: New Images Show Oceans on Extrasolar Planet.

22. Date: July 20, 2007. Headline: Giant Planet Found in Our Solar System Just Beyond Pluto.

23. Date: October 1, 2009. Headline: Astronomers Announce First Detection of an Earth-Size Planet Around Another Star.

24. Date: September 15, 2035. Headline: Sun-Like Star Has Three Planets with Life.

25. Date: March 30, 2027. Headline: More than One-Third of Stars Have Habitable Planets.

26. Date: December 13, 2010. Headline: Planet Orbiting a White Dwarf Has Surface Oceans and Oxygen Atmosphere.

Quick Quiz

Choose the best answer to each of the following. Explain your reasoning with one or more complete sentences.

27. Compared to a star of spectral type K, a star of spectral type A is generally: (a) hotter, more luminous, and more massive; (b) hotter, more luminous, and less massive; (c) cooler, dimmer, and less massive.

28. Stars of types O and B are unlikely to have planets with life because: (a) they have short stellar lives; (b) their intense ultraviolet light would sterilize any planets; (c) they don't have enough heavy elements.

29. How does the habitable zone around a star of spectral type M compare to that around a star of spectral type G? (a) It is larger and farther from its star. (b) It is hotter and much brighter. (c) It is smaller and closer to its star.

30. About how many extrasolar planets have been detected to date? (a) between 10 and 100; (b) between 100 and 1,000; (c) more than 1,000.

31. How have we detected most extrasolar planets discovered to date? (a) transits; (b) Hubble Space Telescope images; (c) the Doppler technique.

32. Which technique will the *Kepler* mission use to search for Earth-size planets around other stars? (a) transits; (b) the astrometric technique; (c) the Doppler technique.

33. Nearly all the extrasolar planets discovered to date are most likely: (a) terrestrial planets; (b) jovian planets; (c) large, icy worlds.

34. What is considered to be the most likely explanation for the close-in orbits of "hot Jupiters"? (a) They formed closer to their stars than Jupiter did. (b) They formed far from their stars, like Jupiter, but then migrated inward. (c) They are actually giant planets made of metal and rock.

35. Jupiter has had an important effect on life on Earth because: (a) Jupiter's heat has helped supply energy to life. (b) Jupiter's gravity helped clear the inner solar system of objects that could cause impacts. (c) Without Jupiter, Earth could not have a stable orbit around the Sun.

36. The main sequence on an H–R diagram represents stars that are: (a) in the final stages of their lives; (b) fusing hydrogen into helium in their cores; (c) all extremely low in mass.

INVESTIGATE FURTHER
In-Depth Questions to Increase Your Understanding

Short-Answer/Essay Questions

37. *Stars with Habitable Planets.* Based on what you've learned about stars in this chapter, make your best estimate of the fraction of all stars around which you'd expect to find planets in habitable zones. Clearly explain how you come up with your estimate. What data will we need to determine whether your estimate is accurate?

38. *Why Not Hubble?* Briefly explain why the Hubble Space Telescope has not been able to obtain images of many planets around other stars. What future technologies and missions should give us such images?

39. *Comparing Methods.* What are the advantages and disadvantages of the Doppler and transit methods? What kinds of planets are easiest to detect in each case? Are there certain planets that each method cannot detect, even if the planets are very large? Explain.

40. *Lots of Big Planets.* Many of the planets discovered so far are more massive than our most massive planet. Does this mean that our solar system is unusual? If so, how or why? If not, why not?

41. *Finding Planets.* Arrange the following planetary systems in order of decreasing detectability. Write a sentence defending each system's position in your list, and describe a scheme that might be able to detect such worlds. (a) Earth-size planets in Earth-size orbits; (b) Jupiter-size planets in Jupiter-size orbits; (c) Earth-size planets in Jupiter-size orbits; (d) Jupiter-size planets in Earth-size orbits; (e) Jupiter-size planets in interstellar space.

42. *Are Earth-Like Planets Common?* Based on what you have learned in this chapter, form an opinion as to whether you think Earth-like planets will ultimately prove to be rare, common, or something in between. Write a one- to two-page essay explaining and defending your opinion.

43. *No Hot Jupiters Here.* How do we think "hot Jupiters" formed? Why didn't one form in our solar system?

44. *Life on a Synchronously Rotating Planet.* Planets in the habitable zones of M stars are likely to rotate synchronously with their orbits, because these habitable zones are so close-in. Computer simulations suggest that on a synchronously rotating planet with a thick atmosphere, winds would carry heat from the side continually facing the star to the back, dark side. If so, there would be a ringlike zone between the light and dark halves that might be habitable. What are some of the adaptations you would expect for life in this zone? Explain.

45. *Ages of Stars on the H–R Diagram.* The giants and supergiants on the H–R diagram are stars in the last stages of their lives. Does this mean they are *older* than most main-sequence stars? Why or why not?

46. *Nightfall.* Read the short story "Nightfall" by Isaac Asimov. If such a planet really exists, do you think the scenario described is realistic? Why or why not? Summarize and defend your opinions in a one- to two-page essay.

Quantitative Problems

Be sure to show all calculations clearly and state your final answers in complete sentences.

47. *Number of Stars with Habitable Planets.* Assume that the Milky Way Galaxy has 500 billion stars (a reasonable estimate). Based on the statistics given in this chapter, approximately how many stars would be Sun-like? How many would be K stars? How many would be M stars? If you assume that an average G star has one planet in its habitable zone, while only one in five K stars and one in ten M stars has such a planet, how many total planets would you expect to find in habitable zones in the Milky Way Galaxy?

48. *Transit of HD209548.* The star HD209548, which has a transiting planet, is roughly the same size as our Sun, which has a radius of about 700,000 kilometers. The planetary transits block 1.7% of the star's light.
    a. Calculate the radius of the transiting planet. (*Hint:* The brightness drop tells us that the planet blocks 1.7% of the *area* of the star's visible disk; the formula for the area of a circle is $\pi \times r^2$.)
    b. The mass of the planet is approximately 0.6 times the mass of Jupiter, and Jupiter's mass is about 1.9×10^{27} kg. Calculate the average density of the planet. Give your answer in grams per cubic centimeter. Compare this density to the average densities of Saturn (0.7 g/cm^3) and Earth (5.5 g/cm^3). (*Hint:* To find the volume of the planet, use the formula for the volume of a sphere: $V = \frac{4}{3} \times \pi \times r^3$; be extremely careful with unit conversions.)

49. *Lost in the Glare.* How hard would it be for an alien astronomer to detect the light from planets in our solar system compared to light from the Sun itself?
    a. Calculate the fraction of the total emitted sunlight that is reflected by Earth. (*Hint:* Imagine a sphere around the Sun the size of the planet's orbit (area = $4\pi a^2$). What fraction of that area does the disk of a planet (area = πr_{planet}^2) take up? Earth's reflectivity is 29%.)
    b. Would detecting Jupiter be easier or harder than detecting Earth? Comment on whether you think Jupiter's larger size or greater distance has a stronger effect on its detectability.

You may neglect any difference in reflectivity between Earth and Jupiter.

50. *Planet Around 51 Pegasi.* The star 51 Pegasi has about the same mass as our Sun. A planet discovered around it has an orbital period of 4.23 days. The mass of the planet is estimated to be 0.6 times the mass of Jupiter. Use Kepler's third law to find the planet's average distance (semimajor axis) from its star. (*Hint:* Because the mass of 51 Pegasi is about the same as the mass of our Sun, you can use Kepler's third law in its original form, $p^2 = a^3$ [Section 2.2]. Be sure to convert the period into years before using this equation.)

51. *Finding a Planetary Mass.* Using the Doppler technique, you discover a planet that is causing its star to move at a maximum speed of 14 meters per second. The planet has an orbital period of 56 days and an average orbital distance of 55 million kilometers from its star. What is the planet's mass? (*Hint:* See Cosmic Calculations 11.1.)

52. *The Doppler Formula.* The amount of Doppler shift for light or radio waves can be calculated from this formula:

$$\frac{\text{wavelength shift}}{\text{rest wavelength}} = \frac{v}{c}$$

The rest wavelength is the wavelength of a particular spectral line in an object that is not moving (relative to us), v is the velocity of the star from which we observe a wavelength shift, and c is the speed of light ($c = 300{,}000{,}000$ m/sec). Suppose that, in a particular star, a spectral line with a rest wavelength of 600 nm is found to be shifted by 0.1 nm (toward the blue). How fast is that star moving toward us, in meters per second? (1 nm $= 10^{-9}$ m.)

53. *Finding a Center of Mass.* In the simple case of a two-body system—for example, a star and a single planet—the position of their center of mass can be determined from:

$$m_{\text{star}}r_{\text{star}} = m_{\text{planet}}r_{\text{planet}}$$

where m_{star} and m_{planet} are the masses of the star and the planet, respectively, and r_{star} and r_{planet} are the distances of the star and the planet from their center of mass. Consider the Sun and Jupiter, which are separated by 780 million kilometers. Using the fact that the Sun's mass is about 1,000 times the mass of Jupiter, about how far is the center of mass from the center of the Sun? How does this distance compare to the Sun's radius of 700,000 kilometers?

Discussion Questions

54. *Future Mission.* A wealthy benefactor has just given you a large grant to search for Earth-like planets around other stars. What would you do? Explain.

55. *The Copernican Principle and Rare Earth.* The Copernican revolution taught us that our planet is not the center of the universe, as had been generally believed before that time. Taking this lesson to heart, we have since assumed that our planet is not "central" or "special" in any way but rather that we are on a fairly typical planet in a fairly typical place in the universe. This principle, often called the *Copernican Principle* or the *Principle of Mediocrity,* has been borne out many times since. For example, we have learned that we are not near the center of our galaxy and that the universe has no center at all. Do you consider this principle to be in conflict with the rare Earth hypothesis? If so, does this make the rare Earth hypothesis any less scientific? Defend your opinions.

WEB PROJECTS

56. *New Planets.* Find the latest information about extrasolar planet discoveries. Create a personal "planet journal," complete with illustrations as needed, with a page for each of at least three recent discoveries of new planets. On each journal page, be sure to note the technique that was used to find the planet, give any information we have about the nature of the planet, and discuss how it does or does not fit in with our current understanding of extrasolar planets in general.

57. *The Kepler Mission.* The *Kepler* mission is the first funded mission designed expressly to look for Earth-size planets around other stars. Go to the *Kepler* Web site and learn more about the mission. Write a one- to two-page summary of the mission's goals and its current status.

58. *Planet-Finding Interferometers.* Other future missions will use interferometry to learn about extrasolar planets. Go to the Web site for one future interferometry mission under consideration, such as *SIM, TPF,* or *Darwin.* For the mission you choose, write a one- to two-page summary of the mission's goals and its current status.

12
The Search for Extraterrestrial Intelligence

LEARNING GOALS

12.1 THE DRAKE EQUATION
- What is the Drake equation?
- How well do we know the terms of the Drake equation?

12.2 THE QUESTION OF INTELLIGENCE
- Even if life is widespread, would intelligence be common?
- Will intelligence inevitably spawn technology?

12.3 SEARCHING FOR INTELLIGENCE
- How did SETI begin?
- How do we search for intelligence today?
- What happens if SETI succeeds?

12.4 THE PROCESS OF SCIENCE IN ACTION: UFOS AND ALIENS ON EARTH
- What have we learned from UFO sightings?
- Have aliens left any verifiable evidence of visitation?
- Is there a case for alien visits?

The probability of success is difficult to estimate;
but if we never search, the chance of success is zero.

P. Morrison and G. Cocconi, *Nature*, 1959

There are approximately 2,500 stars within 50 light-years of Earth. If any of these nearby stellar systems have inhabited planets, the inhabitants could already know that we exist by picking up our high-frequency radio, radar, and television transmissions. These broadcasts—unintentional evidence of our technological society—are moving into space at the speed of light and are currently washing over star systems at the rate of almost one new star system a day. By using sufficiently powerful radio telescopes, others could learn that we're here.

We are only beginning to signal our presence, but other civilizations may have been doing something similar for a long time. In the Milky Way Galaxy alone, there could be tens of billions of Earth-like worlds, and some of these could be filling the interstellar voids with their broadcasts. Using both radio and optical telescopes, scientists are attempting to find such transmissions. These experiments, called the *search for extraterrestrial intelligence* (SETI), are the primary topic of this chapter.

12.1 THE DRAKE EQUATION

The search for extraterrestrial intelligence (SETI) differs in a fundamental way from all the other searches for life that we have discussed in this book. Those searches are concerned not just with finding life itself, but with finding evidence that might point to its existence elsewhere—such as whether habitable planets are common or rare or whether our understanding of the origin of life would allow for life to arise on Mars. In contrast, SETI restricts itself to seeking clear and conclusive evidence of technologically advanced life.

Indeed, if SETI is successful in receiving and perhaps interpreting a message from a distant civilization, it might give us answers to many or most of the other questions we have discussed. To begin with, the discovery of a distant civilization would immediately prove that life is not unique to Earth. Because it is likely that any extraterrestrials we might detect with SETI experiments would be more advanced than we are, there's at least a possibility that we could learn a great deal from their transmissions, if we could understand them.

The principal goal of this chapter is to explore the methods by which scientists are now searching for evidence of other civilizations. First, however, it's worth asking whether our current understanding of life in the universe gives us any good reason to believe that the search may be successful. To some extent, the answer to this question may not matter, as stated so eloquently in the quotation that opens this chapter. That is, our innate scientific curiosity inspires us to search even if we cannot be certain that the search will ever be successful. But although we may not know the probability of success, recent advances in astrobiology allow us to say a lot

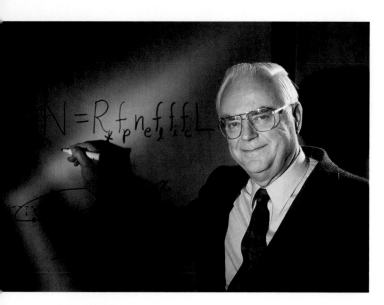

Figure 12.1

Astronomer Frank Drake, with the equation he first wrote in 1961. He is currently a Senior Scientist at the SETI Institute in California.

about the factors that might influence this probability. As in all of science, knowledge about such factors can provide important guidance to research efforts. In this section, we will identify the major factors that relate to the probability of success of SETI efforts—factors that are summarized succinctly by a simple formula known as the *Drake equation.*

• What is the Drake equation?

Early in the search for evidence of other civilizations, scientists realized that it would be useful to try to estimate the chances of finding something. In 1961, the first scientific conference on the search for extraterrestrial intelligence was held in Green Bank, West Virginia, at the radio observatory where a pioneering hunt for an alien signal had recently been conducted. (We will discuss this search, called Project Ozma, in Section 12.3.) There were only about a dozen attendees at the conference—just about the entire world's complement of people with a professional interest in the subject at the time—but their expertise ranged across the disciplines of astronomy, biology, and engineering. In setting the meeting's agenda, astronomer Frank Drake (Figure 12.1) tried to summarize the factors that would determine whether any attempt to detect intelligent extraterrestrials could succeed. He came up with a simple equation that, at least in principle, could be used to calculate the number of civilizations existing elsewhere in our galaxy (or in the universe at large) *from which we could potentially get a signal.* Note that this definition limits what we mean by "civilization" in this context. For example, the ancient Greeks had a remarkable civilization, but theirs doesn't count under this definition because they never developed radio or other technologies that could be used to effectively communicate across space.

The **Drake equation**, as it is now called, does not give us a definitive answer for the number of transmitting civilizations. Rather, it lays out the factors that are important in determining this number. The equation has played a guiding role in research bearing on life in the universe, because much of it deals with life in general and not just the smaller fraction of life that is intelligent enough to produce signals. Many of the factors in the Drake equation have been discussed earlier in this book as part of our search for biology on other worlds. Consequently, we expect that—with improvements in our knowledge of these factors—our estimate of the number of signaling civilizations will get better. However, we needn't await such improvements before we embark on a SETI search.

To keep our discussion of the Drake equation simple, we'll focus only on the number of civilizations in our own galaxy. We can always extend our estimate to the rest of the universe simply by multiplying the result we find for our galaxy by 100 billion, the approximate number of galaxies in the observable universe [Section 3.2]. We limit our focus for practical reasons, too. Any signals from other galaxies will be severely weakened by distance, making them far harder for us to detect. In addition, to stay consistent with key ideas already discussed, we'll consider a slight modification of Drake's original equation. (For the original version, see the box on p. 402.) In modified form, the Drake equation looks like this:

$$\text{Number of civilizations} = N_{\text{HP}} \times f_{\text{life}} \times f_{\text{civ}} \times f_{\text{now}}$$

Let's examine each term to see how this equation tells us the number of civilizations in the Milky Way Galaxy capable of interstellar communication:

- N_{HP} is the number of habitable planets in the galaxy. It is the first term because we assume that a prerequisite to having life or a civilization is having a habitable planet on which that life can evolve.

- f_{life} is the fraction of habitable planets that actually *have* life. For example, if $f_{life} = 1$, it would mean that all habitable planets have life; if $f_{life} = \frac{1}{1,000,000}$, it would mean that only 1 in a million habitable planets has life. Thus, the product $N_{HP} \times f_{life}$ tells us the number of life-bearing planets in the galaxy.

- f_{civ} is the fraction of the life-bearing planets on which a civilization capable of interstellar communication *has at some time* arisen. For example, if $f_{civ} = \frac{1}{1,000}$, it would mean that such a civilization has existed on 1 out of 1,000 planets with life, while the other 999 out of 1,000 have not had a species intelligent enough to build radio transmitters, high-powered lasers, or other devices for interstellar conversation. When we multiply this term by the first two terms to form the product $N_{HP} \times f_{life} \times f_{civ}$, we get the total number of planets on which intelligent beings have evolved and developed a civilization at some time in the galaxy's history.

- f_{now} is the fraction of the civilization-bearing planets that happen to have a civilization *now*, as opposed to, say, millions or billions of years in the past. This term is important because it tells us how many civilizations we could potentially talk to,* since there's no point in listening for civilizations that are long gone. Because the previous three terms told us the total number of civilizations that have *ever* arisen in the galaxy, multiplying by f_{now} tells us how many civilizations we could potentially make contact with today. For example, if the first three terms were to tell us that ten million planets in the galaxy have at some time had a communicating civilization but f_{now} turns out to be one in five million, then only two civilizations could be expected to exist today. As we will see shortly, the value of f_{now} must depend largely on how long civilizations survive once they arise.

To summarize, the Drake equation gives us a way to calculate the number of civilizations capable of interstellar communication that are currently sharing the Milky Way Galaxy with us. As such, it provides a useful way of organizing our thinking about the problem, because it tells us exactly what numbers we need to know to learn the answer. Indeed, it suffers from only one significant drawback: We don't know precise values for any of its terms!

> **THINK ABOUT IT** Try the following sample numbers in the Drake equation. Suppose there are 1,000 habitable planets in our galaxy, that 1 in 10 habitable planets has life, that 1 in 4 planets with life has at some point had an intelligent civilization, and that 1 in 5 civilizations that have ever existed is in existence now. How many civilizations would exist at present? Explain.

---

*For the purposes of the Drake equation, we'll assume that the term f_{now} takes into account the light-travel time for signals from other stars; for example, if a star with a civilization is 10,000 light-years away, it counts in determining f_{now} if the civilization existed 10,000 years ago, because signals broadcast at that time would just now be arriving at Earth.

• How well do we know the terms of the Drake equation?

The only term in the Drake equation for which we can make even a reasonably educated guess is the number of habitable planets, N_{HP}. As we discussed in Chapter 11, the detections of extrasolar planets to date show that planets of some kind are likely to be quite common. While we cannot yet be sure that this also means that *habitable* planets are common, so far we have no good reason to think otherwise. Moreover, the example of our own solar system suggests that it is possible to have more than one habitable planet per system, since we now suspect that Mars was habitable at some point in its past. Given that there are several hundred billion stars in the Milky Way Galaxy, it seems entirely reasonable to suppose that there could be 100 billion or more habitable planets. Nevertheless,

---

SPECIAL TOPIC 12.1: FRANK DRAKE AND HIS EQUATION

If anyone could be called the Father of SETI research, Frank Drake is that person. As a young man, Drake learned electronics in the Navy. He then studied for an advanced degree in astronomy, and on graduation took a job at the new National Radio Astronomy Observatory (NRAO) in Green Bank, West Virginia. In the late 1950s, the Observatory was busy with the construction of a large radio telescope, a project that would take many years. Consequently, the Observatory opted to buy an "off-the-shelf" instrument that could be used right away. This telescope boasted a 26-meter (85-foot) reflector, which at the time was larger than most of the world's operating radio telescopes.

Once this telescope was up and running, the Observatory staff was encouraged to suggest interesting experiments for its use. Drake had already been thinking about the possibility of interstellar communication by radio, and he proposed a simple experiment to search for alien signals from two nearby star systems. This became Project Ozma, the first modern SETI search. A year later, in 1961, Drake organized a conference at Green Bank to discuss the possibility that such an experiment could actually find an extraterrestrial transmission. His "agenda" for that conference became known as the Drake equation.

Drake's original form for his equation is:

$$N = R_* \times f_{planet} \times n_e \times f_{life} \times f_{intell} \times f_{civ} \times L$$

N is the number of transmitting civilizations in our galaxy. R_* is the galactic birthrate of stars suitable for hosting life, in stars per year. For example, because there are roughly 100 billion stars in the Milky Way and the galaxy is approximately 10 billion years old, R_* is approximately 10 per year (making the rather crude assumption that all stars are suitable). The term f_{planet} is the fraction of such stars having planets; n_e is the number of planets per solar system that have an environment favorable for life; f_{life} is the fraction of such planets on which life actually evolved; f_{intell} is the fraction of inhabited worlds that develop intelligent life; f_{civ} is the fraction of planets having intelligent beings that produce a civilization capable of interstellar communication; and L is the lifetime that such civilizations are "on the air," broadcasting signals. You might want to compare these terms with the more compact factors used in the equation on page 400.

Using some admittedly optimistic estimates for the first six terms in the equation, Drake suggested that we could reasonably guess that they would multiply to approximately one per year. This is the "birthrate" of civilized societies in the Milky Way. To find out how many are broadcasting now, we need only multiply this rate by L, the number of years during which they broadcast. This is analogous to determining the number of students attending a college: It's the number per year that enter as freshmen (the entrance rate) times the number of years they spend as students (typically four). If the birthrate for civilizations is taken to be one per year, the Drake equation becomes simply $N = L$; that is, the number of transmitting civilizations is simply the average lifetime (in years) of a transmitting society.

Unfortunately, L is dependent on sociology rather than on astronomy or biology—making it far more difficult to determine its value. Attempts to estimate L usually try to guess what the one technological civilization we know—our own—is likely to do. Only a half-century after inventing radio, we also developed atomic weapons. To some people, this suggests that L might be very short—only a few centuries or less. On the other hand, we can be optimistic and assume that we will survive our own technology and exist as a society for millions of years into the future. Perhaps one of the most important things we could learn from a SETI detection is that not all technologically sophisticated societies are doomed to early self-destruction.

Frank Drake's personalized license plate.

the actual number might be far smaller, and we won't know for sure until we begin to get data about habitable planets—which should begin to happen in the course of the next 10 years, as new technology makes it possible to detect small, rocky worlds and perhaps even to analyze their atmospheres for gases that might betray the presence of life [Section 11.2].

The rest of the terms in the Drake equation present more difficulty. For the moment, we have no rational way to estimate the fraction f_{life} of habitable planets on which life actually arose. The problem is that we cannot generalize when we have only one example to study—our own Earth. Still, we are not completely without guidance. The fact that life apparently arose rapidly on Earth [Section 6.1] suggests that the origin of life was fairly "easy," in which case we might expect that most habitable planets would also have life, making the fraction f_{life} close to 1. However, until we have solid evidence that life arose anywhere else, such as on Mars, it remains possible that Earth was somehow extremely lucky and that f_{life} is so close to 0 that life has never arisen on any other planet in our galaxy.

Similarly, we have little basis on which to guess the fraction f_{civ} of life-bearing planets that eventually develop a civilization. On the one hand, the fact that life flourished on Earth for almost 4 billion years before the rise of humans might suggest that it is very difficult to produce a civilization even when there is life. On the other hand, given that the majority of all stars in the Milky Way are older than our Sun, there has been plenty of time for evolution to work on numerous planets. Any evolutionary drive toward intelligence might inevitably lead to huge numbers of civilizations, even if it takes a long time on any given world. This question of whether intelligence is a rare accident or an inevitable result of evolution is so important to the issue of the search for extraterrestrial civilizations that we will devote the next section to investigating it.

The final term in the equation, f_{now}, is particularly interesting because it is related to the survivability of civilizations. Consider our own example. In the roughly 13 billion years during which our galaxy has existed, we have been capable of interstellar communication via radio for only about 50 years, or only about 1 part in 250 million of our galaxy's history. If we were to destroy ourselves tomorrow (saving students the unpleasantness of a final exam), our technological "lifetime"—the length of time we could make ourselves known to other star systems—would be only 50 years. If this is typical of other civilizations, then our chances of finding a signal from any one of them at any random time would be only 1 in 250 million. In that case, even if there have been hundreds of millions of civilizations in the galaxy's history, no more than a few would be detectable now.

Of course, we have not yet destroyed ourselves, and we may be severely underestimating the fraction f_{now}. For example, suppose civilizations have been arising for the past 10 billion years and routinely survive for 1 billion years. Then the chance of a signal reaching us from any given civilization at a random time is 1 in 10. If $f_{now} = \frac{1}{10}$, then there may be numerous communicating civilizations out there now, even if civilizations arise rather infrequently. To take some numbers, suppose that only 1 in 10 million stars ever gets a planet with a civilization. In a galaxy of 100 billion stars, this would mean that only about 10,000 civilizations ever arise. But if $\frac{1}{10}$ of them are here *now*, then there are some 1,000 civilizations we could potentially find. Four decades ago, considerations like these led Frank Drake to conclude that the typical lifetime of civilizations must be one of the primary factors—perhaps even *the* primary factor—in the potential success of SETI efforts.

The Drake equation is mathematically simple, but its chain of terms is only as strong as its weakest link. If we know one term poorly, there is no way to improve our estimate of the number of civilizations by knowing other terms well. For example, while we might have better estimates of most of the factors in the equation as our knowledge of astronomy and biology improves, f_{now} depends on sociological factors—that is, the behavior of alien civilizations. Do they quickly self-destruct, or do they survive for long periods? The only way we can make realistic estimates of f_{now} is by actually detecting extraterrestrial societies. Thus, as long as we have this uncertainty in f_{now}, we will face a corresponding uncertainty in knowing the total number of signaling worlds even if astrobiology research someday allows us to pin down precise values for terms such as N_{HP} and f_{life}.

As a result of this "weakest link" problem, as well as the fact that all of the terms are still highly uncertain, we cannot draw any definitive conclusions from the Drake equation. Indeed, some of the factors have such a large uncertainty that the numbers we enter into the equation (choosing numbers within the range consistent with our present knowledge) can give us anything from an optimistic view that a large fraction of stars in our galaxy have intelligent, communicating beings to a pessimistic view that we would have to search a large number of galaxies to find even one other example of intelligence. The main value of the Drake equation, then, is in pointing out what factors are important and underscoring the implications of our lack of knowledge about particular factors. Therefore, it can be used to help us recognize what the issues are and where we remain ignorant, and thus steer us to areas in which more research is needed.

> **THINK ABOUT IT** The Drake equation assumes that each transmitting civilization has sprung up independently on its own habitable planet. Is this a reasonable assumption, or might it be too limiting? Defend your opinion.

COSMIC CALCULATIONS 12.1

The Distance Between Signaling Societies

How far is it to the nearest other world with technologically advanced beings? We don't know, of course, but if we use the Drake equation to estimate the number of civilizations, we can then compute their average separation.

We start by estimating the volume V of space available for civilizations in the Milky Way Galaxy, assuming that civilizations are confined to the galaxy's disk:

$$V = \pi R^2 \times T$$

where R is the disk radius and T is the disk thickness.

Suppose the Drake equation tells us that there are N technological civilizations in our galaxy. If we assume that these civilizations are spread randomly, then the average volume of space that contains just one civilization must be the total volume of the galaxy divided by the number of civilizations, V/N. If we consider this volume per civilization to be a cube in which each side measures d light-years, then d is also the distance from the center of one cube to the center of the next, which means it is the average distance between civilizations. The

volume of a cube with side length d is d^3, so $d^3 = V/N$. Solving for d, we find:

$$d = \left(\frac{V}{N}\right)^{1/3} = \left(\frac{\pi R^2 \times T}{N}\right)^{1/3}$$

Example: Suppose that $N = 10{,}000$. What is the average distance between civilizations?

Solution: The disk of the galaxy has radius $R = 50{,}000$ light-years and thickness $T = 1{,}000$ light-years, so its volume is:

$$V = \pi R^2 \times T = \pi \times (50{,}000 \text{ ly})^2 \times (1{,}000 \text{ ly}) \approx 8 \times 10^{12} \text{ ly}^3$$

We use this volume to calculate d:

$$d = \left(\frac{V}{N}\right)^{1/3} = \left(\frac{8 \times 10^{12} \text{ ly}^3}{10{,}000}\right)^{1/3} \approx 900 \text{ ly}$$

If there are 10,000 civilizations in the disk of our galaxy, the average distance between these civilizations is nearly 1,000 light-years.

12.2 THE QUESTION OF INTELLIGENCE

Our discussion of the Drake equation has pointed out several key factors that we must understand better if we want to know how many other civilizations exist. We have already discussed (in earlier chapters) the uncertainties surrounding the number of habitable planets and the origin of life. However, we have not yet discussed the ideas that influence our estimate of f_{civ}, which describes the probability that life will eventually give rise to intelligence and a technologically adept civilization. In this section, we turn our attention to this important topic.

- ### Even if life is widespread, would intelligence be common?

The probability of a SETI success—receiving a signal—depends on how many civilizations are out there and broadcasting, and intelligence is a prerequisite to a civilization. Thus, SETI is likely to be successful only if intelligent life is widespread. But is it? Broadly speaking, there are two opposing schools of thought on this question.

One school considers intelligence that is comparable to our own (in other words, one that is able to develop both science and technology) to be unlikely. From this point of view, biology might be widespread but the evolution of technological intelligence extremely rare. Life has existed on our planet for at least 3.5 billion years. But only in the last few million years has our genus *Homo* developed the capability to understand its environment, and only within the last half-millennium have we come to understand the nature of Earth and begun our exploration of the cosmos. At the very least, the late appearance of *Homo sapiens* on Earth suggests that a long period of evolution must precede the emergence of technologically intelligent creatures.

Moreover, as we discussed in Chapter 6, our existence seems to have resulted from a number of chance events. For example, the Cambrian explosion that gave rise to the "body plans" of all modern animals (including those of our phylum, Chordata) might have been the result of environmental stress introduced by Snowball Earth episodes or a massive asteroid strike, and the rise of mammals might never have occurred if the K–T impact had not wiped out the dinosaurs (Figure 12.2). The chance nature of these events and the many other forks in the evolutionary road suggest to some people that the appearance of technological intelligence on Earth was an enormously improbable event.

The second school of thought holds the opposite view. It proposes that there is evolutionary pressure for intelligence; that is, that various evolutionary mechanisms consistently encourage an increase in intelligence for a wide range of species. If this is true, then some technologically intelligent species would still have evolved on Earth even if a different sequence of past events had prevented the existence of humans. Because we are the only species on our planet that has ever developed an advanced civilization, no other known species directly supports the view that the evolution of technological intelligence is likely. Instead, those who adopt this viewpoint look more generally at the process of evolution for evidence that some of the workings of natural selection promote greater intellectual capability.

Convergent Evolution The evolutionary argument in favor of widespread intelligence is based on the phenomenon of **convergent evolution**, the tendency of organisms of different evolutionary backgrounds

Figure 12.2

The path of evolution on Earth was severely affected by chance events such as the K–T impact 65 million years ago, which wiped out the dinosaurs and most other species. This artist's impression depicts the 10-kilometer-wide asteroid as it approached Earth. If this rock had arrived at Earth's orbit 20 minutes earlier (or 20 minutes later), it would have missed our planet. Would intelligent beings still have eventually arisen?

that occupy similar ecological niches to come to resemble one another. In such cases, natural selection often produces analogous adaptations. One example of convergent evolution is the shape of large marine predators: Dolphins and sharks evolved independently from earlier mammals and fish, respectively, but both have a similar streamlined body form. The obvious reason for this is that being shaped like a torpedo makes for greater speed underwater—and speed has clear survival value for a predator. We say that the evolution of originally quite different animals has *converged* on this optimized underwater shape.

Eyesight offers another example. Vision is a useful adaptation, and most multicellular animals have some ability to see. However, the eye was not a unique evolutionary invention. Studies of evolutionary relationships show that eyes evolved independently at least eight different times. Indeed, the design of the human eye is by no means the best. Compared to the eyes of some other animals, ours have "flaws," such as the fact that the nerves in our eyes come together in a bundle before exiting through the back of our eyeballs, resulting in a small blind spot. The several independent origins of eyes suggest that evolution tends to *converge* toward developing some kind of eye to provide vision.

THINK ABOUT IT All crabs may look quite similar, but DNA studies show that various crab species evolved from very different ancestors; for example, some crab ancestors were shrimplike and others were lobsterlike. Based on this information, how is a crablike body, with its round shape and unusual manner of walking, an example of convergent evolution? Why might this body type be a natural evolutionary development?

Like speed or eyesight, intelligence—the way an animal processes information—is subject to natural selection. If there are ecological roles for which keener intelligence has survival value, then we would expect a convergent evolution to greater brain power for animals in these niches. Intelligence would be just as likely to emerge as any other generally useful adaptation. In this case, evolution would tend to raise the level of intelligence to at least some degree in a great many species (much as eyes developed in many species), which in turn would increase the chance that some of these species would evolve even higher intelligence. With more players in the game, the chance of producing human-style intelligence would be greater. Thus, if we could establish the presence of an evolutionary trend favoring intelligence, we would be encouraged to think that the emergence of technological intelligence elsewhere is not enormously improbable.

Measuring Intelligence

In principle, we can test whether intelligence is evolutionarily favored by measuring the brain power of a variety of animal species over time. But how can we measure the mental ability of animals? Few creatures are eager or able to take IQ tests, but we can resort to a simpler measure of raw brain power based on brain mass.

Figure 12.3 plots brain mass against body mass for a sample of birds and mammals (including primates). There is a clear and expected trend: Heavier animals have heavier brains. By drawing a straight line that fits these data, we define an average brain mass for each body mass. Those animals whose brain mass falls on this line are said to have an **encephalization quotient (EQ)** of 1, which means a typical allotment of mental ability for creatures of their size. They have enough brain mass, we can assume, for a basic set of behaviors. If their brain mass falls above this line, then it seems reasonable to suppose that they are capable of more elaborate behavior. Creatures whose brain mass falls below the line are presumably less mentally agile than average animals in this sample. The EQ, the brain mass relative to the value on the line EQ = 1, serves as an indicator of general intelligence.

Although other indicators of intelligence have been suggested (the amount of brain folding, for example, or the number of neural connections), measuring EQ has the advantage of being fast and easy. For example, notice that the EQ = 1 line shows that the average species with body mass 1 kilogram (read off the horizontal axis) has a brain mass of about 10 grams (along the vertical axis). We'd therefore say that a species with the same body mass but a brain weight of 20 grams has EQ = 2, meaning it has twice as much brain as average for its size. Similarly, a species with body mass 1 kilogram and brain mass 5 grams has EQ = $\frac{1}{2}$, meaning it has only half as much brain as average for its size. Besides this ease of computation, EQ also has the advantage of being something we can estimate for extinct species, since we can often estimate their total body masses and we can gauge their brain masses from the volume of their fossil skulls.

Admittedly, EQ is a simple measure, and it might be likened to judging the computational ability of computers by weighing their CPU chips. Nonetheless, EQ seems to correlate well with complex behavior. Carnivorous animals that need to hunt down their meals generally have higher EQs than leaf eaters, and animals that lavish care on their offspring score higher than those that ignore them.

Exploring the Evolution of Intelligence

What happens when we use EQ to investigate the premise that, on Earth, there has been a trend toward increasing brainpower over time? If we look at contemporary species, we find that although humans don't sport the most massive brains (whales, for instance, have much larger brains than we do), we do

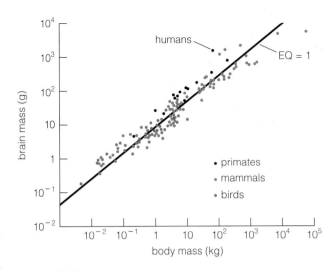

Figure 12.3

This graph shows how brain mass compares to body mass for some mammals (including primates) and birds. The straight line represents an "average" of the ratio of brain mass to body mass, which we define as an encephalization quotient of EQ = 1. Animals that fall above the line have an EQ greater than 1 and are therefore "smarter" than average. Animals that fall below the line have an EQ less than 1 and are "intellectually challenged." Note that the scale uses powers of 10 on both axes. (Adapted from Sagan, 1977.)

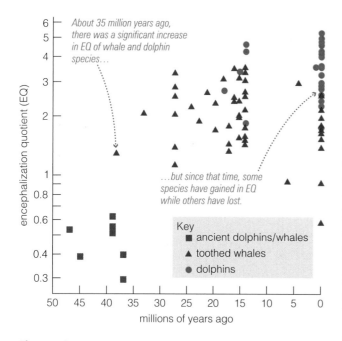

Figure 12.4

This graph shows the change in encephalization quotient (EQ) with time for toothed whales and dolphins. Note that over the course of the last 50 million years, some EQs increased substantially, while others remained relatively unchanged and some even went down. The fact that some dolphins have EQs comparable to our own suggests that high intelligence has survival value in many species, and may be a common evolutionary development. (Adapted from Marino, L., McShea, D., Uhen, M. D. 2004, *The Anatomical Record.* 281A: 1247–1255. Used with permission.)

have the biggest brains in relation to body mass. Our species' encephalization quotient is 7, meaning that our brains are 7 times more massive than would be expected for an "average" mammal of the same body weight. This EQ not only exceeds that of chimps and dolphins, whose EQs are 2.5 and 4 or 5, respectively, but it is higher than that of any other known species, alive or extinct. Thus, these numbers confirm what we already know: We're the cleverest critters on the planet.

However, a look to the past shows that we were not the only creatures evolving toward greater intelligence. Biologists recently measured the EQs for a range of toothed whales and dolphins, and noted how these values changed during the past 50 million years (Figure 12.4). They did this by using computer tomography to determine the brain volumes of long-dead animals whose fossilized skulls were sitting in museums. The animals' weight was estimated by measuring the size of some of the bones around the eye sockets, a parameter known to be strongly correlated with body mass. With data in hand, they then computed the EQs of 200 specimens, representing 62 cetacean species. What they found was that these animals experienced a major improvement in EQ 35 million years ago when they developed the navigation scheme known as echolocation (using audible "pings" as sonar), and some species underwent another big shift 15 million years ago. Not all the EQs went up, and today's cetaceans have a range of EQ from 0.2 to 5. But the smarter ones are not far behind our own EQ of 7. Humans are not closely related to dolphins in an evolutionary sense, so the fact that they have also developed large brains suggests that there is real survival value in cleverness, and that there are many ways that nature can produce it.

Another way to approach the question of the evolution of intelligence is to ask what factors might encourage its appearance and whether those factors are likely to be commonplace. To begin with, a good many "pre-adaptations" seem necessary. High-performance brains like those of birds and mammals need a vigorous metabolism, so intelligence is most likely to arise in warm-blooded animals. In addition, a relatively large body size is necessary to house a large brain (although recent research with insects suggests that, by miniaturizing neurons, nature might in principle be able to pack a lot of intelligence into small packages). A long period of parental care of offspring is probably also a necessary precondition for intelligence, which wouldn't be that useful without a time to learn from parents. These preconditions have existed for a wide range of species on Earth during at least the last 50 million years. Because this is the same period in which intelligence seems to have risen most dramatically, we have some hint that intelligence has indeed been evolutionarily favored.

Still, we might wonder about precisely which evolutionary mechanisms will select for intelligence. After all, having a big brain is metabolically expensive; in humans, for example, roughly one fourth of our metabolism is for the benefit of our brain, even though that brain accounts for only about 2% of body weight. Consequently, big brains will occur only in species in which the cost is rewarded with real survival advantages. One survival benefit might come from the interactions among individuals of social species. Dolphins and primates are highly social animals. Success in a social environment is enhanced by intelligence, because there is survival value in being able to judge the mood and meaning of fellow creatures. Social position in your troop or pod depends on how savvy and canny you are. Moreover, an elevated social position often allows you to have first choice in mates, so clever, high-ranking individuals will tend to produce clever, high-ranking offspring. We might reasonably expect intelligent aliens, if they exist, also to have evolved in interactive social environments.

Competition between species can also ratchet up intellect. For example, large carnivores and their prey encourage improvements in one another's intelligence. When a lioness stalks gazelles for dinner, she's more likely to catch a gazelle that's less aware of its surroundings or less cagey in devising escape maneuvers. The lion gets a meal and in the process inadvertently raises the average intelligence of the surviving gazelles. The rise in gazelle intelligence makes getting a meal tougher for less mentally agile lions, which then preferentially drop out of the gene pool. The less alert, less cunning of both species are weeded out, and the smarter members survive.

In summary, several common circumstances in the animal world naturally select for brain power, and there is evidence that more than one group of animals on our planet has been on the track to high general intelligence. This argues against the idea that the appearance of intelligence is some special, extraordinary accident of evolution on Earth and suggests that, given time and a competitive environment, many intelligent species will arise on any inhabited world. The probability of many candidates means at least some possibility that one or more of these species will develop humanlike brainpower. On the other hand, increased intelligence comes at a cost in terms of resources to support it (such as a high metabolism or carrying around a heavy head), and it is unclear that these resources would not be better utilized by, say, evolving the ability to run faster or fight more fiercely. Moreover, many species *didn't* evolve intelligence over the last 50 or 100 million years. In the end, the example of terrestrial life alone does not tell us whether there is an evolutionary imperative toward humanlike intelligence. And, even if there is such an imperative, we still must ask whether high intelligence necessarily leads to technical competence and the ability to communicate across interstellar distances.

• Will intelligence inevitably spawn technology?

It might seem natural that a species with sufficient brainpower will eventually develop science and technology. However, there are both physiological and sociological counterarguments to this idea.

On the physiological side, suppose dolphins were as intelligent as we are. Their lack of hands and their need to live in water would prevent them from building anything resembling modern technology. They might be smart and have sophisticated social structures, but without telescopes they wouldn't know much about astronomy, and without radios or lasers they wouldn't be able to talk to the dolphins of other worlds. Wolves have mouths, elephants have trunks, and ravens sport beaks and feet, but none of these animals have overall body designs that would allow them to manipulate complex tools—no matter how smart they might be.

On the sociological side, remember that *Homo sapiens* emerged in more or less its current form about 150,000 years ago. Certainly, by 20,000 or 30,000 years ago our ancestors had essentially the same level of intelligence as we do. Yet our ability to communicate through space is less than a century old, and it derives from the emergence of science—itself an endeavor that has developed only gradually over the past 2,500 years. Many different cultures developed mathematics and astronomy to varying degrees, but only the cultural line that emerged from ancient Greece ultimately led to modern science [Section 3.1]. Was this a lucky accident, or was it inevitable that some culture would develop modern technology?

Again, a single example—what happened on our own planet—cannot guarantee that science and technology will be common in the cosmos, even if high intelligence is. The only way we can answer the question of how frequent or rare technological civilizations might be is to find evidence of them on other habitable worlds. We'll now turn our attention to experiments that are trying to do just that.

12.3 SEARCHING FOR INTELLIGENCE

Receiving a message from another civilization could be one of the most important events of human history. We would know that we are not alone in the universe. We could conceivably learn a great deal about science and sociology. We could even learn how other civilizations have successfully survived periods in their history in which they had the power to destroy themselves. How might we receive such a message? In this section, we'll investigate the context in which we engage in SETI efforts. We'll begin with some historical background, then discuss the types of signals that SETI might be able to detect. We'll also explore a few possible ways of detecting extraterrestrial intelligence that don't involve communication signals.

• How did SETI begin?

Shortly after its invention, radio was recognized as a possible means of extraterrestrial communication. After all, radio travels at the speed of light and can easily bridge the airless voids of space. If extraterrestrials are using radio, then we might detect their presence without anyone having to leave the home planet.

Early Claims of Success   As the twentieth century dawned, two pioneers of wireless technology became convinced (wrongly) that they had heard aliens on the airwaves. One of these pioneers was Guglielmo Marconi (1874–1937), generally celebrated as the man who made radio practical. The other was Nikola Tesla (1856–1943).

Tesla was both eccentric and brilliant. He was a prolific inventor (among other things, he made the first fluorescent lamp), but his most enduring legacy was the use of alternating current (AC) for distributing electrical power. AC—produced by having a voltage that cycles positive and negative 60 times a second—is commonplace today, but it was a radical concept when Tesla first proposed its use for a Niagara Falls generating station. The advantage of AC is that transformers can easily raise or lower the voltage, increasing it for low-loss transmission over the great distances from the power plant to your neighborhood and then lowering it for final distribution to your home. If you look at the power poles on any residential street that has aboveground electrical lines, you'll see the "step-down" transformers that provide the 115 volts (in the United States) feeding the wall sockets of the houses.

Tesla promoted AC because of these benefits, but he tried to go even further. He thought it might be possible to bring power to the people with "induction," rather than with copper cables, sending the energy through the air as low-frequency radio waves. To demonstrate his idea, he built a 60-meter-tall transmitting tower, wrapped with wire, in Colorado Springs, Colorado. His device, intended to radiate electrical energy, was an outsized example of what is now known as a Tesla coil. One night in 1901, the inventor noted that his apparatus was picking up electrical disturbances, which he ascribed to interplanetary communication. He wrote, "The feeling is constantly growing on me that I had been the first to hear the greeting of one planet to another."* We now believe that he was listening to an atmospheric phenomenon known as "whistlers," electrical noise created by distant lightning discharges.

That same year, Marconi successfully sent a radio signal across the Atlantic, proving that this invention was useful for communication over large distances. It took little imagination to guess that radio might also serve for sending messages into space. In the early 1920s, Marconi stated that he, too, had picked up signals that came from extraterrestrial sources. These experiments culminated in 1924, during one of the periods when Earth and Mars are closest to each other in their orbits. Marconi and others encouraged anyone with a radio set to listen for martian broadcasts, and even the U.S. Army joined in the search (Figure 12.5). There was considerable optimism that something would be detected, and a cryptographer was standing by in case the signals required decoding. Alas, while signals were picked up, there was no evidence that they were from Mars or any other extraterrestrial source. In retrospect, the signals that Marconi heard may also have been "whistlers" from distant lightning, or possibly garbled U.S. Navy broadcasts of which he was unaware.

In retrospect, these early experiments were doomed. We now know that there are no Martians sophisticated enough to assemble radio transmitters. In addition, we realize that these pioneer listening attempts were all made at the wrong spot on the radio dial. They were conducted at relatively low frequencies that have problems penetrating Earth's ionosphere—a layer of the upper atmosphere that consists of particles ionized by sunlight. The ionosphere acts as a radio "mirror," reflecting low-frequency radio emissions. Indeed, it was the ionosphere that bounced Marconi's 1901 transatlantic signal from England to Canada, thus allowing communication despite Earth's curved surface. The apparatus used by both Tesla and Marconi operated at frequencies too low to penetrate the ionosphere and therefore was insensitive to any cosmic broadcasts.

Although the radio pioneers might have mistaken thunderstorms for Martians, their enthusiasm for radio's use as a long-distance communication medium was ahead of its time. After World War II, when high-frequency radio equipment became widely available, it was possible to listen at frequencies that *could* penetrate the ionosphere. In addition, improvements in antennas and receivers made such experiments enormously more sensitive than the attempts made early in the century. The stage was set for a scientific approach to SETI.

Origins of Modern SETI

The beginning of modern SETI is generally attributed to two physicists working at Cornell University. In 1959, Giuseppe Cocconi and Philip Morrison wondered how difficult it would be to send radio signals over interstellar distances. They made simple calculations showing that communication between nearby stars was possible using technology no more advanced than our own.

Cocconi and Morrison realized that the galaxy is considerably older than our solar system and consequently that there could be civilizations among the stars that have been around far longer than ours—perhaps surviving for many millions of years. For such long-lived societies, the

Figure 12.5

An Army operator listening for martian radio signals, as pictured in *Radio Age*, October 1924.

---

*Original source: Nikola Tesla, "Talking with the Planets," *Collier's Weekly* 26, no. 19, 4 (1901); taken from S. J. Dick, *The Biological Universe*, Cambridge University Press, 1996.

construction of powerful radio beacons that could be used to send "hailing signals" to other star systems should be a simple matter.

Cocconi and Morrison advocated looking for such beacons using large radio telescopes aimed at nearby stars. The exact arrangement of the planets around the stars under scrutiny was irrelevant, because the area of sensitivity of any reasonable-size radio telescope would encompass all of a star's possible planetary environs. The only other technical question concerned which frequencies to monitor with the radio receivers.

The radio portion of the electromagnetic spectrum (see Figure 3.30) extends over a wide range of frequencies. When we build a radio receiver, it is sensitive only to a particular set of frequencies within this wide range, which we refer to as its **band** of operation. For example, if you look at a typical FM radio, you'll see that it is designed to cover the band of frequencies from about 87 to 108 MHz. (Recall that the basic unit of frequency, hertz [Hz], is equivalent to waves [or cycles] per second. Thus, 1 megahertz [MHz] is a frequency of one million cycles per second.)

Of course, radio stations do not broadcast their signals over this entire band—if they did, no matter where you tuned your radio you'd hear the overlapping sounds of all the radio stations that were transmitting. Instead, radio stations limit their signals to only a narrow range of frequencies. For example, if your favorite radio station broadcasts at "97.3 on your FM dial," then you must tune your receiver so that it picks up only a small range of radio frequencies centered on 97.3 MHz. That range of transmitted frequencies is called the **bandwidth** of the signal. As a practical matter, the bandwidth is governed by how much information the broadcast contains. For example, most television signals (Figure 12.6) have a bandwidth 500 times wider than that of an AM radio station, because the TV signal contains picture (video) information in addition to sound (audio).

Terrestrial broadcasts are made with a wide range of bandwidths depending on their purpose, but a hailing signal designed to get the attention of someone light-years away would likely include a very narrow bandwidth signal component, with much of the transmitter energy concentrated at one spot on the radio dial. This would make the signal easier to pick out against the background noise that is naturally produced by interstellar gas, distant galaxies, and the radio receiver itself. But in what part of the radio spectrum would we expect to find such a narrow-band hailing signal? In truth, we cannot say for certain, since a wide range of radio frequencies is serviceable. Because searching all these frequencies would be a hopelessly difficult task, Cocconi and Morrison proposed that SETI experimenters should tune their receivers near a spot on the dial that every scientifically literate society would know: 1,420 MHz. This is the frequency at which neutral hydrogen gas—the major constituent of the thin material that floats between the stars—produces natural radio static. Radio astronomers often use this frequency to study the distribution of interstellar gas in galaxies. Because it is such a useful spot on the dial, astronomers throughout the universe (of whatever species) would have this frequency marked on the receivers of their radio telescopes, and a transmitting beacon tuned near this frequency would be likely to attract the attention of others.

Having set out all the important principles of a radio SETI search, Cocconi and Morrison tried to interest radio astronomers in their idea. They approached those at England's Jodrell Bank Observatory but got little response. However, Frank Drake, then a young astronomer at the Green Bank radio observatory in West Virginia, had independently reached many of the same conclusions as Cocconi and Morrison. In the spring of 1960,

Figure 12.6

The spectrum of a standard-definition television signal, showing how the energy of transmission is spread over a bandwidth of nearly 6 MHz. The large bandwidth is necessary because the signal carries both sound and picture information. However, about one third of the total power in the TV signal is concentrated in the extremely narrow "spike" known as the *video carrier*. This carrier has a bandwidth of less than 1 Hz, or less than one six-millionth of the total bandwidth. Because the video carrier has so much of the total signal power concentrated into such a narrow bandwidth, it would be the easiest part of the signal to detect from a great distance. Hence, it is a good example of the type of narrow-band signal that SETI experiments seek. Note that a high-definition TV signal would, in principle, have even greater bandwidth, although by using digital techniques to compress the video information, this bandwidth bloat can be minimized.

using the 26-meter (85-foot) radio dish at Green Bank, Drake began an experiment much like the one Cocconi and Morrison had in mind (Figure 12.7). He conducted a weeks-long search for 1,420 MHz radio signals from two nearby, Sun-like stars: Epsilon Eridani and Tau Ceti. (These stars are approximately 12 light-years distant.) Drake whimsically named his search Project Ozma, after the fictional princess in the books by Frank Baum. It was the first modern SETI experiment.

Drake's search failed to detect alien signals, but it fired the imagination of many in the science community and ultimately led to a small SETI research program run by the American space agency, NASA. The NASA researchers spent many years studying the feasibility and technology of SETI. Their research led to the construction of specialized receiving systems that could be fitted to large, existing radio telescopes. In 1992, the NASA search began. But a year later, with the data collection barely under way, the program was cancelled by the U.S. Congress. Since then, scaled-down SETI programs have continued with private funding in the United States and as small university research efforts elsewhere in the world.

• How do we search for intelligence today?

Many of the SETI techniques in use today are elaborations on the earlier schemes proposed by Cocconi and Morrison, and first tried by Drake. However, SETI has also branched out to search for signs of intelligence in other ways as well. To understand the modern search for extraterrestrial intelligence, we first need to consider the types of signals we might be able to detect.

Categories of Signals   Drake's Project Ozma searched for a deliberately broadcast signal at a specific frequency that would be known to astronomers anywhere. It was therefore a search for an interstellar hailing signal, or beacon, intentionally sent so that others might detect it. But there are other possible signals we might hope to find. Broadly speaking, alien signals could fall into any one of the following three categories:

1. *Signals used for local communication on the world where intelligent beings live.* Our own radio and television signals fall into this category, because they are designed for our own use and not for interstellar communication. Another local use for radio signals is radar. Advanced civilizations might use radar to locate comets that pose a potential threat to their planet, for example.
2. *Signals used for communication between a civilization's home world and some other site,* such as a colony or spacecraft on another world. We have used relatively weak signals of this type to communicate with our interplanetary spacecraft. Such signals would be far stronger if they were being used, say, to communicate between colonies on planets in different star systems light-years apart.
3. *Intentional signal beacons,* such as the type searched for by Project Ozma, purposefully designed to get the attention of other societies.

In principle, SETI can search for all three types of signals, but in practice, our ability to receive signals depends on the sensitivity of our equipment. To get a rough idea of what we might be able to detect with our current technology, let's consider our own signals as an example.

The first commonplace, high-power, high-frequency transmissions from Earth were our early television broadcasts. These transmissions began in earnest during the 1950s, so they are now some 50 light-years away

Figure 12.7

The 26-meter (85-foot) radio telescope used by Frank Drake in his pioneering 1960 SETI search, Project Ozma. This instrument was the first to be built at the National Radio Astronomy Observatory in Green Bank, West Virginia. Drake scrutinized two nearby star systems for signals near 1,420 MHz in frequency.

in space—followed by all television broadcasts since. Thus, in principle, any civilization within about 50 light-years could watch our old television shows (although it's unclear whether they would *like* them!). However, while television transmitters are fairly powerful, their antenna systems are designed to spread the signal over a wide angle, so that they reach homes in all directions from the transmitting station. This means that the strength of any signal that is, by chance, aimed at any given star is quite weak. And, like all light, these signals continue to weaken with distance (following an inverse square law, as shown in Figure 7.1). If another civilization has the same sort of receiving technology that we do, they could detect our television signals only if they were within about 1 light-year of us, and no stars are that close (the nearest are more than 4 light-years away). Turning the situation around, we see that we are not yet capable of detecting signals in the first category listed above, unless for some reason they are being broadcast with much more power than we use for our own television signals. (Some of our military radars, which employ large antennas to narrowly focus their transmissions, could be detected as far away as a few tens of light-years.) The situation is no better for signals in the second category, at least if such signals are comparable in strength to those we currently use for communicating with our own interplanetary spacecraft.

Our technology is rapidly improving, and in the future we might be able to detect alien signals in any of the three categories. But for the moment, at least, our best chance of detection is with signals in the third category. Beacon signals should be the easiest to detect because they would deliberately be made strong enough to be heard across interstellar distances. They should also be the least difficult to interpret, because they presumably would be designed for easy decoding.

Decoding a Signal   How might an alien civilization design a signal so that we could decode it? In the book and movie *Contact,* author Carl Sagan supposed that the aliens might make it easy for us to recognize their message by playing back to us one of our own television transmissions. However, because our broadcasts have only had enough time to make it a relatively short distance into space (roughly 50 light-years), this strategy could work only if a civilization happened to be near enough to have already found such a signal and returned it to Earth. Fortunately, there are other ways that aliens could make it easy for us.

For example, aliens might choose to broadcast a strong, narrow-band signal. Most natural radio emissions have fairly broad bandwidths, so a signal confined to one narrow spot on the radio dial would immediately offer a hint that it might be artificial (made by intelligent beings). If the signal was also flashing on and off, or switching between two nearby frequencies, we would suspect that it was a coded message. We would undoubtedly record the pattern and try to analyze what was being "said."

The first thing to do would be to look for repetition. If aliens really wanted us to detect and decode their signal, they would repeat the entire broadcast many times, since otherwise we'd have to be listening at just the right moment to catch the signal. Knowing the total length of the message might help us greatly in figuring it out. If the total number of flashes or frequency changes was, for example, 1,679, then we might note that this is the product of two prime numbers, 23 and 73. (A prime number is one that can be divided only by 1 and itself without any remainder. The prime numbers are 1, 3, 7, 11, . . .) We could then arrange the message in a 23-by-73 grid and look for pictures or other figures.

a Arecibo has the world's largest single-dish radio antenna, shown here, with a diameter of 305 meters (1,000 feet).

Figure 12.8

In 1974, the Arecibo radio telescope in Puerto Rico was used to send a 3-minute broadcast toward the globular cluster M13.

b The pictorial message broadcast in 1974, as it looks once you realize that the message is intended to be laid out as a rectangular grid with 73 rows and 23 columns (each of these numbers is a prime number, which should help enable any alien recipients to guess the layout). The colors are shown only to make the components clearer; the actual picture was sent in "black and white."

This simple approach is one we have taken ourselves in one of the few deliberate broadcasts that we have made to the stars. In 1974, the powerful planetary radar transmitter on the Arecibo radio telescope (Figure 12.8a) was fired up and used to send a 3-minute message to the object M13, a globular cluster containing a few hundred thousand stars. The message consisted of 1,679 bits, and each bit was represented by one of two radio frequencies. If someone picks ups the signal and recognizes that 1,679 is the prime number product 23×73, then they could arrange the bits in a rectangular grid with 73 rows and 23 columns. The resulting graphic, shown in Figure 12.8b, represents the Arecibo radio dish, our solar system, a human stick figure, and a schematic of DNA and the eight simple molecules used in its construction. However, because we did not repeat the signal, any aliens living around stars in M13 will have only one chance to receive it as it passes by their planet at the speed of light. That is one reason the signal was sent to a globular cluster: The cluster's several hundred thousand stars would seem to improve the odds of someone receiving the signal. If it is received, it won't be soon. M13 is about 21,000 light-years from Earth, so it will take our signal some 21,000 years to get there and another 21,000 years for any response to make its way back to Earth.

THINK ABOUT IT Some people consider the 1974 broadcast to have been a dangerous exercise that might attract the unwanted attention of hostile aliens. In general, do you think it is "safe" for us to broadcast messages to the stars? Why or why not?

Of course, alien messages could be far more sophisticated and more difficult to decode. In *Contact*, the pictures were three-dimensional, not flat, and in that case we should be looking for messages whose length is the product of three rather than two prime numbers. Alternatively, the message might be encoded in ways similar to the schemes used for sending files on the Internet. Given the enormous variety of possible ways a message could be transmitted, it might be that we would never be able to understand an alien broadcast. However, if advanced societies truly wished to get in touch, then they would undoubtedly go to some trouble to make their messages simple enough to be understandable to any civilization able to build the telescopes necessary to receive them.

Finally, it's important to remember that beacon signals will exist only if other societies make a deliberate decision to broadcast them. If no one does anything more than send occasional short signals like ours in 1974, the chances of detection will be quite small, even if many societies are out there listening. As a result, some people have asked whether we should undertake our own more deliberate effort to send a beacon signal into space. While this question continues to intrigue both researchers and the public, the consensus has been that we should focus on receiving signals first. After all, as we've seen in our consideration of the Drake equation, the chance that anyone will pick up a signal depends on how long broadcasting civilizations are "on the air." There's little point in transmitting signals for only a few weeks or even a few years. A broadcasting project would require long-term investment and a great deal of patience. Perhaps it is too soon for us to consider such a project; we developed radio technology only in the past century. There may be galactic civilizations that have had the ability to transmit signals into space for hundreds of millennia or longer. As the new kids on the block, it might make more sense to listen first.

Radio SETI Today
Let's assume that someone really is broadcasting a signal that we could recognize and potentially decode. How should we go about searching for it? Today, just as in the early days of Drake's Project Ozma, most SETI searches are attempts to detect *radio* transmissions from other worlds.

A large radio antenna—almost always a radio telescope constructed for more conventional research purposes—is used to try to pick up the hoped-for cosmic signals. A low-noise amplifier at the antenna's focus

---

◤ MOVIE MADNESS: *CONTACT*

Do the aliens know we're here? They might, if they're not too far away.

Contact, the movie based on Carl Sagan's novel about getting in touch with our celestial pals, starts out with a nifty sequence in which the camera backs away from Earth, slips by the Moon and planets, and eases out into the galaxy and beyond. During this high-speed countermarch, we hear the sounds of radio programs that have reached each of these cosmic outposts. With every step outward, we move back in time. (OK, there's some cinematic license here: As the camera passes Saturn, we've regressed to 1950s rock and roll. Saturn is never more than 88 light-*minutes* from Earth!)

In fact, less than 100 light-years out, Earth really does go "silent." Easy evidence for the presence of *Homo sapiens* extends only this far—about one tenth of 1% of the distance to the far edges of the galaxy.

Fortunately for the cash-starved SETI researchers in *Contact,* some friendly aliens have an outpost around the bright star Vega, a mere 25 light-years away. They've tuned in one of our early TV broadcasts and, apparently intrigued, have replied. Their response is picked up by the film's heroine, Ellie Arroway, as she uses a pair of earphones

to monitor the cosmic static received by a large radio telescope. (More license here: Radio receivers for SETI sport hundreds of millions of channels. Ellie should either have donned a few hundred million headsets or left the listening to computers.)

The alien reply signal, which sounds like a pile driver hitting a pod of whales, contains an original 1936 broadcast (that way we know that the extraterrestrials are deliberately beaming to us) interweaved with construction details for … well … some sort of large device.

Faced with a SETI detection, the government goes nuts, and so do a lot of the citizenry. Some are ecstatic about the possibility of alien company; others see the news as heralding the apocalypse. Meanwhile, the scientists build the device—a multibillion-dollar machine that looks like the ultimate theme park attraction, but in fact is a wormhole transporter (is there any difference?).

It's nice to think that advanced aliens would want to improve our lot by sending us plans for high-tech hardware. But this seems an unlikely message from space. After all, if we could somehow contact the Neanderthals and give them plans for a personal computer, do you think they could ever, *ever* build it?

---

boosts the signal levels before they are further processed. The processing usually consists of digitally "slicing" the incoming, wide-band signal (typically spanning hundreds of megahertz) into many narrow-band channels. This processing is based on the assumption that some strong narrow-band component will be present in any deliberately broadcast extraterrestrial signal.

During the search, the radio telescope may be either pointed in select directions, such as toward individual stars, or swept across the heavens to study a larger section of the sky. The former technique, known as a **targeted search**, proceeds on the assumption that not all locations in space are equally probable sites for intelligent life. For example, Drake put forth the reasonable hypothesis that civilizations were most likely to exist on planets around Sun-like stars, and consequently he targeted his search at this type of star. The sweep technique, known as a **sky survey**, makes no assumptions about where intelligent aliens might be located.

Several radio SETI projects are currently under way. Table 12.1 summarizes key features of four major projects. The Inner Galactic Plane Survey is conducted on a new telescope, still under construction, called the Allen Telescope Array, which is located in northern California. SERENDIP uses the Arecibo telescope (see Figure 12.8a), while Southern SERENDIP employs the Parkes radio telescope in Australia (Figure 12.9). SETI Italia makes observations concurrently with radio astronomy projects that are being conducted with the Medicina radio telescope. For each project, the search type is indicated, along with the number of narrow-band channels being examined and the overall band of frequencies covered in each search. The last column of the table gives an indication of each experiment's sensitivity: It lists the minimum transmitter power that alien broadcasters would need to be using for us to detect their signals, assuming the aliens are located 100 light-years away and use a transmitting antenna 100 meters in diameter. The Inner Galactic Plane Survey, for example, could detect such an extraterrestrial broadcasting setup if the transmitter had a power of 70 megawatts or more. That's more than most radio and television broadcasts on Earth, but not that much more: Some of our own broadcasting stations have power greater than a megawatt, although they do not use a 100-meter antenna to narrowly focus their transmissions.

TABLE 12.1 *Current Radio SETI Projects*

Name	Institution	Telescope	Type	Total Number of Channels	Width of Single Channel	Band Covered	Detectable Power for 100-Meter Transmitter at 100 Light-Years
Inner Galactic Plane Survey*	SETI Institute	Allen Telescope Array	Survey	450 million	1 Hz	1,390–1,720 MHz	70 megawatts
SERENDIP	University of California, Berkeley**	Arecibo Radio Telescope (305 m)	Sky Survey	168 million	0.6 Hz	1,370–1,470 MHz	1 megawatt
Southern SERENDIP	SETI Australia Centre	Parkes Radio Telescope (64 m)	Sky Survey	58 million	0.6 Hz	1,418–1,421 MHz	1 megawatt
SETI Italia	Istituto di Radioastronomia, Bologna	Medicina radio telescope (32 m)	Sky Survey	24 million	0.6 Hz	Bands centered at 1.4, 2.8, 6.4, and 22.4 thousand MHz	Typically 30 megawatts

* Values given here are for the Allen Telescope Array capabilities at the end of 2006, when it will have 42 of its 6-meter antennas in operation. As more antennas are added (eventually reaching a total of 350), the sensitivity will increase and the observing mode will switch to a targeted search of stars within 1,000 light-years.

** About 2.5% of the data collected by the SERENDIP project is being distributed over the Internet for processing on a downloadable screen saver. This project (called SETI@home) has involved more than five million home computer users.

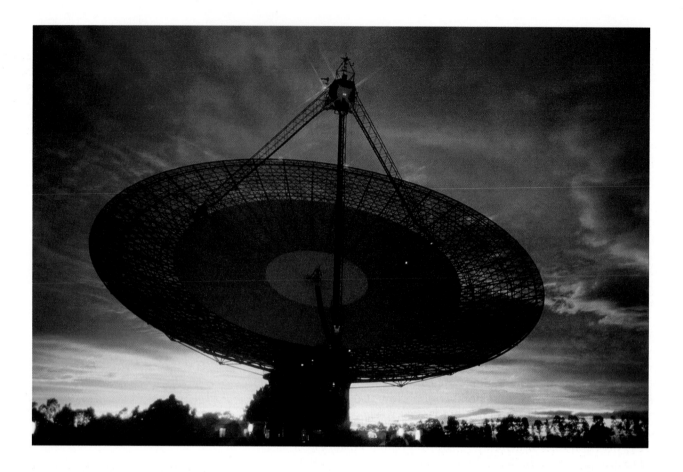

Figure 12.9

The 64-meter Parkes radio telescope in New South Wales, Australia. A SETI experiment "piggybacks" on this telescope while it is engaged in other astronomical research. Piggyback schemes avoid competition for telescope time and therefore provide SETI experiments with a lot of data. On the other hand, the SETI astronomers—who are only "along for the ride"—have no say in where the telescope is aimed. Since there are stars in every direction, however, this type of sky survey could still chance on a transmitting civilization.

While today's radio SETI experiments are vastly more sensitive and comprehensive than Drake's pioneering Project Ozma, they have been limited by two important constraints. First, all projects until now have used telescopes that are also in service for other astronomical research. Thus, they get only limited observing time. In some cases, such as SERENDIP, Southern SERENDIP, and SETI Italia, the SETI studies "piggyback" on other observations, which means that the choice of where the telescope points is dictated by the needs of these other research programs. Second, terrestrial radio interference, especially from radar and orbiting satellites, is becoming an increasing problem for SETI efforts.

The Allen Telescope Array (Figure 12.10) offers a solution to the first problem. Construction is now under way for this telescope in northern California. This instrument will eventually consist of 350 small (6-meter) radio dishes with a combined collecting area equivalent to that of a single 100-meter telescope. One major advantage of an array, as compared to a single-dish radio telescope, is that it can be electronically focused on several stars simultaneously, thus speeding up the search. The Array is a joint venture of the SETI Institute and the University of California, Berkeley, and it will be used full time for SETI observations. It can also be used simultaneously for conventional radio astronomy studies.

Solving the second problem, terrestrial radio interference, is more difficult. This interference greatly hinders SETI searches, because telecommunications satellites, aircraft, and Earth-bound radar produce lots of narrow-band signals—exactly the type being sought. These signals are so strong that they are picked up by radio telescopes no matter which direction they are pointed in. SETI researchers have devised various tricks

Figure 12.10

The Allen Telescope Array, now being constructed by the SETI Institute and the University of California at Berkeley in Hat Creek, California. This array of small (6-meter-diameter) dishes will eventually consist of 350 antennas (inset). The Inner Galactic Plane Survey is being conducted using a small subset of this final number.

to sort out terrestrial signals from possible extraterrestrial ones, but this problem will continue to worsen. Perhaps in the future we will be able to place radio telescopes on the far side of the Moon, where they would be shielded from Earth's noisy radio presence.

Optical SETI and Beyond

The fact that most SETI searches have used radio telescopes may seem logical, since we often think of radio as something that we "listen to," but it is not the only possibility. Remember that radio is a form of light and that we "listen" to it only after the light is converted to sound by a radio receiver. In essence, the radio station uses electronics to convert the information content of the sound (for example, music) into long-wavelength light waves—wavelengths far beyond what the human eye can perceive. These light waves travel through the air to your receiver, where electronics converts the information back into sound. It just so happens that the wavelengths of light that we use for this process are in the radio part of the spectrum (see Figure 3.30), which is why we say we are "listening to the radio." Broadcast television is also encoded in radio waves, but in this case the encoding is used for pictures in addition to sound.

The common use of radio waves for encoding sound or images is a technological choice, not a requirement. In principle, any form of light would do—and often does. For example, if your computer or television is hooked up to a fiber-optic cable network, you are receiving information transmitted in the form of infrared or visible-light waves bouncing through the fiber optics, rather than in the form of radio waves. Thus, it's possible that alien civilizations might choose to communicate with visible light or other forms of light besides radio, and some SETI efforts are designed to search for such signals.

The idea of communicating between worlds via visual signals has a long and interesting history. In fact, scientists considered using light as a communication medium even before radio's invention. During the

COSMIC CALCULATIONS 12.2

Sensitivity of SETI Searches

The distance from which a SETI experiment could detect an alien transmission depends on the sensitivity of the receiver and on the strength of the signal. The inverse square law for light [Section 7.1] tells us that if aliens broadcast a signal with power P, it will have weakened by a factor d^2 by the time it reaches Earth, where d is the distance of the broadcasting civilization. Thus, the strength of the signal S we receive at Earth is this diminished power $\frac{P}{d^2}$ multiplied by the area of the receiving radio dish, A_r, and a constant of proportionality, k:

$$S = k \times A_r \times \frac{P}{d^2}$$

Example: A SETI search using the 300-m-diameter Arecibo radio telescope can pick up a 10-million-watt signal from 1,000 light-years away (if the transmitting antenna is also 300 m in diameter). If we were to detect a similar signal coming from 25,000 light-years away, how powerful would the alien transmitter have to be?

Solution: Since all else is held constant in this problem, the only change is that the second case is 25 times farther away than the first case. Thus, for a given transmitted power P, the signal strength S will be reduced by a factor of $\frac{1}{25}^2 = \frac{1}{625}$. To get the same S, the transmission power from this star would have to be 625 times stronger, or 625×10 million = 6 billion watts.

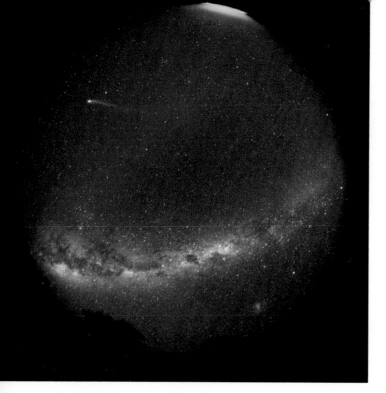

Figure 12.11

A "fish-eye" photograph of the Milky Way in the Australian sky. The dark rift running through the center of the Milky Way is dark because starlight is blocked from our view by interstellar dust. (Comet Hyakutake is visible near the upper left in this 1996 photo.)

nineteenth century, several scientists advanced proposals intended to put us in touch with sophisticated societies that were imagined to live on either the Moon or Mars. The eminent mathematician Karl Gauss is said to have suggested planting trees in the form of a right triangle in Siberia, bounded by clear-cut squares. This greenery graphic presumably would be seen by other solar system inhabitants, proving that Earthlings are at least smart enough to know the theorem of Pythagoras. While this story may be apocryphal, Gauss is definitely known to have proposed using 100 mirrors, each about 1 meter on a side, to shine the light of gas lamps into space. "This would be a discovery even greater than that of America, if we could get in touch with our neighbors on the Moon," he claimed. Another nineteenth-century suggestion was to dig trenches in the Sahara in various geometric shapes, fill them with water and oil, and ignite them at night.

Despite these old ideas, until fairly recently optical SETI did not draw much attention from researchers. Part of the reason had to do with estimates of the energy needed. In general, it requires far less energy to send an interstellar signal via radio than via visible light, so researchers guessed that radio would be the aliens' technology of choice. In addition, while radio waves travel through interstellar space with ease, visible light tends to be absorbed by tiny grains of interstellar dust that float between the stars. This dust creates the dark rift through the center of the Milky Way as we see it in our sky (Figure 12.11). When we look in different directions within the galactic plane, the dust completely blocks our view of visible light beyond a few thousand light-years—a relatively short distance compared to the 100,000-light-year diameter of our galaxy. The fact that visible light can be used to send signals for only limited distances through the galaxy seemed to argue against its use as a beacon for interstellar communication.

Researchers no longer believe that either problem is a serious strike against optical signaling. The limitations of interstellar dust are real, but signals that can travel up to a few thousand light-years could still be quite useful. After all, many millions of stars lie within 1,000 light-years of Earth. Moreover, if longer-distance communication is needed, using infrared rather than visible light would largely circumvent the problem, because infrared light penetrates the interstellar dust. The energy–cost issue argues in favor of radio only if we consider transmitters that beam their signals in all directions into space. For focused communication, such as sending a signal to a particular nearby star, visible light can easily be concentrated into a narrow beam (by using a large lens or mirror). Whereas radio is ideal for a radio station that wants its signal to be picked up by people living in all directions from the station, visible light from a flashlight (or a laser beam) is a better medium if you want to signal someone in a particular place. SETI researchers have concluded that visible light might well be useful for interstellar communication and that optical SETI thus could be a worthwhile enterprise.

How might an alien civilization communicate optically? Simply turning a continuously shining, high-powered laser on someone else's star system isn't particularly effective. A continuous beam could be lost in the glare of starlight from the transmitter's home sun unless it's extremely bright. In addition, a light that is always on doesn't convey much information (think of Morse code). Short bursts of laser light, say a billionth of a second long, would work better, because they can momentarily outshine the starlight even with relatively modest laser power. A series of such bursts—a pulse train—could contain patterns that make up a message. Such arguments have led us to imagine that advanced alien societies might be using automated, high-powered laser transmitters to send pulsed messages to thousands of nearby stars. The messages, perhaps only a few seconds long, could be repeated every several dozen hours. If a civilization in our galactic neighborhood is sending laser messages, then we might hope to find the laser "pings" by monitoring nearby stars.

Optical SETI efforts today are attempting to do just that. Like their radio counterparts, these efforts are passive—that is, they are searching for incoming signals but are not transmitting anything outward—and are systematically checking out the vicinities of local stars. The experiments look for short pulses of light that are bright enough to outshine the background shower of photons naturally produced by the alien world's host star. Photomultiplier tubes, sensitive light detectors able to see short flashes, are affixed to conventional mirror telescopes to hunt for these photon bursts. Figure 12.12 shows the setup for an optical SETI experiment at the Lick Observatory, near San Jose, California. This experiment is quite sensitive: If someone is sending out laser pulses with a transmitter no more sophisticated than we ourselves could build, the Lick experiment could pick it up from as far away as about 500 light-years. Researchers at Harvard University have constructed a telescope whose sole job is to scan the sky for short, bright flashes of light sent from other worlds.

THINK ABOUT IT   It has been suggested that one benevolent use for our nuclear weapons would be to rocket them into space, line them up in geometric patterns, and then detonate them as a "broadcast signal" to distant alien societies. Do you think this would make an effective SETI signal? Why or why not?

Beyond radio and optical SETI, almost any other form of light might also be used for interstellar communication. We've already noted the potential value of using infrared signals in place of visible signals in order to penetrate interstellar dust. Similarly, distant civilizations might choose to communicate with ultraviolet light or X rays. However, these other forms of light have no obvious advantages over radio, infrared, and optical options, and they are more expensive in terms of the energy cost per "bit"

Figure 12.12

An optical SETI experiment being carried out on the 1-meter Nickel telescope at the Lick Observatory, on Mount Hamilton near San Jose, California. The photomultiplier tube detector, which is designed to react to bursts of light, is contained in the white box attached to the back of the telescope.

Figure 12.13

In the movie *2001: A Space Odyssey,* American astronauts dig up a clearly artificial monolith on the Moon. This is a signaling device left behind by an advanced society, and placed where it could be found only when humans had reached a high enough level of technical sophistication.

of information. As a result, SETI researchers feel justified in concentrating at least our current, early searches on radio and optical frequencies.

What about signals using more exotic technologies? For example, some people have suggested that advanced civilizations might send messages via the ghostlike subatomic particles called *neutrinos* or via the so-called *gravity waves* that Einstein predicted but that we have not yet detected directly.* We cannot rule out these possibilities, but we might wonder what the point would be. After all, neither of these schemes would communicate any faster than light, and they would make both transmitters and receivers far harder to build. Of course, it is also possible and perhaps even likely that sufficiently advanced civilizations have developed communication technologies that we could not detect. These technologies might involve physics that we have not yet discovered, or they might use ways of disguising signals so that our detectors cannot distinguish them from natural background signals. Thus, if other civilizations *want* to keep their communications secret from us, they probably can. SETI's hopes rest with civilizations that either are looking for contact or don't care if their communications are intercepted by others. In either of these cases, sticking with the simplest available technologies would seem to make the most sense.

Searching for Artifacts   There might be other ways of detecting extraterrestrial civilizations besides picking up signals traveling through interstellar space. These possibilities apply if civilizations have advanced far beyond our own capabilities and are able either to travel between the stars or to undertake major "astro-engineering" projects that would be visible at a great distance.

If other civilizations have achieved interstellar travel, then they might have visited our solar system. In that case, they may have left artifacts behind, either accidentally or deliberately. Some people claim that we already have such artifacts from UFOs, but as we'll discuss in Section 12.4, these claims do not meet the standards of science. There are more plausible ways to imagine aliens leaving artifacts, and many science fiction writers have considered them. For example, in the classic movie *2001: A Space Odyssey,* based on the story by Arthur C. Clarke, highly advanced aliens buried a monolith on the Moon (Figure 12.13). When humans found the monolith, it notified the extraterrestrials.

It might be a while before we can search the Moon and the planets for artifacts such as buried monoliths, but a few researchers have suggested that aliens wishing to get our attention might leave calling cards in places where we could find them somewhat more easily. In particular, some people believe we should look especially hard at the so-called Lagrange points of the Earth–Moon system, named for the eighteenth-century French mathematician Joseph-Louis Lagrange (1736–1813). At these five positions in space, the effects of gravity from Earth and the Moon "cancel" in such a way that, if you floated weightlessly at one of these five points, you wouldn't be tugged toward either body. Figure 12.14 shows the five Lagrange points for the Earth–Moon system. Note that three of them are on a line passing through the centers of Earth and the Moon and that the other two are located 60° to either side of the Moon. These latter two positions, known as L4 and L5, are of particular interest. Unlike the other three, they are "stable," which means that if you started at L4 or L5 but

---

*However, we have strong indirect evidence for the existence of gravity waves, and a new instrument called the Laser Interferometer Gravitational Wave Observatory (LIGO) may detect them within just a few years.

then drifted slightly away, the competing effects of gravity from Earth and the Moon would bring you back to your starting point. This effect is much like keeping a marble in a shallow bowl; if the marble is moved a bit off-center, it still rolls back to the bottom of the bowl. As a result, L4 and L5 would be obvious places to leave artifacts in cold storage for possible retrieval after long periods of time. Two decades ago, a limited survey of the Lagrange points was made with small telescopes. The hope was to find bright objects that might indicate the presence of parked artifacts. Although the search didn't turn up anything clearly artificial, these locations might be worth a more thorough reconnaissance in the future.

If interstellar travelers have not left artifacts for us to find, we might still be able to detect their powerful, interstellar spacecraft. As we will discuss in Chapter 13, spacecraft capable of traveling at speeds close to the speed of light will likely have enormous engines powered by energy sources such as nuclear fission, nuclear fusion, or matter–antimatter drives. These engines would leave telltale signs of their operation—signs that might be detectable at distances of hundreds or even thousands of light-years away. Although only a limited search for distant rockets has been made, this type of phenomenon might be inadvertently discovered in the course of more conventional astronomical research.

Sufficiently advanced civilizations might also betray their presence through tremendous feats of "astro-engineering." In terms of a civilization's ability to exploit natural energy resources, the twentieth-century Russian physicist Nikolai Kardashev suggested that there might be three distinct categories of civilization:

1. **Planetary (or Type I) civilizations,** which use the resources of their home planet.
2. **Stellar (or Type II) civilizations,** which corral the resources of their home star.
3. **Galactic (or Type III) civilizations,** which employ the resources of their entire galaxy.

We are in the first category, since we exploit (often with abandon) only the meager resources of our home planet. The lights of our cities and the heat from our homes and factories, while considerable, are feeble on a cosmic scale and would be extraordinarily difficult to detect at the distances of the stars. Civilizations in the third category are so advanced that they might be difficult to imagine. As Arthur C. Clarke has stated, "Any sufficiently advanced technology is indistinguishable from magic."

However, we might be able to detect civilizations in the second category. Such a civilization, for example, might decide to capitalize on solar energy in a big way. A star like the Sun puts out a great deal of power. If we could capture just 1 second's worth of the Sun's total energy output, it would be enough to supply today's world demand for energy for approximately the next million years. But nearly all the Sun's energy escapes into space, and Earth intercepts only the tiny bit that heads our way. In principle, a technologically adept civilization could capture *all* of its star's energy by fashioning a large, thin-walled sphere (possibly built from a dismantled outer planet) around its solar system and covering the inner surface with solar cells or their equivalent. Such spheres are called **Dyson spheres**, after physicist Freeman Dyson, who proposed their possible existence (Figure 12.15).

Because a Dyson sphere would absorb all the light of its interior star, you might think it would be invisible; however, the laws of physics dictate that waste heat must escape from the sphere. This heat would be radiated as infrared light that we could detect with specialized telescopes.

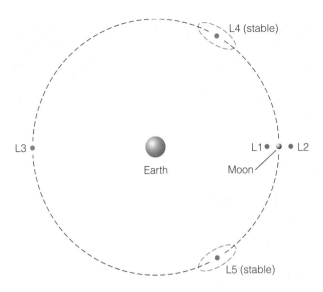

Figure 12.14

The five Lagrange points of the Earth–Moon system. The points L4 and L5, with the stable regions schematically indicated by the dashed ovals, are most attractive for the long-term parking of artifacts or probes. As the Moon orbits Earth, the Lagrange points also orbit, so that they remain in the same relative position with respect to Earth and the Moon.

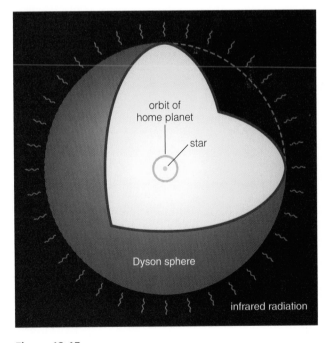

Figure 12.15

Schematic representation of a Dyson sphere, constructed exterior to an inhabited planet, with waste heat escaping from its outer surface in the form of infrared light.

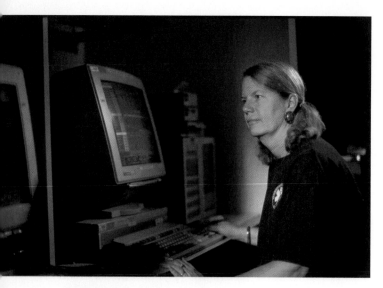

Figure 12.16

A SETI experiment at the Arecibo radio telescope in Puerto Rico. The observer, Jill Tarter (the model for the fictional Ellie Arroway in the novel and movie *Contact*), monitors the progress of the observations on computer workstations. It is the computers that do the "listening," since many millions of channels are monitored simultaneously.

Thus, we could discover the presence of a stellar civilization by finding the infrared signature of a Dyson sphere. Limited searches have already been undertaken, so far to no avail.

• What happens if SETI succeeds?

Until now, no SETI experiment has turned up a confirmed extraterrestrial transmission. Several intriguing signals have been reported, however. Perhaps the best known was found in 1977 by the (now-defunct) automated SETI search at the Ohio State Radio Observatory. When an Ohio State astronomer examined the data from the night's observing, in the form of a computer printout, he found a signal so impressively strong that he wrote "Wow" on the printout margin. The "Wow" signal, made popular by its appealing nomenclature, was never seen again despite repeated attempts at the same frequency and the same sky position. Consequently, SETI researchers do not consider it a detection of extraterrestrial intelligence but instead presume it was some sort of terrestrial interference.

Despite their continued failure to find a persistent, verifiable signal, however, those engaged in SETI research remain optimistic (Figure 12.16). The motivation for this optimism is the rapid improvement in the abilities of the telescopes and detectors used in SETI. Because much of the hardware required for SETI is built with digital electronics, SETI research benefits from the rapid growth in computer capability. The density of transistors placed on silicon chips has been doubling approximately every 18 months for decades (a phenomenon sometimes called *Moore's law*, after Gordon Moore, a cofounder of Intel Corporation). This exponential improvement in electronics motivates the optimism typical of SETI researchers. As an example, Frank Drake has estimated the number of transmitting civilizations, N, as 10,000. If this is so, and if 10% of all stars are suitable for habitation, then one in a million suitable stars will have a transmitting society. The Allen Telescope Array, which is designed to take advantage of technological improvements, will be able to examine approximately a million targets by the year 2025. This suggests that, if Drake's assumptions are correct, a detection might be made within a few decades. Continued progress in astronomy—and in particular, the possibility of learning which extrasolar planets might have atmospheres that indicate the presence of life—could shorten this time even more by improving our choice of star systems to examine.

What will happen if we really do receive an artificial signal from the stars? A signal detection would need to be thoroughly verified by continued observations both at the discovery telescope and at other observatories. The nature of the signal itself will be evidence of artificiality, but scientists need to be sure it isn't due to terrestrial interference, equipment failure, or a college prank. It might take many weeks of careful work by excited astronomers before they are fully confident of a genuine detection.

Once a detection has been verified, the news would be released to the outside world. Given the potential implications of the discovery of an extraterrestrial civilization, this step must be taken with some care. For this reason, SETI research groups have agreed to follow a protocol known as the *Declaration of Principles Concerning Activities Following the Detection of Extraterrestrial Intelligence*. The protocol has no force of law, but it lays out a reasonable course of action. In particular, if a signal has been found and verified as truly extraterrestrial, astronomers around the world would be notified so that they could swing their telescopes in the direction of the signal and learn as much as possible. Governments and the public would also be informed directly. The open nature of research and the necessity to confirm any detection by using other telescopes dictate that the discovery could not be kept secret or covered up. The evidence, after all, is accessible to anyone with access to a large telescope. A recent addition to the protocol addresses the matter of sending a reply to any detected signal. The SETI community has suggested that any deliberate response should represent a consensus of the world's population, not just the wishes of whatever group made the detection.

Suppose we found more than just a signal, but also a message—bits of information. Might we be able to decipher it? In fiction, messages from aliens often have a mathematical bent, because we assume that mathematics will be developed by any technological society. The fictional aliens send us the value of pi or numbers from a well-known mathematical series. This is usually done so that we will recognize that the signal is from intelligent beings. But this sort of labeling wouldn't be needed, because we are seeking narrow-band radio signals or pulsed laser light that clearly would be artificial. In addition, using mathematics as a "language" would be a cumbersome way to express things such as political systems, religious beliefs, art, or even physical appearance. Our own messages to space have been pictorial (see Figure 12.8b and Figure 13.2), and pictures might be a better way to communicate to unfamiliar societies.

Any SETI success in itself would be an astonishing event, because it would prove that we are not alone in the universe. Depending on the nature of the signal, however, we might learn far more. If a civilization is broadcasting to us, it is likely to be hundreds or even thousands of light-years distant, and two-way communication would be tedious at best. A broadcasting society therefore might not expect a reply and might simply send a one-way message, perhaps some version of their encyclopedia or information from their internet. Such a message might take us a while to figure out but would be worth the trouble. From our discussion of the Drake equation, we learned that the chances of detecting another society increase with the average age of technological civilizations. Consequently, if we do find a signal, the chances are great that it comes from a civilization far older than our own. The knowledge we might gain from such an advanced society would be enormous.

Finally, we should also be aware of the possibility that the signal we find will not be a beacon intended for societies like ours, and might contain a message that we can never decode. Even in that case, we would at least have learned that we are neither alone nor particularly special.

12.4 THE PROCESS OF SCIENCE IN ACTION: UFOS AND ALIENS ON EARTH

A fundamental assumption of all SETI efforts is that our best hope of detecting an alien civilization lies with looking beyond Earth for signals or other evidence of their existence. But what if the aliens are already right here on our planet? Remarkably, public opinion polls since the 1960s have consistently shown that about half the American public believes that we are being visited by alien spaceships, and many are convinced that hard evidence of such visits exists. If this were true, then in principle we could learn about extraterrestrial intelligence without ever resorting to the use of expensive telescopes and complex receivers. Unfortunately, when we examine these claims of alien visits using the accepted principles of science, they prove to be far from convincing.

What have we learned from UFO sightings?

The bulk of the claimed evidence for alien visitation consists of UFO sightings—many thousands of UFOs are reported each year. No one doubts that unidentified objects are being seen. Even many astronomers have seen objects in the sky that they cannot identify, and some of the most interesting UFO reports have come from seasoned pilots whose credibility is not in question. However, the mere fact that something is unidentified does not automatically make it an alien spacecraft. The real question is not whether UFOs are seen, but whether these sightings are actually of visitors from the stars.

We can gain some insight into this question by examining both the history of UFO sightings and the history of scientific examination of UFO claims. The first modern report of a UFO as an alien spacecraft was made by businessman Kenneth Arnold in June 1947. Arnold was flying a private plane near Mount Rainier, in the state of Washington, when he saw nine shiny objects having wings, but no tails, that appeared to be streaking across the sky at nearly 2,000 km/hr. A reporter for the United Press wrote up Arnold's experience as a sighting of "flying saucers," and the story became front-page news throughout America.

Within a decade, "flying saucers" had invaded popular culture, if not our planet. They were often seen in books and movies, and the term is still used today. This is ironic, given that Arnold didn't actually describe the objects he saw as saucer shaped, or even round. Three decades after the sighting, he explained that this impression was the result of a newspaperman's error. When, in 1947, the United Press reporter asked him how the objects moved, Arnold answered that they "flew erratic, like a saucer if you skip it across the water." He was describing their motion, not their shape. In fact, several later investigators have suggested that the objects seen by Arnold were meteors, streaks of light caused by small particles entering Earth's atmosphere.

Despite the reporter's misunderstanding, the idea of alien disks buzzing the countryside caught the imagination of the public. It also intrigued the U.S. Air Force, which spent two decades conducting investigations into the nature of UFOs. The military interest was prompted largely by Cold War concerns that UFOs might be new types of aircraft being developed by the Soviet Union.

In the 1950s and 1960s, teams of academics met to study the most interesting of the UFO reports. In the overwhelming majority of cases, these experts were able to plausibly identify the UFOs. They included bright stars and planets, aircraft and gliders, rocket launches, balloons, birds, ball lightning, meteors, atmospheric phenomena, and the occasional hoax. For a minority of the sightings, the investigators could not deduce what had been seen, but their overall conclusion was that there was no reason to believe that UFOs were either highly advanced Soviet craft or visitors from other worlds. The Air Force ultimately dropped its investigation of the UFO phenomenon.

However, some believers felt that the inquiries were either incomplete or part of a ruse organized by the government to put people off the scent of alien visitation. The number of UFO sightings increased (of course, the number of human-made objects in the sky was also growing during these years), and countless books, photos, and film clips were offered as evidence (Figure 12.17). Yet little of the photographic material was compelling. Some of it was ambiguous (is that a distant spacecraft or a nearby bird or bug?), and some was clearly faked. Moreover, the fact that about 90% of

Figure 12.17

A UFO. This photo, which shows far more detail than most pictures made of supposed visitors from other worlds, has the familiar saucer shape made popular by a reporter's error. In fact, this photo is faked, and the saucer is only a lamp shade.

the well-documented UFO cases were explainable as earthly phenomena only encouraged some people to point to the 10% that weren't clearly explained. However, this is not a valid argument for the idea that the unexplained cases represent alien spaceships. After all, if a metropolitan police department solves 90% of the murder cases in a large city—all committed by humans against other humans—it doesn't suggest that the other 10% were committed by, say, aliens.

We can see the problems with UFO "evidence" even more clearly when we evaluate the sightings using the methods of science [Section 2.3]. Recall that modern science seeks natural causes to explain observed phenomena. Although aliens presumably would be "natural" if they really exist, invoking them to explain unidentified objects is no different from invoking ghosts, spirits, or Greek gods and goddesses. Any of these things might be the real explanation for UFOs, but simply saying so does not give us any insight into their true nature. We can learn something only if we can make a model to describe the phenomena of UFOs, and then test the model against additional observations. Reports of UFO sightings do not allow us to do either: Various UFO sighting reports are so different from one another that it's difficult to create any testable hypothesis about the designs of the supposed alien spacecraft, the types of engines that they use to traverse interstellar distances, or the behavior of their occupants. We are left with just individual, eyewitness reports—and as we discussed in Chapter 2, these reports offer no way for other people to evaluate their validity.

The bottom line is that UFO reports tell us nothing more than the fact that people sometimes see things in the sky that they are unable to identify or explain. But the fact that an object has not been identified does not even make it unidentifiable—it might just be that the person making the report did not recognize something that another person might have considered obvious—and it certainly does not make it an alien spacecraft. If we want to establish evidence of alien visitation on Earth, we'll need evidence that's much more solid.

THINK ABOUT IT   Suppose you had a year and an unlimited budget with which to investigate the claim that UFOs represent alien spacecraft. What sorts of experiments would you conduct? What types of evidence might convince you—and the scientific community—that fast-moving lights in the sky are nonterrestrial craft?

• Have aliens left any verifiable evidence of visitation?

If aliens land in Paris, hold press conferences, shake hands with crowds, and provide artifacts from their home world, scientists would surely accept their visit as a fact. But short of that, what could convince scientists that aliens have been here? According to the hallmarks of science, we'd need something that anyone could examine, at least in principle, and that stands up to intense and continued scrutiny. While it's conceivable that photographic evidence might do—for example, if the same UFO was photographed from many different observatories—some sort of artifact or other hard evidence would be even better. Some people claim that such evidence already exists, but to date none of it has withstood serious scientific scrutiny. Let's investigate a few of the more famous claims.

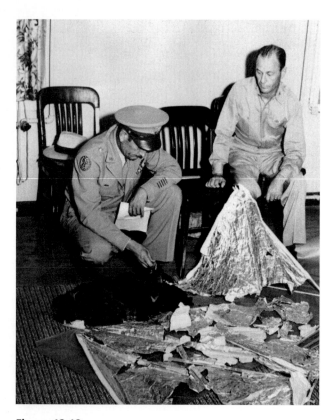

Figure 12.18

Debris recovered in Roswell, New Mexico, in 1947. In this photo, made by newspaper photographer James Johnson, General Roger Ramey is showing the debris to reporters in Fort Worth, Texas.

Crashed Aliens in Roswell   Perhaps the most celebrated case of supposedly alien artifacts concerns the so-called Roswell incident of 1947. The story began in early July 1947, only a few weeks after the nationwide coverage of Kenneth Arnold's "flying saucers." A rancher found some crash remnants in a pasture close to the city of Roswell, New Mexico. He reported the debris to the local sheriff, who then passed on the information to the Roswell Army Air Field, which was nearby. Several military personnel drove out to the ranch, picked up the debris, and explained to the local papers that they had recovered the remains of a "flying disk."

This dismaying story was quickly quashed. Only a day later, an Air Force officer from Fort Worth, Texas, where the debris had been flown, held a press conference in which he dismissed the flying disk notion (Figure 12.18). He stated that the debris was merely a crashed weather balloon, substantially ending interest in the incident at the time. But in 1978, UFO investigator Stanton Friedman began looking into the events at Roswell. Friedman claimed that the debris was from a spacecraft and that alien occupants had been picked up as well. He believed that the government was covering up both the fact of the mishap and its extraterrestrial victims.

The possibility of alien bodies lent this story an appeal beyond the routine sightings of lights in the sky. However, despite the claim of hard evidence, most of the story is based on testimony recorded by Friedman and others—and remember that they conducted their interviews more than *three decades* after the event. Given that long time lag, it's perhaps unsurprising that different witnesses contradicted one another, that some supposed witnesses to the crashed "saucer" had originally claimed not to have seen it, and that a few were caught in flat-out lies. Clearly, the eyewitness testimony by itself could not be considered rigorous enough to prove something as profoundly important as alien visitation.

But what of the crash remains and alien bodies? Something *did* crash at Roswell, and despite the Air Force claim, it was not a weather balloon. Instead, declassified government records tell us that the crash was of a then-secret military balloon experiment designed to detect Soviet nuclear tests. The highly classified operation, known as Project Mogul, used balloon trains to carry the detection devices aloft; the idea was that the balloons would hover at constant altitude near the borders of the Soviet Union and listen for sound waves produced by distant nuclear blasts. Project Mogul was being tested at Roswell in the late 1940s, and the principal scientist behind it has confirmed that the debris recovered by the rancher was from his experiment. Moreover, photos taken of the recovered materials by the Air Force match contemporary photos of Project Mogul balloon trains.

As for the alien bodies or more convincing spacecraft debris, believers can claim only that the government has maintained a tight cover-up, storing the evidence securely in some secret location (such as "Area 51" in Nevada) since the 1947 crash. But consider what that would entail: For some 60 years, through administrations with widely varying political agendas, hundreds of scientists and other military personnel would have had access to such a secret. Yet not a shred of actual evidence has emerged to public view, despite the fact that anyone who produced such evidence would become instantly famous on the talk-show circuit. At minimum, this conspiracy claim seems highly improbable.

In short, claims that the Roswell debris was from a downed alien craft are based on eyewitness accounts made long after the fact, and the supposed evidence of both craft and alien bodies could be real only if you believe the government capable of a highly efficient, six-decade cover-up. The nature of these claims renders them untestable and therefore unscientific, and they seem even more difficult to believe once you realize that we have an alternate explanation—the military experiment—backed by verifiable data in the form of the declassified records and photographs. It might be more interesting to believe that small people from another world just happened to make a navigational error a few weeks after flying saucer mania first swept the country, but the evidence suggests only a crashed Air Force project.

Crop Circles, Abductions, and More

UFOs and crashed aliens can nearly always be readily explained in terms that don't involve visitors from other worlds. The same is true for other highly publicized events interpreted as proof of an extraterrestrial presence in the neighborhood, including abductions, crop circles, and miscellaneous phenomena such as mutilated cattle and goat-eating chupacabras.

Crop circles, geometric patterns made in wheat fields, have generated a great deal of interest among the public (Figure 12.19). For several decades, such patterns have appeared each summer in England. According to some, they are the work of visiting extraterrestrials who are using this method to communicate to us. The proof offered for their nonhuman origin is the speed with which they appear (overnight) and the condition of the matted-down wheat, which is claimed to be inconsistent with simple trampling by humans. But there are many indications that these patterns are merely pranks, not alien messages. The continuing increase in complexity of the patterns over the years suggests that their creators are getting better at their craft—something we would not expect if aliens are using technology brought from distant stars. The fact that the crop patterns are made only at night also suggests human subterfuge. Moreover, many of the crop circles have been acknowledged as pranks, and the human pranksters have demonstrated for news reporters how the crop circles can be made quickly and easily. (Competitions in crop circle making have even been held!) In one interesting case reported in a documentary about crop circles, a self-proclaimed "expert" on the alien origins of crop circles acknowledged that some crop circles were pranks but claimed that not all of them could be. He explained to interviewers why a particular set of circles could not have been made by humans—and then watched in dismay as the interviewers showed him a video of experienced pranksters making the very ones he had just labeled "alien."

Cattle mutilations and the reputed killings of various small animals by short, bipedal creatures called chupacabras have also been attributed to alien activity, despite a centuries-long history of attacks on farm animals by both earthly predators and humans. The only evidence offered to connect these attacks to extraterrestrial beings is uncertainty over who or what is doing the killing. As always, the lack of a confirmed culprit does not mean we are seeing the work of aliens.

A more interesting phenomenon involves claims of alien abduction of humans. Although small in overall percentage of the population, a substantial number of Americans claim to have been alien abductees, often supposedly stolen away by extraterrestrials while asleep (far fewer claim to have been disturbed during waking hours). The victims state that they were either taken aboard spacecraft for observation and unwholesome experiments, or simply watched while they lay immobilized in their beds. These claims may seem difficult to explain away, but a clue to their likely origin comes from research showing that similar phenomena have been described since ancient times in a multitude of cultures. In the past,

Figure 12.19

Crop circles. These relatively simple patterns have been "cropping up" in southern England for several decades and are claimed by some to be manifestations of alien activity. However, the designs are easily produced in a few hours by a small group of motivated students using nothing more than boards and ropes, which suggests a terrestrial origin. In addition, we might wonder why sophisticated extraterrestrials would travel many light-years simply to carve graffiti in our wheat.

the accused culprits have been witches, ghost babies, and goblins. Today, the molester of choice is frequently a visitor from the stars.

According to many psychologists, such abduction experiences are probably attributable to **sleep paralysis,** which occurs during REM (rapid eye movement) sleep. During REM sleep, the body is naturally paralyzed, so that we don't thrash physically along with the movements we make in our dreams. Sometimes, this paralysis persists for a few minutes after the brain has started waking up, and it can give a person the alarming sensation of being awake in a paralyzed body. Visions and other sensations often occur in this state—including strange imaginings that might plausibly include witches, ghosts, goblins, or wide-eyed extraterrestrials.

Surveys suggest that sleep paralysis is experienced by approximately half of all people at some point in their lives. It therefore seems quite reasonable to imagine that a small percentage might believe they had experienced some type of alien abduction. Proponents of alien abduction dispute this idea, noting that a few victims were fully awake and alert during their experience. However, in some cases, daydreams are known to produce sensations similar to those produced by sleep paralysis. Once again, we are left to accept either a simple and earthly explanation (sleep paralysis) or an extraordinary one (abduction by aliens). If you recall our discussion of *Occam's razor* in Chapter 2, you'll understand why scientists prefer the simple explanation, and would accept the extraordinary one only if presented with very strong evidence. As Carl Sagan said, "Extraordinary claims require extraordinary evidence."

Ancient Visitations   The vast majority of those who believe in extraterrestrial visitation claim that the aliens and their craft are among us now. However, some people have suggested that there is evidence for alien visits in the dim past of human history, or even earlier.

The idea that Earth might have been visited seems plausible, especially if civilizations turn out to be common in our galaxy. Our planet might well be of interest to alien scientists, and we'd have no way of knowing whether alien ships studied Earth in the billions of years that preceded the evolution of *Homo sapiens*. Such visits might even have continued after the rise of civilization. But while a few people have claimed that this has happened, the evidence they offer is far from compelling.

The evidence offered for past alien visitation consists of things such as ancient drawings that supposedly show alien visitors or their spacecraft, and archaeological wonders supposedly too sophisticated for our ancestors to have made on their own. These archaeological claims generally make little sense. For example, desert markings on the Nazca plains of Peru (Figure 12.20), claimed by some to be landing strips and markers for alien craft, could just as well be the work of the Nazca Indian culture that occupied the area about 2,000 years ago. No special technology would have been needed to make the markings, and patterns visible only from the air seem more likely to have been intended for gods than for aliens. That some of these drawings look somewhat like astronauts hardly improves the case (Figure 12.21), as they could as easily represent a god or a person wearing a ceremonial headdress.

Claims of alien origin for structures such as the Egyptian pyramids are nothing less than an insult to ancient people, because they seem to suggest that people of these cultures were incapable of constructing such impressive structures on their own. In fact, while the pyramids were a remarkable achievement, their construction was well within the technical expertise of the kingdom of the pharaohs. All other known ancient structures—including Stonehenge, the Mayan pyramids of Central America, and

Figure 12.20

Hundreds of lines and patterns, generally obvious only from the air, are etched in the sand of the Nazca desert in Peru. This aerial photo shows the large figure of a hummingbird. Some people have claimed that these patterns must have been created or inspired by aliens, though it would not have been difficult for the local people to have made them on their own.

the massive stone heads of Easter Island—were also within the technical capability of the cultures that created them.

Perhaps the one thing going for the claims of ancient visitation is that they at least attempt to rely on archaeological artifacts that can in principle be studied by anyone. In that sense, they can be subjected to scientific study. The problem with these claims is that when scientists *have* studied them, the claims have failed to withstand scrutiny.

• Is there a case for alien visits?

Despite the lack of evidence of alien visits past or present, many people continue to believe they have occurred. Champions of alien visitation generally explain away the lack of compelling evidence in one of two ways: government cover-ups or a failure of the mainstream scientific community to take the relevant phenomena seriously. Let's consider both of these arguments in a little more detail.

Government conspiracy and cover-up are popular notions, particularly in the United States. It's certainly conceivable that a secretive government might try to put a lid on the best spacecraft videos and reconnaissance images from orbiting satellites, although the motivation for doing so is unclear. The usual explanations are that the public couldn't handle the news or that the government is taking secret advantage of the alien materials to design (via "reverse engineering") new military hardware. But both explanations are silly. Half the population already believes in alien visitors and would hardly be shocked if newspapers announced tomorrow that aliens were stacked up in government warehouses. As for the reverse engineering of extraterrestrial spacecraft, we should keep in mind the difficulty of traveling from star to star. Any society that could do so would be technologically far superior to our own. Our reverse engineering their spaceships is as unlikely as Neanderthals constructing personal computers just because a laptop somehow landed in their cave. In addition, while a government might successfully hide evidence for a short time, the evidence is unlikely to remain secret for decades (six decades, in the case of the Roswell claims). And, unless the aliens landed only in the United States, can we seriously believe that *every* government would cooperate in hiding the evidence?

The alleged disinterest of the scientific community is also an unimpressive claim. Scientists are constantly competing with one another to be first with a great discovery and thus are aggressive in pursuing any important new phenomenon. Clear evidence that aliens exist and are (or have been) on our planet would be hailed as one of the most important discoveries of all time. Countless researchers would work evenings and weekends, without pay, if they thought such a discovery was possible. The fact that few scientists are engaged in such study reflects not a lack of interest but a lack of evidence worthy of study.

Indeed, lack of physical evidence is the foremost reason given by scientists to justify their skepticism regarding UFOs and alien visits. Airport radars and orbiting satellites have never yielded proof of the passage of an alien spacecraft. No one has ever walked into a research lab with an alien artifact, such as a piece of material that we could not manufacture with our current technology—not even a fragment. In Earth orbit, the U.S. Air Force tracks thousands of pieces of "space junk" from our own satellites and shuttles, including discarded rockets, exploding bolts, a Hasselblad camera, paint chips, and the occasional astronaut glove. But no one has ever found a single piece of space junk that can be attributed to aliens. Most scientists are open to the possibility that we might someday find

Figure 12.21

This 15-meter long etching in the Nazca plains shows an "owl man" that some people claim to be a drawing of an astronaut—presumably one of the astronauts that some people believe used the "runways" also found etched in the desert.

evidence of alien visits, and many would welcome aliens with open arms. But so far, the evidence is simply lacking. Extraterrestrial visitors to Earth remain a routine fixture of cinema and television, but in real life they are like ghosts—a pervasive and attractive idea that the vast majority of scientists treat with skepticism.

Of course, "absence of evidence is not evidence of absence," meaning that a lack of evidence for some claim doesn't make the claim untrue. In the case of alien visitation, it's conceivable that aliens really are here, and that we lack the evidence to prove it because they don't want us to know. If you wish to believe that, no one will stop you—just don't claim that your belief is based on science.

Finally, it is important to distinguish between claims that aliens are visiting us now (or visited Earth in the past) and the possibility that alien civilizations might exist. When, after World War II, space travel moved from the theoretical to the practical, it was only natural to assume that what we were trying to do—travel to other worlds—is routinely done by other civilizations. However, as we'll discuss in the next chapter, there is a great difference between journeys within the solar system and jaunts to the stars. While the former are straightforward, the latter are both enormously difficult and extremely costly in terms of energy. Nonetheless, interstellar travel doesn't violate physics, and if civilizations are common, it is certainly possible and perhaps even likely that some might have been inspired to voyage from star to star.

◯ THE BIG PICTURE
Putting Chapter 12 into Context

In this chapter, we have explored the rationale and the methods of the search for extraterrestrial intelligence. As you continue your study, keep in mind the following "big picture" ideas:

- SETI is both a part of and distinct from other efforts in astrobiology research. Its justifications and methods depend on what we learn more generally about life in the universe. However, whereas other astrobiology research makes slow and steady progress, SETI offers the potential to give us absolute proof in one fell swoop that we are not alone—but only if we receive a clear signal from another civilization.

- We do not yet know enough about life in the universe to make a reasonable estimate either of the number of civilizations that might exist or of our odds of achieving success in SETI efforts. Nevertheless, we have no hope of success unless we try, and contact with another civilization would surely be one of the greatest discoveries in human history.

- The telescopic search for distant civilizations would be rather superfluous if aliens are already here among us. However, despite the many sightings of UFOs and other phenomena supposedly caused by aliens visiting Earth, no compelling evidence for such visits has ever been found. Scientifically, SETI represents our only current hope of detecting other civilizations.

12.1 THE DRAKE EQUATION

• What is the Drake equation?

The **Drake equation** gives us a way to organize our thinking about the question of the number of civilizations in the Milky Way Galaxy. In its modified form, it says that the number of civilizations with whom we could potentially communicate is $N = N_{HP} \times f_{life} \times f_{civ} \times f_{now}$, where N_{HP} is the number of habitable planets in the galaxy, f_{life} is the fraction of habitable planets that actually have life on them, f_{civ} is the fraction of life-bearing planets on which a civilization capable of interstellar communication has at some time arisen, and f_{now} is the fraction of all these civilizations that exist now.

• How well do we know the terms of the Drake equation?

We don't know the values of any of the terms well. We have some data from extrasolar planets that can allow us to make at least an educated guess about the first term, N_{HP}; it seems likely to be quite large, perhaps 100 billion or more habitable planets in our galaxy. For the rest of the terms, we have only the example of Earth to look to, making any guesses far more uncertain.

12.2 THE QUESTION OF INTELLIGENCE

• Even if life is widespread, would intelligence be common?

We really don't know, because we have only the example of Earth. Nevertheless, evolutionary studies indicate at least some drive toward intelligence, so it is at least plausible to imagine intelligence appearing on any planet with life, at least if given enough time.

• Will intelligence inevitably spawn technology?

Certainly, there are some species with physical limitations, such as lack of hands, that would seem to prevent the development of technology. But our own case is ambiguous: It took us a long time to develop technology, but we do not know if this means the development was a fortunate accident or something destined to have happened eventually.

12.3 SEARCHING FOR INTELLIGENCE

• How did SETI begin?

Although there were some attempts at radio contact with aliens in the early twentieth century, in retrospect we know that these efforts were doomed because they used frequencies that are blocked by Earth's iono-

sphere and because they focused on nearby worlds like Mars, where complex life is unlikely to exist. The origin of modern SETI is generally credited to ideas proposed by physicists Giuseppe Cocconi and Philip Morrison. Frank Drake's Project Ozma was the first organized search.

• How do we search for intelligence today?

SETI today is conducted primarily by searching for either radio or optical signals transmitted by distant civilizations. There may be other means of interstellar communication, but it seems reasonable to suppose that radio or optical signals will be used by at least some, if not all, other technological societies. Current signal detection efforts are probably sensitive enough to find only deliberately broadcast beacon signals.

• What happens if SETI succeeds?

The scientific and technological issues of SETI are important but may well pale in comparison to the societal issues if a signal is found. As a result, an important part of SETI work involves thinking about what will happen if the search ultimately proves successful.

12.4 THE PROCESS OF SCIENCE IN ACTION: UFOS AND ALIENS ON EARTH

• What have we learned from UFO sightings?

We've learned that people sometimes see things in the sky that they cannot identify or explain, but we have not found any evidence pointing toward an alien origin for such sightings.

• Have aliens left any verifiable evidence of visitation?

Although many people have made claims of hard evidence from alien visits, none of these claims has ever withstood scientific scrutiny.

• Is there a case for alien visits?

By the tenets of science, there is no current case for alien visits to Earth, either past or present. Keep in mind, however, that absence of evidence is not evidence of absence. If civilizations really are common, then it is conceivable that some aliens have come our way.

REVIEW QUESTIONS
Short-Answer Questions Based on the Reading

1. What is the value of the *Drake equation?* Define each of its terms, and describe the current state of understanding regarding the possible values of each term.

2. What is *convergent evolution?* How does this idea suggest that intelligence would tend to be an evolutionary imperative?

3. Briefly explain the idea of the *encephalization quotient (EQ)*. How does it suggest that humans are indeed intelligent? What does it tell us about intelligence among other animal species?

4. Describe a few physiological and sociological factors that might influence whether an intelligent species can develop technology for interstellar communication.

5. Briefly describe early attempts at interplanetary communication by Marconi and Tesla. Why were these attempts doomed from the start?

6. Briefly discuss early SETI efforts. What do we mean by the *bandwidth* of a signal, and why does SETI concentrate on a search for narrow-bandwidth signals?

7. What are the three general categories of broadcast signals that might be detected at great distance? What are the current prospects for detecting each type of signal through SETI efforts?

8. Why do SETI researchers assume that beacon signals would be designed for easy decoding, and how might we recognize them?

9. Summarize the current techniques of radio SETI and some of the major current projects.

10. Explain why it is reasonable to imagine optical or other signals, and the method behind current optical SETI efforts.

11. Briefly discuss the possibilities of finding other civilizations via artifacts or "astroengineering."

12. What are the three distinct categories of civilization (as outlined by Kardashev)? Which one(s) can we imagine detecting through their use of resources and why?

13. Briefly discuss some of the issues that would surround an actual SETI detection.

14. Discuss several types of claims about alien visitation on Earth. Why, so far at least, do they seem not to reach the level of scientific evidence?

TEST YOUR UNDERSTANDING
Evaluate the Opinions

Each of Problems 15–24 makes a clear statement of opinion. Evaluate each statement and write a few sentences describing why you agree or disagree with it. Explain clearly; not all of these have definitive answers, so your explanation is more important than your chosen answer.

15. Humans are the "crown of creation" and an inevitable result of billions of years of evolution.

16. If, for some reason, we humans were to suddenly wipe out our species, another species—possibly the raccoons—would soon evolve greater intelligence than we possessed.

17. Sea creatures, no matter how clever they are, could never master the technology required to communicate with other worlds.

18. Most of the intelligence in the universe is not biological, but artificial ("machine intelligence").

19. Because SETI researchers are "listening" to star systems that are hundreds of light-years distant, there's a good chance that by the time we hear a signal, the civilization that sent it will have disappeared.

20. No advanced society would ever construct a beacon transmitter, because it would inevitably attract attention and might be dangerous. Similarly, we should not make deliberate transmissions to the stars.

21. We should consider including an "artifact hunt" in the space program that would search on the Moon for objects left behind by advanced extraterrestrial societies.

22. Looking for signals from star systems is a poor approach, because any truly advanced civilization will have moved beyond its home planet and populated interstellar space.

23. If 10,000 people saw the same UFO, scientists would be forced to conclude that an alien visit really occurred.

24. The absence of any scientific evidence for alien visitation on Earth implies that civilizations are rare and that SETI efforts are doomed to failure.

Quick Quiz

Choose the best answer to each of the following. Explain your reasoning with one or more complete sentences.

25. The end result of a calculation with Drake equation is intended to be an estimate of: (a) the number of worlds in the galaxy on which life has arisen; (b) the number of worlds in the galaxy on which intelligence has arisen; (c) the number of worlds in the galaxy on which civilizations are transmitting signals now.

26. Which of the following statements is true about the terms in the Drake equation? (a) Astronomical research will soon give us firm values for all of the terms. (b) Some of the terms depend on sociology, and cannot be determined by astronomers alone. (c) We already know the terms of the equation to an accuracy within a factor of two.

27. The fact that marine predators like dolphins and sharks have similar shapes despite different ancestry is an example of: (a) convergent evolution; (b) narrow bandwidth; (c) spontaneous creation.

28. Which of the following would lead an animal to a higher encephalization quotient (EQ) as it evolved? (a) growth in both body size and brain size; (b) growth in body size but not in brain size; (c) growth in brain size but not in body size.

29. The *bandwidth* of a radio signal is a measure of: (a) its frequency; (b) the range of frequencies that carry information; (c) the amount of power carried by the signal.

30. Why are we more likely to be able to detect a deliberately broadcast "beacon" signal than, say, the television broadcasts of a distant civilization? (a) because we expect beacon signals to be far more common; (b) because our current technology is probably sensitive enough to detect beacons but not much

weaker television transmissions; (c) because television is a sociological phenomenon and beacons are not, so we'd expect all civilizations to have beacons but not all to have television.

31. What is the distinguishing characteristic that those doing radio SETI experiments look for? (a) a signal containing the value of pi and other mathematical constants; (b) a signal that is an echo of an earthly broadcast; (c) a signal that extends over only a narrow band of frequencies.

32. Two-way conversation with other societies is probably unlikely, even if we make contact. This is mainly because: (a) Aliens won't speak our language. (b) It might be dangerous to get in touch. (c) The time it takes for signals to cross the distance to them could be centuries or more.

33. According to the best available evidence, the famous Roswell crash of 1947 involved: (a) an alien spacecraft; (b) an Air Force balloon experiment; (c) There is no evidence that gives us any information about the crash.

34. One reason that scientists doubt that crop circles have alien origin is: (a) They are always beautiful. (b) They can be easily made by humans. (c) Their appearance is not correlated with sightings of bright lights.

INVESTIGATE FURTHER
In-Depth Questions to Increase Your Understanding

Short-Answer/Essay Questions

35. *Drake Values.* Make your own estimates of values for each of the four terms in the modified Drake equation used in this chapter. Explain how you arrived at each estimate, and then use your estimates to calculate N.

36. F_{now}. Suppose that the number of civilizations in the galaxy has been quite large, say, one million. Does that necessarily mean that other civilizations should exist right now? Explain why or why not, and describe the factors that would influence the answer.

37. *Evolution of Intelligence.* Based on your understanding of natural selection and of the evolution of humans as discussed earlier in the book, describe at least three distinct environmental factors that have contributed to the evolution of human intelligence. Explain clearly.

38. *Intelligence on Other Worlds.* Consider again the three factors you identified in Problem 37. For each one, decide whether you think the same factor would be likely to arise and select for intelligence on another world with life, and clearly explain how you reached your conclusions.

39. *Communication.* Imagine that you have to fashion a short message that would tell extraterrestrials something about human society. What would you "say" using only three simple pictures? What would you write if you could use only a half-page of English text?

40. *Talking Back.* Suppose SETI were to find a signal coming from a star system 200 light-years away. Write a one- to two-page essay describing what, if anything, we should do to establish contact. You should think about how quickly we should respond, what the response should be, and what possible dangers might be involved.

41. *Contact.* Watch the movie *Contact,* and pay careful attention to the SETI experiment described in the first third of the film. How

accurately does this experiment reflect any of the current SETI search programs? Did you spot any obvious scientific or technical errors? Write a one- to two-page essay comparing *Contact* to the reality of SETI efforts.

42. *Invasions of Movie Aliens.* Choose a science fiction movie in which aliens are presumed to be visiting Earth. Identify at least three ideas in the movie that either do or do not meet the standards of being testable by science. Describe each in detail.

Quantitative Problems

Be sure to show all calculations clearly and state your final answers in complete sentences.

43. *How Many Stars to Search?* The number of star systems that a SETI search would have to investigate before achieving success depends on how common signaling societies are in the galaxy. This is the number estimated by the Drake equation. Suppose this number is $N = 1$ million. How many star systems must be checked out by SETI in order to find one signal? What if $N = 1,000$? Assume that there are roughly 100 billion stars in our galaxy.

44. *Distance to E.T.* Suppose there are 100,000 signaling societies in our galaxy. What is the average distance between civilizations, assuming that civilizations are spread evenly throughout the galactic disk?

45. *Actual SETI Searches.* Project Phoenix, the largest search of individual star systems for radio signals before 2005, trained its antennas on Sun-like stars up to about 150 light-years away. How many civilizations would there have to be in order for the average distance between civilizations to be 150 light-years? If the actual value of N is 10,000, would we expect Project Phoenix to have made a detection? What if the actual value of N is ten million? Explain.

46. *Power Used by E.T.* A modern SETI search using the 300-meter-diameter Arecibo radio telescope in Puerto Rico can pick up a 10-million-watt signal from 1,000 light-years away (assuming that the broadcasting aliens have a transmitting antenna that's also 300 meters in diameter). Suppose we wish to use Arecibo to search the far side of the Milky Way Galaxy (roughly 80,000 light-years away) under the same assumptions about our setup and the transmitting antenna. What would be the required power of the alien transmitter for us to detect the signal?

47. *Transmitter Used by E.T.* A modern SETI search using the 300-meter-diameter Arecibo radio telescope in Puerto Rico can pick up a 10-million-watt signal from 1,000 light-years away (assuming that the aliens have a transmitting antenna that's also 300 meters in diameter). Suppose an alien civilization is using this same transmitter setup but is on the other side of the Milky Way Galaxy (roughly 80,000 light-years away). How large an antenna would we need to hear the signal?

Discussion Questions

48. *Measuring Intelligence.* In judging the intelligence of animals, we use the *encephalization quotient* (EQ), which depends on the ratio of brain weight to body weight. Can you think of situations for which this might be a poor way to gauge intelligence? For example, could animals have some special processing needs (such as the navigation mechanism of bats) that would make their brains larger without contributing to their intelligence? Alternatively, could animals exist (on Earth or elsewhere) whose brains

were relatively lightweight but who were still highly intelligent? Defend your opinion.

49. *Detecting Signals.* SETI scientists are sometimes criticized for using "old technology" in their search for signals. Perhaps extraterrestrials have moved beyond radio and light signaling and are using something much more sophisticated. Discuss (a) the advantages of radio and light for interstellar communication and (b) any reasonable alternatives you can think of. There is always the possibility that "new physics" will provide faster or more efficient methods for signaling. Do you think this is a reason to limit current SETI efforts?

50. *Societal Reaction.* It is frequently said that the detection of a signal by SETI would revolutionize human society. Does this statement seem reasonable? Some researchers have tried to find historical events, such as the Copernican revolution or the publishing of Darwin's theories of evolution, whose impacts might compare to that of a SETI detection. Are such examples likely to be accurate in predicting how we would react? How likely do you think it is that a SETI discovery would cause either mass panic or an outbreak of universal brotherhood?

51. *Dealing with UFO Claims.* Given the large number of people who claim to have seen a UFO, you are likely to know at least one such person, now or in the future. Perhaps *you* have seen a UFO. Suppose someone who has seen a UFO believes deeply that it was an alien spacecraft. What, if anything, would you say to that person? Why?

WEB PROJECTS

52. *Current SETI Research.* Go to the SETI Institute's Web site and use links listed there to make an inventory of current SETI projects worldwide. Organize these projects according to whether they are radio or optical, and then separate targeted searches from sky surveys. Prepare a one-page summary of this table that discusses how thorough the current searches for extraterrestrial signals are. You should consider how many star systems have been looked at carefully, how wide a band (for radio searches) has been covered, and how sensitive the searches are.

53. *SETI@home.* This is a project organized by researchers at the University of California, Berkeley, to process radio SETI data on home computers. Download the free SETI@home screen saver onto your computer, and use it to analyze data collected by Project SERENDIP. Write a one-page description of the general processing scheme used by SETI@home, as well as the types of signals it is searching for.

54. *What to Do in Case of a Signal Detection?* Download the text of the protocol *Principles Concerning Activities Following the Detection of Extraterrestrial Intelligence* from the SETI Institute's Web site. Write a short discussion of what SETI scientists expect will happen if they detect an extraterrestrial signal and whether this expectation is realistic. In particular, do you think a discovery could be "covered up," or would it leak out to the public before the scientists themselves were sure of the discovery?

55. *Alien Visits.* Learn more about a claim of alien visitation to Earth, past or present. Write a one- to two-page report explaining the claim and the evidence that supports it and discussing the plausibility of the claim.

13
Interstellar Travel and the Fermi Paradox

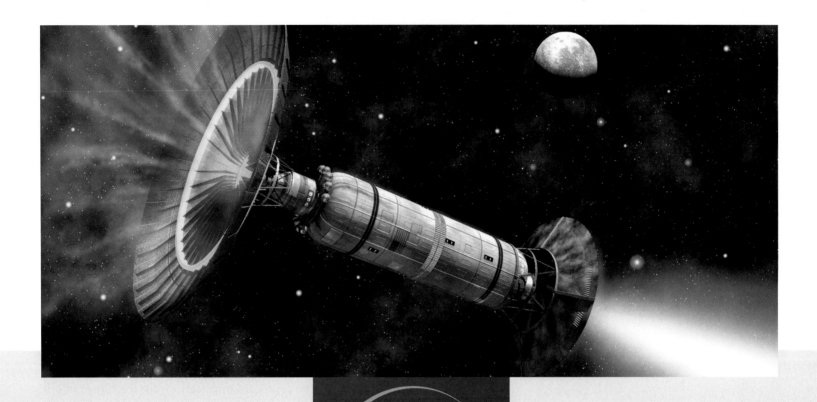

LEARNING GOALS

13.1 THE CHALLENGE OF INTERSTELLAR TRAVEL

- Why is interstellar travel so difficult?
- Could we travel to the stars with existing rockets?

13.2 DESIGNING SPACECRAFT FOR INTERSTELLAR TRAVEL

- How might we build interstellar spacecraft with "conventional" technology?

- How might we build spacecraft that could approach the speed of light?
- Are there ways around the light-speed limitation?

13.3 THE FERMI PARADOX

- Where is everybody?
- Would other civilizations really colonize the galaxy?

- What are possible solutions to the Fermi paradox?
- What are the implications of the Fermi paradox to human civilization?

13.4 THE PROCESS OF SCIENCE IN ACTION: EINSTEIN'S SPECIAL THEORY OF RELATIVITY

- What is "relative" about relativity?
- What evidence supports Einstein's theory?

Provide ships or sails adapted to the heavenly breezes, and there will be some who will not fear even that void ...

Johannes Kepler (in a letter to Galileo), 1593

In an age of rapid technological progress, it may seem inevitable that our rockets will soon reach the depths of interstellar space. The reality, however, is that interstellar travel is much more challenging than bridging the distances to nearby moons and planets. There are engineering and physical constraints, not least of which is the cosmic speed limit—the speed of light. Nevertheless, we can envision at least some ways by which our descendants might someday travel among the stars.

The idea that humans might someday travel throughout the galaxy should make us wonder whether other civilizations have already achieved this ability. Indeed, if civilizations are common, it seems reasonable to expect that some societies—perhaps many—began colonizing the galaxy long before the earliest humans walked the Earth, and maybe even before Earth was born. This idea leads directly to the so-called *Fermi paradox:* If someone could have colonized the galaxy by now, why don't we see any evidence of a galactic civilization?

We will begin this chapter by discussing both the challenges and possibilities of interstellar travel as we understand them today. Then, with that understanding in mind, we will confront the Fermi paradox, and see why, despite the seeming innocence of the question, its solution will undoubtedly have profound implications for the future of our own civilization.

13.1 THE CHALLENGE OF INTERSTELLAR TRAVEL

Science fiction routinely portrays our descendants hurtling through the galaxy, wending their way from one star system to another as easily as we now travel from one country to the next. We have already used rockets to explore other worlds in our solar system. Could future generations travel among the stars just by building larger versions of the rockets we use today? Perhaps surprisingly, the answer is no. The chemical rockets that have sent people to the Moon are wholly inadequate for taking people to the stars.

• Why is interstellar travel so difficult?

The fact that interstellar travel is a daunting enterprise is due to a simple circumstance: the tyranny of distance. The stars are so incredibly remote that only in the nineteenth century did astronomers develop instruments of sufficient precision even to measure the distances of the closest stars besides the Sun. When it was realized just how far away these pinpoints of light are, the French philosopher Blaise Pascal was moved to write that "the eternal silence of infinite spaces" left him terrified.

Spacecraft Bound for the Stars   We've considered the vast distances to the stars in earlier chapters, using the scale model of the solar system introduced in Chapter 3. Here we consider the distances from the point of view of travel. Four of our past interplanetary probes—*Pioneers 10* and *11*, and *Voyagers 1* and *2*—are currently on their way out of the solar system; the *New Horizons* spacecraft will also head toward the stars after it passes by Pluto in 2015. How long will it take these spacecraft to reach the stars?

Let's take the first, *Pioneer 10*, as an example. This spacecraft, launched in the early 1970s, took 21 months to reach its target, the planet Jupiter, before heading out of the solar system. This might seem speedy enough in view of the fact that the giant planet is never closer than 628 million kilometers from Earth. But our nearest stellar neighbor, the Alpha Centauri star system, is 70,000 times farther away than Jupiter. If *Pioneer 10* were to cover the 4.4 light-years to Alpha Centauri at the same average speed at which it traveled to Jupiter, the journey would take 115,000 years. But *Pioneer 10* was not aimed at Alpha Centauri or at any other deliberate target—its trajectory was designed to reach Jupiter and Saturn, not any particular stars beyond. If we plot its trajectory along with the motions of nearby stars, we find that the closest *Pioneer 10* will come to any star in the next million years is 3.3 light-years. In about 2 million years, the probe will reach the general neighborhood of the bright star Aldebaran, in the constellation Taurus.

You can now see why we did not equip *Pioneer 10* or any of these other probes with instruments for studying planets in other star systems. Nevertheless, because the spacecraft themselves should survive unscathed for millions of years in the near-vacuum of interstellar space, we have included messages in case any extraterrestrial beings someday find them. The *Pioneer* probes each carry a small engraved plaque bearing a drawing of a man and a woman as well as diagrams giving the layout of the solar system and our general location in the galaxy (Figure 13.1). The *Voyager* craft, launched about 5 years after the *Pioneer* craft, carry a somewhat more sophisticated message consisting of pictures, multilingual greetings, and two dozen musical selections (ranging from Chuck Berry to Bach) on a gold-plated copper record (Figure 13.2). Although you might wonder how intelligible these earthly calling cards might be to any aliens, the chances that they will ever be found are slim. They are like messages in a bottle thrown into the ocean surf and were intended more as a statement to Earthlings than to extraterrestrials.

THINK ABOUT IT   The *Pioneer* and *Voyager* "messages" are in the form of sounds, music, and pictures. But this assumes that any aliens finding these craft would have sensory organs similar to ours. Is it possible that our messages are too anthropocentric to even be recognized, or are there good reasons to think that E.T. will have eyes and ears, with characteristics similar to ours? Defend your opinion.

The Cosmic Speed Limit   The *Pioneer 10* example makes the problem of interstellar travel quite clear. But it also seems to offer an obvious solution: Build spacecraft that can travel a lot faster. If it would take a little more than a hundred thousand years for *Pioneer 10* to reach the nearest stars, then a spacecraft that travels 100,000 times faster should be able to make the trip in only a little over a year.

However, this seemingly obvious solution is not allowed by the laws of physics. In particular, we know from Einstein's *special theory of relativity*

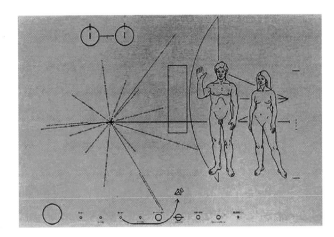

Figure 13.1

The *Pioneer* plaque, carried on both Pioneer 10 and 11 spacecraft, is about the size of an automobile license plate. The human figures are shown in front of a drawing of the spacecraft to give them a sense of scale. The "starburst" to their left shows the Sun's position relative to nearby stellar remnants known as *pulsars*, which are rapidly rotating neutron stars, and Earth's location around the Sun is shown below. Binary code indicates the pulsar periods. Because pulsars slow with time, the periods will allow someone reading the plaque to determine when the spacecraft was launched.

Figure 13.2

Voyagers 1 and *2* carry a phonograph record—a 12-inch gold-plated copper disk containing music, greetings, and images from Earth. One of the etchings on the disk surface gives instructions on how to play it.

that it is impossible to travel through space faster than the speed of light. (We'll discuss this theory and why it imposes a cosmic speed limit in Section 13.4.) This might not seem too limiting, given that light travels incredibly fast—about 300,000 kilometers per second (186,000 miles per second)—fast enough to reach the Moon in barely more than 1 second. But even at this remarkable speed, light takes time to travel the vast distances between the stars, which is why we measure stellar distances in light-years [Section 3.2]. As noted earlier, the nearest star system, Alpha Centauri, is about 4.4 light-years away, which means it takes light 4.4 years to reach us from this system. Because that is the fastest possible speed of travel, the best spacecraft we could hope to build would take longer than 4.4 years for the one-way trip and hence at least 8.8 years for a round trip. To make a trip across our entire galaxy—a distance of 100,000 light-years—would take any spacecraft a minimum of 100,000 years.

Could it be that Einstein's theory is wrong and that we will someday find a way to break this cosmic speed limit? Probably not. Special relativity merits the status of being a scientific theory because it is supported by an enormous body of evidence. Its predictions have been carefully tested and verified in countless experiments, so it cannot simply be "wrong." While it might someday be augmented by a more comprehensive theory, the verified results will not simply disappear; the cosmic speed limit will almost certainly remain in place.

Energy Issues Another challenge of interstellar travel is the tremendous amount of energy it would require, particularly if we wanted to send people and not just lightweight robotic probes to the stars.

Imagine that we wanted to colonize an extrasolar planet. To get a decent-size colony started, we'd need to send a fair number of people with many different sets of skills. For the sake of argument, suppose we wanted to send 5,000 people, meaning we would need a starship of similar capacity to the large starships used in the *Star Trek* television shows and movies. How much energy would such a ship require?

Interestingly, the minimum energy requirement doesn't depend on the fuel source or the ship design at all. Sending a bowling ball flying through the air takes more energy than sending a baseball flying at the same speed, regardless of whether the energy comes from your arm, from a catapult, or from some kind of gas-powered launcher. Similarly, sending either ball flying at a faster speed takes more energy. That is, the energy required to put an object in motion depends on only two things: the object's mass and the speed with which you want it to move.

We can estimate the mass of the starship by comparing it to other ships that transport large numbers of passengers. For example, the *Titanic* weighed about 18,000 kilograms per passenger (although accommodations for most passengers were hardly roomy and it carried provisions for only a couple of weeks, not many years). If we conservatively adopt the same per-person weight for our starship, we would expect its total mass to be about 100 million kilograms. Let's assume further that our starship travels at a modest 10% of light speed, which means it will take more than 40 years to reach the nearest stars. Now that we have estimated both the mass and the speed of our starship, a simple physics formula allows us to calculate the energy needed. (The required formula is the one used to compute kinetic energy, which is equal to $\frac{1}{2}mv^2$, where m is the mass of the moving object and v is its velocity.) The energy needed to get this ship to cruising speed is calculated to be 4.5×10^{22} joules, roughly equivalent to 100 times the world's current annual energy use.

In fact, we should double this value, because the amount of energy required to slow down the ship for a soft landing once we arrive at the colony is the same as that required to accelerate it to cruising speed. Thus, the total energy bill for the trip would be equivalent to at least two centuries' worth of current world energy usage. Let's put this in monetary terms: Using a typical price for home electricity in the United States (10¢ per kilowatt-hour), the energy cost of sending our craft to another star would be about $2,500,000,000,000,000,000. (To this you can add the cost of food and fresh towels for 40 years.) Clearly, unless and until we find a way to produce enormously more energy at vastly lower prices, large-scale interstellar travel will remain out of reach.

• Could we travel to the stars with existing rockets?

Today, when we think of space travel, we invariably think of rockets. But it wasn't always so. The Greeks spoke of celestial travel in the mythological flights of Phaeton and Icarus, neither of which involved rockets and both of which ended badly. As described by the Roman writer Ovid (43 B.C.–18 A.D.), Phaeton was an impetuous youth who borrowed the Sun god's chariot for a joy ride across the sky. Losing control of the horses on this ill-fated mission, Phaeton inadvertently started to set fire to Earth. Humanity's home was saved only when Zeus intervened by killing the cheeky charioteer. Icarus, the son of Daedalus, made a similar ill-advised move and became the first fictional flight victim. The feather wings Icarus used to cruise the heavens were held together with wax, and they underwent a fatal meltdown when, despite his father's warnings, he flew too close to the Sun. These stories suggest that, while they were open to the possibility of other worlds, the Greeks regarded thoughts of human flight as hubris.

The Middle Ages saw a continuation of this less-than-enthusiastic view of human flight. Medieval thinkers were not intrigued by practical travel to the heavens, reserving such jaunts for spiritual ventures. Only after the Renaissance, when the philosophy of modern scientific inquiry took hold, did humankind once again consider the possibility of breaking Earth's bonds. By 1687, Isaac Newton had produced a treatise on universal mechanics that not only described the workings of the heavens but also explained the physics required to reach them.

Newton's third law of motion states that "for every action there is an opposite and equal reaction." Envision the recoil of a gun when it is fired. The bullet moves in one direction, and the gun moves in the opposite direction. Squids, octopuses, and some other mollusks employ a similar technique in their movements. A squid takes in water that it then squirts out at higher speed behind it, thus propelling itself forward. A rocket operates slightly differently, vaporizing on-board fuel that is shot out the back. However, in both cases it is Newton's third law (or, equivalently, the law of conservation of momentum) that accounts for the forward motion.

Development of the Rocket   A rocket has been described as the simplest type of engine, and even scientists as early as Newton realized it could work in empty space. Serious thought about travel to other worlds began in the nineteenth century. In the 1860s and 1870s, the French author Jules Verne wrote influential stories describing travel to the Moon. His propulsion scheme used an oversize artillery shell specially constructed for the task. While this scheme was hardly practical (the enormous acceleration of the shell when fired would turn the passengers to pancakes), Verne's writings stimulated investigation of space travel by three giants of early

COSMIC CALCULATIONS 13.1

The Rocket Equation

The rocket equation tells us how a spacecraft's final velocity, v, depends on the velocity of the exhaust gas expelled out the back, v_e, and the rocket's *mass ratio*. The mass ratio is M_i/M_f, where M_i is the mass of the rocket (including any spacecraft it is carrying) with all its fuel and M_f is the mass of the rocket after the fuel has been burned (that is, the spacecraft and any still-attached but empty fuel tanks). We can write the rocket equation in the following two equivalent forms:

$$v = v_e \ln\left(\frac{M_i}{M_f}\right) \iff \frac{M_i}{M_f} = e^{\left(\frac{v}{v_e}\right)}$$

In the equation at left, "ln" is the natural logarithm; your calculator should have a key for computing this. In the equation at right, e represents a special number with value $e \approx 2.718$; your calculator should also have a key for computing e to any power. If you are familiar with the algebra of logarithms, you can confirm that the two equations are equivalent.

Example: Suppose you want a rocket to achieve escape velocity from Earth (11 km/s) and its engines produce an exhaust velocity of 3 km/s. What mass ratio is required?

Solution: We set the rocket's final velocity to $v = 11$ km/s and its exhaust velocity to $v_e = 3$ km/s, and use the second form of the equation to find the mass ratio:

$$\frac{M_i}{M_f} = e^{\left(\frac{v}{v_e}\right)} = e^{\left(\frac{11 \text{ km/s}}{3 \text{ km/s}}\right)} = e^{\frac{11}{3}} \approx 39$$

The required mass ratio is about 39. As discussed in the text, this mass ratio cannot be achieved with a single-stage rocket, but can be reached with a multi-stage rocket.

rocketry: Konstantin Tsiolkovsky (1857–1935) in Russia, Hermann Oberth (1894–1989) in Germany, and the American Robert Goddard (1882–1945). These fledgling rocket scientists explored many of the theoretical possibilities of this type of propulsion. In particular, they worked out the so-called **rocket equation,** which describes how a vehicle's final speed depends on the propellant's velocity (see Cosmic Calculations 13.1). Both Tsiolkovsky and Goddard realized that it would be difficult for a single rocket to reach **escape velocity,** the speed necessary to overcome gravity and leave Earth behind (about 11 km/s, or 25,000 mi/hr), so they proposed the use of multistage vehicles for space flight. The three pioneers also envisioned space stations, intercontinental ballistic missiles, and ion engines.

At first, few people saw much benefit in turning these ideas into working hardware. In a 1920 technical publication, Goddard mused about the possibility that a sufficiently large rocket could reach the Moon. His speculation was immediately ridiculed by the *New York Times,* which claimed that no lunar-bound rocket could ever work since there is no air between Earth and the Moon, and thus a rocket would have nothing to "push against." However, contrary to the *Times* impression, rockets do not operate by pushing against air—or anything else. They simply employ Newton's third law, firing hot gas in one direction so that the rocket moves in the opposite direction. In fact, atmospheres hinder the performance of rockets, because they create drag that slows rockets down.

A few years later, in 1926, Goddard launched his first liquid-fueled rocket from a field in Auburn, Massachusetts (Figure 13.3a). It reached a height of 13 meters. This heroic, build-it-in-a-garage phase of rocketry was soon surpassed. The 1930s brought larger-scale efforts, particularly in Germany, where the military saw value in guided missiles. The German work culminated in the development of the V-2 rocket, used against England during World War II (Figure 13.3b). The rapid development of rocketry that followed the war was driven largely by German scientists who had been recruited by both the Russians and the Americans. The space age truly began in October 1957 with the launch of the Soviet Union's

Figure 13.3

Early rocketry.

a Robert Goddard stands by his pioneering liquid-fuel rocket in 1926. This craft reached a modest altitude: the height of a four-story building.

b The German V-2 rocket, first launched in 1942, was used against England during World War II. Note that only 16 years separate the V-2 from Goddard's first rocket.

Sputnik I, an 84-kilogram beeping metal ball—the world's first artificial satellite.

Although primarily spurred by national rivalries and military considerations, rocket development over the past five decades has allowed humans and our robot proxies to enter those tantalizing realms that had so long been beyond our grasp. What the Greeks could only dream, we can now do. Today's rockets—the direct descendants of those first envisioned nearly a century ago—are fast enough and powerful enough to allow us to explore the nearby worlds of our solar system. But could they ever be improved enough to take us to the stars?

Limitations of Chemical Rockets   Even today, every rocket we use to launch spacecraft works in basically the same way as Goddard's first rocket. The engines ignite and burn a chemical fuel, such as a mixture of oxygen and kerosene. The chemical burning creates very hot gas, which is expelled through a narrow nozzle, propelling the spacecraft into orbit or to other worlds. These chemical rockets serve our current purposes fairly well (though many people dream of new technologies that would allow us to leave Earth at far lower cost). Unfortunately, they are completely inadequate for interstellar travel.

The largest chemical rocket built to date was the Saturn V, the rocket that carried the *Apollo* astronauts to the Moon (Figure 13.4). This vehicle burned liquid oxygen and kerosene, with water the major combustion product. The hot water vapor was expelled out the back at a speed of about 3 kilometers per second—roughly three times the speed of a rifle bullet. The Saturn V consisted of three separate "stages," that is, three distinct rockets perched atop one another so that each lower stage could drop away after exhausting its fuel. We can see why these multiple stages were useful—and why chemical rockets are limited—by investigating rocket mechanics in a bit more depth.

Let's start by imagining that a rocket like the Saturn V had only a single stage. In order to leave Earth behind, we need to reach Earth's escape velocity of 11 kilometers per second. Using the rocket equation developed by the pioneers of rocketry, we can calculate the **mass ratio** required to attain this speed. The mass ratio is defined as the mass of the fully fueled rocket (including any spacecraft it is carrying) divided by the rocket (and spacecraft) mass after all the fuel is burned. As shown in Cosmic Calculations 13.1, reaching escape velocity requires a mass ratio of 39, meaning that the fueled rocket on the launchpad must weigh 39 times more than the empty rocket and spacecraft alone. That is, the fuel weight would be about 38 times the weight of the spacecraft and the engines. This is clearly a discouraging requirement and one that's just about impossible to meet given the weight of tanks, fuel pumps, fins, and astronauts. Indeed, the best single-stage rockets have mass ratios of only 15 or less.

If it takes a mass ratio of 39 to leave Earth and our best rockets have mass ratios of only 15, how can we ever succeed? As the early rocket pioneers realized, the trick is to use multiple stages. By discarding each stage as its fuel is used up, the upper stages don't need to accelerate the dead weight of those below. The rocket as a whole—with all of its stages—still must weigh 39 times as much as the parts that will actually reach space. But each stage requires a much lower mass ratio, because the weight of the rocket will decrease as stages are discarded. For example, to escape Earth by means of a three-stage rocket (assuming the stages have identical mass ratios and exhaust velocities), each stage need have a mass ratio of only 3.4—well within our capabilities. (See Problem 52 at the end of the chapter.)

Figure 13.4

The Saturn V rocket, which was used to carry the *Apollo* astronauts to the Moon. The most powerful rocket yet built, this now 40-year-old, three-stage design weighed about 30% more than the Space Shuttle at liftoff and was capable of sending a 45,000-kilogram (50-ton) payload to the Moon. With a launchpad mass of 2.8 million kilograms for the Moon trips, its overall mass ratio was 62.

THINK ABOUT IT The Space Shuttle does not use stacked rockets like the Saturn V, but it still uses staging. Explain how. (*Hint:* It should be obvious if you look at a picture of the Shuttle on its launch pad.)

In principle, adding more stages can propel chemical rockets to higher speeds, but not high enough for convenient interstellar travel. For example, an oxygen-kerosene rocket like the Saturn V consisting of a stack of *100* stages (each with a mass ratio of 3.4) would reach a speed of 370 km/sec—33 times faster than the Saturn V but barely more than 0.1% of the speed of light. A trip to Alpha Centauri at this speed would still take some 4,000 years. Using more efficient chemical fuels (such as oxygen and hydrogen, rather than kerosene) can help, but by no more than about a factor of 2. Indeed, no matter what engineering refinements we consider, chemical rockets simply are not powerful enough to deliver large payloads to the stars in a reasonable length of time. Interstellar travel requires a different approach.

13.2 DESIGNING SPACECRAFT FOR INTERSTELLAR TRAVEL

Chemical rockets may be insufficient for travel to the stars, but other technologies hold greater promise. Generally speaking, we can break these technologies into two groups. The first uses "conventional" technology—that is, technology that seems within our grasp (at least if we disregard cost), even if we don't yet have it. The second group involves technologies that are theoretically possible but far beyond our present capabilities. In this section, we'll begin by investigating a few conventional technologies that would allow at least a modest degree of interstellar travel, then move on to explore more far-out ideas.

- **How might we build interstellar spacecraft with "conventional" technology?**

Conventional technologies for interstellar travel are based on the idea that we could adapt existing technologies to the task. None of these technologies would make interstellar travel "easy" — at best, their speeds might reduce the travel time to nearby stars to centuries or decades. Still, if we had unlimited budgets, we could in principle begin work on these technologies today.

Nuclear Rockets Chemical reactions involve shuffling the outer electrons of atoms. While these reactions can seem quite powerful (consider the drama of a Space Shuttle launch), the energy they release is insignificant compared to the amount of energy at least potentially available in the reacting materials. According to Einstein's famous formula, $E = mc^2$, any piece of matter contains an amount of energy E equivalent to its mass m multiplied by the speed of light c squared [Section 3.4], which represents an enormous amount of energy. For example, if you could turn a 1-kilogram (2.2-pound) rock completely into energy, the energy released would be equivalent to that contained in nearly 8 billion liters of gasoline—or as much gasoline as is used by all the cars in the United States in a week. However, while this energy is "there" in any piece of matter, it is very difficult to extract. Chemical reactions extract so little of it that we do not notice any change in the mass of the reacting materials.

Nuclear reactions, in contrast, can noticeably affect the mass of reacting materials. They involve changes in the dense atomic nucleus. Two basic types of nuclear reactions can be used to generate power: fission and fusion. Nuclear **fission** involves the splitting of large nuclei such as uranium or plutonium. When a uranium nucleus is split, approximately 0.07% of its mass is turned into energy. Thus, if 1,000 grams of uranium undergo fission, you'd find that the fission products (the material left over after the fission has occurred) would weigh a total of only 999.3 grams, 0.07% less than the starting weight. Although this mass loss may sound fairly small, the energy it releases dwarfs that released by chemical reactions. Nuclear fission bombs are what destroyed the Japanese cities of Hiroshima and Nagasaki at the end of World War II, and all current nuclear power plants get their energy from fission.

Nuclear *fusion,* the power source of the Sun and other stars [Section 3.2], is about ten times as efficient as fission. Fusion of hydrogen into helium converts about 0.7% of the hydrogen fuel mass into energy. The Sun, for example, fuses 600 million tons of hydrogen into helium *each second.* The resulting helium weighs 0.7% less than the original hydrogen, or about 596 million tons. The other 4 million tons of mass simply "disappears" as it becomes the energy that makes our Sun shine. We humans have managed to achieve nuclear fusion here on Earth, but only in thermonuclear bombs (or "H-bombs") and as yet not in a well-controlled, commercially useful way. This is unfortunate; not only is the fuel for fusion (hydrogen) readily available in water, but the efficiency of fusion is so great—at least compared to current energy sources such as oil, coal, and hydroelectric power—that fusion power would seem almost unlimited if we were able to tap it. For example, if we could somehow hook up a nuclear fusion plant to your kitchen sink, then by continuously fusing the hydrogen in the water flowing from the faucet we could generate more than enough power to meet all the current energy needs of the United States.* That is, with your kitchen faucet fusion plant, we could stop the drilling and importing of oil, dismantle all hydroelectric dams, shut down all coal-burning power stations, get rid of all fission power plants, and still have power to spare. And there'd be no more worries about ongoing contributions to global warming, because fusion does not release any greenhouse gases into the atmosphere.

THINK ABOUT IT   Scientists have been working for decades in hopes of developing the technology for viable nuclear fusion power plants, but so far without success. As a result, governments are reluctant to spend a lot of money on further fusion research, although some is still ongoing. How much effort do you think we should put into fusion power? If we achieved it, how do you think it would change our world?

The tremendous efficiency of nuclear energy over chemical energy was bound to appeal to rocket scientists. In 1955, the U.S. Atomic Energy Commission and the U.S Air Force (and later NASA) embarked on an experiment called *Project Rover* to develop nuclear fission reactors that

---

*Actual attempts to generate fusion power use deuterium (the isotope of hydrogen with one neutron), which is present naturally in the ratio of about 1 part deuterium to 50,000 parts ordinary hydrogen. Thus, with deuterium, the needed water flow would be about 50,000 times greater than that of your kitchen faucet—but this flow (about 130,000 liters per minute) is still only about that of a small stream.

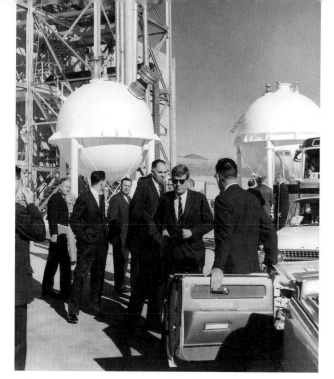

Figure 13.5

President John F. Kennedy departing the Nevada Test Site after viewing a full-scale mock-up of a nuclear-powered engine for Project Rover, December 8, 1962.

could be flown in a rocket. The idea was to use the fission reactor to generate enormous heat, which would be used to bring hydrogen gas to a temperature of millions of degrees before expelling it out the engine nozzles. (Note that the hydrogen was being used as a propellant, not for fusion.) At its peak, Project Rover employed 1,800 people and ultimately tested six fission engines. The program made substantial progress and showed that fission-powered rockets could achieve speeds at least two to three times as great as similar-size chemical rockets. By the late 1960s, NASA officials were confident that the Project Rover rockets could be used to send humans to Mars in what they hoped would be an immediate follow-on to the *Apollo* Moon landings (Figure 13.5). However, the political climate changed, and the United States abandoned its early plans for a human mission to Mars. Project Rover was terminated in 1973.

Another experimental approach, dubbed *Project Orion,* was more radical. Physicists at Nevada's Los Alamos Scientific Laboratory realized that one way to get a rocket up to much higher speed would be to toss small nuclear (fusion) bombs out the rear and let the resulting explosions push the craft forward. The bombs, released at a rate of one every few seconds or more, would drop back about 50 meters and then detonate behind a large metal "pusher plate" affixed to the tail of the rocket (Figure 13.6). This would provide an impulse to move the rocket forward. Despite suffering obvious abuse, the pusher plate wouldn't vaporize because it would be exposed to these searing explosions for only a few milliseconds at a time. The Los Alamos scientists calculated that a spaceship 1 mile long accelerated by the rapid-fire detonation of a million H-bombs could reach Alpha Centauri in just over a century. Thus, Project Orion represented the first true "starship" design to be fashioned by humans. No actual construction ever began, although in principle we could build a Project Orion–type starship with existing technology. However, this kind of starship would be very expensive and would require an exception to the international treaty banning nuclear detonations in space. Project Orion ended in 1965, due both to budget cuts and to the nuclear test ban treaty.

Another nuclear rocket design was developed in the 1970s by the British Interplanetary Society under the name *Project Daedalus* (Figure 13.7).

Figure 13.6

Artist's conception of the Project Orion starship, showing one of the small H-bomb detonations that would propel it. Debris from the detonation impacts the flat disk, called the pusher plate, at the back of the spaceship. The central sections (enclosed in a lattice) hold the bombs, and the front sections house the crew.

Figure 13.7

Artist's conception of a robotic Project Daedalus starship. The front section (upper right) holds the scientific instruments. The large spheres hold the fuel pellets for the central fusion reactor.

The idea was to shoot frozen fuel pellets of deuterium and helium-3 into a reaction chamber where they would undergo nuclear fusion. The fuel pellets, about the size of gravel, would be shot into the chamber at a rate of 250 pellets per second. There they would be encouraged to fuse by electron beams, producing a rapid-fire series of explosions that would propel the ship. Because we cannot yet build nuclear fusion reactors, this design remains beyond our current technological capabilities. Nevertheless, the proponents of Project Daedalus developed a plan for sending a robotic spacecraft to Barnard's star, a dim, type M star 6 light-years distant and the next closest star to Earth beyond the Alpha Centauri system. After 4 years of firing the engine, the craft would reach about one-tenth the speed of light and then spend the next four decades coasting to its destination. Once there, it would deploy probes and sensors to relay back photos and other data, giving us our first close-up view of another stellar system only about 50 years after its launch.

Nuclear-powered rockets are undoubtedly feasible in some form. Still, at best they would achieve speeds of about one-tenth the speed of light. Interstellar journeys would be possible, but it would take decades for them to reach even the nearest stars.

Ions, Sunlight, and Lasers   The propulsion schemes described thus far involve a relatively quick acceleration of the rocket to high speed, after which the engines shut down and the craft cruises for whatever length of time it takes to reach its target. Another approach is to use a low-powered rocket whose engines keep firing continuously. The **ion engine** is an example of this approach. It works something like an old-fashioned television picture tube in that it accelerates charged particles (ions). In a television tube, electrons are fired from the back of the tube to the phosphor screen that faces the viewer. An ion rocket engine does the same, with a nozzle taking the place of the phosphor screen. Both NASA and the European Space Agency (ESA) have already used low-power ion engines successfully. These engines can be started only in space (they don't have enough thrust to lift off Earth and work best in a vacuum), but they can keep firing for long periods of time because the mass expelled per unit of time is small. Moreover, the exhaust ions are shot from the craft at tremendous speeds, and so a powerful ion rocket could in principle reach speeds approaching a percent or so of the speed of light.

Other schemes envision spacecraft that can overcome the limitations of the rocket equation by not taking along the bulky fuel. One possibility that's been considered for nearly a century is to use sunlight as power. Large, highly reflective, very thin (to minimize mass) **solar sails** could be pushed by the pressure exerted by sunlight. This pressure is so slight that we normally don't notice it, but in the vacuum of space, where friction is absent, the steady pressure of sunlight impinging on a mirrored sail could push a spacecraft to impressive speed, particularly with sails hundreds of kilometers in size. Solar sailing might well prove to be a fairly inexpensive way of navigating within the solar system, and it could even be useful for interstellar travel (Figure 13.8). Although the push from the Sun would slowly fade once such craft reached the outer solar system (at Saturn, the light intensity is less than 1% its value near Earth), a solar sailing vehicle that was started very near the Sun might achieve speeds of a few percent of light speed. It could then coast to neighboring stars in less than a century.

The fact that sunlight weakens so much with distance limits the ultimate speed of a solar sailing spacecraft. However, we could get around this problem by using a powerful laser as a power source, instead of sunlight. In principle, the laser could provide a steady and continuous "push" for the solar sail, all the way to its destination if necessary. If building large sails proves too difficult, the laser could be used to vaporize propellant on the rocket that would then move the craft forward in the usual rocket-like manner. As these craft moved light-years away, a large focusing mirror hundreds of kilometers in size would be needed at the laser base to concentrate the beam on the pinpoint target that the spacecraft had become. The primary drawback to these schemes is the power requirement. For example, accelerating a ship to half the speed of light within a few years would require a laser that uses 1,000 times as much power as all current human power consumption. Nevertheless, this approach would allow us to travel to nearby stars in a decade or two rather than many decades.

Figure 13.8

Artist's conception of a spaceship propelled by a solar sail, shown as it approaches a forming planet in a young solar system. The sail is many kilometers across. The scientific payload is at the central meeting point of the four ladder-like structures.

A laser-powered rocket would leave the passengers dependent on the efforts of those back on Earth (or wherever the laser installation was located) to keep the laser shining so they could accelerate to their desired final velocity. This might be somewhat risky given the fact that even the fastest of these transports would be en route and accelerating for decades. What if the laser crew went on strike? In addition, with no laser shining in the *opposite* direction, slowing the spacecraft to a halt at its destination (let alone returning home) would be a problem. One possibility is to use on-board propellant heated by the laser. The propellant could then be fired out the front of the craft to slow it down. Alternative braking schemes that use natural magnetic fields in space have also been suggested.

Interstellar Arks   Another, less demanding approach to interstellar travel is often featured in science fiction. Forget the high-tech rocketry and accept relatively low speeds. Then deal with the resulting long travel times by putting the crew into suspended animation—hibernation, if you will—and letting them doze their way to the stars. A challenging variation on this idea is to somehow allow the travelers to live long enough to cruise the galaxy. A third suggestion is to build enormous craft that can accommodate a very large crew: in essence, an "ark." Many generations would live out their lives aboard this slow-moving vehicle before it finally reached its destination.

The difficulty with the first suggestion is that no one yet knows how to put humans to sleep for hundreds or thousands of years (and then have them wake up). However, genetic researchers have identified genes that control hibernation in animals, so it is possible that genetic engineering techniques could someday allow humans to hibernate as well.

The second suggestion, to allow the crew to live the many thousands of years necessary for interstellar travel at conventional speeds, depends on advances in medical technology. Could we somehow stop the aging process, enabling people to live such long lives that centuries or millennia of travel might seem like a walk to the corner store? We simply don't know.

The idea of building interstellar arks usually gets a skeptical reaction from sociologists. They point out that long voyages on Earth (even those that last only a few months) often end badly. Crews splinter into antagonistic factions and frequently fight for control of the ship. In addition, we might justifiably fear a deterioration in the level of expertise of the crew, with the result that the generation of folk who finally reach the target star system would neither remember why it journeyed there nor have the technical skills required to land on or colonize a world in the system.

• How might we build spacecraft that could approach the speed of light?

We have seen that the conventional approaches to interstellar travel—such as chemical, nuclear, or laser-powered rockets or a solar sail—will not bring us to the stars in anything less than decades. What we really need for interstellar travel are ships that can travel at speeds close to the speed of light. We could then reach the nearest stars in years and explore the space within a few tens of light-years of the Sun in just a few decades. Moreover, such ships would be traveling fast enough for on-board passengers to benefit from some astonishing effects of high-speed travel.

The Role of Relativity   The fact that we cannot exceed the speed of light might at first make distant stars seem forever out of reach. However, the same theory that imposes the cosmic speed limit—Einstein's

TABLE 13.1 *Round-Trip Travel Time to Vega*

This table shows the time that passes on Earth and the time that passes for the crew of a spaceship on round-trip journeys at various speeds to the star Vega, a trip of 25 light-years in each direction. Speeds are given as fractions of the speed of light, $c = 300,000$ km/s. Note that the first-row speed of $0.00005c$ is equivalent to 54,000 kilometers per hour, which is roughly the speed of our fastest chemical rockets today.

Speed	Time Measured on Earth	Time Measured on Ship
$0.00005c$	1,000,000 yrs	1,000,000 yrs
$0.1c$	500 yrs	498 yrs
$0.5c$	100 yrs	86 yrs
$0.7c$	72 yrs	52 yrs
$0.9c$	56 yrs	24 yrs
$0.99c$	50 yrs	7 yrs
$0.999c$	50 yrs	2.2 yrs
$0.9999c$	50 yrs	8 months

special theory of relativity—also tells us that time is different for high-speed travelers than for people who stay at home. We'll discuss the reason this occurs in Section 13.4; here, we'll consider its implications to interstellar travel.

Imagine a trip to the star Vega, about 25 light-years away, in a spaceship traveling at constant speed of 90% of the speed of light ($0.9c$). Because light takes 25 years to travel the distance to Vega, and a ship traveling at $0.9c$ is going 90% as fast as light, the ship's travel time to Vega should be $\frac{25}{0.9} \approx 28$ years. This is indeed the time that would be measured by people staying home on Earth; that is, if the ship made the round trip at this speed, it would return home $28 \times 2 = 56$ years after it left. However, it is *not* the time that would be measured by the ship's crew.

Einstein's theory tells us that when a spaceship (or any other object) travels at close to the speed of light, its length becomes noticeably shorter in the direction of movement, its mass becomes noticeably greater, and time measured aboard proceeds noticeably more slowly than time measured by clocks at rest—a phenomenon called **time dilation**. These changes to time and space are not just speculation—they have all been carefully measured in experiments with subatomic particles that move at speeds close to the speed of light. Note that, even while the ship's time is running slow according to people back on Earth, time would feel perfectly "normal" to the crew. But less time really would pass for them. As shown in Table 13.1 (and calculated in Cosmic Calculations 13.2), only about 24 years would pass on the spaceship during the round-trip voyage to Vega at $0.9c$. In other words, this is what would really happen if a ship left on this journey in the year 2100: The ship would return in the year 2156, but the crew would have aged only 24 years. If a crew member was 20 years old when she left, she'd be 44 on her return, but her twin brother, who stayed home, would be 76.

Table 13.1 shows the benefits of relativistic travel for a hypothetical trip from Earth to Vega at various speeds. Notice that, at low speeds, there's no noticeable difference between ship time and Earth time, but the journey takes a very long time. The closer the ship gets to the speed of light, the less time that passes for the crew. Indeed, from the crew's standpoint, the trip can be made arbitrarily short simply by getting ever closer to the speed of light. But for friends left behind on Earth, the rocket can never return less than 50 years after it left.

If you study the table carefully, you might wonder if the crew of a very-high-velocity rocket would conclude that they're traveling faster than the speed of light—which would violate the cosmic speed limit of relativity. For example, at a speed of 99.99% of the speed of light ($0.9999c$), their trip takes only 8 months for a distance we said was 50 light-years. This would seem to imply a speed some 75 times the speed of light. However, special relativity also tells us that distances shrink at high speed. Once traveling at high speed, the crew would find that the distance to Vega is not 25 light-years as we measure it on Earth but instead has shrunk to a little under 0.4 light-year. Thus, they can cover this short distance in a short time, and they'll never think they are traveling faster than the speed of light.

⌒ **THINK** Suppose you were offered the opportunity to take a trip to Vega
⌒ **ABOUT IT** and back at a speed of $0.9999c$. How long would the round trip take, according to you? How much time would pass on Earth while you were gone? All things considered, are there any circumstances under which you would agree to take such a trip? Explain.

Incredible Journeys In fact, if we could somehow boost our spacecraft to speeds arbitrarily close to the speed of light, we could go anywhere in the universe within a human lifetime. Astronomer Carl Sagan considered a hypothetical rocket that accelerates at a steady 1*g* (or "1 gee")—an acceleration that would feel comfortably like gravity on Earth—to the halfway point of its voyage. This constant acceleration would bring the ship closer and closer to the speed of light, though it would never exceed it. (Note that the acceleration of 1*g* is constant, but the speed is not!) The ship then reverses and decelerates at 1*g* to its destination. During most of the trip, the ship would be traveling at speeds quite close to the speed of light, so time would pass quite slowly on the ship compared to time on Earth. Longer trips would mean longer periods of acceleration, bringing the ship even closer to the speed of light for most of the journey.

Calculations show that such a continuously accelerating ship could make a trip to a star 500 light-years away in only about 12 years according to those on board the ship. However, 500 years would pass on Earth. If a crew of 20-year-olds left Earth in the year 2100, they would be merely 32 years old when they reached their destination; but it would be the year 2600 on Earth (actually a bit later, since they are traveling at not quite the speed of light). If they sent a radio message back to Earth, the message would take 500 years to arrive here across the 500-light-year distance. More than 1,000 years after the crew had left, we'd get a message from people who had aged only 12 years since they'd last been seen on Earth.

Even longer trips would be possible in principle. For example, in a craft with a constant 1*g* acceleration, it would require only about 21 years of ship-board time to bridge the 28,000-light-year distance to the center of the Milky Way Galaxy, where the crew could observe firsthand the mysterious black hole thought to reside there. The 2.5-million-light-year distance to the Andromeda Galaxy could be traveled in only about 29 years of the ship's time. Thus, passengers could travel to the Andromeda Galaxy, spend 2 years studying one of its star systems and taking our first pictures of the Milky Way as it appears from afar, and return only 60 years older than when they had left. However, they would not exactly be returning "home," since 5 million years would have passed on Earth. In this sense, special relativity offers sufficiently fast travelers only a one-way "ticket to the stars." No place is out of reach—but you cannot return home to the same people and places you left behind.

In any event, while such incredible trips are allowed by the laws of physics, the energy costs would be extraordinary. Because special relativity also tells us that an object's mass increases as it approaches the speed of light, the energy cost rises just as much as time slows down. Indeed, that is one explanation for why the speed of light cannot be reached: As the ship gets closer and closer to the speed of light, its mass becomes greater and greater, so the same rocket thrust generates ever less additional speed. The mass heads toward infinity as the ship's speed approaches the speed of light—and no force in the universe can give a push to an infinite mass. That is why the ship can never reach the speed of light, no matter how powerful its engines might be, and even getting close to that speed would require amounts of energy far beyond anything we will be able to muster in the near future. Nevertheless, such practical difficulties can't stop us from speculating, and at least two potential ways of approaching the speed of light are known to exist in principle.

Matter–Antimatter Rocketry

The most efficient energy source we have seen so far is nuclear fusion. But fusion converts only 0.7% of the mass of the fusing hydrogen into energy; 99.3% of the mass still remains,

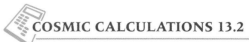

COSMIC CALCULATIONS 13.2

Time Dilation

The effects of time dilation on a fast spaceship can be calculated with a simple formula if we assume that the spaceship travels at constant speed:

$$t_{ship} = t_{Earth}\sqrt{1 - \left(\frac{v}{c}\right)^2}$$

where t_{ship} is the amount of time that passes on the rocket, t_{Earth} is the amount of time that passes on Earth, v is the rocket's velocity (speed), and $c = 3 \times 10^8$ m/s is the speed of light.

Example: Consider a spaceship that travels round-trip to Vega at 90% of the speed of light. As noted in the text, the round-trip travel time measured by people on Earth is 56 years. How much time passes for passengers on the spaceship?

Solution: Because we are given that the ship travels at 90% of the speed of light, or 0.9*c*, we know that $v/c = 0.9$. We plug in this value along with the time that passes on Earth, $t_{Earth} = 56$ yr:

$$t_{ship} = t_{Earth}\sqrt{1 - \left(\frac{v}{c}\right)^2}$$
$$= 56 \text{ yr} \times \sqrt{1 - 0.9^2}$$
$$= 56 \text{ yr} \times \sqrt{1 - 0.81}$$
$$= 56 \text{ yr} \times \sqrt{0.19}$$
$$= 24.4 \text{ yr}$$

This is the approximately 24-year round-trip time shown for the ship in Table 13.1.

as helium. Is there a way to turn more of the mass, or even all of it, into energy? The answer is yes, and it is called **matter–antimatter annihilation.**

Antimatter might sound like the stuff of science fiction, but it really exists. All material things are composed of "ordinary" matter, but physicists have discovered particles that are in some ways the mirror images of normal particles, differing principally in their electrical charge. The first known such particle, christened the *positron*, was discovered in 1932. It is the antimatter twin of the electron; that is, it is identical to an electron except that it has a positive rather than a negative charge. In 1955, the proton's antimatter partner was found: the antiproton. If you were to introduce an antiproton to a positron, they would form an atom of antihydrogen. This antimatter atom would behave chemically just like ordinary hydrogen, except for one thing: You wouldn't want to get near it. When matter and antimatter meet, the result is total annihilation, with 100% of the mass turning into energy. (Note that antimatter still has mass just like ordinary matter; there is no such thing as "antimass.")

The annihilation of matter and antimatter can create energy in a variety of forms. For example, annihilation of positrons with electrons produces energy as a burst of gamma rays. The annihilation of heavier particles, such as antiprotons with protons, produces a gush of particles that soon decay into neutrinos and gamma rays. These might be amenable to powering a rocket, because the flood of reaction particles could be directed out a rearward-facing nozzle. A matter–antimatter rocket could, in principle, achieve speeds of 90% of the speed of light with modest mass ratios.

While such numbers are seductive, the problem lies in rounding up and storing the required antimatter. No practical reservoirs of this material are known. Instead, we would have to manufacture the antimatter, as physicists now do with high-energy particle accelerators. However, current worldwide production of antimatter amounts to only a few billionths of a gram per year, and the energy that would be needed to manufacture larger amounts is staggering. For example, with present technology, manufacturing 1 ton of antimatter—far less than would be needed for an interstellar trip—would take more energy than humankind has used in all of history. Moreover, even if we could make the antimatter, we don't yet know of a good way to store it aboard our rockets, since it would have to be kept in some type of container in which it never touched any ordinary matter at all, since otherwise the result would be premature annihilation.

Interstellar Ramjets   Another approach to achieving relativistic velocities circumvents the problems involved in carrying highly energetic fuel on board. The idea is that a starship could collect its fuel as it goes, using a giant scoop to sweep up interstellar gas. Because this gas would be mostly hydrogen, it could be funneled to a nuclear reactor, fused into helium, and then expelled out the back to propel it. Such propulsion systems are known as **interstellar ramjets** (Figure 13.9). In principle, interstellar ramjets can accelerate continuously by collecting and using fuel nonstop, getting ever closer to the speed of light.

Of course, there are practical difficulties. The typical density of the gas between the stars is only a few atoms per cubic centimeter, so the scoop would need to be hundreds of kilometers across to collect adequate supplies of fuel. As Carl Sagan said, we are talking about "spaceships the size of worlds." Another problem facing an interstellar ramjet—or any ship traveling at relativistic speeds—comes from the interstellar gas and dust itself. At 99% of the speed of light ($0.99c$), a particle the size of a sand grain packs energy equivalent to an explosion of about 100 kilo-

Figure 13.9

Artist's conception of a spaceship powered by an interstellar ramjet. The giant scoop in the front (left) collects interstellar hydrogen for use as fusion fuel.

grams of TNT. Even individual atoms encountered at this speed would be deadly, so the ship would need substantial shielding to protect both its structure and its crew.

Any type of relativistic travel, whether with matter–antimatter engines, interstellar ramjets, or some as-yet-unthought-of method, remains far beyond our current technological capabilities. But we have at least imagined ways by which an extremely advanced civilization *might* be able to travel among the stars. Whether anyone has actually done so remains unknown.

• Are there ways around the light-speed limitation?

If you are a science fiction fan, our discussion of interstellar travel may depress you. Interstellar tourism and commerce seem out of the question, even with ships that travel at speeds close to the speed of light, because of the long times involved (at least as seen from home planets and colonies). If we are ever to travel about the galaxy the way we now travel about Earth, we will need spacecraft that can somehow get us from here to there much faster than the cosmic speed limit would seem to allow. Could such spacecraft be possible?

No one really knows. However, there just might be a "loophole" in the law limiting cosmic speed. In particular, while Einstein's special theory of relativity showed that we can't travel *through* space faster than the speed of light, his general theory of relativity suggests that there might be "shortcuts" that, in effect, let us travel *outside* ordinary space in a way that greatly reduces the distances to be traveled. If so, then we might reach far-off places by taking a shortcut that lessens the distance to them.

Hyperspace In 1916, Einstein enlarged his earlier work (special relativity) to produce the general theory of relativity [Section 2.4]. In this theory, Einstein noted that we live in a four-dimensional universe, with three dimensions of space and one of time; the four dimensions together are usually called **spacetime**. The three spatial dimensions are not rigid and invariant but rather can be warped. Einstein realized that matter can

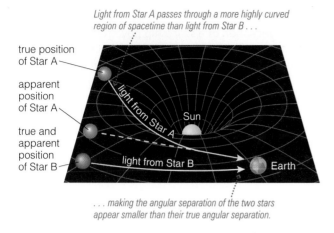

Light from Star A passes through a more highly curved region of spacetime than light from Star B . . .

true position of Star A

apparent position of Star A

true and apparent position of Star B

light from Star A

Sun

light from Star B

Earth

. . . making the angular separation of the two stars appear smaller than their true angular separation.

Figure 13.10

This embedding diagram shows how starlight is bent as it passes near the Sun, causing stars to appear slightly offset from their true positions in space. This effect (exaggerated in this diagram) has been measured during solar eclipses, proving that our space really is curved through hyperspace.

The distance through our universe between Earth and Vega is 25 light-years . . .

our universe

hyperspace

Earth

25 light-years

hyperspace

wormhole

hyperspace

Vega

. . . but the distance would be much shorter if we could travel through a wormhole.

Figure 13.11

Illustration of the idea of a wormhole. Once again, we have reduced our three spatial dimensions to the flat, two-dimensional realm of a rubber sheet. Two distortions of space, one near Earth and one near Vega, have met up in hyperspace, forming a connecting tunnel. Going from Earth to Vega in ordinary space would be a 25-light-year trip. But the wormhole offers a radically shorter route—one that might be traversed in minutes without ever exceeding the speed of light.

provide the distortion. For example, any mass, such as the Sun, will produce space curvature.

How can we tell that space is "curved"? One simple test is to consider the paths of light beams, which travel through space in what we call straight lines. If space had no curvature, then two parallel light beams—for example, from two laser pointers taped side by side—would never cross. However, general relativity insists that the presence of matter can cause a warping of space that will lead these parallel beams to cross. The matter produces a distortion of the ordinary three dimensions of space into other, hypothetical dimensions that we can't see or "get into." These additional dimensions are called **hyperspace.**

To visualize this idea, physicists often resort to "embedding diagrams," such as that depicted in Figure 13.10. Space is reduced from three dimensions to two, and the resulting diagram resembles a rubber sheet; if you inhabited the world shown in the figure, you would be flat, infinitely thin, and incapable of appreciating that any dimension exists on either side of the sheet. Embedded in the sheet is a large mass, such as a star, that causes the sheet to distort into hyperspace. You cannot see the hyperspace dimensions, but you can make measurements on the rubber sheet that will tell you whether or not it is curved. In fact, during solar eclipses, we have measured changes in the apparent positions of stars whose light passes near the Sun. As Figure 13.10 shows, these changes are what we would expect if our space is curved through hyperspace.

The warping of space is usually quite small. Even the bending of starlight passing close to the Sun amounts to only a fraction of a thousandth of a degree. But if space could be distorted more dramatically, the distortion might offer us shortcuts to distant destinations.

Black Holes, Wormholes, and Warp Drive The bizarre objects called black holes are, in fact, holes in spacetime—places where space becomes so distorted that it in effect becomes a bottomless pit. As a result, science fiction writers have sometimes imagined using black holes as shortcuts to other places. Unfortunately, this idea suffers from at least two major drawbacks. First, the only known black holes are themselves extremely far away, so getting to them in the first place would be a problem. Second, we do not know of any way we could survive a close encounter with a black hole.

However, a related phenomenon, called a **wormhole,** might be more useful. Just as a worm might shorten its trip from one side of an apple to the other by tunneling through it, so might a wormhole provide a hyperspace shortcut to a distant part of the universe. Imagine a dense mass floating somewhere near Earth, distorting the surrounding space into hyperspace. Now imagine a similar distortion occurring somewhere else in the cosmos, many light-years away. If these two distortions somehow meet up in hyperspace, they could connect two distant places in ordinary space via a short, hyperspace tunnel (Figure 13.11). Traversing this tunnel might take little time (perhaps minutes) and could short-circuit the necessity of traveling those many light-years.

While this is clearly an appealing idea, could it actually be made to work? In particular, how do we arrange for the wormhole's opportune existence? Quantum physics suggests that on the tiniest scales of the universe, in regions of space far smaller than an atomic particle, spacetime is a seething foam, constantly punctured by distortions into hyperspace. In this highly microscopic world, wormholes might be forming (and self-destructing) all the time. It's conceivable that a highly advanced society might have learned how to capture one of these natural wormholes—one that connected two places of interest—and how to quickly enlarge it to a size that would permit its use for travel. However, such wormhole construction would seem to carry an impossibly large energy cost—estimated to be a thousand times the energy released by an exploding massive star (a supernova). Moreover, even if the energy could be found, we do not yet know of a way to stabilize a wormhole against immediate collapse. All in all, we do not know enough about physics to say for sure whether wormhole travel is even possible. Given this uncertainty, we can but imagine and hope. Carl Sagan's book and movie *Contact* postulated a network of wormhole tunnels permitting fast travel throughout the universe—but even his fictional characters did not know who had built them or how.

Another possible way to travel great distances in a short time might be to exploit the warping of space by placing a dense mass in front of a spacecraft. Hanging like bait on a fishing line, this black hole on a stick would allow the shortening of spatial distances in front of the rocket. Much as you might move across a floor by scrunching a rug in front of you and straightening it out behind you, this highly unusual craft would bend space in front and leave it unaltered behind. You would continually fall into the warped space in front of the craft. This concept comes closest to what we know as the "warp drive" of science fiction. However, it would require either capturing a black hole (difficult, to be charitable) or creating one using enormous amounts of energy. In either case, the black hole would have to be carried along for the ride.

Could there be simpler ways to take advantage of hyperspace? We do not know. This leaves the door wide open for science fiction writers. In the *Star Wars* movies, a simple flip of a lever takes a ship outside our universe and into hyperspace, allowing nearly instantaneous travel to anywhere. In *Star Trek,* a command from the captain sends the ship into warp drive, apparently without the need for a black hole in front of the ship. If such schemes are at all possible, we have no inkling of how they

MOVIE MADNESS: *STAR TREK*

If you want to boldly go where no one has gone before, without spending a few hundred centuries doing it, you need warp drive.

As almost everyone knows, the various incarnations of the *U.S.S. Enterprise, Star Trek*'s famous interstellar transport, high-tail it from one part of the galaxy to another in short order thanks to a futuristic propulsion system. But what *is* warp drive, anyway?

According to the show's technical manuals, the term is merely slang for "continuum distortion propulsion"—a Latinate mouthful that describes a scheme by which powerful fields are used to distort space and allow speeds faster than light. As we discuss in the chapter, rapid travel by warping space is not an entirely nutty idea. It might be possible. And if physics were to allow it in *principle,* could our clever descendants do it in practice?

Maybe yes, maybe no. A major problem is that even if you could warp space, to do so would take enormous amounts of energy. *Star Trek* deals with this small technical detail by fueling the field-generating warp engines with antimatter (anti-hydrogen, to be precise). Combining antimatter with ordinary matter is the most efficient combustion imaginable, as the entire mass of both is converted to energy.

Of course, there's still the problem of making the antimatter, not to mention shipping it to service stations around the galaxy (being careful to keep it out of the hands of pirates—antimatter is costly). But the truly interesting thing about warp drive is the range of speeds attained. At "Warp 1" you're loping along at the speed of light. By "Warp 9"— near the top of the *Enterprise*'s speedometer—you're streaking through space at 1,000 times light speed. This means you can traverse the galaxy in a century, which is short enough to be possible, but long enough to allow you to get effectively stranded and interfere with "Starfleet Command's Prime Directive" to avoid interfering with alien cultures—which, as discussed in Section 13.3, sounds remarkably similar to one of the possible solutions of Fermi's paradox.

Mind you, you could forget warp drive entirely, and stick with the physics we know by building starships that go at 99+ percent of light speed. Special relativity would guarantee that travel times as perceived by the ship's crew would be short. They could cross the galaxy overnight, according to their own watches. But *Star Trek* has opted out of the relativistic approach for good reason, for otherwise the *Enterprise* crew would return home to find all their family and friends long dead and forgotten. Starfleet headquarters would probably be just an archaeological dig.

Better to call up Scotty in the engine room and tell him to put the pedal to the space-bending metal.

Figure 13.12

Enrico Fermi (1901–1954), one of the leading physicists of the twentieth century, received the Nobel Prize in 1938 for work in understanding radioactive decay and predicting the existence of the particles known as neutrinos. By the time he was awarded his Nobel Prize, the Fascists had risen in Italy and the Nazis were in power in Germany. Abhorring these ideologies, he chose not to return to Italy after attending the Nobel Prize ceremony in Sweden. Instead, he moved to the United States, where he became a prominent figure in the Manhattan Project, which was developing the atomic bomb. Element 100 in the periodic table, fermium, was named in his honor.

might work. But who knows what an advanced civilization might have discovered? After all, we have been studying physics in earnest for only a few centuries. Others out there may have been studying it for millions or billions of years.

13.3 THE FERMI PARADOX

We have found that practical interstellar travel is well beyond our capabilities today, but we can envision ways that more advanced civilizations might achieve it. In earlier chapters, we have found good reason to think that habitable planets could be common, some with civilizations, and that many of these civilizations could be much older than ours. These ideas lead to an idea first stated in 1950 by the Nobel Prize–winning Italian-American physicist Enrico Fermi (Figure 13.12). During a lunch at the Los Alamos National Laboratory in northern New Mexico, the conversation drifted to the possibility of extraterrestrial intelligence. The physicists present at the lunch were considering the likelihood that sophisticated cosmic societies might exist in great abundance. Fermi replied to these speculations with a disarmingly simple question: "So where is everybody?" Although serious scientific discussion of his query did not get under way for many years, its central idea is now known as the **Fermi paradox.**

• Where is everybody?

The essence of the Fermi paradox is almost as simple as Fermi's original question. It begins with the idea that neither we nor our planet should be in any way special, in which case other Earth-like planets and other advanced civilizations (meaning civilizations capable of space travel) ought to be fairly common in the galaxy. This is more or less what we conclude from the Drake equation [Section 12.1], unless the rare Earth hypothesis turns out to be correct [Section 11.3]. However, a large number of civilizations would necessarily mean many civilizations with the opportunity to develop advanced technology and interstellar travel long before we came on the scene with our rockets and radio telescopes. In that case, for reasons we'll discuss shortly, it seems that someone else should have colonized the galaxy already. But we see no evidence of such a galactic colonization effort. Thus, we have two seemingly contradictory ideas.

Statement of the Paradox Summarizing the above, we are led to the following two ideas:

1. The idea that neither we nor our planet is in any way special suggests that someone should have colonized the galaxy by now.
2. The idea of a galactic civilization implies that we should be surrounded by evidence of this civilization—but aside from unconvincing claims of extraterrestrial UFOs [Section 12.4], no such evidence exists.

By definition, the existence of two such seemingly contradictory ideas constitutes a *paradox*. But unlike some logical paradoxes (e.g., statements such as "This statement is false"), the Fermi paradox must have some solution. After all, either there is a galactic civilization out there or there isn't.

The Age of Civilizations If you look closely at the first premise of the Fermi paradox, you'll notice that it depends on the idea that, if civilizations are at all common, many should have arisen long before our own arrival.

Recall that the universe is about 14 billion years old [Section 3.2], while Earth is only $4\frac{1}{2}$ billion years old [Section 4.2]. In other words, the universe predates our planet and solar system by some $9\frac{1}{2}$ billion years. Stars began to form quite early in the history of the universe—a fact we know because the ages of the oldest stars are only a few hundred million years short of the age of the universe itself. There's some debate as to whether these early generations of stars had enough heavy elements to make Earth-like planets [Section 11.3], but little doubt that the heavy-element abundance was high enough to make Earth-like planets within a few billion years of the universe's birth. We can play with more precise estimates of these times in a variety of ways, but the bottom line is this: Unless we are misunderstanding some fundamental piece of star and planet formation, it should have been possible for Earth-like planets to have been born starting *at least 5 billion years before* our own planet was born. In other words, some other Earth-like worlds should have had a 5-billion-year head start on ours.

This 5-billion-year head start means that, if intelligent life arose on these planets in the same amount of time that it took intelligent life to arise here on Earth, the first civilizations in our galaxy should have appeared on the scene at least 5 billion years ago. In other words, if these civilizations have survived to the present day, they should be technologically ahead of us by 5 billion years.

We can take the idea a little further by making some guesses about the number of civilizations that have arisen over time. As we discussed in Chapter 12, our current understanding of the Drake equation does not allow any definitive conclusions about the number of civilizations, but it at least seems plausible to imagine that 1 in a million stars would eventually give rise to a civilization on an orbiting planet. Using a conservative estimate of 100 billion stars in the Milky Way—most far older than the Sun—this means there would have been some 100,000 civilizations in our galaxy by now. This is an astonishing idea, and one that becomes even more incredible when we put the number together with the time. If we assume that these 100,000 civilizations have arisen at random times over the past 5 billion years, then on average a civilization arises every 5 billion $\div$ 100,000 = 50,000 years.

Think about what all this means. First, even with the odds of finding a civilization at only 1 in a million stars, there should still have been 100,000 civilizations that would have arisen before we came on the galactic scene. Second, it means that we are almost certainly the youngest civilization in the galaxy at present, and that on average, we'd expect the next youngest civilization to have arisen 50,000 years ago, and the third-youngest to have arisen 100,000 years ago, and so on.

Although the numbers we use for civilizations are essentially wild guesses, things don't change all that much even if we're much more conservative. If we assume that civilizations arise only around 1 in 100 million stars, rather than 1 in 1 million, we still end up with 1,000 civilizations having arisen over the past 5 billion years. And in that case, civilizations would arise about every 5 million years on average, making the next youngest civilization even more advanced relative to us.

THINK ABOUT IT   Consider a couple of other variations on the above theme. How do the numbers change if the fraction of stars with civilizations is higher—say, 1 in 10,000? How do they change if the fraction of stars with civilizations is only 1 in 1 billion? Explain.

These ideas are of course speculative, but you can probably now see why Fermi asked, "So where is everybody?" And you should also be able to see why, if we ever actually meet up with another civilization, it is extremely unlikely that they will be at a technological level of development anywhere near as low as our own.

Machines Deepen the Paradox   So far we have considered the idea that living beings should be out and about in our galaxy, but another possibility is that living beings could send machines outward, much as we have already begun to send robot spacecraft to explore other worlds in our solar system. Consideration of machine technology only deepens the Fermi paradox.

Beyond simply sending out individual robots such as those we can build now, we can imagine that in the future we might build much more sophisticated robots. For example, we might send robots to other worlds with programming instructions to dig up resources on arrival, use the resources to build factories, and use the factories to build spacecraft and more robots. These new robots would then go on to the next world, where they would do the same thing. Thus, these robots would be self-replicating, though in a way quite different from the self-replication of biological beings. The general idea of such self-replicating machines was first proposed by the American mathematician and computer pioneer John Von Neumann (1903–1957), so they are often called **Von Neumann machines.**

The use of Von Neumann machines would allow us to explore much farther and wider than we could by going to other worlds ourselves. Moreover, while interstellar travel poses huge barriers to us due to our limited life spans, these machines could presumably still function after journeys through space that take centuries or millennia. Once we sent the first wave of these machines to a few nearby star systems, they would gradually spread from star system to star system.

In 1981, physicist Frank Tipler used this idea of "colonization" by self-replicating Von Neumann machines to extend the Fermi paradox. In essence, Tipler argued that civilizations could effectively make their presence felt throughout the galaxy even without achieving the ability to send themselves on interstellar journeys. As soon as a civilization reached a level that allowed it to build Von Neumann machines, these machines would begin to spread through the galaxy. Because such colonization would require technology only slightly beyond our own, Tipler argued that if civilizations are common, then the galaxy would already be overrun by self-replicating machines. Because it isn't, Tipler concluded that we are alone and thus that the search for extraterrestrial intelligence (SETI) is a waste of time. Needless to say, other scientists have found plenty of fault with this conclusion, and SETI researchers are still listening hopefully for a signal from the stars. Nevertheless, it's clear that the Fermi paradox is leading us to some deep, philosophical questions about our own civilization and the possible nature of others.

• Would other civilizations really colonize the galaxy?

One obvious question built into the Fermi paradox is that of whether other civilizations really could or would colonize the galaxy. After all, if other civilizations are content to remain quietly on their home planets, or at least in their home star systems, then lots of civilizations could be out there without any having come our way. Thus, to understand the Fermi paradox at a deeper level, we must consider both the capabilities and motives that might lead other civilizations to colonize the galaxy.

Colonization Models   Let's start by assuming that another civilization decided to start sending out spacecraft to colonize other habitable planets. How long would it take this civilization to colonize the entire galaxy?

The answer clearly depends on the civilization's technological capabilities. For example, if it has the technology to build spacecraft that can travel at speeds close to the speed of light, then it could add colonies throughout the galaxy fairly quickly, since trips between nearby stars would take only a few years. Perhaps surprisingly, the conclusion is not that much different if we assume much slower speeds.

Consider a civilization that has nuclear rockets such as the Project Orion or Project Daedalus rockets; as we've discussed, such rockets do not seem that far beyond our own current technological grasp. Recall that such rockets might attain speeds of about 10% of the speed of light ($0.1c$). Given that a typical distance between star systems in our region of the galaxy is about 5 light-years (the average distance is actually slightly lower than this), a nuclear spacecraft traveling at 10% of the speed of light could journey from one star system to the next in about 50 years. This trip would be possible within a human lifetime and might be practical if the colonizers have found ways to hibernate during the voyage or if they have somewhat longer life spans than we do (either naturally or through medical intervention).

After arriving at a new star system, the colonists establish themselves and begin to increase their population. Once the population has grown sufficiently, these colonists send their own pilgrims into space, adding yet more star systems to the growing civilization. Figure 13.13 shows how such colonization would gradually spread through the galaxy. The process starts at the home star system. The first few colonies are located within just a few light-years. These colonies then lead to other colonies at greater distances, as well as at unexplored locations in between. The growth tends to expand the empire around the edges of the existing empire, much like the growth of coral in the sea. For this reason, this type of colonization model is often called a *coral model.*

The overall result is a gradually expanding region in which all habitable planets are colonized. The colonization rate depends on the speed of spacecraft and the time it takes each colony to start sending their own spacecraft to other stars. For travel at 10% of the speed of light, and assuming that it takes 150 years before each colony's population grows enough to send out more colonists, calculations show that the inhabited region of the galaxy expands outward from the home world at about 1% of the speed of light (see Problem 56 at the end of the chapter). Thus, if the home star is near one edge of the galactic disk so that colonizing the entire galaxy means inhabiting star systems 100,000 light-years away, the civilization could expand through the entire galaxy in about 10 million years. The required time would be a few million years less if the home star is in a more central part of the galaxy.

For an even more conservative estimate, suppose the colonists have rockets that travel at only 1% of the speed of light and that it takes each new colony 5,000 years until it is ready to send out additional colonists. Even in this case, the region occupied by this civilization would grow at a rate of roughly $\frac{1}{1,000}$ (0.1%) the speed of light, and the entire galaxy would be colonized in 100 million years. This is still a very short time compared to the time that has been available for civilizations to arise,

Figure 13.13

The coral model of galactic colonization. Colonization begins when the inhabitants of one star system send a few craft to nearby stars. After a time during which the new colonies grow and mature, each new colony sends a few ships with colonists to yet more distant stars, and so on. The colonization "frontier" expands at the edges, much like the growth of coral in the sea.

further deepening the mystery of why we see no evidence that anyone else has done it by now.

Motives for Colonization

In developing a colonization model, we assume that other civilizations would *want* to send out colonists and colonize the galaxy. Is this a reasonable assumption?

We can address this question by considering ourselves as an example, since one of the premises of the Fermi paradox is that we are not special in any way. That is, if we would be likely to colonize the galaxy, then we should assume that others would probably act in the same way. Because we have not yet reached the technological level needed to start interstellar colonization, it's impossible to know with certainty whether we would try it if and when we achieved the capability. However, the history of the human species strongly suggests a predisposition to colonize any new territory available to us.

In many ways our entire history has been one of colonization. Modern humans arose in Africa some 100,000 years ago and almost immediately began expanding around the world. Indeed, the expansion of the human species on Earth probably looked much like the coral model we have described for galactic colonization. Early humans moved outward, gradually encompassing a larger and larger region of our planet. By about 10,000 years ago, our ancestors already lived in almost every place on Earth with a suitable environment for human life. Even after humans had effectively colonized the entire planet, attempts at colonization did not stop. For example, Europeans colonized the Americas, with devastating consequences for the people already living there.

Might our inclination to colonize subside in the future? Possibly, but recent history suggests otherwise. Already there are organizations dedicated to colonizing Mars. It doesn't matter if most people would have no interest in going, because a tiny fraction of the human population would be more than sufficient to start a new colony on another planet. Thus, it seems that if other civilizations are at all like us, they would take advantage of technological opportunities to colonize the galaxy.

Moreover, even if other civilizations don't have an inherited predisposition toward colonization, many other motives might serve to encourage it. For example, some members of an alien civilization might choose to leave their home to escape war or persecution (as did many Europeans and others coming to America, for example). Or a society might deliberately send out colonists in an attempt to make its civilization "extinction proof." For example, while our civilization could easily be wiped out in a variety of ways—from nuclear warfare to environmental catastrophe—a civilization spread among star systems would have a much more difficult time self-destructing, because a particular environmental problem affects only one planet, and the long travel times between stars would make it almost impossible to wage war on multiple planets at once. Finally, if a civilization survived long enough for its star to reach an age at which its home planet was no longer in the habitable zone of its star and life would soon be extinguished [Section 10.4], its members might have no choice but to move on in search of a new home.

THINK ABOUT IT   Consider these and other possible reasons why a civilization might choose to start colonizing other star systems. In general, do you think it is reasonable to assume that other civilizations would attempt galactic colonization? Defend your opinion.

A Hard Look at Motives for Colonization Before we leave this topic, it's worth looking at a few ideas that are sometimes suggested as motives for colonization but that break down on closer examination. The most notable example of such ideas concerns the alleviation of population pressure.

After growing quite slowly for thousands of years, human population began a dramatic upward swing a few hundred years ago. Figure 13.14 shows human population over the past 12,000 years. If the current trend were to continue, today's human population of about six billion (a threshold reached in 1999) would double to 12 billion by about 2050, double again to 24 billion by 2100, and double again to 48 billion by 2150. Indeed, within just a few more centuries of such growth, we would not fit on Earth even if we all stood elbow to elbow (and of course we would have long since exhausted our ability to provide enough food for ourselves by this point). Clearly, our rapid population growth on Earth must stop soon, or we will face an unparalleled catastrophe (see Cosmic Calculations 6.2). So could colonization be the answer to our population problem?

Not a chance. Let's suppose we wanted to stabilize the population at its current size, but through colonization rather than changes in the population growth rate. Currently, we add about 100 million people to Earth each year. Thus, to keep our population stable, we'd need to move 100 million people per year off the planet. A typical Space Shuttle launch costs about $100 million and takes only six or seven people into space (and the Shuttle can't reach other planets, let alone other stars). Even if we somehow found the resources to build enough Space Shuttles or similar craft—and the fuel to launch them—the cost of sending 100 million people into space each year would be more than a quadrillion dollars, some 1,000 times the gross national product of the United States. There are not too many things that we can say are outright impossible, but solving population problems through colonization is one of them.

Another less-than-viable colonization motive is conquest. As we've seen, even if we do find other inhabited planets with intelligent beings, we are likely to be at least either many thousands of years ahead of them or many thousands of years behind them. If we are thousands of years ahead, conquering them would be like the United States conquering cave dwellers—there hardly seems much to gain. If we are thousands of years behind them, we would not be likely to prevail.

Some people suggest that we would colonize not so much through a motivation for direct conquest but rather as the consequence of our species' general tendency toward aggression. This is harder to rule out as a viable motive, but many science fiction writers (including Gene Roddenberry, creator of the *Star Trek* series) have pointed out a potential flaw in this idea: If we continue to be as aggressive and warlike as we have been in the past, we are unlikely to survive long into the future because our capacity for destruction has risen along with our level of technology. Thus, these writers argue that our surviving long enough to achieve the technology for interstellar travel will necessarily mean that we have found ways to overcome our aggressive and warlike tendencies. In that case, colonization will occur because of curiosity and a desire to explore, not because of any desire for empire building.

• What are possible solutions to the Fermi paradox?

We have now seen why the Fermi paradox is real; that is, it really does seem that if civilizations are at all common, then the galaxy should have

Figure 13.14

This graph shows human population over the past 12,000 years. Note the tremendous population growth that has occurred in just the past few centuries.

been colonized long ago. So why don't we see any evidence of a galactic civilization? There are many possible explanations, but broadly speaking, we can group them into three major categories:

1. *We are alone.* There is no galactic civilization because civilizations are extremely rare—so rare that we are the first to have arisen on the galactic scene.
2. *Civilizations are common, but no one has colonized the galaxy.* There are at least three possible reasons why this might be the case:
   i. Technological difficulties. Interstellar travel is much harder or vastly more expensive than we have guessed, so civilizations are unable to venture far from their home worlds.
   ii. Sociological considerations. Our desire to explore is unusual, and other societies choose not to leave their home star systems. Also, for one reason or another, colonizers might run out of steam before they've conquered large tracts of galactic real estate.
   iii. Self-destruction. Many civilizations have arisen, but they have all destroyed themselves before achieving the ability to colonize the stars.
3. *There is a galactic civilization,* but it has deliberately avoided revealing its existence to us.

Let's examine each of these categories in more depth.

We Are Alone   The idea that we are alone is certainly the simplest solution to the Fermi paradox. However, many people object to this solution on philosophical grounds, because it would suggest that our circumstances are very special compared to those that have arisen around any other of the more than 100 billion star systems in the galaxy. If this is true, it would go against almost everything else we have learned since the time of Copernicus. That is, while our ancestors might have imagined our planet to be the center of the universe, more recent astronomical discoveries all seem to suggest that we are not particularly special. Earth is merely a planet orbiting the Sun, our Sun is a rather ordinary star in the Milky Way Galaxy, and our galaxy is much like many other galaxies in the universe.

Of course, while the idea that we are alone might be philosophically unappealing, we cannot rule it out on scientific grounds. Indeed, proponents of the rare Earth hypothesis [Section 11.3] would not be surprised to learn that we are alone. Recall that, according to this hypothesis, the combination of circumstances that allowed for intelligent life to arise on Earth is so rare that we are likely to be the only civilization in the galaxy. The ideas that underlie the rare Earth hypothesis are controversial, but even if they prove incorrect there may be other reasons why we are alone. For example, perhaps some undiscovered law of nature has rendered civilizations impossible until quite recently. In that case, it would not be so strange to imagine that we are the first civilization, even if many others may follow.

Civilizations but No Colonization—Technological Difficulties
The second category of explanations offers the possibility that civilizations are common but colonization is not. Three possible reasons why this might be the case were listed: technological difficulties, sociological considerations, and self-destruction. Let's consider each of these, starting with technological difficulties.

We've seen that interstellar travel seems difficult but not impossible. But could we be underestimating the challenge, and could it actually be so difficult as to be essentially impossible? The energy cost of interstellar travel is sometimes suggested as an impasse. Recall that a large interstellar starship traveling at only about 10% of the speed of light would require energy comparable to what the world currently uses in a hundred years. Clearly, this requirement is prohibitive for us today, and it might be so high compared to the costs of building habitats in our own solar system that migration to other stars might always seem untenable. However, it's certainly possible that an advanced civilization could overcome this problem. The ability to produce power through nuclear fusion, for example, might allow us to generate the needed energy with relative ease. In an extreme case, a civilization capable of building a Dyson sphere [Section 12.3] would have access to all the energy produced by its star. It seems unlikely that the energy requirement alone could preclude all interstellar colonization.

A related possibility is that some other, unknown biological or physical barrier to interstellar travel exists. For example, we have assumed that in the future we'll be able to find a way to keep crews alive for the decades required to go from one star system to the next, but perhaps this is actually much more difficult than we have imagined. Possibly, some type of unknown danger lurking in space prevents intelligent beings from traveling among the stars. Science fiction writers have certainly considered such possibilities—for example, a mysterious effect that causes interstellar travelers to go insane—but they seem far-fetched in light of what we presently know about interstellar space.

It's worth noting that neither energy considerations nor lurking dangers would be enough to stop a civilization from sending out self-replicating Von Neumann machines. But there might be other reasons why no such machines are out there. For example, such machines would tend to grow in number at a rapid (exponential) rate and could in principle use up all the resources in the galaxy in just a few million years. Carl Sagan (in a paper co-written with William Newman) addressed this idea by suggesting that any civilization smart enough to build such machines would also be smart enough to recognize their dangers and therefore would not construct them in the first place.

Civilizations but No Colonization— Sociological Considerations

Let's next turn to sociological considerations. As we've discussed, it seems quite likely that, given the technological and economic opportunity, we will choose to engage in interstellar travel and galactic colonization. But could it be that we are somehow exceptional in having this desire and that other civilizations are perfectly content to stay at home? Like the "we are alone" idea, this idea suggests that we are somehow special rather than typical of intelligent beings and therefore goes against our usual assumption that there's nothing special about our circumstances. After all, we are products of the competitive forces that drive evolution by natural selection, and these forces ought to be similar on any world with life. Moreover, our colonization models show that it would take only *one* other civilization to colonize the entire galaxy in a few million years, so the lack of interest in space travel would have to apply to *every other* civilization that has ever arisen. If civilizations are common—say, if there are the 100,000 civilizations expected if 1 in a million stars has one—it's difficult to believe that not one other civilization has had similar interests to ours.

On the other hand, even with fast rockets, colonization of the entire galaxy would still take millions of years. Perhaps no civilization can maintain enthusiasm for an effort that lasts this long. The individual colonies, separated by many light-years, might evolve along different lines (either biological or cultural), shattering the unity of the empire and bringing further expansion to a halt. While these possibilities are not unreasonable, it still remains true that only one civilization needs to persevere with its colonization efforts in order to bring the entire galaxy under its wing.

One other sociological consideration suggests that advanced societies might start out like us but then "engineer" themselves in such a way so as to shut down their drive to colonize. On our planet, and presumably on others, the development of rocketry occurred at roughly the same time as the invention of nuclear weapons, chemical weapons, and other methods of mass destruction. Societies that remain aggressive are in constant danger of self-destruction. Therefore, civilizations might be motivated to find ways to reduce or channel their aggressive tendencies, perhaps through some type of genetic engineering. The oldest alien cultures, according to this line of reasoning, would have managed to rid themselves of dangerous aggression. It's conceivable that in the process they would have also chosen to focus on improving life on their home planet rather than on moving out into space.

Civilizations but No Colonization—Self-Destruction   A much more sobering possibility for why the galaxy might remain uncolonized even if many civilizations have arisen—one that assumes we are completely typical of intelligent beings—is that societies inevitably self-destruct before attaining the capability for interstellar travel. While this idea is horribly tragic, it is not far-fetched. Nuclear weapons provide clear proof that the technology needed for interstellar travel can also be used to destructive ends. Similarly, any society that learns to tap energy resources would almost certainly use the most accessible energy first, which on any Earth-like planet is likely to be fossil fuels. Thus, like us, other civilizations must face the dangers posed by global warming and other environmental problems. Population growth probably also poses similar problems for all civilizations. Rapid population growth is a natural consequence of biological reproduction for a species that is no longer subject to the whims of predators or childhood diseases, which tend to hold population growth in check. From this perspective, a society can survive long enough to achieve interstellar travel only if it successfully navigates what amounts to a very difficult obstacle course—one in which each obstacle would mean the end of its civilization. Could it be that it simply can't be done?

THINK ABOUT IT   What odds would you give for humanity's surviving long enough to achieve interstellar travel? Defend your choice.

There *Is* a Galactic Civilization   The third category of explanations for the Fermi paradox in essence suggests that there is no paradox at all: The galactic civilization is out there, but we do not yet recognize it. Indeed, UFO buffs might claim that scientists are blind to the obvious proof of this suggestion. While we can't rule out the possibility, no evidence of alien visitation has yet withstood scientific scrutiny. Nevertheless, there are many ways by which a galactic civilization could avoid our detection.

One idea simply assumes that a galactic civilization would have no particular interest in us. After all, to a society that is millions or billions of years ahead of us, we might seem too simple to warrant attention. Of course, if civilizations are communicating or traveling among the stars, we might be able to discover them. Signaling is precisely what SETI experiments look for. The fact that we have not yet received a clearly extraterrestrial broadcast may simply be a consequence of not yet having looked at a sufficient number of stars.

Another possibility is that civilizations are aware of our presence but have deliberately chosen to keep us in the dark. This idea is sometimes called the **zoo hypothesis**, although it might better be called the "wild-life refuge" hypothesis. Just as we set aside nature reserves that are supposed to be left alone as places where wildlife can thrive without our intervention, a galactic civilization might declare planets like ours off-limits to exploration. (*Star Trek* fans might think of this as the "Prime Directive" [to avoid interfering with alien cultures] solution to the Fermi paradox.) One objection to the zoo hypothesis is that even if civilizations wanted to hide from us, we would still be able to intercept their communications among themselves. On the other hand, as we've already noted, SETI searches for signals have so far carefully investigated only a small amount of cosmic real estate and could easily have missed such communications. Alternatively, the communications might involve a technology that we have not yet developed and that therefore is undetectable to us at present.

A closely related idea suggests that a sophisticated galactic civilization might reveal itself to new societies only after they reach a certain level of technology. Perhaps the extraterrestrials place monitoring devices near star systems that show promise of emerging intelligence and patiently wait until these devices record the presence of civilization. This idea is sometimes called the **sentinel hypothesis**, after a science fiction story by Arthur C. Clarke titled "The Sentinel." The story became the basis of the book and movie *2001: A Space Odyssey*, in which a monolith buried on the Moon signals our presence when we finally dig it up (see Movie Madness in Chapter 7). Carl Sagan used a similar idea for the book and movie *Contact* (see Movie Madness in Chapter 12), in which a signaling station around the star Vega amplifies and beams back our own television broadcasts to us, leading us to our first glimpse of a galactic civilization.

• What are the implications of the Fermi paradox to human civilization?

The Fermi paradox may have its origins in a simple question—Where is everybody?—but we have seen that finding an answer is much more complex than asking the question. In fact, if we consider our possible answers in more depth, we find that each leads to astonishing implications for our own species.

Consider the first solution—that we are alone. If this is true, then our civilization is a remarkable achievement. It implies that through all of cosmic evolution, among countless star systems, we are the first piece of the universe ever to know that the rest of the universe exists. Through us, the universe has attained self-awareness. Some philosophers and many religions argue that the ultimate purpose of life is to become truly self-aware. If so, and if we are alone, then the destruction of our civilization and the loss of our scientific knowledge would represent an inglorious end to something that took the universe some 14 billion years to achieve. From this point of view, humanity becomes all the more precious, and the collapse of our civilization would be all the more tragic. Knowing this to be the case might help us learn to put petty bickering and wars behind us so that we might preserve all that is great about our species.

The second category of solutions has much more terrifying implications. If thousands of civilizations before us have all failed to achieve interstellar travel on a large scale, what hope do we have? Unless we somehow think differently than all previous civilizations, we will never go far in space. Given that we have always explored when the opportunity arose, this solution almost inevitably leads to the conclusion that failure will come about because we destroy ourselves. We can only hope that this answer is wrong.

The third solution is perhaps the most intriguing. It says that we are newcomers on the scene of a galactic civilization that has existed for millions or billions of years before us. Perhaps this civilization is deliberately leaving us alone for the time being and will someday decide the time is right to invite us to join it. If so, our entire species might be on the verge of beginning a journey every bit as incredible as that of a baby emerging from the womb and coming into the world.

You can probably now see why the Fermi paradox involves far more than a simple question. No matter what the answer turns out to be, learning it is sure to mark a turning point in the brief history of our species. Moreover, this turning point is likely to be reached within the next few decades or centuries. We already have the ability to destroy our own civilization. If we do so, then our fate is sealed. But if we survive long enough to develop technology that can take us to the stars, the possibilities seem almost limitless.

Imagine for a moment the grand view, a gaze across the centuries and millennia from this moment forward. Picture our descendants living among the stars, having created or joined a great galactic civilization. They will have the privilege of experiencing ideas, worlds, and discoveries far beyond our wildest imagination. Perhaps, in their history lessons, they will learn of our generation—the generation that history placed at the turning point and that managed to steer its way past the dangers of self-destruction and onto the path to the stars.

13.4 THE PROCESS OF SCIENCE IN ACTION: EINSTEIN'S SPECIAL THEORY OF RELATIVITY

In this chapter, we have seen that Einstein's special theory of relativity has important implications to possibilities of interstellar travel. But that is not why Einstein came up with the theory. Rather, as with the development of other scientific theories, Einstein developed the theory of relativity to explain one of the outstanding scientific mysteries of the time, which is why we will take it as this chapter's case study in the process of science in action.

The mystery concerned the speed and nature of light. In 1873, Scottish mathematician and physicist James Clerk Maxwell (1831–1879) published a paper in which he showed that light is an electromagnetic wave [Section 3.4]. His paper included a set of equations—now known as *Maxwell's equations*—that describe the nature of electromagnetic waves. These equations form the basis of our modern understanding of electricity and magnetism, and in essence you are confirming their validity every time you flip a light switch or listen to a radio or television broadcast. However, the idea that light is a wave left a fundamental question: All other types of waves—for example, sound waves, water waves, or waves on a violin string—are carried by some type of medium; what medium carries light waves through "empty" space? Maxwell and other physicists presumed that space must be filled with some medium of unknown composition that could carry the light waves—an idea that had actually been around for some time already—and this medium was known as the *ether*.

If the ether really existed, then Earth's motion around the Sun would make us move in different directions *relative* to the ether at different times of year. In 1887, A. A. Michelson (1852–1931) and E. W. Morley (1838–1923) conducted an elegant experiment—now famous as the *Michelson–Morley experiment*—designed to measure Earth's motion through the ether. The experiment relied on the idea that the speed of light would be slightly faster when it was moving in the same direction as the ether (relative to Earth), slightly slower when moving in the opposite direction, and at speeds in between for other directions. However, their experiment failed to measure any directional difference in the speed of light. In hindsight, the obvious implication was that the ether does not exist. At the time, however, this idea seemed so preposterous that Michelson and Morley went to great lengths to explain how nature might "hide" the ether's existence from human experimenters.

Einstein came up with his theory because he took the results of the Michelson–Morley experiment at face value. Instead of assuming that nature was hiding something from us, he assumed that the speed of light showed no directional variation because none exists; that is, Einstein assumed that the speed of light (through space) is a physical constant, one that will always be measured the same no matter what the motion of the light source or the observer. In doing so, he not only explained the perplexing results of the Michelson–Morley experiment, but he also cleared up some mysteries that had been associated with mathematical implications of Maxwell's equations. In other words, Einstein's theory did not come about by magic or just because he was a smart guy—it came about because there were known problems that needed to be solved, and Einstein was the first to come up with their solution.

Einstein's special theory of relativity is often portrayed as being difficult, but its basic ideas are actually quite easy to understand—though admittedly mind-boggling in their consequences. Because they are so important to our modern understanding of the nature of the universe, let's take a look at the basic ideas and their astonishing implications.

• What is "relative" about relativity?

Imagine a supersonic airplane that flies at a speed of 1,670 km/hr from Nairobi, Kenya, to Quito, Ecuador. How fast is the plane going? At first, this question sounds trivial—we have just said that the plane is going 1,670 km/hr.

But wait. Nairobi and Quito are both nearly on Earth's equator, and the equatorial speed of Earth's rotation is the same 1,670 km/hr speed at which the plane is flying. Moreover, the east-to-west flight from Nairobi to Quito is opposite the direction of Earth's rotation (Figure 13.15). Thus, if you lived on the Moon, the plane would appear to stay put *while Earth rotated beneath it*. When the flight began, you would see the plane lift off the ground in Nairobi. The plane would then remain stationary while Earth's rotation carried Nairobi away from it and Quito toward it. When Quito finally reached the plane's position, the plane would drop back down to the ground.

We have two alternative viewpoints about the plane's flight. People on Earth say that the plane is traveling westward across the surface of Earth. Observers in space say that the plane is stationary while Earth rotates

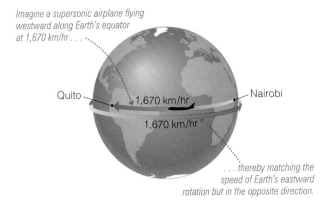

Imagine a supersonic airplane flying westward along Earth's equator at 1,670 km/hr . . .

Quito · · · 1,670 km/hr · · · Nairobi

1,670 km/hr

. . . thereby matching the speed of Earth's eastward rotation but in the opposite direction.

Figure 13.15

A plane flying at 1,670 km/hr from Nairobi to Quito (westward) travels precisely opposite Earth's eastward rotation. Thus, viewed from afar, the plane remains stationary while Earth rotates underneath it.

eastward beneath it. Both viewpoints are equally valid. In fact, there are many other equally valid viewpoints about the plane's flight. Observers looking at the solar system as a whole would see the plane moving at a speed of more than 100,000 km/hr—Earth's speed in its orbit around the Sun. Observers living in another galaxy would see the plane moving at about 800,000 km/hr with the rotation of the Milky Way Galaxy. The only thing all these observers would agree on is that the plane is traveling at 1,670 km/hr *relative to* the surface of Earth.

This example shows that questions like, "Who is really moving?" and "How fast are you going?" have no absolute answers. Einstein's special theory of relativity gets its name because it tells us that measurements of time and space, as well as measurements of motion, make sense only when we describe whom or what they are being measured relative to.

THINK ABOUT IT Suppose you are running on a treadmill and the readout says you are going 8 miles per hour. What is the 8 miles per hour measured relative to? How fast are you going relative to the ground? How fast would an observer on the moon see you going? Describe a few other possible viewpoints on your speed.

The Absolutes of Relativity The theory of relativity tells us that motion is always relative, but it does *not* say that *everything* is relative. In fact, the theory claims that two things in the universe are absolute:

1. The laws of nature are the same for everyone.
2. The speed of light is the same for everyone.

The first absolute, that the laws of nature are the same for everyone, is a more general version of the idea that all viewpoints on motion are equally valid. If they weren't, different observers would disagree about the laws of physics. The second absolute, that the speed of light is the same for everyone, is much more surprising. Ordinarily, we expect speeds to add and subtract. If you watch someone throw a ball forward from a moving car, you'd see the ball traveling at the speed it is thrown plus the speed of the car. But if a person shines a light beam from a moving car, you'd see it moving at precisely the speed of light (about 300,000 kilometers per second), no matter how fast the car is going. This strange fact explains why the Michelson–Morley experiment found no differences in the speed of light, and this has been experimentally verified countless times.

The Speed of Light The cosmic speed limit follows directly from this fact about the speed of light. To see why, imagine that you have just built the most incredible rocket possible, and you are taking it on a test ride. You push the acceleration button and just keep going faster and faster and faster. With enough fuel, you might expect that you'd eventually be moving faster than the speed of light. But can you?

Before we answer this question, the fact that all motion is relative forces us to answer another question: What is your speed being measured relative to? Let's begin with *your* point of view. Imagine that you turn on your rocket's headlights. Because the speed of light is the same for everyone, you must see the headlight beams traveling at the speed of light—which means they are racing away from your rocket at a speed you'll measure to be 300,000 km/s. The fact that you'll see your headlight beams racing away is true no matter how long you have been firing your rocket engines. This shouldn't be too surprising, as it's just another way of saying that you can't catch up with your own light.

Now, however, remember that everyone always measures the same speed of light. This means that people back on Earth—or anyplace else—will also say that your headlight beams are moving through space at 300,000 kilometers per second. In other words, according to anyone watching you from any place in the universe, your headlight beams are traveling at $c = 300,000$ km/s, and they're moving out ahead of you (Figure 13.16). Clearly, this implies that you must be traveling *slower* than the headlight beams, which means slower than the speed of light.

In case you are still not convinced, let's turn the situation around. Imagine that, as you race by some planet, a person on the planet turns on a light beam. Because the speed of light is absolute, you will see the light beam race past you at $c = 300,000$ km/s. The person on the planet will also see the light traveling at $c = 300,000$ km/s and will see the light outrace you. Again, everyone will agree that you are traveling slower than the speed of light.

The same argument applies to any moving object, and it is true with or without headlights. All light travels at the speed of light, including the light that reflects off an object (allowing us to see it) and the infrared light that even cool objects like people emit. As long as the speed of light is absolute, no material object can ever keep up with the light it emits or reflects, which means no material object can reach or exceed the speed of light. Building a spaceship to travel faster than the speed of light is not a mere technological challenge—it simply cannot be done.

The Relativity of Time and Space

Like the cosmic speed limit, the other strange consequences of relativity also follow from the absoluteness of the speed of light. We will not go into details here, but you can understand the general ideas by thinking about the fact that a speed is always equal to a distance divided by a time. For example, a speed of 60 miles per hour means that you travel a distance of 60 miles in a time of 1 hour, and light's speed of 300,000 km/s means that light travels a distance of 300,000 kilometers in a time of 1 second. In our ordinary slow-moving lives, we think of times and distances as absolutes, while speed is relative—as is the case with the speed of the airplane in our Nairobi–Quito example. But Einstein's theory tells us that when it comes to light, speed is absolute—everyone always measures the same speed of light. The consequence is that time and distance must become relative.

We have already seen how time and distance are affected. Time runs slower for high-speed travelers (the phenomenon of *time dilation* that we discussed earlier in the chapter), which is why relativistic starships could allow passengers to make long trips if they traveled fast enough. Distances are also shrunk as measured by the high-speed travelers, which is why high-speed travelers to Vega would see a shorter distance than the distance we measure from Earth. Mass is also affected, so we would measure high-speed objects to have higher mass than they do when they are stationary relative to us.* Another direct consequence of these ideas—

Figure 13.16

The fact that everyone always measures the same speed of light means that you cannot keep up with your own light (top), and therefore no matter what you do, other people must always conclude that you are traveling slower than light. In other words, there is no way for you ever to reach or exceed the speed of light.

---

*You can see why mass must be affected if you think about Newton's second law of motion, which states that force = mass × acceleration (see Figure 2.12). The units of acceleration are a distance divided by a time squared (such as m/s²), so the relativity of time and distance means that even if the force (e.g., from the rocket engines) stays constant, a rocket's acceleration must decrease with increasing speed. The only way to account for this decrease when the force stays the same is to realize that the mass of the rocket must be increasing.

Figure 13.17

If the speed of light were *not* absolute, the speed at which light from a star in a binary system comes toward Earth would depend on its position and velocity toward us in the binary orbit. Thus, we would not see the star as a distinct point of light.

and one that can be derived with nothing more than high school algebra (though we won't do it here)—is Einstein's famous formula, $E = mc^2$.

• What evidence supports Einstein's theory?

Although we haven't gone into details here, we have stated that all the amazing consequences of relativity, including time dilation and $E = mc^2$, follow logically from the absoluteness of the speed of light. However, remember that logic alone is not good enough in science; conclusions must always remain tentative until they pass observational or experimental tests. Does relativity meet the test?

The Absoluteness of the Speed of Light   The first thing we might wish to test is the surprising premise of relativity: the absoluteness of the speed of light. In principle, we can test this premise by measuring the speed of light coming from many different objects and going in many different directions and verifying that the speed is always the same. As we've already discussed, the Michelson–Morley experiment was in essence one such test, and it showed that the speed of light is indeed absolute.

In fact, we have verified this fact many other times. For example, some distant galaxies are moving away from us at speeds close to the speed of light, yet their light still arrives here traveling at the speed of light. Perhaps even more convincingly, if the speed of light were not always the same, we could not see distinct stars in binary star systems. Consider the star moving toward us in Figure 13.17. If the speed of light depended on the star's motion, its light would be coming toward us somewhat faster than the "normal" speed of light. Half an orbit later, when it was moving away from us, its light would travel to us at less than the "normal" speed of light. Thus, the light from the "fast" side of the orbit would tend to catch up with the light emitted earlier from the "slow" side, reaching us at the same time and smearing the star's image so that we'd see it in different positions all at once, instead of seeing it as a distinct star. Thus, the simple fact that we see distinct stars in binary star systems proves that the speed of light does not depend on the stars' motions; that is, light always travels at the same speed. In addition, the speed of light from the two orbiting stars in binary systems, as well as from opposite sides of the rotating Sun, has also been measured, again confirming the same result: The speed of light is always the same.

Experimental Tests of Relativity   Although *we* cannot yet travel at speeds at which the effects of relativity should be obvious, tiny subatomic particles can reach such speeds, thereby allowing us to test the precise predictions of the formulas of special relativity. Let's first consider one way of testing time dilation. In machines called *particle accelerators*, physicists accelerate subatomic particles to speeds near the speed of light and study what happens when the particles collide. The colliding particles have a great deal of kinetic energy, and the collisions convert some of this kinetic energy into mass-energy that emerges as a shower of newly produced particles. Many of these particles have very short lifetimes, at the end of which they decay (change) into other particles. For example, a particle called the π^+ ("pi plus") meson has a lifetime of about 18 nanoseconds (billionths of a second) when produced at rest. But π^+ mesons produced at speeds close to the speed of light in particle accelerators last much longer than 18 nanoseconds—and by precisely the amount predicted by the time dilation formula.

The same experiments also confirm the mass increase predicted by relativity. The amount of energy released when high-speed particles collide depends on the particle masses and speeds. Just as relativity predicts, these masses are greater at high speed than they are at low speed—again by just the amount predicted by Einstein's formulas.

Particle accelerators even offer experimental evidence that nothing can reach the speed of light. It is relatively easy to get particles traveling at 99% of the speed of light in particle accelerators. However, no matter how much more energy is put into the accelerators, the particle speeds get only fractionally closer to the speed of light. Some particles have been accelerated to speeds within 0.00001% of the speed of light, but none have ever reached the speed of light.

Although the effects of relativity are obvious only at very high speeds, modern techniques of measuring time are so precise that effects can be measured even at ordinary speeds. For example, a 1975 experiment compared the amount of time that passed on an airplane flying in circles to the time that passed on the ground. Over 15 hours, the airborne clocks lost a bit under 6 nanoseconds to the ground clocks, matching the result expected from relativity. More recent experiments using the Space Shuttle have confirmed time dilation as well.

Nuclear energy also provides a test of relativity. Remember that $E = mc^2$ is a direct consequence of Einstein's theory. Every time you see film of an atomic bomb, or use electrical power from a nuclear power plant, or feel the energy of sunlight that the Sun generated through nuclear fusion, you are really experiencing direct experimental evidence of relativity.

Beyond Special Relativity   All in all, special relativity is one of the best-tested theories in physics, and it has passed every experimental test to date with flying colors. That is why it merits status as a true, scientific *theory*. Of course, like any scientific theory, it is always open to further testing and refinement. Indeed, Einstein himself realized that special relativity was missing an important ingredient: It dealt successfully with motion through empty space, but not with motion affected by gravity. This missing ingredient provided much of the motivation that led Einstein to press on and publish his general theory of relativity about a decade later.

The general theory expanded on the special theory by including gravity, and in the process Einstein found that it also turned out to be an improvement on Newton's theory of gravity [Section 2.4]. It is likely that the general theory of relativity will someday need its own refinements to make it compatible with the theory of quantum mechanics. If so, it is possible that some of the ideas of special relativity may also change. But the evidence that supports the theory as we know it today is real and cannot be made to disappear. If anyone ever comes up with a better theory, the new theory will still have to explain the many experimental results that support relativity so well.

THE BIG PICTURE
Putting Chapter 13 into Context

In this chapter, we have explored the possibilities for and challenges of interstellar travel, and the surprising Fermi paradox that arises when we think about what other civilizations might already have achieved. As you continue in your studies, keep in mind the following "big picture" ideas:

- Interstellar travel may be a staple of science fiction, but it remains well beyond our current capabilities. Nevertheless, it is possible that the challenges can be surmounted and that other civilizations might already have overcome them.

- The idea that other civilizations might already have achieved interstellar travel leads to the Fermi paradox—the question of why we see no evidence of galactic civilization. This simple question is not easily dismissed, and no matter what its solution turns out to be, it has profound implications to human civilization.

- One key lesson of the Fermi paradox is that we live at a unique moment in the history of the human species. We have the ability to destroy our civilization and perhaps even to drive our species to extinction. But if we survive, our descendants might have a boundless future. From this perspective, no generation has ever borne such great responsibility.

SUMMARY OF KEY CONCEPTS

13.1 THE CHALLENGE OF INTERSTELLAR TRAVEL

- **Why is interstellar travel so difficult?**

Convenient interstellar travel remains well beyond our technological capabilities. Current spacecraft would take more than 100,000 years just to traverse the distance to the nearest stars. The energy requirements for sending people on interstellar trips are enormous, far greater than all current world energy usage.

- **Could we travel to the stars with existing rockets?**
Nearly all rockets built to date are powered by chemical rocket engines. The **rocket equation** shows that these types of engines could not possibly get us to speeds of even 1% of the speed of light, making them impractical for interstellar travel.

13.2 DESIGNING SPACECRAFT FOR INTERSTELLAR TRAVEL

- **How might we build interstellar spacecraft with "conventional" technology?**

Technologies such as nuclear rocket engines, solar sails, and ion engines could in principle allow us to build starships that could travel at speeds up to about 10% of the speed of light—fast enough to reach nearby stars in less than a century.

- **How might we build spacecraft that could approach the speed of light?**
Several technologies that are well beyond us at present but are allowed by the laws of physics could allow starships to reach speeds close to the speed of light. These include **matter–antimatter** engines and **interstellar ramjets** that scoop up fuel as they go.

- **Are there ways around the light-speed limitation?**

No known physical laws prevent hyperspace, **wormholes**, or warp drive from offering "loopholes" that could allow us to get from one place to another in less time than we could by traveling through ordinary space. However, we do not yet know if any of these are really possible.

13.3 THE FERMI PARADOX

- **Where is everybody?**

This seemingly simple question, known as the **Fermi paradox**, comes about because our general assumption that Earth is not unique leads us to expect that many other civilizations should by now have arisen and had the opportunity to colonize the Milky Way Galaxy. Yet we see no evidence of a galactic civilization.

- **Would other civilizations really colonize the galaxy?**

Based on the idea that other beings would have evolved in response to evolutionary pressures similar to those that led to human evolution on Earth, we should expect that other civilizations would have the same inherent drive to colonize that we seem to possess. If so, a civilization should be able to colonize the galaxy in a time that is short compared to the age of the universe, even with technology not much beyond our own.

- **What are possible solutions to the Fermi paradox?**

There are three general categories of solution to the Fermi paradox: (1) We are alone. (2) Civilizations are common, but no one has colonized the galaxy. (3) There is a galactic civilization, but it has deliberately avoided revealing its existence to us.

- **What are the implications of the Fermi paradox to human civilization?**

The first solution implies that we are the only piece of the universe that has ever attained self-awareness. The second suggests that civilizations may either change or destroy themselves before attaining the ability to travel to the stars. The third implies that we might someday meet up with a galactic civilization that predates us by millions or billions of years.

13.4 THE PROCESS OF SCIENCE IN ACTION: EINSTEIN'S SPECIAL THEORY OF RELATIVITY

- **What is "relative" about relativity?**

Special relativity tells us motion is relative, but everyone always agrees on the speed of light. From this it follows that different observers can measure time, distance, and mass differently, and that no material object can reach or exceed the speed of light.

- **What evidence supports Einstein's theory?**

Experiments with light confirm that its speed is always the same. Experiments with subatomic particles in particle accelerators confirm the predictions of **time dilation** and mass increase at speeds close to the speed of light, and time dilation has been verified at relatively low speeds in aircraft and spacecraft. Nuclear power plants and nuclear bombs release energy in accordance with the formula $E = mc^2$, which is also a prediction of special relativity.

EXERCISES AND PROBLEMS

REVIEW QUESTIONS
Short-Answer Questions Based on the Reading

1. Briefly describe the journeys of *Pioneers 10* and *11* and *Voyagers 1* and *2*. How do these spacecraft illustrate the challenge of interstellar travel?

2. How does the speed of light affect the possibility of interstellar travel?

3. About how much energy would it take to send enough people on a trip to start a colony around another star? Explain how you arrive at the answer.

4. How do rockets work? Briefly describe the history of rocketry.

5. What is the *rocket equation* used for? Based on the rocket equation and the *mass ratio*, briefly explain why chemical rockets are inadequate for sending people or large robotic probes to the stars.

6. Describe the proposed fusion-powered starships of Project Orion and Project Daedalus. How quickly could such ships reach the stars?

7. Discuss a few ways of reaching the stars (other than nuclear rockets) that are, at least in principle, within our current technological reach.

8. How would *time dilation* affect space travel at speeds close to the speed of light? Discuss possible ways of achieving such speeds, including matter–antimatter engines and *interstellar ramjets*.

9. Briefly discuss how Einstein's general theory of relativity might allow "shortcuts" by which we could reach distant stars in shorter times than we would expect from their measured distances. Do we know whether these shortcuts are really possible?

10. What is the *Fermi paradox*? What two seemingly contradictory ideas underlie the paradox?

11. Why does it seem that other civilizations, if they exist, should be significantly older than ours? Explain clearly.

12. What are *Von Neumann machines?* How do they affect the Fermi paradox?

13. Describe the *coral model* of galactic colonization. Why do we conclude that civilizations could have colonized the galaxy by now even with technology not much more advanced than ours?

14. Briefly discuss possible motives for galactic colonization, as well as several "motives" that don't hold up.

15. Summarize the three general categories of possible solutions to the Fermi paradox, and discuss each category in some detail.

16. Briefly discuss the profound implications of the Fermi paradox and how the answer to the paradox affects our civilization.

17. What known problems were solved when Einstein discovered the special theory of relativity?

18. Explain how the idea of an absolute speed of light leads automatically to the conclusion that no one can travel faster than light.

19. Besides the idea that you cannot reach the speed of light, what other consequences follow from the absoluteness of the speed of light?

20. Describe at least three tests that have confirmed the validity of the special theory of relativity.

TEST YOUR UNDERSTANDING

Science or Nonscience?

Each of the following describes some futuristic scenario that, while perhaps entertaining, may or may not be plausible. In each case, decide whether the scenario is plausible according to our present understanding of science or whether it is unlikely to be possible. Explain clearly; because not all of these have definitive answers, your explanation is more important than your chosen answer.

21. A brilliant teenager working in her garage discovers a way to build a rocket that burns coal as its fuel and can travel at half the speed of light.

22. Using beamed energy propulsion from a laser powered by energy produced at a windmill farm in the California desert, NASA engineers are able to send a solar sailing ship on a journey to Alpha Centauri that will take only 50 years.

23. Human colonization of the moons of Saturn occurs using spaceships powered by dropping nuclear bombs out the back of the ships.

24. In the year 2750, we receive a signal from a civilization around a nearby star telling us that the *Voyager 2* spacecraft recently crash-landed on its planet.

25. The General Rocket Corporation (a future incarnation of General Motors) unveils a new personal interstellar spacecraft that works as an interstellar ramjet with a scoop about 10 meters across.

26. Members of the first crew of the matter–antimatter spacecraft *Star Apollo,* which left Earth in the year 2165, return to Earth in the year 2450 looking only a few years older than when they left.

27. In the year 2011, we finally uncover definitive evidence of alien visits to Earth when a flying saucer crashes in the Rocky Mountains and its oxygen-kerosene fuel ignites a forest fire.

28. Aliens from a distant star system invade Earth with the intent to destroy us and occupy our planet, but we successfully fight them off when their technology proves no match for ours.

29. Aliens arrive on Earth but virtually ignore our presence, finding the diversity of earthly bacteria to be much more scientifically interesting.

30. A single great galactic civilization exists. It originated on a single planet long ago but is now made up of beings from many different planets, each of which was assimilated into the galactic culture in turn.

Quick Quiz

Choose the best answer to each of the following. Explain your reasoning with one or more complete sentences.

31. The *New Horizons* spacecraft is currently on its way to Pluto and will eventually continue out of our solar system. About how long will it take to travel the distance to the nearest stars? (a) 100 years; (b) 1,000 years; (c) 100,000 years.

32. The amount of energy that would be needed to accelerate a large spaceship to a speed close to the speed of light is (a) about 100 times as much energy as is needed to launch the Space Shuttle; (b) more than the total amount of energy used by the entire world in a year; (c) more than the amount of energy that our Sun emits into space in a year.

33. The rocket engines of our current spacecraft are powered by: (a) chemical energy; (b) nuclear energy; (c) matter–antimatter annihilation.

34. Suppose that a spaceship was launched in the year 2120 on a round-trip journey of 100 light-years, traveling at 99.99% of the speed of light. If one of the crew members was 30 years old when she left, about how old would you expect her to be on her return? (a) 31; (b) 130; (c) 29.

35. Suppose that a spaceship was launched in the year 2120 on a round-trip journey of 100 light-years, traveling at 99.99% of the speed of light. In approximately what year would the ship return to Earth? (a) 2121; (b) 2170; (c) 2220.

36. Which of the following best describes our current understanding of the possibility of fast interstellar travel through hyperspace? (a) Hyperspace travel is the method of choice for all advanced civilizations. (b) We do not know enough to say whether such travel is really possible. (c) The idea of hyperspace is pure fantasy and has no basis in reality.

37. Which of the following questions best represents the Fermi paradox? (a) Why can't we travel faster than the speed of light? (b) Why haven't we found any evidence of a galactic civilization? (c) Why haven't aliens invaded Earth and stolen our resources?

38. According to current scientific understanding, the idea that the Milky Way Galaxy might be home to a civilization millions of years more advanced than ours is: (a) a virtual certainty; (b) extremely unlikely; (c) one reasonable solution to Fermi's paradox.

39. Which of the following is *not* relative in the special theory of relativity? (a) motion; (b) time; (c) the speed of light.

40. What does the famous formula $E = mc^2$ have to do with special relativity? (a) Nothing; it comes from a different theory. (b) It is one of the two starting assumptions of special relativity. (c) It is a direct consequence of the theory, and hence a way of testing the theory's validity.

INVESTIGATE FURTHER
In-Depth Questions to Increase Your Understanding

Short-Answer/Essay Questions

41. *Distant Dream or Near-Reality?* Considering all the issues surrounding interstellar flight, when, if ever, do you think we are likely to begin traveling among the stars? Write a few paragraphs defending your opinion.

42. *What's Wrong with This Picture?* Many science fiction stories have imagined the galaxy divided into a series of empires, each having arisen from a different civilization on a different world, that hold each other at bay because they are all at about the same level of military technology. Is this a realistic scenario? Explain.

43. *Large Rockets.* Suppose we built a rocket that works much like the Space Shuttle, but was 1,000 times as large. Could this rocket get us to speeds close to the speed of light? Explain.

44. *Ticket to the Stars.* In this chapter, we've stated that relativity offers only a one-way "ticket to the stars." Explain why.

45. *Solution to the Fermi Paradox.* Among the various possible solutions we have discussed for the Fermi paradox, which do you think is most likely? (Or, if you have no opinion as to their likelihood, which do you like best?) Write a one- to two-page essay in which you explain why you favor this solution.

46. *Interstellar Travel in the Movies.* Choose a science fiction movie in which aliens (or future humans) are engaged in some type of interstellar travel. In a one- to two-page essay, briefly describe how they supposedly accomplish the travel and evaluate in depth whether the scheme seems plausible.

47. *The "Relative" in Relativity.* Many people have claimed that Einstein showed that "everything is relative." Is this true? Explain.

48. *Relativity as a Theory.* Look back at the hallmarks of science in Chapter 2. Evaluate the special theory of relativity in terms of each of the three hallmarks, showing why it meets the test of being science. Then explain why special relativity qualifies as a scientific *theory,* rather than just a hypothesis.

Quantitative Problems

Be sure to show all calculations clearly and state your final answers in complete sentences.

49. *Cruise Ship Energy.* Suppose we have a spaceship about the size of a typical ocean cruise ship today, which means it has a mass of about 100 million kilograms, and we want to accelerate the ship to a speed of 10% of the speed of light.
    a. How much energy would be required? (*Hint:* You can find the answer simply by calculating the kinetic energy of the ship when it reaches its cruising speed; because 10% of the speed of light is still small compared to the speed of light, you can use the formula that tells us that kinetic energy $= \frac{1}{2} \times m \times v^2$.)
    b. How does your answer compare to total world energy use at present, which is about 5×10^{22} joules per year?
    c. Suppose the cost of energy is 5¢ per 1 million joules. Using this price, how much would it cost to generate the energy needed by this spaceship?

50. *The Rocket Equation I.* Suppose a rocket with mass ratio $M_i/M_f = 15$ has engines that produce an exhaust velocity of 3 km/s. What is its final velocity? Is it sufficient to escape Earth? (*Hint:* See Cosmic Calculations 13.1.)

51. *The Rocket Equation II.* Suppose you want a rocket to achieve escape velocity from Earth (11 km/s) and its engines produce an exhaust velocity of 3 km/s. What mass ratio is required? Briefly explain the meaning of this mass ratio.

52. *The Multistage Rocket Equation.* The rocket equation takes a slightly different form for a multistage rocket:
$$v = n v_e \ln\left(\frac{M_i}{M_f}\right)$$
where n is the number of stages.
    a. Suppose a rocket has three stages with mass ratio $M_i/M_f = 3.4$ and engines that produce an exhaust velocity of 3 km/s. What is its final velocity? Is it sufficient to escape Earth?
    b. Suppose a rocket has 100 stages with mass ratio $M_i/M_f = 3.4$ and engines that produce an exhaust velocity of 3 km/s. What is its final velocity? Compare it to the speed of light.

53. *Relativistic Time Dilation.* Use the time dilation equation from Cosmic Calculations 13.2 to answer each of the following.
    a. Suppose a rocket travels at the escape velocity from Earth (11 km/s). Will time on the rocket differ noticeably from time on Earth? Explain.
    b. Suppose a rocket travels at a speed of $0.9c$. If the rocket is gone from Earth for 10 years as measured on Earth, how much time passes on the rocket?
    c. Suppose a rocket travels at a speed of $0.9999c$. If the rocket is gone from Earth for 10 years as measured on Earth, how much time passes on the rocket?

54. *Testing Relativity.* A π^+ meson produced at rest has a lifetime of 18 nanoseconds (1.8×10^{-8} s). Thus, in its own reference frame, a π^+ meson will always "think" it is at rest and therefore will decay after 18 nanoseconds. Suppose a π^+ meson is produced in a particle accelerator at a speed of $0.998c$. How long will scientists see the particle last before it decays? Briefly explain how an experiment like this helps verify the special theory of relativity. (*Hint:* Use the time dilation equation from Cosmic Calculations 13.2.)

55. *Long Trips at Constant Acceleration.* Consider a spaceship on a long trip with a constant acceleration of $1g$. Although the derivation is beyond the scope of this book, it is possible to show that, as long as the ship is gone from Earth for many years, the amount of time that passes on the spaceship during the trip is approximately:
$$t_{ship} = \frac{2c}{g} \ln\left(\frac{g \times D}{c^2}\right)$$
where D is the distance to the destination and ln is the natural logarithm. If D is in meters, $g = 9.8$ m/s², and $c = 3 \times 10^8$ m/s, the answer will be in units of seconds. Use this formula as needed to answer the following questions. Be sure to convert the distances from light-years to meters and final answers from seconds to years; useful conversions: 1 light-year $\approx 9.5 \times 10^{15}$ m; 1 yr $\approx 3.15 \times 10^7$.
    a. Suppose the ship travels to a star that is 500 light-years away. How much time will pass on the ship? Approximately how much time will pass on Earth? Explain.

b. Suppose the ship travels to the center of the Milky Way Galaxy, about 28,000 light-years away. How much time will pass on the ship? Compare this to the amount of time that passes on Earth.

c. The Andromeda Galaxy is about 2.2 million light-years away. Suppose you had a spaceship that could constantly accelerate at $1g$. Could you go to the Andromeda Galaxy and back within your lifetime? Explain. If you could make the journey, what would you find when you returned to Earth?

56. *The Coral Model of Colonization.* We can estimate the time it would take for a civilization to colonize the galaxy. Imagine that a civilization sends colonists to stars that are an average distance D away and sends them in spacecraft that travel at speed v. The time required for travel, t_{travel}, is then $t_{travel} = D/v$. Suppose that the colonists build up their colony for a time t_{col}, at which point they send out their own set of colonists to other star systems (with the same average distance and same spacecraft speed). Then the speed at which the civilization expands outward from the home star, v_{col} (for the speed of colonization), is $v_{col} = D/(t_{travel} + t_{col})$. However, this is true only if the colonization is always directed straight outward from the home star. In reality, the colonists will sometimes go to uncolonized star systems in other directions, so we will introduce a constant k that accounts for this zigzag motion. Our equation for the speed at which the civilization expands outward from the home star is:

$$v = k \frac{D}{(t_{travel} + t_{col})}$$
$$= k \frac{D}{\left(\dfrac{D}{v} + t_{col}\right)}$$

For the purposes of this problem, assume that $k = \frac{1}{2}$ and that the average distance between star systems is $D = 5$ light-years.

a. How fast (as a fraction of the speed of light) does the civilization expand if its spacecraft travel at $0.1c$ and each colony builds itself up for 150 years before sending out the next wave of colonists? How long would it take the colonists to expand a distance of 100,000 light-years from their home star at this rate?

b. Repeat part (a), but assume that the spacecraft travel at $0.01c$ and that each colony builds itself up for 1,000 years before sending out more colonists.

c. Repeat part (a), but assume that the spacecraft travel at $0.25c$ and that each colony builds itself up for 50 years before sending out more colonists.

Discussion Questions

57. *Seeding the Galaxy.* If interstellar travel is forever impractical, are there other ways an advanced civilization might spread its culture? Clearly, communication is possible, although the speed of light makes conversations between star systems maddeningly tedious. Could a society send the information required to assemble members of its species (its "DNA," for instance) and therefore spread through the galaxy at the speed of light? Can you imagine other ways of spreading a culture without starships? Explain.

58. *Sociology of Interstellar Travel.* Suppose we somehow built a spaceship capable of relativistic travel and volunteers were being recruited for a journey to a star 15 light-years away. Would you volunteer to go? Do you think others would volunteer? In light of the effects of time dilation, discuss the benefits and drawbacks of such a trip.

59. *The Turning Point.* Discuss the idea that our generation has acquired a greater responsibility to future humans than any previous generation. Do you agree with this assessment? If so, how should we deal with this responsibility? Defend your opinions.

WEB PROJECTS

60. *Starship Design.* Find more details about a proposal for starship propulsion or design. How would the starship work? What new technologies would be needed, and what existing technologies could be applied? Summarize your findings in a one- to two-page report.

61. *Advanced Spacecraft Technologies.* NASA supports many efforts to incorporate new technologies into spaceships. Although few of them reach the level of being suitable for interstellar colonization, most are innovative and fascinating. Learn about one such NASA project, and write a short summary of your findings.

62. *Solutions to the Fermi Paradox.* Learn more about someone's pet solution to the Fermi paradox. Write a short summary of that solution, and discuss how it fits with the ideas we have discussed in this chapter.

Epilogue
Contact—Implications of the Search and Discovery

The known is finite, the unknown is infinite; intellectually we stand on an islet in the midst of an illimitable ocean of inexplicability. Our business in every generation is to reclaim a little more land.
—Thomas H. Huxley

We've covered a lot of ground in this book. We've studied life on Earth and learned about the prospects for life elsewhere in our solar system and on planets around other stars. Our scientific discussion mirrors a broader cultural phenomenon—an intense interest in ideas relating to extraterrestrial life. This interest shows up in many aspects of contemporary culture, including TV shows and movies, supermarket tabloids, and a notable public interest in the exploration of space and the search for life beyond Earth.

In this Epilogue, we'll briefly consider a few possible reasons why humans are so deeply interested in the question of whether life is present elsewhere. We'll begin with a short review of why the idea of life beyond Earth seems scientifically reasonable and then explore how the search for it helps us revisit age-old questions about the nature of humanity. We'll also discuss the philosophical and cultural consequences of finding life elsewhere—life of any kind, whether microbial or intelligent or somewhere in between.

IS THERE LIFE ELSEWHERE?

We know of only one example of life existing in the universe—life here on Earth. In this book we have discussed many reasons why it seems reasonable to think that life might exist elsewhere, but we still don't know for sure that it does. Because of this uncertainty, all we can do today is discuss the issues that could determine whether extraterrestrial life is likely or unlikely to exist and how we might search for evidence of it.

Despite these limitations, we are at a unique point in the long debate over the possibility of extraterrestrial life. We have the technological capability to explore Mars and much of the rest of our solar system, and we are rapidly developing technology that should in principle allow us to seek evidence for life on planets around other stars. After millennia of speculation about life beyond Earth, we are finally reaching the point at which we could really discover it, if it exists. This remarkable prospect calls us to discuss the philosophical and cultural consequences of finding life elsewhere. But first let's summarize the key issues in our discussion about the search for life in the universe.

• Why life seems likely

Although we don't yet know whether life exists any place besides Earth, we've seen that current science offers reasons to think it should. We can examine the nature of life on Earth—its building blocks and how it originated—and understand the environmental conditions in which life can exist. We can look at the other planets and moons in our solar system and determine whether the conditions conducive to biology exist there. And we can look for planets around other stars, learn how abundant they are and how they form, and determine whether some of them might be Earth-like planets that could be capable of supporting living things.

When we do these things, we find three key pieces of evidence that point to the idea that life should be common in the universe. We'll list them and then discuss each briefly in turn:

1. The chemical elements that make up life on Earth are common throughout the universe, and complex, carbon-bearing molecules important to life on Earth appear to form easily and naturally under conditions that should be common on many worlds.
2. Life on Earth thrives under a wide range of environmental conditions that we once considered too extreme to be capable of supporting life, and many of these types of environments are likely to be found on other planets in our own solar system and beyond.
3. It seems that life appeared on Earth quite early in its history, making it seem plausible that life is "easy."

Let's consider the first item in our list. The elements from which life is constructed (carbon, hydrogen, oxygen, nitrogen, and almost two dozen other elements) are found nearly everywhere in the universe. Hydrogen is the most abundant element by far, but the other elements used for life exist in at least modest quantities, because they have been created by nuclear fusion in earlier generations of stars and during supernova explosions of those stars. The manufactured elements were ejected by the supernova explosions into interstellar space, where they were incorporated into the gas and dust clouds out of which later stars and planets formed. Moreover, experiments in laboratories on Earth and spectroscopic observations of distant objects show that the elements used by life combine readily into the molecular building blocks of life. We have found molecules as complex as amino acids in meteorites, and we have detected numerous organic molecules even in interstellar space. We therefore expect that organic molecules should form abundantly and naturally under a range of conditions that includes those that were present on the early Earth and are likely to be present on other geologically active worlds. A wide availability of organic molecules would make it likely that the starting points for an origin of life exist on many worlds.

If the starting points for life are commonplace, the next question concerns whether available environments can allow life to arise and thrive. This is the issue addressed by the second point on our list. We have learned that life on Earth is incredibly diverse and that some organisms can thrive in a wide variety of environments that seem "extreme" to humans. For example, we have found life on Earth in the hot water near deep-sea vents, in the dry and frigid deserts of Antarctica, and inside rocks deep underground. The diversity of environments in which we find terrestrial life suggests that life could survive on any world that meets relatively simple environmental requirements—the presence of liquid water (or possibly another liquid), access to the requisite elements and molecules, and an energy source to drive metabolism. These conditions are likely to be met on any geologically active world with a rocky interior and liquid water, whether a planet or a moon. Such worlds would have the necessary elements and the potential for energy to be available via water/rock chemical reactions.

The first two points on our list tell us that the starting points for life are widely available and that life, once started, can survive in a wide range of environments. The third point tells us that getting life to start may not be difficult. Even with the present uncertainties in interpreting the geological record, life must have originated on a time scale that is very short compared to the age of Earth and to the expected ages of other habitable planets. Thus, unless Earth was somehow atypical, it seems that life might be the natural, straightforward consequence of the types of chemical reactions that can occur in planetary environments. When we put all three key pieces of evidence together, it seems reasonable to imagine life existing in at least a few other places in our own solar system and on many similar worlds throughout the universe.

• Prospects for finding life in our solar system

If life is indeed as common and wide-ranging as the evidence suggests it could be, then the first place to look for life is within our own solar system. Here in the Sun's neighborhood, Mars and Europa seem the most likely places besides Earth to harbor life, but we've also found a few other candidates.

Mars shows evidence for liquid water having been present at its surface early in its history and within its crust throughout its history. The martian atmosphere contains several of the key elements of life—notably carbon (in the form of gaseous carbon dioxide), hydrogen and oxygen (in the form of water), and nitrogen. The other necessary elements are found in surface and near-surface rocks. Energy to drive metabolism could come from chemical reactions between the water and the rocks, for example, allowing the possibility of organisms much like those that live within rocks on Earth.

Jupiter's moon Europa may have large amounts of liquid water. Although we are not yet absolutely certain, it seems likely that Europa is even more of a "water world" than Earth, with a global, 100-kilometer-thick ocean lying beneath an ice surface. We also expect that Europa has heavier elements in its rocky interior, so all the elements needed for life should be abundant. Energy for life could be supplied by chemical reactions between the water and the underlying rock, or by chemical compounds created by the impact of high-energy particles onto surface ice (driven by Jupiter's magnetic field). It seems plausible that at least microbial life could exist on Europa if its postulated ocean proves to be real.

Other places in our solar system could also potentially support life. Two other moons of Jupiter, Ganymede and Callisto, show evidence of subsurface liquid oceans similar to Europa's but at greater depth beneath their surfaces. Saturn's moon Titan apparently has liquid methane at least some of the time, and may have liquid water underground in the form of a cold, ammonia/water mixture. Saturn's moon Enceladus also seems likely to have similar liquids in its interior, and Neptune's moon Triton might have the same. These cold worlds are less likely candidates for life because chemical reactions ought to run much more slowly at cold temperatures, but we cannot rule them out.

• Prospects for finding life among the stars

Efforts that will allow us to search for life beyond our own solar system are already under way. We have already discovered almost 200 planets orbiting other stars, and while most of these are closer in size to Jupiter than to our relatively small Earth, we are rapidly developing the technology to detect smaller worlds. Although the characteristics of many of the planets discovered to date have been somewhat surprising in light of our earlier ideas about solar system formation, they do not appear to change the basic theory of planet formation and they validate the idea that planetary systems are common.

We should get more definitive answers about the nature of other planetary systems with new, space-borne telescopes planned for the next couple of decades. The *Kepler* mission should tell us whether planets the size of Earth are common. Later missions such as the *Terrestrial Planet Finder* should help us determine whether Earth-*size* planets are Earth-*like*. With these data, we should be able to answer the question of whether Earth-like planets are rare or common.

If Earth-like planets are indeed common, microbial life could be widespread throughout the galaxy. The possible widespread existence of at least microbial life begs the question of the potential for more complex or intelligent life. Our one example of intelligence here on Earth does not allow us to extrapolate or even hazard a guess as to whether intelligence should be widespread or rare; the only way to find out whether other civilizations exist is to search for them. Searches for radio or light signals from possible extraterrestrial civilizations have been going on for almost five decades. Although these SETI efforts have not yet met with success, only a small sample of the possible homes for civilization has been searched so far. As with the question of extraterrestrial life in general, the question of whether intelligent extraterrestrial life exists remains open.

• Extraterrestrial life and the human condition

We have reviewed the key scientific issues pertaining to life elsewhere. Now we turn our attention to issues that we have generally neglected to this point, including philosophical and societal issues that touch on why the search for life beyond Earth is of such broad intellectual interest to scientists and the public alike.

Why are so many people so interested in the question of whether life exists on other worlds, and what would it mean to find evidence — or a lack of evidence— for such life? We'll discuss a few possible answers to these questions, but you should recognize that the ideas we discuss may not resonate the same way with everyone. Think about what issues might be driving your own interests, which led you to take this course, and especially about how and why these personal issues might differ from the ideas we discuss.

• Our changing perspective on the world

Many people are fascinated by the question of extraterrestrial life because knowing if such life exists would likely have a profound influence on our perspective about our place in the universe. Much of what we have learned about the world around us over the past several thousand years has changed the way we interact with it. For example, a mere 10,000 years ago—a blink of an eye in the history of our planet—humans were primarily a hunter-gatherer society, living out their lives with little knowledge about what was beyond the next mountain or valley. Today, people in nearly every corner of the world recognize our species as part of an enormous cosmos. We have learned that what we do in one place on our planet affects the environment, people, and societies in all other places around the world. In addition, we now see Earth as just one of many worlds in our solar system; our solar system and the Sun as one of more than 100 billion star systems in our galaxy; and our galaxy as just one of billions of similar galaxies in the observable universe. This expansion of our world (or contraction, depending on how you perceive it!) has brought with it a need to rethink our views both of ourselves and of our relationship to the rest of the planet and universe.

Perhaps the first major transition toward a modern scientific view of ourselves (and maybe the most significant one) began nearly 500 years ago with the Copernican revolution. The work of Copernicus, Galileo, Kepler, and Newton allowed us to recognize that Earth is not at the center of the universe but instead orbits the Sun. This displacement of Earth from the center of the universe had profound philosophical and psychological effects. Earth, and by extension humanity, could no longer be viewed as the center of everything, and the world could no longer be viewed as obeying only the laws of providence rather than the laws of physics. This shift in thinking had little impact on people's day-to-day lives—we would be hard-pressed even today to think of a way in which whether Earth goes around the Sun or vice versa matters to our daily activities. However, it made a fundamental difference to our worldview. As a result, the idea initially met widespread resistance in Western society and was not fully accepted for at least a couple of hundred years following Copernicus.

A second major shift occurred in the mid-1800s with the recognition by Charles Darwin and Alfred Russel Wallace of the processes that drive the evolution of species. The basic idea of evolution was not new. By the time of Darwin, scientists already recognized that Earth was old, that fossils represented organisms that had lived in prior times, that older fossils were different from younger ones, and that both were different from current living organisms. These discoveries had already convinced many scientists that the nature of living organisms had changed over time. What Darwin and Wallace discovered was a way to explain why and how species change: through competition for survival and the process we refer to as natural selection. We now recognize that all the species present on Earth represent the product of almost 4 billion years of evolution, traceable back to the earliest history of life on Earth.

Like the Copernican revolution, recognizing the nature of evolution sparked a change in our view of ourselves and of our world. Copernicus and others showed that we are not located at the physical center of the universe, and Darwin and Wallace showed that we are not at the biological center either. Rather, we represent just one of millions of distinct species on Earth, and while we may be "dominant" over many species found in the plant and animal kingdoms, these kingdoms represent just a small part of the great diversity of life on our planet. This shift in perspective has been extremely profound, and may well explain why many people still do not accept the idea of evolution by natural selection. The intensity of the public controversy only underscores the tremendous impact that the theory of evolution has had—and continues to have—on Western society.

A third shift in perspective is taking place today with the discovery of other solar systems and the modern understanding of the potential for life to be widespread in the universe. Although many astronomers had long suspected that planets ought to be common around other stars, the first definitive evidence of extrasolar planets is slightly more than a decade old. If we ultimately confirm our guess that other worlds like ours exist or are common, we will no longer be able to regard Earth as special in any essential way. Throughout the history of scientific thought, new discoveries have increasingly displaced us from the center of the universe, both physically and metaphorically. Finding that life is common throughout the universe would lead to yet another major shift in perspective. In the rest of this Epilogue, we'll briefly explore why.

• The impact of extraterrestrial life on human perspective

People have long speculated about life beyond Earth, and many people—including many scientists—have at times been convinced that life exists on the Moon, on Mars, or on other worlds. Why, then, would the actual discovery of extraterrestrial life have a major impact on human perspective? The answer lies in the difference between guessing and knowing. As long as there is uncertainty about the existence of life on other worlds, people are free to hold a wide range of opinions. People living before the time of Copernicus could continue to believe in an Earth-centered world, but it was quite hard to continue to do so after Galileo, Kepler, and Newton offered convincing proof to the contrary. An actual discovery of life beyond Earth would force us, both as individuals and as a society, to reconsider the place of our planet and our species in the cosmos.

In contemplating the significance of finding life elsewhere, let's begin by considering what would happen if we found microbial life on Mars. The first question we would probably ask is whether the life was genetically related to terrestrial life (suggesting that it had migrated between planets on meteorites) or instead represented an independent origin of life on Mars. We could answer this question by determining the structure of the molecules that make up the martian life. Does it use DNA and RNA molecules similar to those used by terrestrial life? Does it use the same amino acids or the same proteins to carry out enzymatic reactions? Do the molecules that participate in life have the same "handedness" to them? It seems unlikely that there would be only one solution to the problems of containing and passing on the genetic information required for life to reproduce, of catalyzing the chemical reactions that make up life, and of storing and using energy in metabolism. We would therefore expect life that had an origin independent from life on Earth to have a different chemical structure.

While a discovery of microbial life on Mars that was genetically related to terrestrial life would be exciting, it probably would not have as great an impact as a discovery of life that showed evidence of an independent origin. A discovery of life from an independent origin would provide clear proof that the origin of life was not a unique event. With proof that life had originated twice in just our own solar system, we would have every reason to think it has originated many times among the many worlds around other stars and thus that life is widespread in the universe. A discovery of alien microbial life would also help us better understand life in general. We would learn more about the conditions under which life can arise and persist, as well as the conditions under which it can evolve into more complex forms. And we could begin to engage in comparative biology, in which the study of alien life would help us better understand the life on our own world.

While many people who have studied the issue of life in the universe expect that we will find microbial life to be widespread, we cannot truly envision the consequences of an actual discovery unless and until it occurs. Moreover, the history of science tells us to be prepared for surprises. For example, the real discovery of extrasolar planets proved surprising, despite their predicted existence: We learned that other solar systems can be very different from our own. If and when we do find life elsewhere, we should expect to be equally surprised about its nature.

• Extraterrestrial intelligence and the nature of humanity

The discovery of extraterrestrial life of any kind would have a profound effect on our perception of our place in the universe, but it is the potential discovery of extraterrestrial intelligence that generates the greatest public interest. What would finding intelligent life elsewhere mean to us? We discussed some of the philosophical implications in Section 13.3; here we focus on how it might affect our lives more directly.

It is difficult to predict how we would respond to such a discovery, and it is likely that people would exhibit a wide range of responses. At one extreme, some people might look to extraterrestrial intelligence as a source of help for solving our problems. This view has been portrayed in many books and movies, including *Contact* and *2001: A Space Odyssey.* Interestingly, many of these stories suggest, first, that we are not able to solve our own problems without outside help and, second, that extraterrestrials will want to help us! At the other extreme, some people imagine that extraterrestrials would come to Earth and destroy our civilization, either deliberately or by accident. The martians of H. G. Wells's *War of the Worlds* came here bent on our destruction (Figure E.1). In Douglas Adams's *Hitchhiker's Guide to the Galaxy,* aliens destroy Earth as an accidental consequence of their need to build a new "interstellar bypass."

Neither extreme seems especially likely. As we've discussed, intelligent aliens capable of visiting our planet would almost certainly be far more technologically advanced than we are. It's not clear that they would have any interest in us at all (making the first extreme unlikely), and they'd seemingly have little to gain by destroying us (making the second extreme unlikely). If and when intelligent life is discovered, the reality may lie somewhere in between these two extremes—or it might be completely different.

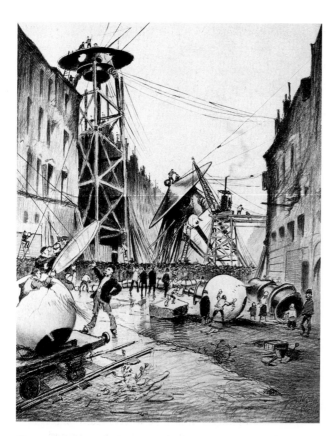

Figure E.1

A 1906 drawing used to illustrate H. G. Wells's book *The War of the Worlds,* in which Earth is invaded by hostile martians.

• Significance of the search itself

A discovery of even microbial extraterrestrial life would undoubtedly bring important practical benefits. For example, studying it will help us understand what characteristics of terrestrial life are unique to Earth and what characteristics apply generally to life anywhere. If we find intelligent life elsewhere, we would learn much more about the nature of intelligence and might be exposed to cultures and societies extremely different from those of humans. If we could communicate with more advanced beings, we could possibly learn the secrets of the nature of the universe, the nature of consciousness and the mind, and technological marvels that could dramatically change our life here on Earth.

MOVIE MADNESS: *E.T.*

When cinema aliens come to Earth, it's usually wise policy to head for the storm cellar. But when wrinkly little E.T. is accidentally left behind by his planetary pals, it turns out to be good news—at least for a few suburban kids.

E.T. is the quintessential alien film—a movie that long held the record for being the most successful picture of all time. There's good reason. Appealing little E.T., who has the stature and gait of a penguin, is every kid's dream. After all, he's a friend (and really useful because with his super powers he can help you outsmart adults), and he's exclusively *your* friend. Other kids will stop kicking verbal sand in your face when you show up at the playground with a guy from another galaxy, even if he has a face like a polished redwood burl.

Aside from this childlike wish fulfillment, *E.T.* encapsulates everything we hope or think is true about intelligent extraterrestrials. To begin with, E.T. is benign and nonthreatening. Unlike evil aliens, who look like reptiles or insects, E.T. resembles a baby, with his short nose, big eyes, and wrinkled skin. He's only 2 feet high, weighs 35 pounds, and has a 25-watt fingertip. He clearly comes from a kinder, gentler planet, since his only interest in Earth's biota is its plants. (No insects

had to die for this film.) Scientists who have given any thought to the true nature of advanced aliens (and those that can come to Earth are clearly advanced) have often equated technological prowess with peaceful behavior. The aliens will be friendly. This doesn't square all that well with our experience on Earth, but one can hope.

In addition, E.T. not only looks a lot like us; he acts like us, functions like us (he can get blotto on beer), and is interested in our personal lives. None of this is likely to be true, of course. He's also well adapted to terrestrial conditions, waddling around without a space suit and, indeed, without any clothes at all. In addition, nefarious federal agents wearing jackets, ties, and drab personalities are busily trying to keep the cuddly creature's visit under wraps, something that many people believe is happening in real life.

All of this may be in keeping with the public's perception of what aliens would be like. But if the search for extraterrestrial intelligence succeeds and we eventually learn the true nature of extraterrestrials, it is far more likely that they will be of a construction and temperament that are far beyond our most fevered imaginings.

However, for many people excited by the scientific search for life in the universe, the search itself is much more than a means to an end. From this viewpoint, the search is just one more critical component in our exploration of the world around us. Other components in astronomy and space science include exploring the planets and moons in our solar system as a way to understand how planets work and exploring stars and galaxies as a way to determine the nature of our universe. Other components in biology include exploring the origin and evolution of life on Earth so that we can understand how we ourselves came to exist.

In all these cases, our exploration does not seem to be driven solely by the desire to find specific answers to the scientific questions we are asking, because in each case we end up asking more questions. Instead, we seem to be driven by our inherent curiosity, our desire to understand the world around us (Figure E.2). Sometimes our curiosity leads to discoveries with practical applications, while at other times it simply helps us understand how or why the world is as it is. Understanding the world around us means learning about the broader-scale environment in which humans exist. Understanding the occurrence of planets orbiting other stars helps us understand the significance of the occurrence of planets orbiting the Sun, including Earth. Understanding the occurrence of life elsewhere allows us to understand the significance of the occurrence of life on Earth. And understanding the potential for intelligent life beyond Earth brings with it an understanding of the meaning of the existence of intelligent life here on Earth. In essence, by learning about the universe around us, we are learning about ourselves and about what it means to be human.

Exploring the universe is but one activity among many that help us learn about ourselves, including searching for new forms of life in extreme environments on Earth, exploring Earth itself, exploring the nature of human consciousness and the human mind, and exploring the human condition through literature and the arts. Each of these activities helps us learn about what it means to be alive and to be human. Thus, the exploration of the universe and the search for extraterrestrial life can be seen as nothing less than an attempt to determine the nature of humanity and our place in the universe.

Figure E.2

Earth as viewed in space from the *Apollo 17* spacecraft. Seen from this perspective, we recognize the strong connections between the terrestrial ecosystem and the planet itself, and we recognize that life, indeed, is a planetary phenomenon. As we explore the universe, we will learn whether the formations of planets around stars and the occurrence of life on planets are rare or commonplace. We may then finally answer the question of whether we are alone.

EXERCISES AND PROBLEMS

DISCUSSION QUESTIONS

1. *Is There Life Elsewhere?* After considering all the evidence to date about the possibility of extraterrestrial life, do you believe it is likely that we'll find microbial life elsewhere? Do you believe it is likely that we'll find intelligent life elsewhere? Defend your opinions, using arguments based on the full range of scientific issues discussed in this book.

2. *Microbial or Intelligent?* Do you think the implications of discovering microbial life elsewhere would be any more or less profound than those of discovering extraterrestrial intelligence? Explain your reasoning.

3. *Extraterrestrial Life and Your Religion.* Would the discovery of extraterrestrial life have any important implications for your own personal religious beliefs? Would it affect the current "official" beliefs (if any) of your religion? Explain.

4. *Extraterrestrial Life and the Debate on Evolution.* Do you think the discovery of extraterrestrial life would affect the current status of the debate over science and religion in the United States? For example, would it alter the controversy surrounding the teaching of evolution in public schools? Why or why not?

5. *Aliens and Everyday Life.* While the discovery of extraterrestrial life would surely be profound, do you think it would alter any aspect of our everyday lives? If so, how? If not, why not?

6. *The Search Itself.* Suppose we spend a fair amount of money and effort searching for life over the next few decades and ultimately find no evidence for life beyond Earth. Will the search have been a waste, a success, or something in between? Defend your opinion.

Appendixes

A | Useful Numbers

ASTRONOMICAL DISTANCES

1 AU $\approx 1.496 \times 10^8$ km

1 light-year $\approx 9.46 \times 10^{12}$ km

1 parsec (pc) $\approx 3.09 \times 10^{13}$ km ≈ 3.26 light-years

1 kiloparsec (kpc) $= 1{,}000$ pc $\approx 3.26 \times 10^3$ light-years

1 megaparsec (Mpc) $= 10^6$ pc $\approx 3.26 \times 10^6$ light-years

ASTRONOMICAL TIMES

1 solar day (average) $= 24^h$

1 sidereal day $\approx 23^h\, 56^m\, 4.09^s$

1 synodic month (average) ≈ 29.53 solar days

1 sidereal month (average) ≈ 27.32 solar days

1 tropical year ≈ 365.242 solar days

1 sidereal year ≈ 365.256 solar days

UNIVERSAL CONSTANTS

Speed of light: $\qquad c = 3 \times 10^5$ km/s $= 3 \times 10^8$ m/s

Gravitational constant: $\qquad G = 6.67 \times 10^{-11} \dfrac{m^3}{kg \times s^2}$

Planck's constant: $\qquad h = 6.626 \times 10^{-34}$ joule $\times$ s

Stefan–Boltzmann constant: $\qquad \sigma = 5.7 \times 10^{-8} \dfrac{watt}{m^2 \times Kelvin^4}$

mass of a proton: $\qquad m_p = 1.67 \times 10^{-27}$ kg

mass of an electron: $\qquad m_e = 9.1 \times 10^{-31}$ kg

USEFUL SUN AND EARTH REFERENCE VALUES

Mass of the Sun: 1 $M_{Sun} \approx 2 \times 10^{30}$ kg

Radius of the Sun: 1 $R_{Sun} \approx 696{,}000$ km

Luminosity of the Sun: 1 $L_{Sun} \approx 3.8 \times 10^{26}$ watts

Mass of Earth: 1 $M_{Earth} \approx 5.97 \times 10^{24}$ kg

Radius (equatorial) of Earth: 1 $R_{Earth} \approx 6{,}378$ km

Acceleration of gravity on Earth: $g = 9.8$ m/s^2

Escape velocity from surface of Earth: $v_{escape} = 11$ km/s $= 11{,}000$ m/s

ENERGY AND POWER UNITS

Basic unit of energy: 1 joule $= 1 \dfrac{kg \times m^2}{s^2}$

Basic unit of power: 1 watt $= 1$ joule/s

Electron-volt: 1 eV $= 1.60 \times 10^{-19}$ joule

B | Useful Formulas

- Universal law of gravitation for the force between objects of mass M_1 and M_2, distance d between their centers:

$$F = G\frac{M_1 M_2}{d^2}$$

- Newton's version of Kepler's third law; p and a are period and semimajor axis, respectively, of either orbiting mass:

$$p^2 = \frac{4\pi^2}{G(M_1 + M_2)}a^3$$

- Escape velocity at distance R from center of object of mass M:

$$v_{escape} = \sqrt{\frac{2GM}{R}}$$

- Relationship between a photon's wavelength (λ), frequency (f), and the speed of light (c):

$$\lambda \times f = c$$

- Energy of a photon of wavelength λ or frequency f:

$$E = hf = \frac{hc}{\lambda}$$

- Stefan–Boltzmann law for thermal radiation at temperature T (in Kelvin):

$$\text{emitted power per unit area} = \sigma T^4$$

- Wien's law for the peak wavelength (λ_{max}) thermal radiation at temperature T (in Kelvin):

$$\lambda_{max} = \frac{2{,}900{,}000}{T}\,\text{nm}$$

- Doppler shift (radial velocity is positive if the object is moving away from us and negative if it is moving toward us):

$$\frac{\text{radial velocity}}{\text{speed of light}} = \frac{\text{shifted wavelength} - \text{rest wavelength}}{\text{rest wavelength}}$$

- Angular separation (α) of two points with an actual separation s, viewed from a distance d (assuming d is much larger than s):

$$\alpha = \frac{s}{2\pi d} \times 360°$$

- Inverse square law for light:

$$\text{apparent brightness} = \frac{\text{luminosity}}{4\pi d^2}$$

(where d is the distance to the object)

C | A Few Mathematical Skills

This appendix reviews the following mathematical skills: powers of 10, scientific notation, working with units, the metric system, and finding a ratio. You should refer to this appendix as needed while studying the textbook.

C.1 POWERS OF 10

Powers of 10 simply indicate how many times to multiply 10 by itself. For example:

$$10^2 = 10 \times 10 = 100$$

$$10^6 = 10 \times 10 \times 10 \times 10 \times 10 \times 10 = 1,000,000$$

Negative powers are the reciprocals of the corresponding positive powers. For example:

$$10^{-2} = \frac{1}{10^2} = \frac{1}{100} = 0.01$$

$$10^{-6} = \frac{1}{10^6} = \frac{1}{1,000,000} = 0.000001$$

Table C.1 lists powers of 10 from 10^{-12} to 10^{12}. Note that powers of 10 follow two basic rules:

1. A positive exponent tells how many zeros follow the 1. For example, 10^0 is a 1 followed by no zeros, and 10^8 is a 1 followed by eight zeros.

TABLE C.1 *Powers of 10*

Zero and Positive Powers			Negative Powers		
Power	**Value**	**Name**	**Power**	**Value**	**Name**
10^0	1	One			
10^1	10	Ten	10^{-1}	0.1	Tenth
10^2	100	Hundred	10^{-2}	0.01	Hundredth
10^3	1,000	Thousand	10^{-3}	0.001	Thousandth
10^4	10,000	Ten thousand	10^{-4}	0.0001	Ten thousandth
10^5	100,000	Hundred thousand	10^{-5}	0.00001	Hundred thousandth
10^6	1,000,000	Million	10^{-6}	0.000001	Millionth
10^7	10,000,000	Ten million	10^{-7}	0.0000001	Ten millionth
10^8	100,000,000	Hundred million	10^{-8}	0.00000001	Hundred millionth
10^9	1,000,000,000	Billion	10^{-9}	0.000000001	Billionth
10^{10}	10,000,000,000	Ten billion	10^{-10}	0.0000000001	Ten billionth
10^{11}	100,000,000,000	Hundred billion	10^{-11}	0.00000000001	Hundred billionth
10^{12}	1,000,000,000,000	Trillion	10^{-12}	0.000000000001	Trillionth

2. A negative exponent tells how many places are to the right of the decimal point, including the 1. For example, $10^{-1} = 0.1$ has one place to the right of the decimal point; $10^{-6} = 0.000001$ has six places to the right of the decimal point.

Multiplying and Dividing Powers of 10

Multiplying powers of 10 simply requires adding exponents, as the following examples show:

$$10^4 \times 10^7 = \underbrace{10,000}_{10^4} \times \underbrace{10,000,000}_{10^7} = \underbrace{100,000,000,000}_{10^{4+7} = 10^{11}} = 10^{11}$$

$$10^5 \times 10^{-3} = \underbrace{100,000}_{10^5} \times \underbrace{0.001}_{10^{-3}} = \underbrace{100}_{10^{5+(-3)}=10^2} = 10^2$$

$$10^{-8} \times 10^{-5} = \underbrace{0.00000001}_{10^{-8}} \times \underbrace{0.00001}_{10^{-5}} = \underbrace{0.0000000000001}_{10^{-8+(-5)} = 10^{-13}} = 10^{-13}$$

Dividing powers of 10 requires subtracting exponents, as in the following examples:

$$\frac{10^5}{10^3} = \underbrace{100,000}_{10^5} \div \underbrace{1,000}_{10^3} = \underbrace{100}_{10^{5-3} = 10^2} = 10^2$$

$$\frac{10^3}{10^7} = \underbrace{1,000}_{10^3} \div \underbrace{10,000,000}_{10^7} = \underbrace{0.0001}_{10^{3-7} = 10^{-4}} = 10^{-4}$$

$$\frac{10^{-4}}{10^{-6}} = \underbrace{0.0001}_{10^{-4}} \div \underbrace{0.000001}_{10^{-6}} = \underbrace{100}_{10^{-4-(-6)} = 10^2} = 10^2$$

Powers of Powers of 10

We can use the multiplication and division rules to raise powers of 10 to other powers or to take roots. For example:

$$(10^4)^3 = 10^4 \times 10^4 \times 10^4 = 10^{4+4+4} = 10^{12}$$

Note that we can get the same end result by simply multiplying the two powers:

$$(10^4)^3 = 10^{4 \times 3} = 10^{12}$$

Because taking a root is the same as raising to a fractional power (e.g., the square root is the same as the $\frac{1}{2}$ power, the cube root is the same as the $\frac{1}{3}$ power, etc.), we can use the same procedure for roots, as in the following example:

$$\sqrt{10^4} = (10^4)^{1/2} = 10^{4 \times (1/2)} = 10^2$$

Adding and Subtracting Powers of 10

Unlike with multiplication and division, there is no shortcut for adding or subtracting powers of 10. The values must be written in longhand notation. For example:

$$10^6 + 10^2 = 1,000,000 + 100 = 1,000,100$$

$$10^8 + 10^{-3} = 100,000,000 + 0.001 = 100,000,000.001$$

$$10^7 - 10^3 = 10,000,000 - 1,000 = 9,999,000$$

Summary

We can summarize our findings using n and m to represent any numbers:

- To *multiply* powers of 10, *add* exponents: $10^n \times 10^m = 10^{n+m}$

- To *divide* powers of 10, *subtract* exponents: $\dfrac{10^n}{10^m} = 10^{n-m}$

- To *raise* powers of 10 to other powers, multiply exponents: $(10^n)^m = 10^{n \times m}$

C.2 SCIENTIFIC NOTATION

When we are dealing with large or small numbers, it's generally easier to write them with powers of 10. For example, it's much easier to write the number 6,000,000,000,000 as 6×10^{12}. This format, in which a number *between* 1 and 10 is multiplied by a power of 10, is called **scientific notation**.

Converting a Number to Scientific Notation

We can convert numbers written in ordinary notation to scientific notation with a simple two-step process:

1. Move the decimal point to come after the *first* nonzero digit.

2. The number of places the decimal point moves tells you the power of 10; the power is *positive* if the decimal point moves to the left and *negative* if it moves to the right.

Examples:

$$3,042 \xrightarrow[\text{3 places to left}]{\text{decimal needs to move}} 3.042 \times 10^3$$

$$0.00012 \xrightarrow[\text{4 places to right}]{\text{decimal needs to move}} 1.2 \times 10^{-4}$$

$$226 \times 10^2 \xrightarrow[\text{2 places to left}]{\text{decimal needs to move}} (2.26 \times 10^2) \times 10^2 = 2.26 \times 10^4$$

Converting a Number from Scientific Notation

We can convert numbers written in scientific notation to ordinary notation by the reverse process:

1. The power of 10 indicates how many places to move the decimal point; move it to the *right* if the power of 10 is positive and to the *left* if it is negative.

2. If moving the decimal point creates any open places, fill them with zeros.

Examples:

$$4.01 \times 10^2 \xrightarrow[\text{2 places to right}]{\text{move decimal}} 401$$

$$3.6 \times 10^6 \xrightarrow[\text{6 places to right}]{\text{move decimal}} 3{,}600{,}000$$

$$5.7 \times 10^{-3} \xrightarrow[\text{3 places to left}]{\text{move decimal}} 0.0057$$

Multiplying or Dividing Numbers in Scientific Notation

Multiplying or dividing numbers in scientific notation simply requires operating on the powers of 10 and the other parts of the number separately.

Examples:

$$(6 \times 10^2) \times (4 \times 10^5) = (6 \times 4) \times (10^2 \times 10^5) = 24 \times 10^7 = (2.4 \times 10^1) \times 10^7 = 2.4 \times 10^8$$

$$\frac{4.2 \times 10^{-2}}{8.4 \times 10^{-5}} = \frac{4.2}{8.4} \times \frac{10^{-2}}{10^{-5}} = 0.5 \times 10^{-2-(-5)} = 0.5 \times 10^3 = (5 \times 10^{-1}) \times 10^3 = 5 \times 10^2$$

Note that, in both these examples, we first found an answer in which the number multiplied by a power of 10 was *not* between 1 and 10. We therefore followed the procedure for converting the final answer to scientific notation.

Addition and Subtraction with Scientific Notation

In general, we must write numbers in ordinary notation before adding or subtracting.

Examples:

$$(3 \times 10^6) + (5 \times 10^2) = 3{,}000{,}000 + 500 = 3{,}000{,}500 = 3.0005 \times 10^6$$

$$(4.6 \times 10^9) - (5 \times 10^8) = 4{,}600{,}000{,}000 - 500{,}000{,}000 = 4{,}100{,}000{,}000 = 4.1 \times 10^9$$

When both numbers have the *same* power of 10, we can factor out the power of 10 first.

Examples:

$$(7 \times 10^{10}) + (4 \times 10^{10}) = (7 + 4) \times 10^{10} = 11 \times 10^{10} = 1.1 \times 10^{11}$$

$$(2.3 \times 10^{-22}) - (1.6 \times 10^{-22}) = (2.3 - 1.6) \times 10^{-22} = 0.7 \times 10^{-22} = 7.0 \times 10^{-23}$$

C.3 WORKING WITH UNITS

Showing the units of a problem as you solve it usually makes the work much easier and also provides a useful way of checking your work. If an answer does not come out with the units you expect, you probably did something wrong. In general, working with units is very similar to working with numbers, as the following guidelines and examples show.

Five Guidelines for Working with Units

Before you begin any problem, think ahead and identify the units you expect for the final answer. Then operate on the units along with the numbers as you solve the problem. The following five guidelines may be helpful when you are working with units:

1. Mathematically, it doesn't matter whether a unit is singular (e.g., meter) or plural (e.g., meters); we can use the same abbreviation (e.g., m) for both.

2. You cannot add or subtract numbers unless they have the *same* units. For example, 5 apples + 3 apples = 8 apples, but the expression 5 apples + 3 oranges cannot be simplified further.

3. You *can* multiply units, divide units, or raise units to powers. Look for key words that tell you what to do.

   - *Per* suggests division. For example, we write a speed of 100 kilometers *per* hour as:

   $$100\frac{km}{hr} \quad \text{or} \quad 100\frac{km}{1\ hr}$$

   - *Of* suggests multiplication. For example, if you launch a 50-kilogram space probe at a launch cost *of* $10,000 per kilogram, the total cost is:

   $$50\ \cancel{kg} \times \frac{\$10,000}{\cancel{kg}} = \$500,000$$

   - *Square* suggests raising to the second power. For example, we write an area of 75 square meters as $75\ m^2$.

   - *Cube* suggests raising to the third power. For example, we write a volume of 12 cubic centimeters as $12\ cm^3$.

4. Often the number you are given is not in the units you wish to work with. For example, you may be given that the speed of light is 300,000 km/s but need it in units of m/s for a particular problem. To convert the units, simply multiply the given number by a *conversion factor:* a fraction in which the numerator (top of the fraction) and denominator (bottom of the fraction) are equal, so that the value of the fraction is 1; the number in the denominator must have the units that you wish to change. In the case of changing the speed of light from units of km/s to m/s, you need a conversion factor for kilometers to meters. Thus, the conversion factor is:

$$\frac{1,000\ m}{1\ km}$$

Note that this conversion factor is equal to 1, since 1,000 meters and 1 kilometer are equal, and that the units to be changed (km) appear in the denominator. We can now convert the speed of light from units of km/s to m/s simply by multiplying by this conversion factor:

$$\underbrace{300,000\frac{\cancel{km}}{s}}_{\substack{\text{speed of light}\\\text{in km/s}}} \times \underbrace{\frac{1,000\ m}{1\ \cancel{km}}}_{\substack{\text{conversion from}\\\text{km to m}}} = \underbrace{3 \times 10^8\frac{m}{s}}_{\substack{\text{speed of light}\\\text{in m/s}}}$$

Note that the units of km cancel, leaving the answer in units of m/s.

5. It's easier to work with units if you replace division with multiplication by the reciprocal. For example, suppose you want to know how many minutes are represented by 300 seconds. We can find the answer by dividing 300 seconds by 60 seconds per minute:

$$300 \text{ s} \div 60 \frac{\text{s}}{\text{min}}$$

However, it is easier to see the unit cancellations if we rewrite this expression by replacing the division with multiplication by the reciprocal (this process is easy to remember as "invert and multiply"):

$$300 \text{ s} \div 60 \frac{\text{s}}{\text{min}} = 300 \, \cancel{s} \times \underbrace{\frac{1 \text{ min}}{60 \, \cancel{s}}}_{\text{invert}} = 5 \text{ min}$$

and multiply

We now see that the units of seconds (s) cancel in the numerator of the first term and the denominator of the second term, leaving the answer in units of minutes.

More Examples of Working with Units

Example 1. How many seconds are there in 1 day?

Solution: We can answer the question by setting up a *chain* of unit conversions in which we start with 1 *day* and end up with *seconds*. We use the facts that there are 24 hours per day (24 hr/day), 60 minutes per hour (60 min/hr), and 60 seconds per minute (60 s/min):

$$\underbrace{1 \, \cancel{\text{day}}}_{\substack{\text{starting} \\ \text{value}}} \times \underbrace{\frac{24 \, \cancel{\text{hr}}}{\cancel{\text{day}}}}_{\substack{\text{conversion} \\ \text{from} \\ \text{day to hr}}} \times \underbrace{\frac{60 \, \cancel{\text{min}}}{\cancel{\text{hr}}}}_{\substack{\text{conversion} \\ \text{from} \\ \text{hr to min}}} \times \underbrace{\frac{60 \text{ s}}{\cancel{\text{min}}}}_{\substack{\text{conversion} \\ \text{from} \\ \text{min to s}}} = 86{,}400 \text{ s}$$

Note that all the units cancel except *seconds*, which is what we want for the answer. There are 86,400 seconds in 1 day.

Example 2. Convert a distance of 10^8 centimeters to kilometers.

Solution: The easiest way to make this conversion is in two steps, since we know that there are 100 centimeters per meter (100 cm/m) and 1,000 meters per kilometer (1,000 m/km):

$$\underbrace{10^8 \text{ cm}}_{\substack{\text{starting} \\ \text{value}}} \times \underbrace{\frac{1 \text{ m}}{100 \text{ cm}}}_{\substack{\text{conversion} \\ \text{from} \\ \text{cm to m}}} \times \underbrace{\frac{1 \text{ km}}{1{,}000 \text{ m}}}_{\substack{\text{conversion} \\ \text{from} \\ \text{m to km}}} = 10^8 \, \cancel{\text{cm}} \times \frac{1 \, \cancel{\text{m}}}{10^2 \, \cancel{\text{cm}}} \times \frac{1 \text{ km}}{10^3 \, \cancel{\text{m}}} = 10^3 \text{ km}$$

Alternatively, if we recognize that the number of kilometers should be smaller than the number of centimeters (because kilometers are larger), we might decide to do this conversion by dividing as follows:

$$10^8 \text{ cm} \div \frac{100 \text{ cm}}{\text{m}} \div \frac{1{,}000 \text{ m}}{\text{km}}$$

In this case, before carrying out the calculation, we replace each division with multiplication by the reciprocal:

$$10^8 \text{ cm} \div \frac{100 \text{ cm}}{\text{m}} \div \frac{1{,}000 \text{ m}}{\text{km}} = 10^8 \text{ cm} \times \frac{1 \text{ m}}{100 \text{ cm}} \times \frac{1 \text{ km}}{1{,}000 \text{ m}}$$

$$= 10^8 \cancel{\text{cm}} \times \frac{1 \cancel{\text{m}}}{10^2 \cancel{\text{cm}}} \times \frac{1 \text{ km}}{10^3 \cancel{\text{m}}}$$

$$= 10^3 \text{ km}$$

Note that we again get the answer that 10^8 cm is the same as 10^3 km, or 1,000 km.

Example 3. Suppose you accelerate at 9.8 m/s^2 for 4 seconds, starting from rest. How fast will you be going?

Solution: The question asks "how fast?" so we expect to end up with a speed. Therefore, we multiply the acceleration by the amount of time you accelerated:

$$9.8 \frac{\text{m}}{\text{s}^2} \times 4 \text{ s} = (9.8 \times 4) \frac{\text{m} \times \cancel{\text{s}}}{\text{s}^{\cancel{2}}} = 39.2 \frac{\text{m}}{\text{s}}$$

Note that the units end up as a speed, showing that you will be traveling 39.2 m/s after 4 seconds of acceleration at 9.8 m/s^2.

Example 4. A reservoir is 2 km long and 3 km wide. Calculate its area, in both square kilometers and square meters.

Solution: We find its area by multiplying its length and width:

$$2 \text{ km} \times 3 \text{ km} = 6 \text{ km}^2$$

Next we need to convert this area of 6 km^2 to square meters, using the fact that there are 1,000 meters per kilometer (1,000 m/km). Note that we must square the term 1,000 m/km when converting from km^2 to m^2:

$$6 \text{ km}^2 \times \left(1{,}000 \frac{\text{m}}{\text{km}}\right)^2 = 6 \text{ km}^2 \times 1{,}000^2 \frac{\text{m}^2}{\text{km}^2} = 6 \cancel{\text{km}^2} \times 1{,}000{,}000 \frac{\text{m}^2}{\cancel{\text{km}^2}}$$

$$= 6{,}000{,}000 \text{ m}^2$$

The reservoir area is 6 km^2, which is the same as 6 million m^2.

C.4 THE METRIC SYSTEM (SI)

The modern version of the metric system, known as *Système Internationale d'Unites* (French for "International System of Units") or **SI**, was formally established in 1960. Today, it is the primary measurement system in nearly every country in the world with the exception of the United States. Even in the United States, it is the system of choice for science and international commerce.

The basic units of length, mass, and time in the SI are:

- The **meter** for length, abbreviated m
- The **kilogram** for mass, abbreviated kg
- The **second** for time, abbreviated s

Multiples of metric units are formed by powers of 10, using a prefix to indicate the power. For example, *kilo* means 10^3 (1,000), so a kilometer is

TABLE C.2 *SI (Metric) Prefixes*

Small Values			Large Values		
Prefix	Abbreviation	Value	Prefix	Abbreviation	Value
Deci	d	10^{-1}	Deca	da	10^{1}
Centi	c	10^{-2}	Hecto	h	10^{2}
Milli	m	10^{-3}	Kilo	k	10^{3}
Micro	μ	10^{-6}	Mega	M	10^{6}
Nano	n	10^{-9}	Giga	G	10^{9}
Pico	p	10^{-12}	Tera	T	10^{12}

1,000 meters; a microgram is 0.000001 gram, because *micro* means 10^{-6}, or one millionth. Some of the more common prefixes are listed in Table C.2.

Metric Conversions

TABLE C.3 *Metric Conversions*

To Metric	From Metric
1 inch = 2.540 cm	1 cm = 0.3937 inch
1 foot = 0.3048 m	1 m = 3.28 feet
1 yard = 0.9144 m	1 m = 1.094 yards
1 mile = 1.6093 km	1 km = 0.6214 mile
1 pound = 0.4536 kg	1 kg = 2.205 pounds

Table C.3 lists conversions between metric units and units used commonly in the United States. Note that the conversions between kilograms and pounds are valid only on Earth, because they depend on the strength of gravity.

Example 1. International athletic competitions generally use metric distances. Compare the length of a 100-meter race to that of a 100-yard race.

Solution: Table C.3 shows that 1 m = 1.094 yd, so 100 m is 109.4 yd. Note that 100 meters is almost 110 yards; a good "rule of thumb" to remember is that distances in meters are about 10% longer than the corresponding number of yards.

Example 2. How many square kilometers are in 1 square mile?

Solution: We use the square of the miles-to-kilometers conversion factor:

$$(1 \text{ mi}^2) \times \left(\frac{1.6093 \text{ km}}{1 \text{ mi}} \right)^2 = (1 \text{ mi}^2) \times \left(1.6093^2 \frac{\text{km}^2}{\text{mi}^2} \right) = 2.5898 \text{ km}^2$$

Therefore, 1 square mile is 2.5898 square kilometers.

C.5 FINDING A RATIO

Suppose you want to compare two quantities, such as the average density of Earth and the average density of Jupiter. The way we do such a comparison is by dividing, which tells us the *ratio* of the two quantities. In this case, Earth's average density is 5.52 grams/cm³ and Jupiter's average density is 1.33 grams/cm³, so the ratio is:

$$\frac{\text{average density of Earth}}{\text{average density of Jupiter}} = \frac{5.52 \text{ g/cm}^3}{1.33 \text{ g/cm}^3} = 4.15$$

Notice how the units cancel on both the top and bottom of the fraction. We can state our result in two equivalent ways:

- The ratio of Earth's average density to Jupiter's average density is 4.15.

- Earth's average density is 4.15 times Jupiter's average density.

Sometimes, the quantities that you want to compare may each involve an equation. In such cases, you could, of course, find the ratio by first calculating each of the two quantities individually and then dividing. However, it is much easier if you first express the ratio as a fraction, putting the equation for one quantity on top and the other equation on the bottom. Some of the terms in the equation may then cancel out, making any calculations much easier.

Example 1. Compare the kinetic energy of a car traveling at 100 km/hr to that of a car traveling at 50 km/hr.

Solution: We do the comparison by finding the ratio of the two kinetic energies, recalling that the formula for kinetic energy is $\frac{1}{2}mv^2$. Since we are not told the mass of the car, you might at first think that we don't have enough information to find the ratio. However, notice what happens when we put the equations for each kinetic energy into the ratio, calling the two speeds v_1 and v_2:

$$\frac{\text{K.E. car at } v_1}{\text{K.E. car at } v_2} = \frac{\frac{1}{2}m_{\text{car}}v_1{}^2}{\frac{1}{2}m_{\text{car}}v_2{}^2} = \frac{v_1{}^2}{v_2{}^2} = \left(\frac{v_1}{v_2}\right)^2$$

All the terms cancel except those with the two speeds, leaving us with a very simple formula for the ratio. Now we put in 100 km/hr for v_1 and 50 km/hr for v_2:

$$\frac{\text{K.E. car at 100 km/hr}}{\text{K.E. car at 50 km/hr}} = \left(\frac{100 \text{ km/hr}}{50 \text{ km/hr}}\right)^2 = 2^4 = 4$$

The ratio of the car's kinetic energies at 100 km/hr and 50 km/hr is 4. That is, the car has four times as much kinetic energy at 100 km/hr as it has at 50 km/hr.

Example 2. Compare the strength of gravity between Earth and the Sun to the strength of gravity between Earth and the Moon.

Solution: We do the comparison by taking the ratio of the Earth–Sun gravity to the Earth–Moon gravity. In this case, each quantity is found from the equation of Newton's law of gravity. (See Section 4.4.) Thus, the ratio is:

$$\frac{\text{Earth–Sun gravity}}{\text{Earth–Moon gravity}} = \frac{G\dfrac{M_{\text{Earth}}M_{\text{Sun}}}{(d_{\text{Earth–Sun}})^2}}{G\dfrac{M_{\text{Earth}}M_{\text{Moon}}}{(d_{\text{Earth–Moon}})^2}} = \frac{M_{\text{Sun}}}{(d_{\text{Earth–Sun}})^2} \times \frac{(d_{\text{Earth–Moon}})^2}{M_{\text{Moon}}}$$

Note how all but four of the terms cancel; the last step comes from replacing the division with multiplication by the reciprocal (the "invert and multiply" rule for division). We can simplify the work further by rearranging the terms so that we have the masses and distances together:

$$\frac{\text{Earth–Sun gravity}}{\text{Earth–Moon gravity}} = \frac{M_{\text{Sun}}}{M_{\text{Moon}}} \times \frac{(d_{\text{Earth–Moon}})^2}{(d_{\text{Earth–Sun}})^2}$$

Now it is just a matter of looking up the numbers (see Appendix E) and calculating:

$$\frac{\text{Earth–Sun gravity}}{\text{Earth–Moon gravity}} = \frac{1.99 \times 10^{30} \text{ kg}}{7.35 \times 10^{22} \text{ kg}} \times \frac{(384.4 \times 10^3 \text{ km})^2}{(149.6 \times 10^6 \text{ km})^2} = 179$$

In other words, the Earth–Sun gravity is 179 times stronger than the Earth–Moon gravity.

D | The Periodic Table of the Elements

Key

12 — Atomic number
Mg — Element's symbol
Magnesium — Element's name
24.305 — Atomic mass*

*Atomic masses are fractions because they represent a weighted average of atomic masses of different isotopes— in proportion to the abundance of each isotope on Earth.

1																	2
H Hydrogen 1.00794																	**He** Helium 4.003
3 **Li** Lithium 6.941	4 **Be** Beryllium 9.01218											5 **B** Boron 10.81	6 **C** Carbon 12.011	7 **N** Nitrogen 14.007	8 **O** Oxygen 15.999	9 **F** Fluorine 18.988	10 **Ne** Neon 20.179
11 **Na** Sodium 22.990	12 **Mg** Magnesium 24.305											13 **Al** Aluminum 26.98	14 **Si** Silicon 28.086	15 **P** Phosphorus 30.974	16 **S** Sulfur 32.06	17 **Cl** Chlorine 35.453	18 **Ar** Argon 39.948
19 **K** Potassium 39.098	20 **Ca** Calcium 40.08	21 **Sc** Scandium 44.956	22 **Ti** Titanium 47.88	23 **V** Vanadium 50.94	24 **Cr** Chromium 51.996	25 **Mn** Manganese 54.938	26 **Fe** Iron 55.847	27 **Co** Cobalt 58.9332	28 **Ni** Nickel 58.69	29 **Cu** Copper 63.546	30 **Zn** Zinc 65.39	31 **Ga** Gallium 69.72	32 **Ge** Germanium 72.59	33 **As** Arsenic 74.922	34 **Se** Selenium 78.96	35 **Br** Bromine 79.904	36 **Kr** Krypton 83.80
37 **Rb** Rubidium 85.468	38 **Sr** Strontium 87.62	39 **Y** Yttrium 88.9059	40 **Zr** Zirconium 91.224	41 **Nb** Niobium 92.91	42 **Mo** Molybdenum 95.94	43 **Tc** Technetium (98)	44 **Ru** Ruthenium 101.07	45 **Rh** Rhodium 102.906	46 **Pd** Palladium 106.42	47 **Ag** Silver 107.868	48 **Cd** Cadmium 112.41	49 **In** Indium 114.82	50 **Sn** Tin 118.71	51 **Sb** Antimony 121.75	52 **Te** Tellurium 127.60	53 **I** Iodine 126.905	54 **Xe** Xenon 131.29
55 **Cs** Cesium 132.91	56 **Ba** Barium 137.34		72 **Hf** Hafnium 178.49	73 **Ta** Tantalum 180.95	74 **W** Tungsten 183.85	75 **Re** Rhenium 186.207	76 **Os** Osmium 190.2	77 **Ir** Iridium 192.22	78 **Pt** Platinum 195.08	79 **Au** Gold 196.967	80 **Hg** Mercury 200.59	81 **Ti** Thallium 204.383	82 **Pb** Lead 207.2	83 **Bi** Bismuth 208.98	84 **Po** Polonium (209)	85 **At** Astatine (210)	86 **Rn** Radon (222)
87 **Fr** Francium (223)	88 **Ra** Radium 226.0254		104 **Rf** Rutherfordium (261)	105 **Db** Dubnium (262)	106 **Sg** Seaborgium (263)	107 **Bh** Bohrium (262)	108 **Hs** Hassium (265)	109 **Mt** Meitnerium (266)	110 **Uun** Ununnilium (269)	111 **Uuu** Unununium (272)	112 **Uub** Ununbium (277)						

Lanthanide Series

57 **La** Lanthanum 138.906	58 **Ce** Cerium 140.12	59 **Pr** Praseodymium 140.908	60 **Nd** Neodymium 144.24	61 **Pm** Promethium (145)	62 **Sm** Samarium 150.36	63 **Eu** Europium 151.96	64 **Gd** Gadolinium 157.25	65 **Tb** Terbium 158.925	66 **Dy** Dysprosium 162.50	67 **Ho** Holmium 164.93	68 **Er** Erbium 167.26	69 **Tm** Thulium 168.934	70 **Yb** Ytterbium 173.04	71 **Lu** Lutetium 174.967

Actinide Series

89 **Ac** Actinium 227.028	90 **Th** Thorium 232.038	91 **Pa** Protactinium 231.036	92 **U** Uranium 238.029	93 **Np** Neptunium 237.048	94 **Pu** Plutonium (244)	95 **Am** Americium (243)	96 **Cm** Curium (247)	97 **Bk** Berkelium (247)	98 **Cf** Californium (251)	99 **Es** Einsteinium (252)	100 **Fm** Fermium (257)	101 **Md** Mendelevium (258)	102 **No** Nobelium (259)	103 **Lr** Lawrencium (260)

E | Planetary Data

TABLE E.1 *Physical Properties of the Sun and Planets*

Name	Radius (Eq[a]) (km)	Radius (Eq) (Earth units)	Mass (kg)	Mass (Earth units)	Average Density (g/cm^3)	Surface Gravity (Earth = 1)	Escape Velocity (km/s)
Sun	695,000	109	1.99×10^{30}	333,000	1.41	27.5	—
Mercury	2,440	0.382	3.30×10^{23}	0.055	5.43	0.38	4.43
Venus	6,051	0.949	4.87×10^{24}	0.815	5.25	0.91	10.4
Earth	6,378	1.00	5.97×10^{24}	1.00	5.52	1.00	11.2
Mars	3,397	0.533	6.42×10^{23}	0.107	3.93	0.38	5.03
Jupiter	71,492	11.19	1.90×10^{27}	317.9	1.32	2.36	59.5
Saturn	60,268	9.46	5.69×10^{26}	95.18	0.70	0.92	35.5
Uranus	25,559	3.98	8.66×10^{25}	14.54	1.22	0.91	21.3
Neptune	24,764	3.81	1.03×10^{26}	17.13	1.64	1.14	23.6
Pluto[b]	1,160	0.181	1.31×10^{22}	0.0022	2.05	0.07	1.25
Eris[b]	1,200	0.19	—	—	—	—	—

[a]Eq = equatorial.

[b]Under IAU definitions approved in August 2006, Pluto and Eris are officially designated "dwarf planets."

TABLE E.2 *Orbital Properties of the Sun and Planets*

Name	Distance from Sun[a] (AU)	Distance from Sun[a] $(10^6 km)$	Orbital Period (years)	Orbital Inclination[b] (degrees)	Orbital Eccentricity	Sidereal Rotation Period (Earth days)[c]	Axis Tilt (degrees)
Sun	—	—	—	—	—	25.4	7.25
Mercury	0.387	57.9	0.2409	7.00	0.206	58.6	0.0
Venus	0.723	108.2	0.6152	3.39	0.007	−243.0	177.3
Earth	1.00	149.6	1.0	0.00	0.017	0.9973	23.45
Mars	1.524	227.9	1.881	1.85	0.093	1.026	25.2
Jupiter	5.203	778.3	11.86	1.31	0.048	0.41	3.08
Saturn	9.539	1,427	29.42	2.48	0.056	0.44	26.73
Uranus	19.19	2,870	84.01	0.77	0.046	−0.72	97.92
Neptune	30.06	4,497	164.8	1.77	0.010	0.67	29.6
Pluto	39.48	5,906	248.0	17.14	0.248	−6.39	112.5
Eris	67.89	10,156	560	44.0	0.438	—	—

[a] Semimajor axis of the orbit.

[b] With respect to the ecliptic.

[c] A negative sign indicates that rotation is backward relative to other planets.

TABLE E.3 *Satellites of the Solar System (as of 2006)[a]*

Planet Satellite	Radius or Dimensions[b] (km)	Distance from Planet[c] (10³ km)	Orbital Period[c] (Earth days)	Mass[d] (kg)	Density[d] (g/cm³)	Notes About the Satellite
Earth						**Earth**
Moon	1,738	384.4	27.322	7.349×10^{22}	3.34	*Moon:* Probably formed in giant impact.
Mars						**Mars**
Phobos	$13 \times 11 \times 9$	9.38	0.319	1.3×10^{16}	1.9	*Phobos, Deimos:* Probable captured asteroids.
Deimos	$8 \times 6 \times 5$	23.5	1.263	1.8×10^{15}	2.2	
Jupiter						**Jupiter**
Small inner moons (4 moons)	8 to 83	128–222	0.295–0.6745	—	—	*Metis, Adrastea, Amalthea, Thebe:* Small moonlets within and near Jupiter's ring system.
Io	1,821	421.6	1.769	8.933×10^{22}	3.57	*Io:* Most volcanically active object in the solar system.
Europa	1,565	670.9	3.551	4.797×10^{22}	2.97	*Europa:* Possible oceans under icy crust.
Ganymede	2,634	1,070.0	7.155	1.482×10^{23}	1.94	*Ganymede:* Largest satellite in solar system; unusual ice geology.
Callisto	2,403	1,883.0	16.689	1.076×10^{23}	1.86	*Callisto:* Cratered iceball.
Irregular group 1 (7 moons)	4–85	7,200–17,000	130–457	—	—	*Themisto, Leda, Himalia, Lysithea, Elara, and 44 others:* Probable captured moons with inclined orbits.
Irregular group 2 (48 moons)	1–30	15,900–29,500	−490 to −983	—	—	*Ananke, Carme, Pasiphae, Sinope, and 46 others:* Probable captured moons in inclined backward orbits.
Saturn						**Saturn**
Small inner moons (11)	3–89	134–212	0.574–1.1	—	—	*Pan, Atlas, Prometheus, Pandora, Epimetheus, Janus, and 5 more:* Small moonlets within and near Saturn's ring system.
Mimas	199	185.52	0.942	3.70×10^{19}	1.17	*Mimas, Enceladus, Tethys:* Small and medium-size iceballs, many with interesting geology.
Enceladus	249	238.02	1.370	1.2×10^{20}	1.24	
Tethys	530	294.66	1.888	6.17×10^{20}	1.26	
Calypso & Telesto	8–12	294.66	1.888	—	—	*Calypso, Telesto:* Small moonlets sharing Tethys's orbit.
Dione	559	377.4	2.737	1.08×10^{21}	1.44	*Dione:* Medium-size iceball, with interesting geology.
Helene & Polydeuces	2–16	377.4	2.737	1.6×10^{16}	—	*Helene & Polydeuces:* Small moonlets sharing Dione's orbit.
Rhea	764	527.04	4.518	2.31×10^{21}	1.33	*Rhea:* Medium-size iceball, with interesting geology.
Titan	2,575	1,221.85	15.945	1.35×10^{23}	1.88	*Titan:* Dense atmosphere shrouds surface; ongoing geological activity possible.
Hyperion	$180 \times 140 \times 112$	1,481.1	21.277	2.8×10^{19}	—	*Hyperion:* Only satellite known not to rotate synchronously.

Iapetus	718	3,561.3	79.331	1.59×10^{21}	1.21	*Iapetus:* Bright and dark hemispheres show greatest contrast in the solar system.
Phoebe	110	12,952	−550.4	1×10^{19}	—	*Phoebe:* Very dark; material ejected from Phoebe may coat one side of Iapetus.
Irregular groups (25 moons)	2–16	11,400–23,400	450–930 / −550 to −1,320	—	—	Probable captured moons with highly inclined and/or backward orbits.
Uranus						**Uranus**
Small inner moons (13 moons)	5–81	49–98	0.4–0.9	—	—	*Cordelia, Ophelia, Bianca, Cressida, Desdemona, Juliet, Portia, Rosalind, Belinda, Puck, 1986 U10, 2003 U1, 2003 U3:* Small moonlets within and near Uranus's ring system.
Miranda	236	129.8	1.413	6.6×10^{19}	1.26	*Miranda, Ariel, Umbriel, Titania, Oberon:* Small and medium-size iceballs, with some interesting geology.
Ariel	579	191.2	2.520	1.35×10^{21}	1.65	
Umbriel	584.7	266.0	4.144	1.17×10^{21}	1.44	
Titania	788.9	435.8	8.706	3.52×10^{21}	1.59	
Oberon	761.4	582.6	13.463	3.01×10^{21}	1.50	
Irregular group (9 moons)	5–95	4,280–21,000	580–2,280	—	—	*Caliban, Sycorax, Stephano, Prospero, Setebos, Trinculo, 2001 U2, 2001 U3, 2003 U3:* Probable captured moons; several in backward orbits.
Neptune						**Neptune**
Small inner moons (5 moons)	29–86	48–74	0.30–0.55	—	—	*Naiad, Thalassa, Despina, Galatea, Larissa:* Small moonlets within and near Neptune's ring system.
Proteus	218 × 208 × 201	117.6	1.121	6×10^{19}	—	
Triton	1,352.6	354.59	−5.875	2.14×10^{22}	2.0	*Triton:* Probable captured Kuiper belt object—largest captured object in solar system.
Nereid	170	5,588.6	360.125	3.1×10^{19}	—	*Nereid:* Small, icy moon; very little else known.
Irregulars (5 moons)	15–27	16,600–48,600	1,870–9,412	—	—	*2002 N1, N2, N3, N4, 2003 N1:* Possible captured moons in inclined or backward orbit.
Pluto						**Pluto**
Charon	593	19.6	6.38718	1.56×10^{21}	1.6	*Charon:* Unusually large compared to its planet; may have formed in giant impact.
Nix and Hydra	32–100	—	—	—	—	Newly discovered moons outside Charon's orbit.

[a] *Note:* Authorities differ substantially on many of the values in this table.

[b] a × b × c values for the dimensions are the approximate lengths of the axes (center to edge) for irregular moons.

[c] Negative sign indicates backward orbit.

[d] Masses and densities are most accurate for those satellites visited by a spacecraft on a flyby. Masses for the smallest moons have not been measured but can be estimated from the radius and an assumed density.

TABLE E.4 *Fifty Extrasolar Planets of Note (listed in order of distance from their star)*

Name	Detection Methods[2]	Minimum Mass (Jupiter Masses)	Semimajor Axis (AU)	Period (days)	Radius (Jupiter radii)	Notes[1]
Gliese 876 d	radial velocity	0.023	0.021	1.94	—	"Hot Jupiter"; smallest known semimajor axis of 2005
OGLE-TR-56 b	radial velocity, transit	1.45	0.023	1.21	1.23	"Hot Jupiter"; first planet discovered by transit
GJ 436 b	radial velocity	0.067	0.028	2.64	—	"Hot Jupiter"; sub-Saturn mass
OGLE-TR-132 b	radial velocity, transit	1.19	0.031	1.69	1.13	"Hot Jupiter"
55 Cnc e	radial velocity	0.045	0.038	2.81	—	"Hot Jupiter"; sub-Neptune mass
TrES-1	transit, radial velocity, eclipse	0.61	0.039	3.03	1.08	"Hot Jupiter"
HD 46375 b	radial velocity	0.25	0.041	3.02	—	"Hot Jupiter"; sub-Saturn mass
OGLE-TR-10 b	transit, radial velocity	0.54	0.042	3.10	1.16	"Hot Jupiter"
HD 149026 b	radial velocity, transit	0.36	0.042	2.88	0.73	"Hot Jupiter"
HD 188753A b	radial velocity	1.14	0.045	3.35	—	"Hot Jupiter"; first planet discovered in triple star system
HD 209458 b	radial velocity, transit, eclipse	0.69	0.045	3.52	1.32	"Hot Jupiter"; first planet and first atmosphere successfully detected by transit
HD 88133 b	radial velocity	0.22	0.047	3.41	—	"Hot Jupiter"; sub-Saturn mass
OGLE-TR-111 b	radial velocity, transit	0.53	0.047	4.02	1.00	"Hot Jupiter"
51 Peg b	radial velocity	0.47	0.052	4.23	—	"Hot Jupiter"; first exoplanet discovered around Sun-like star
Ups And b	radial velocity	0.69	0.059	4.62	—	"Hot Jupiter"; in first multiplanet system discovered around Sun-like star
HD 160691 d	radial velocity	0.044	0.090	9.55	—	"Hot Jupiter"; sub-Neptune mass
55 Cnc b	radial velocity	0.78	0.115	14.7	—	—
HD 190360 c	radial velocity	0.057	0.128	17.1	—	Sub-Saturn mass
Gliese 876 c	radial velocity	0.56	0.130	30.1	—	"Eccentric"[3]
HD 102117 b	radial velocity	0.14	0.149	20.7	—	Sub-Saturn mass
Gliese 876 b	radial velocity	1.94	0.208	60.9	—	—
55 Cnc c	radial velocity	0.22	0.240	43.9	—	"Eccentric"; sub-Saturn mass
HD 16141 b	radial velocity	0.23	0.350	75.6	—	Sub-Saturn mass
HD 80606 b	radial velocity	3.4	0.439	112	—	"Eccentric"
HD 82943 c	radial velocity	0.88	0.730	222	—	"Eccentric"

						Notes
Ups And c	radial velocity	1.89	0.829	242	—	"Eccentric"; in first multiplanet system discovered
HR 810 b	radial velocity	1.94	0.910	311	—	—
HD 210277 b	radial velocity	1.24	1.097	436	—	"Eccentric"; planet mass inferred from surrounding disk
HD 114783 b	radial velocity	0.99	1.200	501	—	—
HD 41004A b	radial velocity	2.3	1.310	655	—	"Eccentric"; planet in a system with two stars and brown dwarf
HD 183263 b	radial velocity	3.7	1.520	634	—	"Eccentric"
HD 114386 b	radial velocity	0.99	1.620	872	—	"Eccentric"
HD 4208 b	radial velocity	0.80	1.670	812	—	—
HD 13189 b	radial velocity	14.0	1.850	472	—	"Eccentric"
Gamma Cephei b	radial velocity	1.59	2.030	902	—	First extrasolar planet discovered in close stellar binary system
47 UMa b	radial velocity	2.5	2.090	1,089	—	—
HD 50554 b	radial velocity	4.9	2.380	1,279	—	"Eccentric"
Ups And d	radial velocity	3.8	2.530	1,284	—	"Eccentric"; in first multiplanet system discovered
HD 202206 c	radial velocity	2.4	2.550	1,383	—	"Eccentric"
HD 12661 c	radial velocity	1.57	2.560	1,445	—	—
HD 168443 c	radial velocity	17.1	2.870	1,740	—	—
HD 37124 d	radial velocity	0.66	3.190	2,295	—	—
Epsilon Eridani b	radial velocity, astrometry	0.86	3.300	2,502	—	"Eccentric"; star surrounded by dust disk
HD 74156 c	radial velocity	6.2	3.400	2,025	—	"Eccentric"
HD 169830 c	radial velocity	4.0	3.600	2,102	—	"Eccentric"
HD 38529 c	radial velocity	12.7	3.680	2,174	—	"Eccentric"
47 UMa c	radial velocity	0.76	3.730	2,594	—	—
HD 50499 b	radial velocity	1.71	3.860	2,583	—	—
HD 72659 b	radial velocity	3.0	4.160	3,177	—	"Eccentric"
55 Cnc d	radial velocity	3.9	5.257	4,517	—	"Eccentric"; largest known semimajor axis as of 2005

[1] The list includes all planets detected by two methods, most planets in multiple systems, most "Hot Jupiters," and a representative sample of other extrasolar planets are not listed.

[2] If two detection methods are listed, the discovery method is given first.

[3] "Eccentric" means eccentricity > 0.25.

F | Selected Astrobiology Web Sites

The Web contains a vast amount of astronomical information. For all your astrobiology Web surfing, the best starting point is the Web site for this textbook:

Life in the Universe Companion Web Site
http://wps.aw.com/aw_bennett_liu_2

The book Web site contains both general links and links organized by chapter. The following tables list some other sites that may be of particular use; if any of the links change, in most cases you can find live links on the book Web site.

PRIMARY ASTROBIOLOGY SITES

General astrobiology sites that serve as great starting points for all kinds of astrobiology information, including the latest news and summaries of key research topics.

Site	Description	Web Address
NASA Astrobiology Institute	NASA's main site for astrobiology	http://nai.arc.nasa.gov
NASA Science News	Great site for all science news related to NASA research, including astrobiology; offers sign-up for e-mail news alerts	http://science.nasa.gov
The SETI Institute	Devoted to the search for life off the Earth	http://www.seti.org
Astrobiology Web	Discovery Channel site dedicated to astrobiology	http://www.astrobiology.com

MORE ASTROBIOLOGY NEWS AND RESOURCES

Additional sites offering the latest astrobiology news and other astrobiology resources.

Site	Description	Web Address
Astrobiology: The Living Universe	Thinkquest student-created site for astrobiology	http://library.thinkquest.org/C003763/
Astronomy Now	General astronomy news, including astrobiology news	http://www.astronomynow.com
The Encyclopedia of Astrobiology, Astronomy, and Spaceflight	Astrobiology news and resources maintained by Dr. David Darling	http://www.daviddarling.info/encyclopedia/ETEmain.html
EurekAlert	The American Association for the Advancement of Science (AAAS) general news site organized by discipline	http://www.eurekalert.org/bysubject/space.php
JPL News	News releases from NASA's Jet Propulsion Laboratory	http://www.jpl.nasa.gov/news
Marsbugs Newsletter	An astrobiology newsletter	http://www.lyon.edu/projects/marsbugs
Science Online	*Science Magazine*'s online site	http://www.sciencemag.org/content/current
Space.com	Lots of space-related news, including areas for SETI and astronomy	http://www.space.com
SpaceScience.com	General space science news	http://science.nasa.gov/SpaceScience.htm
Universe Today	General space news	http://www.universetoday.com

KEY MISSION SITES

The Web pages for major astrobiology-related space missions.

Site	Description	Web Address
NASA's Office of Space Science Missions Page	Direct links to all past, present, and planned NASA space science missions	http://science.hq.nasa.gov/missions/phase.html
Cassini/Huygens	Mission scheduled to arrive at Saturn in 2004	http://www.jpl.nasa.gov/missions/current/cassini.html
Hubble Space Telescope	Latest discoveries, educational activities, and other information from the Hubble Space Telescope	http://hubble.stsci.edu
Mars Exploration Program	Information on current and planned Mars missions	http://mars.jpl.nasa.gov
Kepler	Mission to search for extrasolar planets	http://www.kepler.arc.nasa.gov
Space Interferometry Mission (SIM)	Astronomical observatory that will search for extrasolar planets	http://sim.jpl.nasa.gov/SIM/sim_index.cfm
Terrestrial Planet Finder (TPF)	Proposed mission to search for extrasolar terrestrial planets	http://planetquest.jpl.nasa.gov/TPF/tpf_index.cfm

SITES CONCERNING THE SEARCH FOR EXTRASOLAR PLANETS

Site	Description	Web Address
The Extrasolar Planets Encyclopedia	Information about the search for and discoveries of extrasolar planets	http://cfa-www.harvard.edu/planets
PlanetQuest	NASA site devoted to the search for Earth-like planets	http://planetquest.jpl.nasa.gov/index.html
California and Carnegie Planet Search	Site developed by leading extrasolar planet hunters	http://exoplanets.org

MORE USEFUL WEB SITES

Some of the authors' favorites among many noncommercial resources for astronomy and astrobiology.

Site	Description	Web Address
Astronomical Society of the Pacific	An organization for both professional astronomers and the general public, devoted largely to astronomy education	http://www.astrosociety.org
Astronomy Picture of the Day	An archive of beautiful pictures, updated daily	http://antwrp.gsfc.nasa.gov/apod
AstroWeb	A listing of major resources for astronomy on the Web	http://www.cv.nrao.edu/fits/www/astronomy.html
Canadian Space Agency	Home page for Canada's space program	http://www.space.gc.ca
European Space Agency (ESA)	Home page for this international agency	http://www.esa.int
NASA Home Page	Almost anything you want to learn about NASA	http://www.nasa.gov
The Nine Planets (University of Arizona)	A multimedia tour of the solar system	http://www.nineplanets.org
The Planetary Society	An organization with more than 100,000 members, who are interested in planetary exploration and the search for life in the universe	http://planetary.org
Voyage Scale Model Solar System	The National Mall exhibit discussed in Chapter 3	http://www.challenger.org/teachers/programs/voyage.cfm

ASTROBIOLOGY SOCIETY WEB SITES

The Web sites of large groups or institutions that currently carry out astro-biology-related research, education, and outreach.

Group	Web Address
Arrhenius Laboratory at Scripps Institution of Oceanography	http://www.mrd.ucsd.edu/ga/
Astrobiology Australasia (Australia and New Zealand)	http://www.aao.gov.au/local/www/jab/astrobiology/
Astrobiology: The Living Universe	http://library.thinkquest.org/C003763/
Biosphere 2 Astrobiology Launch Pad	http://www.astro.bio2.edu/astrobiology/
Caltech: Geobiology and Astrobiology	http://www.gps.caltech.edu/options/geobiology/geobio.html
Cardiff Center for Astrobiology	http://www.astrobiology.cf.ac.uk
Center for Astrobiology, Spain (in Spanish)	http://www.cab.inta.es
Arizona State University Astrobiology Institute	http://astrobiology.asu.edu
UCLA Center for Astrobiology	http://astrobiology.ucla.edu
UCLA Astrobiology Society	http://www.studentgroups.ucla.edu/abs/
Johnson Space Center Astrobiology Institute for the Study of Biomarkers	http://ares.jsc.nasa.gov/astrobiology/biomarkers/index.html
Penn State Astrobiology Research Center	http://psarc.geosc.psu.edu
University of Colorado Center for Astrobiology	http://lasp.colorado.edu/life/
University of Washington Astrobiology Program	http://depts.washington.edu/astrobio/
Woods Hole Oceanographic Institute Astrobiology Program	http://astrobiology.mbl.edu
Night Eyes: The Auckland Astronomical Society Junior Section	http://www.astronomy.org.nz/aas/pub/home.asp
SETI and Astrobiology in Australia	http://www1.tpgi.com.au/users/tps-seti/seti.html
Swedish Astrobiology Network	http://www.astrobiologi.se
Southwest Research Institute	http://www.boulder.swri.edu/scinews/astrobio.html
University College London, Astrobiology Institute	http://www.star.ucl.ac.uk/astrobiology
The Astrobiology Society of Britain	http://astrobiologysociety.org/astrouk.html

Glossary

absolute zero The coldest possible temperature, which is $0 \text{ K} = -273.15°\text{C}$.

absorption (of light) The process by which matter absorbs radiative energy.

absorption-line spectrum A spectrum that contains absorption lines.

accelerating universe The possible fate of our universe in which a repulsive force (see *cosmological constant*) causes the expansion of the universe to accelerate with time. Its galaxies will recede from one another increasingly faster, and it will become cold and dark more quickly than a coasting universe.

acceleration The rate at which an object's velocity changes. Its standard units are m/s^2.

acceleration of gravity The acceleration of a falling object. On Earth, the acceleration of gravity, designated by *g*, is 9.8 m/s^2.

accretion The process by which small objects gather together to make larger objects.

adaptive optics A technique in which telescope mirrors flex rapidly to compensate for the bending of starlight caused by atmospheric turbulence.

aerobic organisms Organisms that require molecular oxygen to survive.

albedo Describes the fraction of sunlight reflected by a surface; albedo = 0 means no reflection at all (a perfectly black surface); albedo = 1 means all light is reflected (a perfectly white surface).

Amazonian era The present era on Mars, which began about 1.0 billion years ago.

amino acids The building blocks of proteins. (More technically, an amino acid is a molecule containing both an *amino group* [NH or NH_2] and a *carboxyl group* [COOH].)

anaerobic organisms Organisms that do not require (and may even be poisoned by) molecular oxygen.

Andromeda Galaxy (M13; the Great Galaxy in Andromeda) The nearest large spiral galaxy to the Milky Way.

angular momentum Momentum attributable to rotation or revolution. The angular momentum of an object moving in a circle of radius *r* is the product $m \times v \times r$.

angular resolution (of a telescope) The smallest angular separation that two pointlike objects can have and still be seen as distinct points of light (rather than as a single point of light).

angular size (or **angular distance**) A measure of the angle formed by extending imaginary lines outward from our eyes to span an object (or between two objects).

annihilation See *matter–antimatter annihilation*.

Antarctic Circle The circle on Earth with latitude 66.5°S.

anthropic principle An idea that comes in a variety of forms centering around the fact that our existence is possible only because of a great number of aspects of our universe that are "fine-tuned" for life.

antielectron See *positron*.

antimatter Refers to any particle with the same mass as a particle of ordinary matter but whose other basic properties, such as electrical charge, are precisely opposite.

aphelion The point at which an object orbiting the Sun is farthest from the Sun.

apogee The point at which an object orbiting Earth is farthest from Earth.

apparent brightness The amount of light reaching us *per unit area* from a luminous object; often measured in units of watts/m^2.

apparent magnitude A measure of the apparent brightness of an object in the sky, based on the ancient system developed by Hipparchus.

apparent retrograde motion Refers to the apparent motion of a planet, as viewed from Earth, during the period of a few weeks or months when it moves westward relative to the stars in our sky.

Archaea One of the three domains of life; the others are Eukarya and Bacteria.

arcminutes (or **minutes of arc**) One arcminute is $\frac{1}{60}$ of 1°.

arcseconds (or **seconds of arc**) One arcsecond is $\frac{1}{60}$ of an arcminute, or $\frac{1}{3,600}$ of 1°.

Arctic Circle The circle on Earth with latitude 66.5°N.

Aristotelians Ancient Greek followers of Aristotle, who held that there could be only one Earth and that the heavens were a realm distinct from Earth.

asteroid A relatively small and rocky object that orbits a star; asteroids are sometimes called *minor planets* because they are similar to planets but smaller.

asteroid belt The region of our solar system between the orbits of Mars and Jupiter in which asteroids are heavily concentrated.

astrobiology The study of life on Earth and beyond; emphasizes research into questions of the origin of life, the conditions under which life can survive, and the search for life beyond Earth.

astrometric technique The detection of extrasolar planets through the side-to-side motion of a star caused by gravitational tugs from the planet.

astronomical unit (AU) The average distance (semimajor axis) of Earth from the Sun, which is about 150 million kilometers.

atmosphere A layer of gas that surrounds a planet or moon, usually very thin compared to the size of the object.

atmospheric pressure The surface pressure resulting from the overlying weight of an atmosphere.

atomic mass number The combined number of protons and neutrons in an atom.

atomic number The number of protons in an atom.

atomists Ancient Greek scholars who held that the universe is made from an infinite number of indivisible atoms.

atoms Consist of a nucleus made from protons and neutrons surrounded by a cloud of electrons.

ATP (adenosine triphosphate) The molecule that stores and releases energy for nearly all cellular processes among life on Earth.

aurora Dancing lights in the sky caused by charged particles entering our atmosphere; called the *aurora borealis* in the Northern Hemisphere and the *aurora australis* in the Southern Hemisphere.

autotroph An organism that gets its carbon directly from the atmosphere in the form of carbon dioxide.

autumnal equinox See *fall equinox*.

axis tilt (of a planet in our solar system) The amount by which a planet's axis is tilted with respect to a line perpendicular to the ecliptic plane.

Bacteria One of the three domains of life; the others are Eukarya and Archaea.

band (of sensitivity) The set of frequencies that a particular radio receiver can pick up.

bandwidth (of a transmitted signal) The range of frequencies over which a communication signal is transmitted.

bar The standard unit of pressure, approximately equal to Earth's atmospheric pressure at sea level.

basalt A type of dark, high-density volcanic rock that is rich in iron and magnesium-based silicate minerals; it forms a runny (easily flowing) lava when molten.

Big Bang The name given to the event thought to mark the birth of the universe.

Big Bang theory The scientific theory of the universe's earliest moments, stating that all the matter in our observable universe came into being at a single moment in time as an extremely hot, dense mixture of subatomic particles and radiation.

Big Crunch If gravity ever reverses universal expansion, the universe will someday begin to collapse and presumably end in a Big Crunch.

binary star system A star system that contains two stars.

biochemistry The chemistry of life.

biosphere Refers to the "layer" of life on Earth.

blackbody radiation See *thermal radiation*.

black smokers Structures around seafloor volcanic vents that support a wide variety of life.

blueshift A Doppler shift in which spectral features are shifted to shorter wavelengths, caused when an object is moving toward the observer.

brown dwarf An object that forms much like a star but is too low in mass to sustain nuclear fusion in its core; brown dwarfs have masses much greater than Jupiter but always less than $0.08M_{Sun}$.

Cambrian explosion The dramatic diversification of life on Earth that occurred between about 540 and 500 million years ago.

carbohydrates Molecules such as sugars and starches that provide energy to cells and make important cellular structures.

carbonate rock A carbon-rich rock, such as limestone, that forms underwater from chemical reactions between sediments and carbon dioxide. On Earth, most of the outgassed carbon dioxide currently resides in carbonate rocks.

carbon-based life Life that uses molecules containing carbon for its most critical functions. All life on Earth is carbon-based.

carbon dioxide (CO_2) cycle The process that cycles carbon dioxide between Earth's atmosphere and surface rocks.

catalysis The process of causing or accelerating a chemical reaction by involving a substance or molecule that is not permanently changed by the reaction.

catalyst The unchanged substance or molecule involved in catalysis.

celestial sphere The imaginary sphere on which objects in the sky appear to reside when observed from Earth.

cell The basic structure of all life on Earth, in which the living matter inside is separated from the outside world.

Celsius (temperature scale) The temperature scale commonly used in daily activity internationally; defined so that, on Earth's surface, water freezes at 0°C and boils at 100°C.

center of mass (of orbiting objects) The point at which two or more orbiting objects would balance if they were somehow connected; it is the point around which the orbiting objects actually orbit.

charged particle belts Zones in which ions and electrons accumulate and encircle a planet.

chemical bond The linkage between atoms in a molecule.

chemical element See *element*.

chemical enrichment The process by which the abundance of heavy elements (heavier than helium) in the interstellar medium gradually increases over time as these elements are produced by stars and released into space.

chemoautotroph An organism that gets its carbon directly from the atmosphere and its energy from chemical reactions involving inorganic molecules.

chemoheterotroph An organism that gets both its energy and its carbon by consuming pre-existing organic molecules; all animals are chemoheterotrophs.

chloroplasts Structures in plant cells that produce energy by photosynthesis.

civilization types A way of categorizing civilizations by whether they use resources of their planet, their star, or their galaxy. See also *galactic civilization*, *planetary civilization*, and *stellar civilization*.

clay Any of a variety of common silicate minerals with particular physical structures.

climate Describes the long-term average of weather.

cluster of galaxies A collection of a few dozen or more galaxies bound together by gravity; smaller collections of galaxies are simply called *groups*.

comet A relatively small, icy object that orbits a star.

compound (chemical) A substance made from molecules consisting of two or more atoms with different atomic numbers.

condensates Solid or liquid particles that condense from a cloud of gas.

condensation The formation of solid or liquid particles from a cloud of gas.

conduction (of energy) The process by which thermal energy is transferred by direct contact from warm material to cooler material.

conservation of angular momentum (law of) The principle that, in the absence of net torque (twisting force), the total angular momentum of a system remains constant.

conservation of energy (law of) The principle that energy (including mass-energy) can be neither created nor destroyed, but can change only from one form to another.

conservation of momentum (law of) The principle that, in the absence of net force, the total momentum of a system remains constant.

continental crust The thicker, lower-density crust that makes up Earth's continents. It is made when remelting of seafloor crust allows lower-density rock to separate and erupt to the surface. Continental crust ranges in age from extremely young to as old as about 4 billion years (or more).

continental drift The idea that the continents slowly move around on Earth, now known to be a result of plate tectonics.

continuously habitable zone The region around a star in which conditions could allow for surface habitability throughout the history of the star system.

convection The energy transport process in which warm material expands and rises, while cooler material contracts and falls.

convection cell A small individual region of convecting material.

convergent evolution The tendency of organisms of different evolutionary backgrounds to come to resemble one another because they occupy similar ecological niches.

Copernican revolution The dramatic change, initiated by Copernicus, that occurred when we learned that Earth is a planet orbiting the Sun rather than the center of the universe.

coral model (of colonization) A model of how a civilization might colonize the galaxy, based on growth much like that of coral in the sea.

core (of a planet) The dense central region of a planet that has undergone differentiation.

core (of a star) The central region of a star, in which nuclear fusion can occur.

cosmic microwave background The remnant radiation from the Big Bang, which we detect using radio telescopes sensitive to microwaves (which are short-wavelength radio waves).

cosmic rays Particles such as electrons, protons, and atomic nuclei that zip through interstellar space at close to the speed of light.

cosmos An alternate name for the universe.

crust (of a planet) The low-density surface layer of a planet that has undergone differentiation.

crystal A substance made from atoms arranged in precise geometrical patterns, such as in a mineral.

cultural evolution Changes that arise from the transmission of knowledge accumulated over generations.

cyanobacteria Photosynthetic bacteria thought to have been responsible for making most of the oxygen that gradually built up in Earth's atmosphere.

cycles per second Units of frequency for a wave; describes the number of peaks (or troughs) of a wave that pass by a given point each second. Equivalent to *hertz*.

dark energy Name sometimes given to energy that could be causing the expansion of the universe to accelerate.

dark matter Matter that we infer to exist from its gravitational effects but from which we have not detected any light; dark matter apparently dominates the total mass of the universe.

deuterium A form of hydrogen in which the nucleus contains a proton and a neutron, rather than only a proton (as is the case for most hydrogen nuclei).

differentiation The process in which gravity separates materials according to density, with high-density materials sinking and low-density materials rising.

disequilibrium (chemical) A state in which a mixture undergoing chemical reactions is not in equilibrium.

DNA (deoxyribonucleic acid) The molecule that represents the genetic material of life on Earth.

DNA bases The four DNA bases are adenine (A), cytosine (C), guanine (G), and thymine (T); they can be paired across the two DNA strands only so that A goes with T and C goes with G.

DNA replication The process of copying DNA molecules.

domain (of life) The highest level by which we currently classify life; the three domains are *Eukarya, Bacteria,* and *Archaea.*

Doppler effect (or **shift**) The effect that shifts the wavelengths of spectral features in objects that are moving toward or away from the observer.

Doppler technique The detection of extrasolar planets through the motion of a star toward and away from the observer caused by gravitational tugs from the planet.

Drake equation An equation that lays out the factors that play a role in determining the number of communicating civilizations in our galaxy.

dust (or **dust grains**) Tiny, solid flecks of material; in astronomy, we often discuss *interplanetary dust* (found within a star system) or *interstellar dust* (found among the stars in a galaxy).

Dyson sphere A hypothesized type of large, thin-walled sphere built to surround a star so that an advanced civilization could capture all the energy flowing out from the star; named after physicist Freeman Dyson, who proposed their possible existence.

Earth orbiters (spacecraft) Spacecraft designed to study Earth or the universe from Earth orbit.

eccentricity A measure of how much an ellipse deviates from a perfect circle; defined as the center-to-focus distance divided by the length of the semimajor axis.

ecliptic The Sun's apparent annual path among the constellations.

ecliptic plane The plane of Earth's orbit around the Sun.

ejecta (from an impact) Debris ejected by the blast of an impact.

electrical charge A fundamental property of matter that is described by its amount and as either positive or negative; more technically, it is a measure of how a particle responds to the electromagnetic force.

electromagnetic radiation Another name for light of all types, from radio waves through gamma rays.

electromagnetic spectrum The complete spectrum of light, including radio waves, infrared, visible light, ultraviolet light, X rays, and gamma rays.

electromagnetic wave A synonym for *light,* which consists of waves of electric and magnetic fields.

electron acceptor (in a redox reaction) The chemical (atom or molecule) that gains electrons in an overall chemical reaction.

electron donor (in a redox reaction) The chemical (atom or molecule) that gives up electrons in an overall chemical reaction.

electrons Fundamental particles with negative electric charge; the distribution of electrons in an atom gives the atom its size.

element (chemical) A substance made from individual atoms of a particular atomic number.

ellipse A type of oval that happens to be the shape of bound orbits. An ellipse can be drawn by moving a pencil along a string whose ends are tied to two tacks; the locations of the tacks are the foci (singular, *focus*) of the ellipse.

elliptical galaxies Galaxies that appear round in shape, often longer in one direction, like a football. They have no disks and contain little cool gas and dust compared to spiral galaxies, though they often contain extremely hot, ionized gas.

emission (of light) The process by which matter emits energy in the form of light.

emission line spectrum A spectrum that contains emission lines.

encephalization quotient (EQ) A rough measure of animal intelligence based on the ratio of an animal's brain size to its body mass.

endolith An organism that lives inside of rock.

endospore A special "resting" cell that allows some organisms to remain dormant for long periods of time.

energy Broadly speaking, energy is what makes matter move. The three basic types of energy are *kinetic, potential,* and *radiative.*

enzyme A protein that serves as a catalyst.

eons (geological) The largest divisions of time in Earth's geological history. The four eons are the Hadean, Archaean, Proterozoic, and Phanerozoic.

equilibrium (chemical) A state of balance between the reacting atoms and molecules and the product atoms and molecules in a mixture undergoing chemical reactions.

equinox See *fall equinox* and *spring equinox.*

eras (geological) The second-largest divisions of time in Earth's geological history, after eons. The Phanerozoic eon is subdivided into three eras: the *Paleozoic, Mesozoic,* and *Cenozoic.*

erosion The wearing down or building up of geological features by wind, water, ice, and other phenomena of planetary weather.

eruption The process of releasing hot lava onto the planet's surface.

escape velocity The speed necessary for an object to completely escape the gravity of a large body such as a moon, planet, or star.

Eukarya One of the three domains of life, and the one in which all plants and animals are found; the other domains are Bacteria and Archaea.

eukaryote A living organism that is a member of the domain Eukarya, and therefore is made from one or more eukaryotic cells.

eukaryotic cell A cell that contains a distinct nucleus that is separated from the rest of the cell by its own membrane.

evaporation The process by which atoms or molecules escape into the gas phase from the liquid phase.

evolution (biological) The gradual change in populations of living organisms that is responsible for transforming life on Earth from its primitive origins to the great diversity of life today.

evolutionary adaptation An inherited trait that enhances an organism's ability to survive and reproduce in a particular environment.

expansion (of universe) The idea that the space between galaxies or clusters of galaxies is growing with time.

extrasolar planet A planet orbiting a star other than our Sun.

extraterrestrial life Life that does *not* live on Earth.

extremophile An organism that thrives under conditions that are extreme by human standards.

Fahrenheit (temperature scale) The temperature scale commonly used in daily activity in the United States; defined so that, on Earth's surface, water freezes at 32°F and boils at 212°F.

fall equinox (autumnal equinox) Refers both to the point in Virgo on the celestial sphere where the ecliptic crosses the celestial equator and to the moment in time when the Sun appears at that point each year (around September 21).

false color image An image displayed in colors that are *not* the true, visible-light colors of an object.

fault (geological) A place where rocks slip sideways relative to one another.

feedback processes Processes in which a small change in some property (such as temperature) leads to changes in other properties, which then either amplify or diminish the original small change.

fermions Particles, such as electrons, neutrons, and protons, that obey the exclusion principle.

Fermi's paradox The question posed by Enrico Fermi about extraterrestrial intelligence—"So where is everybody?"—that asks why we have not observed other civilizations even though simple arguments would suggest that some ought to have spread throughout the galaxy by now.

field An abstract concept used to describe how a particle would interact with a force. For example, the idea of a *gravitational field* describes how a particle would react to the local strength of gravity, and the idea of an *electromagnetic field* describes how a charged particle would respond to forces from other charged particles.

fission See *nuclear fission*.

flybys (spacecraft) Spacecraft that fly past a target object (such as a planet), usually just once, as opposed to entering a bound orbit of the object.

force Anything that can cause a change in momentum.

fossil Any relic of an organism that lived and died long ago.

fossil record (or **geological record**) The information about Earth's past that is recorded in fossils (*fossil record*) and rocks (*geological record*). Note that the terms are often used synonymously.

frequency Describes the rate at which peaks of a wave pass by a point; measured in units of 1/s, which are often called *cycles per second* or *hertz*.

frost line The boundary in the solar nebula beyond which ices could condense; only metals and rocks could condense within the frost line.

fundamental forces There are four known fundamental forces in nature: *gravity*, the *electromagnetic force*, the *strong force*, and the *weak force*.

fusion See *nuclear fusion*.

galactic civilization A civilization that employs the resources of its entire galaxy.

galaxy A huge collection of stars, anywhere from a few hundred million to more than a trillion, all bound together by gravity.

galaxy cluster See *cluster of galaxies*.

Galilean moons The four moons of Jupiter that were discovered by Galileo: Io, Europa, Ganymede, and Callisto.

gamma-ray burst A sudden burst of gamma rays from deep space; such bursts apparently come from distant galaxies, but their precise mechanism is unknown.

gamma rays Light with very short wavelengths (and hence high frequencies)—shorter than those of X rays.

gas phase The phase of matter in which atoms or molecules can move essentially independently of one another.

gas pressure Describes the force (per unit area) pushing on any object due to surrounding gas. See also *pressure*.

gene The basic functional unit of an organism's heredity: a single gene consists of a sequence of DNA bases (or RNA bases, in some viruses) that provide the instructions for a single cell function (such as for building a protein).

general theory of relativity Einstein's generalization of his special theory of relativity that makes the general theory also apply when we consider effects of gravity or acceleration.

genetic analysis The analysis of an organism's genes or genome.

genetic code The "language" that living cells use to read the instructions chemically encoded in DNA.

genetic engineering Making deliberate changes to an organism's genome.

genome The complete sequence of DNA bases in an organism, encompassing all of the organism's genes.

genus The next most precise level of classification after *species*; it is a "generic" category to which multiple species may belong.

geocentric model Any of the ancient Greek models of the universe that had Earth at the center of a celestial sphere.

geocentric universe (the ancient belief in) The idea that Earth is the center of the entire universe.

geological activity Processes that change a planet's surface long after its formation, such as volcanism, tectonics, and erosion.

geological processes The four basic geological processes are *impact cratering, volcanism, tectonics*, and *erosion*.

geological record The information about Earth's past that is recorded in both fossils and rocks. See also *fossil record*.

geological time scale The time scale used by scientists to describe major eras in Earth's past. It is divided into four *eons* (the Hadean, Archaean, Proterozoic, and Phanerozoic). The latter (Phanerozoic) eon is subdivided into three *eras* (the Paleozoic, Mesozoic, and Cenozoic), which in turn are subdivided into several *periods*. (The periods are further subdivided into *epochs* and *ages*.)

geology The study of surface features (on a moon, planet, or asteroid) and the processes that create them.

giant impact A collision between a forming planet and a very large planetesimal, such as is thought to have formed our Moon.

giants (among stars) Stars that are near the ends of their lives and that have expanded in radius to extremely large sizes.

global average temperature The average surface temperature of a planet.

global warming An expected increase in Earth's global average temperature caused by human input of carbon dioxide and other greenhouse gases into the atmosphere.

globular cluster A spherically shaped cluster of up to a million or more stars; globular clusters are found primarily in the halos of galaxies and contain only extremely old stars.

granite A light-colored and low-density igneous rock common in mountain ranges on Earth; it gets its name from its grainy appearance and it is composed largely of quartz and feldspar minerals.

gravitation (law of) See *universal law of gravitation*.

gravitational constant The experimentally measured constant G that appears in the law of universal gravitation;

$$G = 6.67 \times 10^{-11} \frac{m^3}{kg \times s^2}$$

gravitational encounter Occurs when two (or more) objects pass near enough that each can feel the effects of the other's gravity and can therefore exchange energy.

gravitational lensing The magnification or distortion (into arcs, rings, or multiple images) of an image caused by light bending through a gravitational field, as predicted by Einstein's general theory of relativity.

gravity One of the four fundamental forces; it is the force that dominates on large scales.

Great Red Spot A large, high-pressure storm on Jupiter.

greenhouse effect The process by which greenhouse gases in an atmosphere make a planet's surface temperature warmer than it would be in the absence of an atmosphere.

greenhouse gases Gases, such as carbon dioxide, water vapor, and methane, that are particularly good absorbers of infrared light but are transparent to visible light.

habitable world A world with environmental conditions under which life could *potentially* arise or survive.

habitable zone The region around a star in which planets could potentially have surface temperatures at which liquid water could exist.

Hadean eon The earliest eon in Earth's history, corresponding to times before about 4.0 billion years ago.

half-life The time it takes for half of the nuclei in a given quantity of a radioactive substance to decay.

halo (of a galaxy) The spherical region surrounding the disk of a spiral galaxy.

handedness The property of some molecules, such as amino acids, that allows them to come in two distinct forms that are mirror images of one another.

heavy bombardment The period in the first few hundred million years after the solar system formed during which the tail end of planetary accretion created most of the craters found on ancient planetary surfaces.

heavy elements In astronomy, *heavy elements* generally refers to all elements *except* hydrogen and helium.

heredity The characteristics of an organism passed to it by its parent(s), which it can pass on to its offspring. The term can also apply to the transmission of these characteristics from one generation to the next.

hertz (Hz) The standard unit of frequency for light waves; equivalent to units of 1/s.

Hertzsprung–Russell (H–R) diagram A graph plotting individual stars as points, with stellar luminosity on the vertical axis and spectral type (or surface temperature) on the horizontal axis.

Hesperian era The middle history of Mars, dating from about 3.8 to 1.0 billion years ago.

heterotroph An organism that gets its carbon by consuming preexisting organic molecules.

hot spot (geological) A place within a plate of the lithosphere where a localized plume of hot mantle material rises.

Hubble's law A mathematical expression of the idea that more distant galaxies move away from us faster.

hydrogen compounds Compounds that contain hydrogen and were common in the solar nebula, such as water (H_2O), ammonia (NH_3), and methane (CH_4).

hydrosphere Refers to the "layer" of water on Earth consisting of oceans, lakes, rivers, ice caps, and other liquid water and ice.

hyperspace Any space with more than three dimensions.

hyperthermophile An organism that thrives under conditions of extremely high temperature compared to what most organisms can tolerate.

hypothesis A tentative model proposed to explain some set of observed facts that has not yet been rigorously tested and confirmed.

ice ages Periods of global cooling during which polar caps, glaciers, and snow cover extend closer to the equator.

ices (in solar system theory) Materials that are solid only at low temperatures, such as the hydrogen compounds water, ammonia, and methane.

igneous rock Rock made when molten rock cools and solidifies.

image A picture of an object made by focusing light.

impact The collision of a small body (such as an asteroid or comet) with a larger object (such as a planet or moon).

impact crater A bowl-shaped depression left by the impact of an object that strikes a planetary surface (as opposed to burning up in the atmosphere).

impactor The object responsible for an impact.

impact sterilization The process of sterilizing a planet as a result of a large impact.

infrared light Light with wavelengths that fall in the portion of the electromagnetic spectrum between radio waves and visible light.

inner solar system Generally considered to encompass the region of our solar system out to about the orbit of Mars.

inorganic Not pertaining to life or the chemistry of carbon molecules.

intensity (of light) A measure of the amount of energy coming from light of a specific wavelength in the spectrum of an object.

interferometry A telescopic technique in which two or more telescopes are used in tandem to produce much better angular resolution than the telescopes could achieve individually.

interstellar cloud A cloud of gas and dust among the stars.

interstellar ramjet A hypothesized type of spaceship that uses a giant scoop to sweep up interstellar gas for use in a nuclear fusion engine.

inverse square law Any quantity that decreases with the square of the distance between two objects is said to follow an inverse square law.

ion engine (rocket) A rocket engine that works by accelerating charged particles and expelling them out its back.

ionization The process of stripping an electron from an atom.

ionization nebula A colorful, wispy cloud of gas that glows because neighboring hot stars irradiate it with ultraviolet photons that can ionize hydrogen atoms.

ionosphere A portion of the thermosphere in which ions are particularly common (due to ionization by X rays from the Sun).

ions Atoms with a positive or negative electrical charge.

isotopes Each different isotope of an element has the *same* number of protons but a *different* number of neutrons.

joule The international unit of energy, equivalent to about $\frac{1}{4,000}$ of a Calorie.

jovian moons The moons of jovian planets.

jovian planets Giant gaseous planets similar in overall composition to Jupiter.

Julian calendar The calendar introduced in 46 B.C. by Julius Caesar and used until it was replaced by the Gregorian calendar.

Kelvin (temperature scale) The most commonly used temperature scale in science, defined such that absolute zero is 0 K and water freezes at 273.15 K.

Kepler's first law States that the orbit of each planet about the Sun is an ellipse with the Sun at one focus.

Kepler's laws of planetary motion Three laws discovered by Kepler that describe the motion of the planets around the Sun.

Kepler's second law States that, as a planet moves around its orbit, it sweeps out equal areas in equal times. This tells us that a planet moves faster when it is closer to the Sun (near perihelion) than when it is farther from the Sun (near aphelion) in its orbit.

Kepler's third law States that the square of a planet's orbital period is proportional to the cube of its average distance from the Sun (semimajor axis), which tells us that more distant planets move more slowly in their orbits. In its original form, it is written $p^2 = a^3$. See also *Newton's version of Kepler's third law*.

kinetic energy Energy of motion, given by the formula $\frac{1}{2}mv^2$.

kingdoms (biological) Except for the three domains, kingdoms (such as the plant kingdom and the animal kingdom) are the highest classification grouping of living organisms.

K–T boundary The thin layer of dark sediments that marks the division between the Cretaceous and Tertiary periods in the fossil record (the "K" comes from the German word for "Cretaceous," *Kreide*).

K–T event (impact) The collision of an asteroid or comet 65 million years ago that caused the mass extinction best known for wiping out the dinosaurs. "K–T" stands for the initials of the geological layers above and below the event.

Kuiper belt The comet-rich region of our solar system that spans distances of about 30–100 AU from the Sun; Kuiper belt comets have orbits that lie fairly close to the plane of planetary orbits and travel around the Sun in the same direction as the planets.

Kuiper belt objects The cometlike objects located in the Kuiper belt.

Lagrange points (of the Earth–Moon system) The five positions in space where the effects of gravity from Earth and the Moon "cancel" in such a way that, if you floated weightlessly at one of these five points, you wouldn't be tugged toward either body.

late heavy bombardment An apparent increase in the impact rate near the end of the heavy bombardment, about 3.9 billion years ago.

latitude The angular north-south distance between Earth's equator and a location on Earth's surface.

light-year The distance that light can travel in 1 year, which is 9.46 trillion kilometers.

lipids Complex molecules in cells, also known as *fats*, that play a variety of functions including being key components of membranes.

liquid phase The phase of matter in which atoms or molecules are held together but move relatively freely.

lithophile See *endolith*. An organism that lives inside of rock.

lithosphere The relatively rigid outer layer of a planet; generally encompasses the crust and the uppermost portion of the mantle.

Local Group The group of about 40 galaxies to which the Milky Way Galaxy belongs.

Local Supercluster The supercluster of galaxies to which the Local Group belongs.

longitude The angular east-west distance between the prime meridian (which passes through Greenwich) and a location on Earth's surface.

luminosity The total power output of an object, usually measured in watts or in units of solar luminosities ($L_{Sun} = 3.8 \times 10^{26}$ watts).

lunar maria The regions of the Moon that look smooth from Earth but actually are impact basins.

magma Underground molten rock.

magnetic field Describes the region surrounding a magnet in which it can affect other magnets or charged particles in its vicinity.

magnetic-field lines Lines that represent how the needles on a series of compasses would point if they were laid out in a magnetic field.

magnetosphere The region surrounding a planet in which charged particles are trapped by the planet's magnetic field.

main sequence The prominent line of points running from the upper left to the lower right on an H–R diagram; main-sequence stars shine by fusing hydrogen in their cores.

main-sequence stars Stars whose temperature and luminosity place them on the main sequence of the H–R diagram. Main-sequence stars release energy by fusing hydrogen into helium in their cores.

mantle (of a planet) The rocky layer that lies between a planet's core and crust.

mantle convection The flow pattern in which hot mantle material expands and rises while cooler material contracts and falls.

martian meteorite This term is used to describe meteorites found on Earth that are thought to have originated on Mars.

mass A measure of the amount of matter in an object.

mass-energy The potential energy of mass, which has an amount $E = mc^2$.

mass extinction An event in which a large fraction of the species living on Earth go extinct, such as the event in which the dinosaurs died out about 65 million years ago.

mass increase (in relativity) Refers to the effect in which an object moving past you seems to have a mass greater than its rest mass.

mass ratio (of a rocket) The ratio of the initial (launch) mass of the rocket M_i, including its fuel and any spacecraft it is carrying, to its mass M after all the fuel is burned.

matter–antimatter annihilation Occurs when a particle of matter and a particle of antimatter meet and convert all of their mass-energy to photons.

membrane (cell) A barrier that separates the inside of a cell (or cell nucleus) from the outside. Cells come in two basic types.

metabolism The many chemical reactions that occur in living organisms.

metals (in solar system theory) Elements, such as nickel, iron, and aluminum, that condense at fairly high temperatures.

metamorphic rock Rock made from igneous or sedimentary rock that gets transformed (but not melted) by high heat or pressure.

meteor A flash of light caused when a particle from space burns up in our atmosphere.

meteorite A rock from space that lands on Earth.

microwaves Light with wavelengths in the range from micrometers to millimeters. Microwaves are generally considered to be a subset of the radio-wave portion of the electromagnetic spectrum.

mid-ocean ridges (on Earth) Long ridges of undersea volcanoes, along which mantle material erupts onto the ocean floor and pushes apart the existing seafloor on either side. These ridges are essentially the source of new seafloor crust, which then makes its way along the ocean bottom for millions of years before returning to the mantle at a subduction zone.

Milankovitch cycles The cyclical changes in Earth's axis tilt and orbit that can change the climate and cause ice ages.

Milky Way Used both as the name of our galaxy and to refer to the band of light we see in the sky when we look into the plane of the Milky Way Galaxy.

Miller–Urey experiment An experiment first performed in the 1950s that was designed to learn how organic molecules might have formed naturally on the early Earth.

mineral A rocky substance with a particular chemical composition and crystal structure.

mitochondria The cellular organs in eukaryotic cells in which oxygen helps produce energy (by making molecules of ATP).

model (scientific) A representation of some aspect of nature that can be used to explain and predict real phenomena without invoking myth, magic, or the supernatural.

moist greenhouse effect A process by which a planet could lose water when the atmospheric circulation allows water vapor to rise high enough to be broken apart by ultraviolet light from the Sun.

molecule Technically the smallest unit of a chemical element or compound; in this text, the term refers only to combinations of two or more atoms held together by chemical bonds.

momentum The product of an object's mass and velocity.

moon An object that orbits a planet.

multiple star system A star system that contains two or more stars.

mutations Errors in the copying process when a living cell replicates itself.

natural selection The process by which mutations that make an organism better able to survive get passed on to future generations.

nebula A cloud of gas in space, usually one that is glowing.

nebular theory The detailed theory that describes how our solar system formed from a cloud of interstellar gas and dust.

neutrons Particles with no electrical charge that are found in atomic nuclei.

newton The standard unit of force in the metric system:

$$1 \text{ newton} = 1 \frac{\text{kg} \times \text{m}}{\text{s}^2}$$

Newton's first law of motion States that, in the absence of a net force, an object moves with constant velocity.

Newton's laws of motion Three basic laws that describe how objects respond to forces.

Newton's second law of motion States how a net force affects an object's motion. Specifically: force = rate of change in momentum, or force = mass × acceleration.

Newton's third law of motion States that, for any force, there is always an equal and opposite reaction force.

Newton's universal law of gravitation See *universal law of gravitation*.

Newton's version of Kepler's third law This generalization of Kepler's third law can be used to calculate the masses of orbiting objects from measurements of orbital period and distance; it is usually written as:

$$p^2 = \frac{4\pi^2}{G(M_1 + M_2)} a^3$$

Noachian era The era on Mars before 3.8 billion years ago.

nonscience As defined in this book, nonscience is any way of searching for knowledge that makes no claim to follow the scientific method, such as seeking knowledge through intuition, tradition, or faith.

nuclear fission The process in which a larger nucleus splits into two (or more) smaller particles.

nuclear fusion The process in which two (or more) smaller nuclei slam together and make one larger nucleus.

nucleus (of an atom) The compact center of an atom made from protons and neutrons.

nucleus (of a cell) The membrane-enclosed region of a eukaryotic cell that contains the cell's DNA.

nucleus (of a comet) The solid portion of a comet, and the only portion that exists when the comet is far from the Sun.

observable universe The portion of the entire universe that, at least in principle, can be seen from Earth.

Occam's razor A principle often used in science, holding that scientists should prefer the simpler of two models that agree equally well with observations. Named after the medieval scholar William of Occam (1285–1349).

Oort cloud A huge, spherical region centered on the Sun, extending perhaps halfway to the nearest stars, in which trillions of comets orbit the Sun with random inclinations, orbital directions, and eccentricities.

orbital energy The sum of an orbiting object's kinetic and gravitational potential energies.

orbital resonance Describes any situation in which one object's orbital period is a simple ratio of another object's period, such as $\frac{1}{2}$, $\frac{1}{4}$, or $\frac{5}{3}$. In such cases, the two objects periodically line up with each other, and the extra gravitational attractions at these times can affect the objects' orbits.

orbiters (of other worlds) Spacecraft that go into orbit of another world for long-term study.

organic chemistry The chemistry of organic molecules (whether or not the molecules are involved in life).

organic molecule This term generally refers to any molecule containing carbon, but with some exceptions among molecules that are most commonly found independent of life. For our purposes in this book, the important exceptions are carbon dioxide (CO_2) and carbonate minerals, which we do not consider to be organic.

outer solar system Generally considered to encompass the region of our solar system beginning at about the orbit of Jupiter.

outgassing The process of releasing gases from a planetary interior, usually through volcanic eruptions.

oxidation Refers to chemical reactions, often with the surface of a planet, that remove oxygen from the atmosphere.

ozone The molecule O_3, which is a particularly good absorber of ultraviolet light.

ozone depletion Refers to the declining levels of atmospheric ozone found worldwide on Earth, especially in Antarctica, in recent years.

ozone hole A place where the concentration of ozone in the stratosphere is dramatically lower than is the norm.

Pangaea A "supercontinent" that existed prior to 225 million years ago, in which all Earth's current continents were linked together.

panspermia The idea that life migrated to Earth from some extraterrestrial location.

paradigm (in science) Refers to general patterns of thought that tend to shape scientific beliefs during a particular time period.

paradox A situation that, at least at first, seems to violate common sense or contradict itself. Resolving paradoxes often leads to deeper understanding.

parallax The apparent shifting of an object against the background, due to viewing it from different positions. See also *stellar parallax*.

perigee The point at which an object orbiting Earth is nearest to Earth.

perihelion The point at which an object orbiting the Sun is closest to the Sun.

periodic table of the elements A table that lists properties of all the known elements in an organized way.

period–luminosity relation The relation that describes how the luminosity of a Cepheid variable star is related to the period between peaks in its brightness; the longer the period, the more luminous the star.

periods (geological) The third-largest divisions of time in Earth's geological history, after eons and eras.

phase (of matter) Describes the way in which atoms or molecules are held together; the common phases are *solid, liquid,* and *gas*.

photoautotroph An organism that gets its carbon directly from the atmosphere and gets its energy from sunlight through photosynthesis; plants are photoautotrophs.

photoheterotroph An organism that gets its carbon by consuming preexisting organic molecules and gets its energy from sunlight through photosynthesis.

photon An individual particle of light, characterized by a wavelength and a frequency.

phyla (singular, *phylum*) The next level of biological classification below kingdoms.

pixel An individual "picture element" in a digital picture.

planet A moderately large object that orbits a star and shines primarily by reflecting light from its star. More precisely, according to a definition approved in 2006, an object can be considered a planet only if it: (1) orbits a star; (2) is large enough for its own gravity to make it round; and (3) has cleared most other objects from its orbital path. An object that is round but has *not* cleared its orbital path, like Pluto, is designated a *dwarf planet.*

planetary civilization A civilization that uses the resources of its home planet; we are a planetary civilization by this definition.

planetary nebula The glowing cloud of gas ejected from a low-mass star at the end of its life.

planetesimals The building blocks of planets, formed by accretion in the solar nebula.

plasma A gas consisting of ions and electrons.

plates (on a planet) Pieces of a lithosphere that apparently float on the denser mantle below.

plate tectonics The geological process in which plates are moved around by stresses in a planet's mantle.

positron The antimatter equivalent of an electron. It is identical to an electron in all respects except that it has a positive rather than a negative electrical charge.

potential energy Energy stored for later conversion into kinetic energy; includes *gravitational potential energy, electrical potential energy,* and *chemical potential energy.*

power The rate of energy usage, usually measured in watts (1 watt = 1 joule/s).

precession The gradual wobble of the axis of a rotating object around a vertical line.

precipitation Condensed atmospheric gases that fall to the surface in the form of rain, snow, or hail.

pressure Describes the force (per unit area) pushing on an object. In astronomy, we are generally interested in pressure applied by surrounding gas (or plasma).

prokaryote A living organism made from cells in which DNA is *not* confined to a distinct, membrane-enclosed nucleus. Most prokaryotes are single-celled. Prokaryotes include all the organisms in two of the three domains of life: Bacteria and Archaea.

prokaryotic cell A cell that lacks a distinct nucleus.

protein A large molecule assembled from amino acids according to instructions encoded in DNA. Proteins play many roles in cells; a special category of proteins, called *enzymes,* catalyzes nearly all of the important biochemical reactions that occur within cells.

protons Particles with positive electrical charge found in atomic nuclei; they are built from three quarks.

protoplanetary disk A disk of material surrounding a young star (or protostar) that may eventually form planets.

protoplanets A term used to describe planetesimals that have grown quite large, to planet-size.

pseudoscience Something that purports to be science or may appear to be scientific but that does not adhere to the testing and verification requirements of the scientific method.

Ptolemaic model The geocentric model of the universe developed by Ptolemy in about A.D. 150.

radar mapping Imaging of a planet by bouncing radar waves off its surface, especially important for Venus and Titan, where thick clouds mask the surface.

radial motion The component of an object's motion directed toward or away from us.

radial velocity The portion of any object's total velocity that is directed toward or away from us. This part of the velocity is the only part that we can measure with the Doppler effect.

radiative energy Energy carried by light; the energy of a photon is Planck's constant times its frequency, or $h \times f$.

radioactive decay The spontaneous change of an atom in which its nucleus breaks apart or a proton turns into an electron, thereby changing into a different element. It also releases heat to a planet's interior.

radioactive element (or **radioactive isotope**) A substance whose nucleus tends to fall apart spontaneously.

radiometric dating The process of determining the age of a rock (i.e., the time since it solidified) by comparing the present amount of a radioactive substance to the amount of its decay product.

radio waves Light with extremely long wavelengths (and hence low frequencies)—longer than those of infrared light.

rare Earth hypothesis A hypothesis holding that the specific circumstances that have made it possible for complex creatures (such as birds or humans) to have evolved on Earth might be so rare that ours may be the only inhabited planet in the galaxy that has anything but the simplest life.

red giant A giant star that is red in color; red giants are a late stage in the life of a star, occurring after the star has exhausted its central supply of core hydrogen.

redox reactions Chemical reactions that involve an exchange or reshuffling of electric charge between the reacting atoms or molecules. A redox reaction always involves the transfer of one or more electrons from an *electron donor* (which becomes oxidized) to an *electron acceptor* (which becomes reduced).

redshift (Doppler) A Doppler shift in which spectral features are shifted to longer wave-

lengths, caused when an object is moving away from the observer.

reduction (chemical) The process of gaining electrons—which reduces the electrical charge (because electrons carry negative charge)—in a chemical reaction.

reference frame (frame of reference) In the theory of relativity, two people (or objects) share the same reference frame if they are *not* moving relative to one another.

resonance See *orbital resonance.*

RNA (ribonucleic acid) A molecule closely related to DNA—but with only a single strand and a slightly different backbone and set of bases—that plays critical roles in carrying out the instructions encoded in DNA.

RNA world The hypothesized period during which life on Earth first evolved and used RNA, rather than DNA, as its genetic material.

rocks (in solar system theory) Material common on the surface of Earth, such as silicon-based minerals, that is solid at temperatures and pressures found on Earth but typically melts or vaporizes at temperatures of 500–1,300 K.

rock cycle The idea that rocks can be transformed between the three basic types igneous, metamorphic, and sedimentary.

rotation The spinning of an object around its axis.

runaway greenhouse effect A positive feedback cycle in which heating caused by the greenhouse effect causes more greenhouse gases to enter the atmosphere, which further enhances the greenhouse effect.

rybozymes RNA molecules that function as catalysts.

sample return mission A space mission designed to return to Earth a sample of another world.

satellite Any object orbiting another object.

science The search for knowledge that can be used to explain or predict natural phenomena in a way that can be confirmed by rigorous observations or experiments.

scientific method An organized approach to explaining observed facts through science.

scientific theory A model of some aspect of nature that has been rigorously tested and has passed all tests to date.

seafloor crust On Earth, the thin, dense crust of basalt created by seafloor spreading.

seafloor spreading On Earth, the creation of new seafloor crust at mid-ocean ridges.

search for extraterrestrial intelligence (SETI) The name given to observing projects designed to search for signs of intelligent life beyond Earth.

second law of thermodynamics The law stating that, when left alone, the energy in a system undergoes conversions that lead to increasing disorder.

sedimentary rock A rock that formed from sediments created and deposited by erosional processes. The sediments tend to build up in distinct layers, or *strata*.

seismic waves Earthquake-induced vibrations that propagate through a planet.

semimajor axis Half the distance across the long axis of an ellipse; in this text, it is usually referred to as the *average* distance of an orbiting object, abbreviated *a* in the formula for Kepler's third law.

sentinel hypothesis A possible solution to the Fermi paradox that suggests that extraterrestrials place monitoring devices near star systems that show promise of emerging intelligence, and patiently wait until these devices record the presence of civilization.

silicate rock A silicon-rich rock.

sleep paralysis The natural paralysis of the body that occurs during REM sleep; it may occasionally persist for a few minutes after the brain has started waking up, giving a person the alarming sensation of being awake in a paralyzed body. Visions and other sensations often occur in this state.

snowball Earth Name given to a hypothesis suggesting that, some 600–700 million years ago, Earth experienced a period in which it became cold enough for glaciers to exist worldwide, even in equatorial regions.

solar luminosity The luminosity of the Sun, which is approximately 4×10^{26} watts.

solar nebula The piece of interstellar cloud from which our own solar system formed.

solar sail A large, highly reflective (and thin, to minimize mass) piece of material that can "sail" through space using pressure exerted by sunlight.

solar system (or **star system**) Consists of a star (sometimes more than one star) and all the objects that orbit it.

solar wind A stream of charged particles ejected from the Sun.

solar wind stripping The stripping away of a planet's atmospheric gas by the solar wind; generally affects only planetary atmospheres that are unprotected by a global magnetic field.

solid phase The phase of matter in which atoms or molecules are held rigidly in place.

solstice See *summer solstice* and *winter solstice*.

special theory of relativity Einstein's theory that describes the effects of the facts that all motion is relative and that everyone always measures the same speed of light.

species A group of organisms that is genetically distinct from other groups; species is the most precise level of biological classification among organisms.

spectral lines Bright or dark lines that appear in an object's spectrum, which we can see when we pass the object's light through a prismlike device that spreads out the light like a rainbow.

spectral resolution Describes the degree of detail that can be seen in a spectrum; the higher the spectral resolution, the more detail we can see.

spectral type A way of classifying a star by the lines that appear in its spectrum; it is related to surface temperature. The basic spectral types are designated by a letter (OBAFGKM, with O for the hottest stars and M for the coolest) and are subdivided with numbers from 0 through 9.

spectroscopy (in astronomical research) The process of obtaining spectra from astronomical objects.

spectrum (of light) See *electromagnetic spectrum*.

speed The rate at which an object moves. Its units are distance divided by time, such as m/s or km/hr.

speed of light The speed at which light travels, which is about 300,000 km/s.

spiral galaxies Galaxies that look like flat, white disks with yellowish bulges at their centers. The disks are filled with cool gas and dust, interspersed with hotter ionized gas, and usually display beautiful spiral arms.

spring equinox (vernal equinox) Refers both to the point in Pisces on the celestial sphere where the ecliptic crosses the celestial equator and to the moment in time when the Sun appears at that point each year (around March 21).

star A large, glowing ball of gas that generates energy through nuclear fusion in its core. The term *star* is sometimes applied to objects that are in the process of becoming true stars (e.g., protostars) and to the remains of stars that have died (e.g., neutron stars).

star system See *solar system*.

stellar civilization A civilization that employs the resources of its home star (that is, not only the resources available on its home planet).

stellar evolution The formation and development of stars.

stellar parallax The apparent shift in the position of a nearby star (relative to distant objects) that occurs as we view the star from different positions in Earth's orbit of the Sun each year.

sterilizing impact An impact large enough that it would have fully vaporized Earth's oceans and killed off any life existing on Earth.

strata (rock) Layers in sedimentary rock.

stromatolites Large bacterial "colonies."

subduction (of tectonic plates) The process in which one plate slides under another.

subduction zones Places where one plate slides under another.

sublimation The process by which atoms or molecules escape into the gas phase from the solid phase.

summer solstice Refers both to the point on the celestial sphere where the ecliptic is farthest north of the celestial equator and to the moment in time when the Sun appears at that point each year (around June 21).

superclusters Superclusters consist of many clusters of galaxies, groups of galaxies, and individual galaxies and are the largest known structures in the universe.

supergiants The largest and brightest of all stars.

supernova The explosion of a star.

symbiotic relationship A relationship in which both an invading organism and a host organism benefit from living together.

synchronous rotation Describes the rotation of an object that always shows the same face to the object that it is orbiting because its rotation period and orbital period are equal.

technological evolution Change driven by the rapid development of technology.

tectonics The disruption of a planet's surface by internal stresses.

temperature A measure of the average kinetic energy of particles in a substance.

terraforming Changing a planet in such a way as to make it more Earth-like.

terrestrial planets Rocky planets similar in overall composition to Earth.

theories of relativity (special and general) Einstein's theories that describe the nature of space, time, and gravity.

theory (in science) See *scientific theory*.

theory of evolution The theory, first advanced by Charles Darwin, that explains *how* evolution occurs through the process of *natural selection*.

thermal energy Represents the collective kinetic energy, as measured by temperature, of the many individual particles moving within a substance.

thermal escape The process in which atoms or molecules in a planet's exosphere move fast enough to escape into space.

thermal radiation The spectrum of radiation produced by an opaque object that depends only on the object's temperature; sometimes called *blackbody radiation*.

thermophile An organism that thrives under conditions of high temperature compared to what most organisms can tolerate.

tidal force A force that is caused when the gravity pulling on one side of an object is larger than that on the other side, causing the object to stretch.

tidal friction Friction within an object that is caused by a tidal force.

tidal heating A source of internal heating created by tidal friction. It is particularly important for satellites with eccentric orbits such as Io and Europa.

time dilation (in relativity) Refers to the effect in which you observe time running slower in reference frames moving relative to you.

transit A transit is an event in which a planet passes in front of a star (or the Sun) as seen from Earth. Only Mercury and Venus can be seen in transit of our Sun. The search for transits of extrasolar planets is an important planet detection strategy.

tree of life An illustration of the biochemical and genetic relationships between different organisms; its three main branches are the three *domains* Bacteria, Archaea, and Eukarya.

troposphere The lowest atmospheric layer, in which convection and weather occur.

turbulence Rapid and random motion.

ultraviolet light Light with wavelengths that fall in the portion of the electromagnetic spectrum between visible light and X rays.

universal law of gravitation The law expressing the force of gravity (F_g) between two objects, given by the formula

$$F_g = G \frac{M_1 M_2}{d^2}$$

$$\left(G = 6.67 \times 10^{-11} \frac{m^3}{kg \times s^2} \right)$$

universe The sum total of all matter and energy.

velocity The combination of speed and direction of motion; it can be stated as a speed in a particular direction, such as 100 km/hr due north.

vernal equinox See *spring equinox*.

viscosity Describes the "thickness" of a liquid in terms of how rapidly it flows; low-viscosity liquids flow quickly (e.g., water), while high-viscosity liquids flow slowly (e.g., molasses).

visible light The light our eyes can see, ranging in wavelength from about 400 to 700 nm.

volatiles Refers to substances such as water, carbon dioxide, and methane that are usually found as gases, liquids, or surface ices on the terrestrial worlds.

volcanism The eruption of molten rock, or lava, from a planet's interior onto its surface.

Von Neumann machines Self-replicating machines first proposed by the American mathematician and computer pioneer John Von Neumann (1903–1957).

watt The standard unit of power in science: 1 watt = 1 joule/s.

wavelength The distance between adjacent peaks (or troughs) of a wave.

weather Describes the ever-varying combination of winds, clouds, temperature, and pressure in a planet's troposphere.

white dwarf The hot, compact corpse of a low-mass star, typically with a mass similar to that of the Sun compressed to a volume the size of the Earth.

winter solstice Refers both to the point on the celestial sphere where the ecliptic is farthest south of the celestial equator and to the moment in time when the Sun appears at that point each year (around December 21).

worldline A line that represents an object on a spacetime diagram.

wormholes The name given to hypothetical tunnels through hyperspace that might connect two distant places in our universe.

X rays Light with wavelengths that fall in the portion of the electromagnetic spectrum between ultraviolet light and gamma rays.

zircons Tiny mineral grains of zirconium silicate, usually found embedded in sedimentary rock.

zoo hypothesis A possible explanation for the Fermi paradox holding that alien civilizations are aware of our presence but have chosen to deliberately avoid contact with us.

References and Bibliography

(*indicates articles or books that should be particularly accessible to students not majoring in science)

General (books covering a wide range of topics relevant to many chapters in this text)

*Bennett, J., Donahue, M., Schneider, N., and Voit, M., *The Cosmic Perspective*, Addison Wesley, 2007.
*Darling, D., *The Extraterrestrial Encyclopedia*, Three Rivers Press, 2000.
*Dick, S. J., *Life on Other Worlds*, Cambridge, 1998.
*Dick, S. J., and Strick, J. E., *The Living Universe*, Rutgers, 2005.
*Grinspoon, D., *Lonely Planets: The Natural Philosophy of Alien Life*, Ecco, 2003.
*Jakosky, B., *The Search for Life on Other Planets*, Cambridge, 1998.
*Sagan, C., *Cosmos*, Random House, 1980.
*Shostak, S., *Sharing the Universe*, Berkeley Hills Books, 1998.
Chyba, C. F., and Hand, K. P., Astrobiology: The study of the living universe, *Annual Review of Astronomy and Astrophysics*, **43,** 31–74, 2005.
Lunine, J. I., *Astrobiology*, Addison Wesley, 2005.
Lunine, J. I., *Earth: Evolution of a Habitable World*, Cambridge, 1999.

Chapter 1

Approaches to Astrobiology

*Morrison, D., and Schmidt, G. K., eds., *Astrobiology Roadmap*, NASA, 1999.
Lederberg, J., Exobiology: Approaches to life beyond the Earth, *Science*, **132,** 393–400, 1960.

Chapter 2

History of Science and Astrobiology

*Asimov, I., *Asimov's Biographical Encyclopedia of Science and Technology*, Doubleday, 1982.
*Dick, S. J., *The Biological Universe*, Cambridge, 1996.
Crowe, M. J., *The Extraterrestrial Life Debate, 1750–1900*, Dover, 1999.
Crowe, M. J., *Theories of the World: From Antiquity to the Copernican Revolution*, Dover, 1990.

The Nature of Science

*Boorstin, D. J., *The Discoverers*, Vintage, 1985.
*Sagan, C., *The Demon-Haunted World: Science as a Candle in the Dark*, Ballantine, 1996.
Cleland, C. E., Historical science, experimental science, and the scientific method, *Geology*, **29,** 987–990, 2001.
Kuhn, T. S., *The Nature of Scientific Revolutions*, University of Chicago Press, 1996.
Popper, K., and Miller, D., eds., *Popper Selections*, Princeton University Press, 1985.

Chapter 3

The Big Bang, Origin of the Elements, Formation of the Solar System

*Bennett, J., *On the Cosmic Horizon*, Addison Wesley, 2001.
*Kirshner, R. P., The Earth's elements, *Scientific American*, **271,** 58–65, 1994.
*Singh, S., *Big Bang*, Harper Perennial, 2005.

*Wood, J. A., Forging the planets, *Sky & Telescope*, January 1999.
Beckwith, S. V. W., and Sargent, A. I., Circumstellar disks and the search for neighboring planetary systems, *Nature*, **383,** 139–144, 1996.
Lissauer, J. J., Planet formation, *Annual Review of Astronomy and Astrophysics*, **31,** 129–174, 1993.
Rees, M. J., Piecing together the biggest puzzle of all, *Science*, **290,** 1919–1925, 2000.

Chapter 4

Radiometric Dating and the Age of the Earth

Dalrymple, G. B., *The Age of the Earth*, Stanford, 1991.

Impact History of the Earth, Hadean Earth

Chyba, C. F., Owen, T. C., and Ip, W. H., Impact delivery of volatiles and organic molecules to the Earth, in *Hazards Due to Comets and Asteroids*, T. Gehrels, ed., University of Arizona Press, 9–58, 1994.
Harrison, T. M., et al., Heterogeneous Hadean hafnium: Evidence of continental crust at 4.4 to 4.5 Ga, *Science*, **310,** 1947–1950, 2005.
Hartmann, W. K., Ryder, G., Dones, L., and Grinspoon, D., The time-dependent intense bombardment of the primordial Earth/Moon system, in *Origin of the Earth and Moon*, R. M. Canup and K. Righter, eds., University of Arizona Press, 493–512, 2000.
Sleep, N. H., Zahnle, K. J., Kasting, J. F., and Morowitz, H. J., Annihilation of ecosystems by large asteroid impacts on the early Earth, *Nature*, **342,** 139–142, 1989.

Geology and Plate Tectonics

Anderson, D. L., Planet Earth, in *The New Solar System*, J. K. Beatty, C. C. Peterson, and A. Chaikin, eds., Sky Publishing Corporation, 111–124, 1999.

Long-Term Climate Change

*Hoffman, P. F., and Schrag, D. P., Snowball Earth, *Scientific American*, **282,** 68–75, 2000.
Kasting, J. F., and Grinspoon, D. H., The faint young sun problem, in *The Sun in Time*, C. P. Sonett, M. S. Giampapa, and M. S. Matthews, eds., University of Arizona Press, 447–462, 1991.

Formation of the Moon

Canup, R., Dynamics of lunar formation, *Annual Review of Astronomy and Astrophysics*, **42,** 441–475, 2004.
Jacobsen, S. B., The Hf-W isotopic system and the origin of the Earth and Moon, *Annual Review of Earth and Planetary Sciences*, **33,** 531–570, 2005.

Chapter 5

Basic Biology and Evolution

*Campbell, N., and Reece, J., *Biology*, Addison Wesley, 2005.
*Zimmer, C., *Evolution: The Triumph of an Idea*, Harper Collins, 2001.
Gould, S. J., Showdown on the Burgess Shale: The reply, *Natural History*, Dec. 1998–Jan. 1999.
Keenan, W. T., The chemistry of life, *Biological Science*, Norton, 47–82, 1980.

Koshland, D. E., Jr., The seven pillars of life, *Science,* **295,** 2215–2216, 2002.

Morris, S. C., Showdown on the Burgess Shale: The challenge, *Natural History,* Dec. 1998–Jan. 1999.

Sagan, C., Life, *Encyclopaedia Britannica,* 1970.

The Three Domains and the Tree of Life

*Doolittle, W. F., Uprooting the tree of life, *Scientific American,* **282,** 90–95, 2000.

Brown, J. F., and Doolittle, W. F., Root of the universal tree of life based on ancient aminoacyl-tRNA synthetase gene duplications, *Proceedings of the National Academy of Sciences,* **92,** 2441–2445, 1995.

Pace, N. R., The early branches in the tree of life, preprint for *Proceedings of a Symposium on Assembling the Tree of Life,* American Museum of Natural History, 2002.

Pace, N. R., A molecular view of microbial diversity and the biosphere, *Science,* **276,** 734–740, 1997.

Woese, C., Interpreting the universal phylogenetic tree, *Proceedings of the National Academy of Sciences,* **97,** 8392–8396, 2000.

Woese, C., The universal ancestor, *Proceedings of the National Academy of Sciences,* **95,** 6854–6859, 1998.

Life in Extreme Environments

*Madigan, M. T., and Marrs, B. L., Extremophiles, *Scientific American,* **276,** 82–87, 1997.

Nealson, K. H., The limits of life on Earth and searching for life on Mars, *Journal of Geophysical Research,* **102,** 23675–23686, 1997.

Sharma, A., et al., Microbial activity at gigapascal pressures, *Science,* **295,** 1514–1516, 2002.

Stevens, T. O., and McKinley, J. P., Lithoautotrophic microbial ecosystems in deep basalt aquifers, *Science,* **270,** 450–454, 1995.

Van Dover, C. L., et al., Biogeography and ecological setting of Indian Ocean hydrothermal vents, *Science,* **294,** 818–823, 2001.

Evolution and Creationism

*National Academy of Sciences, *Science and Creationism: A View from the National Academy of Sciences,* National Academies Press, 1999.

*Pennock, R., *Intelligent Design Creationism and Its Critics: Philosophical, Theological & Scientific Perspectives,* MIT Press, 2001.

*Pennock, R., *Tower of Babel,* MIT Press, 1999.

*Scott, E., *Evolution vs. Creationism,* University of California Press, 2005.

Chapter 6

Early Evidence for Life

Brasier, M., et al., Questioning the evidence for Earth's oldest fossils, *Nature* **416,** 76–81, 2002.

Mojzsis, S. J., Arrhenius, G., McKeegan, K. D., Harrison, T. M., Nutman, A. P., and Friend, C. R. L., Evidence for life on Earth before 3,800 million years ago, *Nature,* **384,** 55–59, 1996.

Mojzsis, S. J., Harrison, T. M., and Pidgeon, R. T., Oxygen-isotope evidence from ancient zircons for liquid water at the Earth's surface 4,300 Myr ago, *Nature,* **409,** 178–181, 2001.

Papineau, D., et al., Composition and structure of microbial communities from stromatolites of Hamelin Pool in Shark Bay, *Applied and Environmental Microbiology,* **71,** 4822–4832, 2005.

Papineau, D., et al., Nitrogen isotopic composition of ammoniated pyllosilicates: Case studies from Precambrian metamorphosed sedimentary rocks, *Chemical Geology,* **216,** 37–58, 2005.

Schopf, J. W., The oldest fossils and what they mean, in *Major Events in the History of Life,* J. W. Schopf, ed., Jones and Bartlett Publishers, 29–63, 1992.

Schopf, J. W., et al., Laser-Raman imagery of Earth's earliest fossils, *Nature,* **416,** 73–76, 2002.

Origin of Life on Earth

*Davies, P., *The Origin of Life,* Penguin Books, 2003.

Cody, G. D., et al., Primordial carbonylated iron-sulfur compounds and the synthesis of pyruvate, *Science,* **298,** 1337–1340, 2000.

Huber, C., and Wachtershauser, G., Activated acetic acid by carbon fixation on (Fe,Ni)S under primordial conditions, *Science,* **276,** 245–247, 1997.

Pace, N. R., Origin of life—Facing up to the physical setting, *Cell,* **65,** 531–533, 1991.

Shock, E. L., Amend, J. P., and Zolotov, M. Y., The early Earth vs. the origin of life, in *Origin of the Earth and Moon,* R. M. Canup and K. Righter, eds., University of Arizona Press, 527–543, 2000.

Miller–Urey Experiment

Miller, S. L., The prebiotic synthesis of organic compounds as a step toward the origin of life, in *Major Events in the History of Life,* J. W. Schopf, ed., Jones and Bartlett Publishers, 1–28, 1992.

Urey, H. C., On the early chemical history of the Earth and the origin of life, *Proceedings of the National Academy of Sciences,* **38,** 351–363, 1952.

Where Did Life Originate?

*Warmflash, D., and Weiss, B., Did life come from another world? *Scientific American,* November 2005.

Baross, J. A., and Hoffman, S. E., Submarine hydrothermal vents and associated gradient environments as sites for the origin and evolution of life, *Origins of Life,* **15,** 327–345, 1985.

D'Hondt, S., Rutherford, S., and Spivak, A. J., Metabolic activity of subsurface life in deep-sea sediments, *Science,* **295,** 2067–2070, 2002.

Ferris, J. P., Huang, C. H., and Hagan, W. J., Jr., Montmorrilonite: A multifunctional mineral catalyst for the prebiological formation of phosphate esters, *Origins of Life and Evolution of the Biosphere,* **18,** 121–133, 1988.

RNA World and Early Life

Cech, T. R., A model for the RNA-catalyzed replication of RNA, *Proceedings of the National Academy of Sciences,* **83,** 4360–4363, 1986.

Doudna, J. A., and Szostak, J. W., RNA-catalyzed synthesis of complementary-strand RNA, *Nature,* **339,** 519–522, 1989.

Joyce, G. F., The rise and fall of the RNA world, *New Biologist,* **3,** 399–407, 1991.

Kerr, R., The story of O_2, *Science,* **308,** 1730–1732, 2005.

Lazcano, A., and Miller, S. L., On the origin of metabolic pathways, *Journal of Molecular Evolution,* **49,** 424–431, 1999.

Morris, S. C., The early evolution of life, in *Understanding the Earth,* G. Brown, C. Hawkesworth, and C. Wilson, eds., Cambridge University Press, 436–457, 1992.

Impacts and Extinction Events

Alvarez, L. W., Alvarez, W., Asaro, F., and Michel, H. V., Extraterrestrial cause for the Cretaceous-Tertiary extinction, *Science,* **208,** 1095–1108, 1980.

Rampino, M. R., and Haggerty, B. M., Extraterrestrial impacts and mass extinctions of life, in *Hazards Due to Comets and Asteroids,* T. Gehrels, ed., University of Arizona Press, 827–857, 1994.

Artificial Life

*Holmes, B., The race to create life from scratch, *New Scientist,* February 12, 2005.

*Stroh, M., Life built to order, *Popular Science,* February 1, 2005.

Chapter 7

Environmental Requirements for Life

Chyba, C. F., and Hand, K. P., Life without photosynthesis, *Science,* **292,** 2026–2027, 2001.

Nealson, K. H., Post-Viking microbiology: New approaches, new data, new insights, *Origins of Life and Evolution of the Biosphere,* **29,** 73–93, 1999.

Pace, N. R., The universal nature of biochemistry, *Proceedings of the National Academy of Sciences,* **98,** 805–808, 2001.

Wald, G., The origins of life, *Proceedings of the National Academy of Sciences,* **52,** 595–611, 1964.

Life in the Solar System

Chang, S., Planetary environments and the conditions of life, *Philosophical Transactions of the Royal Society of London,* **325,** 601–610, 1988.

Soffen, G. A., Life in the solar system, in *The New Solar System,* J. K. Beatty, C. C. Peterson, and A. Chaikin, eds., Sky Publishing Corporation, 365–376, 1999.

Chapter 8

Introduction to Mars

*Squyres, S., *Roving Mars: Spirit, Opportunity, and the Exploration of the Red Planet,* Hyperion, 2005.

Carr, M. H., Mars, in *The New Solar System,* J. K. Beatty, C. C. Peterson, and A. Chaikin, eds., Sky Publishing Corporation, 141–156, 1999.

Zahne, K., Decline and fall of the martian empire, *Nature,* **412,** 209–213, 2001.

Geological and Geochemical Potential for Life on Mars

*Christensen, P. R., The many faces of Mars, *Scientific American,* July 2005.

Baker, V. R., Water and the martian landscape, *Nature,* **412,** 228–236, 2001.

Gladman, B. J., et al., The exchange of impact ejecta between terrestrial planets, *Science,* **271,** 1387–1392, 1996.

Jakosky, B. M., and Phillips, R. J., Mars volatile and climate history, *Nature,* **412,** 237–244, 2001.

Jakosky, B. M., et al., The changing picture of volatiles and climate on Mars, *Science,* **310,** 1439–1440, 2005.

Jakosky, B. M., and Shock, E. L., The biological potential of Mars, the early Earth, and Europa, *Journal of Geophysical Research,* **103,** 19359–19364, 1998.

Weiss, B. P., Yung, Y. L., and Nealson, K. H., Atmospheric energy for subsurface life on Mars? *Proceedings of the National Academy of Sciences,* **97,** 1395–1399, 2000.

Viking Search for Life on Mars

Klein, H. P., Did Viking discover life on Mars? *Origins of Life and Evolution of the Biosphere,* **29,** 625–631, 1999.

Klein, H. P., Horowitz, N. H., and Biemann, K., The search for extant life on Mars, in *Mars,* H. H. Kieffer, B. M. Jakosky, C. W. Snyder, and M. S. Matthews, eds., University of Arizona Press, 1221–1233, 1992.

ALH84001 and Implications for Life on Mars

Barber, D. J., and Scott, E. R. D., Origin of supposedly biogenic magnetite in the martian meteorite Allan Hills 84001, *Proceedings of the National Academy of Sciences,* **99,** 6556–6561, 2002.

Friedmann, E. I., Wierzchos, J., Ascaso, C., and Winklhofer, M., Chains of magnetite crystals in the meteorite ALH84001: Evidence of biological origin, *Proceedings of the National Academy of Sciences,* **98,** 2176–2181, 2001.

Golden, D. C., et al., A simple inorganic process for formation of carbonates, magnetite, and sulfides in martian meteorite ALH84001, *American Mineralogist,* **86,** 370–375, 2001.

McKay, D. S., Gibson, E. K., Jr., Thomas-Keprta, K. L., Vali, H., Romanek, C. S., Clemett, S. J., Chillier, X. D. F., Maechling, C. R., and Zare, R. N., Search for past life on Mars: Possible relic biogenic activity in martian meteorite ALH84001, *Science,* **273,** 924–930, 1996.

Treiman, A. H., A short, critical evaluation of proposed signs of ancient martian life in Antarctic meteorite ALH84001, in *Bioastronomy '99: A New Era in Bioastronomy,* G. A. Lemarchand and K. J. Meech, eds., *Astronomical Society of the Pacific,* 303–314, 2000.

Treiman, A. H., Gleason, J. D., and Bogard, D. D., The SNC meteorites are from Mars, *Planetary and Space Science,* **48,** 1213–1230, 2000.

Searching for Life on Mars

*Zorpette, G., Why go to Mars? *Scientific American,* March 2000.

Carr, M. H., and Garvin, J., Mars exploration, *Nature,* **412,** 250–253, 2001.

DeVincenzi, D. L., Race, M. S., and Klein, H. P., Planetary protection, sample return missions, and Mars exploration: History, status, and future needs, *Journal of Geophysical Research,* **103,** 28577–28585, 1998.

Mojzsis, S. J., and Arrhenius, G., Phosphates and carbon on Mars: Exobiological implications and sample return considerations, *Journal of Geophysical Research,* **103,** 28495–28511, 1998.

Westall, F., et al., An ESA study for the search for life on Mars, *Planetary and Space Science,* **48,** 181–202, 2000.

Chapter 9

Life on Moons

*LePage, A., Habitable moons, *Sky and Telescope,* 51, December 1998.

*Rothery, D. A., *Satellites of the Outer Planets,* Clarendon Press, 1992.

Irwin, L. N., and Schulze-Makuch, D., Assessing the plausibility of life on other worlds, *Astrobiology,* **1,** 143, 2001.

Europa, Ganymede, and Callisto

Chyba, C. F., and Phillips, C. B., Possible ecosystems and the search for life on Europa, *Proceedings of the National Academy of Sciences,* **98,** 801, 2001.

Kivelson, M. G., et al., Galileo magnetometer measurements: A stronger case for a subsurface ocean at Europa, *Science,* **298,** 1340–1343, 2000.

Pappalardo, R. T., Head, J. W., and Greeley, R., The hidden ocean of Europa, *Scientific American,* **281,** 54–63, 1999.

Stevenson, D. J., Jupiter and its moons, *Science,* **294,** 71–72, 2001.

Titan

Lunine, J. I., Does Titan have an ocean? A review of current understanding of Titan's surface, *Reviews of Geophysics,* **31,** 133–149, 1993.

Sagan, C., Thompson, W. R., and Khare, B. N., Titan: A laboratory for prebiological organic chemistry, *Accounts of Chemical Research,* **25,** 286–292, 1992.

Chapter 10

Venus

*Grinspoon, D., *Venus Revealed,* Addison Wesley, 1997.

Habitable Zones and Beyond

Doyle, L., ed., *Circumstellar Habitable Zones,* Travis House, 1996.

Kasting, J. F., and Catling, D., Evolution of a habitable planet, *Annual Review of Astronomy and Astrophysics,* **41,** 429–463, 2003.

Kasting, J. F., Whitmire, D. P., and Reynolds, R. T., Habitable zones around main sequence stars, *Icarus,* **101,** 108–128, 1993.

Stevenson, D. J., Life-sustaining planets in interstellar space? *Nature,* **400,** 32, 1999.

Wetherill, G. W., The formation and habitability of extra–solar planets, *Icarus,* **119,** 219–238, 1996.

Williams, D. M., Kasting, J. F., and Wade, R. A., Habitable moons around extrasolar giant planets, *Nature,* **385,** 234–236, 1997.

Global Warming

*Eilperin, J., World temperatures keep rising with a hot 2005, *Washington Post*, October 13, 2005.

*Kluger, J., Be worried, be very worried, *Time Magazine*, April 3, 2006.

Kerr, R. A., Climate change: A worrying trend of less ice, higher seas, *Science*, **311**, 1698–1701, 2006.

Chapter 11

Discovery of Extrasolar Planets, Implications for Solar System Evolution

*Croswell, K., *Planet Quest*, The Free Press, 1997.

*Doyle, L. R., Deeg, H.-J., and Brown, T. M., Searching for shadows of other Earths, *Scientific American*, **283**, 38, 2000.

*Goldsmith, D., *Worlds Unnumbered*, University Science Books, 1997.

*Malhotra, R., Migrating planets, *Scientific American*, **281**, 56–63, 1999.

Marcy, G. W., and Butler, R. P., A planetary companion to 70 Virginis, *Astrophysical Journal*, **464**, L147–L151, 1996.

Marcy, G. W., and Butler, R. P., Planets orbiting other suns, *Publications of the Astronomical Society of the Pacific*, **112**, 137, 2000.

Marcy, G. W., Cochran, W. D., and Mayor, M., Extrasolar planets around main-sequence stars, in *Protostars and Protoplanets IV*, V. Mannings, A. P. Boss, and S. S. Russell, eds., University of Arizona Press, 1285–1311, 2000.

Santos, N. C., Benz, W., and Mayor, M., Extrasolar planets: Constraints for planet formation models, *Science*, **310**, 251–255, 2005.

How to Search for Evidence of Life on Extrasolar Planets

*Angel, J. R. P., and Woolf, N. J., Searching for life on other planets, *Scientific American*, **274**, 60–66, 1966.

Ford, E. B., Seager, S., and Turner, E. I., Characterization of extrasolar terrestrial planets from diurnal photometric variability, *Nature*, **412**, 885, 2001.

Sagan, C., et al., A search for life on Earth from the Galileo spacecraft, *Nature*, **365**, 715–721, 1993.

Schindler, T. L., and Kasting, J. F., Synthetic spectra of simulated terrestrial atmospheres containing possible biomarker gases, *Icarus*, **145**, 262–271, 2000.

The Rare Earth Hypothesis

*Darling, D., Rare Earths and hidden agendas, in *Life Everywhere*, Basic Books, 2001.

*Gonzalez, G., Brownlee, D., and Ward, P. D., Refuges for life in a hostile universe, *Scientific American*, **285**, 60–67, 2001.

*Ward, P. D., and Brownlee, D., *Rare Earth: Why Complex Life Is Uncommon in the Universe*, Copernicus, 2000.

Stars

*Johnson, G., *Miss Leavitt's Stars: The Untold Story of the Woman Who Discovered How to Measure the Universe*, Norton, 2005.

*Kaler, J. B., Meet the neighbors, *Astronomy*, 46, April 2006.

Chapter 12

Nature of Intelligence and Potential for Its Occurrence

*Calvin, W. H., The emergence of intelligence, *Scientific American*, **271**, 100–107, 1994.

Emery, N. J., and Clayton, N. S., The mentality of crows: Convergent evolution of intelligence in corvids and apes, *Science*, **306**, 1903–1907, 2004.

Jerison, H. J., *Evolution of the Brain and Intelligence*, Academic Press, 1973.

Marino, L., Brain-behavior relations in primates and cetaceans: Implications for the ubiquity of factors leading to the evolution of complex intelligence, in *Astronomical and Biochemical Origins and the Search for Life in the Universe*, C. B. Cosmovici, S. Bowyer, and D. Werthimer, eds., Editrice Compositori, 553, 1997.

Marino, L., McShea, D. W., and Uhen, M. D., The origin and evolution of large brains in toothed whales, *The Anatomical Record*, **281A**, 1247–1255, 2004.

Mayr, E., The probability of extraterrestrial intelligent life, in *Extraterrestrials: Science and Alien Intelligence*, E. Regis, Jr., ed., Cambridge University Press, 23, 1985.

Artifacts

Dyson, F. J., Search for artificial stellar sources of infrared radiation, *Science*, **131**, 1667, 1960.

Freitas, R. A., Jr., and Valdes, F., The search for extraterrestrial artifacts (SETA), *Acta Astronautica*, **12**, 1027–1034, 1985.

Zubrin, R., Detection of extraterrestrial civilizations via the spectral signature of advanced interstellar spacecraft, in *Progress in the Search for Extraterrestrial Life*, ASP Conference Series, vol 74., G. S. Shostak, ed., Astronomical Society of the Pacific, 487, 1995.

SETI

*Chown, M., ET First Contact "Within 20 Years," *New Scientist*, July 23, 2004.

*Davies, P., *Are We Alone?* Basic Books, 1996.

*Drake, F. D., and Sobel, D., *Is Anyone Out There?* Delacorte Press, 1992.

*Harrison, A., *After Contact*, Plenum Press, 1997.

*Mayr, E., The search for extraterrestrial intelligence: Scientific quest or hopeful folly? The improbability of success, *The Planetary Report*, **16**, 4–7, 1996.

*McConnell, B., *Beyond Contact*, O'Reilly and Associates, 2001.

*Sagan, C., The search for extraterrestrial intelligence: Scientific quest or hopeful folly? The abundance of life-bearing planets, *The Planetary Report*, **16**, 8–13, 1996.

*Shostak, S., The Future of SETI, *Sky and Telescope*, 42, April 2001.

*Shostak, S., and Barnett, A., *Cosmic Company*, Cambridge University Press, 2003.

Cocconi, G., and Morrison, P., Searching for interstellar communications, *Nature*, **184**, 844–846, 1959.

Drake, F., Project Ozma, *Physics Today*, **14**, 40, 1961.

Scientific Study of UFOs and Alien Visitation Claims

*Alschuler, W. R., *The Science of UFOs*, St. Martin's Press, 2001.

*Clancy, S., *Abducted: How People Come to Believe They Were Kidnapped by Aliens*, Harvard University Press, 2005.

*Korff, K., *The Roswell UFO Crash: What They Don't Want You to Know*, Prometheus Books, 1997.

*Kristof, N. D., Alien abduction? Science calls it sleep paralysis, *New York Times*, July 6, 1999.

*Sturrock, P. A., *The UFO Enigma*, Warner Books, 1999.

Chapter 13

The Voyager Record

*Sagan, C., Drake, F. D., Dryan, A., Ferris, T., Lomberg, J., Sagan, L. S., *Murmurs of Earth: The Voyager Interstellar Record*, Random House, 1978.

Interstellar Travel

*Mallove, E. F., and Matloff, G. L., *The Starflight Handbook: A Pioneer's Guide to Interstellar Travel*, Wiley, 1989.

Mauldin, J., *Prospects for Interstellar Travel*, Science and Technology Series: A Supplement to Advances in the Astronautical Sciences, Vol. 80, Univelt, 1992.

Relativity, Black Holes and Worm Holes

*Thorne, K. S., *Black Holes and Time Warps: Einstein's Outrageous Legacy*, Norton, 1994.

Will We Try to Colonize Space?

*Bracewell, R., *The Galactic Club*, W. H. Freeman, 1975.

*Dyson, F., *Disturbing the Universe*, Basic Books, 1979.
*Dyson, F., *Infinite in All Directions*, Harper and Row, 1988.
*Heppenheimer, T. A., *Toward Distant Suns*, Stackpole Books, 1979.
*O'Neill, G. K., *The High Frontier: Human Colonies in Space*, Wm. Morrow, 1977.
*Sagan, C., *Pale Blue Dot: A Vision of the Human Future in Space*, Random House, 1994.

Fermi Paradox

*Finney, B., and Jones, E. M., eds., *Interstellar Migration and the Human Experience*, University of California Press, 1985.
*Webb, S., *If the Universe Is Teeming with Aliens . . . Where Is Everybody? Fifty Solutions to Fermi's Paradox and the Problem of Extraterrestrial Life*, Springer, 2002.
Ball, J. A., The zoo hypothesis, *Icarus*, **19,** 347, 1973.
Hart, M., An explanation for the absence of extraterrestrials on Earth,*Quarterly Journal of the Royal Astronomical Society,* **16,** 128, 1975.

Sagan, C., and Newman, W. I., The solipsist approach to extraterrestrial intelligence, *Quarterly Journal of the Royal Astronomical Society,* **24,** 113, 1983.
Tipler, F. J., Extraterrestrial intelligent beings do not exist, *Quarterly Journal of the Royal Astronomical Society,* **21,** 267, 1980.

Epilogue

Philosophical and Societal Issues Related to the Search for Life Elsewhere

Chyba, C. F., and Hand, K. P., Astrobiology: The study of the living universe, *Annual Review of Astronomy and Astrophysics*, 43, 31–74, 2005.
Jakosky, B. M., Philosophical aspects of astrobiology, in *Bioastronomy '99: A New Era in Bioastronomy*, G. A. Lemarchand and K. J. Meech, eds., Astronomical Society of the Pacific, 661–666, 2000.
Jakosky, B. M., and Golombek, M. P., Planetary science, astrobiology, and the role of science and exploration in society, *EOS, Transcripts of the American Geophysical Union*, **81,** 58, February 6, 2000.

Credits

JPL, NASA/JPL/Arizona State University **9.21** NASA/JPL **9.22** NASA/JPL **9.23** NASA/JPL/Space Science Institute, NASA/JPL **9.24** NASA/JPL **9.25** NASA/JPL **9.26** NASA/JPL/Space Science Institute, NASA/JPL **9.27** NASA/JPL

CHAPTER 10 Chapter Opener NASA Earth Observing System

10.2 NASA/JPL, NASA **10.3** NASA/JPL, NASA **10.6** NASA and the Hubble Heritage Team (STScI/AURA) **10.7** "University Corporation for Atmospheric Research Adapted and updated 2006 from Karl and Trenberth (2003), courtesy K. Trenberth, and based on the HADCRUTv3 dataset from the CRU and UKMO" **10.8** NASA

CHAPTER 11 Chapter Opener Lynette R. Cook

11.1 "Dept. of Astronomy, University of Michigan at Cerro Tololo"; 3, Seth Shostak; 4,Data adapted from STARE project: (www.hao.ucar.edu/public/research/stare/hd29458.html; 5, NASA/JPL; 9,"NASA, JPL, Geoffrey Bryden" **11.2** Harvard College Observatory; 1,"Yeshe Fenner, Space Telescope Institute" **11.4** European Southern Observatory **11.5** European Southern University

CHAPTER 12 Chapter Opener Seth Shostak

12.1 Seth Shostak; 0,Seth Shostak; 1,Gordon Garradd; 2,Seth Shostak; 3,MGM / THE KOBAL COLLECTION; 6,Seth Shostak; 7,Seth Shostak; 8,Bettmann/CORBIS; 9,Christopher Cormack/CORBIS **12.2** Seth Shostak; 1,Tom Till/Photographer's Choice/Getty Images **12.5** "Radio Age, October 1924" **12.8** Seth Shostak, "Arecibo Observatory/National Astronomy and Ionosphere Center, operated by Cornell University under a cooperative agreement with the National Science Foundation" **12.9** Seth Shostak **Unnumbered Figure** Seth Shostak

CHAPTER 13 Chapter Opener Joe Bergeron

13.1 NASA/JPL; 2, CORBIS; 7, Joe Bergeron **13.2** NASA/JPL **13.3** NASA **13.4** NASA **13.5** Department of Energy **13.6** NASA/JPL **13.7** Joe Bergeron **13.8** "Joe Bergeron, Addison Wesley" **13.9** NASA/Jet Propulsion Laboratory

EPILOGUE Chapter Opener Graham Johnson/www.fiVth.com

E.1 Bettman/CORBIS **E.2** NASA

FRONT MATTER Graham Johnson/www.fiVth.com

Index

Page references in *italics* refer to figures. Page references preceded by "t" refer to tables. Page references followed by "n" refer to footnotes.

Aberration of starlight, 30n
AC (alternating current), 410
Acceleration of gravity, on Earth, A–3
Accretion
    heat from, 122–123
    life formation and, 237
    in planet formation, 76–77, 79, 237n
Acetylene, on Titan, 314, 317
Acid rain, from giant impacts, 217
Adams, Douglas, 484
Adams, John, 40n
Adenine, 172, *172*
ADP (adenosine diphosphate), 168, *168*
Aerobic organisms, 206–207
Aerobraking, 253
Aether, 24
Agathodaemon, 263
Age
    of ALH84001 martian meteorite, 288, t 288
    of civilizations, 456–458
    of earliest life, 192–195, *192, 193, 194*
    of Earth, 114–116, *114*
    of lunar rocks, 114, 118
    of the Moon, 114–116
    of sedimentary rocks, 194n
    of water on Mars, 273
Ages, geologic, 113–114, *113*
Akilia (Greenland), 194, *194*
Aldebaran, 439
Alexander the Great, 25
ALH84001, 204, *204*, 287–290, *288*, t 288, *289*
Alien visitations. *See also* Fermi paradox
    alien abduction, 429–430
    ancient visitations, 430–431, *430*
    case for, 431–432
    crashed aliens in Roswell, 428–429, *428*
    crop circles and, 429–430, *429*
    in movies, 38
    problems with reports of, 426–427, *426*
    UFO sightings and, 35–36
A-life (artificial life), 160, 226–229
Allen Telescope Array (California), 417–418, t 417, *419, 424*
Almagest (Ptolemy), 21
Alpha Centauri, 55, 56, 367–368
Alpha decay, 106
Alphonso X (king of Spain), 26
Alvarez, Luis and Walter, 215
Amazonian era (Mars), t 272, 272n
Amber, *111*
Amino acids
    defined, 153

genetic code for, 175, *175*
handedness in, 164, *164*
in pre-cells, 201–202, *201*
in the solar system, 238
Amino groups, 153, *163*, 164
Ammonia, 239, t 239, 314
Ammonite fossils, *111*
Amoebas, *161, 165*
Anaerobic organisms, 206–207
Anaximander, 20, 23, 155
The Andromeda Strain (film), 178
Angelou, Maya, *On the Pulse of the Morning*, 100
Angular momentum, conservation of, 75, *75*
Angular separation, A–4
Animals
    cells, *161*
    classification of, 209
    evolution of (*See* Evolution)
    land colonization by, 211–212
    in tree of life, 167, *167*
Antarctica
    extremophiles in rocks in, 180, *180*
    martian meteorites, 204, *204*, 287–290, *288*, t 288, *289*
    melting of ice shelves in, 350, *350*
    subsurface lake in, 307
Anthrax, 180, *181*
Anthropic principle, 68–70
Antimatter–matter rockets, 452
Antiprotons, 452
Apeiron, 23
Aphelion, *27, 28, 28*
Apollo missions
    Earth as viewed from, *485*
    Moon formation models and, 140
    "race" between U.S. and U.S.S.R., 285
    rock samples from, 254
    Saturn V rocket in, 443–444, *443*
Apparent brightness, 389. *See also* Luminosity
Apparent retrograde motion, 21–23, *21, 22*
Archaea
    on early Earth, 207
    as extremophiles, 179, 179n
    metabolic classifications of, 170, t 170
    in the tree of life, 167, *167*
Archean eon, 113, *113*
Arches National Park, path of stars at, *18*
Arcseconds, 373
Arecibo radio telescope (Puerto Rico), 415, *415*, 417, t 417, *424*
Ares Vallis (Mars), 275
Argon, on Titan, 313, 315–316
Argon-potassium dating, 108–110, *108*
Ariel (moon of Uranus), *298*, A–17
Aristarchus, 21–23, *22*
Aristotelians, 24–25

Aristotle
    on evolution, 155
    on gravity, 39
    on planetary motion, 28–29
    on scholarly method, 19
    on a spherical Earth, 20
    as tutor to Alexander the Great, 25
    on uniqueness of life on Earth, 17, 24, 31–32
Arks, interstellar, 449
Armageddon (film), 220
Arnold, Kenneth, 426
Arthropoda, 209
Artificial life (A-life), 160, 226–229
Artificial selection, 157–158, *157*
Asimov, Isaac, 32, 348
A stars, habitability of, 365
Asteroid belt
    formation of, 78, *79*
    location of, 72, 73, *73*
Asteroids
    composition of, 72
    definition of, 51
    formation of, 78
    impacts on Earth from, 215–217, *215, 216*
    location of, 72, 73, *73*
    organic molecules from, 238
    plausibility of life on, 250
    threat of future impacts from, 220–221, *221*
    Trojan, 72n
Astrobiology, 11–12, 38, A–20 to A–22. *See also* Life in the solar system; Life in the universe
"Astro-engineering," 422–424, *423*
Astrometric technique, 372–373
Astronomical Society of the Pacific, A–21
Astronomical unit (AU), 28, A–3
Astronomy
    ancient, 18–23, *20, 21, 22, 23*
    astrobiology, 11–12, 38, A–20 to A–22
    possibilities of extraterrestrial life and, 5
    women in, 362–363, *363*
AstroWeb, A–21
Atmospheres
    on Earth (*See* Atmosphere of Earth)
    on Europa, 309
    on extrasolar planets, 384, *384*
    greenhouse effect, 115, 132–133, *133*, 343
    hydrogen, 331–332
    on Mars, 265, 273, 276–277, 282–283
    on Mercury, 242n
    moist greenhouse effect, 342, 345
    on the Moon, 242n
    role in habitability, 340–341
    runaway greenhouse effect, 244–245, 336–337, *336, 345*

in the solar system, 238
Chemoheterotrophs, 170–171, t 170
Chicxulub crater (Mexico), 216, *216*
Chinese astronomy, ancient, 18
Chirality, 164, *164*
Chloroflexus, 170
Chlorofluorocarbons (CFCs), manufacturing
    on Mars, 286
Chloroplasts, 208, *209*
Chordata, 209
Chromosomes, 173
Chupacabra attacks, 429
Circle, area of, 396
Circular orbits, 21, 26–27, 78
Cities, relative temperatures in, 349
Civilizations
    age of, 456–458
    colonization by, 458–461, *459*, 463
    in the Drake equation, 401, 401n,
        402–404
    evolution of intelligence, 405–410, *406*,
        407, 408
    galactic, 423, 464–466
    planetary, 423
    self-destruction of, 464
    stellar, 423
Clarke, Arthur C., 422, 423, 465
Clay minerals, origin of life and, 201, 201n,
    202
Climate. *See also* Global warming
    axis tilt changes and, 136–137, *136*, 388
    Cambrian explosion and, 210
    carbon dioxide cycle and, 134–135, *134*
    feedback processes, 135–136, *135*
    flotation of solid ice and, 240
    greenhouse effect and, 132–133, *133*, 342
    ice ages, 115, 136–137, *136*
    on Mars, 276–280, *276, 278, 279*
    plate tectonics and, 102
    runaway greenhouse effect, 244–245,
        336–337, *336*, 345
    Snowball Earth, 115, 137–138, *137*, 210
    stability of, 132, 387–388
Cloning, 174
Close encounter hypothesis, 91
Clusters, *52*, 53, 390
Coal, 212
Cocconi, Giuseppe, 399, 411–412
CO_2 cycle. *See* Carbon dioxide cycle
Coin tosses, probability in, 107–108
Colonies
    colonization models, 458–460, *459*, 476
    on extrasolar planets, 440
    on Mars, 285–287
    motives for colonization, 460–461, *461*
    sociological considerations in, 463–464
    technological difficulties in, 462–463
Colonization models, 458–460, *459*, 476
Colors, of stars, 389–390
Columbus, Christopher, 20
Comets. *See also* Impacts
    composition of, 72
    definition of, 51
    deuterium enrichment in, 336
    formation of, 78, *79*
    location of, 73, *73*
    organic molecules from, 238

plausibility of life on, 250
Pluto as, 10
tail formation in, 72
threat of future impacts from, 220–221,
    221
Competition for survival, evolution and,
    156–157, 409
Complementary strands, 172, *173*
Compounds, 83, *83. See also* Organic
    molecules
Conceptual models, 20
Condensation
    implications for life formation, 237, 237n
    in solar system formation, 75–78, t 76,
        78, 79
Conservation of angular momentum, 75, *75*
Conservation of energy, 74, *75*, 85, 153
Conservation of momentum, 75, *75*, 441
Constantinople, fall of, 25
Contact (Sagan; film), 414–415, 416, 455
Contamination
    of meteorites, 289
    from space missions, 284, 285
Continental crust, 115, 125, *125*
Continental drift, 115, 125, *125*
Continental shelf, 126
Continuously habitable zone, 344, *345*
Convection, in interior of Earth, 122, 124
Convection cells, *121*, 122, 126, *126*
Convergent evolution, 405–407
Conversion factors, A–9 to A–11, A–12
Cooling rate, of Earth, 123
Copernican Principle, 397
Copernican revolution, 25–33
    Catholic Church and, 30, 32
    changing perspectives on life from,
        481–482
    Copernicus, 25–26, 26n
    debate on extraterrestrial life in, 31–33,
        50–51
    Galileo, 28–30, *29*
    Kepler, 27–28, *27, 28*
    Tycho Brahe, 27
Copernicus, Nicholas, 25–26, 26n, 32
Coprolites, 112
Coral model of galactic civilization, 459, *459*,
    476
Core
    of the Earth, 121, *121*
    of Europa, 305, *305*
    of Ganymede, 311
    of Mars, 277–278, *278*
    of Mercury, 142
    of the Moon, 140
Coronagraphs, 380
Correlations and causality, 350–351
"Cosmic address," 51–53, *52*
Cosmic calendar, 64–65, *64–65*
Cosmic microwave background, 60
Cosmic rays, 219, 288
Cosmic speed limit, 439–440, 449–450
Cosmos. *See* Universe
Cosmos (Sagan), 261
Crab Nebula, *62*, 63
Creation, biblical accounts of, 184
Creation science, 184
Crop circles, 429–430, *429*

Crust, of Earth, 115, 121–122, *121*, 125, *125*
Cryovolcanism, 316, 317, 318
Crystal structure, 83, *83*
Cubed quantities, A–9
Cultural evolution, 224
Cyanobacteria, 213, *213*
Cytosine, 172, *172*

Dao Vallis (Mars), *274*
Dark energy, 58, 61
Dark matter, 57–58, *58*
Darwin, Charles Robert
    life of, 158–159, *159*
    The Origin of Species, 149, 155, 159
    theory of evolution, 155–158, *156, 157*,
        482
Darwin, Erasmus, 158
Darwin mission, 380, 384
Dating techniques
    radiometric, 106–110, *107, 108*, t 109
    sedimentary strata, 105, *105*
Daughter isotopes, 106
Daylength, increasing, 301
Days of the week, origin of names of, t 20
The Day the Earth Stood Still (film), 69
*Declaration of Principles Concerning Activities
    Following the Detection of Extraterrestrial
    Intelligence,* 424–425
Deep Impact mission, 255, t 255
Deep-sea vents
    chemical disequilibrium near, 320
    comparison with Europa habitability, 307
    energy sources at, 308, *308*
    extremophiles at, 179, *179*
    microfossils from, 193
    origin of life at, 197, 307–308
    pressure at, 239
    sources of organic molecules at, 199
Deimos (moon of Mars), 250, *250*, A–16
Deinococcus radiodurans, 180, 197
Democritus, 24
Density
    of jovian moons, 299
    of Moon *vs.* Earth, 139, 140
    of Sun and planets, A–15
Deuterium, 335–336, 445n, 447
Dialogue Concerning the Two Chief World Systems
    (Galileo), 30
Differentiation, in Earth formation, 115,
    122–123, 122n
Dinosaurs
    extinction of, 215–216, *215, 216*
    fossils of, *111*, 112
    timing of, 116
Dione (moon of Saturn), *298*, A–16
Discovery science, 34n
Diseases, from spacecraft return, 284, 285
Disequilibrium, role in life, 319–321
Disks, galactic, 57, *58*
Distance
    decreases in, at high speeds, 450, 469
    in expanding universe, 60–61, *61*
    within the galaxy, 56–57
    in light-years, 55, A–3
    within the solar system, 53–55, *53, 54*
    to the stars, 29, 55–56
    units of, A–3